MILES PER GALLON

ENGINE A

ENGINE B

20

15

10

5

MPG VS MPH
OF TWO
ENGINES

0   20   40   60   80

MILES PER HR

100

90

80

70

60

50

40

30

20

SPEED (MPH)

1 SECOND

2 SECONDS

3 SECONDS

4 SECONDS

5 SECONDS

FOURTH EDITION

ENGINEERING DESIGN GRAPHICS

James H. Earle

Texas A&M University

Addison-Wesley Publishing Company
Reading, Massachusetts • Menlo Park, California
London • Amsterdam • Don Mills, Ontario • Sydney

Sponsoring Editor: Tom Robbins
Production Editor: Laura R. Skinger
Text Designer: Celine Brandes
Illustrator and Cover Designer: James H. Earle
Art Coordinator: Kristin Belanger
Layout Artist: Lorraine Hodsdon
Production Manager: Sherry Berg

The text of this book was composed in Melior by the Clarinda Company.

Library of Congress Cataloging in Publication Data

Earle, James H.
　Engineering design graphics.

　Includes bibliographies and index.
　1. Engineering design.　2. Engineering graphics.
I. Title.
TA174.E23　1983　　620'.00425　　82-6709
ISBN 0-201-11318-X　　　　　　AACR2

Reprinted with corrections, July 1983

ISBN 0-201-11318-X
DEFGHIJ-HA-8987654

Dedicated to my father,
Hubert Lewis Earle,
October 25, 1900–October 22, 1967

The following supplements are compatible with the following textbooks from Addison-Wesley Publishing Co.: *Drafting Technology, Engineering Design Graphics, Descriptive Geometry,* and *Design Drafting.*

DRAFTING TECHNOLOGY PROBLEMS is a new problem book designed to cover basic graphics, descriptive geometry, and specialty drafting areas. It is designed to accompany a new text, *Drafting Technology,* that is available from Addison-Wesley Publ. Co., Reading, Mass. 01867. 131 pages.

BASIC DRAFTING is a problem book that covers the basics for a one-semester high school drafting course. 67 pages.

CREATIVE DRAFTING is a problem book that covers mechanical drawing and architectural drafting for the high school. 106 pages.

DRAFTING & DESIGN is a problem book for mechanical drawing for a high school or college course. 106 pages.

DRAFTING FUNDAMENTALS 1 is a problem book for a one-year high school course in mechanical drawing. 94 pages.

DRAFTING FUNDAMENTALS 2 is a second version of DRAFTING FUNDAMENTALS 1 for the same level and content. 94 pages.

TECHNICAL ILLUSTRATION is a problem book for a course in pictorial drawing for the college or high school. 70 pages.

ARCHITECTURAL DRAFTING is a problem book for a first course in architectural drafting. 71 pages.

GRAPHICS FOR ENGINEERS 1 is a problem book for a first course in engineering graphics for the college student. 100 pages.

GRAPHICS FOR ENGINEERS 2 is a problem book for a first course in engineering graphics for the college student. 100 pages.

GRAPHICS FOR ENGINEERS 3 is a problem book for a first course in engineering graphics for the college student. 108 pages.

GEOMETRY FOR ENGINEERS 1 is a problem book for a college-level descriptive geometry course. 100 pages.

GEOMETRY FOR ENGINEERS 2 is a problem book for a college-level descriptive geometry course. 100 pages.

GEOMETRY FOR ENGINEERS 3 is a problem book for a college-level descriptive geometry course. 100 pages.

GRAPHICS & GEOMETRY 1 is a problem book for a college course in graphics and descriptive geometry. 119 pages.

GRAPHICS & GEOMETRY 2 is a problem book for a college course in graphics and descriptive geometry. 121 pages.

GRAPHICS & GEOMETRY 3 is a problem book for a college course in graphics and descriptive geometry. 138 pages.

COMPUTER GRAPHICS is a problem book that is supplemented with extensive notes to serve as the basis for a first college-level course in computer graphics. 92 pages.

Creative Publishing Co.

BOX 9292   COLLEGE STATION, TEXAS   77840
PHONE 713-846-7907

PREFACE

This fourth edition of *Engineering Design Graphics* is a major revision of the third edition. There has been considerable reorganization of content and many illustrations and problems have been modified. Much new material has been included to broaden the coverage of the book.

The material on the design process has been grouped in eight sequential chapters (Chapters 2 through 8) to improve the ease of reference. New examples of design projects have been added to these chapters. The list of design projects for class assignment has been expanded to 116 problems.

New chapters with expanded content have been added to improve the coverage of such topics as gears and cams, materials and processes, welding, reproduction methods, pipe drafting, electric/electronics drafting, and computer graphics. All chapters have been based on the latest ANSI standards and the current practices of industry. Metric units are used extensively; however, English units of measurement have not been omitted.

The coverage of engineering graphics, following the chapters on design, is introduced at the basic level, assuming that the student has no prior experience in graphics. This introduction is sufficient to enable the inexperienced student to progress with a minimum of instruction. The progression from the basic concepts to the more advanced ones makes it possible for the book to be used for advanced courses while serving as a source for reviewing fundamentals when needed.

Several techniques of presentation have been used to help the student understand concepts and problem-solving methods. Many problems are solved in a multistep format that separates the steps of the solution. A second color is used throughout the book as a functional means of calling attention to important points and steps. This technique of presentation enables the student to use the text more efficiently when working without the aid of a teacher, and therefore to cover more material than when using a conventional textbook.

More material can be covered by the student if a problem book is used in conjunction with this textbook. Seventeen different problem books are available from Creative Publishing Company, Box 9292, College Station, TX 77840, to aid the student and teacher in covering more material in less time. A list of these can be found on the facing page.

We are grateful for the assistance of many who have influenced the development of this book. We have been significantly aided by the use of many illustrations, especially in the problems, that were developed by the late William E. Street and Carl L. Svensen for their earlier publications. Many industries have furnished photographs and drawings, which have been individually acknowledged. The staff of the Engineering Design Graphics Department of Texas A&M University has been helpful in making suggestions for covering the material in this book.

Professor Tom Pollock was helpful in providing information on various metals and their designations. Professor Bill Zaggle organized the material used in the chapter on computer graphics.

Above all, we are appreciative of the many institutions who have thought enough of our publications to adopt them for classroom use. Since there are so many fine textbooks in the area of engineering graphics, it is indeed an honor for one's publications to be accepted by his colleagues. We are hopeful that this book will fill the needs of the existing engineering graphics programs. As always, comments and suggestions for improvement and revision of this book will be appreciated.

January 1983
College Station, Texas

J. H. E.

CONTENTS

INTRODUCTION TO ENGINEERING AND TECHNOLOGY

1.1 INTRODUCTION

Essentially all our daily activities are assisted by products, systems, and services made possible by the engineer. Our utilities, heating and cooling equipment, automobiles, machinery, and consumer products have been provided at an economical rate to our population by the engineering profession.

The engineer must function as a member of a team composed of other related, and sometimes unrelated, disciplines. Many engineers have been responsible for innovations of life-saving mechanisms used in medicine, which were designed in cooperation with members of the medical profession. Other engineers are technical representatives or salespeople who explain and demonstrate applications of technical products to a specialized segment of the market. Even though there is a wide range of activities within the broad definition of engineering, the engineer is basically a *designer*. This is the activity that most distinguishes him or her from other associated members of the technological team.

This book is devoted to the introduction of elementary design concepts related to the field of engineering and to the application of engineering graphics and descriptive geometry to the design process. Examples are given that have an engineering problem at the core, and that require organization, analysis, problem-solving graphical principles, communication, and skill (Fig. 1.1).

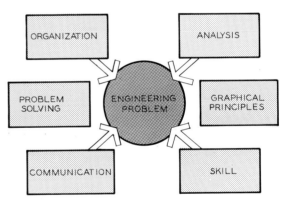

Fig. 1.1 Problems in this text require a total engineering approach with the engineering problem as the central theme.

Creativity and imagination are encouraged as essential ingredients in the engineer's professional activities. A systematic approach to developing a design solution has been used as the format for the entire volume, carrying the process from problem identification to final implementation by successive chapters. Albert Einstein, the famous physicist, said that "Imagination is more important than knowledge, for knowledge is limited, whereas imagination embraces the entire world . . . stimulating progress, or, giving birth to evolution. . . ." (Fig. 1.2).

Fig. 1.2 Albert Einstein, the famous physicist, said "Imagination is more important than knowledge. . . ."

1.2 ENGINEERING GRAPHICS

Engineering graphics is considered to be the total field of graphical problem solving and includes two major areas of specialization, descriptive geometry and working drawings. Other areas that can be utilized for a wide variety of scientific and engineering applications are also included within the field. These areas are nomography, graphical mathematics, empirical equations, technical illustration, vector analysis, data analysis, and other graphical applications associated with each of the different engineering industries.

Engineering graphics is not limited to drafting, since it is more extensive than the communication of an idea in the form of a working drawing. Graphics is the designer's method of thinking, solving, and communicating ideas throughout the design process.

Humankind's progress can be attributed to a great extent to the area of engineering graphics. Even the simplest of structures could not have been designed or built without drawings, diagrams, and details that explained their construction (Fig. 1.3). Gradually, graphical methods were developed to show three related views of an object to simulate its three-dimensional representation. A most significant development in the engineering graphics area was descriptive geometry.

Descriptive Geometry

Gaspard Monge (1746–1818) is considered the "father of descriptive geometry" (Fig. 1.4). Young Monge used this graphical method to solve design problems related to fortifications and battlements while a military student in France. He was scolded by his headmaster for not solving a problem by the usual, long, tedious mathematical process traditionally used for problems of this type. It was only after long explanations and comparisons of the solutions of both methods that he was able to convince the faculty that his graphical methods could be used to solve the problem in considerably less time. This was such an improvement over the mathematical solution that it was kept a military secret for fifteen years before it was allowed to be taught as part of the technical curriculum. Monge became a scientific and mathematical aide to Napoleon during his reign as general and emperor of France.

Descriptive geometry can be defined as the projection of three-dimensional figures onto a two-dimensional plane of paper in such a manner as

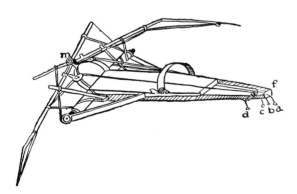

Fig. 1.3 Leonardo da Vinci developed many creative designs through the use of graphical methods.

Fig. 1.4 Gaspard Monge, the "father of descriptive geometry."

to allow geometric manipulations to determine lengths, angles, shapes, and other descriptive information concerning the figures.

1.3 THE TECHNOLOGICAL TEAM

Technology has broadened at a rapid rate during recent years with the introduction of many new processes and fields of specialty unheard of a decade ago. It has become necessary for professional responsibilities to be performed by highly qualified people with specialized training. Thus technology has become a team effort involving the scientist, engineer, technologist, technician, and craftsman (Fig. 1.5). A project may include mechanical design, an advanced electronics system, a structure, and chemical processes, and therefore it may require many engineers, technologists, technicians, and craftsmen to complete the design. The members of the technological team are listed below.

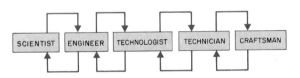

Fig. 1.5 The technological team.

Fig. 1.6 Scientists such as this chemist conduct research to establish fundamental relationships. This chemist is trying to determine the best combination of ingredients to yield a better tire. (Courtesy of Uniroyal, Inc.)

The Scientist

The scientist is primarily a researcher who is seeking to establish new theories and principles through experimentation and testing (Figs. 1.6 and 1.7). Often he or she has little concern for the application of specific principles that are being developed, being primarily interested in isolating significant relationships. Scientific discoveries are used as the basis for related research and for the development of practical applications for future discoveries that may not come into existence until years after the initial discovery.

Fig. 1.7 A researcher solders fine wires to each strain gauge in an experimental model. (Courtesy of the Bureau of Reclamation, U.S. Department of the Interior.)

The Engineer

Engineers are trained in areas of science, mathematics, and industrial processes. This prepares them to apply the basic principles discovered by the scientist to practical problems (Fig. 1.8). They are concerned with the conversion of raw materials and power sources into needed products and services. The emphasis on the practical application of principles distinguishes the engineer from the scientist.

The application of these principles to new products or systems in a creative manner is the process of designing, which is the engineer's most unique function, requiring the highest degree of creativity. Engineers must always be concerned with efficiency and economy in their designs to provide the greatest service to society. In general

Fig. 1.8 One of the modern engineering marvels is the Astrodome of Houston, Texas, shown here in the early stages of construction.

terms, the engineer practices the art of using available principles and resources to achieve a practical end at a reasonable cost.

The Technologist

The technologist is a technically trained person who assists the engineer at a semiprofessional level slightly below the engineer and above the level of the technician. While the engineer is concerned with the application of theoretical concepts, the technologist is more concerned with routine, practical aspects of engineering whether at the planning or production stage (Fig. 1.9). Several technologists working with an engineer will

Fig. 1.9 Engineering technologists combine their knowledge of production techniques with the design talents of the engineers to produce a product. (Courtesy of Omark Industries, Inc.)

greatly increase the engineer's capability of performing the job, since he or she will be relieved of many duties by the technologist. Likewise, the technologist can coordinate the activities of several technicians, thereby sharing supervisory responsibility with the engineer. The technologist usually has a four-year engineering technology background at the college level.

The Technician

Technicians are technically trained individuals who assist the engineer and technologist at a level below the technologist. Their work may vary from conducting routine laboratory experiments to the supervision of craftsmen involved in manufacturing or construction. In general, technicians work as liaisons between the technologist and the craftsman. Technicians must exercise a degree of judgment and imagination in their work and assume supervisory responsibilities beyond those required of the craftsman (Fig. 1.10). They are usually required to have a two-year technical-training background beyond the high-school level to be qualified for their assignment.

The Craftsman

Craftsmen are vital members of the engineering team since they must see that the engineering design is implemented by producing it according to the specifications of the engineer. They may be machinists who fabricate the various components of the product or electrical craftsmen who assemble electrical components. Craftsmen supply technical skill that cannot be provided by the engineer or the technician. The ability to produce a given part in accordance with design specifications is as necessary as the act of designing the part. Craftsmen include electricians, welders, machinists, fabricators, drafters, and members of many other professions (Fig. 1.11).

The Designer

Designers are the people who have special talents for creating solutions to technological problems.

Fig. 1.12 Thomas A. Edison had essentially no formal education, but he gave the world some of its most creative designs, which have never been equaled by another single individual.

The designer may be an engineer, an inventor, or a person who has special talents for devising creative solutions, even though he or she may not have an engineering background. This is often the case in young areas of technology where little precedent has been established by previous experience. Thomas A. Edison (Fig. 1.12) had very little formal education, but he possessed an exceptional ability to design and perfect some of the world's most significant designs. Engineers may find that their formal backgrounds inhibit their design abilities. They may be too quick to label a particular approach as impossible, when in reality it can be solved by a designer who did not know better than to try.

The Stylist

The person who is concerned with the outward appearance of a product rather than the development of a functional design is referred to as a stylist (Fig. 1.13). Stylists may be responsible for the design of an automobile body or the configuration

Fig. 1.10 Technicians are skilled in performing routine jobs requiring technical knowledge and skill. This technician is responsible for inspecting chain assemblies to ensure that quality standards are met. (Courtesy of Omark Industries, Inc.)

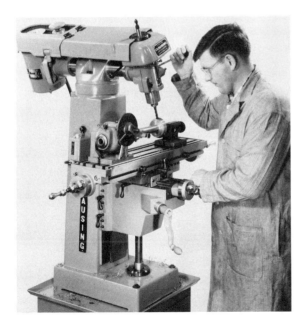

Fig. 1.11 This craftsman is a machinist who is skilled in the making of metal parts in accordance with the specifications that he is given. (Courtesy of the Clausing Corporation.)

Fig. 1.13 The stylist is more concerned with the outward appearance of a product than with the functional aspects of its design. (Courtesy of Ford Motor Corporation.)

of an electric iron, but they are not concerned with the design of the internal details of the product. The stylist must have a high degree of aesthetic awareness plus a feel for the consumer's acceptance of designs.

A stylist who designs an automobile body considers the functional requirements of the body including driver vision, enclosure of passengers, space for the power unit, etc. However, the stylist is not involved with the design of such items as the engine or the steering linkage.

1.4 THE ENGINEERING FIELDS

The following articles briefly outline the more prominent areas of engineering and describe the duties of engineers employed in these fields of the profession. Each of the areas discussed is considerably broader than the description given, since only the more basic activities are listed. The addresses of the professional societies associated with each field are given at the end of this chapter as sources for additional information concerning the opportunities and challenges of each branch.

Significant changes in engineering during recent years have been the emergence of the technologist and technician and the growing number of women who are pursuing engineering careers. More than 3.0 percent of the present freshman en-

gineering class is comprised of women, indicating a growing trend of women in this career field.

Aerospace Engineering

Aerospace engineers in the space exploration branch of this profession work on all types of aircraft and spacecraft, including missiles, rockets, and conventional propeller-driven and jet-powered planes. Their responsibilities include the development of aerospace products from initial planning and design to the final manufacture and testing (Fig. 1.14).

Second only to the auto industry in sales, the aerospace industry contributes immeasurably to national defense and the national economy. The future for aeronautical engineers appears to offer rewarding and challenging opportunities in the years ahead.

Aerospace engineering deals with flight in all aspects—at all speeds and all altitudes. Aerospace engineering assignments range from complex vehicles traveling 350 million miles to Mars to hovering aircraft used in deep sea exploration. This broad field of engineering encompasses many specialized disciplines. As a result, most aerospace engineers usually specialize in a particular area of work. Specialized areas are (1) aerodynamics, (2) structural design, (3) instrumentation, (4) propulsion systems, (5) materials, (6) reliability testing,

Fig. 1.14 Aerospace engineers are responsible for the design and testing of aircraft to determine the most efficient craft possible. (Courtesy of Bell Helicopter Corporation.)

and (7) production methods. They may also specialize in a particular product, such as conventional-powered planes, jet-powered military aircraft, rockets, satellites, or manned space capsules.

In the broadest sense, aerospace engineering can be divided into two major areas: research engineering and design engineering. The research engineer is concerned with the exploration of known principles in search of new ideas and concepts. On the other hand, the design engineer translates the new concepts developed by the researcher into workable applications for the improvement of the existing state of the art. Engineers in each of these two areas must approach the unknowns of their fields with creativity, skill, and knowledge. This approach has elevated the field of aerospace engineering from the Wright brothers' first flight at Kittyhawk, N.C., in 1903 to the penetration of outer space with an unlimited future (Fig. 1.15).

Fig. 1.15 The space shuttle *Columbia* was launched in April, 1981 to introduce a new era for aerospace engineering—the era of the reusable spaceship. (Courtesy of NASA.)

Related engineering specialists are vital members of the aerospace engineering team. A few of the specialty areas are: (1) avionics—the study of electronic, computerized communication and flight-control systems; (2) equipment—the design and installation of equipment for navigation, hydraulic, armament, survival, electrical, and comfort control related to the functional operation of the aircraft; (3) materials and processes—the development, testing, and evaluation of new materials, to determine their applications to aircraft designs; (4) metallurgy—the evaluation and testing of new and specially treated metals to select the most efficient metals for aerospace use.

The professional society of the aeronautical engineers is the American Institute of Aeronautics and Astronautics (AIAA). Student branches of this association are located on most college campuses offering degrees in aeronautical engineering.

Agricultural Engineering

Agricultural engineers are trained to serve the world's largest industry—agriculture. Engineering problems of agriculture deal with the production, processing, and handling of food and fiber. The four major areas of specialization of agriculture are: mechanical power and machinery, farm structures, electrical power and processing equipment, and soil and water control and conservation.

Mechanical Power The agricultural engineer who works with one of the more than 800 manufacturers of farm equipment will be concerned with gasoline and diesel engine equipment such as pumps, irrigation machinery, and tractors. The use of farm machinery designed by agricultural engineers has been responsible for the increased production of agriculture at a higher degree of efficiency. Machinery must be designed for the electrical curing of hay, milk processing and pasteurizing, fruit processing, and artificially heating environments in which to raise livestock and poultry (Fig. 1.16).

Farm Structures The construction of barns, shelters, silos, granaries, processing centers, and other agricultural buildings requires specialists in agricultural engineering. The design of buildings of this type requires that the engineer understand heating, ventilation, and chemical changes that affect the storage of crops. Engineers may also supervise research that will improve methods of design and construction of farm structures.

Fig. 1.16 This walking sprinkler system is an example of the work of agricultural engineers. It distributes 1000 gallons per minute of water to irrigate 140 acres of farm land. (Courtesy of the Bureau of Reclamation.)

Electrical Power A high percentage of the equipment used for agricultural purposes is driven by electrical power. Agricultural engineers design electrical systems and select equipment that will provide efficient operation to meet the requirements of the specific situation. They may serve as consultants or designers for manufacturers or large processors of agricultural products. Their knowledge improves living and working conditions connected with rural activities.

Soil and Water Control The agricultural engineer is responsible for devising systems for improving drainage and irrigation systems, resurfacing fields, and constructing water reservoirs (Fig. 1.17). These activities may be performed in association with the U.S. Department of Agriculture or the Department of Interior, with state agricultural universities, with consulting engineering firms, or with irrigation companies.

Most agricultural engineers are employed in private industry, especially by manufacturers of heavy farm equipment and specialized lines of field, barnyard, and household equipment; electrical service companies; and distributors of farm equipment and supplies. Agricultural engineering is not limited to work in rural areas. Most agricultural engineers live in cities where the leading farm equipment and manufacturing centers are located. Although they need not live on a farm or be

involved in agriculture, it is helpful if they have a clear understanding of agricultural problems, farmwork, crops, animals, and people involved in farming.

Agricultural engineers will be involved in the problems of conservation of resources and the introduction of new agricultural products in the future. Agricultural engineering has been instrumental in raising the farmer's efficiency so much that one farmer today raises enough food for 32 people, whereas he was capable of supplying only four other people 100 years ago. The future offers many opportunities in agricultural engineering.

The professional society of the agricultural engineer is the American Society for Agricultural Engineers.

Fig. 1.17 Agricultural engineers improve production by designing irrigation systems. This irrigation chute is 3¾ miles long. (Courtesy of the Bureau of Reclamation, U.S. Department of the Interior.)

Chemical Engineering

Chemical engineering involves the design and selection of equipment that will facilitate the processing and manufacturing of chemicals in large quantities. These designs are closely related to the principles of chemistry and those of other engineering fields that aid in the economic and efficient production of chemical products (Fig. 1.18).

Chemical engineers design unit operations, including fluid transportation through ducts or pipelines, solid material transportation through pipes or conveyors, heat transfer from one fluid or

Fig. 1.18 The chemical engineer conducts experiments on new refinery methods to improve the quality of products. (Courtesy of Exxon Corporation.)

substance to another fluid or substance through plate or tube walls, absorption of gases by bubbling them through liquids, evaporation of liquids to increase concentration of solutions, distillation under carefully controlled temperatures to separate mixed liquids, and many other similar chemical processes. They may employ chemical reactions of raw products such as oxidation, hydrogenation, reduction, chlorination, nitration, sulfonation, pyrolysis, and polymerization (Fig. 1.19). From these reactions come new materials and products.

Fig. 1.19 The chemical engineer's efforts, in conjunction with efforts of engineers from other fields, are finalized in the construction and operation of a completed refinery producing vital products for our economy and needs. (Courtesy of Houston Oil and Refining Company.)

Process control and instrumentation have developed as important specialties in chemical engineering. The handling and control of large quantities of materials must be possible with a high degree of accuracy and precision. Process control is designed for fully automatic operation with the measurement of quality and quantity by fully automatic instrumentation.

Chemical engineers develop and process chemicals such as acids, alkalies, salts, coal-tar products, dyes, synthetic chemicals, plastics, insecticides, fungicides, and many others for industrial and domestic uses. They are associated with drugs and medicine, cosmetics, explosives, ceramics, cements, paints, petroleum products, lubricants, synthetic fibers, rubber, and detergents. They also design equipment for food preparation and canning plants.

The chemical engineer works with metallurgical and mining engineers in designing processing equipment and in designing and laying out complete plants. They must be well versed in chemistry and must be able to discuss plant design and operation with the plant operator. This combination requires that chemical engineers be well rounded in a number of disciplines, which enables them to select several specialties in which to work.

Approximately 80 percent of chemical engineers work in the manufacturing industries—primarily in the chemical industry. The remainder work for government agencies, independent research institutes, and as independent consulting engineers. New fields that require chemical engineers are nuclear sciences, rocket fuels, and environmental pollution areas. The development of new drugs, fertilizers, paints, and chemicals is expected to increase the demand for chemical engineers in the coming years.

The professional society for chemical engineers is the American Institute of Chemical Engineers (AIChE).

Civil Engineering

Civil engineering, the oldest branch of engineering, is closely related to practically all of our daily activities. The buildings we live in and work in, the transportation facilities we use, the water we drink, and the drainage and sewage systems we use are the results of civil engineering. Civil engineers design and supervise the construction of roads, harbors, airfields, tunnels, bridges, water supply and sewage systems, and many other types of structures. Civil engineers can specialize in a number of areas within the field, with the following being the most prominent: construction, city planning, structural engineering, hydraulic engineering, transportation, highways, and sanitation.

Construction engineers are responsible for the management of resources, manpower, finances, and materials necessary for construction projects. These projects may vary from the erection of skyscrapers to the movement of concrete and earth.

City planners develop plans for the future growth of cities and the various systems related to their operation. Street planning, zoning, and industrial site development are problems encountered in the field of city planning.

Structural engineers are responsible for the design and supervision of the erection of structural systems, including buildings, dams, powerhouses, stadiums, bridges, and numerous other structures. Strength and appearance are considered in designing structures of this type to serve the required needs economically.

Hydraulic engineers work with the behavior of water from its conservation to its transportation. They design wells, canals, dams, pipelines, drainage systems, and other methods of controlling and utilizing water and petroleum products (Fig. 1.20).

Transportation engineers work with the development and improvement of railroads and airlines in all phases of their operations. Railroads are built, modified, and maintained under the supervision of civil engineers. Design and construction of airport runways, control towers, passenger and freight stations, and aircraft hangars are done by civil engineers who specialize in the field of transportation. (Fig. 1.21).

Highway engineers develop the complex network of highways and interchanges for moving automobile traffic. These systems require the design of tunnels, culverts, and traffic control systems.

Sanitary engineers assist in maintaining public health through the purification of water, control of water pollution, and sewage control. Such systems involve the design of pipelines, treatment plants, dams, and related systems.

The activities of the civil engineer are very diversified, with opportunities in a variety of locations from city centers to remote construction sites. Due to their experience in management and the solution of environmental problems, many civil engineers find positions in administration and municipal management. The majority of civil engineers are associated with federal, state, and local government agencies and the construction industry. Many work as consulting engineers for architectural firms and independent consulting engineering firms. The remainder work

Fig. 1.20 Civil engineers design and supervise the construction of such structures as this third power plant of the Grand Coulee Dam of Washington. (Courtesy of the Bureau of Reclamation.)

Fig. 1.21 Civil engineers designed this gravity railroad yard in Houston, Texas to aid in the switching of railroad cars. (Courtesy of Southern Pacific Railroad.)

for public utilities, railroads, educational institutions, steel industries, and other manufacturing industries.

The professional society for civil engineering is the American Society of Civil Engineers (ASCE), founded in 1852, the oldest engineering society in the United States.

Electrical Engineering

Electrical engineers are concerned with the utilization and distribution of electrical energy for the improvement of industrial and domestic functions. The two main divisions of electrical engineering are (1) power and (2) electronics. Power deals with the control of large amounts of energy used by cities and large industries, whereas electronics deals with small amounts of power used for communications and automated operations that have become integral parts of our everyday lives.

These two major divisions of electrical engineering have many areas of specialization. A few of these are given below.

Power generation poses many electrical engineering problems from the development of transmission equipment to the design of generators for producing electricity. Modern methods of power transmission and generation have made it possible for electrical power to be the most economical source of industrial energy (Fig. 1.22).

Power applications are numerous in typical homes where toasters, washers, dryers, vacuum cleaners, and lights are used on a continuing basis. Only about one-quarter of the total energy consumed is used in the home. Industry uses about half of all energy for metal refining, heating, motor drives, welding, machinery controls, chemical processes, plating, and electrolysis.

Fig. 1.22 The design of power transmission systems is the responsibility of the electrical engineer. A helicopter is used to patrol this 1800-mile long Parker-Davis project in Arizona. (Courtesy of the Bureau of Reclamation, U.S. Department of the Interior.)

Transportation industries require electrical engineers to develop electrical systems for automobiles, aircraft, and other forms of transportation. These systems are used for starting ignition, lighting, and instrumentation. Locomotives and ships may power their own generators, which supply electrical power to turn their driving wheels or propellers. The sophisticated signal systems necessary for all forms of transportation require the services of electrical engineers.

Illumination is required at all levels of human activities and environment. The improvement of illumination systems and the economy of illumination energy are challenging areas of study for the electrical engineer.

Computers have become a giant industry that is the domain of the electrical engineer. Computers used in conjunction with industrial electronics have changed manufacturing and production processes of industry, requiring fewer employees and resulting in greater precision. The trend in the production of electronic devices has been toward larger capacities and capabilities, while miniaturizing the circuits and components (Fig. 1.23).

Fig. 1.23 Electrical engineers have made components smaller, lighter, and more efficient. Microminiature components, printed wiring, and advanced encapsulation techniques all combine to reduce complete circuits to a fraction of their former size. (Courtesy of General Dynamics Corporation.)

Communications is the field devoted to the improvement of radio, telephone, telegraph, and television systems, which are the nerve centers of most industrial operations. Communications is vital in the dispatching of a taxicab, the control of ships and aircraft, and in the many other everyday personal and industrial applications.

Instrumentation is the study of systems of electronic instruments used in the precise measuring of industrial processes. Extensive use has been made of the cathode-ray tube and the electronic amplifier in industrial applications and atomic power reactors. Instrumentation has been increasingly applied to medicine for diagnosis and therapy.

Military electronics is utilized in practically all areas of military weapons and tactical systems from the walkie-talkie to the distant radar networks for detecting enemy aircraft. Remote-controlled electronic systems are used for navigation and interception of guided missiles. Many revolutionary advancements are expected to continue in fields of military applications of electrical engineering.

More electrical engineers are employed than any other type of engineer. Electrical engineers are employed in the United States by manufacturers of electrical and electronic equipment, aircraft and parts, business machines, and professional and scientific equipment. The increased need for electrical equipment, for automation, and computerized systems is expected to contribute to a very rapid growth of this field during the coming years.

The professional society for electrical engineers is the Institute of Electrical and Electronic Engineers (IEEE). This is the world's largest technical society, and it was founded in 1884.

Industrial Engineering

Industrial engineering, one of the newer branches of the engineering profession, is defined by the National Professional Society of Industrial Engineers as follows:

Industrial engineering is concerned with the design, improvement, and installation of integrated systems of men, materials, and equipment. It draws upon specialized knowledge and skill in the mathematical, physical, and social sciences together with the principles and methods of engineering analysis and design to specify, predict, and evaluate the results to be obtained from such systems.

Industrial engineering is related to all areas of engineering and business. This field differs from other areas of engineering in that it is more closely related to people and their performance and work-

Fig. 1.24 Industrial engineers lay out and design complex industrial facilities to provide efficient production. This plant produces and assembles automobiles. (Courtesy of Jervis B. Webb Company.)

ing conditions (Fig. 1.24). Consequently, in many instances the industrial engineer is a manager of people, machines, materials, methods, money, and the markets involved.

The industrial engineer may be assigned the responsibility of plant layout, the development of plant processes, or the determination of operating standards that will improve the efficiency of a plant operation. He or she may be responsible for quality control and cost analysis, two operations that are essential to a profitable manufacturing industry.

Several specific areas of industrial engineering are: management, plant design and engineering, electronic data processing, systems analysis and design, control of production and quality, performance standards and measurements, and research. In order for industrial engineers to work in these areas, they must act as members of a team composed of engineers from other branches of the profession. They must view the overall operations of industry and the factors affecting its efficiency rather than be concerned with isolated areas within the total structure.

People-oriented areas include the development of wage incentive systems, job evaluation, work measurement, and the design of environmental systems. Industrial engineers are often involved in management-labor agreements that affect the operation and production of an industry. They design and supervise systems for the improved safety and production of the working forces employed in an industry.

More than two-thirds of industrial engineers are employed in manufacturing industries. Others work for insurance companies, construction and mining firms, public utilities, large businesses, and governmental agencies. With the increasing complexities of industrial operations and the expansion of automated processes, the field of industrial engineering is expected to grow very rapidly.

The professional society of industrial engineers is the American Institute of Industrial Engineers (AIIE), which was organized in 1948.

Mechanical Engineering

The major areas of specialization of the mechanical engineer are: power generation, transportation, aeronautics, marine vessels, manufacturing, power services, and atomic energy. Activities within each of these areas are outlined below.

Power generation requires that prime movers be developed to power electrical generators that will produce electrical energy in stationary power plants. Mechanical engineers design and supervise the operation of steam engines, turbines, internal combustion engines, and other prime movers required in power generation (Fig. 1.25). The storage and handling of fuel used in these systems is also a mechanical engineering problem.

Fig. 1.25 This hydroelectric power plant at Hoover Dam was completed by mechanical engineers. (Courtesy of the Bureau of Reclamation.)

Transportation conveyances are designed and manufactured by mechanical engineers. Automobiles, trucks, buses, locomotives, marine vessels, and aircraft are designed and produced through the efforts of mechanical engineers who must design the power systems of transportation vehicles as well as the structural and fuel systems (Fig. 1.26).

Fig. 1.26 The mechanical engineer is responsible for the design and production of the automobile's power system, suspension system, and much of its entire design. (Courtesy of Chrysler Corporation.)

Aeronautics is a specialized field requiring the mechanical engineer to develop engines to power aircraft. The controls and environmental systems of aircraft are also problems solved by mechanical engineers. The fabrication of the aircraft requires a close coordination between mechanical and aerospace engineers.

Marine vessels that are powered by steam, diesel, or gas-generated engines are designed by mechanical engineers. They are also responsible for supplying power services throughout the vessel, including light, water, refrigeration, and ventilation.

Manufacturing is a challenging field that requires the mechanical engineer to design new products and new factories in which to build them. Mechanical engineers work closely with the industrial engineers in the management of a wide variety of machines. Economy of manufacturing and the achievement of a uniform level of product quality is a major function of this area of mechanical engineering.

Power services include the movement of liquid and gases through pipelines, refrigeration systems, elevators, and escalators. In applying the principles of mechanical engineering, the mechanical engineer must have a knowledge of pumps, ventilation equipment, fans, and compressors.

Atomic energy development has used mechanical engineers for the development and handling of protective equipment and materials. Nuclear reactors, which provide nuclear power for various applications, are constructed as a joint effort with the mechanical engineer playing an important role.

The professional society of the mechanical engineering field is the American Society of Mechanical Engineers (ASME).

Mining and Metallurgical Engineering

Mining and metallurgical engineers are often grouped together as a common profession although their functions are somewhat separate.

Mining engineers are responsible for the extraction of minerals from the earth and for the preparation of minerals for use by the manufacturing industries. They work with geologists to locate ore deposits, which must be exploited through the construction of extensive tunnels and underground operations, necessitating an understanding of safety, ventilation, water supply, and communications.

The metallurgical engineer develops methods of processing and converting metals into useful products. Two main areas of metallurgical engineering are extractive and physical. Extractive metallurgy is concerned with the extraction of metal from raw ores to form pure metals. Physical metallurgy is the development of new products and alloys.

Many metallurgical engineers work on development of machinery for electrical equipment and in the aircraft and aircraft parts industries. With the development of new materials for space flight vehicles, jet aircraft, missiles, and satellites, new lightweight, high-strength materials will increase the need for additional metallurgical engineers at a very rapid rate (Fig. 1.27). They will also be needed to develop economical methods for extracting metal from low-grade ore when the high-grade ore has been depleted.

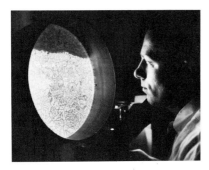

Fig. 1.27 A metallograph is used to show the structure of an alloy that may be used in the construction of a refinery unit. Materials are specially developed for specific applications by the metallurgic engineer. (Courtesy of Exxon Corporation.)

Mining engineers who work at mining sites are usually employed near small communities or in out-of-the-way places, while those in research and consulting are often located in metropolitan areas. The development of new alloys is expected to increase the need for mining engineers for the recovery of relatively little-used ores.

The professional society of this field of engineering is the American Institute of Mining, Metallurgical, and Petroleum Engineering (AIME).

Nuclear Engineering

The field of nuclear engineering has been confined to graduate studies until recent years, when undergraduate programs were developed to provide a more continuous curriculum. This branch of engineering promises to be a spectacular contributor to our future way of life.

The earliest work in the nuclear field has been for military applications and defense purposes. However, the utilization of nuclear power for domestic needs is being developed for a number of areas. Present applications of nuclear energy can be seen in the medical profession, and in various other fields as a power source.

Although nuclear engineering degrees are now offered at the bachelor's level, advanced degrees are recommended for this area of engineering. Advanced degrees can provide the additional specialization especially important in the field. Peaceful applications are divided into two major areas—radiation and nuclear power reactors. Radiation is the propagation of energy through matter or space in the form of waves. In atomic physics

this term has been extended to include fast-moving particles (alpha and beta rays, free neutrons, etc.), gamma rays, and x-rays. Nuclear science is closely associated with botany, chemistry, medicine, and biology.

The production of nuclear power in the form of mechanical or electrical power has become a major area of the peaceful utilization of nuclear energy (Fig. 1.28). For the production of electrical power, nuclear energy is used as the fuel for producing steam that will drive a turbine generator in the conventional manner. This source of power is expected to reduce the expenditure of our depleting supply of coal, oil, and gas presently being used in great quantities for power production.

Most training of the nuclear engineer is centered around the design, construction, and operation of nuclear reactors. Other areas include the processing of nuclear fuels, thermonuclear engineering, and the utilization of various nuclear by-products. The American Nuclear Society is the professional society of nuclear engineers.

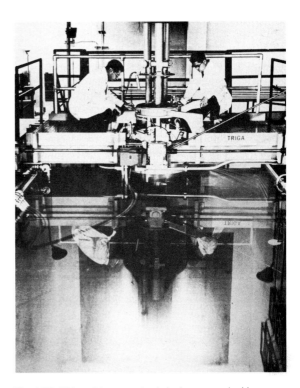

Fig. 1.28 This pulsing reactor is being operated by a nuclear engineer to test the effect of extremely high radiation on delicate equipment. (Courtesy of General Dynamics Corporation.)

Fig. 1.29 The petroleum engineer supervises the operation of offshore platforms used in the exploration for oil. (Courtesy of Marathon Oil Company.)

Petroleum Engineering

Petroleum engineering is the application of engineering to the development and recovery of petroleum resources. The primary concern of petroleum engineers is the recovery of petroleum and gases; however, they must also develop methods for the transportation and separation of various products. They are responsible for the improvement of drilling equipment and its economy of operation when drilling is being carried out at excessive depths, sometimes as great as four miles (Fig. 1.29).

Conservation of petroleum reservoirs has gained increased emphasis with greater consumption each year. New processes have been developed in recent years for recovering increasing amounts of petroleum from dormant oil reservoirs that had previously been abandoned (Fig. 1.30).

The production phase of petroleum engineering requires the cooperation of practically all branches of engineering, due to the wide variety of knowledge and skills demanded. Many new industries have emerged from new petroleum products that have been developed through production methods and research.

The petroleum engineer is assisted by the geologist in the exploration stages of searching for petroleum. Geologists use devices such as the airborne magnetometer, which gives readings on the earth's subsurfaces, indicating where uplifts that could hold oil or gas are located. If these findings are favorable, seismograph crews measure the depths of particular layers of the subsurface. The only sure way of determining whether oil or gas does exist after preliminary surveys have been made is to drill a well.

Oil well drilling is supervised by the petroleum engineer, who also develops the drilling equipment with members of other branches of engineering to offer the most efficient method of oil removal. When oil is found, the petroleum engineer must design piping systems to remove the oil and transport it to its next point of processing. The processing itself is a joint project in which chemical engineers work with petroleum engineers.

Fig. 1.30 Downtown Kilgore, Texas is a reminder of one of the world's richest oil strikes in 1930. Buildings were torn down to make room for more wells.

The future of petroleum engineering appears to be very promising, with the use of petroleum increasing yearly. The daily consumption of petroleum in the United States is 17 million barrels per day, and it is estimated that we will consume over 28 million barrels by the year 2000.

The Society of Petroleum Engineers is a branch of AIME, which includes mining and metallurgical engineers and geologists.

1.5 TECHNOLOGISTS AND TECHNICIANS

As technology has expanded to include more disciplines, materials, and systems, it has become more impractical for an individual to understand and perform all the steps required in the implementation of a design. Instead, it is more efficient to utilize trained specialists who are responsible for specific areas of technology.

With these changes in technology have emerged two new members of the technological team: the technologist and the technician. Both have as their primary mission the assistance of the engineer at a technical level below that of the engineer and above that of the craftsman.

The Technologist

The technologist works under the supervision of an engineer, but is called upon to exercise considerable responsibility to free the engineer for more advanced applications. Most technologists have a four-year college background in a specialty area of engineering technology that enables them to perform semiprofessional jobs with a high degree of skill. They are trained in the basic engineering disciplines, which prepares them to use advanced equipment from computers to testing equipment (Fig. 1.31).

Their interest in the practical aspects of engineering qualifies them to offer valuable advice to the engineer concerning production specifications and on-the-site procedures when new projects are being designed. They may work as designers responsible for converting the preliminary work of the engineer into finished designs that can be efficiently produced.

The Technician

The separation between the technologist and the technician is not always clearly defined, and often these titles are mistakenly interchanged. The technician is a two-year graduate of a technical program who performs tasks that are less technical than those performed by the technologist. Technicians may be repairers, inspectors, production specialists, surveyors (Fig. 1.32), or have similar specialties. They may work under the supervision of an engineer, but ideally they are supervised by a technologist.

Fig. 1.31 This technologist is gathering test data that might detect structural deficiencies in a newly designed tractor. (Courtesy of Ford Motor Company.)

Fig. 1.32 These technicians are making a cartographic survey. Data from the tellurometer are being recorded in a notebook. (Courtesy of the U.S. Forest Service.)

In turn, the technician will coordinate the activities between the technologist and the craftsman. These levels of responsibility make it possible to ensure that the proper skills and qualifications are available throughout the various levels of the chain of command, thereby creating a more efficient and productive use of personnel.

General

Some well-established areas of technology are aerospace technology, chemical technology, civil engineering technology, electronic technology, and design drafting technology. Approximately 700,000 engineering and science technologists are employed in all industries (Fig. 1.33). Twelve percent of these are women. Almost 475,000 technicians are employed in private industry.

The technologist/technician field has been one of the fastest growing occupational groups during recent decades, and a continued increase is expected with the expansion of industry.

Additional information concerning technicians can be obtained from the American Society of Certified Engineering Technicians.

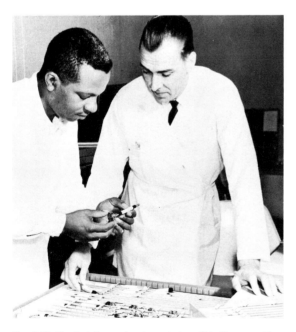

Fig. 1.33 Technicians check parts to within thousandths of an inch to ensure that the part conforms to the specifications. (Courtesy of DoAll Corporation.)

1.6 DRAFTERS

Whereas the drafters of the past spent much of their time in the preparation of ink drawings from which prints could be made, the drafters of today carry a greater responsibility in assisting the engineer and designer. The experienced drafter may be involved in the preparation of complex drawings, selecting materials, detailing designs, and writing specifications.

Design and construction drawings are made by drafters to explain how to fabricate, build, or erect a project or product in fields such as areospace engineering, architecture, machine design, mechanical engineering, and electrical/electronic engineering. Drawings must be prepared with sufficient details and specifications to permit them to be executed by the supervisors of construction and manufacturing with the minimum of supplementary information.

Technical illustration is a type of graphics that is usually prepared as three-dimensional pictorials to illustrate a project or product as realistically as possible. Technical illustration is the most artistic area of engineering design graphics.

Maps, geological sections, and highway plats are used for locating property lines, physical features, strata, right-of-way, building sites, bridges, dams, mines, utility lines, and so forth. Drawings of this type are usually prepared as permanent ink drawings.

Career opportunities in the area of engineering design graphics are based upon education, experience, and competence. The three levels of certification for drafters are: drafters, design drafters, and engineering designers.

Drafters are graduates of a two-year post–high-school curriculum in the area of engineering design graphics.

Design drafters complete two-year programs beyond the high school level in an approved junior college or technical institute. Drafters with this background are technicians.

Engineering designers are drafters who have completed a four-year course at a college offering specialty training in the area of engineering design graphics, which leads to a degree in this area. Graduates of these programs can become certified as technologists.

Professional drafters have opportunities for advancement to positions with titles such as senior drafter, technical illustrator, chief drafter, designer, drafting supervisor, and manager of drafting. With each advancement, the drafter will be given more responsibility of a technical and supervisory nature.

Approximately 485,000 drafters and designers will be employed by 1985, and the need for them is expected to increase in the future. Computerized systems are being adopted by industry to improve the drafter's productivity. Computer graphics systems do not lessen the need for the knowledge of the fundamental graphical principles by the drafter or engineer, but computer graphics offers a different medium of expression. (Fig. 1.34).

The professional society of this field is the American Institute for Design and Drafting (AIDD). Student chapters of AIDD are located on most campuses where programs in engineering design graphics are offered.

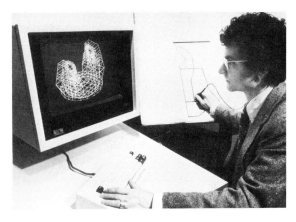

Fig. 1.34 This drafter has constructed a 3-D finite element model of a gun mount on the Applicon display by digitizing a multiview engineering drawing. (Courtesy of Applicon.)

PROBLEMS

1. Write a report of not more than ten typewritten pages that outlines the specific duties and relationships between the scientist, engineer, craftsman, designer, and stylist in an engineering field of your choice. For example, explain this relationship for an engineering team involved in an aspect of civil engineering. Your report should be supported by factual information obtained from interviews, brochures, or library references.

2. Write a report that investigates the employment opportunities, job requirements, professional challenges, and activities of your chosen branch of engineering or technology. Illustrate this report with charts and graphs where possible for easy interpretation. Compare your personal abilities and interests with those required by the profession.

3. Arrange a personal interview with a practicing engineer, technologist, or technician in the field of your interest. Discuss with him or her the general duties and responsibilities of the position to gain a better understanding of this field. Summarize your interview in a written report.

4. Write to the professional society of your field of study for information concerning this area. Prepare a notebook of these materials for easy reference. Include in the notebook a list of books that would provide career information for the engineering student.

ADDRESSES OF PROFESSIONAL SOCIETIES

Publications and information from these societies were used as the basis for the content of this chapter.

Alliance for Engineering in Medicine and Biology
3900 Wisconsin Avenue, N.W., Suite 300, Washington, D.C. 20016

American Ceramic Society
65 Ceramic Drive, Columbus, Ohio 43214

The American Institute of Aeronautics and Astronautics
1290 Avenue of the Americas, New York, N.Y. 10019

American Institute of Chemical Engineers
345 East 47th Street, New York, N.Y. 10017

American Institute for Design and Drafting
3119 Price Road, Bartlesville, Okla. 74003

The American Institute of Industrial Engineers
345 East 47th Street, New York, N.Y. 10017

American Institute of Mining, Metallurgical, and Petroleum Engineering
345 East 47th Street, New York, N.Y. 10017

American Nuclear Society
244A East Ogden Avenue, Hinsdale, Ill. 60521

American Society of Agricultural Engineers
2950 Niles Road, St. Joseph, Mich. 49085

American Society of Civil Engineers
345 East 47th Street, New York, N.Y. 10017

American Society for Engineering Education, Technical Institute Division
11 DuPont Circle, Suite 200, Washington, D.C. 20036

American Society of Mechanical Engineers
345 East 47th Street, New York, N.Y. 10017

The Institute of Electrical and Electronic Engineers
345 East 47th Street, New York, N.Y. 10017

Society of Petroleum Engineers (AIME)
6300 North Central Expressway, Dallas, Tex. 75206

Engineer's Council for Professional Development
345 East 47th Street, New York, N.Y. 10017

National Society of Professional Engineers
2029 K Street, N.W., Washington, D.C. 20006

Society of Women Engineers
United Engineering Center, Room 305,
345 East 47th Street, New York, N.Y. 10017

THE DESIGN PROCESS

2.1 INTRODUCTION

A knowledge of scientific and engineering principles is of little value if these disciplines cannot be harnessed to obtain a tangible end that will fulfill the needs of a given situation. For engineers to function to their fullest extent, they must exercise imagination combined with knowledge and curiosity.

Engineering graphics and descriptive geometry provide methods of solving technical problems, along with other engineering courses. An engineer who is developing a design solution must make many sketches and drawings to develop preliminary ideas before communication with associates is possible. Used in this manner, graphical methods are creative tools.

This chapter will introduce and define the basic steps of the design process.

2.2 TYPES OF DESIGN PROBLEMS

Design problems are numerous and take many forms; however, most fall in one of two categories: **systems design** and **product design.** These two types of design problems will be referred to throughout this textbook. The distinct separation of these types of problems is often difficult, due to an overlap of certain characteristics. We will try, however, to define them for our purposes.

Systems Design

Systems design deals with the arrangement of available products and components into a unique system that yields the desired result. A residential building is a complex system made up of components and products. For example, the typical residence has a heating-cooling system, a utility system, a plumbing system, a gas system, an electrical system, and many other systems that form the overall composite system (Fig. 2.1). These component systems are also called systems because they are composed of a number of individual parts that can be used for other applications. The electrical system involves wiring, insulation, electrical components,

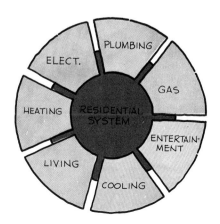

Fig. 2.1 The typical residence is a system composed of many component systems.

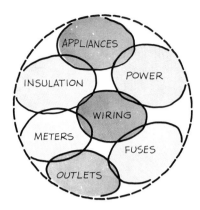

Fig. 2.2 An electrical system of a residence is a composite of related components.

light bulbs, meters, controls, switches, and related items (Fig. 2.2).

An engineering project that requires that a traffic system be developed for a particular need will overlap into other disciplines (Fig. 2.3). The engineering function will be the primary area that will support the project; however, the project will also involve legal problems, economical principles, historical data, human factors, social considerations, scientific principles, and political limitations. The engineer can, of course, design a suitable driving surface, drainage system, overpasses, and other components of the traffic system by application of engineering principles without consideration for

other areas and the limitations imposed by them. However, engineers must adhere to a specific budget in essentially all projects in which they are involved, and the budget is closely related to legal and political problems. Traffic laws, zoning ordinances, right-of-way possession, and liability clearances are other legal areas that must be reconsidered.

Planning for the future is based on past needs and trends, which introduces historical data as a design consideration. Human factors involve driver characteristics, safety features, and other factors that would affect the function of the traffic system. Social problems are associated with traffic systems. Heavily traveled highways will attract commercial establishments, shopping centers, and filling stations, which will affect the adjacent property. Scientific principles developed through laboratory research can be applied by the engineer in building more durable roads, cheaper bridges, and a more functional system. Pressures from special interest groups may conflict with interests of other groups and limit the engineer's freedom of approach.

Example Systems Problem   The following problem is given as an example of a rather simple systems design that can be used to illustrate the various steps of the design process. This particular problem requires a minimum of theoretical engineering principles.

PARKING AREA: Select a building on your campus that is in need of an improved parking lot to accommodate the people who are housed in the building. This may be a dormitory, office building, or classroom building. Design a combination traffic and parking system that will be adequate for the requirements of the building. The solution of this problem must adhere to existing limitations, regulations, and policies of your campus in order for the problem to be as realistic as possible.

Product Design

Product design is concerned with the design, testing, manufacture, and sale of an item that will be mass produced and will perform a specific function. Such a product can be an appliance, a tool, a system component, a toy, or a similar item that can be purchased as a commercial unit. Because of its more limited function, product development is considerably more specific than systems design (Fig. 2.4).

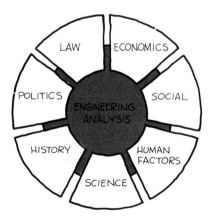

Fig. 2.3 An engineering system may be a complex interaction of many professions, with the engineering function holding the primary area of emphasis. An example of a system of this type is a traffic design problem.

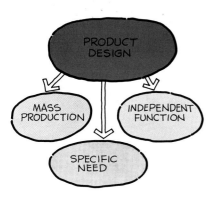

Fig. 2.4 Product design is more limited in scope than systems design and is mass produced.

The distinction between a system and a product is not always clearly apparent. An automotive system has a primary function of providing transportation. However, the automobile must also furnish its passengers with communications, illumination, comfort, and safety, and these would classify it as a system. Nevertheless, the automobile is thought of as a product, since it is mass-produced for a large consumer market. On the other hand, a petroleum refinery is definitely a system, composed of many interrelated functions and components. All refineries will have certain processes in common, but no two can be considered alike in all respects.

Product design is related to the current market needs, cost of production, function, sales, method of distribution, and profit predictions (Fig. 2.5). This concept may be broadened to encompass an entire

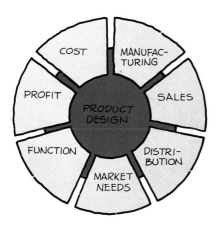

Fig. 2.5 Areas associated with product design are related to the manufacture and sale of completed products.

system that will have sweeping changes of a social and economic nature. An example of this transition from a product to a system is the automobile, whose function has had a significant effect on our way of life. This product has expanded to a system of highways, service stations, repair shops, parking lots, drive-in businesses, residential garages, traffic enforcement, and endless other related components.

Example Product Design Problem   The product problem given below is an example of the type of problem that may be assigned as a class project.

HUNTING SEAT: Many hunters, especially deer hunters, hunt from trees to obtain a vantage point. Sitting in a tree for several hours can be uncomfortable and hazardous to the hunter, which indicates a need for a hunting seat that could be used to improve this situation. Design a seat that would provide the hunter with comfort and safety while hunting from a tree, and meet the general requirements of economy and hunting limitations.

2.3   THE DESIGN PROCESS

The act of devising an original solution to a problem by a combination of principles, resources, and products is design. Design is the most distinguishing responsibility that separates the engineer from the scientist and the technician. The engineer's solutions may involve a combination of existing components in a different arrangement to provide a more efficient result, or they may involve the development of an entirely new product; but in either case the work is referred to as the act of designing.

The design process is the pattern of activities that is followed by the designer in arriving at the solution of a technological problem. This book emphasizes a six-step design process that is a composite of the most commonly employed steps in solving problems. The six steps are (1) problem identification, (2) preliminary ideas, (3) problem refinement, (4) analysis, (5) decision, and (6) implementation (Fig. 2.6). Although designers work sequentially from step to step, they may recycle to previous steps as they progress.

Engineering graphics and descriptive geometry have been integrated into these steps to stress their role in the creative process of designing. These areas are probably more essential to the design process than any other single field of study.

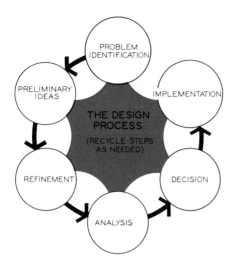

Fig. 2.6 The steps of the design process. Each step can be recycled when needed.

Problem Identification

Most engineering problems are not clearly defined at the outset; consequently they must be identified before an attempt is made to solve the problem (Fig. 2.7). For example, a prominent concern today is air pollution. Before this problem can be solved, you must identify what air pollution is and what causes it. Is pollution caused by automobiles, factories, atmospheric conditions that harbor impurities, or geographic features that contain impure atmospheres?

When you enter a bad street intersection where traffic is unusually congested, do you identify the reasons for it being congested? Are there too many cars, are the signals poorly synchronized, or are there visual obstructions resulting in congested traffic?

Problem identification requires considerable study beyond a simple problem statement like "solve air pollution." You will need to gather data of several types: field data, opinion surveys, historical records, personal observations, experimental data, and physical measurements and characteristics (Fig. 2.7).

Preliminary Ideas

Once the problem has been identified, the next step is to accumulate as many ideas for solution as possible (Fig. 2.8). Preliminary ideas should be sufficiently broad to allow for unique solutions that could revolutionize present methods. All ideas should be recorded in written form. Many rough sketches of preliminary ideas should be made and retained as a means of generating original ideas and stimulating the design process. Ideas and comments should be noted on the sketches as a basis for further preliminary designs.

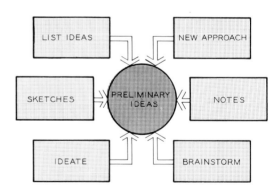

Fig. 2.8 Preliminary ideas are developed after the identification process has been completed. All possibilities should be listed and sketched to give the designer a broad selection of ideas from which to work.

Problem Refinement

Several of the better preliminary ideas are selected for further refinement to determine their true merits. Rough sketches are converted to scale drawings that will permit space analysis, critical measurements, and the calculation of areas and volumes affecting the design (Fig. 2.9). Consideration is given to spatial relationships, angles between planes,

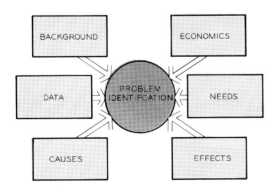

Fig. 2.7 Problem identification requires the accumulation of as much information as possible before a solution is attempted by the designer.

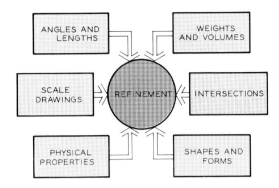

Fig. 2.9 Refinement begins with the construction of scale drawings of the better preliminary ideas. Descriptive geometry and graphical methods are used to find the necessary geometric characteristics.

lengths of structural members, intersections of surfaces and planes.

Descriptive geometry is a very valuable tool for determining information of this type, and it precludes the necessity for tedious mathematical and analytical methods.

An example of a problem of this nature is illustrated in the landing gear of the lunar vehicle shown in Fig. 2.10. It was necessary for the designer to make many freehand sketches of the design and finally a scale drawing to establish clearances with the landing surface. The configuration of the landing gear was drawn to scale in the descriptive views of the landing craft. It was necessary, at this point,

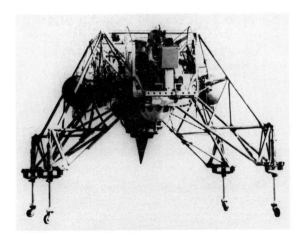

Fig. 2.10 The refinement of the lunar vehicle required the use of descriptive geometry and other graphical methods. (Courtesy of Ryan Aeronautics, Inc.)

to determine certain fundamental lengths, angles, and specifications that are related to the fabrication of the gear. The length of each leg of the landing apparatus and the angles between the members at the point of junction had to be found to design a connector, and the angles the legs made with the body of the spacecraft had to be known in order to design these joints. All of this information was easily and quickly determined with the use of descriptive geometry. The employment of descriptive geometry as a preliminary means of determining this information facilitates the application of analytical principles to convert this information into equations for mathematical solutions.

Analysis

Analysis is the step of the design process where engineering and scientific principles are used most (Fig. 2.11). Analysis involves the evaluation of the best designs to determine the comparative merits of each with respect to cost, strength, function, and market appeal. Graphical principles can also be applied to analysis to a considerable extent. The determination of forces is somewhat simpler with graphical vectors than with the analytical method.

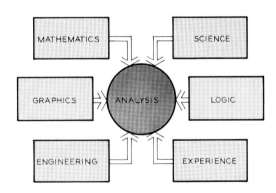

Fig. 2.11 The analysis phase of the design process is the application of all available technological methods from science to graphics in evaluating the refined designs.

Functional relationships between moving parts will also provide data that can be obtained graphically more easily than by analytical methods.

Graphical solutions to analytical problems offer a readily available means of checking the solution, therefore reducing checking time. Graphical methods can also be applied to the conversion of functions of mechanisms to a graphical format that will permit the designer to convert this action into an equation form that will be easy to utilize. Data that would otherwise be difficult to interpret by mathematical means can be gathered and graphically analyzed. For instance, empirical curves that do not fit a normal equation are often integrated graphically when the mathematical process would involve unwieldy equations.

Models constructed at reduced scales are valuable to the analysis of a design to establish relationships of moving parts and outward appearances, and to evaluate other design characteristics. Full-scale prototypes are often constructed after the scale models have been studied for function.

Decision

A decision must be made at this stage to select a single design that will be accepted as the solution of the design problem (Fig. 2.12). Each of the several designs that have been refined and analyzed will offer unique features, and it will probably not be possible to include all of these in a single final solution. In many cases, the final design is a compromise that offers as many of the best features as possible.

The decision may be made by the designer on an independent, unassisted basis, or it may be made by a group of associates. Regardless of the size of the group making the decision as to which design will be accepted, graphics is a primary means of presenting the proposed designs for a decision. The outstanding aspects of each design usually lend themselves to presentation in the form of graphs that compare costs of manufacturing, weights, operational characteristics, and other data that would be considered in arriving at the final decision.

Implementation

The final design concept must be presented in a workable form. This type of presentation refers primarily to the working drawings and specifications that are used as the actual instruments for fabrication of a product, whether it is a small piece of hardware or a bridge (Fig. 2.13). Engineering graphics fundamentals must be used to convert all preliminary designs and data into the language of the manufacturer, who will be responsible for the conversion of the ideas into a reality. Workers must have complete detailed instructions for the manufacture of each single part, measured to a thousandth of an inch to facilitate its proper manufacture. Working drawings must be sufficiently detailed and explicit to provide a legal basis for a contract that will be the document for the contractor's bid on the job.

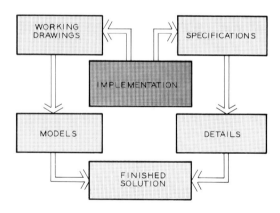

Fig. 2.13 Implementation is the final step of the design process, where drawings and specifications are prepared from which the final product can be constructed.

Designers and engineers must be sufficiently knowledgeable in graphical presentation to be able to supervise the preparation of working drawings even though they may not be involved in the mechanics of producing them. They must approve all plans and specifications prior to their release for production.

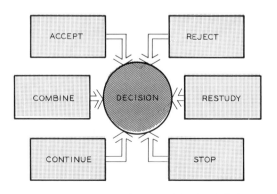

Fig. 2.12 Decision is the selection of the best design or design features to be implemented.

2.4 APPLICATION OF THE DESIGN PROCESS TO A SIMPLE PROBLEM

In order to illustrate the steps of the design process as they would be applied to a simple design problem, the following example is given.

Swing-Set Anchor Problem

A child's swing set has been found unstable during the peak of the swing. The momentum of the swing causes the A-frame to tilt with a possibility of overturning and causing injury. The swing set has swings attached to accommodate three children at a time. Design a device that will eliminate this hazard and have market appeal for owners of swing sets of this type.

Problem Identification   As a first step, the designer writes down the problem statement (Fig. 2.14) and a statement of need. This overt action stimulates flow of thought and assists the designer in attacking the problem systematically. The limitations and desirable features are listed, along with necessary sketches, to enable the designer to develop a better understanding of the problem requirements. Much of the information and notes in the problem identification step may be obvious to the designer, but the act of writing statements about the problem and making freehand sketches helps to get off "dead center," which is a common weakness at the beginning of the creative process.

Preliminary Ideas   A second work sheet is used to sketch preliminary solutions to the problem (Fig. 2.15). This is the most creative part of the process and has the fewest restraints. The designer makes rough sketches and notes to describe preliminary thoughts, without dwelling upon a single design. After ideas have been sketched, the designer goes over the sketches and makes additional notes to indicate the better points of each design, narrowing down the ideas to those with the most merit.

Problem Refinement   The best designs—two or more unless one is highly superior to all others—are drawn to scale as orthographic drawings, as a means of refining preliminary designs. Sufficient notes are used to describe the designs without becoming too involved with details (Fig. 2.16). The refinement provides the physical properties and overall dimensions that must be considered during the earlier stages of the design process.

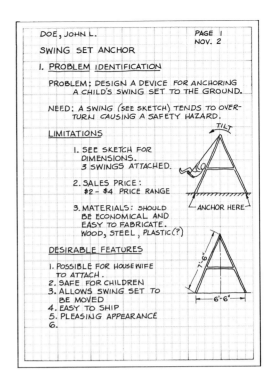

Fig. 2.14 Problem identification work sheet.

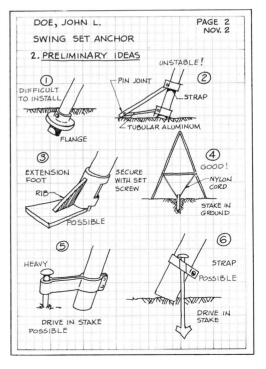

Fig. 2.15 Preliminary ideas work sheet.

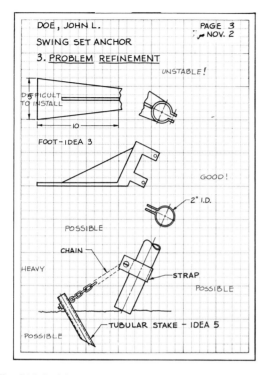

Fig. 2.16 Problem refinement work sheet.

Orthographic projection, working drawing principles, and descriptive geometry may be used, depending upon the particular problem being refined. In this example (Fig. 2.16), simple orthographic views with auxiliary views depict the two designs. These drawings, even though refined, are still subject to change throughout the entire process.

Analysis   Once a preliminary design has been refined to establish fundamental dimensions and relationships, the designs must be analyzed to determine their suitability and other criteria. The maximum angles of swing must be established by observations of a child swinging under average conditions. The force F at the critical angle can be calculated mathematically or estimated by observation (Fig. 2.17). Since three swings may be in use at once, the maximum condition will exist when all three swings are in phase, resulting in a triple pull, or 150 lb in this example.

The danger zones are graphically indicated in the space diagram to show the effects of the foot design and to establish the dimension that it must have in order for it to eliminate the tilting tendency

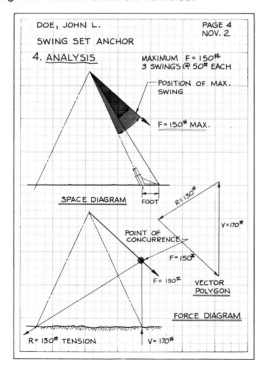

Fig. 2.17 Analysis work sheet.

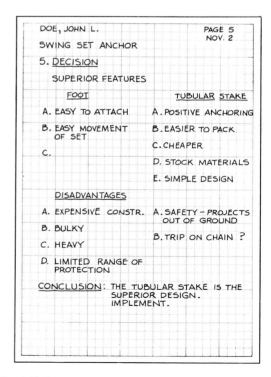

Fig. 2.18 Decision work sheet.

at the maximum angle. The force diagram is drawn to scale to analyze the reaction forces at the extreme condition. A vector polygon is drawn with the vectors parallel to the forces in the force diagram, where the only known force is $F = 150$ lb. The magnitude of resultant R is found to be 130 lb, which is the maximum force that must be overcome at the base of the swing set.

Decision   The designs must be evaluated and the better of the two selected for implementation. In this example (Fig. 2.18), the superior features of each design are listed for easy comparison. The disadvantages of each are listed also to prevent any design weakness from being overlooked. These tabulated lists are reviewed and a final conclusion is reached. A decision is made to implement the tubular stake design.

Implementation   The tubular stake design is presented in the form of a working drawing, in which each individual part is detailed and dimensioned, and from which the parts could be made. All principles of graphical presentation are used, including a freehand sketch illustrating how the parts will be assembled (Fig. 2.19). Note that changes have been made since the initial refinement of this design. These changes were believed to be more operational and economical while serving the desired function.

Standard parts, such as nuts, bolts, and the chain, need not be drawn, but merely noted, since they are parts that will not be specially fabricated. With this drawing, the designer has implemented the design as far as can be done without actually building a prototype, model, or the actual part.

The following chapters will elaborate on each of the steps of the design process in greater detail. The example work sheets shown in Figs. 2.14 through 2.19 illustrate the typical approach to a simple design problem. As simple as this problem is, it would be virtually impossible for it to be designed without the utilization of graphical methods.

A photograph of the actual part is shown in Fig. 2.20 as it would be available for the market. It is shown attached to the swing set in Fig. 2.21 with the stake driven into the ground. No mention has been made of a market analysis or an evaluation of the commercial prospects of the item. This would be an ultimate requirement for the implementation of any device that is produced for consumption by the general market.

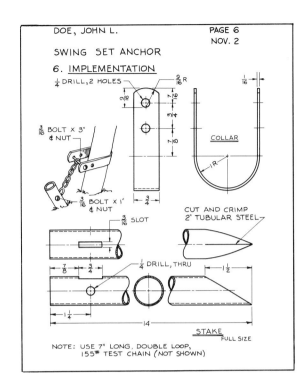

Fig. 2.19 Implementation drawing.

Fig. 2.20 The completed swing-set anchor.

Fig. 2.21 The anchor attached to a swing set.

PROBLEMS

The problems at the end of each chapter are provided to afford students an opportunity to test their understanding of the principles covered in the preceding text. Most problems deal with an understanding of the theoretical concepts rather than with specific applications.

Most problems are to be solved on 8½" × 11" paper, using instruments or drawing freehand as specified. The paper can be printed with a ¼-in. grid to assist in laying out the problems, or plain paper can be used with the layout made with a 16-scale (architect's scale). The grid of the given problems in later chapters represents ¼-in. intervals that can be counted and transferred to a like grid paper or scaled on plain paper.

Each problem sheet should be endorsed. The endorsement should include the student's seat number, and name, the date, and the problem number. Guidelines should be drawn with a straightedge to aid in lettering, using ⅛-in. letters. All points, lines, and planes should be lettered using ⅛-in. letters with guidelines in all cases. Reference planes should be noted appropriately when applicable.

Problems of an essay type should have their answers lettered, using approved, single-stroke, Gothic lettering, as introduced in Chapter 12. Guidelines should be used to assist in alignment and uniformity of lettering. Each page should be numbered and stapled in the upper left corner if turned in for review by the instructor.

You should keep all solved problems for future reference during this (and other) courses.

1. List engineering achievements that have demonstrated a high degree of creativity in the following areas: (a) the household, (b) transportation, (c) recreation, (d) educational facilities, (e) construction, (f) agriculture, (g) power, (h) manufacture.

2. Make an outline of your plan of activities for the weekend. Indicate areas in your plans that you feel display a degree of creativity or imagination. Explain why.

3. Write a short report on the engineering achievement or the person who you feel has exhibited the highest degree of creativity. Justify your selection by outlining the creative aspects of your choice. Your report should not exceed three typewritten pages.

4. Test your creativity in recognizing needs for new designs. List as many improvements for the typical automobile as possible. Make suggestions for implementing these improvements. Follow this same procedure in another area of your choice.

5. List as many systems as possible that affect your daily life. Separate several of these systems into component parts or subsystems.

6. Subdivide the following systems into components: (a) a classroom, (b) a wrist watch, (c) a movie theater, (d) an electric motor, (e) a coffee percolator, (f) a golf course, (g) a service station, (h) a bridge.

7. Indicate which of the items in Problem 6 are systems and which are products. Explain your answers.

8. Make a list of new products that have been introduced within the last five years with which you are familiar.

9. Make a list of products and systems that you would anticipate for life on the moon.

10. Assume that you have been assigned the responsibility for organizing and designing a go-kart installation on your campus. This must be a self-supporting enterprise. Write a paragraph on each of the six steps of the design process to explain how the steps would be applied to the problem. For example, what action would you take to identify the problem?

11. You are responsible for designing a motorized wheelbarrow to be marketed for home use. Write a paragraph on each of the six steps of the design process to explain how the steps would be applied to the problem. For example, what action would you take to identify the problem?

12. List and explain a sequence of steps that you feel would be adequate for the design process, yet different from the six given in this chapter. Your version of the design process may contain as many of the steps discussed here as you desire.

13. Can you design a device for holding a fishing pole in a fishing position while you are fishing in a rowboat? This could be a simple device that will allow you freedom while performing other chores in the boat. Make notes and sketches to describe your design.

14. Assume that you are responsible for designing a car jack that would be more serviceable than present models. Review the six steps of the

design process given in Section 2.3 and make a brief outline of what you would do to apply these steps to your attempt to design a jack. Write the sequential steps and the methods that would be used to carry out each step. List the subject areas that would be used for each step and indicate the more difficult problems that you would anticipate at each step.

15. As an introductory problem to the steps of the design process, design a door stop that could be used to prevent a door from slamming into a wall. This stop could be attached to the floor or the door and should be as simple as possible. Make sketches and notes as necessary to give tangible evidence that you have proceeded through the six steps, and label each step. Your work should be entirely freehand and rapid. Do not spend longer than 30 minutes on this problem. Indicate any information you would need in a final design approach that may not be accessible to you now.

16. List areas that you must consider during the problem identification phase of a design project for the following products: a new skillet design, a lock for a bicycle, a handle for a piece of luggage, an escape from prison, a child's toy, a stadium seat, a desk lamp, an improved umbrella, a hotdog stand.

17. Make a series of rough, freehand sketches to indicate your preliminary ideas for the solution of the following problems: a functional powdered soap dispenser for washing hands, a protector for a football player with an injured elbow, a method of positioning the cross-bar at a pole vault pit, a portable seat for waiting in long lines, a method of protecting windshields of parked cars during freezing weather, a pet-proof garbage can, a bicycle rack, a door knob, a seat to support a small child in a bathtub.

18. Evaluate the sketches made in Problem 4 above and briefly outline in narrative form the information that would be needed to refine your design into a workable form. Use freehand lettering; strive for a neat, readable paper.

19. Many automobiles are available on the market. Explain your decision for selecting the one that would be most appropriate for the activities listed below: a trip on a sightseeing tour in the mountains, a hunting trip in a wooded area for several days, a trip from coast to coast, the delivering of groceries, a business trip downtown. List the type of vehicle, model, its features and why you made your decision to select it.

PROBLEM IDENTIFICATION

3.1 INTRODUCTION

Problem identification is the initial step that a designer takes to solve a problem. Problem identification can be one of two general types: (1) identification of a need, or (2) identification of design criteria (Fig. 3.1).

Identification of a need is the beginning point of the process. A problem, a defect, or a shortcoming in an existing product or situation is recognized, which prompts the designer to investigate the causes and effects that may result in the development of a new product or solution. The need may

Fig. 3.2 Before an automobile is designed, the needs of the market must be identified. (Courtesy of Chrysler-Plymouth Corporation.)

be for an improved automobile safety belt, a solution to air pollution, a special hunting seat, or for a new automobile (Fig. 3.2).

Identification of a need for a new design at this stage is not so thorough as to establish the criteria that must be met in solving the problem. Instead, the designer merely recognizes a need for a better solution.

Identification of design criteria is that part of the problem where the designer conducts an in-depth investigation of the specifications that must be met by a new design. All aspects of the problem's background are reviewed in order to understand clearly the characteristics of a problem.

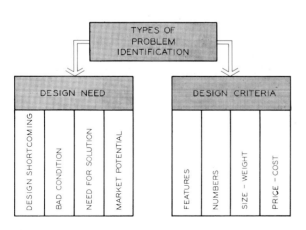

Fig. 3.1 The two basic types of problem identification.

3.2 DESIGN WORKSHEETS

Throughout the design process, designers must make numerous notes and sketches as they search for the solution to the problem. It is essential to their progress that they keep design work—sketches, data, and notes—so they will have a permanent record of progress. Periodically, they must review earlier ideas and notes to avoid overlooking important concepts that were previously identified. From a legal standpoint, a written record of a designer's work on a project will help establish ownership to patentable ideas.

Materials   The following materials will aid designers in maintaining permanent records of their design activities.

1. Sketchpad (8½″ × 11″). A sketchpad can be either grid-lined or plain, depending upon the preference of the individual. Sheets should be punched for insertion in a notebook or file.

2. Pencils. A medium-grade pencil, such as an F grade, is adequate when used with most papers. A colored pencil can be used to emphasize special features and ideas.

3. Binder or envelope. All work sheets should be kept in a binder or envelope for reference.

Format   Each sheet should have the following information (Fig. 3.3):

1. Name of project.
2. Name of designer.
3. Date.
4. Page number of each sheet.

All sketches and notes should be presented in an understandable form. Although precise lettering is not required, make every attempt to develop a fast and readable method of recording your notes.

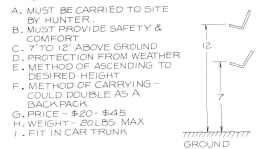

DESIGN

PROBLEM IDENTIFICATION Ch. 3

1. Project title
   SEAT FOR HUNTING FROM A TREE

2. Problem statement
   MANY HUNTERS SIT ON TREE LIMBS WHILE HUNTING WHICH IS UNSAFE AN UNCOMFORTABLE. A TREE'S ELEVATION PROVIDES AN EXCELLENT VANTAGE POINT FOR HUNTING. DESIGN A HUNTING SEAT TO FILL THIS NEED.

3. Requirements and limitations
   A. MUST BE CARRIED TO SITE BY HUNTER.
   B. MUST PROVIDE SAFETY & COMFORT
   C. 7′ TO 12′ ABOVE GROUND
   D. PROTECTION FROM WEATHER
   E. METHOD OF ASCENDING TO DESIRED HEIGHT
   F. METHOD OF CARRYING – COULD DOUBLE AS A BACKPACK
   G. PRICE – $20 - $45
   H. WEIGHT – 20 LBS MAX
   I. FIT IN CAR TRUNK

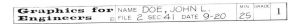

Graphics for Engineers	NAME DOE, JOHN L.		MIN. 25	GRADE	1
	FILE 2 SEC 41 DATE 9-20				

Fig. 3.3 The format of a work sheet for recording the problem identification information concerning a design project.

3.3 THE PROBLEM IDENTIFICATION PROCESS

Problem identification requires the designer to analyze the requirements, limitations, and other background information without becoming involved with the solution to the problem. Guard against trying to solve the problem at this stage; this could result in lost time due to a misunderstanding of the specifications.

The following steps should be used in problem identification (Fig. 3.4).

1. **Problem statement.** Write down the problem statement to begin the thinking process. The statement should be complete and comprehensive, but concise.

2. **Problem requirements.** List the positive requirements that must be achieved in the design. Many of these statements may be questions that will be listed and answered later when data have been gathered.

3. **Problem limitations.** List negative factors that confine the problem to specified limitations. Example: (a) cannot weigh more than 25 lb, (b) must fit in the trunk of a car, etc.

4. **Sketches.** Make sketches of physical characteristics of the problem. Add notes and di-

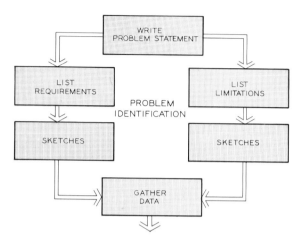

Fig. 3.4 The steps suggested for identifying a problem.

Fig. 3.5 The design of a new automobile is used as an example problem to demonstrate the problem identification step of the design process. (Courtesy of Ford Motor Company.)

mensions that would make these sketches more understandable.

5. **Gather data.** Comprehensive designs may require the study of background data. This data might be population trends, of related designs, physical characteristics, sales records, and market studies. Once collected, this data should be graphed for easy interpretation.

The designer should continually review the identification of the problem throughout the design process. As more is learned about the problem, the new information should be added to the notes.

3.4 AUTOMOBILE DESIGN— PROBLEM IDENTIFICATION

The automobile is a product that we all identify with. It is a necessity, a status symbol, and hobby for many owners.

Suppose that you were involved in the development of a new model automobile that would have a broad market appeal (Fig. 3.5). It would be necessary that you consider the prevailing climate of the marketplace: the need for fuel economy, rigorous emission controls, and expensive safety standards coupled with everchanging lifestyles. Also, you would like to know as much about the consumer as possible as part of problem identification.

The data graphed in Fig. 3.6 were gathered in a national study concerning the backgrounds of

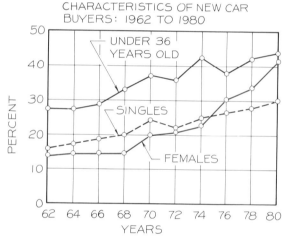

Fig. 3.6 The data plotted here were found by a survey to identify the changing characteristics of new-car buyers since 1962. Data are easier to interpret when graphed.

new-car buyers. It can be seen in the graph that more car buyers are under the age of 36, approximately 45% of the total market. The most rapid growth has been among female car buyers; they buy 40% of all cars sold. Single people buy an increasingly larger portion of new cars.

The data in Fig. 3.7 reveal that 84% of all "person trips" are made in private cars, and the remaining 16% is divided among other forms of transportation. The average trip length is less than nine miles one way. Other data show that Americans drive more than a trillion miles per year: 42%

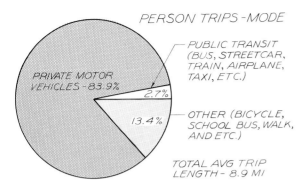

PERSON TRIPS - MODE

PRIVATE MOTOR VEHICLES - 83.9%

PUBLIC TRANSIT (BUS, STREETCAR, TRAIN, AIRPLANE, TAXI, ETC.)

2.7%

13.4%

OTHER (BICYCLE, SCHOOL BUS, WALK, AND ETC.)

TOTAL AVG TRIP LENGTH - 8.9 MI

Fig. 3.7 When a car is to be designed, it is important to know what types of trips it will be used for. This graph shows that 84% of person trips are made in private automobiles.

for business; 33% for recreation and vacations; 19% for family errands; and 5% for educational and civic purposes.

The graph in Fig. 3.8 reflects the boom in the number of men, women, and children who are participating in outdoor sports. This increased interest in sports must be considered when designing a car for this market.

How are the data interpreted? What does it mean? This is where experience and instinct must be used when interpreting the data. Ford Motor

Company concluded that a car was needed that was economical, fun to drive, lively but safe, and sporty but functional. Since younger people and singles were a greater part of the market, there were a number of things the car did not have to be. It did not need a lot of room for passengers; a one or two-seater with room for luggage was adequate for most needs. Likewise, the car did not have to be "hot" to be sporty. Instead, the car needed to be lively and attractive, but one that would give a good overall performance.

The collection of this data, and the knowledge gained from it, resulted in the production of the Ford EXP (Fig. 3.9). Problem identification was essential to the manufacturer in targeting the needs of the market.

Fig. 3.9 The Ford EXP was introduced with the features and characteristics determined by the problem identification phase of their design process. (Courtesy of Ford Motor Company.)

3.5 HUNTING SEAT— PROBLEM IDENTIFICATION

The following example is used to illustrate the problem identification step of designing a hunting seat, which is typical of a problem that might be assigned as a class project.

Hunting Seat

Many hunters, especially deer hunters, hunt from trees to obtain a better vantage point. Design a seat that would provide the hunter with comfort and safety while hunting from a tree and that would meet the general requirements of economy and hunting limitations.

WORK SHEET COMPLETION: The work sheet in Fig. 3.10 is typical of the information that is needed by the designer to understand the background of the problem.

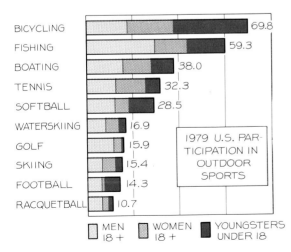

BICYCLING	69.8
FISHING	59.3
BOATING	38.0
TENNIS	32.3
SOFTBALL	28.5
WATERSKIING	16.9
GOLF	15.9
SKIING	15.4
FOOTBALL	14.3
RACQUETBALL	10.7

1979 U.S. PARTICIPATION IN OUTDOOR SPORTS

☐ MEN 18 +   ☐ WOMEN 18 +   ■ YOUNGSTERS UNDER 18

Fig. 3.8 Survey data that show the types of recreational activities in which the public participates are presented in this bar graph. These findings imply that a car design should be designed to meet the needs of a consumer who is active in outdoor sports.

Problem Identification
1. Project title

SEAT FOR HUNTING FROM A TREE

2. Problem statement

MANY HUNTERS SIT ON TREE LIMBS WHILE
HUNTING, WHICH IS UNSAFE AND UNCOM-
FORTABLE. A TREE'S ELEVATION PROVIDES
AN EXCELLENT VANTAGE POINT FOR HUNT-
ING. DESIGN A HUNTING SEAT TO FILL
THIS NEED.

3. Requirements

A. MUST BE CARRIED TO SITE
   BY HUNTER
B. MUST PROVIDE SAFETY &
   COMFORT
C. 7' TO 12' ABOVE GROUND
D. PROTECTION FROM
   WEATHER
E. METHOD OF CARRYING
   – COULD DOUBLE AS A
   BACK PACK
F. METHOD OF ASCENDING
   TO THE DESIRED HEIGHT
G. PRICE: $20 – $45
H. WEIGHT: 20 LBS MAX
I. FIT IN A CAR TRUNK

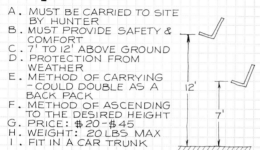

4. Needed information

A. NUMBER OF HUNTERS WHO HUNT FROM
   TREES ? IN STATE ? IN NATION ? – STATE
   GAME COMMISSION
B. WHAT HAPPENS ON A TYPICAL HUNTING ?
   – SURVEY HUNTERS
C. HOW DO HUNTERS HUNT FROM TREES
   WITHOUT SEATS ? INTERVIEW HUNTERS
D. LAWS CONCERNING HUNTING FROM
   TREES – WRITE STATE GAME COMMISSION
E. NUMBER OF BOW & ARROW HUNTERS ?
   – CHECK LIBRARY
F. HUNTING SEASONS ? GAME COMMISSION
G. EQUIPMENT CARRIED BY HUNTER ?
   SURVEY SPORTING GOODS STORES

5. Market considerations

A. WOULD SPORTING GOODS RETAILERS LIKE
   A TREE SEAT ? INTERVIEW DEALERS
B. WHAT IS THE COMPETITION ? REVIEW ADS
   IN HUNTING MAGAZINES. LIBRARY
C. WHAT WOULD BE A GOOD PRICE RANGE ?
   – INTERVIEW DEALERS
D. HOW MUCH DO HUNTERS SPEND PER
   SEASON ? – INTERVIEW DEALERS
E. FEATURES DESIRED BY HUNTERS ?
   – INTERVIEW HUNTERS
F. POSSIBLE MARKET OUTLETS ?

Fig. 3.10 A work sheet for the problem identification step of the design process for the hunting seat design.

Title and Problem Statement   The title of the project is recorded along with a brief problem statement that could be easily understood by anyone.

Requirements and Limitations   The requirements and limitations are listed along with any sketches that would aid in a better understanding of the problem. In some cases, it might be necessary to list the requirements as questions for which you have no answer at the time. Later, after further investigation, you should list the limits such as "must cost between $60 and $100." It is better to give a price or weight range rather than attempting to be exact in your estimates.

A source for information such as sales prices, weights, and sizes are catalogs dealing with similar products. If catalogs must be written for, the correspondence should be mailed as early as possible to give the necessary time for a response, so as not to delay your progress.

Needed Information   You will wish to have data that would tell you more about your problem, as listed in Fig. 3.10. How many hunters are there? How many hunt from trees or elevated blinds? You could get this and similar information from your state game office.

What is the average income of the hunter? How much do they spend on their hobby per year? Sporting-goods dealers could help you here by sharing their experiences, and perhaps they could direct you to other sources for answers to these questions. These dealers could also suggest the best price range for your design.

Market Considerations   The designer must think about cost control at all stages of the project, even

DATA

NUMBER OF HUNTING LICENSES SOLD STATE-
WIDE AND AN ESTIMATE OF THE NUMBER OF
HUNTERS WHO HUNT FROM TREES

YR	TOTAL	FROM TREES
1965	467,000	305,000
1970	481,000	370,000
1975	520,000	380,000
1980	542,000	392,000

CONSUMER SURVEY

QUESTIONNAIRE GIVEN TO 50 HUNTERS
RANKING: 1-HIGH, 4-LOW

OPINION OF SEAT IDEA

RANK	NO	%
1	20	40
2	15	30
3	8	16
4	7	14
	50	

RETAILER SURVEY
4 SPORTING GOODS DEALERS WERE POSI-
TIVE ABOUT THE CHANCES OF THE PROD-
UCT. SUGGESTED PRICE—$30

ANNUAL EXPENDITURES BY HUNTERS

AMMUNITION	$50
CLOTHING	70
LEASES	100
GUN	100
TRAVEL	70
	$390

Fig. 3.11 A work sheet for collecting data.

in the problem identification step. Figures 3.10 and 3.11 are typical examples of work sheets that list data gathered by the designer that pertains to market considerations.

By referring to Fig. 3.12, you can see how an item is priced from wholesale to retail. The percentages will vary by product: There is less profit in retail food sales than in furniture sales. If an item is to retail for $50, it cannot cost more than about $20 to produce for there to be the necessary margins. It is customary for the designer to think in terms of what the product should retail for, and thus work backward to the necessary wholesale price.

Another method of collecting information would be to conduct a survey among hunters you know to gather firsthand opinions of the merits of introducing a hunting seat on the market. The results of a survey of this type are listed on the work

sheet in Fig. 3.11. This shows that 40% of fifty people surveyed gave the market potential of the seat a high ranking. A dealer survey shows that about $390 is spent by each hunter per season, and a listing of the number of hunters is also given.

Graphs All tabular data is easier to interpret and to understand if presented in the form of a graph. For example, the number of hunters is compared with those who hunt from trees in Fig. 3.13. A thorough problem identification will include graphs, sketches, and schematics that improve the communication of the findings and conclusions of the designer.

The problem identification of this problem is not yet complete; it is difficult to decide when enough information has been gathered in any pro-

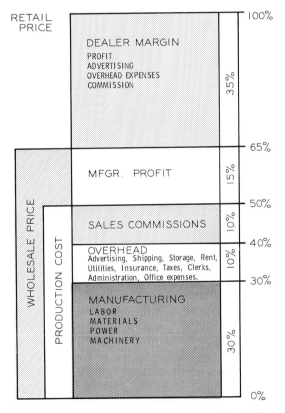

Fig. 3.12 An example model that shows the breakdown of expenses and costs involved in arriving at the retail price of a product.

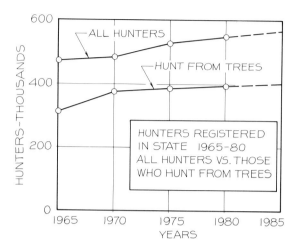

Fig. 3.13 Survey data is plotted in this graph to describe the population trends of hunters who are potential customers for the hunting seat.

ject. However, you should be able to follow the method in this example and incorporate your own personal innovations to arrive at a satisfactory problem identification.

3.6 ORGANIZATION OF EFFORT

The proper organization and scheduling of design effort and related activities is essential to productive results. The designer who has anticipated the various activities that must be performed in achieving a solution is much better prepared than the designer who has neglected these steps.

A schedule of design efforts should be prepared immediately after the identification of a need has been sufficiently established to warrant proceeding with the design process. This may come after the approval of a design proposal by an administrator, engineer, or teacher if performed as a class assignment.

A technique of scheduling project work is "Project Evaluation and Review Technique" (PERT), which was developed by industries participating in certain governmental projects where coordination of many activities within a prescribed time limit was essential to the successful completion of the project. PERT provides a means of scheduling the activities in their appropriate sequence and reviewing the progress being made in their completion.

The **critical-path method** of scheduling project activities evolved from PERT, and it is used in conjunction with PERT in most cases. The critical-path method is concerned with the determination of the chain of events that depend upon other activities. Some jobs cannot be performed until previous tasks have been completed. The critical path is a sequence of tasks that will require the *longest period* of time with the minimum of flexibility before the project can be completed. Other activities, not in the critical path, can be scheduled to receive secondary emphasis, since they are not critical to the completion of the project.

3.7 PLANNING DESIGN ACTIVITIES

A student design team will find that it can function more effectively if a few of the basic principles of PERT and critical-path scheduling are applied to its planning. Although scheduling is important to an individual in planning his or her own activities, it is increasingly valuable to a group effort as the size of the group increases.

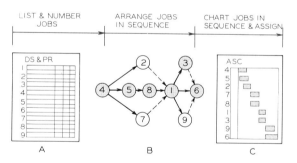

Fig. 3.14 The three steps of planning and scheduling a project.

Figure 3.14 is a flow chart showing the suggested steps for the best results in planning your project. The steps of planning are: (a) List the jobs that must be performed on a form called a Design Schedule and Progress Record. (b) Prepare an Activities Network to place the jobs in sequence. (c) Prepare an Activity Sequence Chart that graphs the jobs in the same sequence as shown in the network.

Design Schedule & Progress Record

TEAM __7__ PROJECT __PRODUCT DESIGN__

WORK PERIODS __13__ MAN-HOURS __200__ FINISHING DATE __12-10__ SECTION __500__

JOB	ASSIGNMENT	ESTIMD HOURS	ACTUAL HOURS	PER CENT COMPLETE 0 50 100
1	WRITE REPORT - ALL	15		
2	BRAINSTORM - ALL	0.5		
3	WRITE LETTERS - JHE	2		
4	GRAPH DATA - BLW	2		
5	MARKET ANAL - ALL	2		
6	COL. DATA - JWS	2		
7	ORGANIZATION - ALL	1		

Fig. 3.15 A Design Schedule and Progress Record and typical entries shown to identify example design tasks.

Design Schedule and Progress Record A suggested form for this chart is given in Fig. 3.15. Each team member should participate in the listing of the jobs that must be completed to arrive at a final solution to a project. Each job is broken into reasonably small tasks, which are listed on the Design Schedule and Progress Record (DS&PR) form in the order in which they come to mind, with no concern for their sequence. When all of these are listed on the form, an estimate of the time required for each job is listed in the second column. The sum of the times for each job should be adjusted to approximate the total time allotted for the project. Extra time may be left unassigned for emergencies. Each job is given a number to identify it. This number has no relation to the sequence in which the jobs will be completed.

Activities Network The numbered jobs listed on the DS&PR must now be arranged in the proper sequence so that they can be completed in the most efficient manner. The method of graphically arranging the jobs is shown in Fig. 3.16. The note

form (Fig. 3.16A) lists the events by name. The symbol form (Fig. 3.16B) uses numbers of the jobs from the DS&PR as the events. An *event* is some specific point in the Activities Network. Events do not require time, but are considered to be milestones along the sequence of activities.

Events are connected by arrows that represent the *activities* that consume time from one event to the next. For example, in Fig. 3.16A, it requires 0.5 hour to assemble preliminary notes after beginning the technical report. The time is marked on the activity arrow, and the length of the arrow has no relation to the amount of time.

Dummy activities are often used in completing an Activities Network to indicate a connection between activities even though the dummy activity requires "zero time." All events must be connected by activity arrows that come to the event and also leave the event to connect it with the next successive event, even though no time is required in this step. Two dummy activities are used in the Activities Network in Fig. 3.17 to show a sequential connection but no expenditure of time. Dashed lines are used for dummy activities.

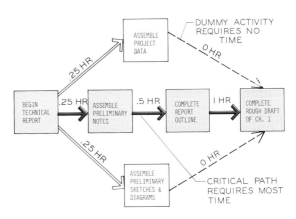

Fig. 3.17 The critical path is the path that requires the most time to arrive at the last event of a project.

The *critical path* is the longest time path from the first event to the completion of the project. This path is the sequence of jobs that must be concentrated on throughout the project to remain on schedule. The critical path is marked in the portion of the Activity Network shown in Fig. 3.17 where 1.75 hours are required to complete the rough draft of the first chapter of a report after beginning.

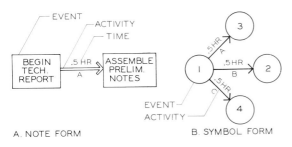

Fig. 3.16 Two forms of preparing an Activities Network, which graphically places the jobs in sequence.

A more sophisticated method of estimating the activity times of an Activity Network is shown in Fig. 3.18. Each activity is given an optimistic time, a most likely time, and a pessimistic time. These three times are written on each activity arrow and are used in arriving at critical paths based on each of these estimates.

An Activities Network for a project is shown in Fig. 3.19. Different teams will prepare different networks for the same set of activities. However, the more important jobs should be scheduled in generally the same order, and the identification of these is the most important function of the Activities Network.

Activity Sequence Chart   The jobs of the Activities Network can be listed on the Activity Sequence Chart (ACS) (Fig. 3.20). The job numbers are taken from the Network and are listed in sequence and assigned to members of the team. A bar graph is used to graph the time for each job. For example, the first half hour of the project will be used to perform job 7, which is "organization," listed on the DS&PR. If your team has four members, this is equivalent to two man-hours of work. This will require the cooperation of the entire team and this must be completed before the next job can begin. Continue until all the jobs are listed. The project hours across the top of the ASC can be used to schedule the events and to estimate when each event should be reached during the project.

As progress is made with the completion of the jobs, the status of each assignment can be graphed on the DS&PR (Fig. 3.21). When a job is

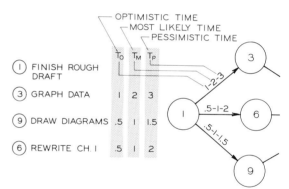

Fig. 3.18 An Activities Network can be constructed to show optimistic, most likely, and pessimistic times on the activity arrows to give a tolerance in the schedule.

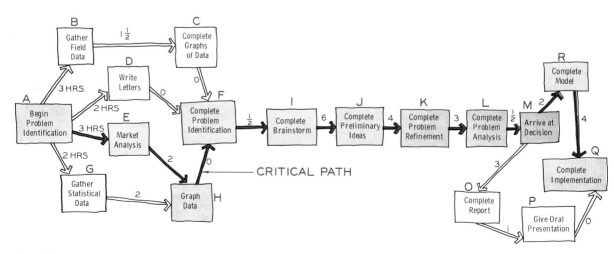

Fig. 3.19 An Activities Network that is used to schedule design effort where the critical path is noted.

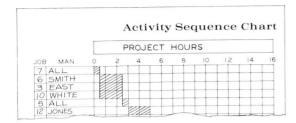

Fig. 3.20 The jobs listed on the DS&PR are converted into an Activities Network and are then graphed on an Activity Sequence Chart (ASC) to assign each job to a fixed time schedule in the proper order.

completed, the completion date and the hours required are listed in the last two columns. If more time was required than was scheduled, it is helpful to know this as soon as possible so that adjustments can be made in subsequent jobs to compensate for this loss.

You should always refer to the critical path of the ASC to be sure that the jobs on this path re-

ceive priority since they will require the most time.

This method of planning and scheduling is closely related to PERT, but it has been simplified to introduce the basic principles in a direct fashion. You may wish to make modifications of this approach that would be more helpful to a specific situation.

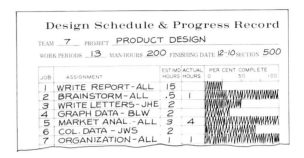

Fig. 3.21 The Design Schedule and Progress Record is completed by graphing the actual time required for each job.

PROBLEMS

Problems should be presented on 8½″ × 11″ paper, grid or plain. All notes, sketches, drawings, and graphical work should be neatly prepared in keeping with good practices, as introduced in this volume. Written matter should be typed or lettered using ⅛″ guidelines.

General

1. Identify a need for a design solution that could be used as a short design problem for a class assignment (less than three man-hours for complete solution). This can be a system or a product. Submit a proposal that briefly outlines this need and your general plan for solution. Limit the proposal to two typewritten pages.

2. Assume that you were marooned on a deserted island with no tools, supplies, or anything. Identify major problems that you would be required to solve. List factors that would identify the problem in detail. *Example:* Need for food: determine (a) available sources of food on island, (b) method of storing food supply, (c) method of

cooking, (d) method of hunting and trapping, etc. Although you are not in a position to gather data or supply answers to these questions, list factors of this type that would need to be answered before a solution could be attempted.

3. While walking to class, what irritations or discomforts did you recognize? Identify the problem causing these irritations, using work sheets and writing out the problem statement, recognition of a need, requirements, and limitations.

4. Apply the criteria given in Problem 3 above to your living quarters, your classroom, recreation facilities, dining facilities, and other environments with which you come in contact.

Product Design Problems

5. Assume that you are attempting to reconstruct the designer's approach to the development of the self-opening can. Even though the problem has been solved and completed, follow the identification steps with which the designer was concerned. Using the procedure outlined in Section 3.3, list these on work sheets. After identifying the

problem, do you feel that the solution is the most appropriate one or does your identification suggest other designs?

6. Follow the same procedure outlined in Problem 5 in identifying the problem of designing a travel iron for pressing clothes. Retain your notes and sketches on work sheets.

7. Identify the problems of designing a motorized wheelbarrow. List your ideas on work sheets.

8. You have noticed the need for a device that, attached to a bicycle, would allow the bicycle to ride over street curbs to sidewalk level. Identify the problem to determine its application to the general market.

9. Identify the problem and need for the development of a portable engineering travel kit that would provide the engineer with on-the-road facilities to make engineering calculations, notes, sketches, and drawings. This may take the form of a case that includes a calculator, drawing instruments, paper, reference material, etc. Investigate this problem and identify the needs of such a product.

Systems Design Problems

10. Water pollution is a complex systems problem. Use the work-sheet approach to identifying the problem. State the problem and list the need, requirements, and limitations. Data need not be gathered, but indicate the type of data needed and their probable sources.

11. Assume that you are assigned the responsibility of planning for the expansion of your campus during the next twenty years. This applies to all aspects of the campus, including classroom buildings, dormitories, traffic, libraries, etc. Although this comprehensive problem would require the assistance of many experts, test your logic by identifying the major aspects of the problem. Without gathering field data, indicate the type of data needed and where they would be found. Prepare work sheets as a record of your effort.

12. Identify the problems of designing a water supply that would serve your campus, assuming that the only source was rainfall over the campus. Gather any data that would contribute to problem identification. Prepare work sheets.

13. Identify the problems of designing an outdoor, drive-in movie installation. Use the worksheet approach and identify all areas of the problem, including traffic, engineering, and economics.

14. Assume that you are assigned the responsibility of correcting the existing campus system that is in most critical need of modification. List the several systems that you recognize as needing improvement. Select the most critical problem and identify it on work sheets. Gather any necessary data to assist in this identification. Submit a proposal for design if the identification supports the need for a solution.

15. Identify the problems of constructing, equipping, and operating a commercial pistol range. Use work sheets to retain your ideas.

PRELIMINARY IDEAS

4.1 INTRODUCTION

Coming up with preliminary ideas is the most creative step of the design process. Graphical methods, and freehand sketching in particular, are very helpful to the designer in developing and communicating preliminary solutions of a problem.

The relationship between creativity and information accumulation can be seen in Fig. 4.1. During the process of generating preliminary ideas, you have no limitations; be as creative as possible, even if your ideas are "wild." As the design process progresses toward the final production of the finished design, the need for creativity diminishes and information accumulation increases.

Fig. 4.2 A unique design is the "instant spoon," which is used as a package for instant coffee. Bouillon, dry-soup mixes, tea, hot chocolate, and fruit juice concentrates are a few product possibilities for the air-formed packages. (Courtesy of Alcoa).

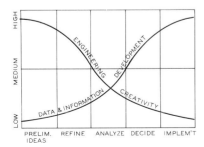

Fig. 4.1 Engineering creativity is highest during the initial stages of the design process, while data and information development increase during the final stages.

The "instant spoon" in Fig. 4.2 is an example of a product that began as a radical idea. The spoon contains instant coffee for use with hot water. Dry soup, tea, chocolate, and fruit juices could also be packaged in a dehydrated form in this air-formed container.

This chapter will be devoted to presenting the methods that can be used to develop preliminary ideas.

4.2 INDIVIDUAL VERSUS TEAM

Designers work as individuals and as members of design teams. Each approach is covered below.

Individual Approach

Designers who work alone must make notes and sketches to communicate, not with others, but with themselves. Their primary goal is to obtain as many ideas as possible on the assumption that the better ideas will be more likely to come from a long list rather than from a short list.

Ideas are sketched as possible solutions with explanatory notes and schematics. These sketches are not working drawings, but rapid freehand sketches used to retain ideas that might otherwise be lost. All sketches and notes are kept in the designer's file for future reference.

Team Approach

Due to the complexity of technology, the team approach is necessary to solve many of today's problems. Many specialists in various fields must work together toward a common solution. As more people become involved in a single project, the more serious become the problems of management and human relations. A team must overcome personality differences and individual ego problems.

A design team should alternate between individual and group work. For example, team members could individually develop a series of preliminary ideas, which would then be presented and discussed with their team.

Most teams perform best if a leader is selected by the members to be responsible for moderating and guiding the activities of the team. The strength of the team approach is the opportunity for the whole team to utilize the special talents of the individuals on the team.

4.3 PLAN OF ACTION

A systematic approach should be used in gathering preliminary ideas for your design problem. The following sequence (Fig. 4.3) of steps is suggested: (1) hold brainstorming session, (2) prepare sketches and notes, (3) research existing designs, and (4) conduct surveys.

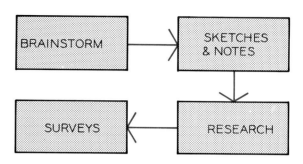

Fig. 4.3 A suggested plan of action for gathering preliminary ideas for a design solution.

You should also periodically review your notes and work sheets from the previous step, problem identification, to ensure that your efforts are directed toward the "target."

4.4 BRAINSTORMING

To "brainstorm" is defined as the practice of a conference technique by which a group attempts to find a solution for a specific problem by amassing all the ideas spontaneously contributed by its members. Brainstorming is used to take advantage of the combined ideas of a group of people.

Rules of Brainstorming

The basic guidelines of a brainstorming session are:*

1. *Criticism is ruled out.* Adverse judgment of ideas must be withheld until later.
2. *"Free-wheeling" is welcomed.* The wilder the idea, the better; it is easier to tame down than to think up.
3. *Quantity is wanted.* The greater the number of ideas, the more likelihood of useful ideas.
4. *Combination and improvement are sought.* Participants should seek ways of improving the ideas of others.

Organization of a Brainstorming Session

The organizational steps of a brainstorming session are (1) selection of the panel, (2) preliminary

---

*From Alex F. Osborn, *Applied Imagination.* New York: Scribner, 1963, p. 156.

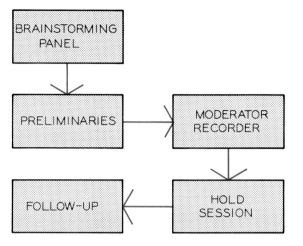

Fig. 4.4 The organizational steps of planning a brainstorming session.

is assigned the job of recording the ideas as they are verbalized by the panel.

The Session   The leader tosses out the first problem and recognizes the first member who holds up his or her hand. That person responds with an idea given in as few words as possible. As rapidly as possible, the moderator recognizes the next person who holds up a hand, and the process continues in this manner.

Hopefully, a suggestion made by one member might stimulate a related idea in another panel member. If this occurs, that member should hold up his or her hand and snap his fingers to signify that he wishes to "hitch-hike" on the previous idea, adding a supplementary idea to it. This expansion of concepts is the crux of the brainstorming session, and when successful, it can cause a chain reaction of developing a single concept.

The moderator should not permit individuals to give long responses that would slow down the flow of ideas.

Length   A session should move at a brisk pace and should be called to a halt when ideas slow to an unproductive rate. An effective session can last anywhere from a few minutes to an hour, but a 25-minute session is considered to be about the best length.

The Follow-up   The recorder should reproduce the list of ideas gathered during the session for distribution to the participants. Approximately 100 ideas are usually gathered during a 25-minute session. The list should be pared down to the ideas believed to have the most merit—probably 10 or 12.

Topics to be Brainstormed

Brainstorming can be used to determine possible solutions to a problem, or to identify problems that need to be solved. This technique could be used by a class to gather a list of ideas for design projects in need of solution. These areas of brainstorming for specific problems could be:

1. Products to improve study environment,
2. Improvements for classrooms,
3. Improvements for vehicular and pedestrian traffic,
4. Shortcoming in recreational facilities,
5. New products needed for various applications.

group work, (3) selection of a moderator and recorder, (4) the session, and (5) the follow-up (Fig. 4.4).

Panel Selection   The optimum number of participants in a brainstorming session is about 12 panel members. To encourage a variety of ideas, the group should be composed of people with and without knowledge of the subject to be brainstormed. Supervisors and superiors often restrict a flow of ideas; consequently, it is desirable that panels be composed of people with a similar professional status.

Preliminary Work   A one-page outline of information pertaining to the session should be prepared and given to the panel about two days before the session to allow the idea process to "incubate" during this period. The rules of brainstorming should be listed on the sheet if the participants are unfamiliar with the process.

The Problem   The problem to be brainstormed should be concisely defined. For example, instead of posing the problem as "how to improve our campus," it should be presented more specifically such as "how to improve instruction" or "how to improve student parking."

The Moderator and Recorder   A moderator is selected to be in charge of the session. The recorder

Fig. 4.5 Designers communicate with themselves, as well as with others, through sketches and notes. This graphical technique enables designers to develop their preliminary ideas. (Courtesy of Ford Motor Company.)

4.5 SKETCHING AND NOTES

Sketching is the designer's most important medium for developing preliminary ideas (Fig. 4.5). By sketching, the designer's ideas take form in three dimensions as a pictorial (Fig. 4.6) or as two-dimensional orthographic views. When properly used, sketching becomes an extension of the thinking process that is both rapid and visual.

Since designers must keep all of their design work, the standard work sheet format that was introduced in Chapter 3 should be used. Notes and sketches should be recorded with sufficient clarity and technique to be readable by others as well as the designer.

A series of sketches is shown in Fig. 4.7 that was used in finalizing the styling features of an

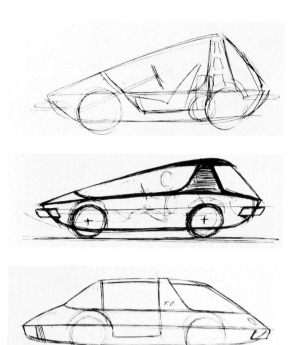

Fig. 4.7 The final design of the American Motors Pacer was the evolutionary process of developing conceptual sketches into final drawings. (Courtesy of American Motors Corporation.)

Fig. 4.6 Sketching is used by the designer and stylist to develop the exterior design of an automobile. (Courtesy of Ford Motor Company.)

automobile. Additional sketches were used to develop preliminary designs for a rearview mirror for an automobile (Fig. 4.8).

The Transportable Uni-Lodge (Fig. 4.9) illustrates the value of sketches in developing and communicating ideas. These sketches are preliminary ideas for a transportable lodge that may someday become a reality. The Uni-Lodge could be transported by helicopter to previously unreachable areas.

Retractable legs with pontoons permit it to float on water. Special features that are noted on the drawing are accommodations for six people for two weeks, water for drinking and bathing, solar

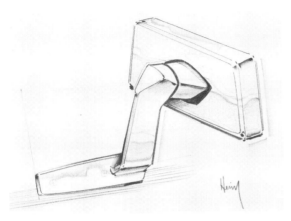

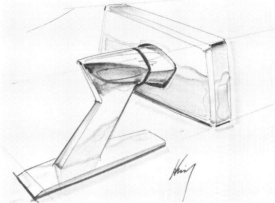

Fig. 4.8 Preliminary sketches of a rearview mirror design. (Courtesy of Ford Motor Company.)

power for electrical power, and other conveniences. Most of these ideas were indicated on the sketches by notes. Pictorials and view drawings were used in combination to depict the system. Without sketching and graphics, these ideas would have been almost impossible to develop.

Sketching techniques were also used to develop the new concept shown in Fig. 4.10, a check-out system for a grocery store. The shopper sets the

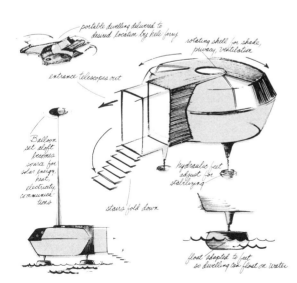

Fig. 4.9 Preliminary sketches of a Transportable Uni-Lodge for the future, a mobile dwelling. (Courtesy of Lippincott and Margulies, Inc., and Charles Bruning Company.)

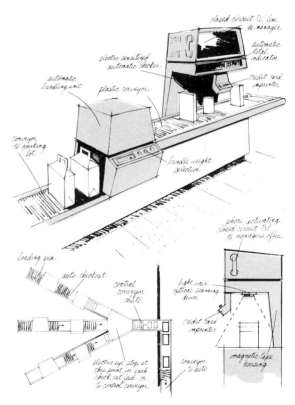

Fig. 4.10 The conceptualization of an automatic checkout/packaging unit for a supermarket is shown in a sequence of sketches. (Courtesy of Lester Beall, Inc., and Charles Bruning Company.)

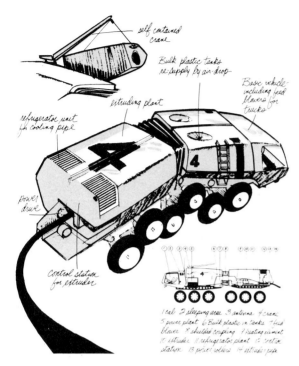

Fig. 4.11 A designer's preliminary design of a self-contained pipelayer, which is intended to lay pipe for the irrigation of large tracts of desert. (Courtesy of Donald Desky Associates, Inc., and Charles Bruning Company.)

machine in operation by inserting a credit card in a slot. After the card is scanned, the customer receives an order number tag, and the conveyor moves the items under a scanner that totals the prices of the items. If the customer has any questions concerning the purchases, the machine can be stopped for communication with the store's employees by lifting the phone.

The totaled items are conveyed into a packaging unit where they are packaged in plastic containers marked with the customer's number. Large orders are transported by a central conveyor to an exterior pick-up point near where the customer is parked.

Another new concept, a self-contained pipelayer, is shown in Fig. 4.11. The pipelayer can be used to lay pipe to provide irrigation of desert areas. The first unit is the tractor, which includes a cab, sleeping accommodations, radio equipment, power plant, and bulk storage tanks for plastic. The second unit, the van, consists of an extrusion machine, a

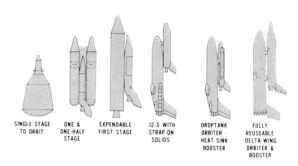

Fig. 4.12 A comparison of six preliminary design concepts for the space shuttle project. (Courtesy of NASA.)

refrigeration unit, and a control station. The unit is capable of transporting bulk plastic and machinery to extrude and lay approximately two miles of plastic pipe from each pair of storage tanks. The tanks are discarded when empty and are replaced by new tanks that are air-dropped into the area.

Many different preliminary ideas should be developed at this stage of the design process. A number of concepts were developed before arriving at the final design for the space shuttle that was developed by NASA (Fig. 4.12). Graphics were also used to present such concepts as the method of launching and retrieving the space shuttle as shown in Fig. 4.13.

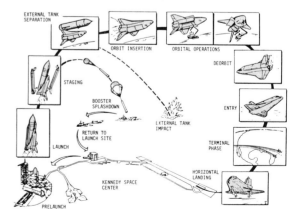

Fig. 4.13 Graphical techniques are used to illustrate the stages of the space shuttle mission. (Courtesy of NASA.)

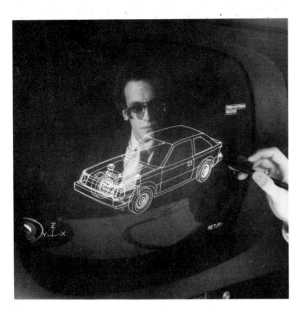

Fig. 4.14 After a number of freehand sketches have been made to develop several concepts, these preliminary ideas can be developed further by computer graphics techniques. (Courtesy of Ford Motor Company.)

Computer graphics can be used for modifying and developing a number of ideas for consideration (Fig. 4.14), but the computer is used only after much freehand sketching has been done.

4.6 RESEARCH METHODS

Preliminary ideas can be obtained through research of similar products and designs in use. The process of applying known principles to new applications and designs is called **synthesis.** Some reference sources that are available for this research are technical magazines, general magazines, manufacturer's brochures, patents, and consultants.

Technical Magazines   Articles in technical magazines often give detailed explanations of unique designs, complete with sketches and photographs. Advertisements in these magazines can furnish information on materials and innovations that may be helpful in obtaining ideas.

General Magazines   Significant design developments are usually reported in general periodicals subscribed to by most families. Magazines that are

several years old can also be helpful in finding ideas.

Patents   Patents from the U.S. Patent Office can be used to good advantage by the designer. Patents are available to anyone at a cost of 50¢ each. Figure 4.15 is a reproduction of Thomas Edison's patent illustration for the phonograph.

Consultants   When designers must investigate areas with which they are unfamiliar, they may find it necessary to employ a consultant who is a specialist in this field. A comprehensive design may require a team of specialists in structures, electronics, power systems, and instrumentation. Manufacturers' representatives are also available for consultation to assist with many problems that are related to their products.

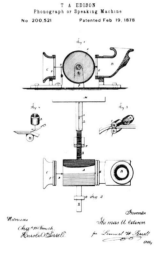

Fig. 4.15 Patent illustrations and specifications are available from the U.S. Patent Office for over 3,000,000 patents. This patent drawing illustrates Thomas Edison's phonograph. (Courtesy of the U.S. Patent Office.)

4.7 SURVEY METHODS

Survey methods are used to gather opinions and reactions to a preliminary design or a completed design. This is especially important when a product is being designed for the general market.

Opinions

Designers need to know the consumer's attitude toward their preliminary designs. Is the potential buyer excited about the design, or in desperate need of it, or does he show little interest in it? This will have an impact on the future of the project.

Agencies are available to conduct surveys to determine market reactions. Before conducting the survey, the particular group whose opinions would be of most value should be targeted for survey by one of the following methods: (1) personal interview, (2) telephone interview, or (3) mail questionnaire.

The personal interview should be organized to provide reliable and unbiased opinions. Questions that require a minimum of explanation should be listed on a sheet that will permit the interviewer to tabulate the replies as they are given. Questions should be true–false or multiple choice to make conclusions easier to determine.

Interviewers should introduce themselves, give the purpose of the interview, and ask for permission to proceed. They should then tabulate the responses as they are given and thank the respondent for his or her participation at the close of the interview.

The telephone interview can be conducted by randomly picking names from the directory if the opinions of a general cross section of the public are desired. If the opinions of a select group— sporting goods retailers, for example—would be more valuable, the Yellow Pages of a city telephone directory would supply a list of prospects. The telephone interviews should be conducted and tabulated in the same manner as the personal interview.

The mail questionnaire is an economical method of contacting large numbers of people in a wide range of locations. The questionnaire should be tested by sending it to a small group as a pilot test to identify questions that need to be revised. The final questionnaire should have multiple-choice questions for ease of tabulation and completion. When mailed, a self-addressed, stamped envelope should be included to improve response to the inquiry. At least three times as many questionnaires should be mailed as the number of responses that you wish to receive.

Larger samples are more meaningful and reliable than opinions received from a small group. Likewise, opinions must be obtained from a typical cross section of the population being consid-

ered, since opinions often vary geographically and with the economic status of the interviewee.

What types of opinions would be helpful to designers? First, they would like to know if there is a need for the product they are working on. If the consumer is not positive toward the product, what features does he like and dislike? What price would he be willing to pay for it? If he is a retailer, what does he think it will sell for? Does size matter? Color? Has he bought similar products before?

All findings and data should be graphed where appropriate for study and analysis.

4.8 HUNTING SEAT— PRELIMINARY IDEAS

The design of the tree-borne hunting seat that was introduced in Chapter 3 is used as an example to illustrate the preliminary-ideas step of the design process. The problem is restated below.

Preliminary Ideas
1. Brainstorming ideas

A. USE LAWN CHAIR
B. CHAIR ON STILTS
C. INFLATABLE CHAIR
D. PROVIDE FOOT REST
E. SHELTER FROM RAIN
F. PADDED SEAT
G. HEAD REST
H. RIFLE REST
I. AMMUNITION COMPARTMENT
J. REFRESHMENT COMPARTMENT
K. ENTERTAINMENT COMPARTMENT
L. TV ACCESSORY
M. RADIO
N. CB RADIO SYSTEM
O. HOIST SYSTEM
   1. PULLEYS & CABLES
   2. TREE CLIMBER
   3. LADDER
   4. STEPS
P. PLATFORM FOR STANDING
Q. PLATFORM FOR SLEEPING
R. LIGHTS FOR NIGHT
S. SAFETY BELT
T. SAFETY BELT FOR RIFLE
U. SAFETY BELT FOR BOW & ARROW
V. DOUBLES AS BACK PACK
W. DOUBLES AS CAMP CHAIR
X. DOUBLES AS TENT
Y. CARRYING CASE FOR SEAT
Z. MOTORIZED SEAT
A1. SEAT ON WHEELS
B1. ETC.

Fig. 4.16 A worksheet that lists the brainstorming ideas that were recorded by a design team.

HUNTING SEAT: Many hunters hunt from a sitting position in trees to obtain a better vantage point. Design a seat that would provide the hunter with comfort and safety while hunting from a tree and that would meet the general requirements of economy and hunting limitations.

Brainstorming ideas are gathered from a brainstorming session with classmates. The ideas, all of them, are listed on a work sheet (Fig. 4.16). Remember, wild ideas are encouraged to stimulate the act of forming ideas.

The better ideas are listed on a second work sheet (Fig. 4.17) that summarizes the features that would be desirable in the final design of your project. You may list more features than would be possible to include in a single design, but be sure that no ideas are forgotten or lost at this stage.

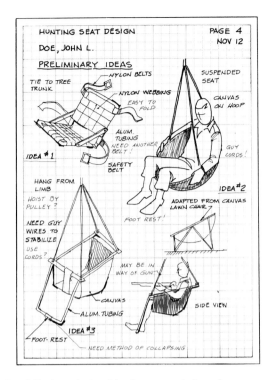

Fig. 4.18 A work sheet for the presentation of preliminary ideas for the development of a hunting seat. Notes and sketches are used to supplement the sketches.

2. Description of best ideas

A. SEAT WITH FOOT REST

B. NEED METHOD OF ASCENDING TREE

C. ACCESSORIES

    1. STANDING PLATFORM

    2. RIFLE RACK

    3. BOW RACK

    4. CLIMBING ACCESSORY

D. SAFETY FEATURES

E. SEAT SHOULD FOLD UP

F. WEIGH UNDER 12 LBS

G. PROTECTION FROM RAIN

3. Attach sketches

Fig. 4.17 A work sheet that lists the better ideas that were selected from the original brainstorming ideas.

Sketches of preliminary ideas are drawn on additional work sheets, using rapid freehand techniques. Orthographic views and pictorial methods are used in combination. Thoughts or questions that come to mind during this process of sketching should be noted on the drawings. Lettering and sketching techniques need not be highly detailed or precisely executed. However, the techniques should be readable and understandable.

In Fig. 4.18, notice how ideas have been adapted from various types of known chairs, lawn chairs in particular. Each idea is numbered for identification purposes.

A fourth work sheet (Fig. 4.19) gives a fourth idea. Idea #2 is modified on this sheet to include a footrest and to suggest a method of guying the seat while suspended. The upper part of the sheet shows sketches that identify the need for tilting the seat designs for comfort. Idea #1 is revised to include a V-frame and the seat is anchored and tilted.

Many other ideas of this type need to be developed and sketched before leaving this step of the design process. These four work sheets are representative of the initial steps that should be taken to complete this step of the process. No ideas are discarded; all are retained as part of the work-sheet file.

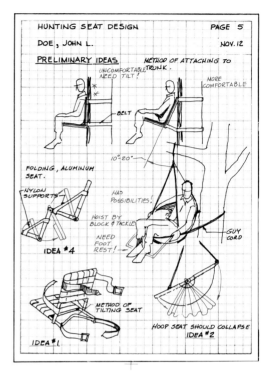

Fig. 4.19 Additional preliminary ideas are recorded to illustrate design concepts for the hunting seat problem.

PROBLEMS

Problems should be presented on 8½″ × 11″ paper, grid or plain. Each grid square represents ¼″. All notes, sketches, drawings, and graphical work should be neatly prepared in keeping with good practices as introduced in this volume. Written matter should be legibly lettered using ⅛″ guidelines.

1. Assume that you had been given access to the following items. Select one or several of these and list as many uses as possible. Look for applications that would be unique or unusual. The items are: empty vegetable cans (3″ diameter × 5″ tall), 2000 sheets of 8½″ × 11″ bond paper, one cu yd of dirt, three empty oil drums (24″ diameter × 36″ tall), a load of egg cartons, 25 bamboo poles (10″ long), 10 old tires, old newspapers.

2. If you were going to select an ideal team to develop an engineering problem, what would you look for? List the characteristics with your explanations. Letter your response.

3. What are the advantages and the disadvantages of working independently on a project? What are the advantages and the disadvantages of working as a member of a design team? List your reasons and give examples of the types of problems where each approach would have the greater advantage.

4. Conduct a brief research of materials available to you to accumulate information on one of the following design problems or on one of your own selection. You are concerned with costs, methods of construction, dimensions, existing models, estimates of need, and other information of a general nature that will assist you in understanding the problem and deciding whether or not it is a feasible project. List the references you used. Prepare your research in a presentable form that could be reviewed by your instructor. The design problems are: a one-person canoe, a built-in car jack, an automatic blackboard eraser, a built-in coffee maker for an automobile, a self-opening

door to permit a pet to leave or enter the house, an emergency fire escape for a two-story building, a rain protector for persons attending sporting events and other spectator activities, a new household appliance, a home exerciser.

5. List and describe the type of consulting services on an engineering or professional level that would be required in the following design projects: a zoning system for a city of 20,000 population, a shopping-center development, a go-cart, a water purification facility, a hydroelectrical system, a nuclear fallout disaster plan, a processing plant for refining petroleum products, a drainage system for residential and rural areas.

6. Develop a questionnaire that could be given to the general public to determine its attitude toward a particular product. Select a product and prepare the questionnaire to measure reponse to its unique features. The questionnaire should be simple and brief and should require mostly multiple-choice answers, with the minimum of sub-jective answers. Indicate how you would tabulate the information received from your questionnaire.

7. A brainstorming session is a good method of loosening one's imagination and releasing latent ideas. After reviewing the brainstorming session techniques, beginning with Section 4.4, organize a group of associates to brainstorm a selected problem or to determine a problem in need of solution. List all ideas as they are suggested. Write a brief review of the results of this session.

8. Prepare preliminary sketches and notes of ideas to develop the problems of a systems nature that were identified in Chapter 3, Problems 10 through 15.

9. Select a product-design problem similar to the one covered in Section 4.8 and sketch rough ideas that might be possible solutions to the problem. Some example problems are a stadium seat, a trailer hitch, a handle for a filing cabinet, a portable clothes rod for installation in a closet, or other products of this nature.

5

DESIGN REFINEMENT

5.1 INTRODUCTION

Descriptive geometry is the graphical discipline that has the greatest application to the refinement step of the design process. In this step, it is necessary to make scale drawings with instruments to check the critical dimensions that cannot be accurately shown in sketches (Fig. 5.1).

Refinement is the first departure from unrestricted creativity and imagination. Practicality and function must now be given primary consideration. Therefore, several better ideas should be

Fig. 5.1 The designer's first step in the refinement step of the design process is to draw preliminary ideas as scale drawings with instruments. (Courtesy of Chrysler Corporation.)

selected and refined in order to make a comparison among them during the analysis and decision steps of the design process.

5.2 PHYSICAL PROPERTIES

One of the important concerns of the design's refinement is the determination of the physical properties of the proposed solutions. For example, scale drawings of six configurations of the space shuttle were made in Fig. 5.2. These scale drawings evolved from many preliminary sketches and ideas that were developed during the preliminary ideas step of the design process. As simple as these scale drawings are, they give a good comparison of the physical dimensions of the different design studies.

Many additional refinement drawings must be drawn to scale to determine such properties as sizes, volumes, and the inside configurations that will house the astronauts and the gear that is necessary to operate the space shuttle. All mechanisms and functional parts of the shuttle must be drawn and refined by scale drawings to understand better the practicality of these features.

An example of a refinement drawing of the exterior of an automobile with the appropriate overall dimensions is shown in Fig. 5.3. This series of orthographic views illustrates the overall appearance and gives the sizes of the design. Ad-

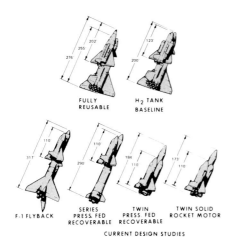

FULLY REUSABLE H₂ TANK BASELINE

F-1 FLYBACK SERIES PRESS. FED RECOVERABLE TWIN PRESS. FED RECOVERABLE TWIN SOLID ROCKET MOTOR

CURRENT DESIGN STUDIES

Fig. 5.2 Several refinement drawings of the space shuttle are shown here with only the major dimensions given for comparison. (Courtesy of NASA.)

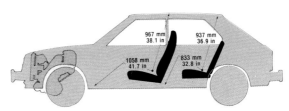

HEAD ROOM AND LEG ROOM DIMENSIONS

Fig. 5.4 This scale drawing gives the important dimensions and clearances necessary for a comfortable interior of an automobile. (Courtesy of Chrysler Corporation.)

ditional scale drawings can also be used to present graphically dimensions and functions that are essential to good automobile design, such as seating space as shown in Fig. 5.4.

5.3 APPLICATION OF DESCRIPTIVE GEOMETRY

Descriptive geometry is the study of points, lines, and surfaces in three-dimensional space. The calculation of practically any given properties begins with basic geometric elements—points, lines, areas, volumes, and angles.

Before descriptive geometry can be applied, a series of orthographic views must be drawn to scale from which auxiliary views can be projected with accuracy. An example problem has been solved in Fig. 5.5 where the clearance between a hydraulic cylinder and the fender of an automobile has been determined with descriptive geometry.

Similarly, the angle between the planes of a windshield design (Fig. 5.6) can be found by descriptive geometry when the top and front views of the windshield are first drawn to scale as orthographic views. You should consider how a problem of this type would be solved without the use of descriptive geometry. Additional views could be constructed to determine the area and shape of the windshield when laid out in a flat plane.

Descriptive geometry is used in Fig. 5.7 to determine the opening size for an automobile fender to allow a clearance between the fender and the wheel. The wheel is turned to its maximum steering angles to locate lines of interference, which will determine the minimum opening of the front fender.

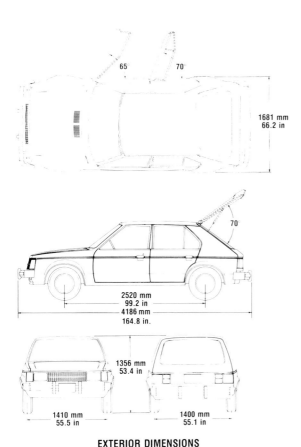

EXTERIOR DIMENSIONS

Fig. 5.3 A series of orthographic views were drawn to describe the features and dimensions of a car design. (Courtesy of Chrysler Corporation.)

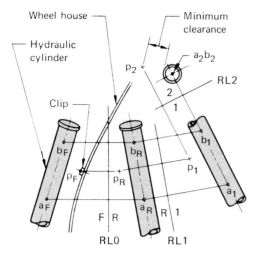

Fig. 5.5 Descriptive geometry is an effective means of determining clearances between components as shown in this example where a hydraulic cylinder's clearance with a fender is determined. (Courtesy of General Motors Corporation.)

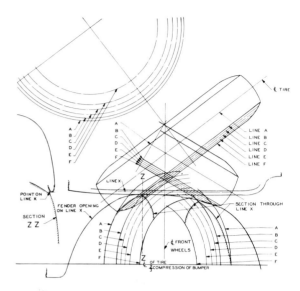

Fig. 5.7 The clearance between a fender and a tire must be found to determine the fender opening of an automobile. The clearance is found by descriptive geometry in this example. (Courtesy of Chrysler Corporation.)

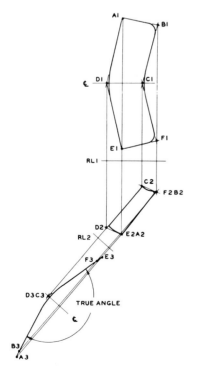

Fig. 5.6 The angle between the planes of a windshield can be determined by descriptive geometry. (Courtesy of Chrysler Corporation.)

The design of a surgical light (Fig. 5.8) is a problem that involves the application of geometry at a sophisticated level. The light fixture had to be designed in such a manner so as to provide the maximum of light on the operating area. A scaled refinement drawing is shown in Fig. 5.8(a) where you can observe the converging beams of light that are emitted from the reflectors. The beams are very narrow at their centers and are positioned at shoulder level to minimize shadows cast by interference of the surgeon's shoulders, arms, and hands.

Additional scale drawings are drawn in Fig. 5.8(b) so that the geometry of the fixture can be studied in detail. Measurements, angles, areas, and other geometry can be determined from these scale drawings (Fig. 5.9).

Figure 5.10 is an example of a refinement drawing that shows critical dimensions of the surgical lamp. Eventually this geometry will be tested by construction of a working model to confirm the information found in refinement drawings.

Another example of a problem refined by descriptive geometry is the structural frame for a 10-foot-diameter underwater sphere (Fig. 5.11). Before working drawings can be made, the physical properties and dimensions of the spherical pentagons must be determined through a series of auxiliary

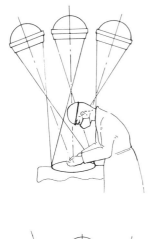

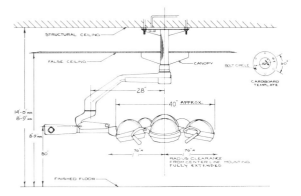

Fig. 5.10 The overall dimensions of the final design of the surgical lamp are shown in this refinement drawing. (Courtesy of Sybron Corporation.)

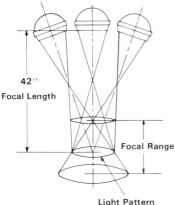

Fig. 5.8 A well-adapted surgical lamp emits light that passes around the surgeon's shoulders with the mimimum of shadow. The focal range of this surgical lamp is between 30″ and 60″. (Courtesy of Sybron Corporation.)

Fig. 5.11 This 10-foot diameter underwater sphere could not have been designed without utilizing descriptive geometry methods. (Courtesy of the U.S. Navy.)

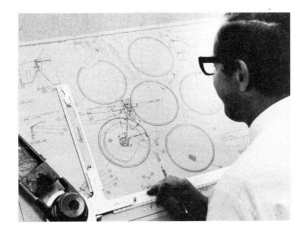

Fig. 5.9 The geometry of a surgical lamp can be studied by using scale drawings developed in the refinement step of the design process. (Courtesy of Sybron Corporation. Photograph by Brad Bliss.)

views. The angles between the members had to be found before the joints could be detailed to fit properly (Fig. 5.12). Also, the geometry had to be determined for designing the jigs that were necessary for holding the structural parts in position during assembly.

5.4 REFINEMENT CONSIDERATIONS

Advanced designs, such as the development of a new model automobile, have numerous features that must be refined and optimized at this stage. The car's interior must be developed and presented in

Fig. 5.12 The compound joints of the structural members of the underwater sphere were designed and fabricated with the use of descriptive geometry. (Courtesy of the U.S. Navy.)

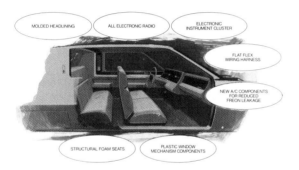

Fig. 5.13 The various features of an automobile's interior are indicated in this illustration. (Courtesy of Ford Motor Company.)

an understandable form (Fig. 5.13). The various features are shown pictorially or are listed in the surrounding balloons when they are not visible in the drawing. The same approach is used to present the various details of the chassis design (Fig. 5.14).

The overall dimensions of the automobile are shown in the orthographic view in Fig. 5.15. Also shown are the weight reductions over the 1977 models. The suspension system design of an automobile is shown graphically after it has been refined (Fig. 5.16) to explain its installation. In Fig. 5.17, the arrangement of the exhaust system is shown pictorially. The exhaust was designed to conform to the California standards imposed on the manufacturer.

You can easily understand that many intermediate refinement drawings were required before these general pictorials were drawn, which show only the overall locations of the various components. For example, descriptive geometry had to be used to determine the bend angles and clearances in the exhaust pipe to fit a particular chassis (Fig. 5.18).

Computer graphics is a powerful tool that can be used to refine a preliminary idea. In this example, the designer is using a "light pen" on a cathode-ray tube to refine a windshield wiper system (Fig. 5.19). The image on the CRT can be shown in either two or three dimensions.

Similar to the automobile design, the Space Shuttle Orbiter was refined using a combination of orthographic drawings and notes. A profile view of the Orbiter (Fig. 5.20) gives the overall size of the craft plus a series of notes that show the locations

Fig. 5.14 As a design of an automobile is refined, the various features of its chassis are noted on the drawing. (Courtesy of Ford Motor Company.)

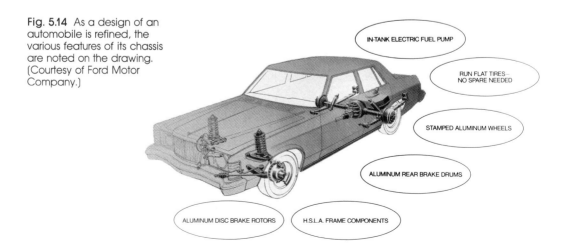

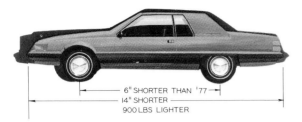

Fig. 5.15 The side view of an automobile design with the significant dimensions given. (Courtesy of Ford Motor Company.)

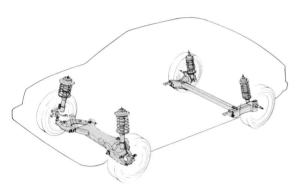

SUSPENSION SYSTEM
PLYMOUTH HORIZON — DODGE OMNI

Fig. 5.16 A pictorial drawing that illustrates the suspension system of an automobile. (Courtesy of Chrysler Corporation.)

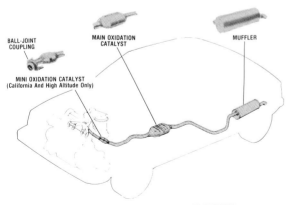

CALIFORNIA OXIDATION CATALYST SYSTEM

Fig. 5.17 The exhaust system of an automobile must be developed by using descriptive geometry to determine the lengths and angles necessary to clear the structural members of the chassis. (Courtesy of Chrysler Corporation.)

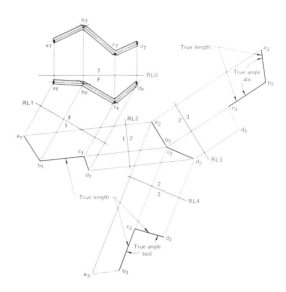

Fig. 5.18 An application of the descriptive geometry is shown in this example where the lengths and angles between pipe segments are found. (Courtesy of General Motors Corporation.)

Fig. 5.19 A windshield system can be refined by plotting it on a cathode-ray tube, using computer graphics. (Courtesy of Ford Motor Company.)

of various payload accommodations. Each of these parts of the design had to be developed with extensive drawings early in the design process.

More detailed physical properties of the vehicle system design are shown in the orthographic views of Fig. 5.21. Although not fully detailed, the overall dimensions and weights are given, even if they are

Scope of Orbiter Payload Accommodations

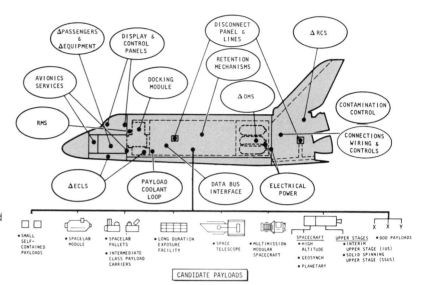

Fig. 5.20 A scale drawing and a number of features are identified in this example of a refined drawing of the space shuttle orbiter. (Courtesy of NASA.)

Fig. 5.21 A refined drawing of the space shuttle is shown here with its overall dimensions. (Courtesy of NASA.)

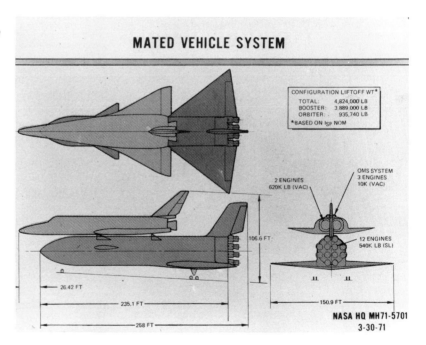

Orbiter Vehicle Dimensions

GEOMETRY	WING	VERTICAL STAB.
AREA	2690 FT^2	413.25 FT^2
ASPECT RATIO	2.265	1.675
AIRFOIL Y_o 199	0010 MOD	WEDGE
SWEEP (LEADING EDGE)	45 DEG	45 DEG
(WING GLOVE)	81 DEG	
M.A.C.	474.81 IN.	199.81 IN.
DIHEDRAL (TRAILING EDGE)	3 DEG 30 MIN	

CONTROL SURFACE AREA & MAX DEFLECTION

AREA		MAX DEFLECTION
ELEVON (ONE SIDE)	206.57 FT^2	-35 TO +20 DEG
RUDDER	97.15 FT^2	±22.8 DEG
SPEED BRAKE	97.15 FT^2	0 TO 87.2 DEG (TOTAL)
BODY FLAP	135.75 FT^2	-11.7 TO +22.55 DEG

Fig. 5.22 Additional information and specifications for the orbiter vehicle are given in this refinement drawing. (Courtesy of NASA.)

general estimates at this point. The more complete specifications of the Orbiter vehicle are given in Fig. 5.22.

5.5 HUNTING SEAT—PROBLEM IDENTIFICATION

To illustrate the method of refining a preliminary product design, the hunting seat problem is continued as an example.

HUNTING SEAT: Many hunters, especially deer hunters, hunt from trees to obtain a better vantage point. Design a seat that would provide the hunter with comfort and safety while hunting from a tree and that would meet the general requirements of economy and hunting limitations.

Refinement Preliminary ideas for this design were developed in Chapter 4. For an example of refinement, we have selected two preliminary concepts for development, but in practice, several should be refined. A list of the design features that are to be incorporated into the design are listed on a work sheet (Fig. 5.23).

Refinement
1. Description of design

A. SUPPORT 350 LBS

B. WEIGH UNDER 10 LBS

C. FIT TREES 6" TO 18" IN DIA

D. WITH ELEVATING (CLIMBING) APPARATUS

E. HAS SAFETY SEAT BELT

F. HAS COMFORTABLE SEAT

G. FOLDS FOR EASY STORAGE

H. STANDING PLATFORM (PERHAPS)

2. Attach scale drawings

Fig. 5.23 A list of a design's specifications and desirable features are listed on a work sheet of this type. In this example, the hunting seat is refined.

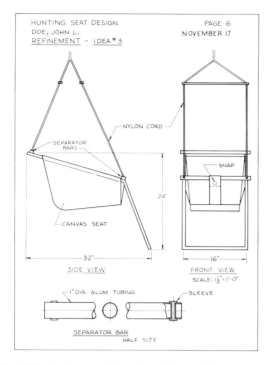

Fig. 5.24 A refinement drawing of idea #3 for a hunting seat previously developed in Chapter 4. Only general dimensions are given on the scale drawing.

Idea #3 is refined in Fig. 5.24 where a scale drawing of the seat is given in orthographic views. Structural components are blocked in, but no attempt was made to detail each part. Tubular parts, such as the separator bars, are blocked in to expedite the drawing process. Also, some hidden lines are omitted.

The important requirement of the refinement drawings is that they be made to scale to give an accurate proportion of the design and to serve as a basis for finding angles, lengths, shapes, and other geometric specifications. Overall dimensions are given on the drawing to give general overall sizes. Specific details are shown for the separator bar to explain an idea for a sleeve to protect the nylon cord from being cut.

The canvas seat is developed as a flat pattern in Fig. 5.25. Stitch lines are shown to explain the details of assembly. The lengths of the nylon cords

are found using descriptive geometry by projecting from the orthographic views. A means of adjusting and securing the footrest is refined in Detail A, where a wing nut is used to tighten the tubular members against a friction washer to lock the footrest in the desired position.

A note is given in colored pencil to indicate that additional details are required to clarify this joint. Also, a method of attaching the nylon cord near the footrest junction is needed.

Another design concept is shown developed as a refinement drawing in Fig. 5.26. Only the major dimensions are given on the scaled instrument drawing.

These two example sheets do not represent a complete refinement of the design; they are merely examples of the type of drawings required in this step of the design process.

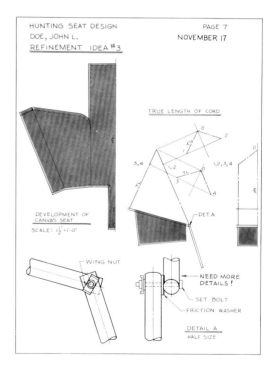

Fig. 5.25 The hunting seat design is refined using descriptive geometry and working drawings. Many additional drawings of this type are required to completely refine a preliminary concept.

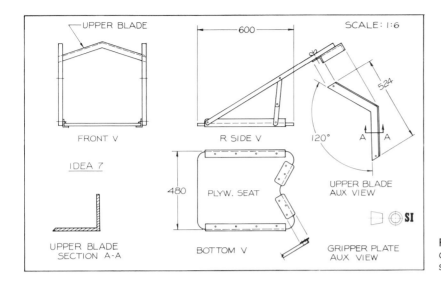

UPPER BLADE

600

SCALE: 1:6

524

120°

FRONT V

R SIDE V

A A

IDEA 7

UPPER BLADE
AUX VIEW

480

PLYW. SEAT

SI

UPPER BLADE
SECTION A-A

BOTTOM V

GRIPPER PLATE
AUX VIEW

Fig. 5.26 Another design concept for a hunting seat is shown as a refinement drawing.

PROBLEMS

Problems should be presented on 8½ × 11 inch paper, grid or plain. Each grid square represents ¼″. All notes, sketches, drawings, and graphical work should be neatly prepared in keeping with good practices as covered in this volume. Written matter should be legibly lettered using ⅛″ guidelines.

1. When refining a design for a folding lawn chair, what physical properties would a designer need to determine? What physical properties would be needed for the following items: a TV-set base, a golf cart, a child's swing set, a portable typewriter, an earthen dam, a shortwave radio, a portable camping tent, a warehouse dolly used for moving heavy boxes?

2. Why should scale drawings be used in the refinement of a design rather than freehand sketches? Explain.

3. List five examples of problems that involve spatial relationships that could be solved by the application of descriptive geometry. Explain your answers.

4. Make a freehand sketch of two oblique planes that intersect. Indicate by notes and algebraic equations how you would determine the angle between these planes mathematically.

5. What is the difference between a working drawing and a refinement drawing? Explain your answer and give examples.

6. How many preliminary designs should be refined when this step of the design process is reached? Explain.

7. Prepare refinement drawings of problems that were identified in the problem sections at the end of Chapter 3, Problems 5 through 9 and 10 through 15. Keep all of these drawings together as they accumulate throughout the design process. It may be necessary to postpone preparing these refinement drawings until you read some of the succeeding chapters and understand sufficient theory.

8. Make a list of refinement drawings that would be needed to develop the installation and design of a 100-foot radio antenna. Make rough sketches indicating the type of drawings needed with notes to explain their purposes.

9. Make a list of refinement drawings that would be needed to refine a preliminary design for a rearview mirror that will attach to the outside of an automobile. Refer to Fig. 4.8.

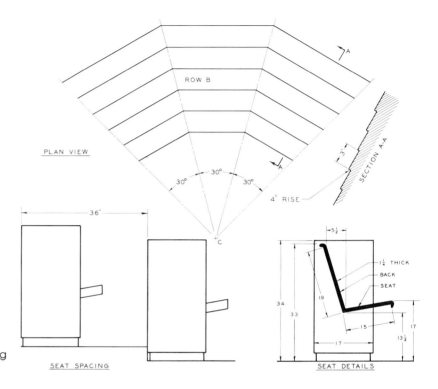

ROW B

PLAN VIEW

SECTION A-A

30° 30° 30°

4" RISE

3"

C

36"

SEAT SPACING

5½

1¼ THICK

BACK

SEAT

34

33

19

15

17

13½

17

SEAT DETAILS

Fig. 5.27 Auditorium seating refinement (Problem 14).

10. After a refinement drawing has been made, the design is found to be lacking in some respects so that it is eliminated as a possible solution. What should be the designer's next step? Explain.

11. Would a pictorial be helpful as a refinement drawing? Explain your answer.

12. List several design projects that an engineer or technician in your particular field of engineering would probably be responsible for. Outline the type of refinement drawings that would be necessary in projects of this type.

13. For the hunting seat design covered in Section 4.8, what refinement drawings are necessary that were not given on the example work sheets? Make freehand sketches of the drawings needed, with notes to explain what they would reveal.

14. A seating layout is shown in Fig. 5.27 for continuous pews, which are shown in detail. Prepare the necessary refinement drawings to determine the following by descriptive geometry.

a) The angle between the seats of row B.
b) The angle between the pew backs of row B.
c) The miter angles of the seats and the backs.
d) A scale drawing of each seat and back.

Select the appropriate scale and sheet size for solving this problem.

Refinement Problems

The following sketches show products that a designer has developed as preliminary ideas. You are to prepare scaled refinement drawings with instruments of each to understand better how the products are to be made and detailed. The types of refinement drawings that you can make are:

a) Orthographic views of each individual part of the design.
b) Orthographic views of assemblies and sub-assemblies of the products.
c) Pictorials to explain relationships of parts.

You must develop and design the details as they are drawn since the preliminary sketches are just that—preliminary. Devise solutions that will work by adding your own inventiveness to the refinement drawings.

15. Sit-up device. A device that clamps to an interior door to hold one's feet while doing sit-ups for physical fitness (Fig. 5.28).

Fig. 5.28 Sit-up device refinement (Problem 15).

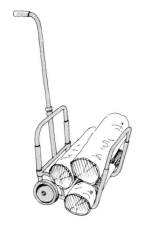

Fig. 5.29 Fireplace caddy refinement (Problem 16).

16. Fireplace caddy. A small hand cart for carrying firewood to an indoor fireplace that is decorative enough to be used as a holder of the wood once it is brought inside (Fig. 5.29).

17. Woodworking clamp. A clamp that is to be permanently attached to a workbench for holding down wood pieces up to 6″ (150 mm) in thick-

ness to aid the woodworker (Fig. 5.30). The design involves the refinement of the mechanism and the collar that attaches to the workbench to permit height adjustment and easy removal.

18. Hold-down clamp. This hold-down clamp is designed to attach to a workbench by drilling and counterboring a hole for a mounting bolt that fits in the T-slot of the clamp. When the clamp is removed, the bolt will drop below the surface of the workbench. (Fig. 5.31).

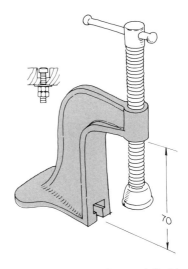

Fig. 5.31 Hold-down clamp refinement (Problem 18).

19. Pointer mount. A mount that connects on the top of a drawing table to hold a rotational pencil pointer (Fig. 5.32).

20. Sharpener guide. A device for holding a chisel edge at a constant angle while being sharpened on a whetstone (Fig. 5.33).

21. Luggage carrier. A portable cart for carrying luggage that will fold up to as small a size as possible (Fig. 5.34).

22. Side view mirror. A fully adjustable mirror that is to be mounted on the side of an automobile (Fig. 5.35).

23. Rotary pump. A rotary pump that operates by squeezing a liquid through a flexible tube (Fig. 5.36).

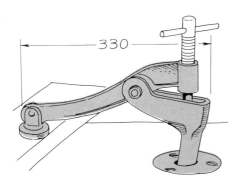

Fig. 5.30 Woodworking clamp refinement (Problem 17).

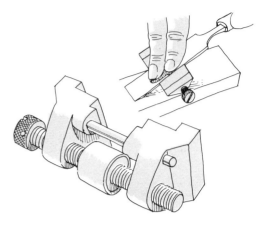

Fig. 5.33 Sharpener guide refinement (Problem 20).

Fig. 5.32 Pointer mount refinement (Problem 19).

Fig. 5.34 Luggage carrier refinement (Problem 21).

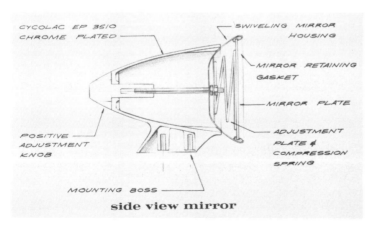

CYCOLAC EP 3510
CHROME PLATED

SWIVELING MIRROR HOUSING

MIRROR RETAINING GASKET

MIRROR PLATE

POSITIVE ADJUSTMENT KNOB

ADJUSTMENT PLATE & COMPRESSION SPRING

MOUNTING BOSS

side view mirror

Fig. 5.35 Side view mirror refinement (Problem 22).

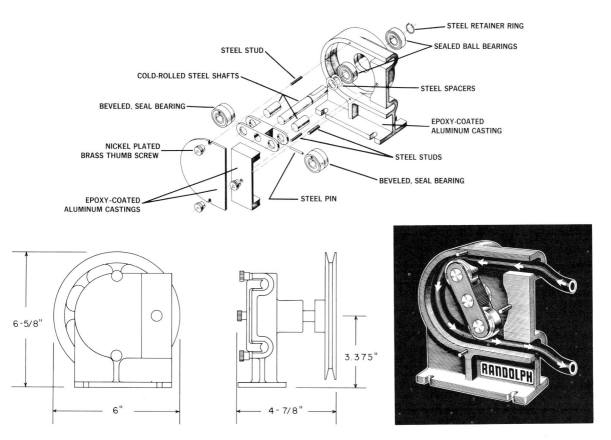

STEEL RETAINER RING

SEALED BALL BEARINGS

STEEL STUD

COLD-ROLLED STEEL SHAFTS

STEEL SPACERS

BEVELED, SEAL BEARING

EPOXY-COATED
ALUMINUM CASTING

NICKEL PLATED
BRASS THUMB SCREW

STEEL STUDS

BEVELED, SEAL BEARING

EPOXY-COATED
ALUMINUM CASTINGS

STEEL PIN

6-5/8"

6"

3.375"

4-7/8"

RANDOLPH

Fig. 5.36 Rotary pump refinement (Problem 23).

DESIGN ANALYSIS

6.1 INTRODUCTION

Analysis is the process most commonly associated with traditional engineering courses. For example, when a bridge is designed to span a river and support a given load, the proposed design must be analyzed to select the proper size structural materials and components for it to be attractive, strong, economical, and functional.

Analysis is the process of evaluation of a proposed design. This stage is characterized by objective thinking and the application of technical knowledge. Less and less creativity is employed as the design process progresses.

6.2 TYPES OF ANALYSIS

The general areas of analysis are (1) functional analysis, (2) human engineering, (3) market and product analysis, (4) specifications analysis, (5) strength analysis, (6) economic analysis, and (7) model analysis.

Functional Analysis   Functional analysis is the most important characteristic of a design; if a design does not function it is a failure regardless of its other desirable features (Fig. 6.1). A doorknob that will not open a door is an unacceptable design even if it is attractive, strong, or economic.

Seldom does a design not function at all. The question is usually deciding what degree of func-

Fig. 6.1 Experimental automobile designs are analyzed and tested to arrive at the most functional and efficient designs. (Courtesy of Ford Motor Company.)

tion is provided by a given design. A bathroom faucet handle may regulate the flow of water, but does it work as well as other designs?

Human Engineering   All designs must serve humans in some respect. They will use the product, travel on it or in it, or profit from its existence. Therefore, the designer must consider the human needs as related to the physical, mental, and emotional characteristics of the user of the product.

Market and Product Analysis   The market for which a product is designed is usually studied in detail during the initial stages of its development (Chapters 3 and 4) and prior to its production. The

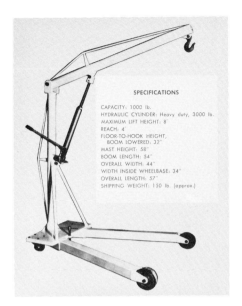

Fig. 6.2 Designs must be analyzed to determine their physical properties, including dimensions, weights, capacities, and other data of this nature. (Courtesy of Air Technical industries.)

initial market survey is performed to determine consumer attitude toward a proposed concept. Product analysis at this later stage seeks to determine the consumer's attitude toward a *specific product design*.

Physical Specfications Analysis In the previous chapter, various specifications were obtained such as lengths, areas, shapes, angles, and so forth. This information needs to be expanded to include such specifications as weights, volumes, capacities, velocities, and ranges of operation. Completed designs are accompanied by specifications as shown in Fig. 6.2.

Strength Analysis A proposed design must be sufficiently strong to support the maximum design load that can be anticipated. Strength is closely associated with function, since a weak design is not a functional design. Strength analysis is a major area of engineering analysis.

Economic Analysis Designs that are unduly expensive have little chance of being profitable in a competitive marketplace. For this reason, the designer must consider economy and type of fabrication as the design nears the completion stage.

Model Analysis A proposed design is seldom produced before a model or prototype has been constructed for analysis and evaluation. Extensive tests may be run on a functional design to gather data on its performance to support its acceptance or rejection.

6.3 GRAPHICS AND ANALYSIS

Engineering graphics and descriptive geometry are valuable in the analysis step of the design process. Systems of forces can often be solved graphically in less time than would be required by the analytical methods employed in engineering techniques.

Empirical data obtained from laboratory experiments and field data can be transformed into algebraic equations by graphical techniques (Fig. 6.3). Mathematical evaluation can also be handled graphically. Graphical calculus must be used when the data does not fit an algebraic equation, but instead plots as an irregular curve.

Clearances between parts and linkage systems (Fig. 6.4) can be analyzed graphically. Even if analysis problems are solved mathematically, it is advantageous to solve them first by graphics to make the mathematical solution easier.

Data such as market surveys, populations, trends, and technical data can be presented and analyzed graphically (Fig. 6.5). A graphical format of this type is a definite aid to the analysis and evaluation of pertinent information.

6.4 FUNCTIONAL ANALYSIS

The analysis of a design's functional characteristics can be performed using mathematics, graphics, and engineering disciplines. Above all, the analysis of function is an area that requires the application of judgment and experience in order to balance the economy, durability, appearance, and marketability of a product's design.

For example, in times of high fuel costs, there is a need for an economical automobile, but simply designing a car that gets good mileage is not the only consideration. Most buyers are willing to give up some degree of luxury, but few would give up all convenient extras on a car just for better gasoline economy. Judgment must be exercised to obtain the optimal blend of economy and comfort.

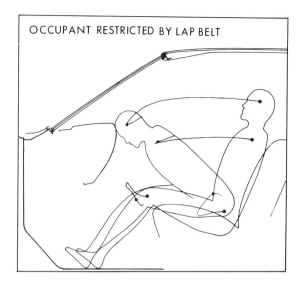

OCCUPANT RESTRICTED BY LAP BELT

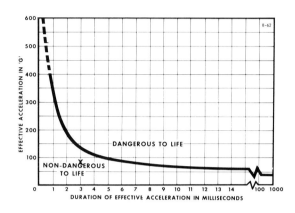

IMPACT ACCELERATION - TIME
TOLERANCE FOR THE HUMAN BRAIN IN FOREHEAD
IMPACTS AGAINST PLANE, UNYIELDING SURFACES

Fig. 6.3 Empirical data obtained from laboratory experiments can be analyzed more efficiently by graphical techniques than by mathematical methods. (Courtesy of General Motors Corporation.)

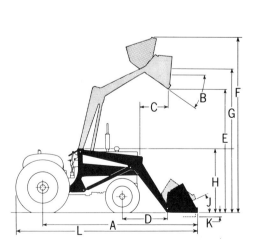

Fig. 6.4 Clearance between functional parts and linkage systems can be analyzed efficiently by using graphical methods. (Courtesy of International Harvester Company.)

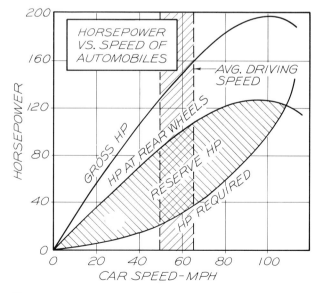

Fig. 6.5 An automobile's power system is analyzed by plotting experimental data on a graph of this type.

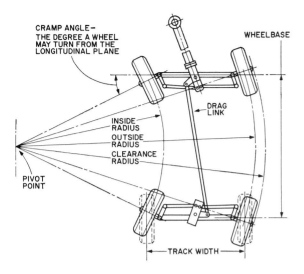

Fig. 6.6 This linkage ensures that an in-plant cargo trailer will tow properly within the available traffic lanes. (Courtesy of *Plant Engineering.*)

Certain functional characteristics can be analytically evaluated when function is the only consideration. The in-plant cargo trailer was designed and its function analyzed graphically in Fig. 6.6. Similarly, the function of a hand-operated clamp is analyzed for its function of clamping a part in position while the part is machined or drilled (Fig. 6.7). In cases of this type, function is limited to a narrow range of requirements: provide a rapid method of manually holding a part in position with a plunger that travels a maximum of 1.5 inches.

6.5 HUMAN ENGINEERING

Human engineering is a significant area of design that is defined by Woodson as follows:

"The design of human tasks, man-machine systems, and specific items of man-operated equipment for the most effective accomplishments of the job, including displays for presenting information to the human senses, controls for human operation, and complex man-machine systems. In the design of equipment, human engineering places major emphasis upon efficiency, as measured by speed and accuracy, of human performance, in the use and operation of equipment. Allied with efficiency are safety and comfort of the operator."

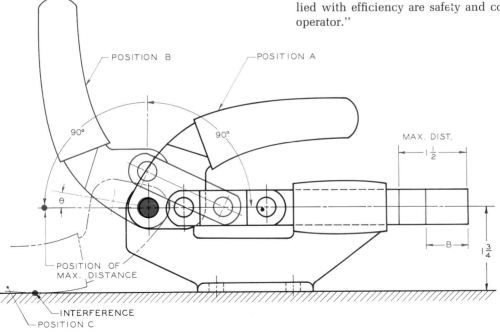

Fig. 6.7 Graphical analysis of a hand-operated clamping device of this type is an efficient means of determining operating limits.

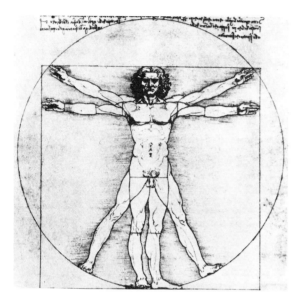

Fig. 6.8 Leonardo da Vinci analyzed body dimensions and proportions in the year 1473.

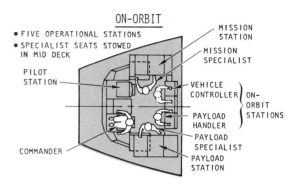

Fig. 6.10 Human factors and living environments had to be analyzed when the crew cabin was designed for the space shuttle. (Courtesy of Rockwell International.)

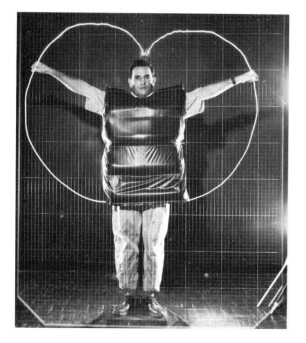

Fig. 6.9 Body dimensions and movements are measured to analyze human mobility that is restricted by a radiation protection vest used by astronauts in space travel. (Courtesy of General Dynamics Corporation.)

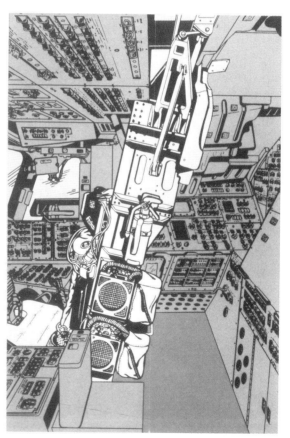

Fig. 6.11 The location of the controls and instruments in the cockpit of the space shuttle was an extensive problem of human engineering. All controls are in triplicate for the maximum of safety. (Courtesy of Rockwell International.)

Leonardo da Vinci analyzed body dimensions and proportions in about 1473 (Fig. 6.8). A close similarity can be seen in Fig. 6.9, where human factors are measured to determine the restriction of human mobility by a radiation protection garment used by astronauts.

Human factors are very important to a successful space program since astronauts must function in a weightless atmosphere where the simplest tasks are quite different than when performed on earth. The spacecraft's cabin must be designed to provide the proper life systems and a flight deck work area in which the on-orbit assignments can be performed (Fig. 6.10). Likewise, the location of the controls and instruments is determined by the reach and movement of the astronauts in the space shuttle (Fig. 6.11). Each lever is designed to require the most natural movement possible for operation. For safety, all controls and systems are in triplicate.

Dimensions   A design must take into account the dimensions, ranges of manipulations, and the senses of the person who will be using the finished

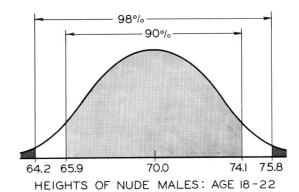

HEIGHTS OF NUDE MALES: AGE 18-22

Fig. 6.13 The distribution of the average heights in inches of American males 18–22 years of age. (Based on data gathered by B. D. Corpinos, *Human Biology* 30:292.) Fifty percent of American males in this age range are taller than 70 inches, and 50 percent are shorter.

product (Fig. 6.12). Variations in physical characteristics tend to conform to the normal distribution curve shown in Fig. 6.13. Note that the average male is 5' 10" tall. Statistical analysis tells us that 90% of American males have a height between 5'6" and 6'2".

Body dimensions of the average American male are shown in Figs. 6.14 and 6.15. The measurements were collected by Henry Dreyfuss to be used for industrial designs that are closely related to human dimensions. The average female body dimensions are given in Figs. 6.16 and 6.17.

Two car seats are shown in Figs. 6.18 and 6.19. The safety bar in Fig. 6.18 can easily be raised to clear the child's head and thus permits her to dismount from the seat with a minimum of assistance. The car seat shown in Fig. 6.19 has a safety bar that cannot be lifted as easily, which makes it difficult for the child to dismount without help. Consequently, it would be more desirable to protect an unattended child riding in a back seat.

Motion   The study of body motion includes the amount of space required to function comfortably, safely, and efficiently. A great deal of study is devoted to the analysis of the interior of an automobile (Fig. 6.20) in order to adapt it to the human body. The seat belt system within the automobile is also related to body dimensions and motion in order to balance freedom of motion with the required restraint in case of a collision (Fig. 6.21).

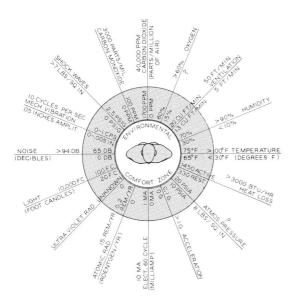

Fig. 6.12 The environmental comfort zone is represented by the inner circle, and the outer circle is the bearable limit zone of human environment. (Courtesy of Henry Dreyfuss, *The Measure of Man*, New York: Whitney Library of Design, 1967.)

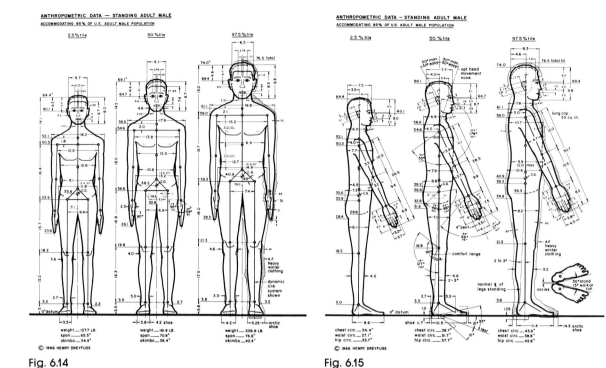

Fig. 6.14

Fig. 6.15

Fig. 6.16

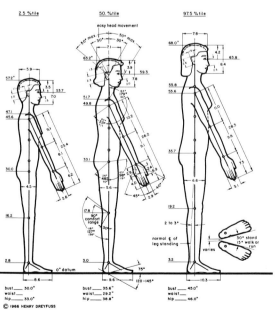

Fig. 6.17

Figs. 6.14 to 6.17 Front and profile body measurements of the adult male and female. These measurements describe 95% of the U.S. adult population. (Courtesy of Henry Dreyfuss, *The Measure of Man,* New York: Whitney Library of Design, 1967.)

Fig. 6.18 This child's seat has a safety bar that encircles the child's body. The safety bar lifts over the child's head with generous clearance.

Fig. 6.19 A child's seat designed for the back seat has a smaller safety bar. The safety bar does not lift over the child's head, which makes it difficult for her to get out of the seat without help.

Fig. 6.20 The design of an automobile is based on the dimensions of the consumers who will use it. This is an application of human engineering. (Courtesy of General Motors Corporation.)

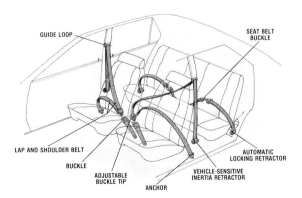

GUIDE LOOP

SEAT BELT BUCKLE

LAP AND SHOULDER BELT

BUCKLE

ADJUSTABLE BUCKLE TIP

ANCHOR

VEHICLE-SENSITIVE INERTIA RETRACTOR

AUTOMATIC LOCKING RETRACTOR

Fig. 6.21 A properly designed passenger-restraint system for an automobile is an example of human engineering. (Courtesy of the Chrysler Corporation.)

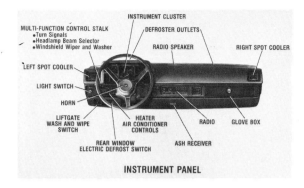

Fig. 6.22 The design of efficient and safe instrument panels in automobiles is a problem involving human engineering. (Courtesy of Chrysler Corporation.)

Vision Designs with gauges and controls are developed to take advantage of the most visually effective means of aiding the operator. You will notice that most gauges are somewhat standardized, and even the same colors are often used among designs to make them as universal as possible. The dashboard configuration of the automobile shown in Fig. 6.22 is very similar to those of other models. Lights are used to make it easy for the driver to manipulate the controls and to obtain instant visual information from the instruments, night or day.

Sound Sound must be received within specified frequencies to be audible. Preliminary studies indicate that sound affects the stress level and, consequently, the efficiency of a person's productivity.

Environment Working environments may include the total industrial plant layout, the conditions in a particular work station, or a specialized location, such as the cockpit of an airplane. The environment includes (1) temperature, (2) lighting, (3) color scheme, (4) sound control, and (5) comfort of operation. Attractive pleasant surroundings are more likely to contribute to efficiency than cluttered, distasteful, or noisy environments.

6.6 MARKET AND PRODUCT ANALYSIS

A product must be analyzed to determine its acceptance by the market before it is released for production. Areas of product analysis are: (1) po-

tential market evaluation, (2) market outlets, (3) advertising methods, and (4) sales features.

Potential Market Analysis General information about the market should be determined by including predicted age groups, income brackets, and geographical locations of the consumers who will be prospective purchasers of the product. Market information of this type will also be helpful in planning advertising campaigns to reach the customer. This step of market evaluation is an in-depth continuation of the market study done in the problem identification step of the design process.

Market Outlets A product may be introduced to the market through existing distribution outlets, such as retail outlets; or unique dealerships might need to be established. For example, large-size electronic computers are not suitable for distribution through general retail outlets; instead, technical representatives work as consultants with clients on an individual basis. On the other hand, a camping seat or other products used by the sportsman can be sold effectively through department and sporting goods outlets.

Unique Features A listing of unique features incorporated in new designs should be made and kept throughout the design process to support the feasibility of the proposed design. Unique features are likely to stimulate interest in a product and attract consumers. Economy of purchase and ease of operation are vital features to any design.

An example is the many innovations that have occurred in writing instruments. We have gone from the quill pen, to the fountain pen, to the ball-point pen, to the felt-tip pen, and so on. Each unique writing innovation has opened vast markets that have attracted consumers away from existing writing instruments.

Advertising Advertising to introduce a product can be done through several media including personal contact, direct mail, radio, television, newspapers, and periodicals. A product for the homemaker can be advertised in magazines subscribed to by homemakers, but a product developed for a business executive must be advertised in a different format. Selection of the appropriate medium must be measured with regard to the comparative advertising costs. Timing is also important for products that tend to be used on a seasonal basis.

6.7 PHYSICAL SPECIFICATIONS ANALYSIS

The physical specifications of a product must be analyzed to finalize the design.

Sizes   A product's overall size and dimensions must be evaluated to ensure that it meets the standards of size (if any) that it should, such as the permissible widths in the case of automobile design. Is its weight within the prescribed range? What is the size of the product when extended, contracted, and when positioned differently?

Ranges   Many products have ranges of operation, capacities, and speeds that need to be analyzed to finalize a design. The designer must calculate or estimate ranges such as seating capacity, miles per gallon, pounds of laundry per cycle, flow in gallons per minute, or electrical power required for operation.

Shipping Specifications   Finally, the designer must be concerned with the packaging and shipping of the product. Will it be shipped by air, by mail, by rail, or by truck? Will it likely be shipped one at a time or in quantities? Will it be shipped assembled, partially assembled, or disassembled. How much will the shipping crate cost and how much will shipping cost? What would be the size of the product when crated for shipment?

6.8 STRENGTH ANALYSIS

Much of engineering is devoted to the analysis of the strength of designs to support dead loads, withstand shocks, and to endure repetitive usage at a variety of motions ranging from slow to fast. A proven technique of measuring a product's strength is by testing in the laboratory or field. The steering wheel of an automobile (Fig. 6.23) is tested in the laboratory to ensure that it is sufficiently durable to function under its design limitations.

In other cases, data is gathered from experiments and is plotted graphically to establish strength characteristics. An example is the strength of clay tile as related to its absorption characteristics (Fig. 6.24). Graphical analysis can be used to analyze the strength of a design from

Fig. 6.23 The strength of a steering wheel can be analyzed by submitting it to fatigue testing in the laboratory. (Courtesy of Ford Motor Company.)

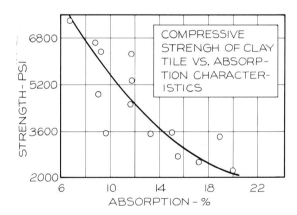

Fig. 6.24 An example of an approximate curve that represents the strength of a structural clay material related to its percent of absorption. (Courtesy of the Structural Clay Products Institute.)

which the proper structural members can be specified.

In Fig. 6.25 the shear and bending moments have been determined by graphical calculus. The resulting load calculations are necessary to select the materials and their sizes required by the design.

You have more instinct for strength of a design than you probably realize. Most people can recognize a chair that is too weak to sit in, a bridge

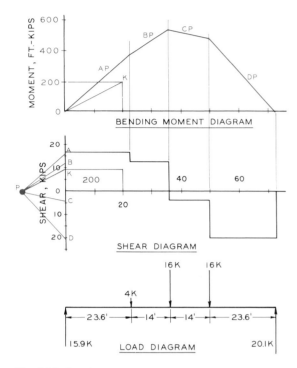

Fig. 6.25 Graphical calculus can be used as an analysis technique to develop a finished design.

that is in need of support, and structural members that are undersize. Designers and engineers use their instincts also in the design of routine structures and products that conform to standards and strengths obtained from previous experience.

6.9 ECONOMIC ANALYSIS

Before a product is released for production, a cost analysis must be performed to determine the item's production cost and the margin of profit that can be realized from it. Two methods of pricing a product or project for the average student assignment are: (1) itemizing, and (2) comparative pricing.

Itemizing   Industrial firms staffed with experienced estimators can determine construction or production cost by itemizing each expense throughout the process of producing a design solution or product. A number of costs that must be itemized are shown in Fig. 6.26. The percentages in this figure vary for different areas of manufac-

turing and retailing. The more specialized the product, the greater the markup, since competition and sales volume will be less.

By studying the detailed working drawings of a project, the cost analyst estimates the costs for materials, manufacturing, labor, overhead, and related production costs. Finally, the production cost for the product, or the construction cost for the project, is determined. Then the profit can be added to arrive at the wholesale price.

Comparative Pricing   The cost of a product can be approximated by comparing the proposed product with the cost of similar products on the market. Such an example is the comparative pricing of the power tools shown in Fig. 6.27. Each is priced at $33.

These tools are very similar; all have the same power sources, the same materials, and the same styling. And most importantly, each has identical market potentialities. Approximately the same

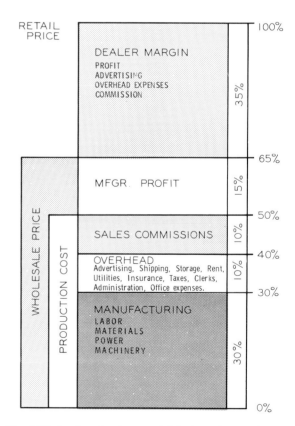

Fig. 6.26 A price structure model that can be used as a guide in the economic analysis of a product.

Fig. 6.27 These three products retail for $33 each. You can see that each is similar in design and will have essentially identical market potential. This is an example of comparative pricing. (Courtesy of Sears, Roebuck and Company.)

numbers of drills are sold as sanders and saws; consequently, production costs and retail price will be similar for each.

An example of using comparative pricing is the price of the hunting seat (Fig. 6.28) compared with that of the exercise bench sold by Best Company (Fig. 6.29). Both the manufacturing requirements and the market volumes are similar; consequently, both sell for about $90. On the other hand, the baby stroller sells for a proportionately

Fig. 6.29 This exercise apparatus is priced at $100, which is similar in manufacture and market appeal as the hunting seat shown in the previous figure. (Courtesy of Best Products Co., Inc.)

Fig. 6.28 This hunting seat can be compared with the exercise apparatus in Fig. 6.29. Each is comparatively priced in the $90–$100 range since both have similar manufacturing problems and similar market potentials. (Courtesy of Baker Manufacturing Company, Valdosta, Georgia.)

smaller price ($45–$65) because it has a larger market volume (Fig. 6.30). The cost of shipping must be considered also.

Comparative pricing is used by manufacturers by determining factors such as cost per square foot, cost per mile, cost per cubic foot, or cost per day. These factors enable them to formulate rough estimates of costs prior to performing more detailed studies.

Miscellaneous Expenses It is easy to overlook some expenses that are usually incurred in developing a new product. Shipping and packaging costs must be considered in pricing a product, and this may be a sizable expense. The designer must investigate shipping charges by the available carriers in order to select the best means.

Fig. 6.30 These baby strollers are priced from $45 to $65, which is less than the hunting seat and the exercise apparatus. The market potential for the strollers is greater than that of the other two products, which makes the strollers cheaper because they can be mass marketed. (Courtesy of Best Products Co., Inc.)

Warehousing and storage of an inventory of finished products may have to be absorbed in the price of a product when an extensive inventory space must be maintained. Coupled with warehousing is the expense of insurance, temperature control, shelving, forklifts, and employees that must be funded. The failure to recognize an expense of this type can affect the profitability of a product; it can even result in a loss.

6.10 MODEL ANALYSIS

Models are effective aids in analyzing a design in final stages of its development. A three-dimensional model can be used to study the design's proportion, operation, size, function, and efficiency. The types of models are:

1. Conceptual models,
2. Mock-up models,
3. Prototype models, and
4. System layout models.

Conceptual Models A conceptual model is a rough model made by designers at any stage of the design process to help them analyze a design or a feature. Conceptual models are for designers' own use rather than as a means of presenting ideas to others in a formal manner. Models of this type may illustrate only a single feature of the product, rather than the entire design.

Mock-ups Mock-ups are full-size dummies of the finished design that give the proper appearance of the product (Fig. 6.31). Mock-ups are constructed more for the presentation of size, shape, appearance, and component relationships than for operational movements.

Prototypes A prototype is a full-size working model that follows the final specifications in all respects. The only exceptions may be in the use of materials. Since a prototype is made mostly by hand, materials that are easier to fabricate by hand may be used instead of those used in final production.

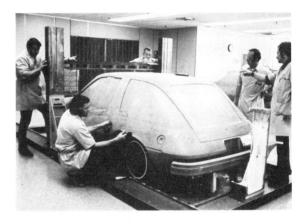

Fig. 6.31 A mock-up model is a full-size dummy of a product made to give a representation of the final appearance, but mock-ups often are not built to function or operate. (Courtesy of American Motors Corporation.)

Fig. 6.32 A system layout model is used to analyze the details of construction of refinery design. (Courtesy of E. I. du Pont de Nemours and Company.)

System Layout Models   System layout models are models used to show the relationships between large layouts such as manufacturing systems, architectural developments, and traffic systems. Models of refineries are often constructed to supplement working drawings during the construction of the facility (Fig. 6.32). Models of this type are used to determine overall layouts and clearance between components within the system.

Model Construction

Model Materials   Model supply dealers will be able to furnish most of the materials required for the construction of student-made models. Balsawood is commonly used in model construction because it is easy to shape with a minimum of specialized tools. Standard parts such as wheels, tubing, figures, dowels, and other structural shapes can be purchased, rather than made, to save time and effort. In some cases, the designer can find the needed

components, such as wheels, on an existing toy that can be purchased at a nominal cost.

Clay and plaster are effective materials for forming a shape with contoured surfaces (Fig. 6.33). Clear plastic can be used to construct models that illustrate both inside and outside design features (Fig. 6.34).

Models should be finished to give a realistic impression of the completed design when they are to be used for presentation to others. Student models can be effectively finished by painting the model to simulate the materials that will be used in the final product.

Fig. 6.34 A model built of clear plastic permits both inside and outside details to be observed. (Courtesy of the Chrysler Corporation.)

Fig. 6.33 This working model of the Hoover Dam was built to show the final appearance of the dam and the surrounding terrain. It was built of plaster and clay. (Courtesy of the U.S. Department of the Interior.)

Model Scale   A model that is used to analyze moving parts of a functional product should be scaled so the smallest moving part will operate. For example, a student model of a portable home caddy is shown in Fig. 6.35. This balsawood model is constructed to demonstrate the function of a linkage system that permits the wheels to be collapsed for convenience of storage. Although the model is small, the linkage system can be operated in the same way as in the completed product.

In general, the model should be constructed to be at least 12'' in overall size. Some models, due to their subject, must be much larger.

Model Testing   The analysis process can be aided by testing models that have been built to the specifications of the final design. The aerodynamic characteristics of the rear styling of two automobiles can be evaluated by wind tunnel tests as

Fig. 6.35 A student model of a portable home caddy was designed to demonstrate how the device folds flat for ease of storage.

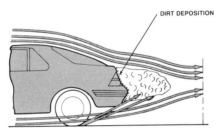

DIRT DEPOSITION

DIRT DEPOSITION

Fig. 6.36 The aerodynamic characteristics of an automobile can be determined by testing a scale model in a wind tunnel. (Courtesy of Ford Motor Company.)

FASTBACK TYPE BACKLIGHT
- Reduces wake size — lowers drag.
- Attached flow is maintained over rear glass — results in clean glass.

SQUAREBACK TYPE BACKLIGHT
- Large wake size — higher drag level.
- Flow separation off roof results in dirt depositing on rear glass.

Fig. 6.37 Aerodynamic tests on automobile models provide information as to how design features should be modified. (Courtesy of Ford Motor Company.)

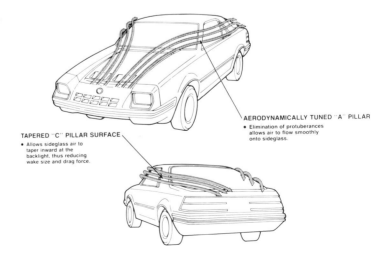

AERODYNAMICALLY TUNED "A" PILLAR
- Elimination of protuberances allows air to flow smoothly onto sideglass.

TAPERED "C" PILLAR SURFACE
- Allows sideglass air to taper inward at the backlight, thus reducing wake size and drag force.

shown in Fig. 6.36. The information obtained from these tests will suggest methods of improvement of the exterior styling to make the automobiles more aerodynámically efficient (Fig. 6.37).

Physical relationships and the functional workings of movable components, such as the hatch and storage area of a car, can be tested in a proto-type model (Fig. 6.38). Models used in this manner permit the most realistic form of evaluation and testing that is possible. Models also are used to test consumer reactions to new products before proceeding with production.

Fig. 6.39 The computer is a powerful analytical tool. This computer links the United States and England for transmitting engineering data. (Courtesy of Ford Motor Company.)

Fig. 6.38 The physical relationships of design components can be observed and tested with the use of full-size models. (Courtesy of Chrysler Corporation.)

6.11 ANALYSIS BY COMPUTER

Increasingly more computer systems are being used to assist the designer, especially in the analysis step of the design process. The data link system shown in Fig. 6.39 is used by Ford Motor Company to instantaneously transmit engineering data and de-sign information from the United States to England in one-twentieth of a second.

Computers can be used to develop and analyze designs in a graphical form for visual study. The designer creates designs for analysis by using an electronic pen, tablet, and a typewriter keyboard to obtain a display on a CRT scope (Fig. 6.40). This important area of engineering analysis is covered in· more detail in Chapter 37.

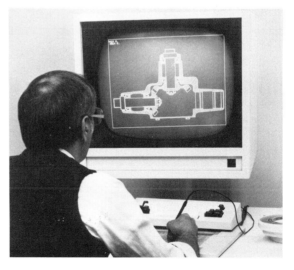

Fig. 6.40 Computer graphics enables the designer to make and analyze drawings by use of an electronic pen, tablet, and a typewriter keyboard. (Courtesy of Applicon.)

Analysis
1. Function

    A. PROVIDES METHOD OF CLIMBING TREE
    B. PROVIDES COMFORTABLE SEATING
    C. PROVIDES DECK FOR STANDING

2. Human factors

    A. SAFETY BELT
    B. 360° VISION
    C. FOOT REST FOR COMFORT
    D. EASE OF CLIMBING
    E. CONVENIENCE ACCESSORIES
    F. PORTABLE : 10-15 LBS

3. Market acceptance

    A. POTENTIAL MARKET
      1. STATE - 40,000
      2. NATION - 1.8 MILLION
    B. CHEAPER THAN DEER STAND
    C. ECONOMICAL: $100
    D. ADVERTISE IN SPORTING MAGAZINES
    E. RETAIL THROUGH SPORTING GOODS DEALERS

4. Physical description

    A. PLATFORM 19" X 24" (456 SQ IN) STAINED
    B. FITS TREES 5" TO 18" DIA
    C. STRAPS FOR CARRYING ON BACK
    D. HAND CLIMBER-SEAT INCLUDED
    E. FOLDS FLAT
    F. WEIGHT - 10 LBS
    G. FOLDS TO 20" X 32"
    H. SHIPPED IN CARDBOARD BOX

5. Strength

    A. SHOULD SUPPORT OVER 500 LBS
    B. SAFETY BELT SUPPORTS OVER 500 LBS
    C. SHOULDER STRAPS SUPPORT OVER 100 LBS
    D. HAND CLIMBER SUPPORTS 400 LBS

6. Production procedures

    A. STRUCTURAL MEMBERS CUT FROM STANDARD ALUMINUM CHANNELS
    B. JOINTS CONNECTED WITH NUTS & BOLTS
    C. EDGES SMOOTHED BY GRINDING
    D. SEAT MADE OF .75" PLYWOOD -FINISHED BY STAINING
    E. SEAT FRAME COVERED WITH VINYL WHICH IS BUCKLED ON

7. Economics

MATERIALS	$10
LABOR	21
SHIPPING	3
WAREHOUSING	2
MISCELLANEOUS	2
TOTAL	$38
COMMISSION	5
PROFIT	20
WHOLESALE PRICE	63
RETAIL PRICE	$90

Fig. 6.41 A typical work sheet used to analyze the various features of a design. It is used here to analyze the hunting seat.

6.12 HUNTING SEAT ANALYSIS

To illustrate a method of analyzing a product design, the hunting seat problem is continued as an example.

HUNTING SEAT: Many hunters, especially deer hunters, hunt from trees to obtain a better vantage point. Design a seat that would provide the hunter with comfort and safety while hunting from a tree, and that would meet the general requirements of economy and hunting limitations.

Analysis

The major areas of analysis are listed on the three work sheets in Fig. 6.41. Additional sheets should be used to elaborate on each of these as required to cover each category of design thoroughly.

To determine the strength of the hunting seat's support system of one design, the loads are deter-

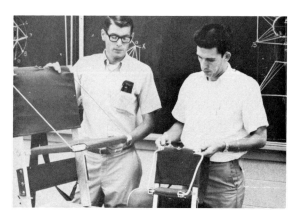

Fig. 6.43 Full-size and half-size scale models were built by Keith Sherman and Larry Oakes to aid them in analyzing their design.

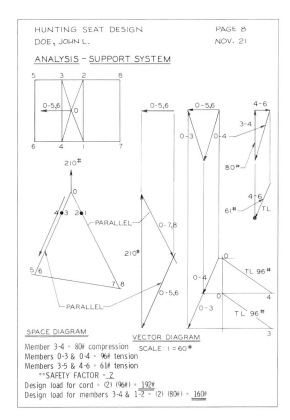

Fig. 6.42 A work sheet on which the support system of a proposed hunting seat is analyzed with graphical vectors.

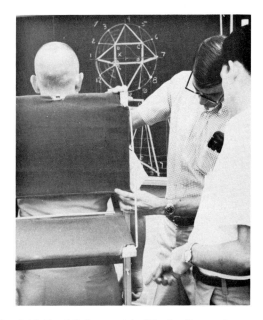

Fig. 6.44 The full-size model of the hunting seat was tested for its adaptation as a backpack.

mined by using graphical vectors (Fig. 6.42). The loads in each support cable are found from which the proper size cable will be selected. The use of vectors is covered in detail in Chapter 30.

For further analysis, models are constructed at a reduced scale and full size (Fig. 6.43). The adaptability of the seat as a backpack can be tested to evaluate comfort and other human engineering factors (Fig. 6.44).

Fig. 6.45 The hunting seat of Baker Manufacturing Company

(a)

(b)

(c)

(d)

(e)

(f)

(a) The hunter hugs the tree and lifts the hunting seat with his feet.
(b) The hunting seat used as a platform for standing.

(c) The hunting seat used for sitting.
(d) An accessory can be used to assist the hunter in climbing the tree.
(e) The hunter pulls upward and lifts

the seat, repeating this until the proper height is attained.
(f) The hunting seat can be used for sitting or standing.

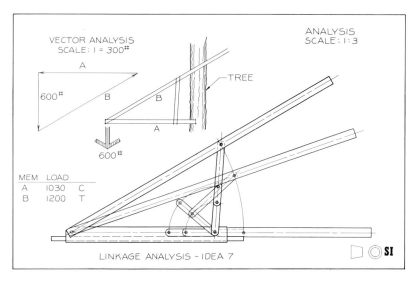

Fig. 6.46 The drawing is used to analyze the linkage system of the hunting seat and a vector diagram is used to determine the forces in the members when the seat is loaded to maximum.

This is the Original Patented Baker Tree Stand

BAKER TREE STAND FEATURES:
1. Platform 19″ x 24″ (456 sq. in.) stained.
2. Back pack wt. 10 lbs.
3. Tested to hold 560 lbs.
4. Fits trees 5″ to 18″ diameter.
5. Riveted assembly folds flat for carrying.
6. Hand Climber fits inside frame.
7. Back packs with Strap Assembly.
8. Safety Belt with Extension and a Tie Down (for safety strap) included.

Fig. 6.47 A summary of the features and physical properties of the Baker Tree Stand. (Courtesy of Baker Manufacturing Company.)

A commercial version of the seat is illustrated in Fig. 6.45, where it is tested to measure its functional features including the method of using it to climb a tree.

An analysis drawing of the hunting seat shown in Fig. 6.46 illustrates the operation of the linkage system that permits the seat to collapse into a single plane for carrying ease. The forces in the members are also found graphically by using vector analysis.

The overall features and the physical properties of the Baker Tree Stand are shown in Fig. 6.47. These features are helpful to a consumer in making a purchase.

PROBLEMS

The following problems should be solved on 8 1/2″ × 11″ paper, accompanied by the necessary drawings, notes, and text. Answers to essay problems can be typed or lettered. All sheets should be stapled together or included in a binder or folder.

General

1. Make a list of human factors that must be considered in the design of the following items: a canoe, a hairbrush, a water cooler, an automobile, a wheelbarrow, a drawing table, a study desk, a pair of binoculars, a baby stroller, a golf course, the seating in a stadium.

2. What physical quantities would have to be determined in the designs listed in Problem 1?

3. Select one of the items given in Problem 1 and make an outline of the various steps that should be taken to satisfy the following areas of analysis: (1) human engineering, (2) market analysis, (3) prototype analysis, (4) physical quantities, (5) strength, (6) function, and (7) economy.

Human Engineering

4. Using your body as the average, make a drawing to indicate the optimum working areas for you when in a sitting position at a drawing table. Assume that you are to use your measurements as a basis for designing a drawing table to satisfy the needs of your class. This table will be marketed to schools similar to yours. Your reach, posture, and vision will have considerable effect upon its dimensions. Your finished drawing should give three views of the ideal working area for you while drawing; the drawing should also show the most efficient positioning of instruments for working. Experiment with the angle of tilt of the table top to determine the most comfortable position for working.

5. Using the dimensions for the average person given in this chapter, design a stadium seating arrangement that will serve the optimum needs of the spectators. Determine the dimensions shown in Fig. 6.48. A primary consideration will be the slope of the stadium seating, which should be designed to provide an adequate view of the playing field. The comfort of the average spectator and provision for traffic between seats must also be

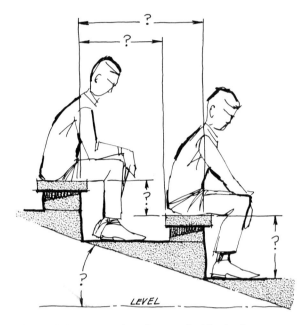

Fig. 6.48 Human engineering applied to stadium seating.

considered. It is desirable to get as many people in the stadium as possible while providing adequate comfort. Use the average dimensions given in this chapter and your own body dimensions to simulate these conditions in your classroom.

6. Compare the dimensions of the students in your class with the standards given in the Section 6.5. For example, compare the average height of your class with the national standard height.

7. Design a backpack that will be used on a camping trip. Decide what are the minimum belongings a camper should carry, and use their weights and volumes in establishing the design criteria. Make sketches of the pack and the method of attaching it to the body to provide the optimum in mobility, comfort, and capacity. Determine the optimum load a camper could carry on his or her back during a hike lasting several hours.

8. Establish the dimensions, facilities, and other provisions that would be needed in a one-person bomb shelter to provide protection for a period of 48 hours. Make sketches of the interior in relationship to a person and supplies. How will ventilation, water, food, and other vital resources be provided? Explain your design as it relates to human engineering needs.

9. As an engineer you must design a manhole access to an underground facility (Fig. 6.49). What must the diameter of the manhole be to permit a person to climb a ladder for a distance of 10′ with freedom of movement? Make a sketch of your design and explain your method of solving your problem.

10. Assume that you are assigned to design a one-person facility for temporary observation service in the Arctic. This facility is to be absolutely as compact as possible to provide for the needs of a single person during 72-hour periods while he or she serves as an observer and operates a radio. Determine the facilities and provisions that would be needed, including heat, ventilation, and insulation. Make sketches of your design and explain items that you consider to be essential to the human engineering aspects of the problem.

11. Design a configuration for an automobile steering wheel that would differ from present design, but be just as functional. Base your design on human factors such as arm position, grip, and vision. Make sketches of your design and list items that you considered.

12. Make sketches to indicate safety features that could be built into your automobile to reduce the seriousness of injury caused by accidents. Explain your ideas and the advantages of your designs. Primary consideration should be given to the human aspects of the designs.

13. Assume that you prefer to alternate between a sitting and a standing position when working at a drawing table. Determine the ideal height of the table top for work in each position. Indicate how a table could be devised to permit instant conversion from the height for standing to the height for sitting.

14. Identify some human engineering problems that you recognize as being in need of solving. Present several of these to your instructor for approval. Solve the approved problems. Make a series of sketches and notes to explain your approach.

Market Analysis

15. Assume that you are responsible for conducting a market analysis of the drill shown in Fig. 6.27. Include in your analysis all of the areas covered in Section 6.6. Assume that this product is new and has never been introduced in an electrically powered form before. Outline the steps you would take in conducting a product and market analysis.

16. Make a product and market analysis of the car seat shown in Fig. 6.18, or the one shown in Fig. 6.19, following the steps suggested in Section 6.9. Arrive at a market value that you feel would be satisfactory, and determine the outlets and other information of this type that would be important to your analysis.

17. Assume that the following cost estimates of producing hunting seats were given: 100 seats, $35 each; 200 seats, $20 each; 400 seats, $10 each; 1000 seats, $8.50 each. Using these figures, determine the price at which you could introduce the seats to the market on a trial basis and still have some financial protection. Explain your plan.

18. List as many unique features of the hunting seat as you possibly can that would be important to a sales campaign and to advertising. Make sketches and notes to explain these features.

Models

19. Give examples of items for which it would be necessary to build full-size prototypes for de-

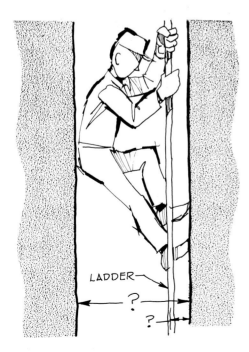

Fig. 6.49 Optimum size for access to a manhole.

tailed analysis and testing before the product was produced. Give examples of products or designs that would not require a full-scale prototype, since a small-scale model could serve for analysis. Explain your answers.

20. List the scales and the materials that you would use to construct models of the items given in Problem 1. Explain your choices.

21. List several types of designs that you think would be most effectively presented in model form to a group of stockholders for possible financial help. Explain your choices. Give examples of projects whose positions would not be improved by presentation of a model. Explain.

DECISION

7

7.1 INTRODUCTION

Once the design has been conceived, developed, refined, and analyzed, a decision must be made to determine which design is the most worthy of implementation. The decision step is based on facts and data; but at best, it is still subjective and must be made by experienced individuals.

In this chapter, we discuss the details of organizing and planning an oral presentation, the preparation of technical reports, and the process of making a decision.

7.2 TYPES OF PRESENTATIONS

The presentation for decision can be given to a few people or to a large group. The groups can vary from a single project associate to laymen who are unfamiliar with the project and its objectives.

Informal Presentation The most fundamental type of presentation is that given to several immediate associates or to a supervisor. This presentation is very informal, without specially prepared visual aids of the type used for a large group. However, pictorials, schematics, sketches, and models should be used to convey an understanding of the design concept.

Ideas may be sketched and discussed when only a few people are involved (Fig. 7.1). In a discussion that includes a few more individuals who

Fig. 7.1 A decision may be the outcome of a presentation to a single individual, where ideas and designs are discussed and sketched informally. (Courtesy of Ford Motor Company.)

may be working jointly on a project, ideas, schematics, and sketches can be drawn on a blackboard (Fig. 7.2). Informal presentations are usually preliminary presentations, made during the beginning stages of a design development; the same material will be formally presented in more detail at a later date.

Formal Presentation The formal presentation receives the most emphasis in this chapter, because it is usually more crucial than the informal presentation.

Fig. 7.2 An informal presentation may be given to professional associates by means of preliminary notes and blackboard sketches. (Courtesy of Exxon Research of New Jersey.)

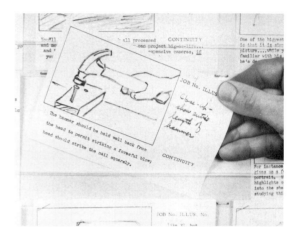

Fig. 7.3 A presentation should be planned by using a series of 3″ × 5″ cards on which the various topics and visual aids can be noted. (Courtesy of Eastman Kodak.)

Fig. 7.4 Planning cards can be used to prepare the sequence of the presentation by using a planning board (as shown here) or by merely arranging the cards on a table top. This method permits easy modification in the sequence by repositioning the cards. (Courtesy of Eastman Kodak.)

The group to whom the presentation is given may be composed exclusively of professional associates, administrators, or laymen, or it may be mixed. Function and acceptability of the design from an engineering standpoint are the primary considerations of the engineering associates. Administrators will be concerned with the economic feasibility of a design and its estimated profit return. The group of laymen could be the clients for whom the project is designed, stockholders, or members of the general public who may vote on the approval of a design.

7.3 ORGANIZING A PRESENTATION

An acceptable method of planning a formal, oral, or written report is the use of 3″ by 5″ cards (Fig. 7.3). Separate ideas that are to be illustrated graphically are first written on separate cards. A rough sketch is made to indicate the type of illustration required and the method of reproduction to be used, and brief notes are added to suggest a general outline of the discussion that will be given orally to accompany the visual aid. The total sequence of the presentation can be easily reviewed by displaying the cards on a table, bulletin board, or planning board (Fig. 7.4).

The sequence or content of the cards can be changed easily at this point. The completed cards

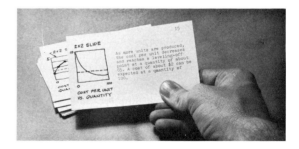

Fig.7.5 An example of the layout of a 3″ × 5″ card that shows a sketch of the visual and its accompanying text.

(Fig. 7.5) should contain the following information:

1. *Number.* The card's position in the sequence of visual aids.
2. *Illustration.* A sketch of the illustration that must be prepared. Notes can be included to indicate special effects, types of visual aids, and other specifications that will be helpful in preparation.
3. *Text.* A brief outline of the oral presentation that will be given with the visual aid.

If a variety of visual aids will be used during the presentation such as flip charts, slides, transparencies, or combinations of these, the material should be divided into sections that will allow a smooth transition from one type of presentation to another.

After the 3″ by 5″ cards have been properly sequenced and related to the presentation, they should be grouped according to similarities in illustration or production. All graphs that are to be illustrated by transparencies should be grouped for preparation and production at the same time. Grouping of similar graphs and illustrations that will be photographed to make slides will facilitate production and reduce set-up time and preliminary preparations for a particular process.

7.4 VISUAL AIDS FOR PRESENTATION

Communication of any type of information can be enhanced by the utilization of visual aids. Visual aids help to ensure an understanding of the material.

The visual aids most commonly used are flip charts, photographic slides, overhead projector transparencies, and models. General rules that will improve the effectiveness of any type of visual aid selected are:

1. Each slide or chart should convey one thought only.
2. Lengthy statements should be reduced to key phrases or words that will communicate the thought intended.
3. Tabular data should be presented in the form of a graph for easy comprehension.
4. Each slide or chart should be clearly readable by the group for whom the presentation is intended.
5. Illustrations, color, and attention-getting devices should be incorporated into the slides or charts for appropriate emphasis.
6. A sufficient number of slides or graphs should be prepared in order to ensure that no notes are needed by the presenter.

Flip Charts

Flip charts consist of a series of illustrations prepared on medium-weight paper and mounted on a backing board that can be placed on an easel for presentation before a group (Fig. 7.6). Each successive chart is flipped over the backing board to proceed to the next during the discussion. Flip charts are used for small conferences or presentations in an area no larger than an average-sized classroom. A common format for a flip chart is 30″ by 36″; smaller sizes tend to be too small for most groups.

Paper   Brown or white wrapping paper is very suitable for a series of flip charts that will be used

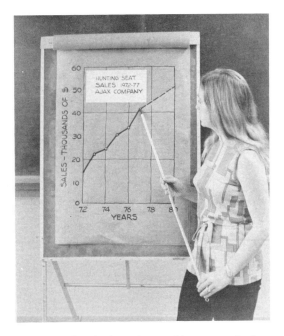

Fig. 7.6 The flip chart can be used to communicate with small groups. This chart was drawn lightly in pencil and finished using India ink applied with a brush.

Fig. 7.7 Felt-tip markers are well suited to the preparation of flip charts and other types of visual aids.

for a presentation to be given only once. A corrugated cardboard can be sized to serve as a backing board to which the flip charts can be attached. The sheets are cut to a size that will match the backing-board dimensions. The card sequence is used as a guide in laying out each chart.

Light guidelines can be ruled with a straight edge to improve layout.

Lettering Materials The final lettering can be done with felt-tip markers, chalk, ink, tempera or sign paints, or overlay film. Felt-tip markers, used extensively for fast bold lines in a variety of colors, give very sophisticated effects if properly used. Both freehand and straight-edge drawings can be made with them (Fig. 7.7). India ink is an

effective medium for lettering and for adding emphasis to a chart. It can be applied freehand with a brush or by means of special lettering pens for bold lines (Fig. 7.8). The designer should develop a freehand technique for lettering presentations, since it is faster than mechanical lettering and is equally attractive. Tempera or sign paints come in many colors and can be applied with a brush for an attractive layout (Fig. 7.9).

Transfer Films These commercially produced products save a great deal of time in the preparation of artwork of a repetitive nature while giving a highly professional appearance. The application of a specially produced sheet is shown in Fig. 7.10.

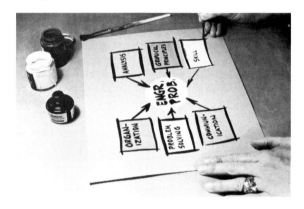

Fig. 7.9 Tempera (a bottled type of water color) and other water-base paints are excellent for the preparation of colorful visual aids.

Fig. 7.8 This flip chart was drawn on brown wrapping paper using India ink and a brush. A title is being applied with rubber cement. Colored construction paper can be used to highlight flip charts.

Fig. 7.10 Transfer films are available for the preparation of visual aids. In this example, an emblem is being transferred onto the working surface. (Courtesy of Instantype® Incorporated.)

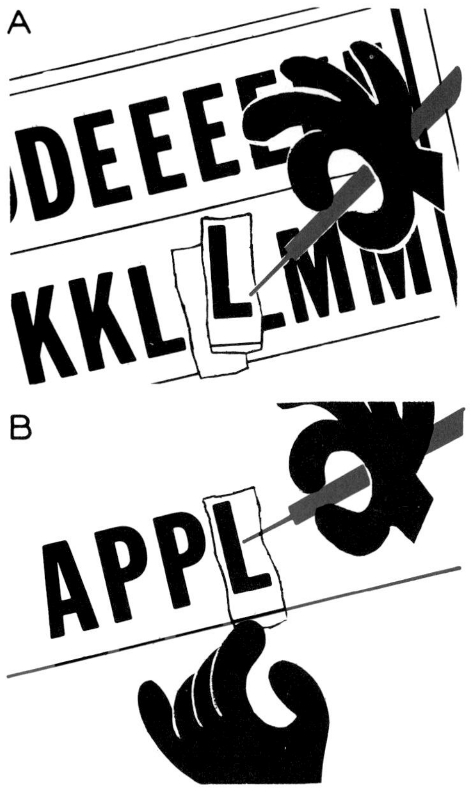

Fig. 7.11

Step 1 Remove desired letter from the sheet by cutting the film.

Step 2 Align the printed line on the film with your guide line. Burnish. (Courtesy of Artype® Incorporated.)

Another useful application of this type of material is shown in Fig. 7.11. In this case, the letters are printed permanently on the film and the desired letters are removed from the sheet by cutting a portion of the film, as shown in Step 1. A letter is removed and the line printed under the letter on the film is aligned with the horizontal guideline (Step 2). The letter is burnished to secure letter and film to the paper. The guideline is then cut away.

Sheets of this type may have a glossy surface or a mat finish that is hardly visible. They are available with dot patterns, symbols, letters, numbers, and in various colors for many applications to the presentation for decision.

Color A chart will be more attractive and effective if color is used to add variety. Any color selected should be readable from the audience and compatible with the other colors used.

Color can be added to a chart very easily through the use of construction paper and rubber cement. A graph or schematic can be laid out as in the initial steps of any layout. Instead of shading or painting areas to be colored, construction paper can be cut and pasted into position with rubber cement. In general, a chart of this type should be bold and should have no tedious or time-consuming details. This technique is especially useful for bar graphs, as illustrated in Fig. 7.12.

Assembly The finished charts are assembled in order, with a title chart covered by a blank sheet of paper. All sheets are stapled or attached to the backing board. The blank cover sheet conceals the theme of the flip charts and eliminates any anticipation of the presentation.

Presentation In advance of the meeting, the flip charts should be mounted on an easel or secured in the most suitable position for viewing. It may be appropriate to give a few introductory remarks concerning the presentation before revealing the title page. The charts are flipped in sequence and referred to with a pointer to ensure that the speaker is not blocking the view of the audience. Written notes can be kept to a minimum since the flip charts serve as notes. But it may be helpful to write certain key points lightly (visible only to the presenter) on the charts. As in the technical report, there should be concluding charts that emphasize and review key points.

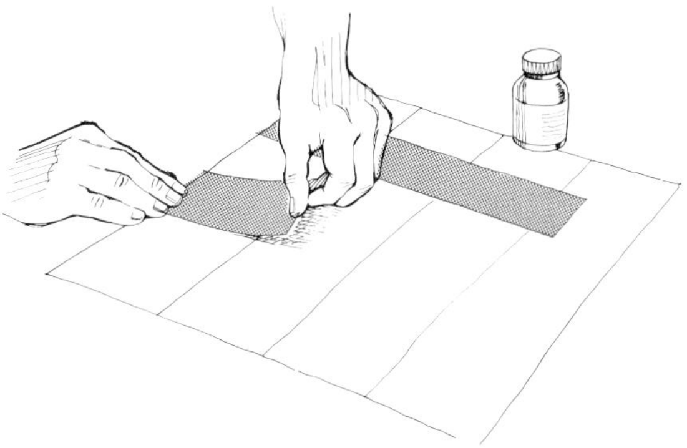

Fig. 7.12 Colored construction paper can be rubber cemented onto flip charts and other visuals to give an attractive and colorful appearance with the minimum of effort and expense.

Photographic Slides

Photographic slides are very effective in situations where flip charts are too small for good visibility, as in presentations to large groups. They are a necessity if photographs are required to depict actual scenes or examples. Photographic slides, usually 2″ by 2″ in size, can be easily filed and used in other presentations at relatively low cost.

Layout of Artwork for Photographic Slides The film portion of the average 2″ by 2″ slide has a ratio of 2:3, and the same ratio should be used to lay out a surface on which to work. Figure 7.13 illustrates a method for sizing the layout in correct proportion for a 35-mm slide. A format size of 8″ by 12″ is appropriate for most information that will be presented by slides.

Fig. 7.14 Artwork for photographic slides can be prepared by using felt-tip markers, tempera paints, ink, and other techniques.

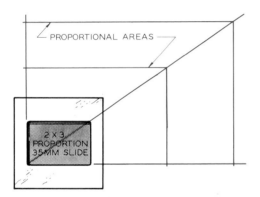

Fig. 7.13 This method of sizing artwork to proportional sizes can be used to ensure that the artwork will properly fill a photographic slide.

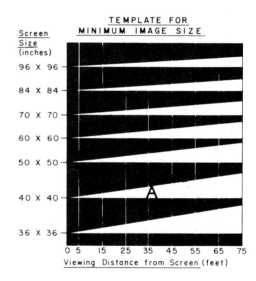

Fig. 7.15 This template can be used to determine the minimum size for lettering that is to be projected onto a screen with an overhead projector. The letter A is located on the chart for a screen size of 40″ × 40″ and a viewing distance from the screen of 35 feet. (This template has been reduced to half size; consequently, the heights of the letters must be doubled.

A combination of different colors on each slide will add variety to the sequence and maintain a higher level of interest. Colored construction papers, mat board, and other poster materials can be used in laying out the artwork for a slide. Light guidelines are drawn with a pencil, which will not leave distracting traces on the finished slide.

A margin of at least 1″ on all sides of a layout should be allowed to ensure that no edges will show when the art is photographed. Colored chalk, ink, tempera, or other media used in the preparation of flip charts can be used for preparing photographic-slide layouts (Fig. 7.14). Selection of the proper size for lettering is critical to ensure

readability from any point in the audience. A template for selecting the proper size lettering is given in Fig. 7.15. Only capitals should be used, with space between lines equivalent to the height of the letters. The lettering specifications in the artwork in Fig. 7.16 are suggested for an illustration of 5 3/4″ by 8 1/2″ for good projection.

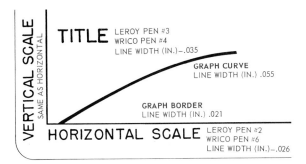

Fig. 7.16 The line weights for preparing photographic slides. Lettering or lines should be at least this bold. (Courtesy of Eastman Kodak.)

Fig. 7.18 A light meter and a 35-mm reflex camera are required for photographing artwork for photographic 2″ × 2″ slides.

Special effects can be achieved with three-dimensional letters (Fig. 7.17) or unusual backgrounds. Paper cutouts add color and interest to the slide. Photographs, schematics, and diagrams may be photographed directly from magazines, textbooks, and other references if details are bold enough to be legible when the slide is projected.

Copying the Layouts   The equipment needed for producing the slides consists of a camera and a copy stand, light meter and lights (Fig. 7.18). A 35-mm reflex camera is recommended because the reflex camera has a through-the-lens viewfinder, and the photographer can accurately focus and align each slide before photographing it. The copy stand holds the camera in the proper position during the photographing and thereby reduces problems of focusing and movement of the camera (Fig. 7.19).

Fig. 7.19 Artwork and charts can be reproduced with a 35-mm reflex camera mounted onto a copy stand, which facilitates the positioning of the camera.

Fig. 7.17 Three-dimensional letters are available for the preparation of photographic charts.

In many cases, however, the camera can be hand held.

If all layouts are drawn the same size, the camera can be left in the same position during the entire photographing process. If slides are made indoors, photographic lights that match the type of film being used are needed. When weather permits, copy work can be photographed in natural light. Book illustrations that are too small for regular copying can be photographed with a close-up lens that will fill the slide with the area being photographed.

Fig. 7.20 Photographic slides should be sorted and arranged in proper sequence prior to loading in the slide tray.

The developed and mounted slides should be reviewed and sorted to determine their quality prior to the presentation (Fig. 7.20). If the information on a slide will be needed at more than one interval during the presentation, duplicate slides should be made rather than attempting to return to a previously shown slide. The mounts of all slides should be numbered in their final sequence.

The Slide Script A slide presentation that will remain on permanent file for repetitive use should have a script to serve not only in the presentation

but also as a review guide. An example of a slide script is shown in Fig. 7.21.

A script can be prepared on standard 8 1/2″ by 11″ paper and maintained in a binder or a notebook. The left-hand side of the script sheets contain black and white photographs of the slides used with the narration entered on the right-hand side. The narration is not intended to be read but to serve as a guide for the person who will give the presentation. The photographs allow the speaker to be aware of the slides that will be shown in sequence and enable him or her to key the narration to the slides without getting out of phase during the presentation.

Overhead Transparencies

Overhead transparencies are reproduced on 8 1/2″ by 11″ transparent materials by the heat-transfer process or the diazo process, and are attached to mounts (Fig. 7.22A). Tracing paper is the most

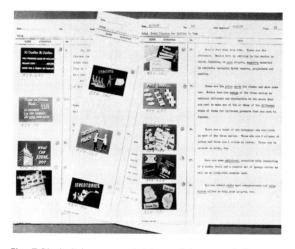

Fig. 7.21 A slide manuscript for a slide presentation should be prepared for an important presentation or for a presentation that will be given on a repetitive basis. (Courtesy of Eastman Kodak.)

Fig. 7.22 (A) The transparency used on an overhead projector is comprised of an 8 1/2″ × 11″ transparency mounted on a 10″ × 12″ frame. The projection area within the frame is about 7 1/2″ × 9 1/2″. (B) Different colors can be added as overlay flips to add interest and permit the user to show information in steps.

common material used in the preparation of transparencies since it can be used in either process without special equipment.

Line work should be prepared in black India ink for best reproduction. Overlay materials and graphing tapes can be used to give a professional appearance. Lettering should be legible (at least 1/4″ high) and brief for easy reading. The finished tracing-paper drawing is transferred to the transparencies in much the same way a diazo print is made. Both sheets are placed in contact and run through the duplicator.

Color Overlays   Transparency materials are available in a variety of colors. Several overlays of different colors can be hinged to the basic transparency for a sequential presentation that shows the development of an idea or problem. The artwork for overlays is prepared on tracing paper that has been positioned over the basic layout. Areas or lines that will be reproduced in a color during the duplicating process are drawn with black India ink. Register marks (Fig. 7.22B) are used to align the transparencies with one another during the artwork presentation and mounting phase.

Presentation with Transparencies   The overhead projector can be placed at the front of the room, much nearer the screen than a slide projector can be placed. The presenter can stand or sit near the projector in a semilighted room and refer to the transparencies while facing the audience (Fig. 7.23). With a small pointer, the presenter can indicate important points on the stage of the projector, and the image of the pointer will project on the screen. Just as several overlays can be hinged into position to develop an idea in sequential steps, opaque paper overlays can be hinged to the mount with tape to conceal portions of the overlay and focus audience attention on a single topic at a time.

A transparency should not be left on the screen during any long period of time in which it is not germane to the presentation. Instead the projector should be turned off to direct audience attention to the speaker and turned on again only when attention should be redirected to the screen.

Models

A model is an excellent means of communicating the final design concept in its most realistic form.

Fig. 7.23 The overhead projector can be used in a semilighted room while the presenter faces the audience.

Fig. 7.24 A full-size prototype model of a design can be used as an effective presentation to a small group. (Courtesy of the Chrysler Corporation.)

The general configuration of a design, as well as its major working parts, can be shown. Models are especially helpful in explaining three-dimensional concepts to a group unfamiliar with the interpretation of drawings.

A model that is an actual-size prototype of the completed design (Fig. 7.24) gives the most accurate impression of the finished design. However, in many cases the model is considerably smaller.

The model should be at least 12″ in size to be visible from a distance of more than 12 feet. Photographic slides can effectively supplement the model during a group presentation. A series of close-ups taken from appropriate angles can help to give each member of the audience a clear view of a relatively small model.

7.5 THE GROUP PRESENTATION

Preparation

Planning for the presentation should be done in advance and should include consideration of seating arrangement, location of visual aid equipment, and facilities available in the room.

Most conference rooms, classrooms, and auditoriums are arranged to afford good viewing when visual aids are used, but conditions should be checked by viewing the presentation materials from extreme locations in the audience area prior to the meeting.

The room where the presentation will be given should be checked for good ventilation, lighting, sound, and electrical control. Arrangements should be made for control of lights during the presentation by an assistant or by a remote control device. Sound amplification may be necessary if room and audience are large.

All projectors and other visual-aid equipment should be positioned and focused before the audience arrives. The screen or flip charts should be located where they afford the best view from all positions. Remote controls for slide projectors should be ready for use near the speaker's position.

Slide trays should be loaded and on the projector for immediate operation. Overhead transparencies should be grouped in sequence near the projector. Flip charts should be secured to the easel and a pointer provided for calling attention to specific points during the presentation.

The speaker should take care not to block the view of the audience when employing visual aids. A rehearsal with an assistant in the seating area to call attention to any blocking movements will help the speaker avoid them.

The Presentation

The speaker should give the presentation at a moderate pace. Visual aids should be used only when they are directly related to the narration so as not to be a distraction (Fig. 7.25).

The speaker should thoroughly familiarize the audience with the objectives of the presentation in an introduction. The order of presentation should conform very closely to the technical report, and in the discussion emphasis should be placed on significant points.

A positive approach in selling ideas should not be confused with high-pressure salesmanship that may be deceiving. The presenter should be the first to point out design weaknesses in his presentation, but these weaknesses should be offset by alternatives that compensate for them. A presentation should not be given for decision unless the designer feels that he or she has a worthy solution to propose. Strong and weak points should be discussed in an objective manner. Conclusions and recommendations should be given in light of available research and analysis. If after thorough analysis a designer feels that he or she cannot recommend a design, the reasons for that conclusion should be stated.

A period for questions and answers is usually provided at the end of the presentation to clarify technical points that may not be fully understood. If possible, the technical report (discussed in the next section) should be available to members of the audience to accompany the oral presentation. In answering specific questions, the speaker may refer to the technical report where detailed supplementary data are available.

Presentation Critique

The evaluation of a presentation in a classroom situation is important to the student, since it is through an evaluation that the student improves.

The evaluation form in Fig. 7.26 has been designed to aid the student in planning as well as evaluating a presentation. The names of each team member are given at the top of the sheet and their percent contribution is computed by the team as a group. The total of this column must add up to 100 percent. The F-factor is found for each member

Fig. 7.25 A presentation should make full use of graphical aids and models to assist the speaker in communicating ideas. (Courtesy of Bendix Corporation.)

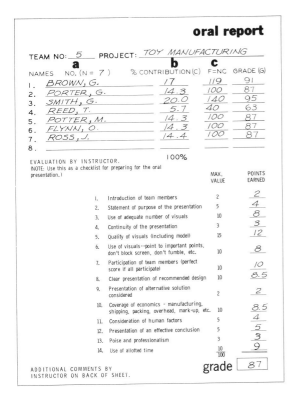

Fig. 7.26 An evaluation form for grading a team's oral presentation. Individual grades are found by using the chart in Appendix 50.

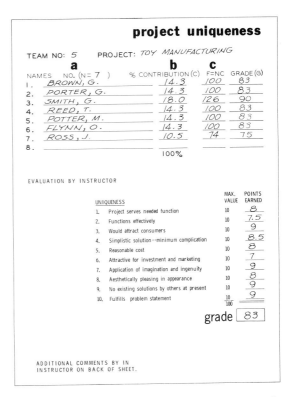

Fig. 7.27 An evaluation for grading the uniqueness of a team's project. Individual grades are found by using the chart in Appendix 50.

by multiplying the number of members on the team by the percent contribution of each.

The grade of each individual can then be found by using the chart in Appendix 50. This method ensures that each member of the team is awarded a grade commensurate with his or her contribution to the presentation.

Another form (Fig. 7.27) is used to evaluate the creativity of the project being presented. Project uniqueness is that part of the evaluation process that considers the level of creativity that was exercised in arriving at the solution to a design problem.

This second form allows for the division of the team's grade among the members of the team based on the contribution of each. The instructor grades the project and then uses the chart in Appendix 50 and the F-factors determined by the team as a whole to arrive at an individual grade for each team member.

7.6   THE TECHNICAL REPORT

Engineers, technologists, and technicians at all levels must know how to prepare a written report, since this is the universal means of transmitting information.

The report can be written for one of the following purposes: (1) proposal for a project, (2) progress report, and (3) final report (Fig. 7.28).

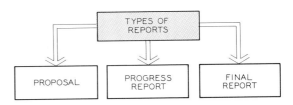

Fig. 7.28 The three basic types of technical reports.

The Proposal

The proposal is a report written to substantiate the need for a given project that will require the authorization of funds and the utilization of manpower within the organization. It could also be a report to be submitted to a client, outlining a recommendation that will necessitate an expenditure of funds. Since a project will not become a reality unless approved, the proposal is usually written in a very thorough, formal form that reflects sound analysis and thinking.

The proposal is organized to include data, costs, specifications, time schedules, personnel requirements, completion dates, and any specific information that will aid the reader in understanding the project. Above all, the purpose and importance of the proposed project must be clearly stated with emphasis on the benefit of the project to the client or organization.

The proposal must be written in the language of the reader. The businessperson will be more interested in profits and benefits that a project promises, whereas the chief engineer will be concerned with its feasibility from an engineering standpoint. A proposal may begin as a technical report that is reviewed by the people directly affected by the proposed project. After the report has been approved at this level, the proposal is rewritten to present the overall picture in less technical terms for the investor, client, stockholder, or businessperson who may be indirectly involved.

A typical proposal (Fig. 7.29) should contain the following major elements:

- *Statement of the problem*. The problem is clearly identified to present the purpose of the project. In most cases, about one typewritten page or less will suffice.
- *Method of approach*. The procedures for attacking the problem are outlined and explained in detail. This can be done in outline form for easy review (usually the longest part of a proposal).
- *Personnel needs and facilities*. Requirements for equipment, space, and personnel are itemized.
- *Time schedule*. A time schedule gives an estimate of the completion dates for the various phases of the project. This schedule should be coordinated with any other related activity that may be affected.
- *Budget*. The funds required should be presented in sufficiently detailed form to permit analysis by those who review the proposal.
- *Summary*. The report is summarized to emphasize the important points that are the basis of the proposal. The importance of the project and its contribution to the reviewer should be strongly stated.

The Progress Report

The progress report is used to review periodically the status of a project or an assignment. Some progress reports may be in the form of a letter or a memorandum that will be circulated among those interested in the project. More comprehensive projects require a more detailed formal report.

It usually gives a projection of an increase or decrease in expenditures or time schedules to permit a revision of project plans. These reports are essential to keep top management in touch with any variation or deviation in the operation of a project.

Reference to Project Evaluation and Review Techniques (PERT) introduced in Chapter 3 suggests effective methods of reporting the status of a design project. Essentially the same forms used to schedule activities could be submitted as part of a progress report to give an accurate picture of the progress that has been made. A weekly progress report for student projects is shown in Fig. 7.30.

Fig. 7.29 The general divisions of a proposal.

LETTER OF TRANSMITTAL – SECOND PAGE OF
YOUR REPORT. ADDRESS TO YOUR TEACHER.

May 8, 1977

Professor J. T. Coppinger
Engineering Design Graphics Department
Texas A&M University
College Station, Texas 77840

Dear Professor Coppinger:

Attached is our report which outlines our recommendations for
the establishment of a manufacturing operation to produce an
educational toy at a rate of 5,000 per month.

We have researched and studied this problem very closely, and
feel that our conclusions and recommendations are based on
sound judgment. We are hopeful that our findings meet with
your approval.

Sincerely,

Gerald Brown

Gerald Brown, Chief Designer
Team 5, EDG 105, Section 145

Fig. 7.33 A letter of transmittal for a technical report.

DOUBLE SPACE

TABLE OF CONTENTS

Fig. 7.35 The table of contents for a technical report.

LIST HOME ADDRESSES
ON TITLE PAGE.

ORGANIZATION AND DESIGN
OF TOY MANUFACTURING OPERATION

by

TEAM 5

Instructor: J. T. Coppinger
Engineering Design Graphics 105
Texas A&M University
May 8, 1977

TEAM MEMBERS

Gerald Brown, 3305 NW 60 Street, Oklahoma City, Oklahoma

Greg Porter, 330 Veda Mae, San Antonio, Texas

Gary Smith, 918 Sutton, San Antonio, Texas

Tom Reed, 388 Highway 36, Caldwell, Texas

Clifton Potter, 212 East Avenue F, Robstown, Texas

Olivaree Flynn, 1002 North Flores, Rio Grande, Texas

James R. Ross, 108 Ridgeway, San Marcos, Texas

Fig. 7.34 The title page of a technical report prepared
by a student team.

OPTIONAL IN INFORMAL REPORTS

TABLE OF ILLUSTRATIONS

Fig. 7.36 The table of illustrations. This page can be
omitted in informal reports.

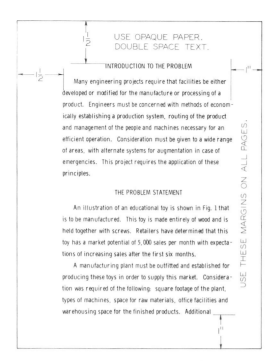

Fig. 7.37 A typical page of text of a technical report. Notice the margins that must be used.

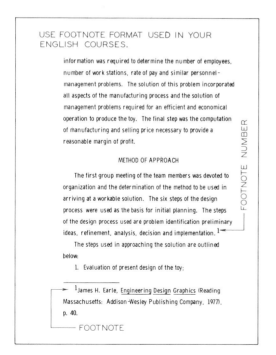

Fig. 7.38 A page of text with a footnote.

for all pages of a report, including those where illustrations are given. Use the third person; do not say, "We investigated the problem and I suggested. . . ."

The text should refer to a figure that follows. Example: "The number of boats sold between 1950 and 1976 is shown in Figure 6." The figure should be placed immediately after the text that refers to it or on the following page. Figure. 7.39 shows a figure that is referred to in Fig. 7.37.

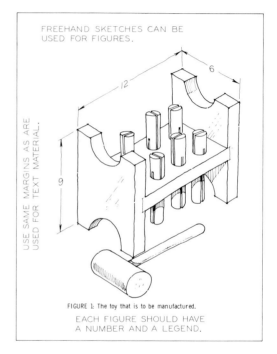

Fig. 7.39 A freehand sketch can be used as a figure. Figures must adhere to the same margins as pages of text.

8. *Method.* (Use a heading appropriate for your report rather than this general term.) This section should cover the general method that was used in solving the problem. It can be written in outline or essay form.

9. *Body.* (Use a heading appropriate for your report rather than this general term.) This portion elaborates on the solution of the problem. Subheadings should be used to make the parts of the report stand out. Where possible, illustrate your report with sketches, instrument drawings, graphs, or photographs to commu-

nicate its contents. Examples of illustrations are shown in Figs. 7.39, 7.40, 7.41, and 7.43.

Illustrations should be drawn on opaque paper or on tracing paper, and a print should be made for inclusion in the report. Ink illustrations are preferable. If a drawing must be turned lengthwise on the sheet, the top of the figure should be placed toward the left side of the sheet so that it can be read from the right side of the page.

Graphs and illustrations can be drawn proportional to a photographic slide or to an overhead transparency if these will be used in an oral presentation. This will eliminate duplication of effort.

Large drawings that must be folded for insertion in a report should be folded to $8\frac{1}{2}'' \times 11''$ to allow them to be conveniently unfolded by the reader.

10. *Findings.* (Use a heading appropriate for your report rather than this general term.) The results of the report should be tabulated and presented graphically with explanatory text. The findings should be evaluated and presented for easy interpretation. The data pre-

Fig. 7.41 Photographs can be attached to a page with rubber cement and used as figures.

sented in Fig. 7.42 are easier to interpret when graphed as shown in Fig. 7.43.

11. *Conclusions.* The conclusions should summarize the entire report very briefly and end with specific conclusions and recommendations, either positive or negative. If there are several conclusions, they may be emphasized by a listing in numerical order.

12. *Bibliography.* This is a list of references—books, magazines, brochures, conversations—that were used in the preparation of the report. They are usually listed alphabetically by author (Fig. 7.44). Refer to an English textbook for the format suitable for your report; several formats are acceptable for footnotes and bibliographies.

13. *Appendix.* The appendix should include less important drawings, sketches, raw data, brochures, letters, and other general information that supports but is not appropriate for inclusion in the main part of the report (Figs. 7.45 and 7.46). Preliminary ideas and student progress reports should be included here.

Drawings and illustrations that are important to the development of the report should be included

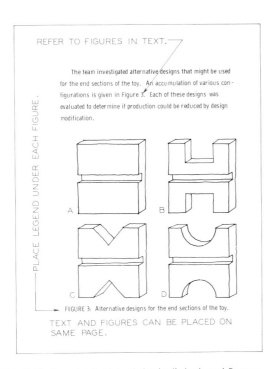

Fig. 7.40 A page that contains both text and figures.

TABLES OF DATA ARE EASIER TO INTERPRET IF GRAPHED.

OPERATIONAL COSTS

The monthly cost of materials required for the toy totals $1,423.42. The equipment, building, wages, insurance, materials, utilities and taxes total $57,860 for one year. For all practical purposes, this is considered as $58,000 for the first year. Although materials and insurance will cost $23,172 for one year, this is listed below as $23,000.

If the product is sold for 50¢ per toy, gross income will be $30,000 each year excluding any expansion. On a yearly basis, this process will pay for the equipment and building space after 5 years (Fig. 23). In actuality, the payments would be made on a monthly basis and over a number of years everything would be paid for. The yearly expenditures and income are shown below.

	Gross Income	Payments Necessary	Balance
First Year	$30,000	$58,000	−$28,000
Second Year	$30,000	$51,000	−$21,000
Third Year	$30,000	$44,000	−$14,000
Fourth Year	$30,000	$37,000	−$ 7,000
Fifth Year	$30,000	$30,000	$ 0,000
Sixth Year	$30,000	$23,000	+$ 7,000
Seventh Year	$30,000	$23,000	+$14,000
Eighth Year	$30,000	$23,000	+$21,000

Fig. 7.42 Tabular data can be given in a report, but they are easier to interpret if graphed.

BIBLIOGRAPHY

Alford, Leon Pratt. Production Handbook. New York: The Ronald Press, 1944.

Anderson, E.A. The Science of Production. New York: J. Wiley and Sons, 1938.

Bethel, Lawrence L. Production Control. New York: McGraw Hill Book Company, 1942.

Coventon, Walter. Woodwork Tools and Their Use. New York: Hutchinson's, 1953.

De Cristofoso, R. J. Modern Power Tool Woodworking. New York: Raymond, 1967.

Douglas, James Harvey. Woodworking With Machines. Bloomington, Illinois: McKnight and McKnight Publishing Co., 1960.

Earle, James H. Engineering Design Graphics. Reading, Mass: Addison-Wesley Publishing Co., 1977.

Groneman, Chris Harold. General Woodworking. New York: McGraw-Hill, 1952.

Groneman, Chris Harold. Exploring the Industries for the General Shop and Laboratory of Industries. Austin, Texas:

Fig. 7.44 One form of giving the bibliography of references used in the report.

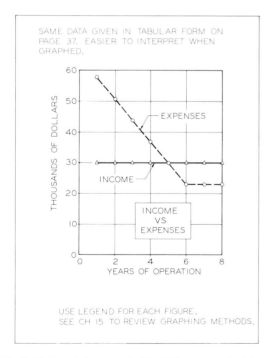

SAME DATA GIVEN IN TABULAR FORM ON PAGE 37. EASIER TO INTERPRET WHEN GRAPHED.

EXPENSES

INCOME

INCOME VS EXPENSES

THOUSANDS OF DOLLARS

YEARS OF OPERATION

USE LEGEND FOR EACH FIGURE. SEE CH 15 TO REVIEW GRAPHING METHODS.

Fig. 7.43 The data shown in Fig. 7.42 are graphed in this example for easier interpretation.

APPENDIX MATERIALS INCLUDE RAW DATA, NOTES, SKETCHES, PROGRESS REPORTS, ETC.

SUMMARY SHEET
OF OPINION SURVEY
Number of Votes Per Section

Team Member	1	2	3	4	5	6
#1	0	2	7	1	0	0
#2	0	3	3	2	1	1
#3	0	2	6	2	0	0
#4	1	1	5	3	0	0
Totals	1	8	21	8	1	1

Each of the four members interviewed ten people of his choice. Each team member had a data sheet on which he recorded the opinions of the persons interviewed. This data was combined into the chart as shown below.

Piece number 3 received the most votes (21) as the most attractive design.

Fig. 7.45 Data and survey information can be placed in the appendix.

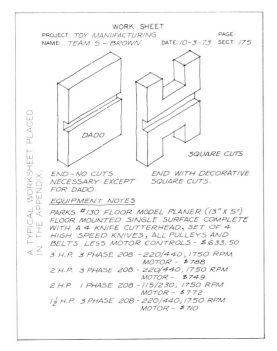

WORK SHEET

PROJECT: *TOY MANUFACTURING*   PAGE:
NAME: *TEAM 5 - BROWN*   DATE: *10-3-73* SECT: *175*

DADO

SQUARE CUTS

END - NO CUTS   END WITH DECORATIVE
NECESSARY EXCEPT   SQUARE CUTS.
FOR DADO.

EQUIPMENT NOTES

PARKS #130 FLOOR MODEL PLANER (13"X 5")
FLOOR MOUNTED SINGLE SURFACE COMPLETE
WITH A 4 KNIFE CUTTERHEAD, SET OF 4
HIGH SPEED KNIVES, ALL PULLEYS AND
BELTS LESS MOTOR CONTROLS - $633.50

3 H.P. 3 PHASE 208 -220/440, 1750 RPM
MOTOR - $788
2 H.P. 3 PHASE 208 -220/440, 1750 RPM
MOTOR - $749
2 H.P. 1 PHASE 208 -115/230, 1750 RPM
MOTOR - $772
1½ H.P. 3 PHASE 208 -220/440, 1750 RPM
MOTOR - $710

A TYPICAL WORKSHEET PLACED IN THE APPENDIX.

Fig. 7.46 A typical work sheet that is worthy of inclusion in the appendix, showing preliminary ideas considered. Sloppy, hard-to-read sketches should be omitted from the report.

in the main portion of the report as numbered figures, with legends to describe them and to relate them to the text.

Common Omissions

Elements that are often omitted from technical reports that deal with product design, but should be included, are cost estimates, overhead expenses, human and psychological factors, shipping costs, packing specifications, method of advertising, sales considerations, and summarizing recommendations. As you prepare your report, assume that you are the reader and that you are unfamiliar with the purpose of the report. Doing this will assist you in critically evaluating its contents and completeness.

7.9 DECISION

The purpose of oral and technical reports is to present the basis for a decision on a design; a decision of whether or not a recommended design should be implemented. No venture is one hundred percent safe, but the more facts you have in your favor, the more confident you can be regarding a decision.

Acceptance   A design may be accepted in its entirety. This is a compliment to the designer and his or her efforts in researching and solving the problem and winning its acceptance.

Rejection   A design recommendation can be rejected in its entirety. This may be because of changes in the economic climate or moves by competitors that would make your design unprofitable. On the other hand, some designs may be rejected because of a poor presentation of their merits.

Compromise   A design may not be approved in its entirety, but a compromise might be suggested in one or more areas. The initial quantity that will be produced might be increased or decreased; or several features might be modified to make a design more attractive.

Decision—Hunting Seat

The design of a hunting seat that was introduced in Chapter 3 is used as an example to illustrate the decision step of the design process. The problem is restated below.

HUNTING SEAT: Many hunters hunt from trees to obtain a better vantage point. Design a seat that would provide the hunter with comfort and safety while hunting from a tree and that would meet the general requirements of economy and hunting limitations.

Decision Chart   The chart shown in Fig. 7.47 can be used to compare the various designs that are available to choose from for implementation. Each idea is listed and given a number such as Design 1, Design 2, and so forth for identification.

Next, maximum values for the various factors of analysis are assigned in order for the total of all factors to be 10 points. You must use your judgment to determine these values since they will vary from product to product. Using the maximum values as guide, you can now evaluate each factor of the competing designs.

Decision table

DESIGN 1: BUCKET SEAT
DESIGN 2: FOLDING SEAT
DESIGN 3: PLATFORM SEAT
DESIGN 4:
DESIGN 5:
DESIGN 6:
DESIGN 7:
DESIGN 8:

MAX VALUE	FACTORS FOR ANALYSIS	1	2	3	4	5	6	7	8
3.0	FUNCTION	2.0	2.5	2.3					
2.0	HUMAN FACTORS	1.6	1.7	1.4					
.5	MARKET ANALYSIS	.4	.4	.4					
1.0	STRENGTH	1.0	1.0	1.0					
.5	PRODUCTION PR PROCEDURES	.3	.4	.2					
1.0	COST	.7	.8	.6					
1.5	PROFITABILITY	1.1	1.3	1.0					
.5	APPEARANCE	.3	.4	.4					
10	TOTALS	7.4	8.5	7.3					

Fig. 7.47 A work sheet with a decision table that is used to evaluate the developed design alternatives.

The vertical columns of numbers are summed to determine the design with the lowest total, the best design, perhaps. However, your instincts may disagree with your numerical analysis. If this is the case, you should have enough faith in your judgment not to be restricted by your numerical decision. Highly successful individuals are often those who place more weight on their instincts than on the facts.

In short, the availability of facts, an outstanding presentation, and a well-analyzed design will not ensure a profitable product. Decision will always remain the most subjective part of the design process.

Conclusion Once a decision has been made, it should be clearly stated, along with the reasons of

its acceptance (Fig. 7.48). Additional information may be given such as: number to be produced initially; selling price per unit; expected profit per unit; expected sales during the first year, second, etc.; number that must be sold to break even; and most marketable features.

It is possible that you would recommend that a design not be implemented for a number of reasons. If this is the case, the design process should not be considered a failure. A negative decision could save an investor from large losses by recognizing a poor venture at the outset.

If a design is recommended for implementation, the next step of the design process, implementation, will conclude the design cycle.

Conclusions

IMPLEMENT AND PRODUCE THE FLAT FOLDING SEAT. THIS DESIGN IS BELIEVED TO BE THE BEST SOLUTION AND ONE THAT CAN BE PROFITABLY MARKETED THROUGH SPORTING GOODS DEALERS.

SALES PRICE	$ 90
SHIPPING EXPENSES	5
NUMBER TO SELL TO BREAK EVEN	1500
MANUFACTURED BY CONTRACTORS AT OUTSET	
ESTIMATED PROFIT PER SEAT	$20

Fig. 7.48 The decision is summarized on this work sheet to give the designer's conclusion and recommendation concerning the next step, implementation.

PROBLEMS

1. Prepare a checklist that could be used to evaluate an oral presentation of one of your classmates. List important items that should be considered, and determine a point system that could be assigned to each. Keep the form simple, yet thorough enough to be of value to the presenter for improving his or her technique of presentation. Devise a means of tabulating the evaluation to arrive at an overall rating. Present this grading sheet in final form that can be used in an actual class presentation.

2. Prepare a series of 3" × 5" cards to plan a flip-chart presentation that will last no more than five minutes. The subject of your flip-chart presentation may be one of your choosing or one assigned by your teacher. Example topics: your career plans for the first two years after graduation, the role of this course in your total educational program, the importance of effective communications, the identification of a need for a design project that you are proposing, a comparison of the engineering profession with another profession of your choice.

3. Prepare graphical aids for an oral presentation, using the methods and materials indicated in this chapter to familiarize yourself with these techniques.

4. For a technique of your choice or one assigned by your instructor, prepare a five-minute briefing using the planning cards developed in Problem 2. Give this briefing to your class as assigned.

5. Assume that you are an engineer responsible for representing your firm in the presentation of a proposal for a sizable contract. Make a list of instructions that you could give to your assistants to coordinate the preparation of your presentation for a group of 20 persons ranging in background from bankers to engineers. The topic is not so important as the method of presentation, since most topics will require much the same preparation. Consequently, use a topic of your choice, one assigned by your teacher, or one suggested in Problem 2. Your instructions should outline the materials you need, method of preparation of graphical aids, number required, method of projection or presentation, assistance needed in presenting materials, room seating arrangements, and other factors. Your outline should be sufficiently complete to cover the entire program of an ideal presentation within the time you think most desirable.

6. Prepare a model that can be used effectively to communicate an engineering graphics principle to your classmates and that might be used for instructional purposes. Demonstrate it in a class presentation.

7. Prepare a series of photographic slides of the model constructed in Problem 6. Present these to your class for their evaluation.

8. Prepare a slide manuscript for the series of slides developed in Problem 7. Insert these in a binder for permanent filing.

9. Prepare an overhead transparency to illustrate a descriptive-geometry principle, using hinged overlays to present the problem in sequential steps. Demonstrate the use of this transparency to your class.

IMPLEMENTATION

8.1 INTRODUCTION

Implementation is the final step of the design process during which the design becomes a reality. Implementation is the last phase of the design process, but it is one that never completely ends.

Graphical methods are particularly important during the initial steps of implementation, since all products are constructed from working drawings with specifications. Drawings that illustrate how a series of parts is put together to form the final product are assembly drawings.

8.2 WORKING DRAWINGS

Working drawings are drawn using orthographic views that have been dimensioned and noted to show how the parts are to be made as individual pieces. It would be almost impossible to obtain the desired results by any method other than by using graphics. Since dimensions and notes must be clearly understood when the drawings are read, the designer should be sure that good lettering practices are used.

An example of a working drawing is given in Fig. 8.1, where a base plate mount is described by three views with the necessary dimensions to give its measurements. Notes are used to specify the materials and the sizes of the holes. A properly drawn working drawing, when followed correctly, will result in the same product, regardless of the shop in which it is made.

Several parts can be drawn on the same sheet when making working drawings, with no attempt to arrange the different parts in relationship to each other. The names of the parts, their identifying numbers, and their materials are given near the views. Each sheet is numbered in the title block along with other required information.

8.3 SPECIFICATIONS

Specifications are notes and instructions that are used to supplement working drawings. Specifications may be given in as typed documents when graphical representations are unnecessary.

For example, instructions of this type can be given effectively as written specifications:

"Metallurgical inspection is required before machining"

or

"Paint with two coats of flat black paint (NO. 780) after finishing."

It is more convenient to give specifications on the working drawing by the drafter when space permits.

8.4   ASSEMBLY DRAWINGS

Drawings that illustrate how parts are put together after they have been made are called **assembly drawings.** Assembly drawings can be drawn as pictorial or orthographic views that are fully assembled, fully exploded, or partially exploded. Figure 8.2 is a partially exploded orthographic assembly where sections have been used to clarify internal details. A parts list is usually given along with the assembly drawing to provide a bill of materials. The preparation of working drawings and assembly drawings will be discussed in Chapter 23.

8.5   MISCELLANEOUS CONSIDERATIONS

Several miscellaneous areas of the implementation step must be considered in addition to the preparation of the necessary drawings and specifications

from which the product will be constructed. These are packaging, storage, shipping, and marketing of the product.

Packaging   Practically all products are packaged in some manner prior to shipment to the final destination from the manufacturer. In some cases, such as in the toy industry, the packaging is very elaborate and may be as expensive as the product. Designers must be aware of the packaging needs as they develop a design, since a product that is difficult to package will cost more for this reason.

Many products are shipped partially disassembled to make them easier to package and, therefore, more economical. Likewise, more compact packages require less storage space.

Storage   Most manufacturers maintain a backlog of products for shipment as orders are received. Otherwise, an unexpected flow of orders may cause delays in shipments while waiting for the

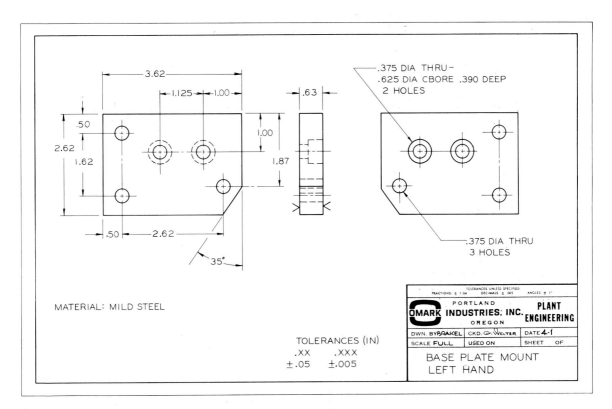

Fig. 8.1 A working drawing of a single part that was made as the implementation step of the design process. (Courtesy of Omark Industries, Inc.)

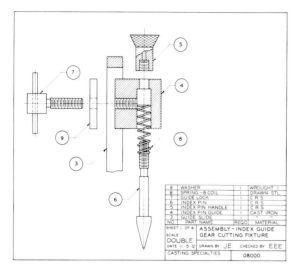

9	WASHER	1	WROUGHT I
8	SPRING - 8 COIL	1	DRAWN STL
7	GUIDE LOCK	1	C R S
6	INDEX PIN	1	C R S
5	INDEX PIN HANDLE	1	C R S
4	INDEX PIN GUIDE	1	CAST IRON
3	GUIDE SLIDE	"	"
NO	PART NAME	REQD	MATERIAL

SHEET 1 OF 4	ASSEMBLY - INDEX GUIDE
SCALE	GEAR CUTTING FIXTURE
DOUBLE	
DATE 11-5-72	DRAWN BY JE	CHECKED BY EEE
CASTING SPECIALTIES	08000

Fig. 8.2 A partially exploded orthographic assembly drawing of an index guide to a cutting fixture, which shows how the individual parts of an assembly are put together.

products to be manufactured; and consequently, this delay may result in the loss of business. The cost of storage of inventories must be figured as part of the product's final selling price. Some products, such as electronic gear, must be stored in temperature controlled environments, which increases the expense of warehousing.

Shipping   The cost of shipping products to their outlets must be evaluated to select the most economical and satisfactory means of serving the market. Some industries locate their shipping facilities in the middle of their market areas to reduce shipping costs. On the other hand, companies at the extreme ends of their market's geographical region just add additional cost to their products to cover shipping.

Marketing   Good designers will be concerned with all aspects of a product after it leaves their company and goes to the marketplace. They will be concerned with its marketability and acceptance by the consumer. Complaints about the product's effectiveness and function will be of importance to good designers since they will want to modify any defect in future versions of the product. Good designers realize that their product must be at least as good as that of the competition to survive, and better to succeed.

8.6 IMPLEMENTATION— HUNTING SEAT

The design of a hunting seat that was introduced in Chapter 3 is used as an example to illustrate the implementation step of the design process. The problem is restated below.

HUNTING SEAT: Many hunters hunt from trees to obtain a better vantage point. Design a seat that would provide the hunter with comfort and safety while hunting from a tree and that would meet the general requirements of economy and hunting limitations.

Four working drawing sheets (Figs. 8.3 through 8.6) have been prepared to present the details of the hunting seat design that was selected for implementation. The fifth sheet, Fig. 8.7, is an assembly drawing with a parts list that illustrates how the parts are assembled once they have been made as individual pieces. (This particular design was developed, patented, and is marketed by Baker Manufacturing Company, Valdosta, Georgia. It is the Baker Favorite Seat, Patent No. 3460649.)

All parts have been dimensioned using metric units, which are millimeters. Standard parts that are purchased from suppliers are not drawn, but are itemized on the drawing, given parts numbers, and listed in the parts list on the assembly drawing.

Each part shown in the working drawings is fully dimensioned and noted to explain the details of construction. Various scales are used when necessary to clarify details that cannot be shown at the same scale.

The assembly drawing (Fig. 8.7) is a pictorial with the different parts identified by balloons attached to leaders. Each part is listed in the parts list by number with general information to describe it. When more information is needed than is given in the parts list, you should refer to the working drawings or the specifications where this information must be given in full.

The drawings given in this example design give only the details for the basic Baker Favorite Seat. Additional accessories are available as shown in Fig. 8.8.

Packaging   The Baker Favorite Seat is packaged in a corrugated cardboard box that is 20″ × 38″ × 2.5″ in size, and the package weighs approximately 10 lbs when it contains the hunting seat (Fig. 8.9). When shipped by motor freight, the shipping costs

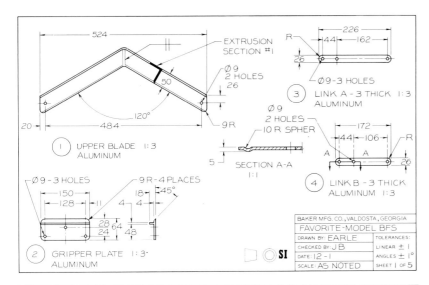

Fig. 8.3 A working drawing sheet of parts of a hunting seat design, sheet 1 of 5.

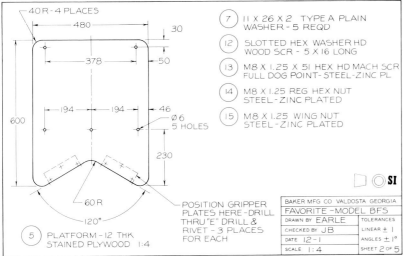

Fig. 8.4 A working drawing sheet of parts of a hunting seat design, sheet 2 of 5.

Fig. 8.5 A working drawing sheet of parts of a hunting seat design, sheet 3 of 5.

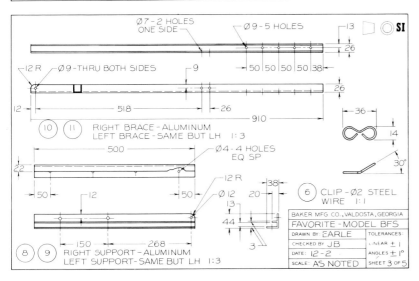

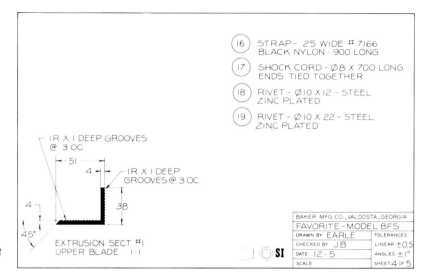

Fig. 8.6 A working drawing sheet of parts of a hunting seat design, sheet 4 of 5.

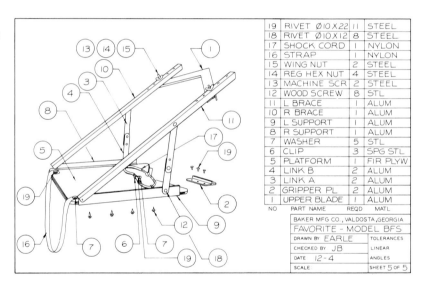

Fig. 8.7 An assembly drawing that demonstrates how the parts of the hunting seat design are assembled. Sheet 5 of 5.

are computed by hundred-pound multiples (Cwt). Consequently, shipping costs are reduced when shipping in volume. The box has been customized to advertise the product and for easy recognition of each model by the shipping personnel (Fig. 8.10).

Storage   An inventory of seats must be maintained to meet the regular seasonal orders that are anticipated throughout the year. The periods prior to hunting seasons will require more inventory than during the rest of the year.

An inventory of seats that are waiting to be sold adds to the cost of overhead in the form of

interest payments, warehouse rent, warehouse personnel, and loading equipment. Costs of this nature are just as real as those expended in the product's manufacture.

Shipping   Shipping costs must be evaluated for all types of carriers that are available: rail, motor freight, air delivery, and mail services. The shipping cost for a Baker Favorite Seat with its accessories is $5.75 when shipped one at a time by United Parcel Service. The cost per unit is reduced by about 50% when they are shipped in bundles of ten to the same destination by truck.

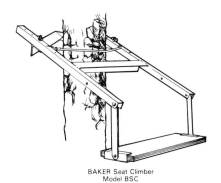

BAKER Seat Climber
Model BSC

A Conversion Kit, Model CK, will
convert Hand Climbers to Seat
Climbers.

An accessory to the Seat Climber
is the Padded Pouch — Model
PP.

Secure Seat Climber and Tree
Stand with tie down for added
safty while hunting.

Fig. 8.8 Examples of accessories
that have been designed to
accompany the basic hunting seat.
(Courtesy of Baker Manufacturing
Co., Valdosta, Georgia.)

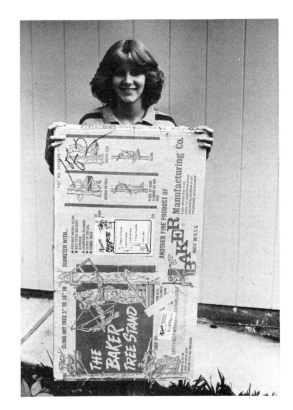

Fig. 8.9 The hunting seat is shipped in a cardboard
box to the retail outlet, or to the consumer.

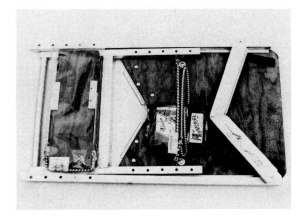

Fig. 8.10 The basic hunting seat is folded into a flat
position for ease of packaging prior to shipment.

Accessories As a product of this type gains acceptance and is used by the consumer, the need
for accessories becomes apparent. Examples are:
fold-down seats, hand climbers, and add-on seats
to name a few (Fig. 8.8). Additional accessories of
this type provide for the special needs of the hunters, increase the marketability of the product, and
increase sales volume. Each accessory must be designed by the same process that was used to develop the basic seat.

The retail price for the Baker Seat is about $90 when purchased by the consumer. The retail price is about five or six times higher than the materials and labor cost necessary to manufacture the seat. Retailers are given approximately a 40% margin, and the area distributors who take the orders from the retailers earn about 10%. The remainder of the overhead is advertising expenses and the other areas of overhead mentioned in this article. All expenses, such as 20% interest rates, must be absorbed by the consumer who ultimately purchases the product.

8.7 PATENTS

A designer who has developed an original and novel solution to a design problem should investigate the possibilities of obtaining a patent from the U.S. Patent Office (Fig. 8.11). This must be done prior to disclosing the invention, since premature disclosure may forfeit the patent right. There are many benefits in having a successful patent.

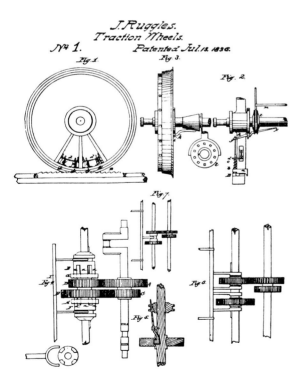

Fig. 8.11 The first patent, which was issued by the U.S. Patent Office in 1836.

A helpful pamphlet (available free of charge from the Patent Office) is *General Information Concerning Patents*. It is a good source for more specific details of the patent procedure. (This publication was the primary reference for the following text.)

General Patent Requirements

Designers must have a general understanding of the patent requirements. They need to understand what can be patented and who is eligible to apply for a patent.

What Can Be Patented In the language of the patent law now in effect, any person who "invents or discovers any new and useful process, machine, manufacture, or composition of matter, or any new and useful improvement thereof, may obtain a patent," subject to the conditions and requirements of law. Essentially, these categories include everything made by humans and the processes for making them (Fig. 8.12).

Court interpretations of this statute have excluded specific categories of inventions from the field of patentable items. For example, inventions used solely for the development of nuclear and atomic weapons for warfare are not patentable since these are not considered to be "useful." Another example, a design for a functional mechanism that will not operate in keeping with its intended purpose is not patentable. An *idea* for a new invention or machine is not patentable. The specific design and description of the machine must be available before it can be eligible for consideration as a patentable item.

Who Can Apply for a Patent Only the inventor of a device may apply for a patent. A patent given to a person who was not the inventor would be void, and the person would be subject to prosecution for committing perjury. Application for a patent *can* be made by the executor of a deceased inventor's estate. Two or more persons may apply for a patent as joint inventors, but the person or firm making a financial contribution cannot be joined in the application as an inventor.

Patent Rights A patent granted to an inventor gives him or her the right to exclude others from making, using, or selling the invention throughout the United States for a period of 17 years. After

T. A. EDISON.
Electric-Lamp.

No. 223,898.   Patented Jan. 27, 1880.

Fig. 8.12 Thomas Edison's patent drawing for the electric lamp, 1880.

expiration of the 17-year patent term, the invention may be made, used, or sold by anyone without authorization from the holder of the patent. The term may not be extended except by special act of Congress.

Patented articles must be marked with the word "Patent" and the number of the patent. Failure to mark an item in this manner may forfeit rights to damages if a person infringes on an invention not properly marked and continues to infringe after the application of the proper marking. Markings using the terms "Patent Pending" have no legal effect, since protection does not begin until the actual grant is made.

Application for a Patent

An inventor applying for a patent must include the following:

1. a written document that comprises a petition, a specification (description and claims), and an oath or declaration;

2. a drawing in those cases in which a drawing is possible; and

3. the filing fee.

The patent application will not be accepted for consideration unless it is complete and complies with the rules of the Patent Office.

Petition, Oath, or Declaration The petition and oath are usually combined in one form; on this form, the inventor petitions or requests that he or she be given a patent on the invention. The oath or declaration is a statement declaring that the inventor believes himself to be the original and first inventor of the invention described in the application.

Specification of the Patent The specifications or description of a patent must be attached to the application in written form describing the invention in full detail so that a person skilled in the field to which the invention pertains can produce the item. These descriptions should specifically point out features that distinguish the invention from similar patents. Drawings should be referred to in the text by figures and part numbers (Figs. 8.13 through 8.16).

The following format is suggested for the specifications:

1. Title of the invention, or a preamble stating the name, citizenship, and residence of the applicant and the title of the invention.

2. Brief summary of the invention.

3. If there are drawings, a brief description of the several views given.

4. Detailed description.

5. Claims.

Claims are brief descriptions of the details of the invention that distinguish new features from features of already patented material. The claims are the most significant parts of the patent, since they will be used as the basis to ascertain the novelty and patentability of an invention. A single invention may incorporate a number of features that will be stated as separate claims in the specification.

Fee The application for a patent must be accompanied by the filing fee. The basic filing fee is $65. An additional charge of $2 per claim is made for each claim in excess of ten on any one application. After the application has been accepted, a notice

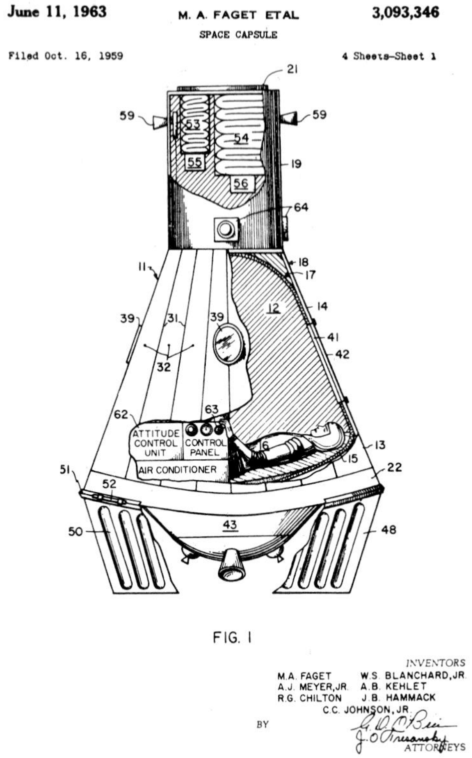

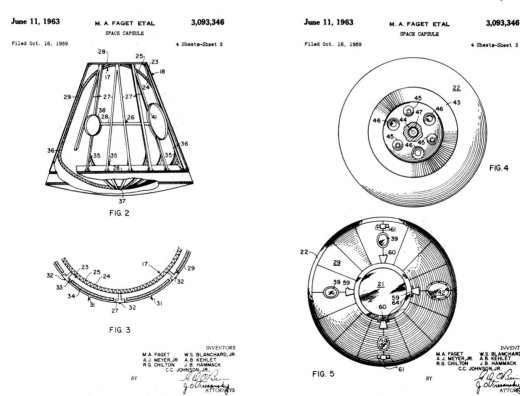

Fig. 8.13 The patent drawing of a space capsule developed by the National Aeronautics and Space Administration (NASA). (Courtesy of the U.S. Patent Office.

Fig. 8.14 Structural details of the space capsule. (Courtesy of the U.S. Patent Office.)

Fig. 8.15 The top and bottom views of the space capsule. (Courtesy of the U.S. Patent Office.)

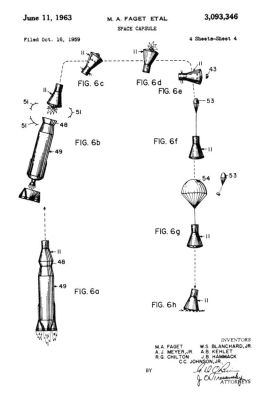

June 11, 1963   M. A. FAGET ETAL   3,093,346

SPACE CAPSULE

Filed Oct. 16, 1959   4 Sheets-Sheet 4

FIG. 6c   FIG. 6d   FIG. 6e

FIG. 6b   FIG. 6f

FIG. 6g

FIG. 6a   FIG. 6h

INVENTORS
M.A. FAGET   W.S. BLANCHARD, JR.
A.J. MEYER, JR.   A.B. KEHLET
R.G. CHILTON   J.B. HAMMACK
C.C. JOHNSON, JR.

BY

ATTORNEYS

Fig. 8.16 The sequence of events involving the space capsule from launch to landing. (Courtesy of the U.S. Patent Office.)

will be sent to the applicant giving him or her three months from that date to remit an issue fee of $100. An additional fee of $10 is charged for each page of the specification printed, and $2 for each sheet of drawing.

8.8 THE PREPARATION OF PATENT DRAWINGS

When drawings are necessary to describe an invention, the applicant for a patent must submit these drawings with the application. In some cases it may be desirable to prepare flow diagrams, schematics, and similar diagrams. Due to the many thousands of patents that are processed by the Patent Office, it is imperative that all drawings conform to established standards and rules.

A booklet, *Guide for Patent Draftsmen*, is available from the U.S. Government Printing Office. It outlines the procedures for the preparation of patent drawings. Most of the rules and illustrations included in this booklet are presented in the following articles. Specific rules concerning sheet size, spacing, and notation are of special importance, since even a well-prepared drawing will be rejected unless the format rules are observed. If the inventor cannot furnish his or her own drawings, the Patent Office will refer him to a drafter who can prepare the drawings. This service will be at the expense of the inventor.

Patent Drawing Standards

When the patent is issued, the completed drawing is printed and published. Drawing sheets are reduced about one-third in size; this requires that the original drawings be prepared uniformly to 150 percent. The quality of the drawings should be excellent. The following rules must be followed as closely as possible to prevent rejection of the application.

Paper and Ink Drawings must be made on pure white paper of thickness corresponding to a two-ply or three-ply Bristol board. The surface must be calendered and smooth to permit erasure and correction. Only India ink will secure perfectly black solid lines. The use of white pigment to cover lines is not acceptable.

Sheet Size and Margins The sheet size must be exactly 8 1/2″ by 14″ (21.6 by 35.6 cm) or exactly 21.0 by 29.7 cm. All sheets in a particular application must be the same size. One of the shorter sides is regarded as the top of the sheet. On 8 1/2″ by 14″ sheets, the top margin must be 2″ and the side and bottom margins 1/4″. Margin border lines cannot be drawn on the sheets, but all work must be included within the margins. The sheets may be punched with two 1/4″ DIA holes with their centerlines spaced 11/16″ below the top edge and 2 3/4″ apart and centered from the sides of the sheet.

The margins for 21.0 by 29.7 cm sheets are 2.5 cm from the top, 2.5 cm from the left, and 1.5 cm from the right, and 1 cm from the bottom. A significant patent is shown in Fig. 8.17: Goddard's patent on the rocket.

Character of Lines All lines and lettering must be absolutely black regardless of how fine the lines may be. All lines should be drawn with instru-

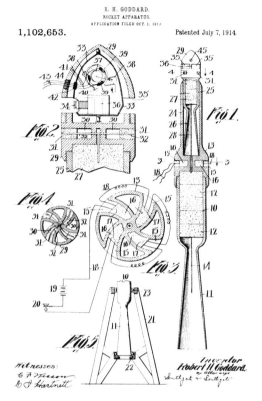

R. H. GODDARD.
ROCKET APPARATUS.
APPLICATION FILED OCT. 1, 1913.

1,102,653.   Patented July 7, 1914.

Fig. 8.17 R.H. Goddard's patent on a rocket, issued in 1914. (Courtesy of the U.S. Patent Office.)

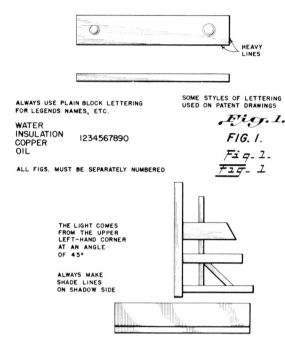

Fig. 8.18 Typical examples of lines and lettering recommended for patent drawings.

ments. Freehand work should be avoided. Lines should not be crowded.

Hatching and Shading Hatching lines, used to shade the surface of an object, should be parallel lines not less than about 1/20'' apart (Fig. 8.18). Heavy lines are used on the shade side of the drawing; however, they should not be used if they are likely to confuse the drawing. The light is assumed to come from the upper left-hand corner at an angle of 45°. Examples of this form of shading are shown in Fig. 8.19. Types of surface delineation are given in Figs. 8.20 and 8.21.

Scale The scale should be large enough to show the mechanism without crowding when the drawing is reduced for reproduction. Certain portions of the mechanism may be drawn at a larger scale to show additional details. Additional patent drawings can be used if necessary, but no more should be used than is necessary.

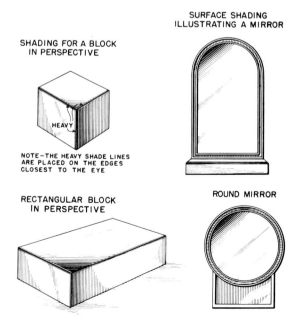

Fig. 8.19 Techniques of shading patent drawings.

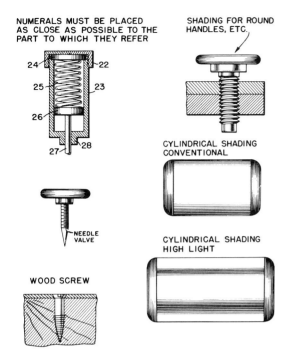

Fig. 8.20 Methods of numbering parts and rendering details for patent drawings.

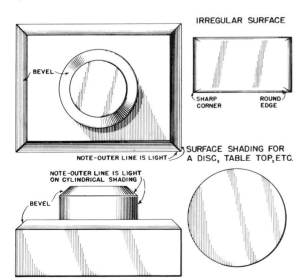

Fig. 8.21 Techniques of representing surfaces and beveled planes on patent drawings.

Reference Characters The different views of a mechanism should be identified by consecutive figure numbers, using plain, legible, and carefully prepared numerals. They should be at least 1/8″ in height, not encircled, and placed close to the parts to which they apply without confusing the drawing, as shown in Fig. 8.20. A leader is used to indicate the parts to which they refer. Numbers should not be placed on hatched surfaces unless a blank space is provided for them. The same part appearing in more than one view of the drawing should be designated by the same character, and this character should never be used to designate other parts.

Symbols Symbols used to represent various materials in sections, electrical components, and mechanical devices are suggested by the Patent Office, but these conform to the usual conventional engineering drawing standards covered in Chapter 16. All symbols used must be adequately identified in the specifications. Legends may be used on the drawing to explain the symbols used.

Signature and Names The signature of the applicant, or the name of the applicant and the signature of the attorney or agent, may be placed in the lower right-hand corner of each sheet within the marginal lines or below the lower marginal lines.

Views The patent drawing should contain as many figures as are necessary to explain an invention. If possible, the figures should be numbered consecutively in order of their appearance. Figures may be plan, elevation, section, perspective, or detail views (Figs. 8.22 and 8.23). Shading is used to show the shape of the components and the details of each part. Exploded views such as those of the Colt Revolving Gun (Fig. 8.24) can be used to advantage to describe the assembly of a number of parts. Large parts may be broken into sections and drawn on several sheets if this approach does not confuse the matter. Removed sections can be used, provided that the cutting plane is labeled to indicate the section by number. Views should not be connected by projection lines. All sheet headings and signatures will be placed in the same position on the sheet whether the drawing is arranged to read from the bottom of the sheet or from the right side of the sheet. It is desirable that the drawing be positioned so that it can be read when the sheet

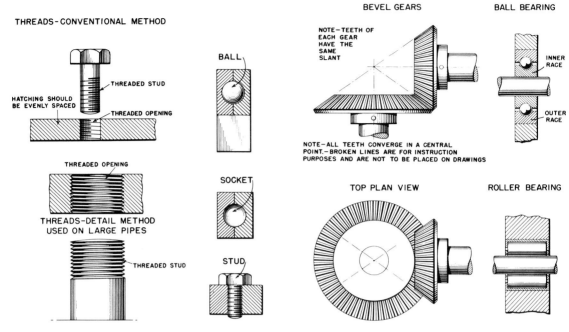

Fig. 8.22 Techniques of representing threads and small components on patent drawings.

Fig. 8.23 Representation of gears and ball bearings on patent drawings.

Fig. 8.24 An assembly of parts used to describe the workings of a patent. (Courtesy of the U.S. Patent Office.)

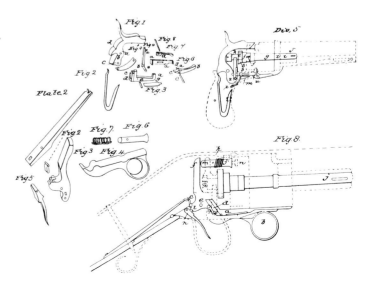

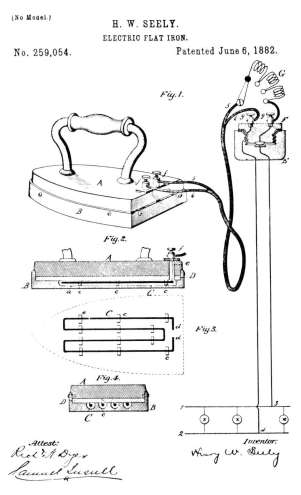

H. W. SEELY.
ELECTRIC FLAT IRON.

No. 259,054.

Patented June 6, 1882.

Fig. 1.

Fig. 2.

Fig. 3.

Fig. 4.

Attest:

Inventor:

Fig. 8.25 The complete patent drawing for the electric iron. (Courtesy of the U.S. Patent Office.)

is held upright; however, in some cases this may not be practical.

No extraneous matter, such as an agent's or attorney's stamp or address, is permitted to appear on the face of the drawing. The completed drawings should be sent flat, protected by heavy board or rolled in a suitable mailing tube. Folded or mutilated drawings must be redrawn. Drawings used for an accepted patent will not be returned to the applicant, but will remain on permanent record.

An example of patent drawings and specifications for an electric iron are shown in Figs. 8.25 and 8.26. These drawing specifications represent the published record of the patent in the U.S. Patent Office.

8.9 PATENT SEARCHES

A patent can be granted only after the Patent Office examiners have searched existing patents to determine whether the invention has been previously patented. With over 3,000,000 patents on record, a patent search is the most time-consuming portion of the process of obtaining a patent. Many inventors employ patent attorneys or agents to conduct a preliminary search of existing patents to discover whether an invention infringes on another. These preliminary searches serve to establish whether or not a patent application has a chance of success; however, the Patent Office examiner may find prior patents covering the same invention that were not found during the preliminary search.

Patents are filed in the Search Room of the Patent Office by classes and subclasses according to subject matter. The searcher will review the subclass that covers patents in the field of the application he or she is working on. However, other seemingly unrelated patents may cover the invention being submitted, thereby disallowing the patent.

8.10 QUESTIONS AND ANSWERS ABOUT PATENTS

For best results, patent applications must be handled with the assistance of a qualified patent attorney or patent agent. As a further help to the potential inventor, the Patent Office has published a pamphlet, *Questions and Answers About Patents*. Most of these questions and answers are listed here to supplement the previous articles on patents.

Nature and Duration of Patents

1. Q. *What is a patent?*

   A. A patent is a grant issued by the U.S. Government giving an inventor the right to exclude all others from making, using, or selling his or her invention within the United States, its territories and possessions.

2. Q. *For how long a term of years is a patent granted?*

   A. Seventeen years from the date on which it is issued; except for patents on ornamental

Fig. 8.26 The complete patent specifications for the electric iron. Copies are available from the U.S. Patent Office. (Courtesy of the U.S. Patent Office.)

UNITED STATES PATENT OFFICE.

HENRY W. SEELY, OF NEW YORK, N. Y., ASSIGNOR OF TWO-THIRDS TO RICHARD N. DYER AND SAMUEL INSULL, OF SAME PLACE.

ELECTRIC FLAT-IRON.

SPECIFICATION forming part of Letters Patent No. 259,054, dated June 6, 1882.

Application filed December 8, 1881. (No model.)

To all whom it may concern:

Be it known that I, HENRY W. SEELY, a citizen of the United States, residing at New York, in the county and State of New York, 5 have invented a new and useful Electric Flat-Iron, of which the following is a specification.

The object of my invention is to utilize electric currents derived from any suitable source of electric energy for the purpose of heating 10 flat-irons, fluting-irons, and other similar utensils. To accomplish this object I place within the iron and close to its face a resistance, preferably of carbon, and of such size and shape that it will heat the face of the iron suf-15 ficiently and equally. This resistance has terminals, by means of which it may be connected in an electric circuit, preferably a multiple-arc circuit of an electric lighting system.

In the accompanying drawings, Figure 1 is 20 a perspective view of a flat-iron connected with a multiple-arc system of electric lighting; Fig. 2, a vertical longitudinal section of the iron; Fig. 3, a plan view of the heating-resistance, and Fig. 4 a transverse vertical section of the 25 iron.

Similar letters of reference refer to corresponding parts in all these figures.

The base of the flat-iron is made in two parts, A B, the upper part, A, fitting into the lower 30 one, B. In the interior of B is formed a groove, *a*, whose shape corresponds to that of the carbon resistance C, which is laid in the groove. This resistance is preferably molded or formed as one continuous piece of carbon, though, in-35 stead of this, a number of carbon sticks could be laid parallel in grooves connected together by wires electroplated to their ends. To prevent contact between the carbon and the metal below and around it, it is laid in supporting 40 saddles *c c*, of some suitable non-conducting and non-combustible material.

Above the resistance is placed a layer, D, of an insulating substance, which is also both non-combustible and a poor conductor of heat. 45 This substance is preferably one which can be put in its place while in a soft or plastic condition and then allowed to harden—as, for instance, plaster-of-paris. Before pouring in such substance the grooves and resistance should be covered with a sheet of paper or similar mate-50 rial, in order that the plastic substance may not penetrate between the carbon and the iron, and thus impair the conduction of heat between them. The upper part, A, of the iron is set directly upon the top of the insulating sub-55 stance D, and is secured to the lower part by rivets, or in any other suitable manner.

If desired, a packing of felt or other substance which is a non-conductor of heat may be placed in the joint between A and B, so 60 that all the heat will be retained in the lower part of the iron.

The ends *d d* of the resistance C are electro-plated or otherwise attached to wires which pass up through an aperture, *e*, (being insulated 65 from the iron where they pass through it,) to binding-posts *f f*, attached to a plate of insulating material fastened to the top of the base. By means of these binding-posts connection is made with the wires from any suit-70 able source of electricity.

In Fig. 1 the flat-iron is shown in connection with a multiple-arc system of electric lighting.

1 2 are floor-mains of the system in derived circuits, from which are placed incandescent 75 electric lamps, (represented at *x x*.)

3 4 is a multiple-arc circuit leading to the interior terminals of an ordinary lamp-socket, E. From this the lamp has been removed, and instead a plug, F, having exterior termi-80 nals corresponding to the socket-terminals, is placed in the socket. The plug-terminals are connected to binding-posts *g g*, from which flexible conducting-wires 5 6, of sufficient length to allow the iron to be moved back and forth, 85 lead to the binding-posts *f f*.

An adjustable resistance, G, may, if desired, be placed in the circuit between the socket and the iron, in order that the heat of the latter may be properly regulated. 90

A safety-catch should be provided, preferably located within the plug, F, to protect the system in case of a short circuit occurring.

It is evident that my invention could be applied to fluting-irons in which a curved corru-95 gated iron bears on a corrugated base by placing a heating-resistance in the base, or in both the base and the moving iron.

2   259,054

What I claim is—

1. The combination, with a flat-iron or similar utensil, of an electrical resistance located within the same, the face of said iron being 5 heated by radiation from said resistance, substantially as set forth.

2. A chambered flat-iron or similar utensil, in combination with an electrical resistance inclosed entirely thereby, whereby all the heat 10 radiated from such resistance will be utilized, substantially as set forth.

3. A chambered flat-iron or similar utensil, in combination with an electrical resistance inclosed thereby, and a layer of non-heat-conducting material to confine the heat to the face 15 of the iron, substantially as set forth.

This specification signed and witnessed this 6th day of December, 1881.

HENRY W. SEELY.

Witnesses:
RICHD. N. DYER,
SAMUEL INSULL.

designs, which are granted for terms of 3 1/2, 7, or 14 years.

3. Q. *May the term of a patent be extended?*

A. Only by special act of Congress, and this occurs very rarely and only in most exceptional circumstances.

4. Q. *Does the patentee continue to have any control over the use of the invention after the patent expires?*

A. No. Anyone has the free right to use an invention covered in an expired patent, so long as he does not use features covered in other unexpired patents in doing so.

5. Q. *On what subject matter may a patent be granted?*

A. A patent may be granted to the inventor or discoverer of any new and useful process, machine, manufacture, or composition of matter, or any new and useful improvement thereof, or on any distinct and new variety of plant, other than a tuber-propagated plant, which is asexually reproduced, or on any new, original, and ornamental design for an article of manufacture.

6. Q. *On what subject matter may a patent not be granted?*

A. A patent may not be granted on a useless device, on printed matter, on a method of doing business, on an improvement in a device which would be obvious to a person skilled in the art, or on a machine which will not operate, particularly on an alleged perpetual motion machine.

Meaning of Words "Patent Pending"

7. Q. *What do the terms "patent pending" and "patent applied for" mean?*

A. They are used by a manufacturer or seller of an article to inform the public that an application for patent on that article is on file in the Patent Office. The law imposes a fine on those who use these terms falsely to deceive the public.

Patent Applications

8. Q. *I have made some changes and improvements in my invention after my patent appli-*cation was filed in the Patent Ooffice. May I ammend my patent application by adding a description or illustration of these features?

A. No. The law specifically provides that new matter shall not be introduced into the disclosure of a patent application. However, you should call the attention of your attorney or patent agent promptly to any such changes you may make or plan to make, so that he or she may take or recommend any steps that may be necessary for your protection.

9. Q. *How does one apply for a patent?*

A. By making the proper application to the Commissioner of Patents, Washington, D.C., 20231.

10. Q. *What is the best way to prepare an application?*

A. As the preparation and prosecution of an application are highly complex proceedings, they should preferably be conducted by an attorney trained in this specialized practice. The Patent Office therefore advises inventors to employ a patent attorney or agent who is registered in the Patent Office.

11. Q. *Of what does a patent application consist?*

A. An application fee, a petition, a specification, and claims describing and defining the invention, an oath or declaration, and a drawing if the invention can be illustrated.

12. Q. *What are the Patent Office fees in connection with filing of an application for patent and issuance of the patent?*

A. A filing fee of $65 plus certain additional charges for claims, depending on their number and the manner of their presentation, are required when the application is filed. A final or issue fee of $100 plus certain printing charges are also required if the patent is to be granted. The final fee is not required until your application is allowed by the Patent Office.

13. Q. *Are models required as a part of the application?*

A. Only in the most exceptional cases. The Patent Office has the power to require that a model be furnished, but rarely exercises it.

14. Q. *Is it necessary to go to the Patent Office in Washington to transact business concerning patent matters?*

A. No; most business with the Patent Office is conducted by correspondence. Interviews regarding pending applications can be arranged with examiners if necessary, however, and are often helpful.

15. Q. *Can the Patent Office give advice as to whether an inventor should apply for a patent?*

A. No. It can only consider the patentability of an invention when this question comes regularly before it in the form of a patent application.

16. Q. *Is there any danger that the Patent Office will give others information contained in my application while it is pending?*

A. No. All patent applications are maintained in the strictest secrecy until the patent is issued. After the patent is issued, however, the Patent Office file containing the application and all correspondence leading up to issuance of the patent is made available in the Patent Office Search Room for inspection by anyone, and copies of these files may be purchased from the Patent Office.

17. Q. *May I write to the Patent Office directly about my application after it is filed?*

A. The Patent Office will answer an applicant's inquiries as to the status of the application and inform him or her whether the application has been rejected, allowed, or is awaiting action by the Patent Office. However, if you have a patent attorney or agent, the Patent Office cannot correspond with both you and the attorney concerning the merits of your application. All comments concerning your invention should be forwarded through your patent attorney or agent.

18. Q. *What happens when two inventors apply separately for a patent on the same invention?*

A. An "interference" is declared and testimony may be submitted to the Patent Office to determine which inventor is entitled to the patent. Your attorney or agent can give you further information about this if it becomes necessary.

19. Q. *Can the six-month period allowed by the Patent Office for response to an office action in a pending application be extended?*

A. No. This time is fixed by law and cannot be extended by the Patent Office, but it may be reduced to not less than thirty days. The application will be abandoned unless proper response is received in the Patent Office within the time allowed.

20. Q. *May applications be examined out of their regular order?*

A. No. all applications are examined in the order in which they are filed, except under certain very special conditions.

When to Apply for a Patent

21. Q. *I have been making and selling my invention for the past 13 months and have not filed any patent application. Is it too late for me to apply for a patent?*

A. Yes. A valid patent may not be obtained if the invention was in public use or on sale in this country for more than one year prior to the filing of your patent application. Your own use and sale of the invention for more than a year before your application is filed will bar your right to a patent just as effectively as though this use and sale had been done by someone else.

22. Q. *I published an article describing my invention in a magazine 13 months ago. Is it too late to apply for a patent?*

A. Yes. The fact that you are the author of the article will not save your patent application. The law provides that the inventor is not entitled to a patent if the invention has been described in a printed publication anywhere in the world more than a year before his patent application is filed.

Who May Obtain a Patent

23. Q. *Is there any restriction as to persons who may obtain a United States patent?*

A. No. Any inventor may obtain a patent regardless of age or sex, by complying with the provisions of the law. A foreign citizen may obtain a patent under exactly the same conditions as a United States citizen.

24. Q. *If two or more persons work together to make an invention, to whom will the patent be granted?*

A. If each had a share in the ideas forming the invention, they are joint inventors and a patent will be issued to them jointly on the basis of a proper patent application filed by them jointly. If, on the other hand, one of these persons has provided all of the ideas of the invention, and the other has only followed instructions in making it, the person who contributed the ideas is the sole inventor and the patent application and patent should be in his or her name only.

25. Q. *If one person furnishes all of the ideas to make an invention and another employs him or her or furnishes the money for building and testing the invention, should the patent application be filed by them jointly?*

A. No. The application must be signed, executed, sworn to, and filed in the Patent Office in the name of the inventor. This is the person who furnishes the ideas, not the employer or the person who furnishes the money.

26. Q. *May a patent be granted if an inventor dies before filing his application?*

A. Yes; the application may be filed by the inventor's executor or administrator.

27. Q. *While in England this summer, I found an article on sale which was very ingenious and has not been introduced into the United States or patented or described. May I obtain a United States patent on this invention?*

A. No. A United States patent may be obtained only by the true inventor, not by someone who learns of an invention of another.

Ownership and Sale of Patent Rights

28. Q. *May the inventor sell or otherwise transfer his or her right to his patent or patent application to someone else?*

A. Yes. He or she may sell all or part of the interest in the patent application or patent to anyone by a properly worded assignment. The application must be filed in the Patent Office as the invention of the true inventor, however, and not as the invention of the person who has purchased the invention from him.

29. Q. *Is it advisable to conduct a search of patents and other records before applying for a patent?*

A. Yes. If it is found that the device is shown in some prior patent it is useless to make application. By making a search beforehand the expense involved in filing a needless application is often saved.

Patent Searching

30. Q. *Where can a search be conducted?*

A. In the Search Room of the Patent Office in the Department of Commerce Building, 14th and E Street, NW, Washington, D.C. Classified and numerically arranged sets of United States and foreign patents are kept there for public use.

31. Q. *Will the Patent Office make searches for individuals to help them decide whether to file patent applications?*

A. No. But it will assist inventors who come to Washington by helping them to find the proper patent classes in which to make their searches. For a reasonable fee it will furnish lists of patents in any class and subclass, and copies of these patents may be purchased for 50¢ each.

Technical Knowledge Available from Patents

32. Q. *I have not made an invention but have encountered a problem. Can I obtain knowledge through patents of what has been done by others to solve the problem?*

A. The patents of the Patent Office Search Room in Washington contain a vast wealth of technical information and suggestions, organized in a manner that will enable you to review those most closely related to your field of interest. You may come to Washington and review these patents, or engage a patent practitioner to do this for you and to send you copies of the patents most closely related to your problem.

33. Q. *Can I make a search or obtain technical information from patents at locations other than the Patent Office Search Room in Washington?*

A. Yes. Libraries have sets of patent copies numerically arranged in bound volumes, and these patents may be used for search or other information purposes as discussed in the answer to Question 34.

34. Q. *How can technical information be found in a library collection of patents arranged in bound volumes in numerical order?*

A. You must first find out from the *Manual of Classification* in the library the Patent Office classes and subclasses which cover the field of your invention or interest. You can then, by referring to microfilm reels or volumes of the Index of Patents in the library, identify the patents in these subclasses and, thence, look at them in the bound volumes. Further information on this subject may be found in the leaflet *Obtaining Information from Patents*, a copy of which may be requested from the Patent Office.

Infringement of Others' Patents

35. Q. *If I obtain a patent on my invention, will that protect me against the claims of others who assert that I am infringing their patents when I make, use, or sell my own invention?*

A. No. There may be a patent of a more basic nature on which your invention is an improvement. If your invention is a detailed refinement or feature of such a basically protected invention, you may not use it without the consent of the patentee, just as no one will have the right to use your patented improvement without your consent. You should seek competent legal advice before starting to make or sell or use your invention commercially, even though it is protected by a patent granted to you.

Enforcement of Patent Rights

36. Q. *Will the Patent Office help me to prosecute*

others if they infringe the rights granted to me by my patent?

A. No. The Patent Office has no jurisdiction over questions relating to the infringement of patent rights. If your patent is infringed, you may sue the infringer in the appropriate United States court at your own expense.

Patent Protection in Foreign Countries

37. Q. *Does a United States patent give protection in foreign countries?*

A. No. The United States patent protects your invention only in this country. If you wish to protect your invention in foreign countries, you must file an application in the Patent Office of each such country within the time permitted by law. This may be quite expensive, both because of the cost of filing and processing the individual patent applications, and because of the fact that most foreign countries require payment of taxes to maintain the patents in force.

How to Obtain Further Information

38. Q. *How does one obtain information as to patent applications, fees, and other details concerning patents?*

A. By ordering a pamphlet entitled *General Information Concerning Patents*

39. Q. *How can I obtain information about the steps I should take in deciding whether to try to obtain a patent, in securing the best possible patent protection, and in developing and marketing my invention successfully?*

A. By ordering a pamphlet entitled *Patents and Inventions, An Information Aid for Inventors.*

PROBLEMS

Working Drawings

1. Prepare working drawings as the implementation step of the design process of one of the problems that was refined at the end of Chapter 5.

Draw the working drawings and assembly on 11″ × 17″ sheets of tracing vellum or film.

2. Prepare working drawings of one of the problems assigned or selected from those at the end of Chapter 23. Draw the working drawings and assembly on 11″ × 17″ sheets of tracing vellum or film.

Patents

3. Write for a copy of a patent that would be of interest to you. Make a list of the features that were used as a basis for obtaining the patent.

4. Suggest modifications that could be used to modify the patent mentioned in the previous problem. Make sketches of innovations that would be possible improvements of the patented mechanism.

5. Write the U. S. Patent Office for patent application forms. Prepare a patent application for a simple invention that has been previously patented, such as a fountain pen, drafting instrument, or similar item. Determine what drawings and materials are needed to complete your application.

6. Prepare patent drawings in accordance with the standards established in Section 8.8 to depict a simple patented object, such as those mentioned in Problem 5. Strive for a finished technique that would make your drawing acceptable as a patent drawing.

7. Make a list of ideas that you believe to be patentable. These may be ideas that you have developed during work on design problems assigned in class.

8. Write a technical report investigating the history and significance of the patent system and its role in our industrial society. Consult your library and available government publications on patents. Give information and data that will improve your understanding of patents.

DESIGN PROBLEMS

9.1 GENERAL

Engineering problems are seldom compartmentalized by subject area in a clearly defined structure that suggests a single solution. It is more common for problems to contain a blend of several areas, ranging from psychological and social factors to different fields of engineering.

This chapter will introduce problems that can be used for class assignments, team projects, and other combinations of approaches to provide an experience in applying all of the previously covered principles and techniques and those that follow in this volume.

9.2 THE INDIVIDUAL APPROACH

The short problem—requiring one or two hours of work—will probably be assigned for solution by students working individually, with responsibility for each step of the design process and the solution of the problem. An advantage to the individual approach is unity of control with the authority to make decisions without consulting associates. No time or effort need be expended in the management of team members.

The procedure for solving a design problem is the same for all types of problems. A simple design problem may involve fewer details and less

depth in each step, but all design steps are applied as in the more sophisticated, comprehensive problem.

9.3 TEAM APPROACH

An effectively organized team working on a single problem has access to more talent than does the typical individual who devotes the same number of hours to the project. The application of this talent will be a problem if the team is not properly organized toward a unified effort. Team management and working effectively with others are valuable traits that should be developed by the student; they are requirements in professional practice.

Team Size A student design team should have from three to seven members. Three is considered a minimum number for a valid team experience in the group approach. The optimum size, four, lessens the possibility of domination by one or more members. Teams larger than seven approach an unmanageable size.

Team Composition Teams need not be composed of close friends or persons with particular relationships. In actual practice, the engineering team may be composed of professionals from different firms who may be total strangers. This arrangement can be advantageous in that it reduces the impact of personalities and individual traits that may affect an association of close friends.

Team Leader A leader is necessary for most teams to function effectively. He or she must take the responsibility for making assignments, seeing that deadlines are met, and acting as arbitrator in cases of disagreement.

9.4 THE SELECTION OF A PROBLEM

The best problem for a student design project is one that involves familiar and accessible conditions. Evaluation of a situation by means of first-hand observation is always preferable.

A design for a water-ski rack for an automobile is more feasible than one for a support bracket for an airplane. The average student is familiar with water skis and has access to an automobile to establish the limiting factors of the design problem.

Student-proposed Problem The best design problem is one that has been recognized and proposed by the students who will be working toward its solution. The mere ability to recognize the need for a new design or a modification of an existing design is the first step toward the development of a creative attitude.

When the members of a design team decide on a design project, they should prepare a written proposal to identify the problem and to outline its limits. The format for the proposal and a description of its contents are given in Section 7.6.

Instructor-assigned Problems Problems may be assigned by the instructor to student teams. Assigned problems are appropriate for classroom situations, and they are analogous to the method of assignment used in industry.

9.5 PROBLEM SPECIFICATIONS

The following specifications are those an individual or design team must consider when preparing a design proposal or outlining their assignments.

The Short Problem A typical short problem can be completed during one class period or as an outside assignment by a single student. A problem of this type will be solved as indicated in Section 2.3. The specifications for a short problem could include all or part of the following.

1. Completed work sheets illustrating the development of the design process (Chapter 2).
2. Freehand sketches of the design for implementation (Chapter 4).
3. An instrument drawing of the proposed design (Chapters 8 and 23).
4. A dimensioned instrument drawing of the proposed design (Chapter 23).
5. A pictorial sketch (or one made with instruments) illustrating the design (Chapters 3 and 25).
6. Visual aids, flip charts, or other media to present the design to a group (Chapter 7).

The short problem may be presented entirely by means of freehand sketching, which reduces time but requires a high degree of sketching skill on the part of the designer.

The Comprehensive Problem Comprehensive problems can vary in time, but a typical one takes an average of 80 work-hours. More complex problems can require 120 work-hours, or about 20 hours per person when solved by a team of six.

The following specifications will apply in total or in part to a comprehensive problem of the systems- or product-development type. For example, a market survey would not be appropriate for a systems problem involving the design of a parking lot.

1. A proposal outlining the problem, the method of approach, and the specifications used in solving the problem (Chapter 7).
2. Completed work sheets illustrating the development of the design process (Chapter 2).
3. Schematic diagrams, flow charts, or other symbolic methods of illustrating the design and its function (Chapter 32).
4. An opinion survey determining interest for the proposed design (Chapter 6).
5. A market survey evaluating the product's possible acceptance and estimated profit (Chapter 3).

6. A model or prototype for analysis or presentation (Chapter 6).

7. Pictorials explaining features of the final design solution that are not clearly shown in other drawings (Chapter 25).

8. Dimensioned engineering drawings giving details and specifications of the design (Chapter 23).

9. A technical report, fully illustrated with charts and graphs, explaining the activities leading to the solution and ending with conclusions and recommendations (Chapter 7).

9.6 SCHEDULING TEAM ACTIVITIES

The following can be used by a team as guidelines to assist members in working toward the completion of their project. (In addition, references are provided to those portions of the text where more detailed instructions may be found.)

A progress report should be prepared after each project period; Figure 7.30 is an example of a progress report format. The periods of this team project assignment are spread over 14 weeks, beginning with the second week of the semester and ending with the 15th week. The instructor may choose to have the team work on their team assignments during regular class periods throughout the semester, or he or she may ask the students to meet as a team outside of class periods.

WEEK 1: Team organization: Each team should discuss each part of the project as a group, but it will probably be necessary to appoint a project chief. This assignment can be for the duration of the project or it can be rotated among other team members.

WEEK 2: First meeting: The following activities should be discussed and assigned to specific team members.

1. After the project has been selected, each team member should review these text references before the first group meeting: Chapters 2, 3, 4, and 9.

2. Do not attempt to solve the problems confronting you during the first meeting. Instead, try to identify the problems and the factors af-

fecting them. Use work sheets (Section 3.2) to list all facts and information as a permanent record.

3. Determine the data that must be gathered to identify your problems—dimensions, trends, surveys, and similar information (Chapter 3).

WEEK 3: Second meeting: Problem identification.

1. List jobs that must be performed on the DS & PR form (Chapter 3).

2. Prepare an activities network (Chapter 3) to determine the sequence of activities necessary to complete the project.

3. List the activities on an Activity Sequence Chart (ASC) (Chapter 3), and assign each job to a team member.

4. Each team member should make a copy of the ASC for reference.

5. Prepare a progress report to cover this period's effort (see Fig. 7.30).

WEEK 4: Third meeting: Preliminary ideas.

1. Team members may make more progress if, at the beginning of each meeting, they work individually to develop as many ideas as possible. Notes and sketches should be prepared legibly and kept for group discussion (Chapter 4).

2. Review individual ideas as a group (Chapter 4).

3. Conduct a brainstorming session (Chapter 4). List all ideas.

4. Make assignments to team members.

5. Discuss the need for a survey (Chapter 4).

6. Prepare a progress report for your instructor (see Fig. 7.30).

WEEK 5: Visiting engineers.

1. Invite engineers from industry or representatives from other departments on your campus to serve as consultants to your class during one class period. These consultants can discuss your problems with you and offer advice that will be of help.

2. The consultants cannot be expected to solve your problems, but instead, they will serve as counselors and as coaches to guide each team.

3. Each team should make a brief progress report to each consultant at the outset of the meeting.

4. Questions can be asked each consultant concerning his or her job, company, or any other aspect of the profession that might improve your understanding of career opportunities.

WEEK 6: Fourth meeting: Preliminary ideas.

1. Continue collecting ideas. Keep all work sheets and notes for inclusion in the appendix of your written report (Chapter 4).

2. Prepare a progress report for your instructor.

WEEK 7: Fifth meeting: Design refinement.

1. Begin making scale drawings and determining specific characteristics of your project (Chapter 5).

2. Prepare graphs, diagrams, and schematics that can be used in your report and also as visual aids in the oral presentation (Chapter 32).

3. Apply more critical judgment at this stage than in the previous examples.

WEEK 8: Sixth meeting: Analysis.

1. Review all collected data, preliminary ideas, and refined designs. Consider weights of materials, strength, economy, market prospects, and human factors as they might affect your design (Chapter 6).

2. Apply engineering fundamentals learned in other courses.

3. Prepare a progress report for your instructor.

WEEK 9: Seventh meeting: Decision and implementation.

1. Your team must select or reject the best solution at this point. If a solution is not feasible, your report should clearly state why this is the case. Lack of a solution is not considered a team failure if you have carefully studied the problem (Chapter 7).

2. Implementation is the final step of the design project. All findings will be presented in a technical report; a model may be required for product designs (Chapter 8).

3. If your team has kept data, sketches, and graphs, and has accumulated a record of all activities, a great portion of the report will be completed except for editing (Chapter 7).

4. Assign responsibilities for the preparation of the final drawings and the text of the report.

5. Prepare a progress report.

WEEK 10: Eighth meeting: Implementation and report.

1. Make individual assignments and continue with the various parts of the written report (Chapter 8).

2. Refine final figures and illustrations. Prepare a letter of transmittal.

3. Prepare a progress report.

WEEK 11: Ninth meeting: Complete report.

1. Prepare an evaluation form that reports the contribution of each team member (Chapter 7).

2. Complete the first draft of final report.

WEEK 12: Tenth meeting: Preparation for presentation.

1. Organize your presentation. Determine the types of visuals that are needed (Chapter 7).

2. Prepare 3 × 5 cards for each visual (Chapter 7).

3. Assign responsibilities for the preparation of visuals.

4. Refer to Chapter 32 for good practices of preparing graphs and visuals.

WEEK 13: Eleventh meeting: Preparation for presentation.

1. Conduct a practice presentation (Chapter 7).

2. Use the same techniques of presentation that will be used in the final presentation.

3. Refer to oral report evaluation forms.

4. Have as many team members share in the presentation as possible.

WEEK 14: Twelfth meeting: Final presentation.

1. Give final presentation to the returning consultants.

2. Complete evaluation forms for oral presentation. Include percentage contributed by each team member to the oral presentation (Section 7.5).

3. Classes will evaluate each team presentation using the forms provided. Teams will not grade their own (Fig. 7.26).

9.7 SHORT DESIGN PROBLEMS

The following are short problems that can be completed in less than two periods. They are usually assigned as individual assignments.

1: Lamp bracket. Design a simple bracket to attach a desk lamp to a vertical wall for reading in bed. The lamp should be easily removable so that it can be used as a conventional desk lamp.

2: Towel bar. Design a towel bar for a kitchen or bathroom. Determine optimum size, and consider styling, ease of use, and method of attachment. This design will be a modification of those already available on the market.

3: Pipe aligner for welding. Pipes are often welded together in sections. The initial problem of joining pipes with a butt weld is the alignment of the pipes in the desired position. Design a device with which to align pipes for on-the-job welding. For ease of operation, a hand-held device would be desirable. Assume that the pipes will vary in diameter from 2″ to 4″.

4: Film reel design. The film reel used on projectors is often difficult to thread due to the limited working space. A typical 12″ diameter movie reel is shown in Fig. 9.1. Redesign this type of reel to allow more space for threading. Your design should also be attractive, since the reel is a consumer product that must compete against other designs.

5: Side view mirror. In most cars, rearview mirrors are attached to the side of the automobile

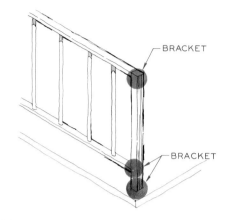

Fig. 9.2 Problem 6. Porch railing system.

to improve the driver's view of the road. Design a side view mirror that is an improvement over those with which you are familiar. Consider the aerodynamics of your design, protection from inclement weather, and other factors that would affect the function of the mirror.

6: Railing post mount. An ornamental iron railing is to be attached to a wooden porch surface supported by several 1″ square tubular posts. Design the mounting piece that will attach the post to the surface. Figure 9.2 shows a typical railing to clarify the problem requirements.

A second attachment is needed to assemble the railing with the support post at each end. If two screw holes are to be used, design the part that can be used to secure the two perpendicular members.

7: Nail feeder. Workers lose time in covering a roof with shingles if they have to fumble for nails. Design a device that can be attached to a worker's chest and will hold nails in such a way that these will be fed in a lined-up position ready for driving. Determine the number of nails that should be held in this device at any one time.

8: Cupboard door closer. Kitchen cupboard and cabinet doors are usually not self-closing and thus are safety hazards and are unsightly. Design a device that will close doors left partially open. It would be advantageous to provide a means for disengaging the closer when desired.

9: Paint-can holder. Paint cans are designed with a simple wire bail that, when held, makes it difficult to get a paint brush in the can. Wire bails are also painful to hold for any length of time. De-

Fig. 9.1 Problem 4. Typical movie projector reel.

sign a holding device that can be easily attached and removed from a gallon size paint can (6.5″ diameter × 7.5″ height). Consider the human factors such as comfort, grip, balance, and function.

10: Self-closing or opening hinge. Interior doors of residential dwellings tend to remain partially open instead of staying in a completely open position against the wall. This introduces a problem when a door opens into the end of a hall, since the edge of the door that is ajar in the middle of the hall can be a hazard to the occupants.

Design a hinge that will hold the door in a completely open position. This can be used to replace door stops and other devices used to hold doors.

11: Tape cartridge storage unit. Tape players are used frequently in automobiles. Design a storage unit that will hold a number of these cartridges in an orderly fashion, so that the driver can select and insert them with a minimum of motion and distraction. Determine the best location for this unit and the method of attachment to an automobile. Provision should be made to protect the cartridges from theft.

12: Slide projector elevator. Most commercial slide and movie projectors have adjustment feet used to raise the projector to the proper position for casting an image on a screen (Fig. 9.3). The range of variation of this adjustment is usually less than 2″. Study a slide projector available to you to determine the specific needs and the limitations of an adjustment that would provide a greater variation in elevation. Design a device to serve this purpose

Fig. 9.3 Problem 12. Movie projector. (Courtesy of Eastman Kodak.)

as part of the original design or as an accessory that could be used on existing projectors.

13: Bookholder for reading in bed. As a student, you may often desire to read while lying in bed. Design a holder that can be used for supporting a book in the desired position while providing comfortable conditions.

14: Table leg design. Do-it-yourselfers build a variety of tables using slab doors, plywood, and commercially available legs. Table tops come in a number of sizes, but table heights are fairly standard. Determine what the standard table heights are and design a family of legs that can be attached to table tops with screws. For improved appearance, the legs should not be vertical, but should slant from the table top. Design the legs and attachments, indicating the method of manufacture, size, cost, and method of attachment.

15: Canoe mounting system. Canoes and light boats are often transported on the top of automobiles on a luggage rack or similar attachment. Design an accessory that will enable a person to remove and load a boat on top of an automobile without additional help. This attachment should accommodate aluminum boats varying from 14′ to 17′ in length and weighing 100 lbs to 200 lbs. Give specifications for a method of securing the boat after it is in its final position on top of the automobile.

16: Toothbrush holder. Design a toothbrush holder that can be attached to a bathroom wall and can hold a drinking cup and two toothbrushes.

17: Napkin holder. Design a device that will hold 25 paper napkins on a dining room table for easy access.

18: Book holder. Design a holder that will support your textbook on your drawing table in a position that will make it more readable and accessible.

19: Clothes hook. Design a clothes hook that can be attached to a closet door for hanging clothes. It should be easy to manufacture and simple to use.

20: Automobile ashtray. Design an ashtray that can be attached to the dashboard of any automobile. It should be easy to attach and to remove for emptying.

21: Pencil holder. Design a holder that can be attached to the interior of the car for holding pencils or pens.

22: Door stop. Design a door stop that can be attached to a vertical wall or floor to prevent the door knob from bumping the wall.

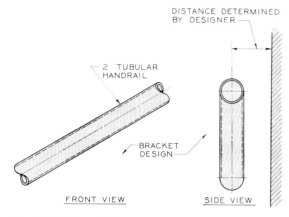

Fig. 9.4 Problem 27. A handrail bracket.

Fig. 9.5 Problem 28. A latch-pole hanger.

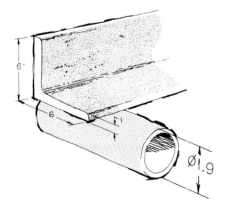

Fig. 9.6 Problem 29. A pipe clamp.

23: Teaching aid. Design an apparatus that can be used by a teacher to illustrate the basic principles of orthographic projection. Investigate the market potential of such a product.

24: Cup dispenser. Design a paper cup dispenser that can be attached to a vertical wall. This dispenser should hold a series of cups 2″ in diameter that measure 6″ tall when stacked together.

25: Drawer handle. Design a handle that would be satisfactory for a standard file cabinet drawer.

26: Paper dispenser. Design a dispenser that will hold a 6″ diameter by 24″ wide roll of wrapping paper. The paper will be used on a table top for wrapping packages.

27: Handrail bracket. Design a bracket that will support a tubular handrail that will be used on a staircase (Fig. 9.4). Consider the weight that the handrail must support.

28: Latch-pole hanger. Design a hanger that can be used to support a latch-pole from a vertical wall. It should be easy to install and use (Fig. 9.5).

29: Pipe clamp. A pipe with a 4″ diameter must be supported by angles (Fig. 9.6) that are spaced eight feet apart. Design a clamp that will support the pipe without drilling holes in the angles.

30: TV yoke. Design a yoke that can support a TV set from the ceiling of a classroom and permit it to be adjusted at the best position for viewing.

31: Flag pole socket. Design a socket for flags that is to be attached to a vertical wall. Determine the best angle of inclination for the flag pole.

32: Crutches. Design a portable crutch that could be used by a person with a temporary leg injury.

33: Cup holder. Design a holder that will support a soft drink can or bottle on the interior of an automobile.

34: Gate hinge. Design a hinge that could be attached to a 3″ diameter tubular post to support a 3′ wide gate.

35: Safety lock. Design a safety lock that will hold a high voltage power switch in either the "off" or "on" positions to prevent an accident (Fig. 9.7).

36: Tubular hinge. Design a hinge that can be used to hinge 2.5″ OD high strength aluminum pipe in the manner shown in Fig. 9.8. A hinge of this type is needed for portable scaffolding.

37: Miter jig. Design a jig that can be used for assembling wooden frames at 90° angles. The stock for the frames is to be rectangular in cross-sections

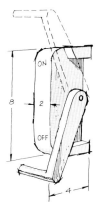

Fig. 9.7 Problem 35. A safety lock.

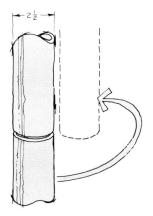

Fig. 9.8 Problem 36. A tubular hinge.

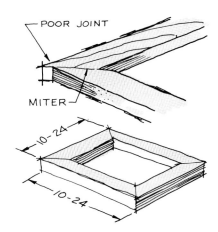

Fig. 9.9 Problem 37. A miter jig.

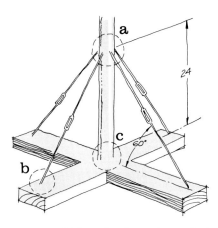

Fig. 9.10 Problem 38. Base hardware for a volleyball net.

that vary from 0.75″ × 1.5″ to 1.60″ × 3.60″. Outside dimensions vary from 10″ to 24″ (Fig. 9.9).

38: Base hardware. Design the hardware needed at the points indicated for a standard volleyball net. The 7′ pipes are supported by crossing two-by-fours. Design the hardware needed at points *a*, *b*, and *c* (Fig. 9.10).

39: Conduit connector/hanger. Design a support that will attach to a 0.75″ conduit that will support a channel that is used as an adjustable raceway for electrical wiring. Your design should permit ease of adjustment. (Fig. 9.11).

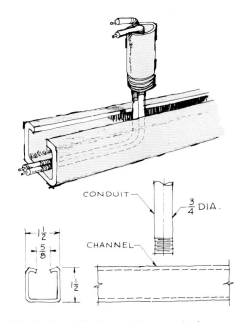

Fig. 9.11 Problem 39. A conduit connector hanger.

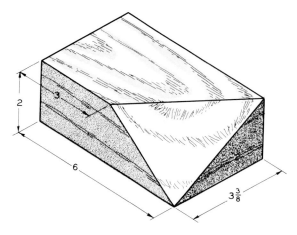

Fig. 9.12 Problem 40. A fixture design.

40: Fixture design. Design a fixture that will permit a small-scale manufacturer to saw the corner of the block as shown in Fig. 9.12.

41: Drum truck. Design a truck that can be used for moving a 55-gallon drum of turpentine (7.28 pounds per gallon). The drum will be kept in a horizontal position, but it would be advantageous to incorporate a feature into the truck that would permit it to be set in an upright position (Fig. 9.13).

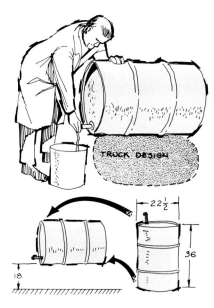

Fig. 9.13 Problem 41. A drum truck.

9.8 SYSTEMS DESIGN PROBLEMS

The systems design problem is a broad engineering problem that requires the study of the interrelationships of various components—social, economic, physical, and management considerations. The end result of a systems problem may be a plan, a conclusion, or a recommendation—as opposed to a hardware design for product development. An engineering-management problem is the problem of locating an industrial plant in a given area. The engineer must analyze the requirements of both company and community, and the benefits to be derived by each. Engineering problems are studied concurrently with management needs. Predictions are based on the available information, obtained from like situations previously tried or from entirely unproven criteria.

42: A multi-purpose utility meter. Today's residential units have separate meters for electricity, water, and gas that are checked each month by separate companies. Hence there is considerable duplication of effort and expense. Consider the feasibility of combining all meters into a single unit that could be read by a meter-reading service. A unit of this type would reduce the number of meter readers to one-third. Outline how such a system could be organized and implemented.

43: Portable bleachers. Portable bleachers are required for some outdoor and indoor activities that attract only small crowds of 100 or less. Design a portable bleacher system that can be assembled and disassembled with a minimum of effort, and that can be stored. Consider the structure, size, materials, and method of construction. Identify a variety of uses for these bleachers that would justify their production for the general market.

44: Archery range. Determine the feasibility of providing an archery range for beginning and experienced archers that can be operated at a profit. Your problem is to investigate the need for such a facility, its potential market, its location, the equipment needed, method of operation, costs and fees, and other factors that would bear upon your decision regarding the feasibility of the project. Consider all engineering aspects of site preparation, utilities, concessions, and parking.

45: Car rental system. Investigate the possibilities and feasibility of a student-operated rental car system. Data must be collected to determine student interest, cost factors, number of cars needed, rates, personal needs, storage, maintenance, and other factors affecting the feasibility of the operation. Specify the details of your system, including the location, the cost of operation, and the expected level of income.

46: Model airplane field. Model airplane hobbyists who build and fly gasoline engine models often have inadequate facilities for pursuing their hobby. Analyze their needs, the space required, the type of surfaces, control of sound, safety factors, method of operation, and come up with a total evaluation of the system. Select a site on or near your campus that you feel is adequate for this facility. Evaluate the equipment, utilities, and site preparation that would be required. Determine the degree to which a facility of this sort would be utilized and the income, if any, that could be expected from it.

47: Overnight campsite. Analyze the feasibility of converting a local lot near a major highway into a complex of campsites for overnight campers. Determine the facilities that would be desired by the campers and the expenditures that would be required for operation. Determine the profit margins for your particular design.

48: Swimming pool study. A group of students wishes to construct a swimming pool on your campus that would be self-supporting. Determine the cost of the pool, the area in which it would be located, the students it would serve, and the equipment and labor necessary to operate it. Determine whether or not it would be feasible as the conclusion of your problem.

49: Outdoor shower. You are the owner of a weekend cottage and you do not have a hot and cold water system—you have only cold water. Design a system that can take advantage of the sun's heat in the summer to heat the water used for bathing and kitchen purposes. Devise a means of using the same system in the winter with heat provided from some other source (a wood fire, kerosene, or other method). Explain your design, the operating temperature, and the cost of operating the system when heated artificially in the winter. Can you design a totally portable shower that can be used on camping trips?

50: Information center. Visitors to a typical college campus often find it difficult to locate buildings, parking lots, and campus facilities. In most cases, there is a need for a system that would be available to provide general information to off-campus visitors.

Analyze your college campus to determine the most logical location for a drive-in information center. Develop a design for this center that could provide an easy traffic flow to accommodate the expected number using the system. Determine the informational material that should be included to assist a visitor. Consider using slides, photographs, maps, sound, and other audiovisual aids to accomplish your goal.

Your design should include the location of the system, its plot plan, equipment, housing and a detailed description of its operation.

51: Golf driving-range ball return system. Golf driving ranges are usually designed in such a way that the balls must be retrieved by hand or by means of a specially designed vehicle that collects them from the range. Design a system capable of automatically returning the balls to the driving area. Consider all possibilities of this system from site modification to selection of a special site. Evaluate the market potential of your design.

52: Instructional system. Classroom instruction could be improved by providing two-way communication between the teacher and the students. For example, the teacher could proceed at a more efficient rate if he or she had some idea as to whether or not the class was understanding the points being made. Determine whether a system could be developed that would give the student some means of signaling understanding, or lack of understanding, of the lecture without having to ask questions.

53: An instructional console system. You have been assigned the responsibility for designing the ideal classroom. Determine the optimum class size, the room layout, lighting requirements, choice of furniture, and other factors of this type that affect environment. Study the possibility of developing a visual-aid console that incorporates the latest projectors, recorders, and other devices that would improve instruction and learning (Fig. 9.14).

54: Pedestrian transport system. Your campus has considerably more pedestrian traffic in some areas than in others, causing traffic congestion. Consider the possibility of developing a system that would provide a more even flow of pedestrian traffic during rush periods. Much time is lost between classes when a student must leave the upper floor of a building, descend to ground level with many other students, travel to another building at ground level, and climb a series of

Fig. 9.14 Problem 53. An instructional console system.

stairs to an upper story of a building. Consider a solution to this problem of pedestrian traffic.

55: Instant motel. Many communities have periodic needs for more housing than is available on a regular basis for events like ball games and celebrations. Investigate the different methods of providing an "instant motel" that could fulfill this need for a few days.

Consider such options as tents, vans, trailers, train cars, etc. Determine the profit margins for your proposed solution.

56: Helicopter service. You have been assigned the responsibility of planning a helicopter passenger service that will connect with the local airlines in your community. Helicopters will transport passengers from the terminal to your campus on a regular schedule to reduce time loss due to excessive automobile traffic. Analyze the needs of your campus to determine whether such a system could be feasible and self-supporting. Determine the helicopter landing area on your campus and the flight schedule.

57: Fallout shelters. Recent emphasis has been placed on the importance of shelters for protection against nuclear fallout. Many new urban buildings have been constructed to include shelters as an integral part of the structure. Referring to available data from local and governmental agencies concerning the need for nuclear protection, determine the general needs of your campus

for fallout protection ten years from now. Estimate the basic needs, length of protection, provisions, etc., as a means of arriving at the space requirements for protection.

58: Campus planning. Assume that your college campus was to begin planning for full-capacity utilization. Full utilization means the total use of your present facilities on a 12-month, 24-hour basis. Determine how many students could be accommodated in your classrooms, dormitories, and in other facilities without adding new buildings.

Analysis of teaching schedules, faculty, and service personnel needed should be studied to identify possible problems. Evaluate the changes that will occur in parking systems, pedestrian traffic, and the management of the total system. A considerable portion of this problem is the identification of the critical areas that must be studied.

59: Diazo machine operation. Blueline prints are made on a diazo machine by putting the tracing paper drawing in contact with the diazo paper and feeding it through the diazo machine. A reproduction department in a company will have to reproduce many prints; hence a full-time operator will be required. Assume that you have been given the assignment of establishing a system to provide the most efficient utilization of the equipment and the operator's time to meet the specifications outlined below.

The machine will accept individual drawings or groups of drawings along its 40″ belt at a rate of 10′ per minute. The drawing size used most frequently by your company is 11″ × 17″. The diazo paper must be run through the developing chamber of the machine directly above the intake at a speed of 10′ per minute. The machine is 60″ long, 24″ deep, and 48″ high. Determine the equipment and tables and their optimum arrangement for most efficient operation. Analyze the working space required and the sequence of activities.

What is the cost per drawing of the operator working at peak efficiency? Consider the other duties of the operator, such as stapling sets of drawings together and gathering original drawings to be returned with their prints.

60: Drive-in theater. You have been commissioned to develop an area for a drive-in motion picture theater. As the chief designer you must determine the optimum size of the drive-in to provide adequate parking and viewing from the audience, as well as an adequate profit. Consideration must be given to traffic flow, screen size, and position, electrical problems, drainage, utilities, concessions, and other facilities commonly found in a treater of this type. Determine the overall layout of the drive-in, its traffic system, and major components. Detail a typical parking space for a single car, indicating the contour of the surface to provide the proper viewing angle for the car.

List the areas that would require specialists such as electrical engineers, civil engineers, etc. Outline your plan for involving consultants who could assist you in preparing your final design. Refer to the Yellow Pages of a telephone directory for names of individuals or firms that could help you.

61: Boat launching facility. You are to design a boat launching area at a lake where boat trailers could be positioned. Assume that as many as 500 boats may be unloaded during the peak hours of a day. Analyze the requirements of a workable system that will control this traffic with a minimum of confusion. Thought must be given to the unloading area, the space required, the parking area, and the type of terrain required for the facility. Determine the charges required to maintain this facility and pay for help if any is needed.

62: Football stadium expansion. Study the attendance figures of your football stadium to determine what the future holds for it. Will a larger stadium be required? If so, when and how much extra seating will be required. Search for historical data and information that would help you make your recommendation.

63: Shopping checkout system. An acute problem of the shopping center and grocery market is that of checking out goods, payment, and delivery of purchases to the customers' automobiles. Your project is to develop a system for expediting the transfer of goods selected by the customer, from the shelf to the customer's car. This includes the process of packaging and payment. Under the present method, a number of sackers and delivery people are required to package the groceries and cart them to each customer's car. A more efficient and economical method may be warranted.

64: Injury-proof playground. Most playgrounds are unsafe and many injuries occur to children who use them. Study the activities of children at play. Design a playground system that permits the greatest degree of participation by the children with the least risk of injury.

65: Loading dock system. The trucking in-

dustry provides a sizable portion of the transportation of goods and supplies. It is important that these materials be unloaded and stored as efficiently as possible to keep the trucks en route with a minimum of layovers. Unloading is usually performed by a fork-lift truck at a loading dock that is level with the truck's floor. Design a method ensuring that the dock can be adjusted to level with the truck's floor. The height of trucks may vary; consequently, an adjustment may be needed to position the floor.

Design a system that will allow the truck to be enclosed or protected from the cold weather during its loading or unloading while taking advantage of the warehouse's heating. The dimensions of average trucks are given in Fig. 9.15 to serve as a guide. In addition to the dock design, make a layout of the plan of your design.

66: Modification of a drive-in theater. Assume that you are the owner of a drive-in movie theater that has fallen on bad times, and you wish to convert this facility into a different operation that would be profitable. Consider the possibilities that are open to you, with the least expense and the greatest possibility for profit return. You might select an actual theater in your neighborhood to use as an example.

67: Modification of an existing facility. Isolate an area on your campus or in your community that that is inadequate for the present demands made on it. This could be a traffic intersection, parking lot, recreational area, or a classroom. Identify the

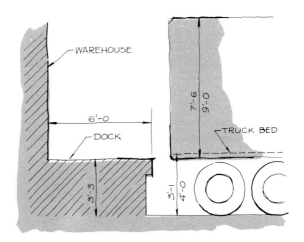

Fig. 9.15 Problem 65. The dimensions of an average truck.

problem and the deficiencies that should be corrected. Propose modifications that would improve the existing facility.

68: Tape-recording system. In the majority of classes, the student must spend most of his or her time taking notes; this activity interferes with concentration on the concepts being presented. Design a system that would provide students with a tape recording of class lectures, which they could take to their rooms and play for review purposes. Consider the possibility of providing an automatic system that would record illustrations and diagrams presented by the instructor in class, eliminating manual copying. Estimate the cost of such a system and determine its value to the educational program. Approach this design as though your company were planning to produce an experimental system of this type for the educational market.

69: Service station modification. The service station system used today differs little in arrangement and function from the first filling stations to come into existence. One major change, however, is the conversion to the credit card system. A ticket must be prepared and signed by the customer, requiring more effort than the cash payment does. The method of servicing automobiles—cleaning windshields, checking oil, and other routine chores—is approached in the same manner, with little improvement in technique.

Consider the usual operations performed in servicing an automobile to develop a more efficient system. This may involve a new layout for the station and the utilities used in servicing cars.

70: Recreational facility. Analyze the various recreational activities on your campus that could be improved with the construction of a multipurpose facility to accommodate these activities. These activities might include such things as outdoor movies, plays, sports, meetings, dances, and etc. The facility should be analyzed as an outdoor installation with the minimum of conventional structures. Develop an open, multipurpose area that could be converted to as many uses as possible.

71: Educational toy. Assume that you are assigned the responsibility of establishing the production system for manufacturing the educational toy shown in Fig. 9.16 at a rate of 5000 per month. You must determine the square footage needed, types of machines required, space for raw materials, office facilities, and warehousing necessary to sustain this operation. Determine the number of employees needed, their rate of pay, and the num-

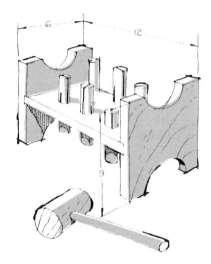

Fig. 9.16 Problem 71. An educational toy.

ber of work stations required. Compute the expense to produce the item and the selling price necessary to provide the required profit margin.

72: Child's furniture. Proceed as in Problem 71 but assume that the child's furniture shown in Fig. 9.17 is the product to be produced. Determine the number to be manufactured per month to break even and establish the selling price. Establish scales for selling prices in quantities in excess of the break-even level.

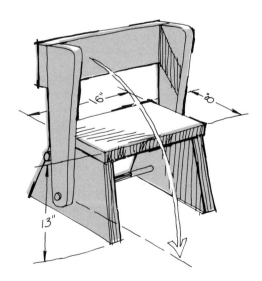

Fig. 9.17 Problem 72. Child's furniture.

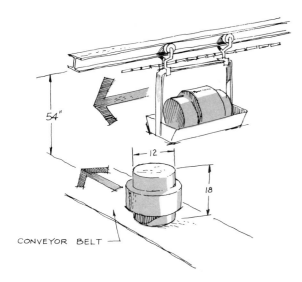

Fig. 9.18 Problem 74. Conveyor system.

73: Simple product. Select a simple product and establish the production system and requirements for its manufacture, proceeding as in Problem 71. Have the product approved by your instructor before going ahead.

74: Conveyor system. You have been assigned the responsibility of designing a conveyor system for delivery of green tire carcasses to the vulcanizing press. The carcasses from the overhead chain conveyor must be transferred to the belt in an upright position as shown (Fig. 9.18). Each carcass weighs approximately 30 lbs and is 12″ in diameter and 18″ tall. Your solution will probably involve the incorporation of standard conveyor components.

9.9 PRODUCT DESIGN

A product design involves the development of a device that will perform a specific function and will be mass-produced and sold to a broad market. It is necessary to be concerned with the tastes and needs of the market, methods of manufacturing and marketing, patent considerations, human factors, shipping, and so on.

75: Hunting blind. Hunters of geese and ducks must remain concealed while hunting. Design a portable hunting blind to house two hunters. This blind should be completely portable so

that it can be carried in separate sections by each of the hunters. Specify its details and how it is to be assembled and used.

76: Convertible drafting table. Many laboratories are equipped with drawing tables that have drafting machines attached to them. This arrangement clutters the table top surface when the tables are used for classes that do not require drafting machines. Design a drafting table that can enclose a drafting machine, thereby concealing it and protecting it when it is not needed.

77: Writing tablet for a folding chair. Design a writing tablet arm for a folding chair that could be used in an emergency or when a class needs more seating. To allow easy storage, the tablet arm must fold with the chair (Fig. 9.19). Use a folding chair available to you for dimensions and specifications.

Fig. 9.19 Problem 77. Folding chair with a writing tablet attached.

78: Portable truck ramp. Delivery trucks need ramps to load and unload supplies and materials at their destinations (Fig. 9.20). Assume that the bed of the truck is 20″ from ground level. Design a portable ramp that would permit the unloading of goods with the use of a hand dolly and reduce manual lifting.

79: Sensor retaining device. The Instrumentation Department of the Naval Oceanographic Office uses underwater sensors to learn more about the ocean. These sensors are submerged on cables from a boat on the surface. The winch used to retrieve the sensor frequently over-runs (continues pulling when it has been retrieved), causing the cable to break and the sensor to be lost. Design a safety device that will retain the sensor if the cable is broken when the sensor reaches a pulley. The

Fig. 9.20 Problem 78. Delivery ramp for unloading goods.

problem is illustrated in Fig. 9.21. The sensor weighs 75 lbs.

80: Flexible trailer hitch. In combat zones, vehicles must tow trailers where terrain may be very uneven and hazardous. Design a trailer hitch that will provide for the most extreme conditions possible. Study the problem requirements and limitations to identify the parameters within which your design must function.

81: Worker's stilts—human engineering. Workers who apply gypsum board and other types of wallboards to the interiors of buildings and homes must work on scaffolds or wear some type of stilts to be able to reach the ceiling to nail the 4′ × 8′ boards into position.

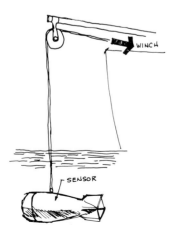

Fig. 9.21 Problem 79. Underwater sensor.

Design stilts that will provide the worker with access to a ceiling 8′ high while permitting him or her to perform the job of nailing ceiling panels with comfort. The stilts should be adjustable to accommodate workers of various weights, sizes, and heights. Analyze the needs of the workers and the requirements for stilts that will provide comfort, traction, safety, and adapt to the human body.

82: Pole-vault standards. Many pole vaulters are exceeding the 18 foot height in track meets, which introduces a problem for the officials of this event. The pole vault uprights must be adjusted for each vaulter by moving them forward or backward plus or minus 18″. Also, the crossbar must be replaced with great difficulty at these heights by using forked sticks and ladders, which are crude and inefficient. Develop a more efficient set of uprights that can be easily repositioned and will allow the crossbar to be replaced with greater ease.

83: Sportman's chair. Analyze the need for a sportman's chair that could be used for camping, fishing from a bank or boat, at sporting events, and for as many other purposes as you can think of. The need is not for a special-purpose chair, but for a chair that is suitable for a wide variety of uses to fully justify it as a marketable item. Make a list of as many applications as possible and use this as a basis for your design.

84: Portable toilet. Design a portable toilet unit for the camper and outdoorsperson. This unit should be highly portable, with consideration given to the method of waste disposal. Evaluate the market potential for this product.

85: Child carrier for a bicycle. Design a seat that can be used to carry a small child as a passenger on a bicycle. Assume that the bicycle will be ridden by an older youth or an adult. Determine the age of the child who would probably be carried as a passenger. Design the seat for safety and comfort.

86: Lawn sprinkler control. Design a sprinkler that can be used to water irregularly shaped yards while giving a uniform coverage. This sprinkler should be adjustable so that it can be adapted, within its range, to yards of any shape. Also consider a method of cutting the water off at certain sprinkler positions to prevent the watering of patios or other areas that are to remain dry.

87: Power lawn fertilizer attachment. The rotary-power lawn mower emits a force through its outlet caused by the air pressure from the rotating blades. This force might be used to distribute fertilizer while the lawn is being mowed. Design an attachment for a power mower that could spread fertilizer while the mower is performing its usual cutting operation.

88: Car and window washer. The force of water coming from a hose provides a source of power that could operate a mechanism that could be used to wash windows or cars. Design an attachment for the typical garden hose that would apply water and agitation (for optimum action) to the surface being cleaned. Consider other applications of the force exerted by water pressure in the performance of yard and household chores that involve water and require agitation that could be provided by the water force.

89: Projector cabinet. Many homes have slide projectors, but each showing of the family slides must be preceded by the time-consuming effort of setting up the equipment. Design a cabinet that could serve as an end table or some other function while also housing a slide projector ready for use at any time. The cabinet might also serve as storage for slide trays. It should have electrical power for the projector. Evaluate the market for a multipurpose cabinet of this type.

90: Heavy appliance mover. Design a device that can be used for moving large appliances, such as stoves, refrigerators, and washers, about the house. This product would not be used often—only for rearrangement, cleaning, and for servicing of the equipment.

91: Car jack. The conventional car jack is a somewhat dangerous means of changing tires on any surface other than a horizontal one. The average jack does not attach itself adequately to the automobile's frame or bumper, introducing a severe safety problem. Design a jack that would be an improvement over existing jacks and possibly employ a different method of applying a lifting force to a car. Consider the various types of terrain on which the device must serve.

92: Map holder. The driver of an automobile who is traveling alone in an unfamiliar part of the country must frequently refer to a map. Design a system that will give the driver a ready view of the map in a convenient location in the car. Provide a means of lighting the map during night driving that will not distract the driver. Consider all possibilities of making the map's information more available to the driver.

93: Bicycle-for-two adapter. Design the parts and assembly required to convert the typical bicycle into a bicycle built for two (tandem) when mated with another bicycle of the same make and size. Work from an existing bicycle, and consider, among other things, how each rider can equally share in the pedaling. Determine the cost of your assembly and its method of attachment to the average bicycle. Use existing stock parts when possible to reduce special machining.

94: Automobile unsticker. All drivers have had the experience of getting their car stuck in soft sand, mud, or snow and are familiar with the lack of traction and the sound of spinning wheels. Design a kit to be carried in the car trunk in a minimum of space that will contain the items required by the driver who must get his or her car "unstuck" when no other help is available. This kit can be composed of one or several items. Investigate the need for such a kit and the main factors that lead to the loss of traction.

95: Stump remover. Assume that a number of tree stumps must be removed from the ground to clear land for construction. The stumps are dead with partially deteriorated root systems and require a force of approximately 2000 pounds to remove them.Design an apparatus that could be attached to the bumper of a car that could be used to remove the stumps by either pushing or pulling. The device should be easy to attach to the stump and to the car.

96: Gate opener. An aggravation to farmers and ranchers is the necessity of opening and closing gates when driving from one fenced area to the next. It would be desirable to design a gate that could be opened and closed without getting out of the vehicle. Design a manually operated gate that would appeal to this market.

Study all aspects of the problem to determine the features that would make your design as marketable as possible.

97: Paint mixer. Paint purchased from the shelf of a paint store must be mixed by stirring or some other form of agitation that will bring it to a consistent mixture. Design a product that could be used by the paint store or the paint contractor to quickly mix paint in the store and on the job. Determine the standard size paint cans for which your mixer will be designed. Consider all possibilities and methods available for this operation.

98: Mounting for an outboard motor on a canoe. Unlike a square-end boat, the pointed-end canoe does not provide a suitable surface for attaching an outboard motor. Design an attachment that will adapt an outboard motor to a canoe. Indicate how the motor will be controlled by the operator in the canoe.

99: Automobile coffee maker. More and more accessories and conveniences are being incorporated into the design of automobiles. Adequate heat is available in the automobile's power system to prepare coffee in minimum time. Design an attachment as an integral system of an automobile that will serve coffee from the dashboard area. Consider the type of coffee to be used, instant or regular, method of changing or adding water, the spigot system, and similar details.

100: Baby seat (cantilever). Design a child's chair that can be attached to a standard table top and will support the child at the required height. The chair should be designed to ensure that the child cannot crawl out of or detach it from the table top. A possible solution could be a design that would cantilever from the table top, using the child's body as a means of applying the force necessary to grip the table top. The design would be further improved if the chair were collapsible or suitable for other purposes. Determine the age group that would be most in need of the chair and base your design on the dimensions of a child of this age.

101: Miniature-TV support. Miniature television sets for close viewing are available with a screen size of 5″ × 5″. An attachment is needed that would support sets ranging in size from 6″ × 6″ to 7″ × 7″ for viewing from a bed. Determine the placement of the set with respect to a viewer for best results. Provide adjustments on the support that will be used to position the set properly. Analyze the method of concealing electrical wires within the apparatus.

102: Panel applicator. A worker who applies 4′ × 8′ gypsum board or paneling must be assisted by a helper who holds the panel in position while it is being nailed to the ceiling. This helper is only partially efficient and adds to the cost of labor.

Design a device that could be used in this capacity; it should be collapsible for easy transportation, economical, and versatile. Consider the most efficient way in which the operator could control the mechanism and list the features that would be advantageous to the marketing of the device. The average ceiling height is 8′, but provide adjustments that would adapt the device to lower or higher ceilings.

103: Backpack. Design a backpack that can be used by the outdoorsperson who must carry supplies while hiking. Your design should be based on the analysis of the supplies that would be carried by the average outdoorsperson. A major portion of your design effort should be devoted to adapting the backpack to the human body for maximum comfort and the best leverage for carrying a load over an extended period of time. Can other uses be made of your design?

104: Automobile controls. Design driving controls that can be attached to the standard automobile that will permit an injured person to drive a car without the use of his or her legs. This device should be easy to attach and to operate with the maximum of safety.

105: Bathing apparatus. Design an apparatus that would assist a wheelchair-bound person, who does not have use of his or her legs, to get in and out of a bathtub without assistance from others.

106: Adjustable TV base. Design a TV base to support full-size TV sets that would allow the maximum of adjustment: up and down, rotation about a vertical axis and about a horizontal axis. Design the base to be as versatile as possible.

107: Trailer jack. Design a trailer jack that can be used to repair flat tires that may occur on a boat trailer. It may be possible to build in jack devices as permanent features of the trailer.

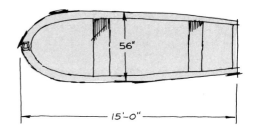

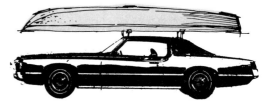

Fig. 9.22 Problem 109. Boat specifications.

108: Projector cabinet. Design a portable cabinet that could be left permanently in a classroom that would house a slide projector and a movie projector. The cabinet should provide both convenience and security from vandalism and theft.

109: Boat loader. Design a rack and a system whereby a single person could load a boat on top of a car for transporting from site to site. Use the boat specifications in Fig. 9.22; the boat weighs 110 lbs.

110: Door opener. Design a method whereby a trucker at a loading dock could open the warehouse door without having to get out of the truck. The doors are dimensioned in Fig. 9.23, and the dock extends 8' from the doors.

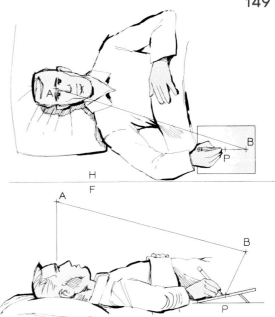

Fig. 9.25 Problem 116. A writing tablet for the handicapped.

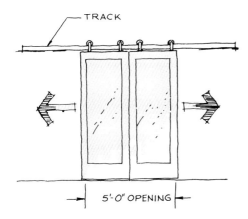

Fig. 9.23 Problem 110. Door dimensions.

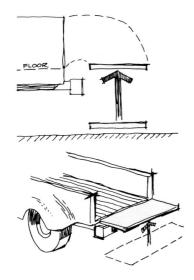

Fig. 9.24 Problem 115. Pickup truck hoist.

111: Projector eraser. Design a device that would erase grease pencil markings from the acetate roll of an overhead projector as the acetate is cranked past the stage of the projector.

112: Cement mixer. Design a portable and simple cement mixer that can be operated manually by a home owner. The mixer should be sufficiently simple and economical so it could be justified as a seldom-used product.

113: Boat trailer. Design a trailer that supports the boat under the trailer rather than on top of it. Develop this design so that a boat can be launched in more shallow water than is required now.

114: Washing machine. Design a manually operated washing machine that can be used by underdeveloped countries without power. This could be considered as an "undesign" of a powered washing machine.

115: Pickup truck hoist. Design a tail gate that could be attached to the tail gate of a pickup truck that can be used for raising and lowering loads from the ground to the floor of the truck (Fig. 9.24). Design it to be operated without a motor.

116: Writing tablet for the handicapped. Design a tablet that will permit a person to write from a prone position as shown in Fig. 9.25 using a series of mirrors. This problem involves application of human engineering considerations.

10

DRAWING INSTRUMENTS

10.1 INTRODUCTION

This chapter is devoted to the discussion of drafting instruments and their uses. With drafting instruments, a person with limited artistic ability can produce drawings of professional quality.

Instruments are designed to help the drafter produce drawings in a minimum of time with the least amount of effort. Although anyone can draw a straight line with a straightedge, skill is required to produce a straight line that is uniform in darkness and thickness. This skill must be developed through practice.

10.2 THE PENCIL

Since any drawing begins with the pencil, the proper pencil must be selected and sharpened correctly to yield the desired results. Pencils may be the conventional wood pencil or the lead holder, which is a mechanical pencil (Fig. 10.1). Both types

Fig. 10.1 The mechanical pencil (lead holder) or the wood pencil can be used for mechanical drawing. The ends of the lead and the wood pencil are labeled to indicate the grade of the pencil lead.

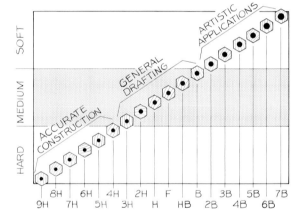

Fig. 10.2 The hardest pencil lead is 9H and the softest is 7B. Note the diameter of the hard leads is smaller than the soft leads.

are identified by a number and/or letter at the end. Sharpen the end opposite these markings so you will not sharpen away the identity of the grade of lead.

Pencil grades are shown graphically in Fig. 10.2, ranging from the hardest, 9H, to the softest, 7B. The pencils in the medium grade range of 3H–B are the pencils most often used for drafting work.

After you have selected your pencil, it is important that it be properly sharpened. This can be done with a small knife or a drafter's pencil sharpener, which removes the wood and leaves ap-

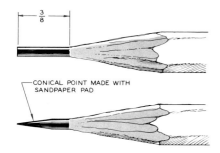

Fig. 10.3 The drafting pencil should be sharpened to a tapered conical point (not a needle point) with a sandpaper pad or other type of sharpener.

Fig. 10.5 The professional drafter will often use a pencil pointer of this type to sharpen pencils.

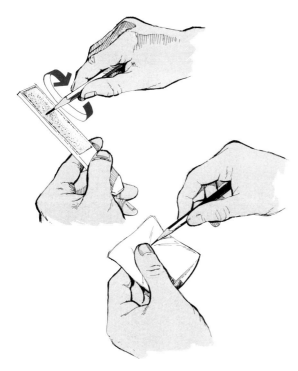

Fig. 10.4 The drafting pencil is revolved about its axis as you stroke the sandpaper pad to form a conical point. The graphite is wiped from the sharpened point with a tissue or a cloth.

proximately ⅜ inch of lead exposed (Fig. 10.3). The point can then be sharpened with a sandpaper pad to a conical point, by stroking the sandpaper with the pencil point while the pencil is being revolved between the fingers (Fig. 10.4). The excess graphite is wiped from the point with a cloth or tissue.

A pencil pointer that is used by professional drafters (Fig. 10.5) can be used to sharpen either

wood or mechanical pencils. Insert the pencil in the hole and revolve it to sharpen the lead to a conical point. Other types of small hand-held point sharpeners that work on the same principle are available.

10.3 PAPERS AND DRAFTING MEDIA

Sizes   The surface on which a drawing is made must be carefully selected to yield the best results for a given application. This usually begins by selecting the sheet size for making a set of drawings. Sheet sizes are specified by letters such as Size A, Size B, and so forth. These sizes are multiples of either the standard 8½″ × 11″ sheet or the 9″ × 12 ″ sheet. These sizes are listed below:

Size A	8½″ × 11″	9″ × 12″
Size B	11″ × 17″	12″ × 18″
Size C	17″ × 22″	18″ × 24″
Size D	22″ × 34	24″ × 36″
Size E	34″ × 44	36 × 48

Detail Paper   When drawings are not to be reproduced by the diazo process, (which is a blue-line print) an opaque paper, called *detail paper*, can be used as the drawing surface.

The higher the rag content (cotton additive) of the paper, the better will be its quality and durability because it contains cotton rather than just wood pulp.

Preliminary layouts can be drawn on detail paper and then traced onto the final tracing surface.

Tracing Paper A thin translucent paper that is used for making detail drawings is *tracing paper* or *tracing vellum*. These papers are translucent to permit the passage of light through them so drawings can be reproduced by the diazo process (blue-line process). The highest quality tracing papers are very translucent and yield the best reproductions.

Vellum is tracing paper that has been treated chemically to improve its translucency. Vellum does not retain its original quality as long as does high-quality, untreated tracing paper.

It is advantageous to lay out a drawing on detail paper, overlay it with tracing paper, and trace the final reproducible drawing.

Tracing Cloth *Tracing cloth* is permanent drafting medium that is available for both ink and pencil drawings. It is made of cotton fabric that has been covered with a compound of starch to provide a tough, erasable drafting surface that yields excellent blue-line reproductions. This material is more stable than paper, which means that it does not change its shape with variations in temperature and humidity as much as does tracing paper.

Erasures can be made on tracing cloth repeatedly without damaging the surface. This is especially important when drawing with ink.

Polyester Film An excellent drafting surface is polyester film. It is available under several trade names such as *Mylar film*. This material is highly transparent (much more so than paper and cloth), very stable, and is the toughest medium available. It is waterproof and is very difficult to tear.

Mylar film is used for both pencil and ink drawings. The drawing is made on the matte side of the film; the other side is glossy and will not take pencil or ink lines. Some films specify that a plastic-lead pencil be used and others adapt well to standard lead pencils. India ink and inks especially made for Mylar film can be used for ink drawings.

Ink lines will not wash off with water and will not erase with a dry eraser; but erasures can be made with a dampened hand-held eraser. An electric eraser is not recommended for use with this medium unless it is equipped with an eraser of the type recommended by the manufacturer of the film.

10.4 THE T-SQUARE AND BOARD

The T-square and drafting board are the basic pieces of equipment used by the beginning drafter (Fig. 10.6). The T-square can be moved with its head in contact with the edge of the board for drawing parallel horizontal lines. The drawing paper should be attached with drafting tape to the drawing board parallel to the blade of the T-square (Fig. 10.7).

The drafting board is made of basswood, which is lightweight but strong. Standard sizes of boards are $12'' \times 14''$, $15'' \times 20''$, and $21'' \times 26''$. The working edge of a drawing board is the edge where the T-square head is held firmly in position when each horizontal line is drawn. Some drawing boards have built-in steel working edges for a higher degree of accuracy.

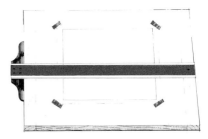

Fig. 10.6 The T-square and drafting board are the basic tools used by the student drafter. The drawing is taped to the board with drafting tape.

Fig. 10.7 The drafting tape placed at each corner of a drawing should be square and be cut prior to taping the drawing.

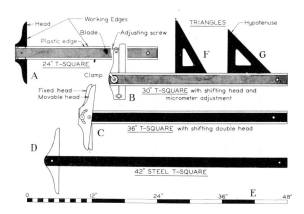

Fig. 10.8 A variety of types of T-squares and the two most commonly used triangles.

Types of T-squares are illustrated in Fig. 10.8. They are made of many materials including plastic, wood, and metal. Care should be taken not to nick the upper working edge of a T-square because this ruins the instrument.

10.5 DRAFTING MACHINES

Although the T-square is used in industry and in the classroom, most professional drafters prefer the mechanical *drafting machine*, which is attached to the drawing board or table top (Fig. 10.9). These machines have fingertip controls for drawing lines

Fig. 10.9 The drafting machine is often used instead of the T-square and drafting board in both industry and school laboratories. (Courtesy of Keuffel & Esser Co., Morristown, N.J.)

at any angle, and can be easily returned to their original position. Drafting machines provide a considerable degree of convenience to the drafter, but the same drawing can be made with the T-square and triangle in the hands of a skilled drafter.

For large drawings such as those made by architects, a parallel blade is available (Fig. 10.10). The blade is attached to a cable at each end that keeps it parallel in any position. The angle of the blade can be changed by adjusting the cables.

Fig. 10.10 The parallel blade is advantageous to the drafter who makes very large drawings. It is guided by the cables attached to the board. (Courtesy of Keuffel & Esser Co., Morristown, N.J.)

10.6 ALPHABET OF LINES

The type of line produced by a pencil depends upon the hardness of the lead, the drawing surface, and the technique of the drafter. Examples of the standard lines, or the *alphabet of lines*, is shown in Fig. 10.11, along with the recommended pencils for drawing the lines. These pencil grades may vary greatly with the drawing surface being used.

Guidelines that are used to aid in lettering and in laying out a drawing are very light lines, just dark enough to be seen. A 4H pencil is recommended for drawing most guidelines.

Except for guidelines, all other lines should be drawn dark and black so they will reproduce well. The important characteristic of pencil lines is their relative widths, as shown in Fig. 10.11. When drawing these lines, assume that you are drawing them in ink where their blackness is uniform and the only variable is their widths.

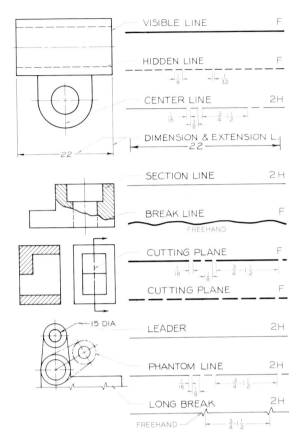

VISIBLE LINE	F
HIDDEN LINE	F
CENTER LINE	2H
DIMENSION & EXTENSION L.	
SECTION LINE	2 H
BREAK LINE	F
CUTTING PLANE	F
CUTTING PLANE	F
LEADER	2H
PHANTOM LINE	2H
LONG BREAK	2H

Fig. 10.11 The alphabet of lines varies in width to produce a finished mechanical drawing of an object. The full-size lines are shown in the right column along with the recommended pencil grades for drawing the lines.

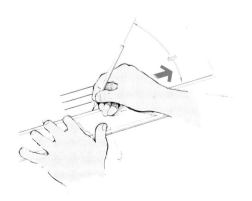

Fig. 10.12 Horizontal lines are drawn with a pencil held in a plane perpendicular to the paper and at 60° to the surface. These lines are drawn left to right along the upper edge of the T-square.

10.7 HORIZONTAL LINES

A horizontal line is drawn using the upper edge of your horizontal straightedge, and drawing the line from left to right, for the right-handed person (Fig. 10.12). Your pencil should be held in a vertical plane to make a 60° angle with the drawing surface.

As horizontal lines are drawn, the pencil should be rotated about its axis to allow its point to wear evenly (Fig. 10.13). If necessary, lines can be darkened by drawing over them one or more times. A small space should be left between the straightedge and the pencil point for the best line (Fig. 10.14).

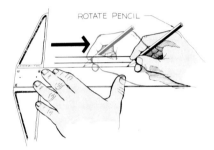

Fig. 10.13 As the horizontal lines are drawn, the pencil should be rotated about its axis so that the point will wear down evenly.

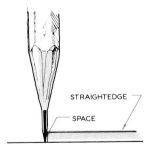

Fig. 10.14 The pencil point should be held in a vertical plane and inclined 60° to leave a space between the point and the straightedge being used.

10.8 VERTICAL LINES

A triangle is used in conjunction with the blade of a T-square for drawing vertical lines since all drawing triangles have one 90° angle. While the T-square is held firmly with one hand, the triangle can be positioned where needed and the vertical lines drawn (Fig. 10.15).

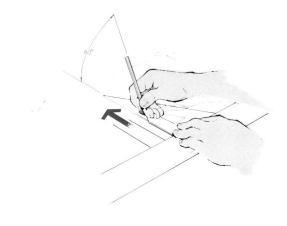

Fig. 10.15 Vertical lines are drawn along the left side of a triangle in an upward direction with the pencil held in a vertical plane at 60° to the surface.

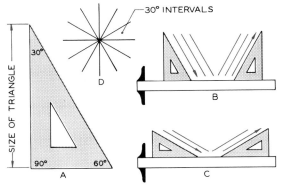

Fig. 10.16 The 30°–60° triangle can be used to construct lines at 60° and 30° to the horizontal. Lines can be spaced at 30° intervals throughout 360°.

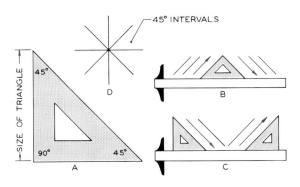

Fig. 10.17 The 45° triangle can be used to draw lines at 45° intervals throughout 360°.

Vertical lines are drawn upward along the left side of the triangle while holding the pencil in a vertical plane at 60° to the drawing surface.

10.9 DRAFTING TRIANGLES

The two most often used triangles are the 45° triangle and the 30°–60° triangle. The 30°–60° triangle is specified by the longer of the two sides adjacent to the 90° angle (Fig. 10.16). Standard sizes of 30°–60° triangles range in 2-inch intervals from 4″ to 24″. The variety of lines that can be drawn with this triangle and T-square are shown in Fig. 10.16D.

The 45° triangle is specified by the length of the sides adjacent to the 90° angle. These range in size from 4″ to 24″ at 2-inch intervals, but the 6″ and 10″ sizes are adequate for most classroom applications. The various angles that can be drawn with this triangle are shown in Fig. 10.17.

By using the 45° and 30°–60° triangles in combination, angles can be drawn at 15° intervals throughout 360° (Fig. 10.18).

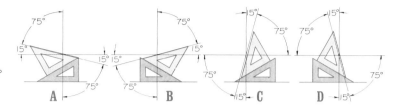

Fig. 10.18 By using the 30°–60° triangle in combination with the 45° triangle, angles can be drawn at 15° intervals without the use of a protractor.

10.10   THE PROTRACTOR

When lines must be drawn or measured at angles other than at multiples of 15°, a protractor is used (Fig. 10.19). Protractors are available as semicircles (180°) or as circles (360°).

An adjustable triangle also serves as a protractor and a drawing edge at the same time (Fig. 10.19B). Most drafting machines have built-in protractors that are easier to use and more convenient than traditional protractors.

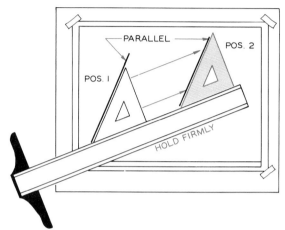

Fig. 10.20 A T-square and a 45° triangle can be used for drawing a series of parallel lines. The T-square is held firmly in position and the triangle is moved from position 1 to position 2.

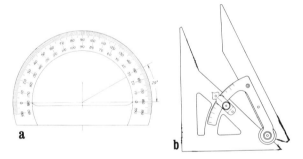

Fig. 10.19 The semicircular protractor can be used to measure angles. The adjustable triangle can be used as a drawing edge in addition to being used to measure angles.

10.11   PARALLEL LINES

A series of lines can be drawn parallel to a given line by using a triangle and a straightedge (Fig. 10.20).

The 45° triangle is placed parallel to a given line and is held in contact with the straightedge (which may be another triangle). By holding the straightedge in one position, the triangle can be moved to various positions for drawing series of parallel lines.

10.12   PERPENDICULAR LINES

Perpendicular lines can be constructed by using either of the standard triangles. A 30°–60° triangle is used with a T-square or another triangle to draw line 3–4 perpendicular to line 1–2 (Fig. 10.21). One edge of the triangle is placed parallel to line 1–2 in position 1 with the T-square in contact with the triangle. By holding the T-square in place, the triangle is rotated and moved to position 2 to draw the perpendicular line.

10.13   IRREGULAR CURVES

Curves that are not arcs must be drawn with an instrument called an *irregular curve*. These plastic curves come in a variety of sizes and shapes, but

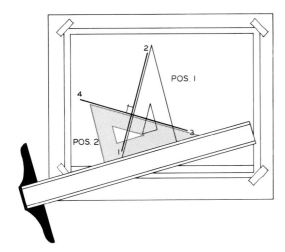

Fig. 10.21 A 30°–60° triangle and a T-square can be used to construct a line perpendicular to line 1–2. The triangle is aligned with 1–2 in position 1 and is then rotated to position 2 to construct line 3–4.

the one shown in Fig. 10.22 is typical of those that are used.

The use of the irregular curve is shown in this figure where a series of points is connected to form a smooth curve. Note that the plastic curve must be repositioned several times to draw the complete curve.

The *flexible spline* is an instrument used for drawing long irregular curves (Fig. 10.23). The

spline is held in position by weights while the curve is drawn.

10.14 ERASING

Erasers are manufactured to match the lines and papers that are used for drafting. Surfaces may vary from soft paper to polyester film, and the lines may be drawn in ink or pencil.

Erasing should be done with the softest eraser that will serve the purpose. For example, ink erasers should not be used to erase pencil lines, because ink erasers are coarse and may damage the surface of the paper.

When it is necessary to erase in small areas, an erasing shield can be used to prevent accidental erasing of adjacent lines (Fig. 10.24). Place the shield over the line, hold it firmly, and erase the desired line. All erasing should be followed by brushing away the "crumbs" from the drawing with a brush. Do not use your hands for this purpose; this may smudge your drawing.

Electric erasers are available. A cordless model is shown in Fig. 10.25. This eraser is recharged by its desk stand that is left plugged into a wall outlet. Other models are electric and must remain connected to an electrical outlet. The erasers that are used in these rotating machines are available in several grades for erasing ink and pencil lines.

Fig. 10.22 Use of the irregular curve

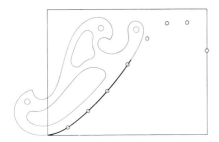
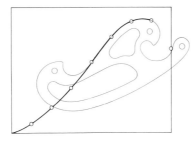
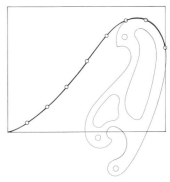

Step 1 To connect points with a smooth curve, the irregular curve is positioned to pass through as many points as possible and a portion of the curve is drawn.

Step 2 The irregular curve is positioned for drawing another portion of the connecting curve.

Step 3 The last portion is drawn to complete the smooth curve. Most irregular curves must be drawn in separate steps, in this manner.

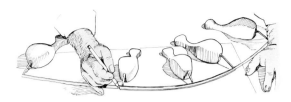

Fig. 10.23 A flexible spline can be used for drawing large irregular curves. The spline is held in position by weights.

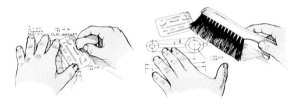

Fig. 10.24 The erasing shield is used for erasing in tight spots without removing the wrong lines. The dusting brush is used to remove the erased material when finished. Do not brush with the palm of your hand; this will smear the drawing.

Fig. 10.25 This cordless electric eraser is typical of those used by professional drafters.

10.15 SCALES

All engineering drawings require the use of scales to measure lengths, sizes, and other measurements. The types of scales may be flat or triangular as shown in Fig. 10.26, and they are made of wood, plastic, and metal. Triangular engineers' and architects' scales are shown in Fig. 10.27.

Most scales are either 6″ or 12″ long. The scales covered in this section are the architects', engineers', mechanical engineers', and metric scales.

The Architects' Scale

The architects' scale is used to dimension and scale features encountered by the architect such as cabinets, plumbing, and electrical layouts. Most indoor measurements are made in feet and inches with an architects' scale.

The basic form of indicating on a drawing the scale that is being used is shown in Fig. 10.28. This form should be used on the drawing in the title block or in some prominent location so that the scale of the drawing can be known.

Since the dimensions made with the architects' scale are in feet and inches, it is very difficult to handle the arithmetic associated with these dimensions. It is necessary to convert all dimensions to decimal equivalents (all feet or all inches) before the simplest arithmetic can be performed.

Scale: Full Size The 16 scale is used for measuring full-size lines (Fig. 10.29a). An inch on the 16 scale is divided into sixteenths to match the ruler used by the carpenter. This example is measured to be $3\frac{1}{8}″$. Note that when the measurement is less than one foot, a zero may be used to precede the inch measurements, and the inch marks are omitted in all cases.

Scale: 1 = 1′ − 0 In Fig. 10.29b, a line is measured to its nearest whole feet (2 ft in this case) and the remainder is measured in inches at the end of the scale ($3\frac{1}{2}″$) for a total of $2′−3\frac{1}{2}$. At the end of each architects' scale, a foot has been divided into inches for measuring dimensions that are less than a foot.

The scale 1″ = 1′−0 is the same as saying 1″ is equal to 12″, or a $\frac{1}{12}$th size.

Scale: $\frac{3}{8}$ = 1′−0 When this scale is used $\frac{3}{8}″$ is used to represent 12″ on a drawing. Figure 10.29c is measured to be 7′−5.

Scale: $\frac{1}{2}$ = 1′−0 A line is measured to be $5′−8\frac{1}{2}$ in Fig. 10.29d.

Scale: Half Size The 16 scale is used to measure or draw a line that is half size. This is sometimes specified as Scale: 6 = 12 (inch marks omitted). The line in Fig. 10.29e is measured to be $0′−6\frac{3}{8}$.

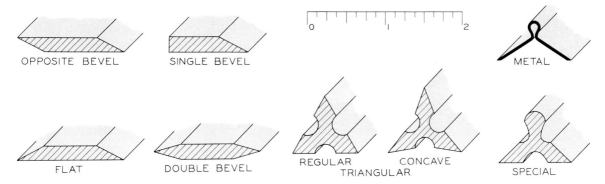

OPPOSITE BEVEL SINGLE BEVEL METAL

FLAT DOUBLE BEVEL REGULAR TRIANGULAR CONCAVE SPECIAL

Fig. 10.26 Scales are made of wood, plastic, bamboo, and metal, and are available in any of the shapes shown here.

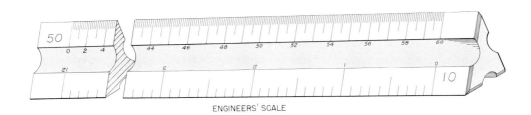

ENGINEERS' SCALE

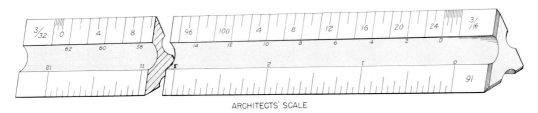

ARCHITECTS' SCALE

Fig. 10.27 The architects' scale is used to measure in feet and inches, whereas the engineers' scale measures in decimal units.

ARCHITECTS' SCALE

BASIC FORM $SCALE: \frac{X}{X} = 1'-0$

┌─FROM END OF SCALE

Fig. 10.28 The basic form of indicating the scale on the architects' scale, and the variety of scales available.

TYPICAL SCALES

SCALE: FULL SIZE (USE 16-SCALE)

SCALE: HALF SIZE (USE 16-SCALE)

SCALE: $3 = 1'-0$ SCALE: $1\frac{1}{2} = 1'-0$

SCALE: $1\frac{1}{2} = 1'-0$ SCALE: $\frac{3}{4} = 1'-0$

SCALE: $\frac{1}{2} = 1'-0$ SCALE: $\frac{3}{8} = 1'-0$

SCALE: $\frac{3}{16} = 1'-0$ SCALE: $\frac{1}{8} = 1'-0$

SCALE: $\frac{3}{32} = 1'-0$

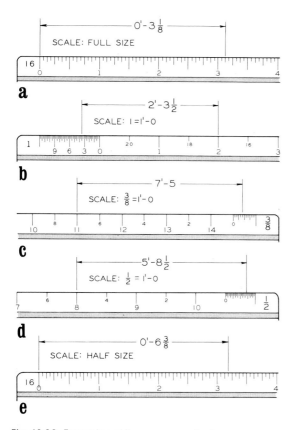

a SCALE: FULL SIZE — 0'-3 1/8

b SCALE: 1=1'-0 — 2'-3 1/2

c SCALE: 3/8 =1'-0 — 7'-5

d SCALE: 1/2 = 1'-0 — 5'-8 1/2

e SCALE: HALF SIZE — 0'-6 3/8

Fig. 10.29 Examples of lines measured using an architects' scale.

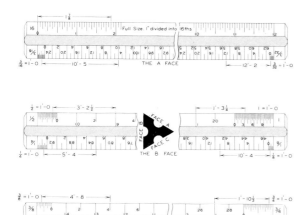

Fig. 10.30 Examples of lines measured on a three-sided architects' scale.

Other lines measured on the architects' scale are shown in Fig. 10.30. When locating dimensions using any scale, hold your pencil in a vertical position for the greatest accuracy when marking measurements (Fig. 10.31).

When indicating dimensions in feet and inches, they should be in the form shown in Fig. 10.32. Notice that the fractions are twice as tall as the whole numerals.

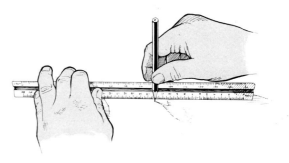

Fig. 10.31 When marking off measurements along a scale, be sure to hold your pencil vertically for the most accurate measurement.

2'-7 1/2 ; 4'-0 1/4 ; 0'-1 1/2
OMIT INCH MARKS — ZERO HERE — ZERO OPTIONAL

Fig. 10.32 Inch marks are omitted according to current standards, but foot marks are shown. A leading zero is used when the inch measurements are less than a whole inch. When representing feet, a zero is optional if the measurement is less than a foot.

ENGINEERS' SCALES

BASIC FORM SCALE: 1= XX
FROM END OF ENGR. SCALE

EXAMPLE SCALES

10 SCALE: 1=10'; SCALE: 1 = 1,000'
20 SCALE: 1=200'; SCALE: 1=20LB
30 SCALE: 1=0.3; SCALE: 1=3,000'
40 SCALE: 1=4'; SCALE: 1 = 40'
50 SCALE: 1=50'; SCALE: 1 = 500'
60 SCALE: 1=6; SCALE: 1 = 0.6'

Fig. 10.33 The basic form for indicating the scale when the engineers' scale is used, and the variety of scales that are available on this scale.

The Engineers' Scale

The engineers' scale is a decimal scale on which each division is a multiple of 10 units. It is used for making drawings of engineering projects that are located outdoors, such as streets, structures, land measurements, and other large dimensions associated with topography. For this reason, it is sometimes called the civil engineers' scale.

Since the measurements are in decimal form, it is easy to perform arithmetic operations without the need of converting feet and inches as when the architects' scale is used. Areas and volumes can be found easily.

The form of specifying scales on the engineers' scale is shown in Fig. 10.33, such as Scale: 1 = 10′. Each end of the scale is labeled 10, 20, 30, etc. This indicates the number of units per inch on the scale. Many combinations may be obtained by moving the decimal places of a given scale, as indicated in Fig. 10.33.

10 Scale   In Fig. 10.34a, the 10 scale is used to measure a line at the scale of 1 = 10′. The line is 32.0 feet long.

20 Scale   In Fig. 10.34b, the 20 scale is used to measure a line drawn at a scale of 1 = 200.0′. The line is 540.0 feet long.

30 Scale   A line of 10.6 (inch marks omitted) is measured using the scale of 1 = 3.0 in Fig. 10.34c.

The format for indicating measurements in feet and inches is shown in Fig. 10.35. It is customary to omit zeros in front of decimal points when dimensioning an object using English (Imperial) units and inch marks are always omitted if the dimensions are given in inches.

Mechanical Engineers' Scale

The mechanical engineers' scale is used to draw small parts (Fig. 10.36). This scale is used to represent drawings in inches using common fractions. These scales are available in ratios of half size, one-quarter size, and one-eighth size. For example, on the half-size scale, 1 inch is used to represent 2 inches. On a quarter-size scale, 1 inch would represent 4 inches.

The Metric System—SI Units

The English system (Imperial system) of measurements has been used in the United States, Britain,

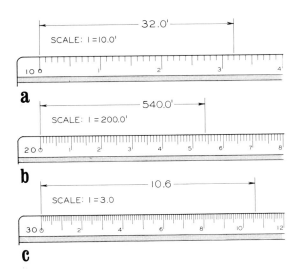

Fig. 10.34 Examples of lines measured with the engineers' scale.

Fig. 10.35 When using English units (inches), decimal fractions do not have leading zeros and inch marks are omitted. Be sure to provide adequate space for decimal points between the numbers. Foot marks are shown.

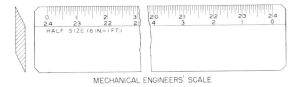

Fig. 10.36 The mechanical engineers' scales are used for measuring small parts at scales of half size, quarter size, and one-eighth size. These units are in inches with common fractions.

and Canada since these countries were established. Presently a movement is underway to convert to the more universal metric system.

The English system was based on arbitrary units of the inch, foot, cubit, yard, and mile (Fig. 10.37). There is no common relationship between these units of measurement; consequently the system is cumbersome to use when simple arithmetic is performed. For example, finding the area of a

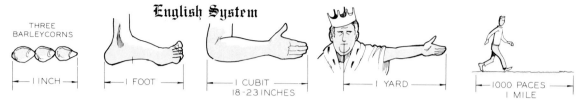

Fig. 10.37 The units of the English system were based on arbitrary dimensions.

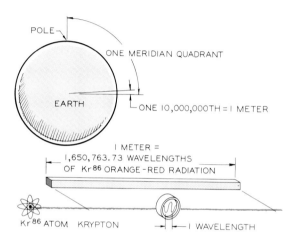

Fig. 10.38 The origin of the meter was based on the dimensions of the earth, but it has since been based on the wavelength of krypton-86. A meter is 39.37 inches.

SI UNITS			DERIVED UNITS		
LENGTH	METER	m	AREA	SQ METER	m²
MASS	KILOGRAM	kg	VOLUME	CU METER	m³
TIME	SECOND	s	DENSITY	KILOGRAM/CU MET	kg/m³
ELECTRICAL			PRESSURE	NEWTON/SQ MET	N/m²
CURRENT	AMPERE	A			
TEMPERATURE	KELVIN	K			
LUMINOUS					
INTENSITY	CANDELA	cd			

Fig. 10.39 The basic SI units and their abbreviations. The derived units are units that have come into common usage.

basic SI units are shown in Fig. 10.39 with their abbreviations. It is important that lowercase and uppercase abbreviations be used properly as shown.

Several practical units of measurement have been derived (Fig. 10.40) to make them easier to use in many applications. These unofficial SI units are widely used. Note that degrees Celsius (centigrade) is recommended over the official temperature measurement, Kelvin. When using Kelvin, the freezing and boiling temperatures are 273.15°K and 373.15°K respectively. Pressure is measured in bars, where one bar is equal to 0.1 megapascal or 100,000 pascals.

Many SI units have prefixes to indicate placement of the decimal. The more common prefixes and their abbreviations are shown in Fig. 10.41.

rectangle that is 25 inches by 6¾ yards is a complex problem.

The metric system was proposed by France in the fifteenth century. In 1793, the French National Assembly agreed that the meter (m) would be one ten-millionth of the meridian quadrant of the earth (Fig. 10.38). Fractions of the meter were expressed as decimal fractions. Debate continued until an international commission officially adopted the metric system in 1875. Since a slight error in the first measurement of the meter was found, the meter was later established as equal to 1,650,763.73 wavelengths of the orange-red light given off by krypton-86 (Fig. 10.38).

The international organization charged with the establishment and the promotion of the metric system is called the *International Standards Organization (ISO)*. The system they have endorsed is called *Système International d'Unités* (International System of Units) and is abbreviated SI. The

PARAMETER	PRACTICAL UNITS		SI EQUIVALENT
TEMPERATURE	DEGREES CELSIUS	°C	0°C = 273.15 K
LIQUID VOLUME	LITER	l	l = dm³
PRESSURE	BAR	BAR	BAR = 0.1 MPa
MASS WEIGHT	METRIC TON	t	t = 10³ kg
LAND MEASURE	HECTARE	ha	ha = 10⁴ m²
PLANE ANGLE	DEGREE	°	1° = π/180 RAD

Fig. 10.40 These practical metric units are a few of those that are widely used because they are easier to deal with than the official SI units.

VALUE		PREFIX	SYMBOL
1,000,000	$= 10^6 =$	MEGA	M
1,000	$= 10^3 =$	KILO	k
100	$= 10^2 =$	HECTO	h
10	$= 10^1 =$	DEKA	da
1	$= 10^0 =$		
.1	$= 10^{-1} =$	DECI	d
.01	$= 10^{-2} =$	CENTI	c
.001	$= 10^{-3} =$	MILLI	m
.000 001	$= 10^{-6} =$	MICRO	μ

Fig. 10.41 The prefixes and abbreviations used to indicate the decimal placement for SI measurements.

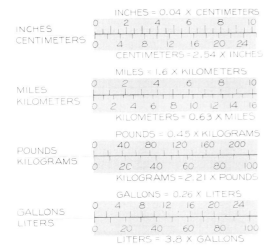

Fig. 10.42 A comparison of metric units with those used in the English system of measurement.

Several comparisons of English and SI units are given in Fig. 10.42. Other conversion factors are given in Appendix 2.

10.16 METRIC SCALES

The basic unit of measurement on an engineering drawing is the millimeter (mm), which is one-thousandth of a meter, or one-tenth of a centimeter. These units are understood unless otherwise specified on a drawing. The width of the fingernail of your index finger can serve as a convenient gage to approximate the dimension of a centimeter, or ten millimeters (Fig. 10.43).

Metric scales are indicated on a drawing in the form shown in Fig. 10.44. A colon is placed between the numeral 1 and the ratio of the drawing size. The units are not specified since the mil-

limeter is understood to be the unit of measurement.

Decimal fractions are unnecessary on most metrically dimensioned drawings; consequently the dimensions are usually rounded off to whole numbers except for these measurements that are dimensioned with specified tolerances. For metric units less than 1, a leading zero is placed in front of the decimal. In the English system, the zero is omitted from inch measurements (Fig. 10.45).

In some industries (e.g., the automotive), it is recommended that all millimeter dimensions be given with one number after the decimal point, or

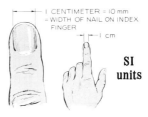

Fig. 10.43 The width of the nail on your index finger is approximately equal to a centimeter or ten millimeters.

Fig. 10.44 The basic form for indicating scales when the metric scale is used, and the variety of scales that are available.

Fig. 10.45 When decimal fractions are shown in metric units, a zero is used to precede the decimal. Be sure to allow adequate space for the decimal point when numbers with decimals are lettered.

$$1\,dm = \frac{m}{10}\,; \quad 1\,cm = \frac{m}{100}\,; \quad 1\,mm = \frac{m}{1\,000}\,; \quad 1\,\mu m = \frac{m}{1\,000\,000}$$

Fig. 10.46 The dekameter is one-tenth of a meter; the centimeter is one-hundredth of a meter; a millimeter is one-thousandth of a meter; and a micrometer is one-millionth of a meter.

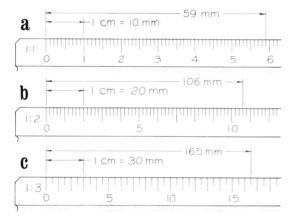

Fig. 10.47 Examples of lines measured with metric scales.

Other Scales

Many other metric (SI) scales are used: 1:250, 1:400, 1:500, and so on. The scale ratios mean that one unit represents the number of units on the right of the colon. For example, 1:20 means that one millimeter equals 20 mm, or one centimeter equals 20 cm, or one meter represents 20 m.

Metric Symbols

When drawings are made in metric units, this can be noted in the titleblock or elsewhere using the SI symbol (Fig. 10.48). The large SI indicates Système International. The two views of the partial cone are used to denote whether the orthographic views were drawn in the U.S. system (third-angle projection) or the European system (first-angle projection).

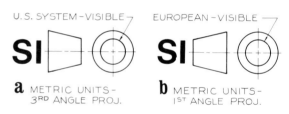

Fig. 10.48 The large letters SI indicate that the units of measurement are in metric units. The partial cones indicate that the views are arranged using the third-angle projection (the U.S. system) or the first-angle projection (the European system).

Scale Conversion

Tables for converting inches to millimeters are given in Appendix 2; however, this conversion can be performed by multiplying decimal inches by 25.4 to obtain millimeters. For example, 1.5 inches would be $1.5 \times 25.4 = 38.1$ mm.

To convert an architect's scale to an approximate metric scale, the scale must be multipled by 12. For example, Scale: $\frac{1}{8}$ = 1′–0 is the same as $\frac{1}{8}$ inch = 12 inches, or 1 inch = 96 inches. This scale closely approximates the metric scale of 1:100. Many of the scales used in the metric system cannot be converted to exact English scales. The scale of 1 = 5′ converts exactly to the metric scale of 1:60.

a zero if there is no fraction. They recommend that toleranced dimensions be carried out to three decimal places when the unit is the millimeter.

Scale 1:1 The full-size metric scale (Fig. 10.46) shows the relationship between the metric units of the dekameter, centimeter, millimeter, and the micrometer. There are 10 dekameters in a meter; 100 centimeters in a meter; 1000 millimeters in a meter; and 1,000,000 micrometers in a meter. A line of 59 mm is measured in Fig. 10.47.

Scale 1:2 This scale is used when 1 mm is equal to 2 mm, 20 mm, 200 mm, etc. The line in Fig. 10.47b is 106 mm long.

Scale 1:3 A line of 165 mm is measured in Fig. 10.47c where 1 mm is used to represent 30 mm.

OMIT COMMAS AND GROUP INTO THREES
1 000 000 000 NOT 1,000,000,000
USE RAISED DOT FOR MULTIPLICATION
N·M NOT NM
INDICATE DIVISION BY EITHER
kg/m OR kg·m⁻¹
USE ZERO PRECEDING DECIMALS
0.72 mm NOT .72 mm
INDICATE SI SCALES AS
SCALE: 1:2 SI

Fig. 10.49 General rules to be used with the SI system.

Expression of Metric Units

The general rules for expressing SI units are given in Fig. 10.49. Commas are not used between sets of zeros, but instead a space is left between them.

10.17 THE INSTRUMENT SET

A basic set of drawing instruments is shown in Fig. 10.50 and the name of each part is given. These can be purchased separately, but they are available assembled as a set in a case similar to the one shown in Fig. 10.51. A more elaborate set of instruments is shown in Fig. 10.52.

This section will cover the care and use of these instruments.

The Compass

The *compass* is used to draw circles and arcs in ink and in pencil (Fig. 10.53). In order to obtain

RULING PEN
PEN POINT
EXTRA HANDLE
LARGE BOW COMPASS
NEEDLE
NEEDLE BOX
SMALL BOW COMPASS
PEN POINT
LEAD BOX
DIVIDERS
EXTENSION BAR COMPASS

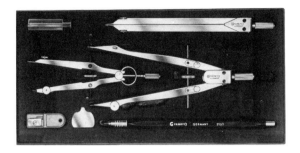

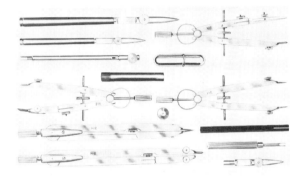

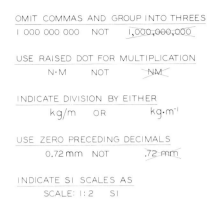

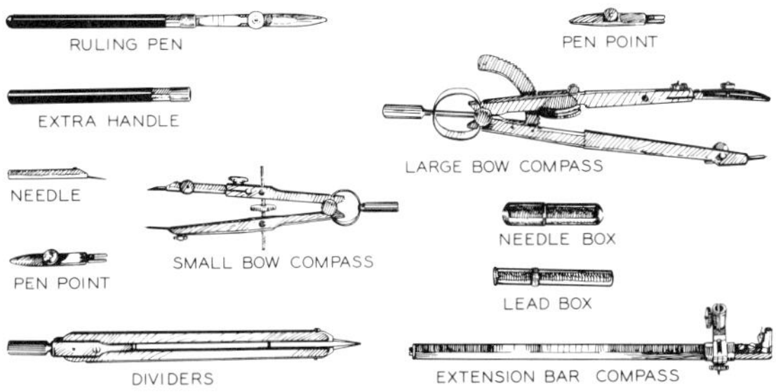

Fig. 10.50 The parts of a set of drafting instruments.

Fig. 10.51 Instruments usually come as a cased set. (Courtesy of Gramercy Guild.)

Fig. 10.52 A set of instruments with three bows for the advanced student. (Courtesy of Keuffel & Esser Co., Morristown, N.J.)

2.250
4 3/8
2.156
4.312

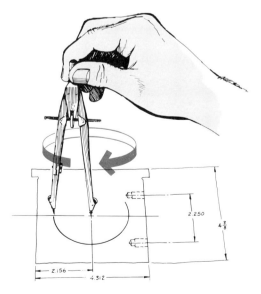

Fig. 10.53 The compass is used for drawing circles.

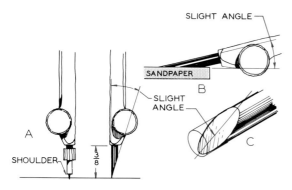

Fig. 10.54 The compass lead should be sharpened from the outside on a sandpaper pad at B, to a wedge point (C). The pencil point should be about the same length as the compass point at A.

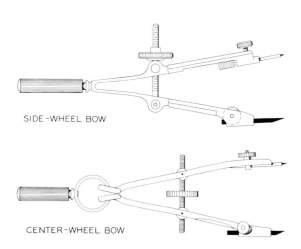

Fig. 10.55 Two types of small bow compasses for drawing circles of about one-inch radius.

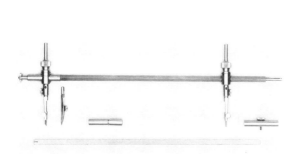

Fig. 10.56 Large circles can be drawn with the beam compass. Note that ink attachments are available also.

good results with the compass, its pencil point must be sharpened on its outside with a sandpaper board (Fig. 10.54). A bevel cut of this type gives the best all-round point for drawing a circle. Note that the needle point of the compass must be adjusted to match the length of the lead. When the compass point is set in the drawing surface, it should not be inserted to the shoulder of the point but just enough for a firm set.

When the table top has a hard covering, several sheets of paper should be placed under the drawing to provide a seat for the compass point. A center tack (the circular part shown in Fig. 10.52) can be placed over the center point and used for setting the compass point for drawing circles and for preventing the enlargement of the center hole when used repetitively.

Bow compasses are provided in some sets (Fig. 10.55) for drawing small circles with ink and pencil. For large circles, extension bars are provided to extend the range of the large-bow compass. If the circles are much larger, a beam compass can be used (Fig. 10.56).

Small circles and ellipses can be effectively drawn with a circle template that is aligned with the perpendicular center lines of the circle. The circle or ellipse is drawn with a pencil to match the other lines of the drawing (Fig. 10.57).

The Dividers

The *dividers* looks much like a compass but is used for laying off and transferring dimensions onto a drawing. For example, equal divisions can

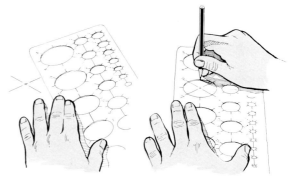

Fig. 10.57 The circle template can be used for drawing circles without the use of a compass. The circle or ellipse is aligned with the center lines.

Fig. 10.58 The dividers are used to step off measurements and to transfer dimensions on a drawing.

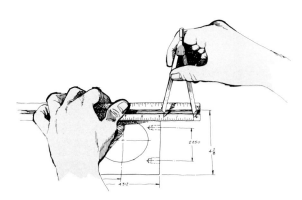

Fig. 10.59 Dividers are used to transfer dimensions from a scale to a drawing.

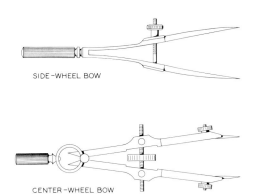

Fig. 10.60 Two types of bow dividers for transferring small dimensions, such as the spacing between guidelines for lettering.

be stepped off rapidly along a line (Fig. 10.58). A slight impression is made in the drawing surface with the points as each measurement is made.

Dividers can be used to transfer dimensions from a scale to a drawing (Fig. 10.59). Another use for the dividers is the division of a line into a number of equal parts. This is done by trial and error. You begin by estimating the spacing and stepping off the space until the correct spacing is found.

Small bow dividers can be used for transferring smaller dimensions such as the spacing between the guidelines for lettering. Two types of bow dividers are shown in Fig. 10.60.

Proportional Dividers

Dimensions can be transferred from one scale to another by using a special type of dividers, the *proportional dividers*. The central pivot point can be moved from position to position to vary the ratio of the spacing at one end of the dividers to the ratio at the other end (Fig. 10.61). This instrument is very helpful in enlarging and reducing dimensions on a drawing.

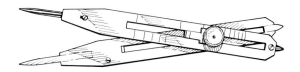

Fig. 10.61 Proportional dividers can be used for making measurements that are proportional to other dimensions. The pivot can be set to give the desired ratio.

10.18 INK DRAWING

Although the majority of drafting and design work is done in pencil, inking is required for many applications, especially for drawings that will be reproduced in publications and reports. Pencil drawings have a tendency to lose their sharpness as instruments are moved about the drawing surface during preparation. Ink drawings remain dark and distinct without danger of losing their quality.

Ink lines give the best reproductions when reproduced by the printing press, diazo process, or microfilming.

Materials for Ink Drawing

An average good grade of tracing paper can be used for ink drawings, but erasing errors may result in holes in the paper and the loss of your time. Therefore, tracing film or tracing cloth should be used, which will withstand many erasings and corrections.

When using drafting film, the drawing should be made on the matte surface in accordance with the manufacturer's directions. A cleaning solution is available that can be used to prepare the surface for ink and to remove spots that might not take the ink properly. These films can also be cleaned with a damp cloth.

Tracing cloths need to be prepared for inking by applying a coating of powder or pounce to absorb oily spots that will otherwise repel an ink line. These oily spots can be left on a drawing by fingerprints.

The drawing ink used for engineering drawings is called India ink and is available under numerous trade names. This dense black, carbon ink is much thicker and faster-drying than regular fountain pen ink. Some drafters prefer to "season" their ink by leaving the top of the bottle open for several days to thicken it.

Ink should be removed from instruments before drying to prevent clogging, which restricts an easy flow of ink from the pen to the paper. Inking instruments will not work well unless kept clean of dried ink. Ultrasonic pen cleaners are available to aid in the cleaning of pen points that have been used for inking. The points are immersed in a cup of pen cleaner solution that is vibrated at about 80,000 cycles per second to free the dried particles of ink. Several manufacturers produce these cleaners.

Ruling Pen

Two types of ruling pens are shown in Fig. 10.62. Both have set screws for varying the widths of the lines that are drawn.

The ruling pen should be inked with the spout on the cap of the ink bottle (Fig. 10.63). Do not overload the pen with ink since this will cause the lines to be wet and perhaps run on the drawing. Experiment with your particular pen to learn the proper amount of ink to apply to the nibs.

When ruling horizontal lines, the ruling pen is held in the same position as a pencil (Fig. 10.64), maintaining a space between the nibs and

Fig. 10.63 The pen is inked between the nibs with the spout on the ink bottle cap.

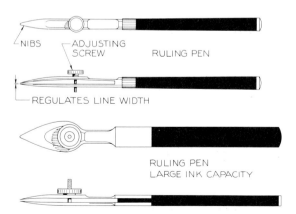

Fig. 10.62 The ruling pen has two nibs that are adjusted by a set screw to vary the width of the ink lines drawn with the pen.

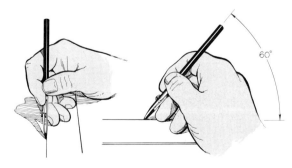

Fig. 10.64 The ruling pen is held in a vertical plane at 60° to the drawing surface for drawing ink lines.

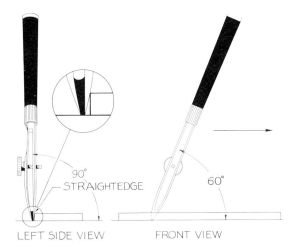

Fig. 10.65 Two views showing the position for holding the ruling pen for drawing ink lines.

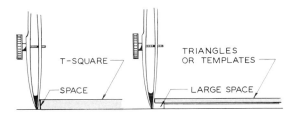

Fig. 10.66 The ruling pen should be held in a vertical plane so that there will be a space between the T-square and the nibs. A triangle or template can be placed under the straightedge for a greater margin of safety.

Fig. 10.67 An India ink technical pen that can be used for making ink drawings.

the straightedge. Two views of this position are shown in Fig. 10.65.

An extra margin of safety can be obtained by placing a triangle or template under the straightedge, as shown in Fig. 10.66. This prevents wet ink from coming in contact with the straightedge and avoids smearing.

An alternative pen that can be used is the technical ink fountain pen as illustrated in Fig. 10.67. These pens come in sets (Fig. 10.68) with

various-sized pen points that are used for the alphabet of lines (Fig. 10.69). Lines drawn with fountain pens of this type dry much faster than those drawn with ruling pens because the ink is applied in a thinner layer.

Examples of poorly drawn ink lines are shown in Fig. 10.70. These mistakes can be avoided by learning the technical faults that cause them.

Fig. 10.68 A set of inking pens available in various line weights.

Available in 18 "Kolor-Koded" line widths *

	6 x 0
	5 x 0
	4 x 0
	3 x 0
	00
	0
	1
	2
	2½
	3
	4
	6
	7
	8
	9
	10
	12
	14

*Approximate only. (Line widths will vary, depending on type of surface, type of ink, speed at which line is drawn, etc.)

Fig. 10.69 This chart shows the variety of technical pens that is available for drawing lines of graduated widths. (Courtesy of Koh-I-Noor Rapidograph, Inc.)

A. GOOD-EVEN LINE

B. POOR-INK RAN UNDER STRAIGHTEDGE

C. POOR-TOO MUCH INK; SLOW AT ENDS

D. POOR-NIBS TOUCH IMPROPERLY

Fig. 10.70 Examples of poor ink lines are shown at B, C, and D.

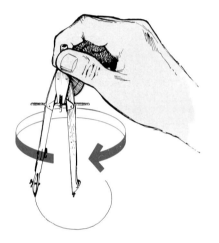

Fig. 10.71 An inking attachment can be used with a large bow compass for drawing circles and arcs in ink.

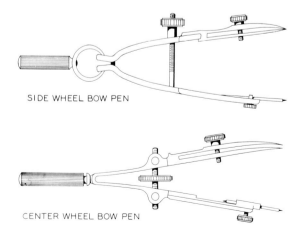

SIDE WHEEL BOW PEN

CENTER WHEEL BOW PEN

Fig. 10.72 Two types of small bow compasses for drawing arcs in ink up to a radius of about one inch.

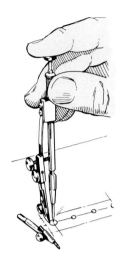

Fig. 10.73 The spring bow compass is used for drawing small circles of about one-quarter inch diameter. A pencil attachment is available for pencil circles.

Inking Compass

The inking compass is usually the same compass used for the circles drawn by pencil, with the inking attachment inserted in place of the pencil attachment. The inking compass may have an elbow in its legs to allow the points to be approximately perpendicular to the drawing surface. The circle can be drawn with one continuous line, as shown in Fig. 10.71.

Two types of compasses for drawing smaller circles are shown in Fig. 10.72. These can be used to draw circles up to a radius of one inch. The spring bow compass (Fig. 10.73) can be used to

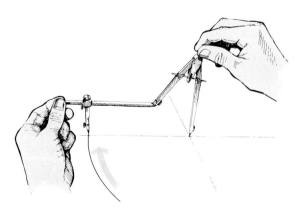

Fig. 10.74 An extension bar can be used with a bow compass for drawing large arcs.

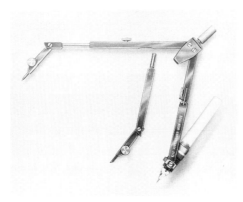

Fig. 10.75 Special compasses with adapters are available for using technical ink pens for drawing arcs. (Courtesy of Koh-I-Noor Rapidograph, Inc.)

draw pencil and ink circles that are very small, usually about one-eighth inch in radius.

Larger circles can be drawn using the extension bar with the large compass (Fig. 10.74). Much larger circles can be drawn with the beam compass (as shown in Fig. 10.56) using the inking attachment. Special compasses are available for drawing circles with a technical fountain pen (Fig. 10.75). These pens screw into an adapter that fits the compass.

Order of Inking

When a drawing is to be inked, begin by locating the center lines and tangent points associated with the arcs as shown in Step 1 of Fig. 10.76. This construction should be laid out in pencil prior to inking.

Ink the arcs and circles first, being careful to stop all arcs at their points of tangency, Step 2. Connect the arcs with straight lines at the points of tangency to complete the drawing.

When a drawing is composed mostly of straight lines, begin at the top of the drawing and draw all of the horizontal lines as you progress from the top of the sheet to the bottom. In this manner, you are moving away from the wet lines, which are left to dry as you progress. After allowing the last horizontal line to dry, the right-handed person should ink the vertical lines by beginning with the far left and moving across your drawing to the right, away from the wet lines. The direction may be reversed for the left-handed drafter.

The proper methods of connecting ink lines are shown in Fig. 10.77. Ink lines should make good junctions with their connecting lines. Ink lines should be centered over the previously drawn pencil guidelines.

Templates

A wide variety of templates is available for preparing drawings in both pencil and ink. These are available for drawing nuts and bolts, circles and

Fig. 10.76 Order of inking

CONSTRUCTION LAYOUT

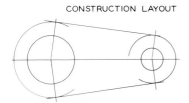

INK ARCS

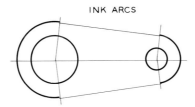

INK STRAIGHT LINES

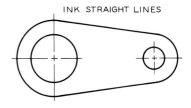

Step 1 The drawing is laid out with light pencil construction lines. All centers and tangent points are accurately located.

Step 2 The arcs and circles are always inked first. Arcs should stop at their points of tangency.

Step 3 Straight lines are drawn to match the ends of the arcs. Centerlines are shown to complete the drawing.

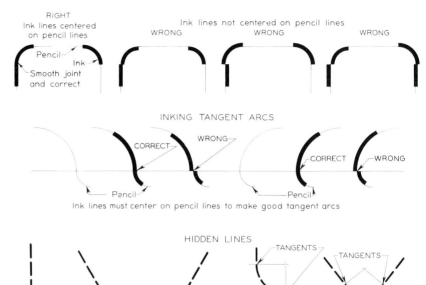

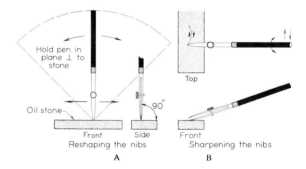

Fig. 10.77 Techniques of matching and joining ink lines.

ellipses, architectural symbols, and many other applications. They work best when used with technical fountain pens rather than with the traditional ruling pen. The wide range of templates is shown in Fig. 10.78.

Sharpening the Ruling Pen

When the ruling pen does not begin to draw when placed in contact with the drawing surface, it is in need of either cleaning or sharpening. It can be easily cleaned with water and a cloth. Sometimes it must be scraped to remove the dry ink with a

Fig. 10.79 To sharpen the nibs of a ruling pen, begin by closing the nibs until they touch. Sharpen the outside of nibs, but not the inside.

Fig. 10.78 Many types of templates are available to aid drafters in their work. (Courtesy of Rapidesign.)

pen knife or a razor blade. If the pen still does not draw without effort, the nibs are in need of sharpening.

The pen is sharpened by closing the nibs with the set screw until the points close. The nibs are then passed across the oil stone (whetrock) as shown in Fig. 10.79A to even the points. This should require only a few strokes.

Avoid nibs that are pointed; they should be slightly rounded for best results. Open the nibs to shape them (Fig. 10.79B). Be sure that the nibs are sharpened to the same lengths so both will touch the paper when an ink line is drawn. Do not sharpen the inside of the nibs.

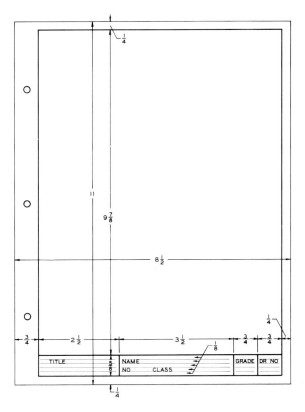

Fig. 10.80 The format and title strip for a Size A sheet (8½" × 11") suggested for solving the problems at the end of each chapter. When the sheet is in the vertical format it will be called a Size AV.

10.19 SOLUTIONS OF PROBLEMS

Problems at the end of each chapter are given to test your understanding of principles covered in the chapter.

The following formats are suggested for the layout of problem sheets. Most problems will be drawn on 8½" × 11" sheets as shown in Fig. 10.80. A title strip is suggested in this figure, with

a border as shown. Guidelines should be drawn very lightly to be only faintly visible. The 8½" × 11" sheet in the vertical format is called Size AV throughout the remainder of this textbook. When this size sheet is in the horizontal format, as shown in Fig. 10.81, it will be called Size AH.

The standard sizes of sheets from Size A through Size E are shown in Fig. 10.81. An alternative title strip for Sizes B, C, D, and E is shown in Fig. 10.82. Guidelines should always be used for lettering title strips.

Another title block and parts list is given in Fig. 10.83. These are placed in the lower right-hand corner of the sheet against the borders. When both are used on the same drawing, the parts list is placed directly above and in contact with the title block or the title strip as the case may be.

You should refer to this section when solving problems from other chapters as you progress through the textbook.

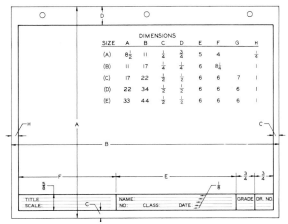

Fig. 10.81 The format for Size AH (an 8½" × 11" sheet in a horizontal position) and the sizes of other sheets. The dimensions under columns A through H give the various layouts.

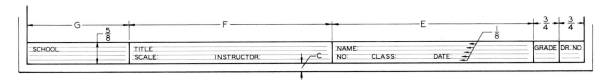

Fig. 10.82 An alternative title strip that can be used on sheet Sizes B, C, D, and E instead of the one given in Fig. 10.81.

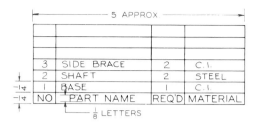

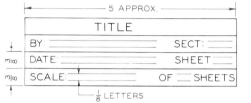

Fig. 10.83 A title block and parts list that can be used on some problem sheets if needed.

PROBLEMS

The problems (Figs. 10.84–10.88) are to be constructed on Size AH (8½″ × 11″) paper, plain or with a printed grid, using the format shown in Fig. 10.80. Two problems can be constructed per sheet. Use pencil or ink as assigned by your instructor.

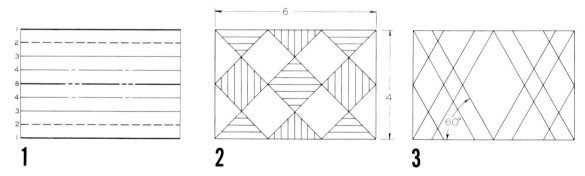

Fig. 10.84 Each of these problems is to be drawn on a Size AV sheet. Begin by lightly laying out a 4″ × 6″ rectangle. Problem 1: construct the following lines: 1—visible line, 2—hidden line, 3—dimension line, 4—center line, and 8—cutting plane line. Problems 2 and 3: construct the patterns shown using line weights equal to that of visible lines.

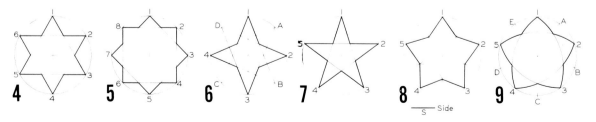

Fig. 10.85 Study the figures closely and draw one per sheet of Size AV paper. The circles are 4″ in diameter. The dimension S in Problem 8 is 1.2 inches.

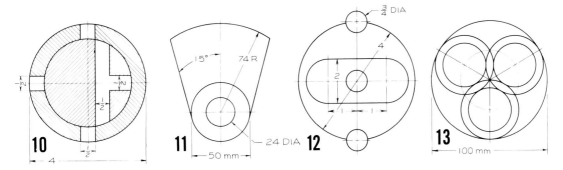

Fig. 10.86 Draw the problems on a Size AV sheet, one per sheet, using the given dimensions and your instruments.

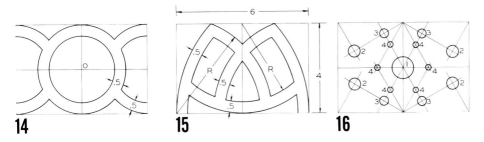

Fig. 10.87 Construct the problems inside of the 4" × 6" rectangels, using the given dimensions. Problem 16: the hole diameters are as follows: No. 1—1", No. 2—.5", No. 3—.4", and No. 4—.24".

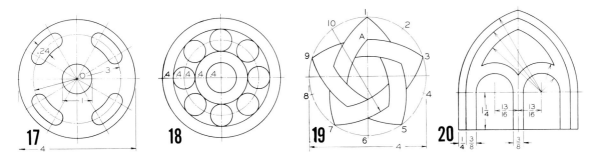

Fig. 10.88 Construct the drawings in problems 17–20 using the given dimensions (the large circles are 4" in diameter), one problem per sheet. Use a Size AV sheet.

11

LETTERING

11.1 LETTERING

All drawings are supplemented with notes, dimensions, and specifications that must be lettered. Consequently, the ability to construct legible freehand letters is a very important skill to develop since it affects the usage and interpretation of a drawing.

The drawing in Fig. 11.1 illustrates the variety of lettered notes and dimensions that must be

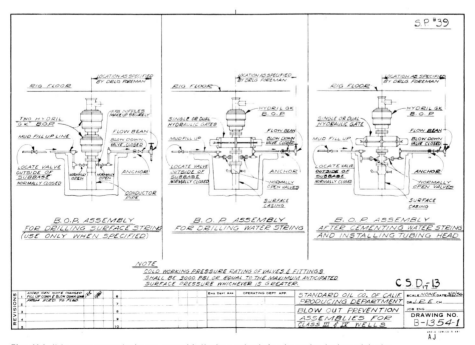

Fig. 11.1 It is necessary to learn good lettering principles in order to be able to show the notes and dimensions required on this working drawing. (Courtesy of Unisorb Machinery Installation Systems.)

used. You can see that the words are easy to read, and the dimensions can be clearly understood.

A measure of the professionalism of drafters, technologists, or engineers is their technique of lettering. Good lettering results from a good attitude and the willingness to put forth one's best efforts. On the other hand, poor lettering indicates an indifferent attitude and a lack of pride in one's work.

11.2 TOOLS OF LETTERING

Good lettering requires the use of proper instruments. The best grade pencil for lettering on most surfaces is a pencil in the H–HB grade range, with an F pencil being the most commonly used grade. Some papers and films are coarser than others and may require a harder pencil lead. The point of the pencil should be slightly rounded to give the desired line width (Fig. 11.2). A needle point will break off when pressure is applied.

Fig. 11.3 When lettering a drawing, use a protective sheet under your hand to prevent smudges. Your lettering will be best when you are working from a comfortable position; you may wish to turn your paper for the most natural strokes.

Fig. 11.2 Good lettering begins with a properly sharpened pencil point. The point should be slightly rounded, not a needle point. The F pencil is usually the best grade for lettering.

When lettering, the pencil should be revolved slightly between your fingers as the strokes are being made so the lead will wear down gradually and evenly. Bear down firmly to make letters black and bright for good reproduction. To prevent smudging your drawing during the lettering process, it is helpful to place a sheet of paper under your hand to protect the drawing (Fig. 11.3).

For freehand lettering with ink, a number of pen points are available to use with a pen staff. You will notice in Fig. 11.4 that points come in various graduations to give lines of varying widths. These pens should be used with India ink for best results.

Another type of inking pen that is widely used is the India ink technical pen (Fig. 11.5).

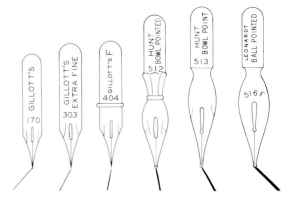

Fig. 11.4 Inking pens are available to give a variety of line widths for freehand lettering. These are used with pen staffs and India ink.

These have tubular points that are kept clear by a movable plunger inside the tubes. The point is kept clear by gently shaking the pen up and down to activate the plunger. These pens are available in a range of tubular point sizes for making lines of varying widths. These pens can hold a supply of ink sufficient for hours of use without refilling.

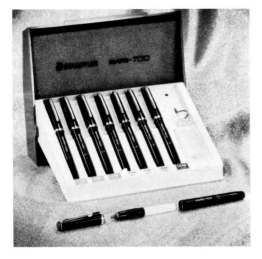

Fig. 11.5 Ink fountain pens for the drafter have been widely used by professionals who work with ink. The line widths vary with each pen, each of which has its own number. (Courtesy of J. S. Staedtler, Inc.)

11.3 GOTHIC LETTERING

The standard type of lettering that is recommended for engineering drawings is *single-stroke Gothic lettering*. This form of lettering is given this name because the letters are made with a series of single strokes and the letter form is a variation of Gothic lettering. The strokes are made uniformly with no variation in line weight.

Two general categories of Gothic lettering are *vertical* and *inclined* lettering (Fig. 11.6). Each is equally acceptable; however, these types should not be mixed on the same drawing.

VERTICAL GOTHIC
INCLINED GOTHIC

Fig. 11.6 Two types of Gothic lettering recommended by engineering standards are vertical and inclined lettering.

11.4 GUIDELINES

The most important rule of lettering is: *Use guidelines at all times*. This applies whether you are lettering a paragraph or a single letter or numeral. The method of constructing and using guidelines can be seen in Fig. 11.7. Use a sharp pencil in the 3H–5H grade range and draw these lines very lightly, just dark enough for them to be seen.

Most lettering is done with the capital letters $\frac{1}{8}''$ high (3 mm high). The spacing between the lines of lettering should be no closer than half the height of the capital letters, $\frac{1}{16}''$ in this case.

Vertical guidelines should be drawn at random to serve as a visual guide in addition to the horizontal lines. These guidelines will be slanted for inclined lettering.

Lettering Guides

The two most-used instruments for drawing guidelines for lettering are the *Braddock-Rowe lettering triangle* and the *Ames lettering instrument*.

The Braddock-Rowe triangle is pierced with sets of holes for spacing guidelines (Fig. 11.8). The numbers under each set of holes represent thirty-

Fig. 11.7 Lettering guidelines

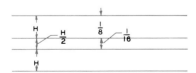

Step 1 Letter heights, *H*, are laid off and thin construction lines are drawn with a 4H pencil. The spacing between the lines should be no closer than *H*/2, or $\frac{1}{16}''$ when $\frac{1}{8}''$ letters are used.

Step 2 Vertical guidelines are drawn as light, thin lines. These are randomly spaced to serve as visual guides for lettering.

Step 3 The letters are drawn with single strokes using a medium-grade pencil, H-HB. The guidelines need not be erased since they are drawn lightly.

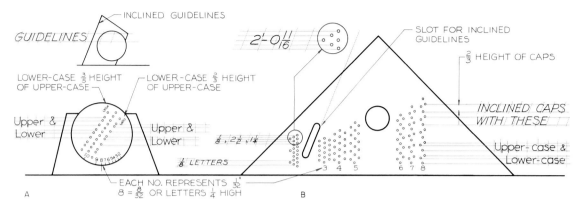

Fig. 11.8 A. The Ames lettering guide can be used for drawing guidelines for uppercase and lowercase letters, vertical or inclined. The dial is set to the desired number of thirty-seconds of an inch for the height of uppercase letters. B. The Braddock-Rowe triangle can be used as a 45° triangle as well as an instrument for constructing guidelines. The numbers designating the guidelines represent thirty-seconds of an inch. For example, the number 4 represents ⁴/₃₂ or ⅛ inch for the height of uppercase letters.

seconds of an inch. For example, the numeral 4 represents ⁴/₃₂″, or guidelines that are placed ⅛″ apart for making uppercase (capital) letters. Some triangles are marked for metric lettering in millimeters. Note in Fig. 11.8 that intermediate holes are provided for guidelines for lowercase letters, which are not as tall as the capital letters.

The Braddock-Rowe triangle is used in conjunction with a horizontal straightedge, such as a T-square, held firmly in position with the triangle placed against its edge. A sharp 4H pencil is placed in one hole of the desired set of holes to contact the drawing surface, and the pencil point is guided across the paper to draw the guideline while the triangle slides against the straightedge. This process is repeated as the pencil point is moved successively to each hole until the desired number of guidelines are drawn.

A slanted slot for drawing guidelines for inclined lettering is cut in the triangle. These slanting guidelines are spaced randomly by eye.

The Ames lettering guide is a very similar device with a circular dial for selecting the proper spacing of guidelines. Again, the numbers around the dial represent thirty-seconds of an inch. The number 8 represents ⁸/₃₂″, or guidelines for drawing capital letters that are ¼″ tall. Metric guides are labeled in millimeters.

This instrument is used with a pencil and straightedge, as previously explained for using the Braddock-Rowe triangle. Be sure to keep the guidelines very light so that they will not interfere with the legibility of the lettering.

11.5 VERTICAL LETTERS

Vertical Capital Letters

Capital letters (uppercase letters) are commonly used on working drawings. They are very legible and result in a word or phrase that is easy to read.

The capital letters for the *single-stroke Gothic* alphabet are shown in Fig. 11.9. Each letter is drawn inside a square box of guidelines to help you learn their correct proportions. Some letters require the full area of the box; some require less, and a few require more space. Each straight-line stroke should be drawn as a single stroke without stopping. For example, the letter A is drawn with three single strokes. Letters composed of curves can best be drawn in segments. The letter O can be drawn by joining two semicircles to form the full circle.

The shape and proportion of letters is important to good lettering. Memorize the shape of each letter given in this alphabet. Small wiggles in your strokes will not detract from your lettering if the letter forms are correct.

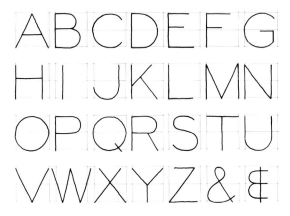

Fig. 11.9 The uppercase letters used in single-stroke Gothic lettering. Each is drawn inside a square to help you learn the proportions of each letter.

Fig. 11.10 There are many ways to letter poorly. A few of them, and the reasons why the lettering is inferior, are shown here. *Do not* make these mistakes.

Examples of poor lettering are shown in Fig. 11.10. Observe the reason given for the lettering being poor in each example.

Vertical Lowercase Letters

An alphabet of lowercase letters is shown in Fig. 11.11. Lowercase letters are either two-thirds or three-fifths as tall as the uppercase letters that they are used with. Both of these ratios are labeled on the Ames guide. Only the two-thirds ratio is available on the Braddock-Rowe triangle.

Some lowercase letters have ascenders that extend above the body of the letter such as the letter b; and some have descenders that extend below the body of the letters such as the letter y. The ascenders are the same length as the descenders.

The guidelines in Fig. 11.11 form perfect squares about the body of each letter to illustrate the proportions. A number of these letters have bodies that are perfect circles that touch all sides of the squares.

Capital and lowercase letters are used together in Fig. 11.12 as in a title. You can see the difference between the lowercase letters that are two-thirds the height of capitals and those that are three-fifths the height of capitals.

Vertical Numerals

Vertical numerals are shown in Fig. 11.13 where each number is enclosed in a square box of guide-

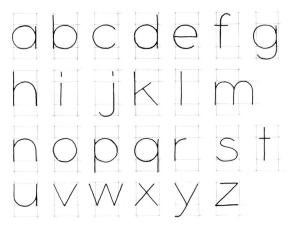

Fig. 11.11 The lowercase alphabet used in single-stroke Gothic lettering. The body of each letter is drawn inside a square to help you learn the proportions.

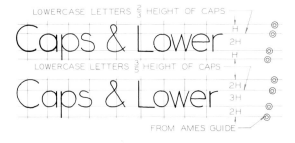

Fig. 11.12 Uppercase and lowercase letters are sometimes used together. The ratio of the lowercase letters to the uppercase letters will be either two-thirds or three-fifths. The Ames guide has both, and the Braddock-Rowe triangle has only the three-fifths ratio.

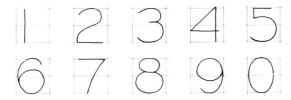

Fig. 11.13 The numerals for single-stroke Gothic lettering. Each is drawn inside a square to help you learn the proportions.

Fig. 11.14 The uppercase alphabet for single-stroke inclined Gothic lettering.

lines. As with lettering, you must learn the proportions of the numerals in order to use them properly. Each number is made the same height as the capital letters being used; usually ⅛″ high. The numeral zero is an oval and the letter O is a perfect circle in vertical lettering.

11.6 INCLINED LETTERS

Inclined Capital Letters

Inclined uppercase letters (capitals) have the same heights and proportions as vertical letters, the only difference being their inclination of 68° to the horizontal. The inclined alphabet is shown in Fig. 11.14.

Inclined guidelines should be drawn using the Braddock-Rowe triangle or the Ames guide, as illustrated in Fig. 11.8. When these are not available, a 2 × 5 angle can be constructed and parallel guidelines drawn as shown in Fig. 11.15.

Lettering features that appear as circles in vertical lettering will appear as ellipses when inclined lettering is used.

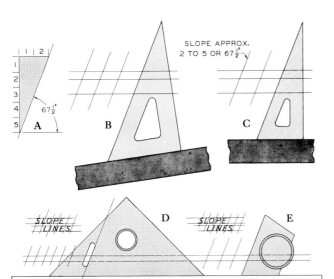

Fig. 11.15 Inclined guidelines can be constructed by any of the following: (A) draw a 2 × 5 right triangle for establishing the angle of inclination of 67.5°; (C) use a specially designed lettering triangle that has an angle of 67.5°; (D) use the slot in the Braddock-Rowe triangle; or (E) use the angle of the Ames guide.

Inclined Lowercase Letters

Lowercase inclined letters are drawn in the same manner as the vertical lowercase letters. This alphabet is shown in Fig. 11.16. Ovals (ellipses) are used instead of the circles used in vertical lettering. The angle of inclination is 68°, the same as is used for uppercase letters.

Inclined Numerals

The inclined numerals that should be used in conjunction with inclined lettering are shown in Fig. 11.17. Except for the inclination of 68° to the horizontal, they are drawn the same as vertical numbers.

The use of inclined letters and numbers in combination is seen in Fig. 11.18. The guidelines in this example were constructed using the Braddock-Rowe triangle (Fig. 11.8).

Fig. 11.16 The lowercase, single-stroke alphabet for inclined Gothic lettering. The body of each letter is drawn inside a rhombus to help you learn the proportions.

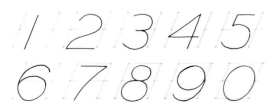

Fig. 11.17 The numerals for single-stroke Gothic inclined lettering. Each number is drawn in a rhombus to help you learn the proportions.

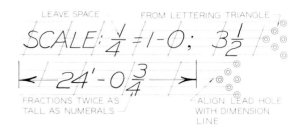

Fig. 11.18 Inclined common fractions are twice as tall as single numerals. Inch marks are omitted when numerals are used to show dimensions.

11.7 SPACING NUMERALS AND LETTERS

Common fractions are twice as tall as single numerals (Fig. 11.19). The fractions will be ¼″ tall when they are used with ⅛″ lettering. A separate set of holes for common fractions is given on the Braddock-Rowe triangle and on the Ames guide. These are equally spaced ¹⁄₁₆″ apart with the center line being used for the fraction's crossbar. Examples of these holes can be seen in Fig. 11.19 and Fig. 11.8.

When numbers are used with decimals, space should be provided for the decimal point (Fig. 11.20). Common mistakes of spacing decimal fractions are shown at B and C. The correct method of drawing common fractions is illustrated at D, and several of the often-encountered errors are shown at E, F, and G.

When letters are grouped together to spell words, the areas between the letters should be approximately equal for the most pleasing result (Fig. 11.21). The incorrect use of guidelines and

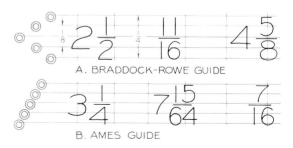

Fig. 11.19 Common fractions are twice as tall as single numerals. Guidelines for these can be drawn by using the Ames or the Braddock-Rowe triangle.

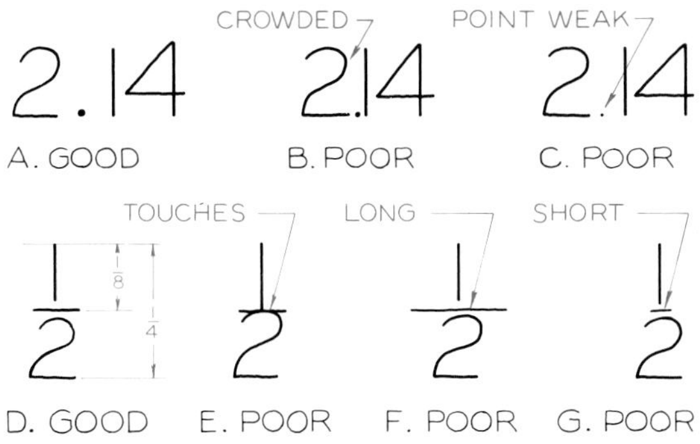

CROWDED ⌐ POINT WEAK ⌐

2.14   2.14   2.14

A. GOOD   B. POOR   C. POOR

TOUCHES ⌐   LONG ⌐   SHORT ⌐

$\frac{1}{2}$ $\frac{1}{2}$ $\frac{1}{2}$ $\frac{1}{2}$

D. GOOD   E. POOR   F. POOR   G. POOR

Fig. 11.20 Examples of poor spacing of numerals that result in poor lettering.

A. GOOD – EQUAL AREAS BETWEEN LETTERS

B. POOR - EQUAL SPACING

C. POOR - EQUAL SPACING

Fig. 11.21 Proper spacing of letters is necessary for good lettering and good appearance. The areas between letters should be approximately equal.

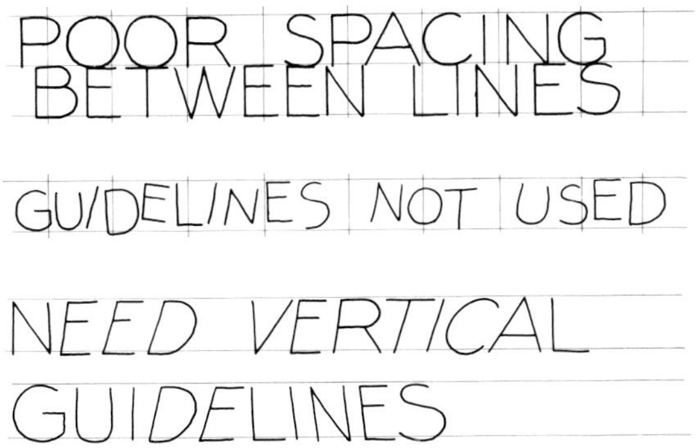

POOR SPACING BETWEEN LINES

GUIDELINES NOT USED

NEED VERTICAL GUIDELINES

Fig. 11.22 Always leave space between lines of lettering. After constructing guidelines, *use them.* Use vertical guidelines to improve the angle of your vertical strokes.

other violations of good lettering practice are shown in Fig. 11.22. Avoid making these errors when you are lettering.

11.8 MECHANICAL LETTERING

Drawings and illustrations that are to be reproduced by a printing process are usually drawn in India ink, and the lettering must also be in ink. Several mechanical aides for lettering are available.

The Wrico lettering template (Fig. 11.23) can be placed against a fixed straightedge for aligning the letters. You move the template from position to position while drawing each letter (with a lettering pen) through the raised portion of the template where the holes form the letters.

A slightly different template and pen are used by the Rapid-O-Graph system as illustrated in Fig. 11.24. The method of lettering with this system is the same as the Wrico system. A variety of pen sizes is available from each manufacturer for

Fig. 11.23 A Wrico lettering template can be used for mechanical lettering. (Courtesy of Wood-Reagan Instrument Company.)

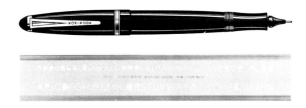

Fig. 11.24 A typical India ink fountain pen and template that can be used for mechanical lettering. (Courtesy of Koh-I-Noor Rapidograph, Inc.)

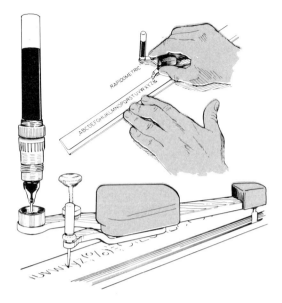

Fig. 11.25 Templates and scribes of this type are available for mechanical lettering.

Fig. 11.26 This portable typewriter can be placed on the drawing surface so that notes and numerals can be typed on a drawing. (Courtesy of Grintzner, Inc.)

drawing different sized letters and numbers with thin or bold lines.

Another system of mechanical lettering makes use of a grooved template that is used in conjunction with a scriber that follows the grooves and inks the letters on the drawing surface. Note that a standard India ink technical pen can be unscrewed from its barrel and attached to the scriber (Fig. 11.25). Many templates for varying styles of lettering and symbols are available for this system of mechanical lettering.

11.9 LETTERING BY TYPING

Some drafting departments type many of their notes and specifications to reduce drafting time and to improve the readability of a drawing. Large typewriters are available with long carriages that will accept extremely large drawings, and the notes are typed on the drawings in the conventional manner.

A portable typewriter (Fig. 11.26) can be placed on top of a drawing for typing without the necessity of a typewriter carriage. Numbers and letters are then typed after the drawing has been made.

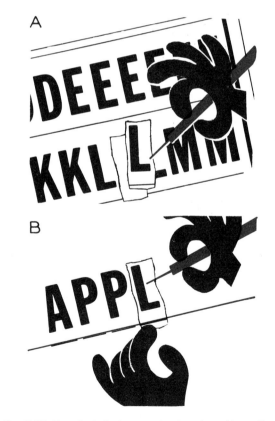

Fig. 11.27 Transfer lettering can be transferred from film to the drawing surface. Transfer lettering comes in many sizes and styles. (Courtesy of Artype, Incorporated.)

11.10 TRANSFER LETTERING

Transfer lettering comes printed on transparent sheets with adhesive backings on one side (Fig. 11.27). The letters are cut from the sheet, aligned with drawn guidelines, and then burnished down permanently. Successive letters are applied in the same manner until the desired word is completed.

Some types of transfer lettering are burnished directly from the transparent sheet to the drawing surface. Each letter is rubbed firmly and evenly to make it transfer. This is done repetitively until the word is complete.

A multitude of letter forms and sizes is available from many manufacturers of transfer letters. Although many symbols and patterns are available in addition to lettering, it is possible to have custom transfer sheets produced for trademarks, title blocks, and other often-used applications.

PROBLEMS

Lettering problems are to be presented on Size AV (8½″ × 11″) paper, plain or grid, using the format shown in Fig. 11.28.

1. Practice lettering the vertical uppercase alphabet shown in Fig. 11.9. Construct each letter four times: four A's, four B's, etc. Use a medium-weight pencil—H, F, or HB.

2. Practice lettering vertical numerals and the lowercase alphabet as shown in Fig. 11.29. Construct each letter and numeral three times: three 1's, three 2's, etc. Use a medium-weight pencil—H, F, or HB.

3. Practice lettering the inclined uppercase alphabet shown in Fig. 11.14. Construct each letter

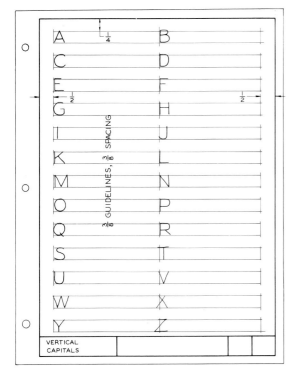

Fig. 11.28 Problem 1. Construct each vertical uppercase letter four times.

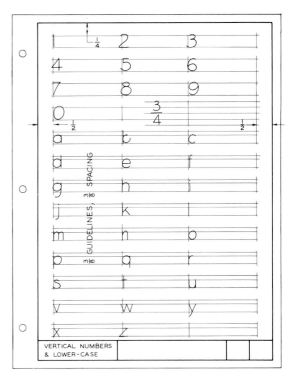

Fig. 11.29 Problem 2. Construct each vertical numeral and lowercase letter three times.

four times: four A's, four B's, etc. Use a medium-weight pencil—H, F, or HB.

4. Practice lettering the vertical numerals and the lowercase alphabet shown in Figs. 11.11 and 11.13. Construct each letter three times: three 1's, three 2's, etc. Use a medium-weight pencil—H, F, or HB.

5. Construct guidelines for $\frac{1}{8}$" capital letters starting $\frac{1}{4}$" from the top border. Each should end $\frac{1}{2}$" from the left and right borders. Using these guidelines, letter the first paragraph of the text of this chapter. Use all vertical capitals. Spacing between the lines should be $\frac{1}{8}$".

6. Repeat Problem 5, but use all inclined capital letters. Use inclined guidelines to assist you in slanting your letters uniformly.

7. Repeat Problem 5 but use vertical capitals and lowercase letters in combination. Capitalize only those words that are capitalized in the text.

8. Repeat Problem 5 but use inclined capitals and lowercase letters in combination. Capitalize only the words that are capitalized in the text.

GEOMETRIC CONSTRUCTION

12.1 INTRODUCTION

Many problems in drafting and graphics can be solved only by the application of geometry and geometric construction. Engineering and technical problems are often solved by geometric construction during the design process.

Mathematics was an outgrowth of graphical construction; consequently, there is a close relationship between the two areas. The proofs of many principles of plane geometry and trigonometry may be developed by using graphics. Graphical methods can be applied to algebra and arithmetic, and virtually all problems of analytical geometry can be solved graphically.

An understanding of the principles covered in this chapter will improve your understanding of mathematics and geometry.

12.2 ANGLES

A fundamental requirement of geometric construction is the construction of lines that join at specified angles with each other. You should learn the terminology associated with angular measurements. The definitions of various angles are given in Fig. 12.1.

The unit of angular measurement is the degree, and a circle has 360 degrees. A degree (°) can be divided into 60 parts called minutes ('), and a

minute can be divided into 60 parts called seconds ("). An angle of 15°32'14" is an angle of 15 degrees, 32 minutes, and 14 seconds.

12.3 TRIANGLES

The *triangle* is a three-sided polygon (or figure) that is named according to its shape. The four types of triangles are the *scalene, isosceles, equilateral,* and *right triangles* (Fig. 12.2). The sum of the angles inside a triangle will always be equal to 180°.

12.4 QUADRILATERALS

A *quadrilateral* is a four-sided figure of any shape. The sum of the angles inside a quadrilateral is 360°.

The various types of quadrilaterals and their respective names are shown in Fig. 12.3, along with the equations for the areas of these figures.

12.5 POLYGONS

A *polygon* is a multi-sided plane figure of any number of sides. (The triangle and quadrilateral are polygons.) If the sides of the polygon are equal in length, the polygon is a *regular polygon*. Four

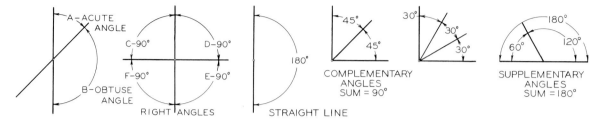

Fig. 12.1 Standard types of angles and their definitions.

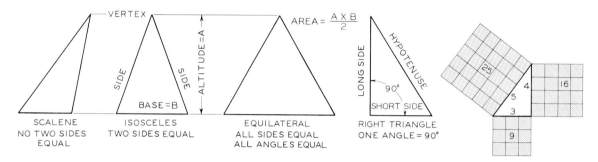

Fig. 12.2 Types of triangles and their definitions. The hypotenuse of the right triangle is equal to the square root of the sum of the squares of the other two sides. This is the Pythagorean theorem.

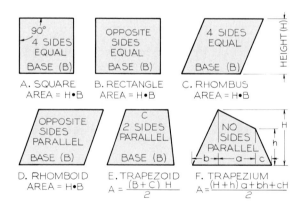

Fig. 12.3 Types of quadrilaterals (four-sided plane figures).

Fig. 12.4 Regular polygons inscribed in circles.

types of regular polygons are shown in Fig. 12.4. Note that a regular polygon can be inscribed in a circle and all of the corner points will lie on the circle.

Other regular polygons not pictured are: the *heptagon* with seven sides, the *nonagon* with nine sides, the *decagon* with ten sides and the *dodecagon* with 12 sides.

The sum of the angles inside any polygon can be found by the equation: Sum = $(n - 2) \times 180°$; where n is equal to the number of sides of the polygon.

12.6 ELEMENTS OF A CIRCLE

A circle of 360° can be divided into a number of parts, each of which has its own special name (Fig. 12.5). The equation for finding the area of a circle is $A = \pi r^2$; where r is the radius and π is equal to 3.14 (pi). The equation for finding the circumference is $C = \pi d$; where d is the diameter.

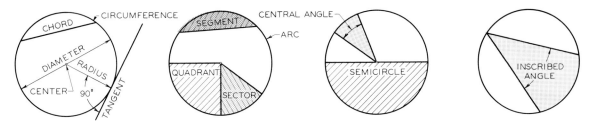

Fig. 12.5 Definitions of the elements of a circle.

12.7 GEOMETRIC SOLIDS

The various types of solid geometric shapes are shown in Fig. 12.6 along with their names and definitions.

Polyhedra   A multi-sided solid formed by intersecting planes is called a *polyhedron*. If the faces of a polyhedron are regular polygons, it is called a *regular polyhedron*. The five regular polyhedra are the *tetrahedron* with four sides, the *hexahedron* with six sides, the *octahedron* with eight sides, the *dodecahedron* with 12 sides, and the *icosahedron* with 20 sides.

Prisms   A *prism* is a solid that has two parallel bases that are equal in shape. The bases are connected by sides that are parallelograms. The line from the center of one base to the other is called the *axis*. If the axis is perpendicular to the bases, the axis is called the *altitude* and the prism is a *right prism*. If the axis is not perpendicular to the base, the prism is an *oblique prism*.

A prism that has been cut off to form a base that is not parallel to the other is called a *truncated prism*.

A *parallelepiped* is a prism with a base that is either a rectangle or a parallelogram.

Pyramids   The *pyramid* is a solid with a polygon as a base and triangular faces that converge at a point called the *vertex*. The line from the vertex to the center of the base is called the *axis*. If the axis is perpendicular to the base, it is the *altitude* of the pyramid, and the pyramid is a *right pyramid*. If the axis is not perpendicular to the base, the pyramid is an *oblique pyramid*. A truncated pyramid is called a *frustum* of a pyramid.

Cylinders   The *cylinder* is formed by a line or element, called a generatrix, that moves about the circle while remaining parallel to its axis.

The axis of a cylinder connects the centers of each end of a cylinder. If the axis is perpendicular to the bases, it is the *altitude* of the cylinder. When the axis is perpendicular to the bases the cylinder is a *right cylinder*. When the axis does not make a 90° angle with the base, the cylinder is an *oblique cylinder*.

Cones   A *cone* is formed by a line or element, called a generatrix, with one end that moves about the curved base while the other end remains at a fixed point called the *vertex*. The line from the center of the base to the vertex is called the *axis*. If the axis is perpendicular to the base, it is called the *altitude* and the cone is a *right cone*. A truncated cone is called a *frustum* of a cone.

Spheres   The *sphere* is generated by the plane of a circle that is revolved about one of its diameters to form a solid. The ends of an axis through the center of the sphere are called *poles*.

The equations for finding the volumes of geometric solids are given in Fig. 12.7.

12.8 CONSTRUCTING A TRIANGLE

When three sides of a triangle are given, the triangle can be constructed by using a compass as shown in Fig. 12.8. Only one triangle can be found when the sides are given.

A right triangle can be constructed by inscribing it inside a semicircle as shown in Fig. 12.9. Any triangle inscribed in a semicircle will always be a right triangle.

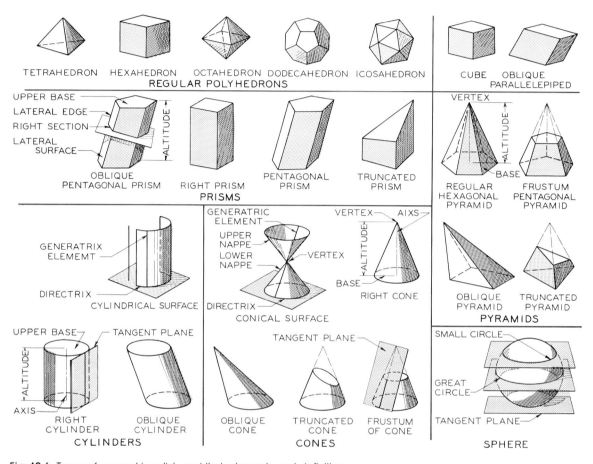

TETRAHEDRON HEXAHEDRON OCTAHEDRON DODECAHEDRON ICOSAHEDRON

REGULAR POLYHEDRONS

CUBE OBLIQUE
PARALLELEPIPED

UPPER BASE
LATERAL EDGE
RIGHT SECTION
LATERAL SURFACE
ALTITUDE

OBLIQUE PENTAGONAL PRISM RIGHT PRISM PENTAGONAL PRISM TRUNCATED PRISM

PRISMS

VERTEX
ALTITUDE
BASE

REGULAR HEXAGONAL PYRAMID FRUSTUM PENTAGONAL PYRAMID

GENERATRIX ELEMEMT
DIRECTRIX
CYLINDRICAL SURFACE

GENERATRIC ELEMENT
UPPER NAPPE
LOWER NAPPE
VERTEX
DIRECTRIX
CONICAL SURFACE

VERTEX AIXS
ALTITUDE
BASE
RIGHT CONE

OBLIQUE PYRAMID TRUNCATED PYRAMID

PYRAMIDS

UPPER BASE TANGENT PLANE
ALTITUDE
AXIS
RIGHT CYLINDER OBLIQUE CYLINDER

CYLINDERS

TANGENT PLANE
OBLIQUE CONE TRUNCATED CONE FRUSTUM OF CONE

CONES

SMALL CIRCLE
GREAT CIRCLE
TANGENT PLANE

SPHERE

Fig. 12.6 Types of geometric solids and their elements and definitions.

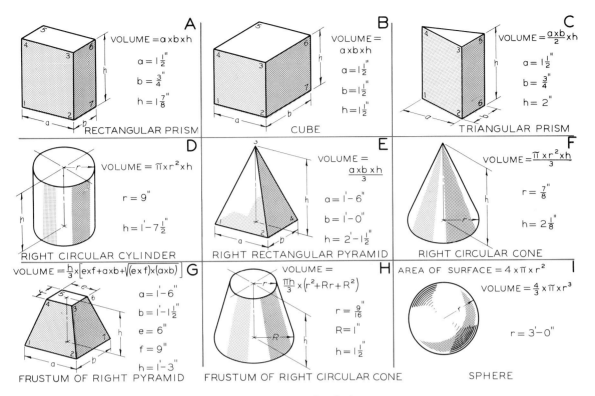

A RECTANGULAR PRISM

$$\text{VOLUME} = a \times b \times h$$

$a = 1\frac{1}{2}''$

$b = \frac{3}{4}''$

$h = 1\frac{7}{8}''$

B CUBE

$$\text{VOLUME} = a \times b \times h$$

$a = 1\frac{1}{2}''$

$b = 1\frac{1}{2}''$

$h = 1\frac{1}{2}''$

C TRIANGULAR PRISM

$$\text{VOLUME} = \frac{a \times b}{2} \times h$$

$a = 1\frac{1}{2}''$

$b = \frac{3}{4}''$

$h = 2''$

D RIGHT CIRCULAR CYLINDER

$$\text{VOLUME} = \pi \times r^2 \times h$$

$r = 9''$

$h = 1' - 7\frac{1}{2}''$

E RIGHT RECTANGULAR PYRAMID

$$\text{VOLUME} = \frac{a \times b \times h}{3}$$

$a = 1' - 6''$

$b = 1' - 0''$

$h = 2' - 1\frac{1}{2}''$

F RIGHT CIRCULAR CONE

$$\text{VOLUME} = \frac{\pi \times r^2 \times h}{3}$$

$r = \frac{7}{8}''$

$h = 2\frac{1}{8}''$

G FRUSTUM OF RIGHT PYRAMID

$$\text{VOLUME} = \frac{h}{3} \times \left[e \times f + a \times b + \sqrt{(e \times f) \times (a \times b)} \right]$$

$a = 1' - 6''$

$b = 1' - 1\frac{1}{2}''$

$e = 6''$

$f = 9''$

$h = 1' - 3''$

H FRUSTUM OF RIGHT CIRCULAR CONE

$$\text{VOLUME} = \frac{\pi h}{3} \times \left(r^2 + Rr + R^2 \right)$$

$r = \frac{9}{16}''$

$R = 1''$

$h = 1\frac{1}{2}''$

I SPHERE

$$\text{AREA OF SURFACE} = 4 \times \pi \times r^2$$

$$\text{VOLUME} = \frac{4}{3} \times \pi \times r^3$$

$r = 3' - 0''$

Fig. 12.7 Standard geometric solids and the equations for finding their volumes.

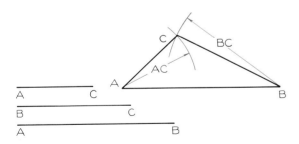

Fig. 12.8 When three sides are given, a triangle can be drawn with a compass.

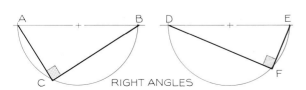

RIGHT ANGLES

Fig. 12.9 Any angle that is inscribed in a semicircle will be a right angle.

12.9 CONSTRUCTING POLYGONS

A regular polygon (having equal sides) can be inscribed in a circle or circumscribed about a circle. When inscribed, all the corner points will lie along the circle. This makes it possible to divide the circle into the desired number of sectors to locate the points (Fig. 12.10).

For example, a 12-sided polygon is constructed by dividing the circle into 12 sectors and connecting the points to form the polygon.

12.10 THE HEXAGON

Examples of inscribed and circumscribed hexagons are shown in Fig. 12.11. These are drawn with 30°–60° triangles either inside or outside the circles. Note that the circle represents the distance from corner to corner when inscribed, and from flat to flat when circumscribed.

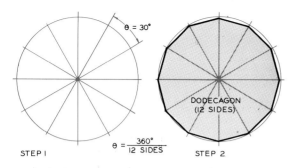

Fig. 12.10 The regular polygon

Step 1 To construct a regular polygon, divide the circumference of a circle into the same number of divisions as the sides of the polygon, 12 in this case.

Step 2 Connect the divisions along the circumference with straight lines to form the sides of the polygon.

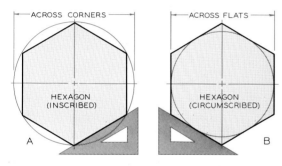

Fig. 12.11 A circle can be inscribed or circumscribed to form a hexagon by using a 30°–60° triangle.

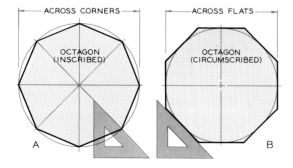

Fig. 12.12 A circle can be inscribed or circumscribed to form an octagon with a 45° triangle.

12.11   THE OCTAGON

The octagon, an eight-sided regular polygon, can be inscribed in or circumscribed about a circle (Fig. 12.12) by using a 45° triangle. A second method inscribes the octagon inside a square (Fig. 12.13).

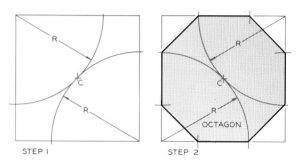

Fig. 12.13 Octagon in a square

Step 1 Construct a square and locate its center, C. Construct two arcs from opposite corners that will pass through C.

Step 2 Repeat this step by using the other corners. Connect the points located on the square to form the octagon.

12.12   THE PENTAGON

Since the pentagon is a five-sided regular polygon, it can be inscribed in or circumscribed about a circle, as previously covered. Another method of constructing a pentagon is shown in Fig. 12.14. This construction is performed with the use of a compass and a straightedge.

12.13   BISECTING LINES AND ANGLES

Finding the midpoint of a line, or the perpendicular bisector of a line, is a basic technique of geometric construction. Two methods are illustrated in Fig. 12.15.

The first method involves the use of a compass to construct a perpendicular to a line. The second method uses a standard triangle. The compass method can be used to find the midpoint of an arc as well as a straight line.

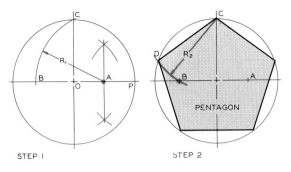

Fig. 12.14 The pentagon

Step 1 Bisect radius OP to locate point A. With A as the center and AC as the radius R_1, locate point B on the diameter.

Step 2 With point C as the center and BC as the radius R_2, locate point D on the arc. Line CD is the chord that can be used to locate the other corners of the pentagon.

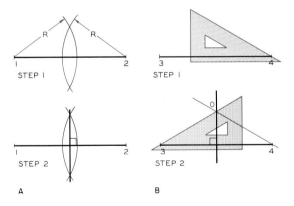

STEP I   STEP I

STEP 2   STEP 2

A   B

Fig. 12.15 Bisecting a line

A line can be bisected by using a compass and any radius or a standard triangle and a straightedge.

The angle in Fig. 12.16 can be bisected with a compass by drawing three arcs.

12.14 REVOLUTION OF A FIGURE

Rotating a triangle about one of its points is demonstrated in Fig. 12.17. In this case, the triangle is rotated about point 1 of line 1–4. Point 4 is rotated to its desired position using a compass. Points 2

and 3 are found by triangulation to complete the rotated view.

This same principle will work on any plane regardless of its shape.

12.15 ENLARGEMENT AND REDUCTION OF A FIGURE

In Fig. 12.18, the small figure is enlarged by using a series of radial lines from the lower left corner. The smaller figure is completed as a rectangle and the larger rectangle is drawn proportional to the small one. The upper right notch is located using the same technique.

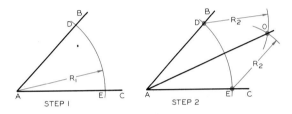

STEP I   STEP 2

Fig. 12.16 Bisecting an angle

Step 1 Swing an arc of any radius to locate points D and E.

Step 2 Using the same radius, draw two arcs from D and E to locate point O. Line AO is the bisector of the angle.

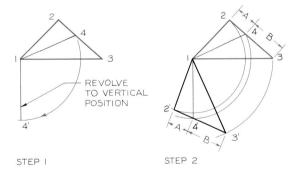

STEP I   STEP 2

Fig. 12.17 Rotation of a figure

Step 1 A plane figure can be rotated about any point. Line 1–4 is rotated about point 1 to its desired position with a compass.

Step 2 Points 2' and 3' are located by measuring distances A and B from 4'.

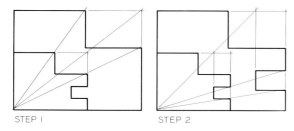

Fig. 12.18 Enlargement of a figure

Step 1 A proportional enlargement can be made by using a series of diagonals drawn through a single point, the lower left corner in this case.

Step 2 Additional diagonals are drawn to locate the other features of the object. This process can be reversed for reducing an object.

The smaller notch is found by using three construction lines projected from the lower left corner. This is based on the principle of similar and proportional triangles.

This method could have been used to reduce the larger drawing to the smaller one.

12.16 DIVISION OF A LINE

It is often necessary to divide a line into a number of equal parts when a convenient scale is not available for this purpose. For example, suppose a

one-inch line is to be divided into seven equal parts. No scale is available that divides an inch into sevenths, and mathematical units involve hard-to-measure decimals.

The method shown in Fig. 12.19 is an efficient way to solve this problem.

An application of this principle is used for locating lines on a graph that are equally spaced (Fig. 12.20). A scale with the desired number of units (0 to 5) is laid across from left to right on the graph. Vertical lines are drawn through these points to divide the graph into five equal divisions.

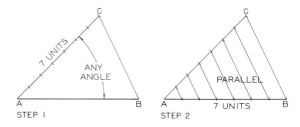

Fig. 12.19 Division of a line

Step 1 Line *AB* is divided into 7 equal divisions by constructing a line through *A* and dividing it into 7 known units with your dividers. Point *C* is connected to point *B*.

Step 2 A series of lines are drawn parallel to *CB* to locate the divisions along line *AB*.

Fig. 12.20 Division of a space

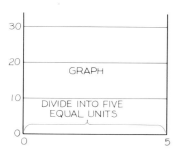

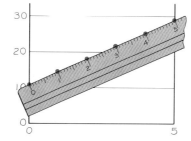

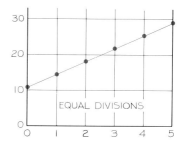

Given It is desired to divide a graph into five equal divisions along the *x*-axis.

Step 1 A convenient scale with five units of measurement that approximate the horizontal distance is laid across the graph aligning the 0 and 5 markings with the lines.

Step 2 Construct vertical lines through the points that were found in step 1. This method could have been used to calibrate the divisions along the *y*-axis.

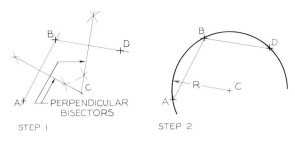

Fig. 12.21 An arc through three points

Step 1 Connect the points with two lines and find their perpendicular bisectors. The bisectors will intersect at the center, C.

Step 2 Using the center, C, and the distance to the points as the radius, construct the arc through the points.

12.17   A CIRCLE THROUGH THREE POINTS

An arc can be drawn through any three points by connecting the points with two lines (Fig. 12.21). Perpendicular bisectors are found for each line to locate the center at point *C*. The radius is drawn, and the lines *AB* and *BD* become chords of the circle.

This system can be reversed to find the center of a given circle. Draw two chords that intersect at a point on the circumference and bisect them. The perpendicular bisectors will intersect at the center of the circle.

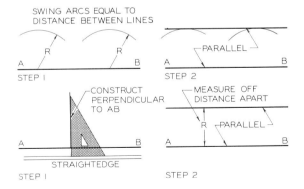

Fig. 12.22 Construction of parallel lines

Either of the above methods can be used for constructing one line parallel to another. The first method uses a compass and a straightedge; the other uses a triangle and T-square.

12.18   PARALLEL LINES

A line can be drawn parallel to another by using either of the methods shown in Fig. 12.22.

The first method involves the use of a compass to draw two arcs to locate a parallel line that is the desired distance away.

The second method requires the construction of a perpendicular from a given line and the measurement of the distance *R* to locate the parallel, which is drawn with a T-square.

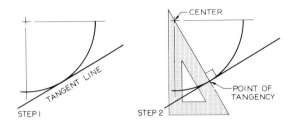

Fig. 12.23 Locating a tangent point

Step 1 Align your triangle with the tangent line while holding it firmly against a straightedge.

Step 2 Hold the straightedge in position, rotate the triangle, and construct a line through the center that is perpendicular to the line, to locate the point of tangency.

12.19   POINTS OF TANGENCY

A point of tangency is the theoretical point where a straight line joins an arc or where two arcs join making a smooth transition. In Fig. 12.23, a line is tangent to an arc. The point of tangency is located by constructing a perpendicular to the line from the center of the arc. A thin line is drawn from the center through the point to mark the point of tangency.

The conventional methods of marking points of tangency are shown in Fig. 12.24. Thin lines are drawn through the points from centers of the arcs. This method should be used to mark all points of tangency when solving tangency problems.

12.20   LINE TANGENT TO AN ARC

Although you can approximate the point of tangency between a line and an arc, the method of

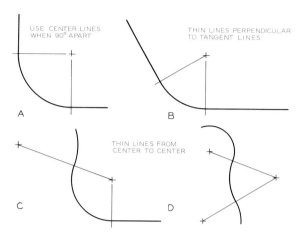

Fig. 12.24 Thin lines that extend beyond the curves from the centers are used to mark points of tangency. These lines should always be shown in this manner.

Fig. 12.25 Line from a point tangent to an arc

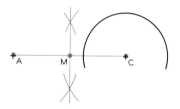 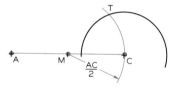 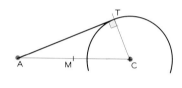

Step 1 Connect point A with center C. Locate point M by bisecting AC.

Step 2 Using point M as the center and MC as the radius, locate point T on the arc.

Step 3 Draw the line from A to T that is tangent to the arc of point T.

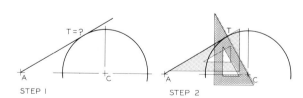

Fig. 12.26 Line from a point tangent to an arc

Step 1 A line can be drawn from point A tangent to the arc by eye.

Step 2 By rotating your triangle, the point of tangency can be located at the 90° angle with the line that passes through the center.

finding the exact point of tangency is shown in Fig. 12.25. Point A and the arc are given. Point A is connected to the center in step 1, AC is bisected in step 2, and T is located in step 3. This is the exact point of tangency.

The point of tangency could also have been found by using a standard triangle, as shown in Fig. 12.26. One edge of the triangle is aligned with TA while the T-square is held firmly. The triangle is rotated to construct a line through the center that is perpendicular to AT, locating the point of tangency, T.

12.21 ARC TANGENT TO A LINE FROM A POINT

If an arc is to be constructed tangent to line CD at T (Fig. 12.27) and pass through point P, a perpendicular bisector of TP is drawn. A perpendicular to CD is drawn at T to locate the center at O.

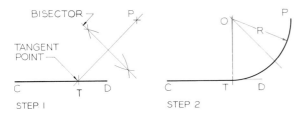

Fig. 12.27 An arc through two points

Step 1 If an arc must be tangent to a given line at a certain point and pass through P, find the perpendicular bisector of line *TP*.

Step 2 Construct a perpendicular to the line at *T* to intersect the bisector. The arc is drawn from center O with radius *OT*.

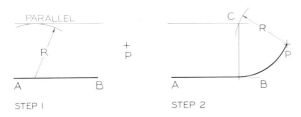

Fig. 12.28 An arc tangent to a line and a point

Step 1 When an arc of a given radius is to be drawn tangent to a line and through point P, draw a line parallel to *AB* and R from it.

Step 2 Draw an arc from P with radius R to locate the center at C. The arc is drawn with radius R and center C.

A similar problem in Fig. 12.28 requires you to draw an arc of a given radius that will be tangent to line *AB* and pass through point P. In this case the point of tangency on the line is not known until the problem has been solved.

12.22 ARC TANGENT TO TWO LINES

An arc of a given radius can be constructed tangent to two nonparallel lines if the radius is given. This construction may be used to round a corner of a product or to design a curb at a traffic intersection. This method is shown in Fig. 12.29 where two lines form an acute angle.

The same steps are used to find an arc that is tangent to two lines that form an obtuse angle (Fig. 12.30). In both cases the points of tangency are located with thin lines drawn from the centers through the points of tangency.

A different technique can be used to find an arc of a given radius that is tangent to perpendicular lines (Fig. 12.31). This method will work only for perpendicular lines.

12.23 ARC TANGENT TO AN ARC AND A LINE

When a radius is given, an arc can be drawn that is tangent to an arc and a line. These steps of construction are given in Fig. 12.32.

Fig. 12. 29 Arc tangent to two lines

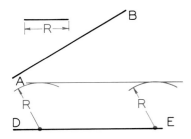

Step 1 Construct a line parallel to *DE* with the radius of the specified arc *R*.

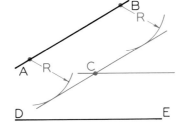

Step 2 Draw a second construction line parallel to *AB* to locate the center, C.

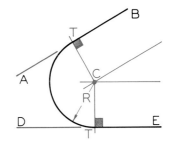

Step 3 Thin lines are drawn from C perpendicular to *AB* and *DE* to locate the points of tangency. The tangent arc is drawn using the center C.

Fig. 12.30 An arc tangent to two lines

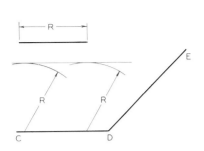
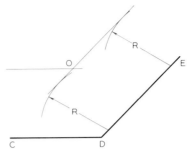
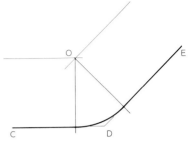

Step 1 Using the specified radius R, construct a line parallel to CD.

Step 2 Construct a line parallel to DE that is distance R from it to locate center C.

Step 3 Construct thin lines from center O perpendicular to lines CD and DE to locate the points of tangency. Draw the arc using radius R and center O.

Fig. 12.31 An arc tangent to perpendicular lines

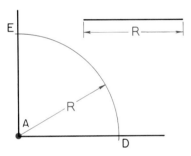
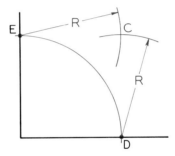
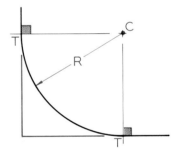

Step 1 Using the specified radius, R, locate points D and E by using center A.

Step 2 Swing two arcs using the radius R that was used in step 1 to locate point O.

Step 3 Locate the tangent points with lines from the center C. Draw the arc with radius R and center C.

Fig. 12.32 An arc tangent to an arc and a line

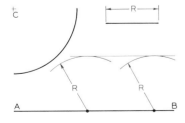
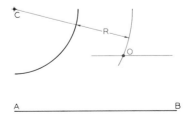
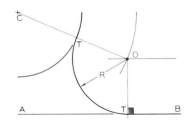

Step 1 Construct a line parallel to AB that is R from it. Use thin construction lines.

Step 2 Add radius R to the extended radius through point C. Use this large radius to locate point O.

Step 3 Lines OC and OT are drawn to locate the tangency points. The arc with radius R and center O is drawn.

Fig. 12.33 An arc tangent to an arc and a line

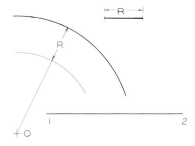 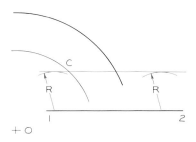 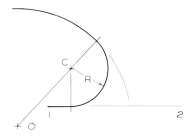

Step 1 The specified arc R is subtracted to the extended radius through the arc's center at O. A concentric arc is drawn with the shortened radius.

Step 2 A line parallel to 1–2 is drawn a distance of R from it to locate the center, point C.

Step 3 The tangent points are located with lines from O through C, and through C perpendicular to 1–2. Draw the tangent arc with radius R.

Fig. 12.34 An arc tangent to two arcs

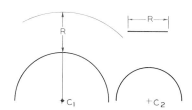 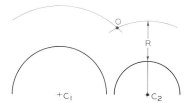 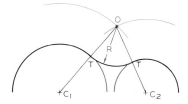

Step 1 The radius of one circle is extended and the radius R is added to it. The extended radius is used for drawing a concentric arc.

Step 2 The radius of the other circle is extended and the radius R is added to it. The extended radius is used to construct an arc and to locate point O, the center.

Step 3 The centers are connected with point O to locate the points of tangency. The arc is drawn tangent to the two arcs with radius R.

A variation of this principle of construction is shown in Fig. 12.33 where the arc is drawn parallel to an arc and line with the arc in a reverse position.

12.24 ARC TANGENT TO TWO ARCS

A third arc is drawn tangent to two given arcs in Fig. 12.34. Thin lines are drawn from the centers to locate the points of tangency. This tangent arc is concave from the top.

A convex arc can be drawn tangent to the given arcs if the radius of the arc is greater than the radius of either of the given arcs (Fig. 12.35).

A variation of this problem is shown in Fig. 12.36 where an arc of a given radius is drawn tangent to the top of one arc and the bottom of the other. The two arcs are of different sizes.

A similar problem is shown in Fig. 12.37 where an arc is drawn tangent to a circle and a larger arc.

In all cases, the points of tangency are located and marked by thin lines drawn from center to center of the tangent arcs.

Fig. 12.35 Arc tangent to two arcs

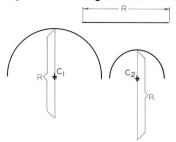

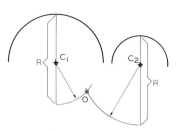

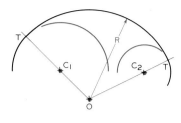

Step 1 The radius of each arc is extended from the arc past its center and the specified radius R is laid off from the arcs along these lines.

Step 2 The distance from each center to the ends of the extended radii are used for two concentric arcs to locate the center O.

Step 3 Thin lines from O through centers C_1 and C_2 locate the points of tangency. The arc is drawn using point O as the center.

Fig. 12.36 An arc tangent to two circles

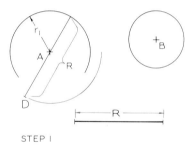

STEP 1

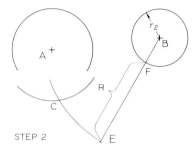

STEP 2

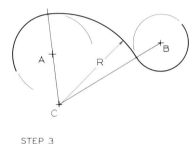

STEP 3

Step 1 The specified radius R is laid off from the arc along the extended radius to locate point D. Radius AD is used to construct a concentric arc.

Step 2 The radius through center B is extended and the radius R is added to it from point F. Radius BE is used to locate the center C.

Step 3 The tangent arc is drawn from center C with radius R. The points of tangency are located with thin lines from C through the given centers.

Fig. 12.37 Arc tangent to two arcs

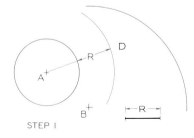

STEP 1

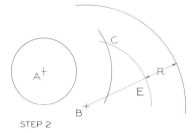

STEP 2

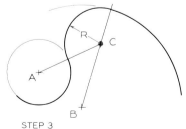

STEP 3

Step 1 Radius R is added to the radius through center A. Radius AD is used to draw a concentric arc with center A.

Step 2 Radius R is subtracted from the radius through B. Radius BE is used to construct a concentric arc and locate point C.

Step 3 The points of tangency are located with thin lines BC and AC. The tangent arc is drawn with center C.

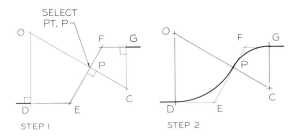

Fig. 12.38 An ogee curve

Step 1 To draw an ogee curve between two parallel lines, draw line *EF* at any angle. Locate a point of your choosing along *EF*, *P* in this case. Find the tangent points by making *FG* equal to *FP* and *DE* equal to *EP*. Draw perpendiculars at *G* and *D* to intersect the perpendicular bisector of *EF*.

Step 2 Using radii *CP* and *OP*, draw the two tangent arcs to complete the ogee curve.

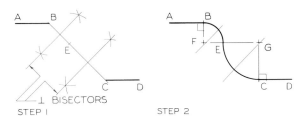

Fig. 12.40 The unequal ogee curve

Step 1 Two parallel lines are to be connected by an ogee curve that passes through *B* and *C*. Draw line *BC* and select point *E* on the line. Bisect *BE* and *EC*.

Step 2 Construct perpendiculars at *B* and *C* to intersect the bisectors to locate centers *F* and *G*. Locate the points of tangency and draw the ogee curve using radii *FB* and *GC*.

12.25 THE OGEE CURVE

The ogee curve can be thought of as a double curve formed by tangent arcs. The ogee curve in Fig. 12.38 was found by constructing two arcs tangent to three intersecting lines.

An ogee curve can be drawn between two parallel lines (Fig. 12.39) from points *B* to *C* by geometric construction.

An alternative method of drawing an ogee curve that passes through points *B*, *E*, and *C* is illustrated in Fig. 12.40. The method of drawing an ogee curve through two nonparallel lines is shown in Fig. 12.41.

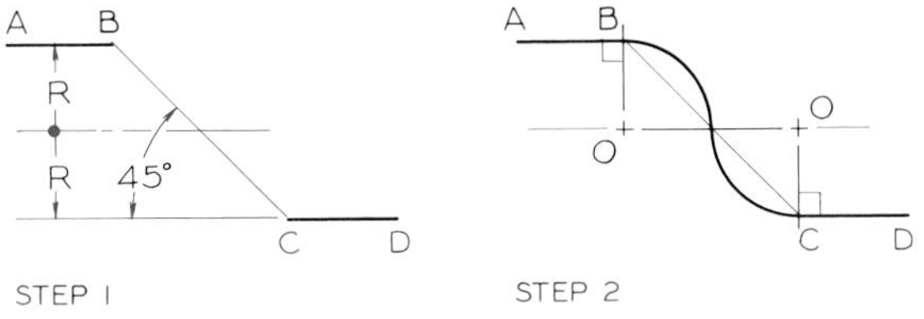

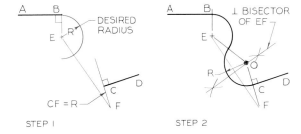

Fig. 12.41 Ogee curve between nonparallel lines

Step 1 Draw a perpendicular to *AB* at *B* and draw the arc with the desired radius from center *E*. Draw a perpendicular to line *CD* at *C* and make *CF* equal to the radius *R*. Connect *E* and *F*.

Step 2 Draw a perpendicular bisector of *EF* to locate point *O*. Use radius *OC* (which may be different from the first radius) to complete the curve. Mark points of tangency.

Fig. 12.39 An ogee curve

Step 1 To draw an ogee curve formed by two equal arcs, draw a line at 45° between the lines. Divide the distance between *AB* and *CD* in half to find the radius *R*.

Step 2 Construct perpendiculars at *B* and *C* to locate the centers along the line between *AB* and *CD*. Draw the arcs to complete the ogee curve.

12.26 A CURVE OF ARCS

An irregular curve formed with tangent arcs can be constructed as shown in Fig. 12.42. In this case the radii of the arcs are selected to give the desired curve by moving from one set of points to the next.

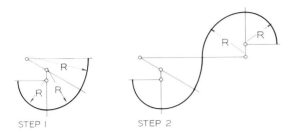

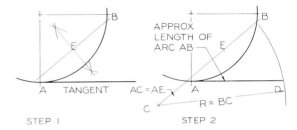

Fig. 12.42 A curve of arcs

Step 1 A series of arcs can be joined to form a smooth curve. Begin with the small arc, extend the radius through its center, draw the second and then the third.

Step 2 The curve can be reversed by extending the radius in the opposite direction and repeating the same process.

Fig. 12.44 To rectify an arc—compass method

Step 1 If you wish to rectify an arc from A to B, draw chord AB and find its midpoint.

Step 2 Extend AB to make AC equal to AE. The length of arc AB is approximated to be the distance from A to D found by swinging arc BC.

12.28 CONIC SECTIONS

Conic sections are plane figures that can be described graphically as well as mathematically. They are formed by passing imaginary cutting planes through a right cone (Fig. 12.45).

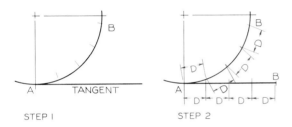

Fig. 12.43 To rectify an arc

Step 1 An arc has been rectified when its length has been laid out along a straight line. Construct a line tangent to the arc and divide the arc into a series of equal divisions from A to B.

Step 2 The chordal distances, D, along the arc are laid out along the straight line until point B is located.

12.29 THE ELLIPSE

The *ellipse* is a conic section formed by passing a plane through a right cone at an angle (Fig. 12.45). The ellipse is mathematically defined as the path of a point that moves in such a way that the sum of the distances from two focal points is a constant.

The construction of an ellipse is found by revolving the edge view of a circle as shown in Fig. 12.46. These points are connected by a smooth curve. This ellipse could have been drawn using the ellipse template as shown in Fig. 12.47. The angle between the line of sight and the edge view of the circle is the angle of the ellipse template that should be used (or the one closest to this size).

The largest diameter of an ellipse is always the true length and is called the *major diameter*. The shortest diameter is perpendicular to the major diameter and is called the *minor diameter*. The crossing diameters are used to align the ellipse template.

Ellipse templates are available in intervals of 5° and in variations in size of the major diameter of about ⅛″ (Fig. 12.48).

12.27 RECTIFYING AN ARC

An arc is rectified when its true length is laid out along a straight line. A method of rectifying an arc is illustrated in Fig. 12.43.

A second method of rectifying an arc is given in Fig. 12.44 by using a different form of geometric construction.

In addition to these methods, an arc can be rectified by using the mathematical equation for finding the circumference of the circle. Since a circle has 360°, the arc of a 30° sector is ¹⁄₁₂ of the full circumference. Therefore, if the circumference is 12 inches, the arc of 30° is equal to one inch.

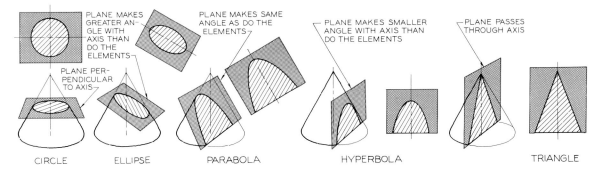

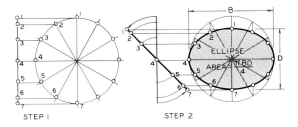

Fig. 12.45 The conic sections are formed by passing cutting planes at various angles through right cones. The conic sections are the circle, ellipse, parabola, hyperbola, and triangle.

Fig. 12.46 An ellipse by revolution

Step 1 When the edge view of a circle is perpendicular to the projectors between its adjacent view, the view will be a true circle. Mark equally spaced points along the arc and project them to the edge.

Step 2 Revolve the edge of the circle to the desired position and project the points to the circular view, which will now appear as an ellipse. Note that the points are projected vertically downward to their new positions.

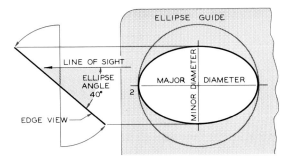

Fig. 12.47 The ellipse template

When the edge view of a circle is revolved so the line of sight between the two views is not perpendicular to the edge view, the circle will appear as an ellipse. The major diameter remains constant but the minor diameter will vary. The angle between the line of sight and the edge view of the circle will give the angle of the ellipse guide template.

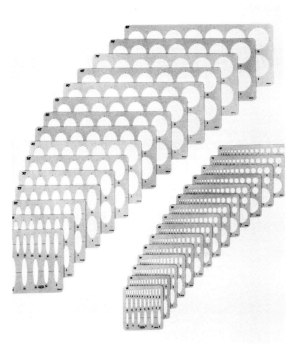

Fig. 12.48 Ellipse templates come in a variety of sizes. Most are calibrated at 5° intervals from 15° up to 60°. (Courtesy of Timely Products, Inc.)

The mathematical equation of the ellipse is

$$\frac{x^2}{a^2} + \frac{y^2}{b^2} = 1, \quad \text{where } a, b \neq 0.$$

Letters a and b are constants, and x and y are variables. This equation can be plotted on graph paper.

The ellipse can be constructed inside a rectangular box or a parallelogram, as illustrated in

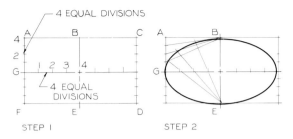

Fig. 12.49 Ellipse—parallelogram method

Step 1 An ellipse can be drawn inside of a rectangle or a parallelogram by dividing the horizontal center line into the same number of equal divisions as the shorter sides, *AF* and *CD*.

Step 2 The construction of the curve in one quadrant is shown by using sets of rays from *E* and *B* to plot the points.

Fig. 12.49, where a series of points is plotted to form an elliptical curve.

Two circles can be used for constructing an ellipse by making the diameter of the large one equal to the major diameter and the diameter of the small one equal to the minor diameter (Fig. 12.50).

The *conjugate diameters* of an ellipse are diameters that are parallel to the tangents at the ends

of each, as illustrated in Fig. 12.51. A single ellipse has an infinite number of sets of conjugate diameters.

When the ellipse and a pair of conjugate diameters are given, the major and minor diameters of the ellipse can be found by using the method illustrated in Fig. 12.52. If the conjugate diameters are not given, they can be constructed as shown in Fig. 12.51, and the major and minor diameters

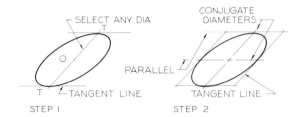

Fig. 12.51 Conjugate diameters

Step 1 Many sets of conjugate diameters can be constructed for an ellipse. A conjugate diameter is one that is parallel to the tangents at the ends of the other conjugate diameter. A diameter is selected and parallel tangents are drawn.

Step 2 The horizontal conjugate diameter is drawn parallel to the horizontal tangents, and the inclined tangents are drawn parallel to the conjugate diameter found in step 1.

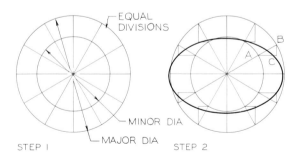

Fig. 12.50 Ellipse—circle method

Step 1 Two concentric circles are drawn with the large one equal to the major diameter and the small one equal to the minor diameter. Divide them into equal sectors.

Step 2 Plot points on the ellipse by projecting downward from the large curve to intersect horizontal construction lines drawn from the intersections on the small circle.

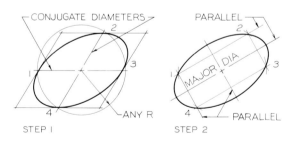

Fig. 12.52 Finding the axes of an ellipse

Step 1 When the conjugate diameters of an ellipse are given, you can find the major and minor diameters of the ellipse by drawing a circle of any radius from the intersection of the diameters.

Step 2 The circle cuts four points along the ellipse. The points are connected to form a rectangle. The major and minor diameters are parallel to the sides of the rectangle.

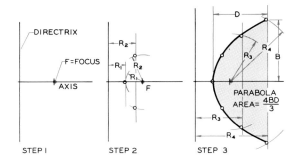

Fig. 12.53 The parabola—mathematical method

Step 1 Draw an axis perpendicular to a line called a directrix. Choose a point for the focus, F, on the axis.

Step 2 Locate points by using a series of selected radii to plot points on the curve. For example, draw a line parallel to the directrix and R_2 from it. Swing R_2 from F to intersect the line and plot the point.

Step 3 Continue the process with a series of arcs of varying radii until an adequate number of points have been found to complete the curve.

found. The major and minor diameters are necessary for using the ellipse template.

12.30 THE PARABOLA

The parabola is mathematically defined as a plane curve, each point of which is equidistant from a directrix (a straight line) and its focal point. The parabola is formed when the cutting plane makes the same angle with the base of a cone as do the elements of the cone.

The construction of a parabola by using its mathematical definition is shown in Fig. 12.53.

A parabolic curve can be constructed by geometric construction as shown in Fig. 12.54 by dividing the two perpendicular lines into the same number of divisions. The parabola is drawn through the plotted points with an irregular curve.

A third method of construction is illustrated in Fig. 12.55 using the parallelogram method.

The mathematical equation of the parabola is

$$y = ax^2 + bx + c, \qquad \text{where } a \neq 0.$$

Letters a, b, and c are constants, and x and y are variables. This equation can be written by interchanging the y's and x's. The equation can be plotted on graph paper.

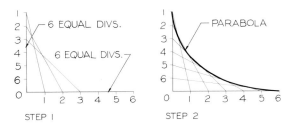

Fig. 12.54 The parabola—tangent method

Step 1 Construct two lines at a convenient angle and divide each of them into the same number of divisions. Connect the points with a series of diagonals.

Step 2 When finished, construct the parabolic curve to be tangent to the diagonals.

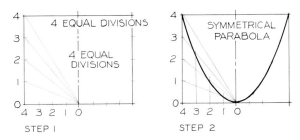

Fig. 12.55 The parabola—parallelogram method

Step 1 Construct a rectangle or a parallelogram to contain the parabola, and locate its axis parallel to the sides through O. Divide the sides into equal divisions. Connect the divisions with point O.

Step 2 Construct lines parallel to the sides (vertical in this case) to locate the points along the rays from O. Draw the parabola.

12.31 THE HYPERBOLA

The hyperbola is a two-part conic section that is mathematically defined as the path of a point that moves in such a way that the difference of its distances from two focal points is a constant. This definition is used to construct the hyperbola in Fig. 12.56.

A second method of construction is shown in Fig. 12.57. Two perpendicular lines are drawn through point B as asymptotes. The hyperbolic curve becomes more nearly parallel and closer to the asymptotes as the hyperbola is extended, but the curve never merges with the asymptotes.

Fig. 12.56 The hyperbola

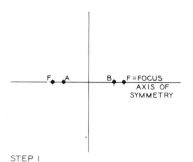

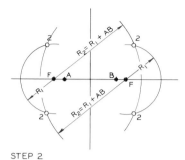

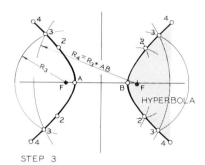

STEP I

STEP 2

STEP 3

Step 1 A perpendicular is drawn through the axis of symmetry, and focal points F are located equidistant from it on both sides. Points on the curve, A and B, are located equidistant from the perpendicular at a location of your choice but between the focal points.

Step 2 Radius R_1 is selected to draw arcs using focal points F as the centers. R_1 is added to AB (the distance between the nearest points on the hyperbolas) to find R_2. Radius R_2 is used to draw arcs using the focal points as centers. The intersections of R_1 and R_2 establish points 2 on the hyperbola.

Step 3 Other radii are selected and added to distance AB to locate additional points in the same manner as described in step 2. A smooth curve is drawn through the points to form the hyperbolic curves.

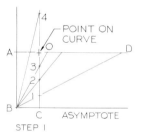

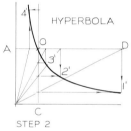

STEP I

STEP 2

Fig. 12.57 The equilateral hyperbola

Step 1 Two perpendiculars are drawn through B and any point O on the curve is located. Horizontal and vertical lines are drawn through O. Line CO is divided into equal divisions and rays from B are drawn through them to the horizontal line.

Step 2 Horizontal construction lines are drawn from the divisions along line OC, and lines from AD are projected vertically to locate points 1' through 4' on the curve.

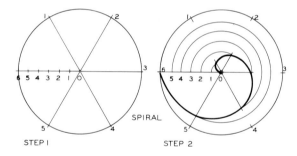

STEP I

STEP 2

Fig. 12.58 The spiral

Step 1 Draw a circle and divide it into equal parts. The radius is divided into the same number of equal parts—six in this example.

Step 2 By beginning on the inside, draw arc 0–1 to intersect radius 0–1. Then swing arc 0–2 to radius 0–2 and continue until the last point is reached at 6, which lies on the original circle.

12.32 THE SPIRAL OF ARCHIMEDES

A spiral is a coil that begins at a point and becomes larger as it travels around the origin. A spiral lies in a single plane. The steps of constructing a spiral are shown in Fig. 12.58.

12.33 THE HELIX

A helix is a curve that coils around a cylinder or a cone at a constant angle of inclination. Examples of helixes are corkscrews or threads on a screw. A helix is constructed about a cylinder in Fig. 12.59, and about a cone in Fig. 12.60.

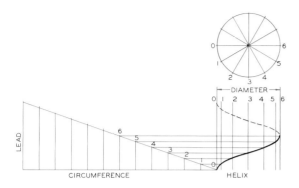

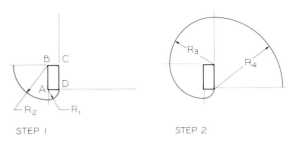

STEP 1   **STEP 2**

Fig. 12.61 The involute

Step 1 Side *AD* is used as a radius for drawing arc *AD.* *AB* is added to *AD* to form radius R_2. Draw a second arc using center *B.*

Step 2 The two remaining sides are used to unwind the involute back to its point of origin.

12.34 THE INVOLUTE

The involute is the path of the end of a line as it unwinds from a line or a plane figure. In Fig. 12.61, an involute is formed by unwinding a line from a rectangle. Successively different radii that are equal in length to the sides are used to develop the involute.

12.35 THE CYCLOID

The cycloid is a plane curve formed by a point on a circle as the circle rolls along a straight line.

    In Fig. 12.62, the distance from 1 to 9 must be equal to the circumference of the circle. The circle is located at the center point 5, and it is rolled to the left to locate points A_4, A_3, A_2, and A_1 as the center moves from O_4 to O_1. These points are connected with a smooth curve to complete the left side of the cycloid.

    The same construction is repeated as the circle moves to the right to complete the symmetrical curve.

12.36 THE EPICYCLOID AND HYPOCYCLOID

The *epicycloid* is a curve formed by a point on a circle as the circle rolls along the convex side of a larger circle (Fig. 12.63). The circumference of the rolling circle is divided into equal divisions.

Fig. 12.59 The helix

Divide the top view of the cylinder into equal divisions and project them to the front view. Lay out the circumference and the height of the cylinder, which is the lead. Divide the circumference into the same number of equal parts by taking the measurements from the top view. Project the points along the inclined rise to their respective elements to find the helix.

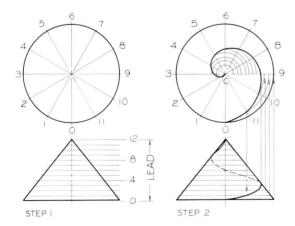

STEP 1   **STEP 2**

Fig. 12.60 A conical helix

Step 1 Divide the cone's base into equal parts. Pass a series of horizontal cutting planes through the front view of the cone. Use the same number as the divisions on the base, 12 in this case.

Step 2 Project all of the divisions along the front view of the cone to line C9, and draw a series of arcs from center C to their respective radii in the top view to plot the points. Project the points to their respective cutting planes in the front view.

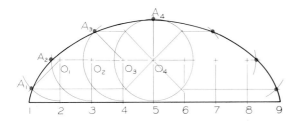

Fig. 12.62 The cycloid

Step 1 A circle is centered on a horizontal tangent line. It is divided into a number of equal divisions. The circumference of the circle is laid off along line 1–9, which is divided into the same number of parts as the circle.

Step 2 The center of the arc is moved from O_4 to O_3 to locate point A_3, then to O_2, etc. The points are connected to form the cycloidal curve.

These same units are measured along the circumference of the large arc. These will be the contact points as the circle rolls along the arc. The plotted points are connected to form the epicycloid.

The *hypocycloid* is a curve formed by a point on a circle as the circle rolls along the concave side of a larger arc. As in the epicycloid, the circumference of the small circle is divided into equal divisions that are laid off along the larger arc. The curve is plotted and drawn as a smooth curve as the circle is rolled left and right.

The elliptical-appearing curve in step 3 of Fig. 12.63 is a curve formed by connecting the epicycloid and hypocycloid curves by rolling a circle of the same size on each side of a given arc.

Fig. 12.63 The hypocycloid and epicycloid

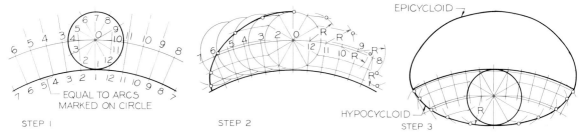

Step 1 Divide the circle into a number of equal parts. Measure the same lengths along the large arc. Locate the positions of the center O along an arc drawn through the center O.

Step 2 Move the circle to the positions 1 through 6 to find the points along the epicycloid. Repeat this process at the right side and draw the curve.

Step 3 The hypocycloid is found in the same manner, but the circle rolls along the inside of the large arc. This figure shows both the epicycloid and the hypocycloid drawn together.

PROBLEMS

The following problems are similar to those solved as examples in the chapter. These are to be solved on Size AV paper similar to the one shown in Fig. 12.64 where problems 1–5 are laid out. Each inch on the grid is equal to 0.20 inches; therefore, use your engineers' 10 scale to lay out the problems. By equating each grid to 5 mm, you can use your full-size metric scale to lay out and solve the problems.

Show your construction and mark all points of tangency, as discussed in the chapter.

1. Draw triangle *ABC* using the given sides.

2–3. Inscribe an angle in the semicircles with the vertexes at point *P*.

4. Inscribe a nine-sided regular polygon inside the circle.

5. Circumscribe a ten-sided regular polygon about the circle.

6. Circumscribe a hexagon about the circle.

7. Inscribe an octagon in the circle.

8. Circumscribe an octagon about the circle.

9. Construct a pentagon inside the circle using the compass method.

10–11. Bisect the lines.

12–13. Bisect the angle.

14. Rotate the triangle 80° in a clockwise direction about point *A*.

15. Enlarge the given shape to the size indicated by the diagonal.

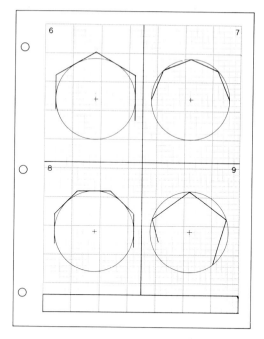

Fig. 12.65 Problems 6–9: Construction of regular polygons.

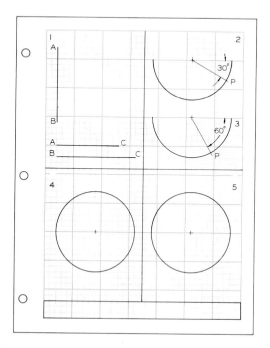

Fig. 12.64 Problems 1–5: Basic constructions.

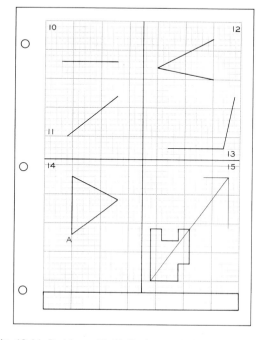

Fig. 12.66 Problems 10–15: Basic constructions.

16. Divide *AB* into seven equal parts. Draw the construction line through *B* for your construction.

17. Divide the two vertical lines into four equal divisions. Draw three equally spaced vertical lines at the divisions.

18. Construct an arc with radius *R* that is tangent to the line at *J* and passes through point *P*.

19. Construct an arc with radius *R* that is tangent to the line and passes through *P*.

20. Construct a line from *P* that is tangent to the semicircle. Locate the points of tangency. Use the compass method.

21–23. Construct arcs with the given radii tangent to the lines.

24–27. Construct arcs that are tangent to the arcs and/or lines. The radii are given for each problem.

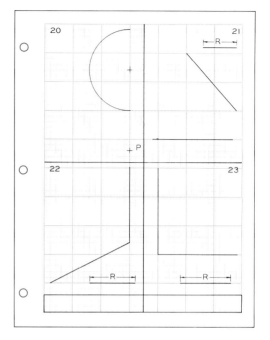

Fig. 12.68 Problems 20–23: Tangency construction.

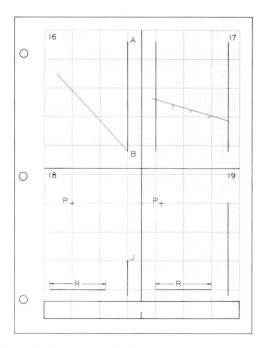

Fig. 12.67 Problems 16–19: Tangency construction.

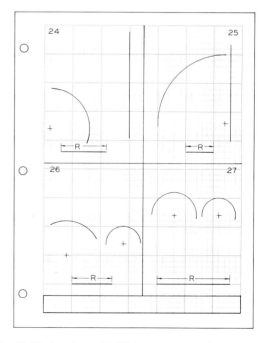

Fig. 12.69 Problems 24–27: Tangency construction.

28–31. Construct ogee curves that connect the ends of the given lines and pass through points *P* where given. In Problem 31, the radii for the arcs are given.

32–33. Using the given radii, connect the given arcs with a tangent arc as indicated in the freehand sketches.

34. Rectify the arc along the given line by dividing the circumference into equal divisions and laying them off with your dividers.

35. Rectify the arc by using the compass method as shown in Fig. 12.44.

36. Construct an ellipse inside the rectangular layout.

37. Construct an ellipse inside the large circle. The small circle represents the minor diameter.

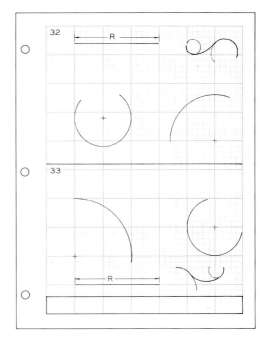

Fig. 12.71 Problems 32–33: Tangency construction.

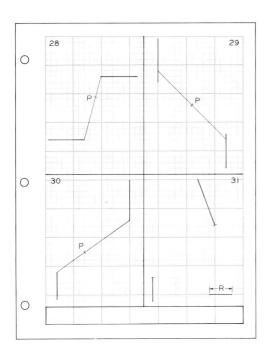

Fig. 12.70 Problems 28–31: Ogee curve construction.

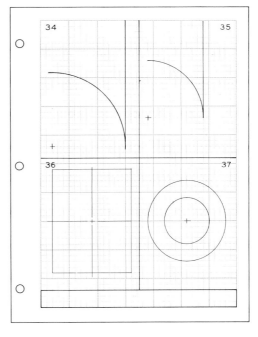

Fig. 12.72 Problems 34–37: Rectifying an arc, ellipse construction.

38. Construct an ellipse inside the circle when the edge view has been rotated 45° as shown.

39. Using the focal point, F, and the directrix, plot and draw the parabola formed by these elements.

40. Construct a parabola using perpendicular lines by either of the methods shown in Figs. 12.54 and 12.55.

41. Using the focal points, F, points A and B on the curve, and the axis of symmetry, construct the hyperbolic curve.

42. Construct a hyperbola that passes through O. The perpendicular lines are asymptotes.

43. Construct a spiral of Archimedes by using the four divisions that are marked along the radius.

44–45. Construct a helix that has a rise equal to the heights of the cylinder and cone. Show construction and the curve in all views.

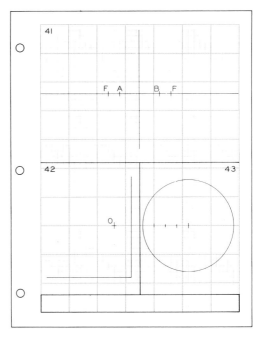

Fig. 12.74 Problems 41–43: Hyperbola and spiral construction.

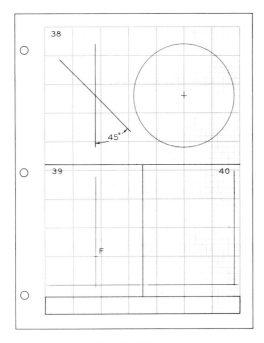

Fig. 12.73 Problems 38–40: Ellipse and parabola construction.

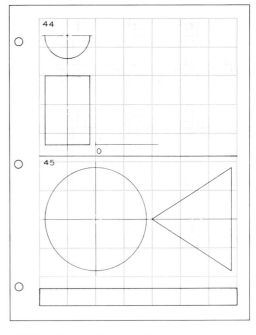

Fig. 12.75 Problems 44–45: Helix construction.

46–47. Construct involutes by unwinding the triangle and rectangle in a clockwise direction, beginning with point A in each.

48. Construct an epicycloid by rolling the circle along the curve whose center is at point C.

49. Construct a cycloid by rolling the circle about the horizontal line.

50–62. Construct these problems (Figs. 12.77–12.89) on Size A sheets, one problem per sheet. Select the proper scale that will best fit the problem to the sheet. Mark all points of tangency and strive for good line quality.

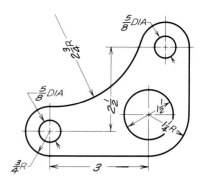

Fig. 12.78 Problem 51: Lever crank.

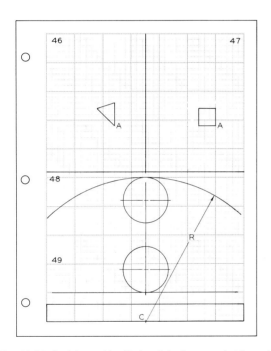

Fig. 12.76 Problems 46–49: Involute, hypocycloid, and epicycloid construction.

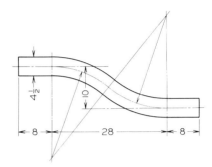

Fig. 12.79 Problem 52: Road tangency.

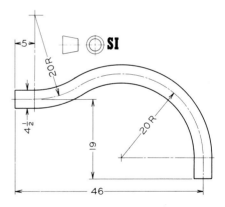

Fig. 12.80 Problem 53: Road tangency.

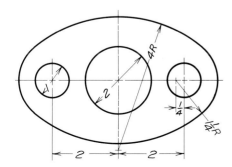

Fig. 12.77 Problem 50: Gasket.

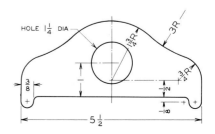

Fig. 12.81 Problem 54: Gasket.

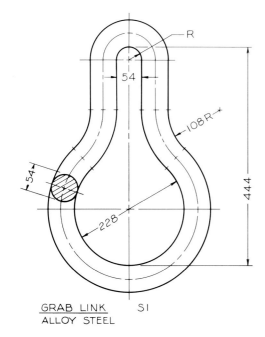

GRAB LINK   SI
ALLOY STEEL

Fig. 12.82 Problem 55: Grab link.

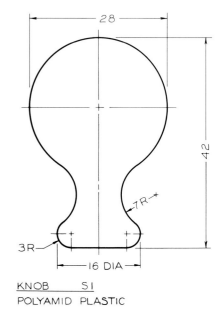

KNOB   SI
POLYAMID PLASTIC

Fig. 12.84 Problem 57: Knob.

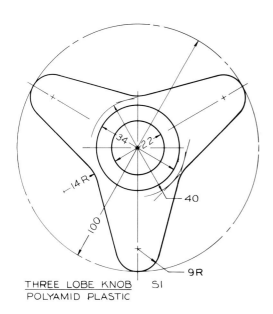

THREE LOBE KNOB   SI
POLYAMID PLASTIC

Fig. 12.83 Problem 56: Three-lobe knob.

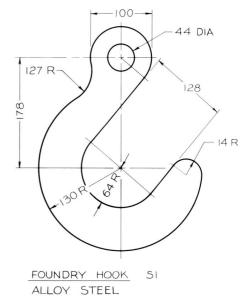

FOUNDRY HOOK   SI
ALLOY STEEL

Fig. 12.85 Problem 58: Foundry hook.

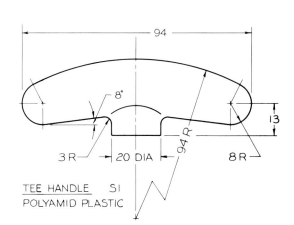

Fig. 12.86 Problem 59: Tee handle.

94

8°

3 R — 20 DIA — 94 R — 8 R

13

TEE HANDLE   SI
POLYAMID PLASTIC

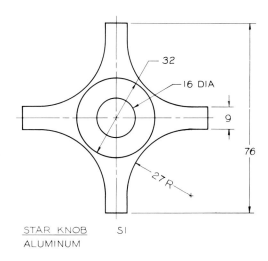

Fig. 12.88 Problem 61: Star knob.

32

16 DIA

9

76

27 R

STAR KNOB   SI
ALUMINUM

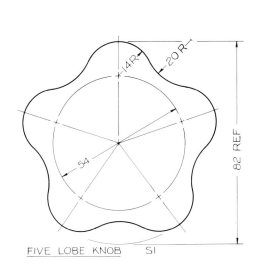

Fig. 12.87 Problem 60: Five-lobe knob.

14 R   20 R

54

82 REF

FIVE LOBE KNOB   SI

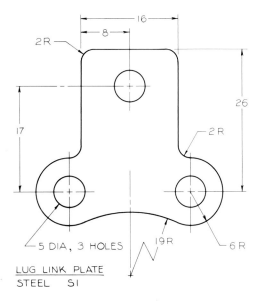

Fig. 12.89 Problem 62: Lug link plate.

16

8

2R

17

26

2 R

5 DIA, 3 HOLES   19 R   6 R

LUG LINK PLATE
STEEL   SI

13

MULTIVIEW SKETCHING

13.1 THE PURPOSE OF SKETCHING

Sketching is a thinking process as well as a technique of communication. Designers must develop their ideas by making many sketches, revising them, and finally arriving at the desired solution. Later, these sketches are converted into instrument drawings.

Sketching should be a rapid method of drawing. If speed is not developed with a degree of skill in sketching, this technique is not effective. If you develop your sketching skills, you can assign drafting work to assistants who can then prepare the finished drawings by working from your sketches. If your sketches are not sufficiently clear to communicate your ideas to someone else, then it is likely that you have not thought out the solution well enough, even for your own understanding. The ability to communicate by any means is a great asset, and sketching is one of the more powerful techniques of communication.

This chapter will introduce you to sketching techniques, and methods of developing this ability.

13.2 SHAPE DESCRIPTION

A pictorial of an object is shown in Fig. 13.1 with three arrows that indicate the directions of sight that will give top, front, and right-side views of the object. Each view is a two-dimensional view rather than a three-dimensional pictorial.

This is called *orthographic projection* or *multiview projection*. In multiview projection, it is important that the views be located as shown in Fig. 13.1. The top view is placed over the front view, since both views share the dimension of width. The side view is placed to the right of the front view, since these views share the dimension of height. The distance between the views can vary, but they must be positioned so that the views project from each other as shown here.

Several examples of poorly arranged views are shown in Fig. 13.2. You can see that although

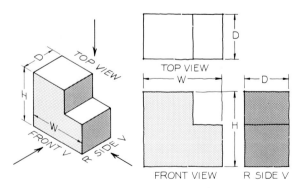

Fig. 13.1 Three views of an object can be found by looking at the object in this manner. The three views— the top, front, and right side—describe the object.

these views are correct, they are hard to interpret because the views are not placed in their standard positions.

13.3 SIX-VIEW DRAWINGS

Six principal views may be found for any object by using the rules of orthographic or multiview projection. This is the maximum number of principal views.

The directions of sight for the six views are shown in Fig. 13.3 where the views are drawn in their standard positions as illustrated. The width dimension is common to the top, front, and bottom views. Height is common to the right-side, front, left-side, and rear views.

Seldom will an object be so complex as to require six orthographic views, but if six views are needed, they should be arranged as shown in this example.

13.4 SKETCHING TECHNIQUES

Sketching means freehand drawing without the use of instruments or straightedges. The best pen-cil grades for sketching are medium weight pencils such as H, F, or HB grades. The standard lines used in multiview drawing and their respective line weights are shown in Fig. 13.4. Using the correct line weight improves the readability of a drawing.

You will be able to use the same grade of pencil for all lines when you are sketching, by sharpening the pencil point to match the desired line width. The different point sizes are shown in Fig. 13.5. A line that is drawn freehand should have a freehand appearance; no attempt should be made to give the line the appearance of one drawn by instruments, since these two techniques of drawing are completely different.

Sketching technique can be improved by using a printed grid on sketching paper, or by overlaying a printed grid with translucent tracing paper (Fig. 13.6) so the grid can be seen through the paper.

When a freehand sketch is made, some lines will be vertical, others horizontal or angular. If you do not tape your drawing to the table top, you will be able to position the sheet for the most comfortable strokes, which are (for the right-handed drafter) left to right (Fig. 13.7).

Fig. 13.2 Arrangement of views

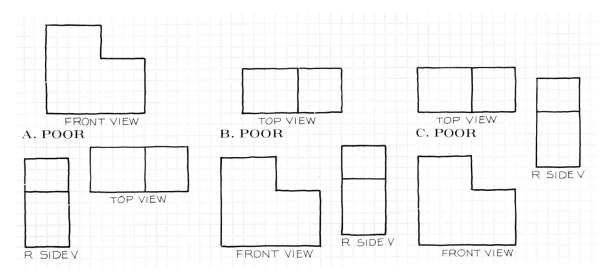

A. These views are sketched incorrectly. The views are scrambled.

B. These views are nearly correct, but they do not project from view to view.

C. The top and front views are correctly positioned, but the right-side view is incorrectly positioned.

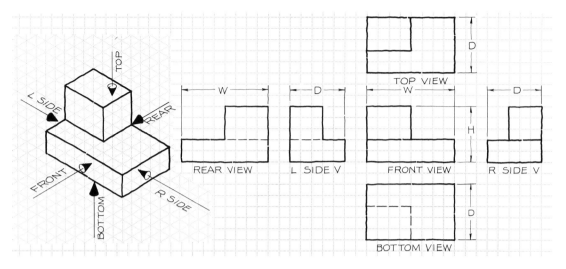

Fig. 13.3 Six principal views can be sketched by looking at the object in the directions indicated by the lines of sight. Note how the dimensions are placed on the views. Height (*H*) is shared by all four of the horizontally positioned views.

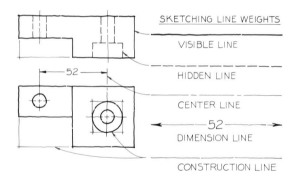

Fig. 13.4 These lines are examples of those that you should sketch with an F or an HB pencil when drawing views of an object. Note that some lines are thin and others are wider, but all are black except construction lines.

The lines that are sketched to form the various views should intersect as indicated in Fig. 13.8 for the best effect.

13.5 THE THREE-VIEW SKETCH

The steps of drawing three orthographic views on a printed grid are shown in Fig. 13.9. The most commonly used combination of views are the front, top, and right-side views as in this example. The overall dimensions of the object are sketched in Step 1. The slanted surface is drawn in the top view and projected to the other views. The final lines are darkened, the views labeled, and the

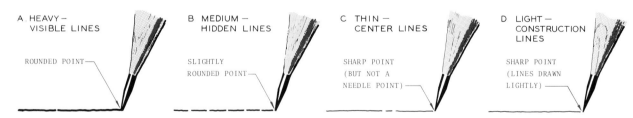

Fig. 13.5 The alphabet of lines that are sketched freehand and are all made with the same pencil grade (F or HB). The variation in the lines is achieved by varying the sharpness of the pencil point.

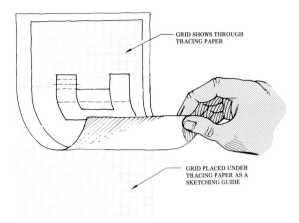

GRID SHOWS THROUGH
TRACING PAPER

GRID PLACED UNDER
TRACING PAPER AS A
SKETCHING GUIDE

Fig. 13.6 A grid can be placed under a sheet of tracing paper to aid you in freehand sketching. The grid can be used as guidelines for your sketching.

Fig. 13.7 Freehand sketching techniques

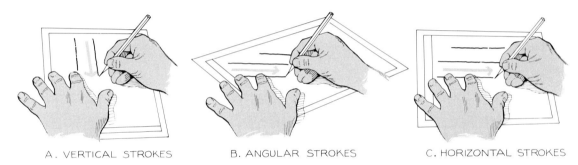

A. VERTICAL STROKES

B. ANGULAR STROKES

C. HORIZONTAL STROKES

A. Vertical lines should be sketched in a downward direction.

B. Angular strokes can be sketched left to right, if you rotate your sheet slightly.

C. Horizontal strokes are made best in a left-to-right direction. Always sketch from a comfortable position, and turn your paper if necessary.

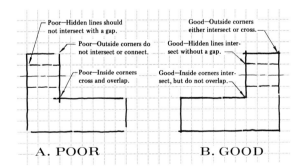

Poor—Hidden lines should not intersect with a gap.

Poor—Outside corners do not intersect or connect.

Poor—Inside corners cross and overlap.

Good—Outside corners either intersect or cross.

Good—Hidden lines intersect without a gap.

Good—Inside corners intersect, but do not overlap.

A. POOR

B. GOOD

Fig. 13.8 For good sketches, follow these examples of good technique. Compare the good drawing (B) with the drawing made when these rules were not followed (A).

overall dimensions of height, width, and depth are applied to the views.

When surfaces are slanted, they will not appear true shape in the principal views of orthographic projection (Fig. 13.10). Surfaces that do not appear true size are either *foreshortened*, or they appear as *edges*. In Fig. 13.10C, two planes of the object are slanted, thus both of them appear foreshortened in the right-side view.

A good exercise for analyzing the given views is to find the missing view when two views are given (Fig. 13.11). In Step 1, the front view can be blocked in by projecting from the top and side views. In Step 2, the notch is located, and in Step 3 the final view is completed and darkened to match the other views.

Fig. 13.9 Three-view sketching

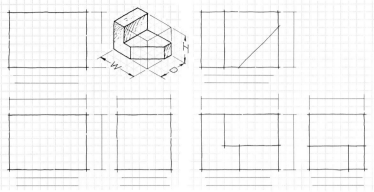
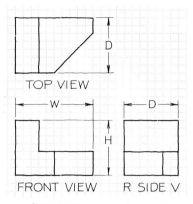

Step 1 Block in the views by using the overall dimensions. Allow proper spacing for labeling and dimensioning the views.

Step 2 Remove the notches and project from view to view as shown.

Step 3 Check your layout for correctness; darken the lines and complete the labels and dimensions.

Fig. 13.10 Views of planes

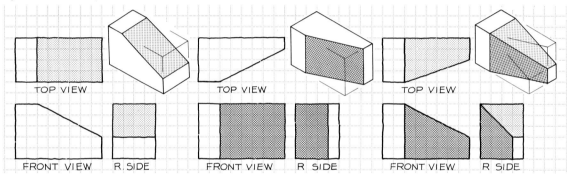

A. The plane appears as an edge in the front view and it is foreshortened in the top and side views.

B. The plane is an edge in the top view and it is foreshortened in the front and side views.

C. These two planes appear foreshortened in the right-side view. Each appears as an edge in either the top or front views.

Fig. 13.11 Missing views

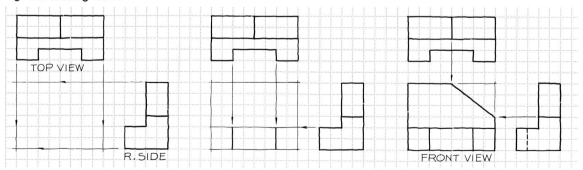

Step 1 When sketches of two views are given and the third is required, begin by projecting the overall dimensions from the top and right-side views.

Step 2 The various features of the object are sketched using construction lines.

Step 3 The features are completed, the view is checked for correctness, and the lines are darkened to the proper line quality.

The right-side view is found in Fig. 13.12 where the top and front views are given. The right-side view has the depth dimension in common with the top view, and height in common with the front view. Knowing this enables us to block in the side view in Step 1. The side view is developed in Step 2, and is completed in Step 3. A pictorial of the part is shown in Step 3 to help you visualize the object. It is better if you learn to analyze the views together rather than having to rely upon a pictorial prepared by someone else.

Another exercise is that of completing the views when some or all of them have missing lines (Fig. 13.13). A pictorial is provided to help you analyze the given views. Remember that depth is common to the top and side views as shown in Step 2.

13.6 CIRCULAR FEATURES

The circle is often used in the design of a part. For example, the top view of a drilled hole will be a circle, which means that it is a cylinder.

Fig. 13.12 Missing views

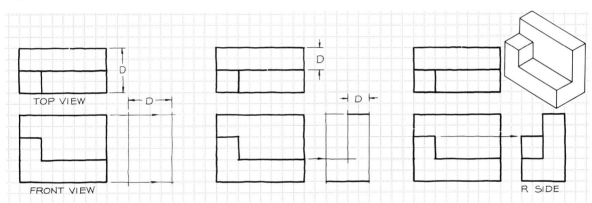

Step 1 To find the right-side view when the top and front views are given, block in the view with the overall dimensions.

Step 2 Develop the features of the view by analyzing the views together. Use light construction lines.

Step 3 Check the view for correctness and darken the lines to their proper line weight.

Fig. 13.13 Missing lines

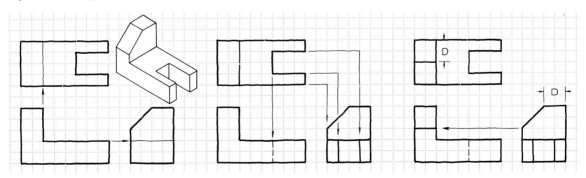

Step 1 Lines may be missing in all views in this type of problem. The first missing line is found by projecting the edges of the planes from the front to the top and side views.

Step 2 The notch in the top view is projected to the front and side views. The line in the front view is a hidden line.

Step 3 The line formed by the beveled surface is found in the front view by projecting from the side view.

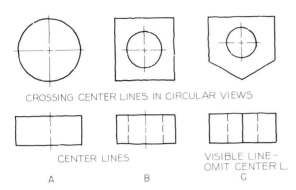

CROSSING CENTER LINES IN CIRCULAR VIEWS

CENTER LINES

A   B

VISIBLE LINE—
OMIT CENTER L.
C

Fig. 13.14 Center lines are used to indicate the centers of circles and the axes of cylinders. These are drawn as very thin lines. When they coincide with visible or hidden lines, center lines are omitted.

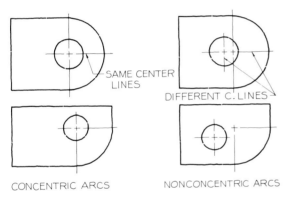

SAME CENTER LINES

DIFFERENT C. LINES

CONCENTRIC ARCS   NONCONCENTRIC ARCS

Fig. 13.15 The center line should extend beyond the last arc that has the same center. When the arcs are not concentric, separate center lines should be drawn.

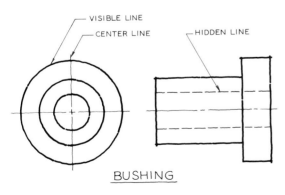

VISIBLE LINE

CENTER LINE

HIDDEN LINE

BUSHING

Fig. 13.16 Here you can see the application of center lines of concentric cylinders, and the relative weight of hidden, visible, and center lines.

The lines that are used with circles and cylinders to indicate that the features are true circles or cylinders are called *centerlines*. Examples of these are shown in Fig. 13.14.

Centerlines cross in the circular views to indicate the position of the center of the circle. Centerlines are thin lines with short dashes spaced at intervals about every inch along the line. The short dashes should cross in the circular views. Refer to Fig. 13.4 for several examples of center lines applied to a drawing.

If a centerline coincides with an object line—visible or hidden—the centerline should be omitted since the object lines are more important (Fig. 13.14C).

The application of centerlines is shown in Fig. 13.15 where they indicate whether or not circles and arcs are concentric (share the same centers). The centerline should extend beyond the arc by about one-eighth of an inch. The correct manner of applying centerlines is shown in Fig. 13.16. The circular view clearly indicates that the cylinders are concentric since each shares the same centerlines.

Sketching Circles

Circles can be sketched by using light guidelines in conjunction with centerlines (Fig. 13.17). The arcs are drawn using the guidelines and a series of short arcs. It is difficult to draw a freehand circle in one continuous line. Arcs of less than a full cir-

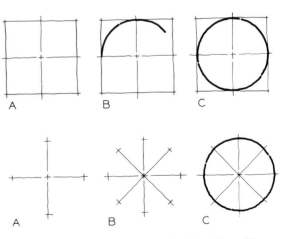

A   B   C

A   B   C

Fig. 13.17 Circles can be sketched using either of the construction methods shown here. The use of guidelines is essential to freehand sketching of circles and arcs.

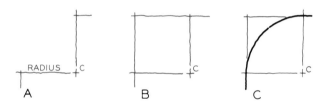

Fig. 13.18 Partial circles (arcs) should be drawn using guidelines.

Fig. 13.19 Circular features in orthographic views

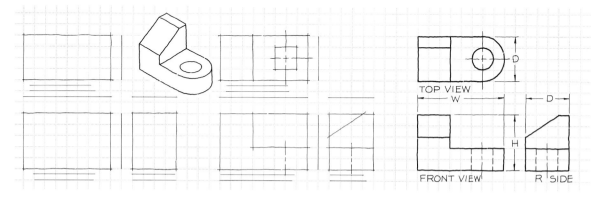

Step 1 To draw orthographic views of the object shown in this pictorial, begin by blocking in the overall dimensions. Leave room for the labels and dimensions.

Step 2 Construct the center lines and the squares about the center lines in which the circles will be drawn. Show the slanted surface in the side view.

Step 3 Sketch the arcs, and darken the final lines of the views. Label the views and show the dimensions of W, D, and H.

cle are drawn as shown in Fig. 13.18 by using light guidelines and centerlines.

If you fail to become reasonably skilled at sketching circles, use a circle template to draw the circle or the arc lightly, then darken the line freehand to match the other lines of your sketch. A compass can be used in the same manner, and the lines darkened freehand.

The construction of three orthographic views with circular features is shown in Fig. 13.19. Note that the circular features are located with centerlines and light guidelines in Step 2; and then they are sketched in and darkened in Step 3.

A similar example is given in Fig. 13.20 where the part is composed of circular features and arcs. The circles should be drawn first so that their corresponding rectangular views (such as the hidden hole in the top view) can be found by projecting from the circular view.

13.7 ISOMETRIC SKETCHING

Another type of pictorial is the *isometric drawing*, which may be drawn on a specially printed grid. You will notice that the grid in Fig. 13.21 is composed of a series of lines making 120° angles with each other to form the axes for drawing the pictorials.

The squares in the orthographic views can be laid off along the isometric grids as shown in Step 1. The notch is located in the same manner in Steps 2 and 3 to complete the isometric pictorial.

Isometric pictorials are helpful in communicating an idea in three dimensions rather than by a series of two-dimensional orthographic views.

Angles cannot be measured with a protractor in isometric pictorials; they must be drawn by measuring coordinates along the three axes of the printed grid. In Fig. 13.22, the sloping surface is

Fig. 13.20 Circular features in orthographic views

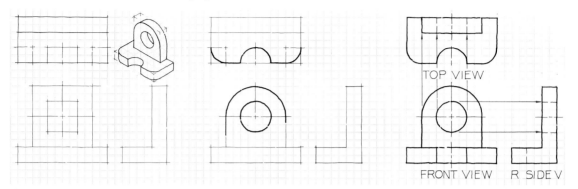

Step 1 When sketching orthographic views with circular features, you should begin by sketching the center lines and guidelines first.

Step 2 Using the guidelines, sketch the circular features. These can be darkened as they are drawn if they will be final lines.

Step 3 The corresponding outlines of the circular features are found by projecting from the views found in step 2. All of the lines are darkened.

Fig. 13.21 Isometric pictorial sketching

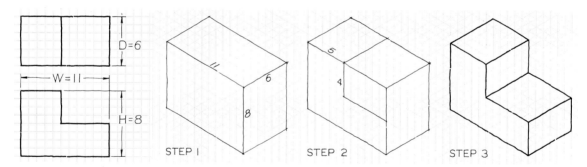

Step 1 When orthographic views of a part are given, an isometric pictorial can be sketched by using a printed isometric grid. Begin by constructing a box using the overall dimensions from the given views.

Step 2 The notch can be located by measuring over 5 squares as shown in the orthographic views. The notch is measured 4 squares downward by counting the units.

Step 3 The pictorial is completed and the lines are darkened. This is a three-dimensional pictorial, whereas the orthographic views are each two-dimensional views.

Fig. 13.22 Angles in isometric pictorials   225

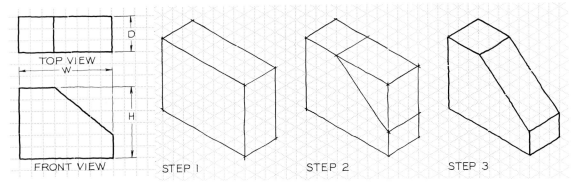

Step 1 Begin by drawing a box using the overall dimensions given in the orthographic views. Count the squares and transfer them to the isometric grid.

Step 2 Angles cannot be measured with a protractor. Angles will be either larger or smaller than their true measurement. Find each end of the angle by measuring along the axes.

Step 3 The extreme ends of the angles are connected to give the pictorial view of the object with a slanting surface. Note that dimensions can be measured only in directions parallel to the three axes.

Fig. 13.23 Double angles in isometric pictorials

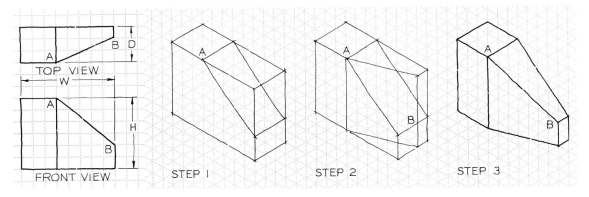

Step 1 When part of an object has a double angle, begin by constructing the overall box and finding one of the angles.

Step 2 Find the second angle that locates point B. Point A will connect to point B to give the intersection line.

Step 3 The final lines are darkened. Line AB is the line of intersection between the two sloping planes.

located by measuring in the direction of width and height to locate the ends of the angular slope.

When an object has two sloping planes that intersect (Fig. 13.23), it is necessary to draw the sloping planes one at a time to find point B. The line from A to B is the line of intersection between the two planes.

Circles in Isometric Pictorials

Circles will appear as ellipses in isometric pictorials. These can be sketched by locating their cen-

terlines as shown in Step 1 of Fig. 13.24. The center must be located equidistant from the top, bottom, and end of the front view (Step 1).

The circles are blocked in with guidelines, which will appear as rhombuses. The approximate elliptical views of the circular features will pass through the points where the centerlines intersect the guidelines (Step 2). The rear of the object is found in the same manner to complete the pictorial in Step 3.

The steps of constructing elliptical views of circles by two methods are shown in Fig. 13.25.

Fig. 13.24 Circles in isometric pictorials

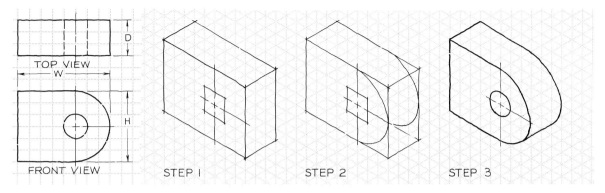

TOP VIEW
W

FRONT VIEW
D
H

STEP 1   STEP 2   STEP 3

Step 1 Begin by constructing a box using the overall dimensions given orthographically and omit the arcs and circles. Draw the center lines and a square (rhombus) of guidelines around the circular hole.

Step 2 Draw the pictorial views of the arcs tangent to the boxes formed by the guidelines. These arcs will appear as ellipses instead of circular arcs.

Step 3 Construct the small hole and darken the lines. Hidden lines are normally omitted in pictorial sketches.

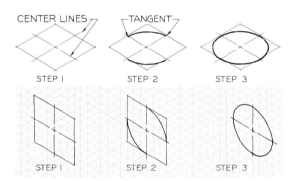

CENTER LINES   TANGENT

STEP 1   STEP 2   STEP 3

STEP 1   STEP 2   STEP 3

Fig. 13.25 Sketching ellipses

Two methods of sketching ellipses are shown: one without a grid, and one with a grid. When a grid is not used, the center lines are drawn at 30° to the horizontal (for a horizontal circle). A rhombus is drawn about them and the ellipse is sketched inside the guidelines. When there is a grid, the same technique is used except that the lines of the grid become the guidelines.

The first method is sketched without the use of guidelines and the second method utilizes guidelines. The ellipses are sketched tangent to the sides of the rhombus formed by the guidelines or the grid.

Techniques of sketching circular and cylindrical shapes are shown in Fig. 13.26. When drawing cylinders, the outside elements are drawn parallel to the axis of the cylinder and tangent to the elliptical ends of the cylinder.

A complex part composed of a number of cylindrical forms is shown in Fig. 13.27a. These three views are used as the basis for an isometric pictorial shown in the steps of construction. When an isometric grid is not used, the axes of the isometric sketch are positioned 120° apart. In other words, the height dimension is vertical, and the width and depth dimensions make 30° angles with the horizontal direction on your paper.

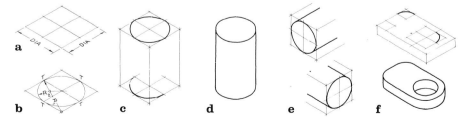

a

b

c   d   e   f

Fig. 13.26 Sketching circular features

A variety of examples of sketching circles and cylinders are shown here. Note that in all cases, guidelines are used to aid in proportional sketching.

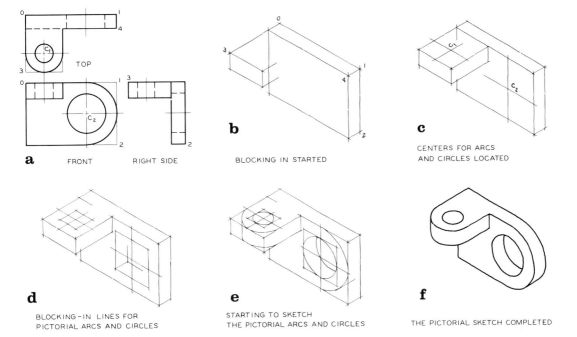

a   FRONT   RIGHT SIDE   **b**   BLOCKING IN STARTED   **c**   CENTERS FOR ARCS AND CIRCLES LOCATED

d   BLOCKING-IN LINES FOR PICTORIAL ARCS AND CIRCLES   **e**   STARTING TO SKETCH THE PICTORIAL ARCS AND CIRCLES   **f**   THE PICTORIAL SKETCH COMPLETED

Fig. 13.27 Sketching an object isometrically

The five steps of constructing an isometric pictorial of a complex part are shown here. Since a grid is not given, the guidelines are drawn vertical and at 30° to the horizontal. Center lines are used for locating the circular features at c. The object is developed at d and e and darkened at f.

PROBLEMS

These sketching problems should be drawn on Size A (8½″ × 11″) paper with or without a printed grid. A typical format for this size sheet is shown in Fig. 13.28 where a one-quarter–inch grid is given. (This grid can be converted to an approximate metric grid by equating each square to five millimeters.) All sketches and lettering should be neatly executed by applying the principles covered in this chapter. Figures 13.29–13.31 contain the problems and instructions.

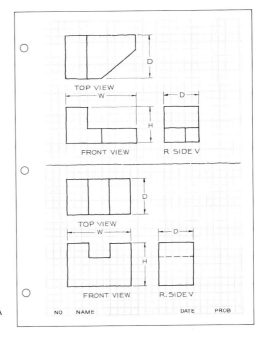

Fig. 13.28 The layout of a Size A sheet for sketching problems.

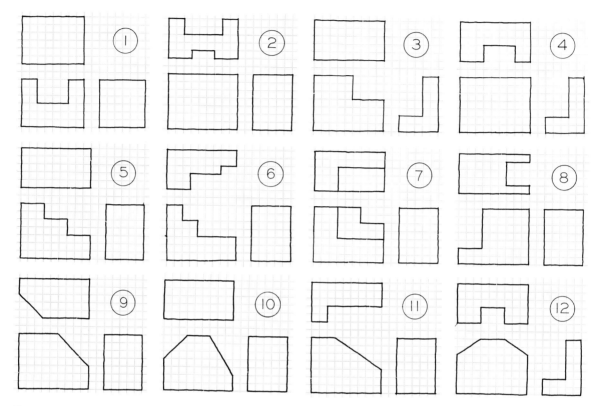

Fig. 13.29 On Size A paper sketch top, front, and right-side views of the problems assigned. Two problems can be drawn on each sheet. Give the overall dimensions of *W, D,* and *H* and label each view.

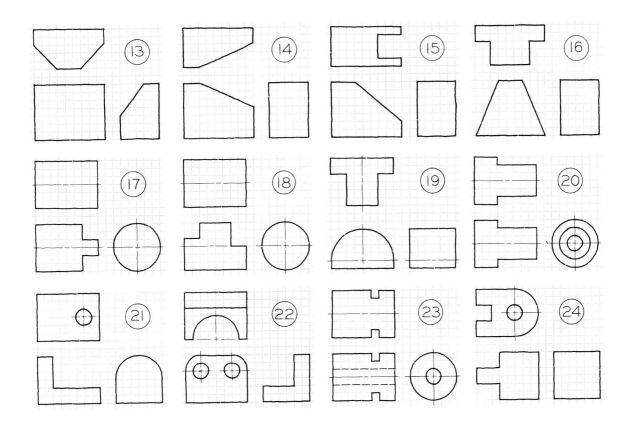

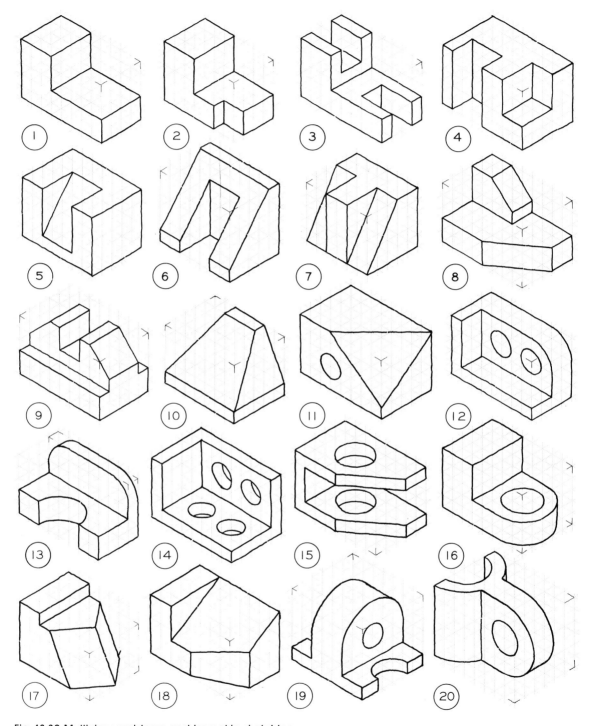

Fig. 13.30 Multiview problems and isometric sketching

On Size A paper, sketch the top, front, and right-side views of the problems assigned. Supply the lines that may be missing from all views. Then sketch isometric pictorials of the object assigned, two per sheet.

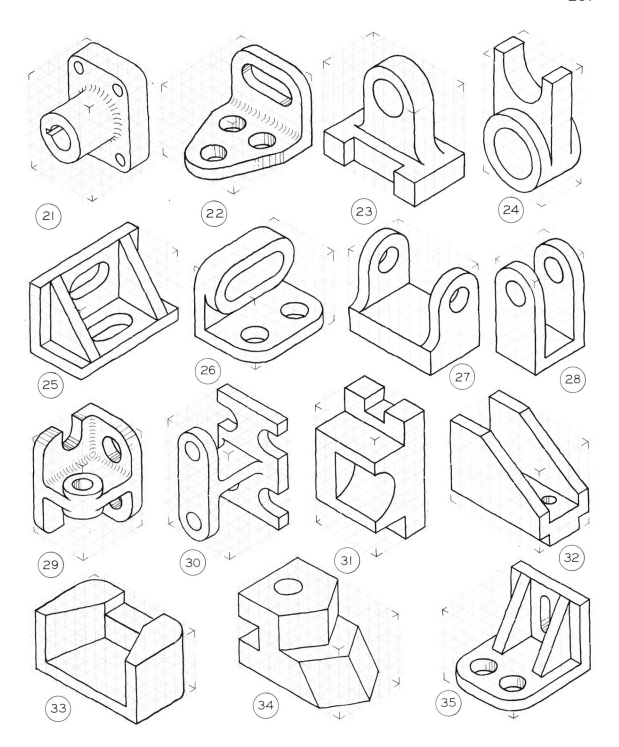

21

22

23

24

25

26

27

28

29

30

31

32

33

34

35

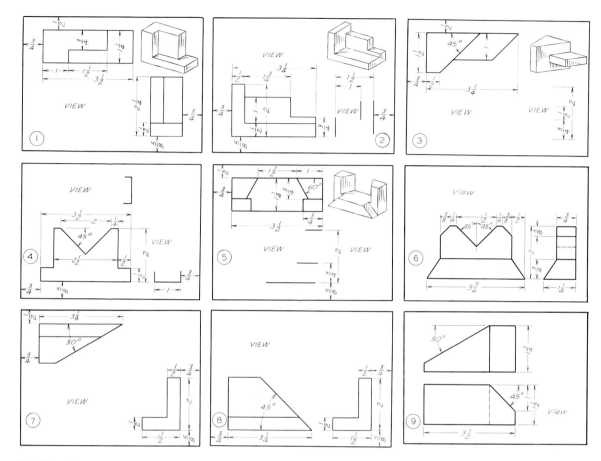

Fig. 13.31 Multiview problems and isometric pictorials

Sketch the problems assigned by your instructor, two per sheet on Size A paper. Use the given dimensions to locate and draw the views. Draw the missing views. Then sketch isometric pictorials of the objects assigned, two per sheet on Size A paper.

14

MULTIVIEW DRAWING WITH INSTRUMENTS

14.1 INTRODUCTION

Multiview drawing is the system of representing three-dimensional objects on a sheet of paper by separate views arranged in a standard manner that is familiar to the members of the engineering and technological fields. These drawings are usually executed with the instruments and drafting aids that were discussed in Chapter 2; consequently, these drawings are often referred to as **mechanical drawings.** When dimensions and notes have been added to complete the specifications of the parts that have been drawn, they are called **working drawings** or **detail drawings.**

This system of constructing multiview drawings is called **orthographic projection.** This method has evolved over the years into a system that, when the rules of the system are followed, is readily understood by the technical community. This chapter will cover the rules and conventional practices of orthographic drawing, which is truly the language of the engineer.

14.2 ORTHOGRAPHIC PROJECTION

Whereas the artist is likely to draw pictorials to represent objects in an impressionistic manner, drafters must be more precise and give more attention to detail in order for their drawings to be un-

derstood clearly. The method of preparing this type of drawing is *orthographic projection* or *multiview drawing.*

In this system, the views of an object are projected perpendicularly onto projection planes with parallel projectors. The process of finding one orthographic view, the front view, is illustrated in Fig. 14.1. The front view is projected on a vertical frontal plane with parallel projectors. The resulting front view is two dimensional since it has no depth and lies in a single plane described by two dimensions, width and height.

The top view of the same object is projected onto a horizontal projection plane that is perpendicular to the frontal projection plane in Fig. 14.2A. The right-side view is projected onto a vertical profile plane that is perpendicular to both the horizontal and frontal planes (Fig. 14.2B).

Imagine that the same object has been enclosed in a glass box composed of the frontal, horizontal, and profile projection planes. While in the glass box, the views are projected onto the projection planes (Fig. 14.3), and then the box is opened into the plane of the drawing surface. This gives the standard positions for the three orthographic views.

A similar example of the same principle is the object in the projection box in Fig. 14.4. Again, the three views are positioned in the same manner: the top view over the front, and the right-side view to the right of the front view.

Fig. 14.1 Orthographic projection

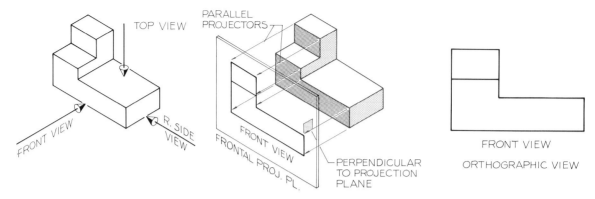

Step 1 Three mutually perpendicular lines of sight are drawn to obtain three views of the object.

Step 2 The frontal plane is a vertical plane on which the front view is projected with parallel projectors that are perpendicular to the frontal plane.

Step 3 The resulting view is the front view of the object. This is a two-dimensional orthographic view.

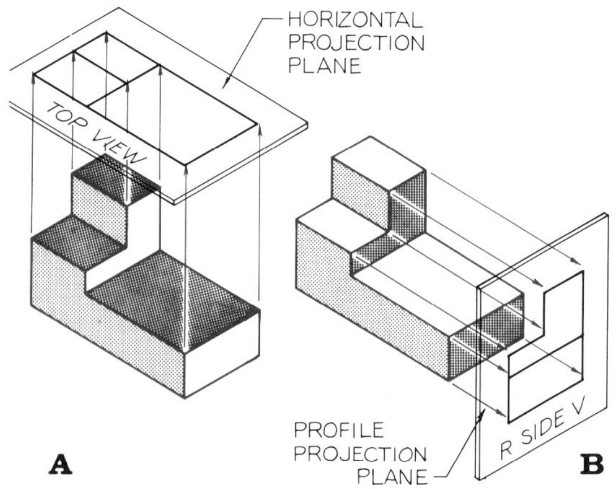

> The three projection planes that we have been dealing with in these examples are principal planes referred to as the *horizontal (H), frontal (F),* and *profile (P)* planes.

Any view that is projected onto one of these principal planes is called a **principal view.** The three dimensions of an object that are used to show its three-dimensional form are *height, width,* and *depth.*

14.3   ALPHABET OF LINES

As mentioned in the previous chapter, the use of proper line weights in a drawing will greatly improve the drawing's readability and appearance. All lines should be drawn dark and dense as if they were drawn with ink, varying the lines only by their width. The only lines that are exceptions are guidelines and construction lines, which are drawn very lightly for lettering and laying out a drawing. Skill in drafting is necessary to draw lines that have the proper width and that contrast with other types of lines.

The lines of an orthographic view are labeled along with the suggested pencil grades for drawing

Fig. 14.2 The top view is projected onto a horizontal projection plane. The right-side view is projected onto a vertical profile plane that is perpendicular to the horizontal and frontal planes.

Fig. 14.3 Glass box theory

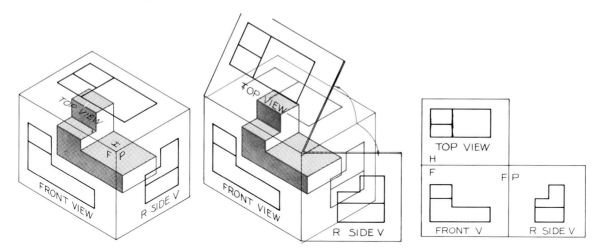

Step 1 Imagine that the object has been placed inside a box formed by the horizontal, frontal, and profile planes on which the top, front, and right-side views have been projected.

Step 2 The three projection planes are then opened into the plane of the drawing surface.

Step 3 The three views are positioned with the top view over the front view and the right-side view to the right. The planes are labeled H, F, and P at the fold lines.

Fig. 14.4 The principal projection planes

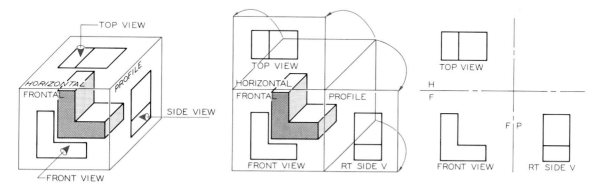

A. The three principal projection planes of orthographic projection can be thought of as planes of a glass box.

B. The views of an object are projected onto the projection planes that are opened into the plane of the drawing surface.

C. The outlines of the planes are omitted. The fold lines are drawn and labeled.

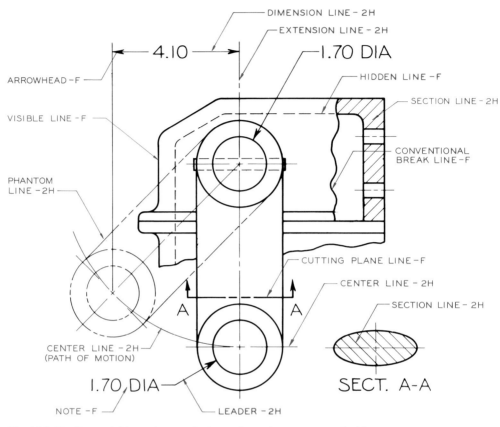

DIMENSION LINE – 2H
EXTENSION LINE – 2H
1.70 DIA
HIDDEN LINE – F
ARROWHEAD – F
SECTION LINE – 2H
4.10
VISIBLE LINE – F
CONVENTIONAL BREAK LINE – F
PHANTOM LINE – 2H
CUTTING PLANE LINE – F
CENTER LINE – 2H
SECTION LINE – 2H
A   A
CENTER LINE – 2H (PATH OF MOTION)
1.70 DIA
SECT. A-A
NOTE – F
LEADER – 2H

Fig. 14.5 The line weights and suggested pencil grades recommended for orthographic views.

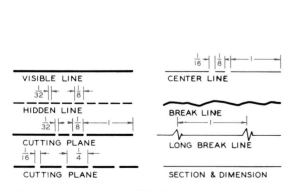

VISIBLE LINE
$\frac{1}{32}$ $\frac{1}{8}$
HIDDEN LINE
$\frac{1}{32}$ $\frac{1}{8}$
CUTTING PLANE
$\frac{1}{16}$ $\frac{1}{4}$
CUTTING PLANE

CENTER LINE
$\frac{1}{16}$ $\frac{1}{8}$ 1
BREAK LINE
LONG BREAK LINE   1
SECTION & DIMENSION

Fig. 14.6 A comparison of the line weights for orthographic views. These dimensions will vary for different sizes of drawings and should be approximated by eye.

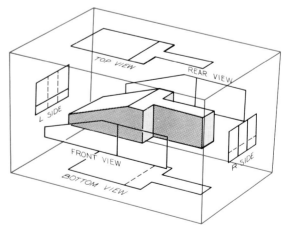

TOP VIEW
REAR VIEW
L SIDE
FRONT VIEW
R SIDE
BOTTOM VIEW

Fig. 14.7 Six principal views of an object can be drawn in orthographic projection. You can imagine that the object is in a glass box with the views projected onto the six planes of projection.

them in Fig. 14.5. The lengths of dashes in hidden lines and centerlines are drawn longer as the size of a drawing increases. Additional specifications for these lines are given in Fig. 14.6.

14.4   THE SIX-VIEW DRAWING

A view that is projected onto a principal plane—the horizontal, frontal, or profile plane—is called a principal view. Again, if you visualize an object placed inside a glass box, you will see that there are two horizontal planes, two frontal planes, and two profile planes (Fig. 14.7). Consequently, the maximum number of principal views that can be used to represent an object is six.

The top and bottom views are projected onto horizontal planes, the front and rear views are projected onto frontal planes, and the right- and left-side views are projected onto profile planes.

To draw the views on a sheet of paper, the glass box is imagined to be opened up into the plane of the drawing paper (Fig. 14.8). When in a fully opened position, the views will appear as shown in Fig. 14.9.

Note the order and arrangement of the six views and you will see the logic behind the system of orthographic projection. The top view is placed over the front view, the bottom view under the front view, the right-side view to the right, the left-side view to the left, and the rear view is placed to the left of the left-side view.

The three dimensions of an object that are necessary to give its size are *height*, *width*, and *depth*. The standard arrangement of the six views allows some of the views to share dimensions by projection. For example, the height dimension applies to the four views that are arranged horizontally, and this dimension is shown only once between the front and right-side views. The width dimension is placed between the top and front views, but it also applies to the bottom view, which is located under the front view.

Note that the projectors align the views both horizontally and vertically about the front view in Fig. 14.9. This is one of the reasons why this system of drawing is referred to as a system of projection.

Each side of the fold lines of the glass box is labeled with the letters *H*, *F*, and *P*. These letters identify the projection planes on a given side of the fold lines.

14.5   THE THREE-VIEW DRAWING

The most commonly used orthographic arrangement is the three-view drawing composed of the front, top, and right-side views. This is because three views are usually adequate to describe an object.

The same object used in the previous example is shown placed in a glass box in Fig. 14.10, which is opened onto the plane of the drawing surface.

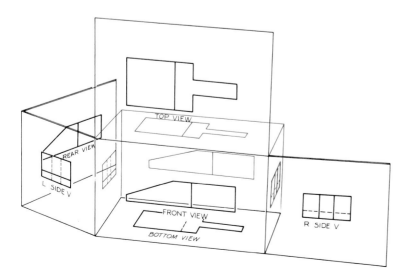

Fig. 14.8 The glass box is opened onto the plane of the drawing surface, which locates the views in their standard positions.

TOP VIEW

REAR VIEW

L. SIDE V

FRONT VIEW

BOTTOM VIEW

R SIDE V

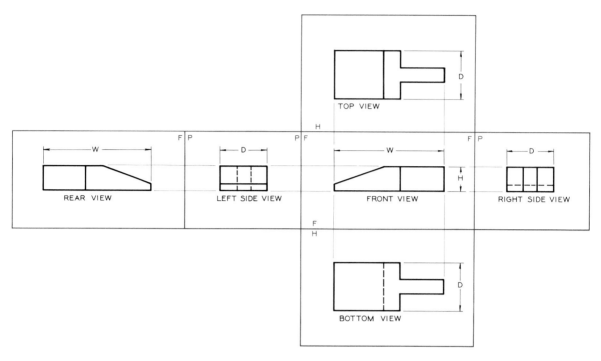

Fig. 14.9 Once the box is completely opened into a single plane, the six views are arranged to describe the object. The outlines of the planes are usually omitted. They are shown here to assist you in relating this figure to the previous one.

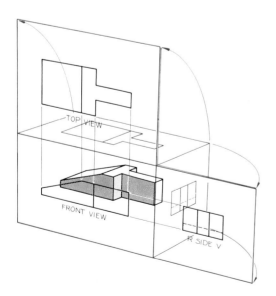

Fig. 14.10 Three-view drawings are commonly used for describing machine parts and small parts. The glass box is used to illustrate how the views are projected onto their projection planes.

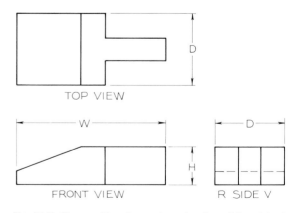

Fig. 14.11 The resulting three-view drawing of the object from the previous figure.

Fig. 14.12 Positioning orthographic views

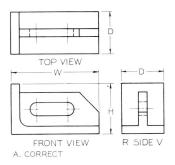

FRONT VIEW
A. CORRECT

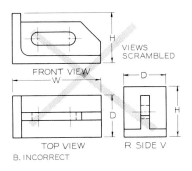

B. INCORRECT

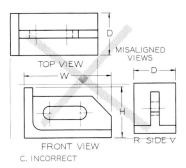

C. INCORRECT

A. This is a correct arrangement of views, labels, and dimensions. The views project from each other in proper alignment.

B. These views have been scrambled into unconventional positions, making it hard to interpret them. The dimensions have been unnecessarily repeated.

C. These views have been misaligned where they do not project from one to the other. This is incorrect.

The resulting three-view arrangement is shown in Fig. 14.11 where the views are labeled and dimensioned.

Since the views are understood to be projections from one view to the next, the dimensions are placed between the adjacent views and are not repeated. A single dimension will apply to all views that project along a single direction.

14.6 ARRANGEMENT OF VIEWS

The proper arrangement of orthographic views is essential to their understanding and interpretation. The standard positions for a three-view drawing consisting of the top, front, and right-side views are shown in Fig. 14.12A. The top and side views are projected directly from the front view. The views are properly labeled and dimensioned.

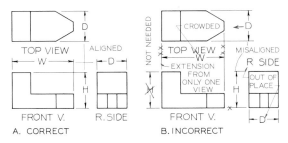

A. CORRECT

B. INCORRECT

Fig. 14.13 A. These views are properly aligned and the dimensions are correctly located. B. A number of errors are indicated in this incorrect arrangement.

You can see the problems that you would cause by rearranging the views in a nonstandard sequence as shown in Fig. 14.12B. Similarly, views that do not project from view to view are improperly drawn, as in the example in Fig. 14.12C. These rules of arrangement are emphasized in Fig. 14.13.

14.7 SELECTION OF VIEWS

When drawing an object by orthographic projection, you should select the views with the fewest hidden lines. That is why the right-side view is preferred over the left-side view in Fig. 14.14A.

Although the three-view arrangement of top, front, and right-side views is the most commonly used, the arrangement of the top, front, and left-side views is equally acceptable (Fig. 14.14B) if this view has fewer hidden lines than the right-side view.

Some objects have standard views that are considered to be the front view, top view, and so forth. For example, a chair has front and top views that are recognized as such by everyone; therefore the accepted front view should be used as the orthographic front view.

Fig. 14.14 Selection of views

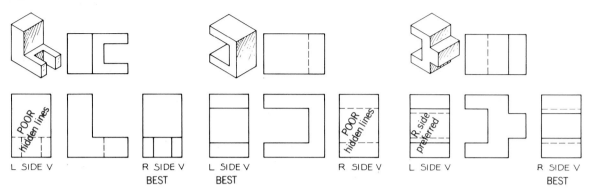

L SIDE V   R SIDE V
BEST

L SIDE V   R SIDE V
BEST

L SIDE V   R SIDE V
BEST

A. In orthographic projection, you should select the sequence of views with the fewest hidden lines.

B. The left-side view has fewer hidden lines; and therefore, this view is selected over the right-side view.

C. When both views have an equal number of hidden lines, the right-side view is traditionally selected.

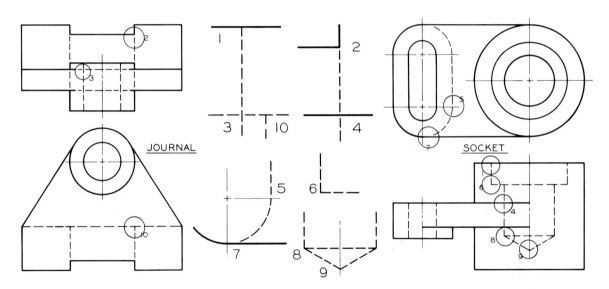

Fig. 14.15 When lines intersect in orthographic projection, they should intersect as shown here.

14.8 LINE TECHNIQUES

As drawings become more complex, you will encounter more instances where lines will overlap and intersect in a variety of ways similar to those shown in Fig. 14.15. This illustration shows the standard techniques of handling intersecting lines of most types.

The methods of constructing hidden lines composed of straight lines and curved segments is shown in Fig. 14.16.

You should become familiar with the order of precedence (priority) of lines (Fig. 14.17). The most important line that dominates all others is the visible object line. It will be shown regardless of any other type of line that lies behind it. The

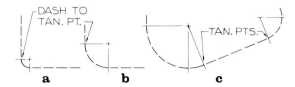

Fig. 14.16 Hidden lines in orthographic projection that are composed of curves should be drawn in this manner.

hidden object line is of next importance, and it takes precedence over the centerline.

14.9 POINT NUMBERING

It will be helpful to you in constructing orthographic views if you become familiar with the method of numbering points and lines of an object. The rules of point numbering are introduced in Fig. 14.18.

An object is shown in Fig. 14.19 that has been numbered to aid in the construction of the missing front view when the top and side views are given. By projecting a selected point from the top and side views, the front view of the point can be found.

This method is recommended when you are having difficulty in interpreting given views.

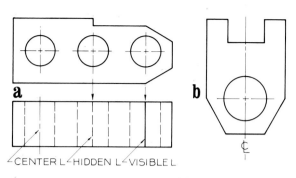

Fig. 14.17 The order of importance (precedence) of lines is: visible, hidden, and center lines, part A. The symbol made of the letters C and L in part B is used to label a center line when this is needed on symmetrical parts.

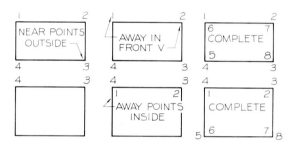

Fig. 14.18 When numbering points, the near points are labeled on the outside of the view, and away points are labeled inside the view.

Fig. 14.19 Point numbering

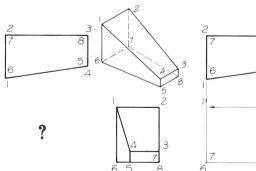

Step 1 When a missing orthographic view is to be drawn, it is helpful to number the points in the given views.

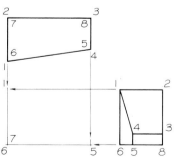

Step 2 Points 1, 5, 6, and 7 are found by projecting from the given views of these points.

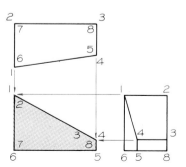

Step 3 The plotted points are connected to form the missing front view.

Fig. 14.20 Lines and planes

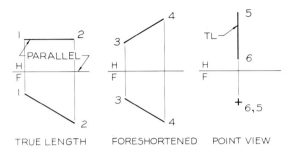

TRUE LENGTH   FORESHORTENED   POINT VIEW

A line will project in orthographic projection as true length, foreshortened, or a point.

TRUE SIZE   FORESHORTENED   EDGE VIEW

A plane in orthographic projection will appear as true size, foreshortened, or an edge.

14.10   LINE AND PLANES

An orthographic view of a line can appear true length, foreshortened, or as a point (Fig. 14.20). When a line appears true length, it must be parallel to the reference line in the previous view.

A plane in orthographic projection can appear true size, foreshortened, or as an edge (Fig. 14.20).

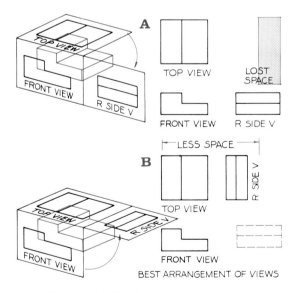

Fig. 14.21 The right-side view can be projected from the top view and positioned as shown in part B to save space when an object has a much larger depth than height.

14.11   ALTERNATE ARRANGEMENT OF VIEWS

Although the right-side view is usually placed to the right of the front view as shown in Fig. 14.21A, the side view can be projected from the top view as shown in part B. This is advisable when the object has a large depth dimension as compared to its height.

Different methods of positioning and arranging views are shown in Fig. 14.22. All of these arrangements are acceptable.

14.12   LAYING OUT THE THREE-VIEW DRAWING

The depth dimension applies to both the top and side views, but these views are usually positioned where this dimension does not project between them (Fig. 14.23). The depth dimension can be graphically transferred to the two views by using a 45° line, an arc, or a pair of dividers. It is preferable that you learn to use your dividers for this purpose.

These and similar methods of transferring dimensions from view to view are illustrated in Fig. 14.24.

The method of laying out a three-view drawing in a 10″ × 15″ space is shown in Fig. 14.25. The views are blocked in with light construction lines using the overall dimensions of the views. Centerlines and tangent points are drawn next, and then the straight lines. Once the layout has

Fig. 14.22 Arrangement of views

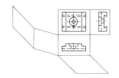

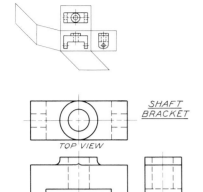

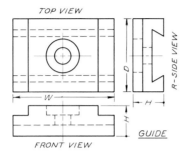

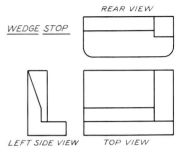

A. This is the most standard arrangement of views for a three-view drawing.

B. The right-side view is projected off the top view in this arrangement.

C. This is an unconventional but acceptable arrangement of three views

Fig. 14.23 Transferring depth dimensions

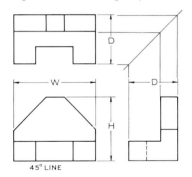

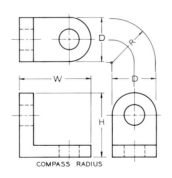

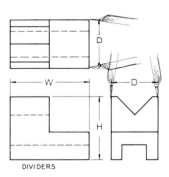

A. The depth dimension can be projected from the top view to the right-side view by constructing a 45° line positioned as shown.

B. The depth dimension can be projected from the top view to the side view using a compass and a center point.

C. The depth dimension can be transferred from the top view to the side view by using dividers.

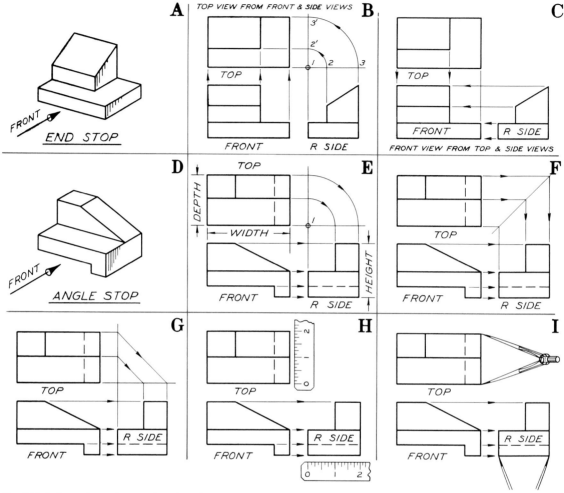

Fig. 14.24 A number of examples of methods of transferring dimensions from the views are shown here.

been checked, the lines are darkened to their proper weight.

When overall dimensions and labels are required, the three-view sketch should be positioned as shown in Figs. 14.25 through 14.28.

14.13 THE TWO-VIEW DRAWING

Some objects can be adequately explained in two views, and it is good economy of time and space to use only the views that are necessary to depict an object. Two parts that require only two views are shown in Fig. 14.29.

Cylindrical parts need only two views as shown in Fig. 14.30. It is preferable to select the views with the fewest hidden lines; consequently the right-side view is the best view in this case.

The layout of a two-view drawing is shown in Fig. 14.31 in a 10″ × 15″ space. The overall dimensions are used to block in and center the views. Second, the centerlines are located and drawn using light construction lines with a 2H–4H pencil. When the views have been completed, they are checked, and then the lines are darkened to complete the two-view drawing.

Views that involve arcs and tangent lines

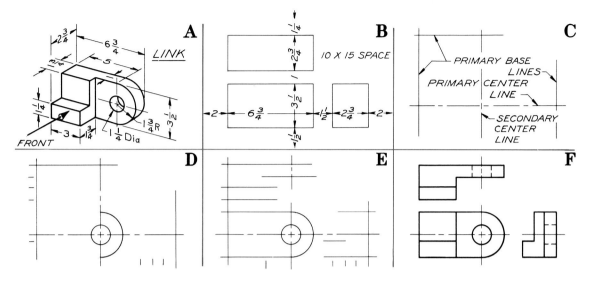

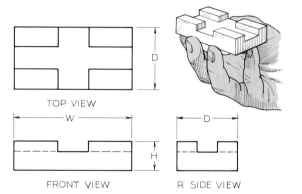

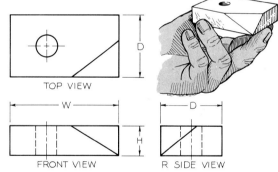

Fig. 14.25 Lay out a three-view drawing in this order: Center the views, draw the center lines, draw the arcs, draw the straight lines, and darken your layout.

Fig. 14.26 A three-view drawing of an object.

Fig. 14.28 A three-view drawing of an object.

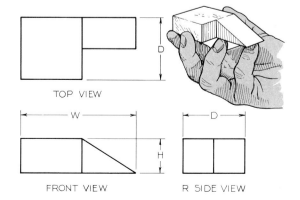

Fig. 14.27 A three-view drawing of an object.

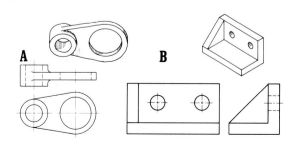

Fig. 14.29 These objects can be adequately described with two orthographic views.

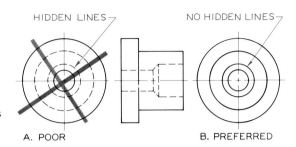

Fig. 14.30 Cylindrical objects can be depicted with two views. Always select views with the fewest hidden lines.

A. POOR

NO HIDDEN LINES

B. PREFERRED

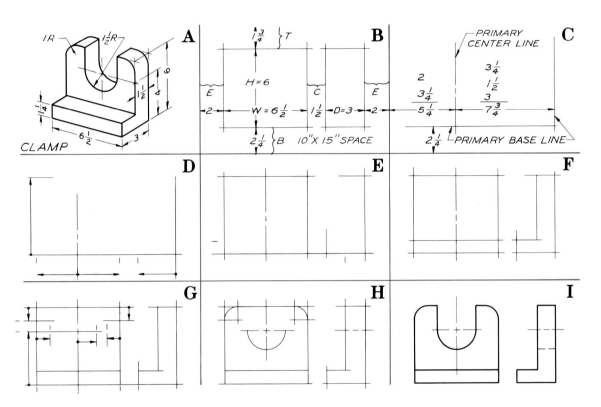

Fig. 14.31 Lay out two-view drawings in this order: Position the views, locate the center lines, block in the views, locate the centers, draw the arcs, draw the straight lines, and darken the lines of the finished views.

should be laid out by locating the centers and centerlines of the arcs as the first step (Fig. 14.32). The tangent points are found using light construction lines. Begin by drawing the arcs first. The completed views are checked and the lines are darkened to their proper weights. It is important that the arcs are drawn to stop at the points of tangency. This is especially necessary when the drawing is being inked.

14.14 ONE-VIEW DRAWINGS

Cylindrical parts and those with a uniform thickness can be described in one view (Fig. 14.33).

In both cases, notes are used to explain the missing feature or dimension. The note, DIA, is placed after the diameter dimension for the cylindrical part. The thickness of the washer and the shim is specified by a note.

Fig. 14.32 Views with circular features

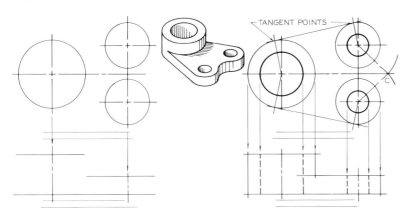

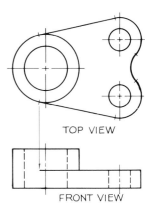

TOP VIEW

FRONT VIEW

Step 1 Begin the layout by locating the centers and drawing the center lines and arcs. Draw these lines lightly.

Step 2 Draw the tangent lines and locate the points of tangency. Block in the front view.

Step 3 After checking the views for correctness, darken the lines to their proper widths. Draw the arcs first, being careful to stop them at the tangent points.

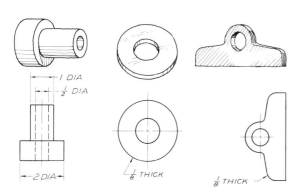

Fig. 14.33 Objects of this type can be described with only one orthographic view with the use of supplementary notes.

Fig. 14.34 Unnecessary and confusing hidden lines have been omitted in the side views to improve their clarity.

14.15 INCOMPLETE AND REMOVED VIEWS

The right- and left-side views of the part in Fig. 14.34 would be very complex and hard to interpret if all hidden lines were shown as specified by the rules of orthographic projection. Therefore, it is best to omit lines that confuse a clear understanding of the views. The two side views that are shown are easier to understand than the views drawn in their entirety.

In many cases, it is difficult to show a feature because of its location. Standard views can be

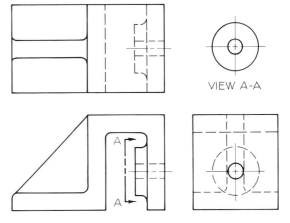

VIEW A-A

Fig. 14.35 A removed view, indicated by the directional arrows, can be used to draw views in hard-to-view locations.

Fig. 14.36 Plotting curved lines

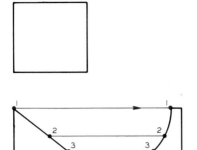

Step 1 Curves in orthographic projection can be plotted by locating and numbering the points in two views.

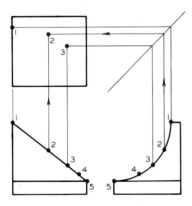

Step 2 Two views of each point are projected to the third view, where the projectors intersect.

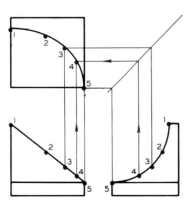

Step 3 All points are projected in this manner. The points are connected with an irregular curve.

confusing when lines overlap from other features. The view in Fig. 14.35 is more clearly shown when the view indicated by the lines of sight is removed to an isolated position.

14.16 CURVE PLOTTING

An irregular curve can be drawn by following the rules of orthographic projection. Such an example is shown in Fig. 14.36.

The process of plotting points is begun by locating a series of points along the curve in two given views. By following the rules of projection, these points are projected to the top view where each point is located and the points are then connected by a smooth curve.

Figure 14.37 is a similar example where an ellipse is plotted in the top view by projecting from the front and side views.

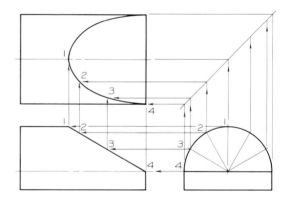

Fig. 14.37 The ellipse in the top view was found by numbering points in the front and side views, and then projecting them to the top view.

You will find it helpful to number points that are to be plotted when curves are being located by projection.

14.17 PARTIAL VIEWS

A partial view can be used to save time and space when the parts are symmetrical or cylindrical. By omitting the rear of the circular top view in Fig. 14.38, space can be saved without sacrificing clarity.

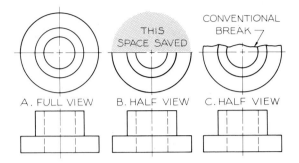

Fig. **14.38** To save space and drawing time, the top view of a cylindrical part can be drawn as a partial view using either of these methods.

A partial break may be used to make it more apparent that a portion of the view has been omitted. Either method is correct and acceptable.

14.18 CONVENTIONAL REVOLUTIONS

The readability of an orthographic view may be improved if the rules of projection are violated.

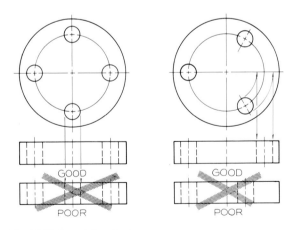

Fig **14.39** Revolving holes

A. A true projection of equally spaced holes gives a misleading impression that the center hole passes through the center of the plate.

B. A conventional view is used to show the true radial distances of the holes from the center by revolution. The third hole is omitted.

Established violations of rules that are customarily made for the sake of clarity are called **conventional practices.**

When holes are symmetrically spaced in a circular plane as shown in Fig. 14.39, it is conventional practice to show them at their true radial distance from the center of the plane. This requires an imagined revolution of the holes in the top view.

This same principle of revolution applies to symmetrically positioned features such as the three lugs on the outside of the part in Fig. 14.40. The conventional view is better than the true orthographic projection.

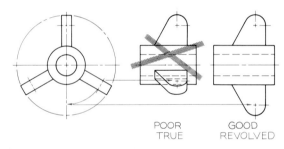

Fig. **14.40** Symmetrically positioned external features, such as these lugs, are revolved to their true-size positions for the best views.

The conventional and desired method of drawing holes and ribs in combination in the same view is shown in Fig. 14.41.

Another conventional practice is illustrated in Fig. 14.42, where an inclined feature is revolved to a frontal position in the top view, so it can be drawn as true size in the front view. The revolution of the part is not drawn since it is an imagined revolution. A similar object with a revolved feature is shown in Fig. 14.43.

Other parts whose views are improved by revolution are those shown in Fig. 14.44. Unless there is a reason for them to be positioned otherwise, it is desirable to show the top view features at 45° so they will not coincide with the centerlines. The front views are drawn by imagining that the features have been revolved.

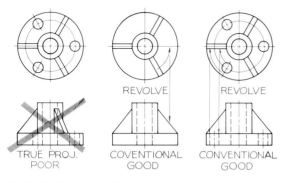

Fig. 14.41 The conventional methods of revolving holes and ribs in combination for improved clarity.

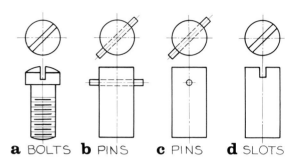

a BOLTS **b** PINS **c** PINS **d** SLOTS

Fig. 14.44 Parts of this type are drawn at 45° angles in the top views, but the front views are drawn to show the details as revolved views.

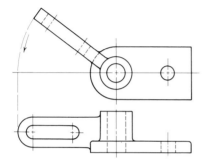

Fig. 14.42 It is conventional practice to imagine that features of this type have been revolved in order for them to appear true shape in the front view.

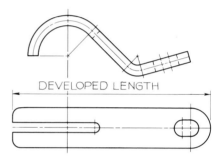

DEVELOPED LENGTH

Fig. 14.45 Objects that have been shaped by bending thin stock can be shown as developed views as if the views have been flattened out.

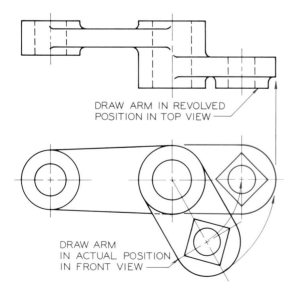

DRAW ARM IN REVOLVED POSITION IN TOP VIEW

DRAW ARM IN ACTUAL POSITION IN FRONT VIEW

Fig. 14.43 The arm in the front view is imagined to be revolved so its true length can be drawn in the top view. This is an accepted conventional practice.

A closely related type of conventional view is the developed view where a bent piece of material is drawn as if it were flattened out (Fig. 14.45).

14.19 INTERSECTIONS

In orthographic projection, the intersection between planes results in a line that describes the object. In Fig. 14.46 examples of views are shown where lines may or may not be required.

The standard types of intersections between cylinders are shown in Fig. 14.47. Those at A and B are conventional intersections, which means that they are approximations, and they are drawn in this manner for ease of construction. The example at C is a true intersection where cylinders of equal diameters intersect. Similar intersections are shown in Fig. 14.48.

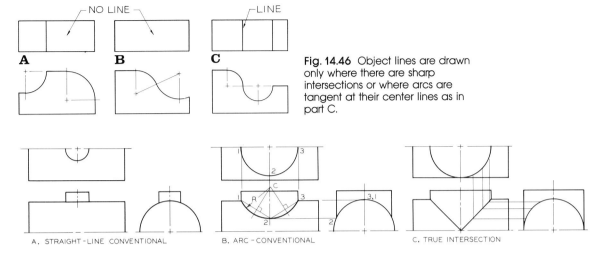

Fig. 14.46 Object lines are drawn only where there are sharp intersections or where arcs are tangent at their center lines as in part C.

A. STRAIGHT-LINE CONVENTIONAL   B. ARC-CONVENTIONAL   C. TRUE INTERSECTION

Fig. 14.47 The conventional methods of showing intersections between cylinders. These are approximations except for part C.

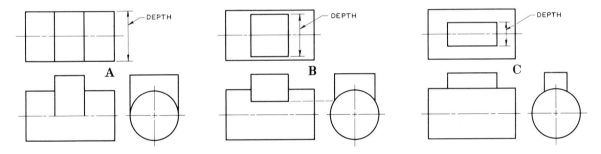

Fig. 14.48 Conventional intersections between cylinders and rectangular shapes. The intersection at C is an approximate intersection.

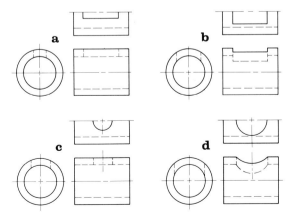

Fig. 14.49 Conventional intersections between cylinders and holes piercing them.

The types of intersections that are formed by holes in cylinders are shown in Fig. 14.49. These are conventional intersections that are sufficient for depicting these features.

14.20 FILLETS AND ROUNDS

Fillets and **rounds** are rounded corners that are used on castings such as the body of the Collet Index Fixture shown in Fig. 14.50. A fillet is an inside rounding, and a round is an external rounding. The radii of fillets and rounds may be many sizes but they are usually about ¼″; they are used on castings for added strength and improved appearance.

Fig. 14.50 The edges of this Collet Index Fixture are rounded to form fillets and rounds. Note that the surface of the casting is rough except where it has been machined. (Courtesy of Hardinge Brothers Inc.)

Fig. 14.51 Fillets and rounds

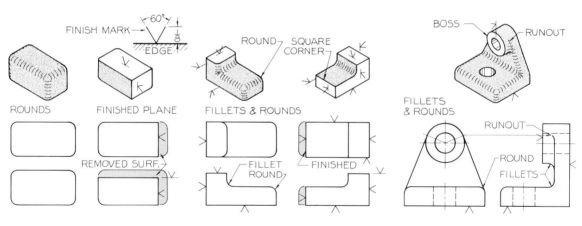

A. A casting has rounded corners on the outside called rounds. When a surface has been finished by machining, the rounds are removed and the corners are squared. The finish mark is a V that is placed on the edge view of the surfaces that are finished.

B. A fillet is a rounded inside corner. The rounds are removed when the outside surfaces are finished. The fillets can be seen only in the front view.

C. The views of an object with fillets and rounds must be drawn in such a way as to call attention to them.

The orthographic views of a part with fillets and rounds must be drawn in such a manner that these detailed features can be seen (Fig. 14.51).

A casting will have square corners only when its surfaces have been finished, which is the process of machining away a portion of the surface to a smooth finish (Fig. 14.51B).

> The finished surface is indicated on a view by placing a **finish mark** (V) on the outside edge views of the surface that is to be finished in all views whether the edges are *visible* or *hidden*.

You can see in Fig. 14.51. how fillets and rounds are shown, as well as the square corners without rounded corners due to finishing. Note

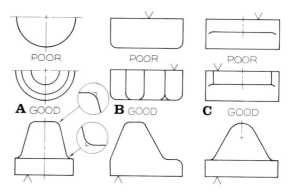

Fig. 14.53 Examples of conventionally drawn fillets and rounds.

that a boss is a raised cylindrical feature that is thickened to receive a shaft or to be threaded.

An alternate finish mark that can be used is the one in Fig. 14.52.

The techniques of showing fillets and rounds on orthographic views are given in Fig. 14.53.

The curve formed by a fillet at a point of tangency is called a **runout.** A comparison of intersections and runouts of parts with and without fillets and rounds is shown in Fig. 14.54. Large runouts are constructed as an eighth of a circle with a compass as illustrated in Fig. 14.55. When the runouts are small, they can be drawn with a circle template.

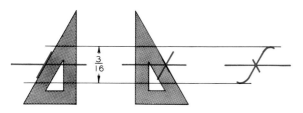

Fig. 14.52 An alternate type of finish mark is the traditional f that is applied to the edge views of the finished surface, whether hidden or visible.

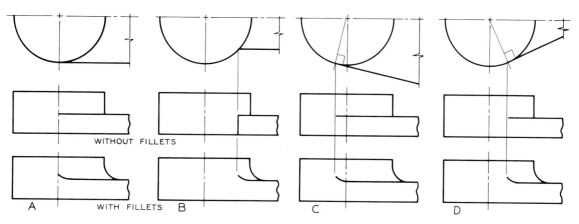

Fig. 14.54 Intersections between features of types of objects. These are called runouts.

Fig. 14.55 Plotting runouts

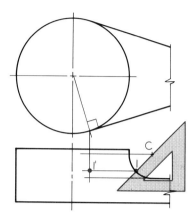 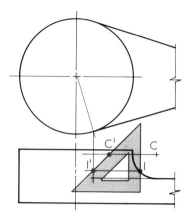 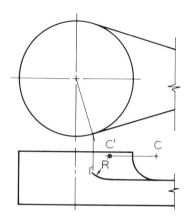

Step 1 Find the point of tangency in the top view and project it to the front view. A 45° triangle is used to find point 1, which is projected to locate point 1'.

Step 2 A 45° triangle is used to locate point C', which is on the horizontal projector from the center of the fillet, C.

Step 3 The radius of the fillet is used to draw the runout with C' as the center. The runout arc is equal to one-eighth of a circle.

Fig. 14.56 Runouts are shown for differently shaped ribs. One has fillets, and the one at B is a rib with rounded edges.

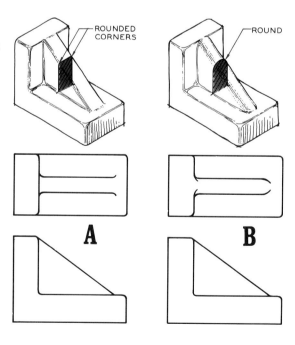

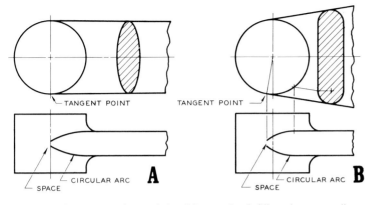
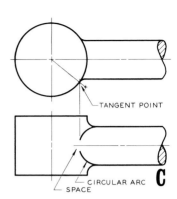

Fig. 14.57 Conventional runouts involving parts of different cross sections.

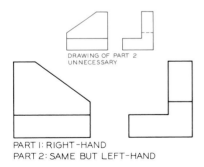

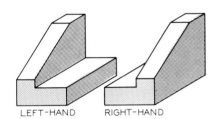

Fig. 14.58 Left-hand and right-hand parts

A. Some parts are required to be similar except that one is a left-hand part and the other is a right-hand part.

B. One of the parts can be drawn and labeled, right-hand in this case. The other part need not be drawn, but merely indicated by a note.

When properly drawn, the runouts on orthographic views will tell much about the details of an object (Fig. 14.56). For example, the runouts in the top views tell us that the ribs at A have rounded corners, and the rib at B is completely round.

Methods of showing intersections of other types are shown in Fig. 14.57.

14.21 LEFT-HAND AND RIGHT-HAND VIEWS

Two parts are often required that are very similar to each other, but one part is actually a "mirror image" of the other. Two parts of this type are shown in Fig. 14.58. Your first impression is that the parts are interchangeable, but the parts are actually as different as a pair of shoes.

The drafter can reduce drawing time by drawing views of only one of the parts and labeling these views as shown in Fig. 14.58. A note can be added to indicate that the other matching part has the same dimensions.

14.22 FIRST-ANGLE PROJECTION

The problems of this chapter have been presented as third-angle projections where the top view is placed over the front view, and the right-side view to the right of the front view. This method is used extensively in America, Britain, and Canada. Most

Fig. 14.59 First-angle projection

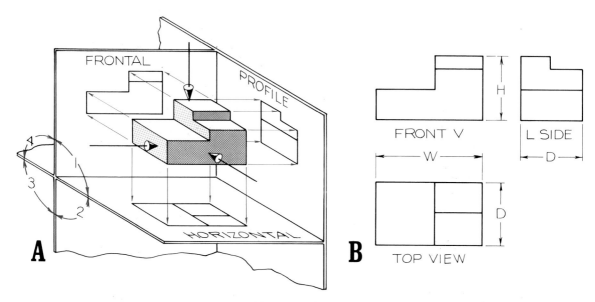

A. The first angle of projection is used by many of the countries that use the metric system. You imagine that the object is placed above the horizontal and in front of the frontal plane.

B. The views are drawn in this location, which is different from the third angle of projection that is usually used in America.

a METRIC UNITS –
3RD ANGLE PROJ.

b METRIC UNITS –
1ST ANGLE PROJ.

Fig. 14.60 The angle of projection that is used to prepare a set of drawings is indicated by this truncated cone. It is placed in or near the title block of a drawing.

of the rest of the industrial world uses the *first-angle* of projection.

The first-angle system is illustrated in Fig. 14.59 where an object is placed above the horizontal plane and in front of the frontal plane. When these projection planes are opened onto the surface of the drawing paper, the front view is projected over the top view, and the left-side view is placed to the right of the front view.

It is important that the angle of projection be indicated on a drawing to aid in the interpretation of the views. This is done by placing a truncated cone in or near the title block (Fig. 14.60). When metric units of measurements are used, the cone and the symbol, SI, are placed together on the drawing.

PROBLEMS

The following problems are to be drawn as orthographic views on Size A or Size B paper as assigned by your instructor. (Refer to Section 2.33 for instructions on laying out a sheet.)

1–11 (Figs. 14.61–14.71) Draw the given views using the dimensions provided, and then construct the missing view; either the top, front, or right-side view. Use Size A sheets and draw one or two problems per sheet as assigned by your instructor.

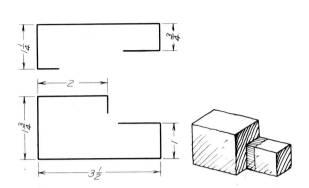

Fig. 14.61 Problem 1: Guide block.

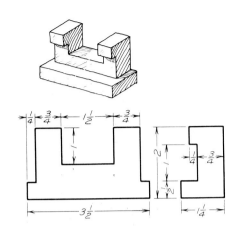

Fig. 14.64 Problem 4: Lock catch.

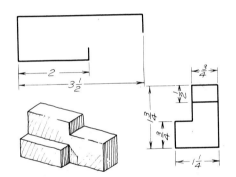

Fig. 14.62 Problem 2: Double step.

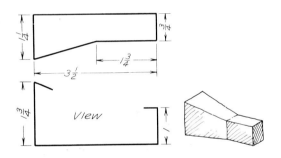

Fig. 14.65 Problem 5: Two-way adjuster.

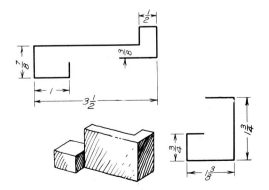

Fig. 14.63 Problem 3: Adjustable stop.

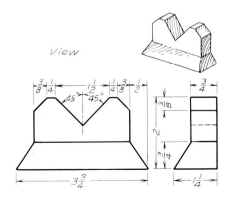

Fig. 14.66 Problem 6: Vee block.

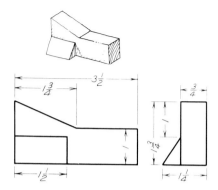

Fig. 14.67 Problem 7: Filler.

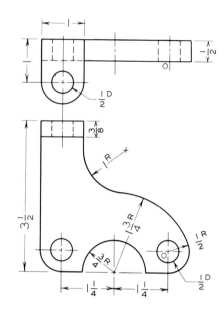

Fig. 14.70 Problem 10: Support brace.

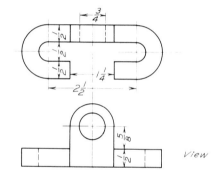

Fig. 14.68 Problem 8: Slide stop.

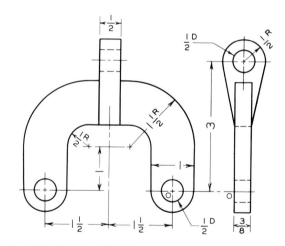

Fig. 14.71 Problem 11: Yoke.

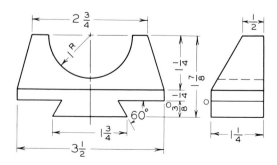

Fig. 14.69 Problem 9: Shaft support.

12–52 Construct the necessary orthographic views to describe the objects from Fig. 14.72 through Fig. 14.112. Draw these on Size A or Size B sheets. Label the views and show the overall dimensions of W, D, and H on the appropriate views.

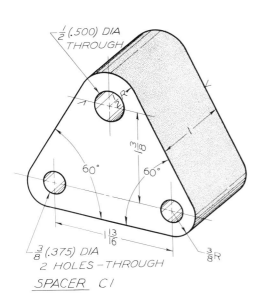

Fig. 14.72 Problem 12: Spacer.

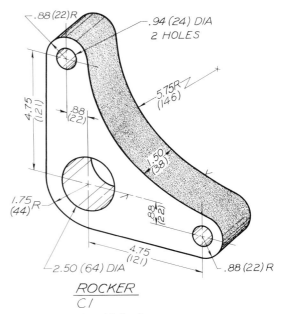

Fig. 14.74 Problem 14: Rocker.

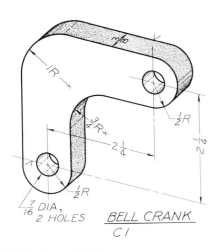

Fig. 14.73 Problem 13: Bell crank.

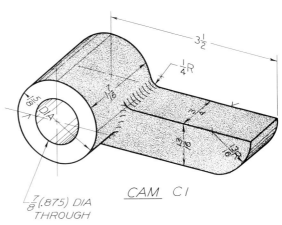

Fig. 14.75 Problem 15: Rod guide.

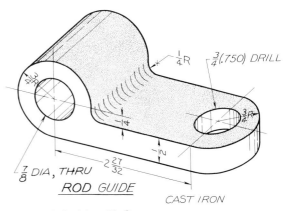

Fig. 14.76 Problem 16: Cam.

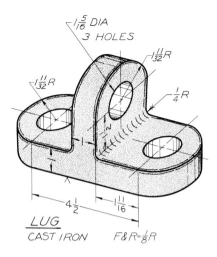

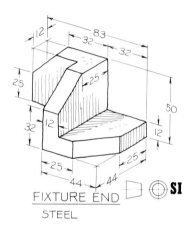

Fig. 14.77 Problem 17: Lug.

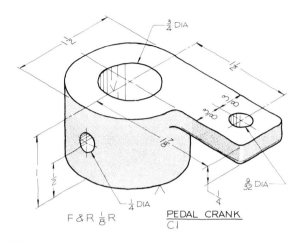

Fig. 14.80 Problem 20: Pedal crank.

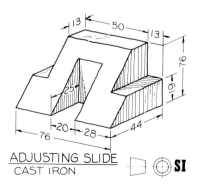

Fig. 14.78 Problem 18: Fixture end.

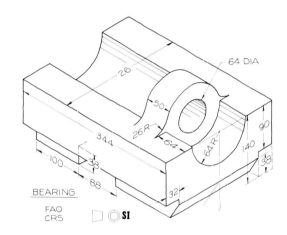

Fig. 14.81 Problem 21: Bearing.

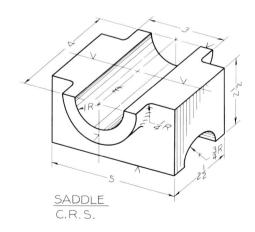

Fig. 14.79 Problem 19: Adjusting slide.

Fig. 14.82 Problem 22: Saddle.

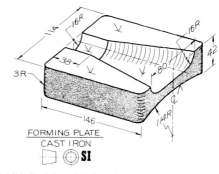

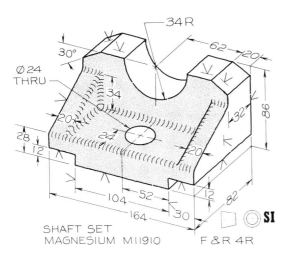

Fig. 14.83 Problem 23: Shaft set.

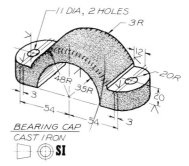

Fig. 14.86 Problem 26: Forming plate.

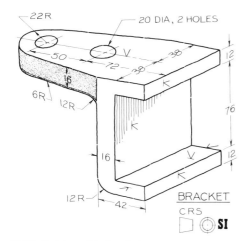

Fig. 14.84 Problem 24: Bracket.

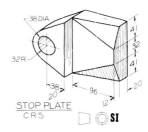

Fig. 14.87 Problem 27: Bearing cap.

Fig. 14.88 Problem 28: Stop plate.

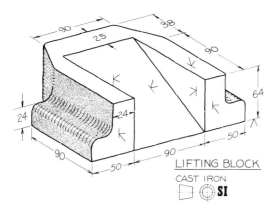

Fig. 14.85 Problem 25: Lifting block.

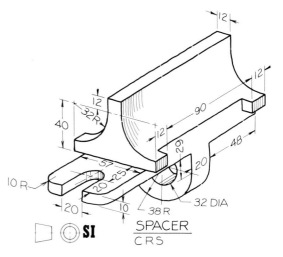

Fig. 14.89 Problem 29: Spacer.

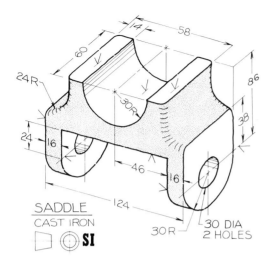

SADDLE
CAST IRON

Fig. 14.90 Problem 30: Saddle.

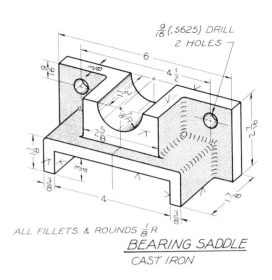

$\frac{9}{16}$ (.5625) DRILL
2 HOLES

ALL FILLETS & ROUNDS $\frac{1}{8}$ R

BEARING SADDLE
CAST IRON

Fig. 14.93 Problem 33: Bearing saddle.

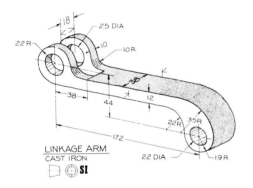

LINKAGE ARM
CAST IRON

Fig. 14.91 Problem 31: Linkage arm.

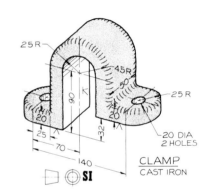

25 R

45 R

25 R

20 DIA
2 HOLES

CLAMP
CAST IRON

Fig. 14.94 Problem 34: Clamp.

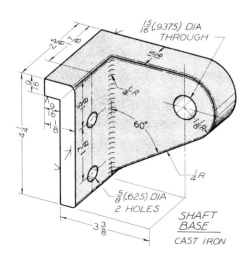

$\frac{15}{16}$ (.9375) DIA
THROUGH

60°

$\frac{5}{8}$ (.625) DIA
2 HOLES

SHAFT
BASE
CAST IRON

Fig. 14.92 Problem 32: Shaft base.

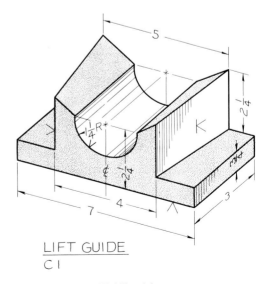

LIFT GUIDE
C I

Fig. 14.95 Problem 35: Lift guide.

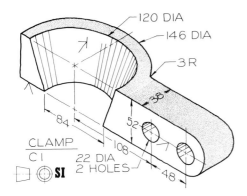

120 DIA
146 DIA
3 R
38
84
52
108
CLAMP
C1
22 DIA
2 HOLES
48
SI

Fig. 14.96 Problem 36: Clamp.

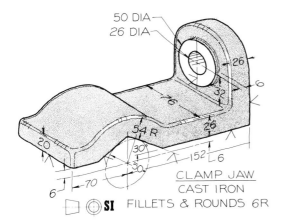

50 DIA
26 DIA
26
6
76
32
20
54 R
26
30°
152 6
CLAMP JAW
CAST IRON
6
70
SI   FILLETS & ROUNDS 6R

Fig. 14.97 Problem 37: Clamp jaw.

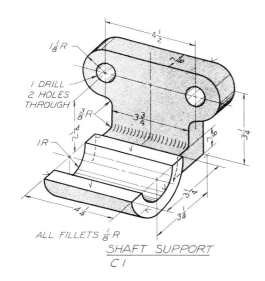

$4\frac{1}{2}$
$1\frac{1}{8}R$
1 DRILL
2 HOLES
THROUGH
$\frac{3}{8}R$
$3\frac{3}{4}$
$3\frac{1}{4}$
1R
$3\frac{1}{4}$
$4\frac{1}{4}$
$3\frac{1}{4}$
ALL FILLETS $\frac{1}{8}$ R
SHAFT SUPPORT
C1

Fig. 14.98 Problem 38: Shaft support.

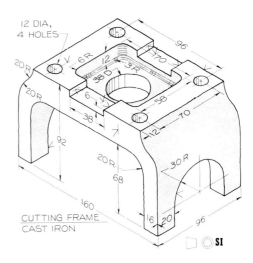

12 DIA,
4 HOLES
96
20R
6R
12
70
20 R
38 D
3R
6
58
38
70
92
12
20 R
30 R
68
160
16 20
96
CUTTING FRAME
CAST IRON
SI

Fig. 14.99 Problem 39: Cutting frame.

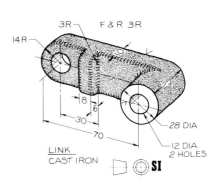

3R   F & R 3R
14R
8
6
30
28 DIA
70
12 DIA
2 HOLES
LINK
CAST IRON   SI

Fig. 14.100 Problem 40: Link.

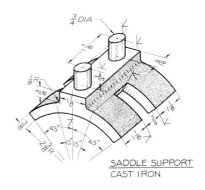

$\frac{3}{4}$ DIA
$\frac{1}{8}$R
$\frac{3}{8}$
$1\frac{1}{8}$
45°
15° 15°
45°
$\frac{3}{4}$
$1\frac{1}{8}$
SADDLE SUPPORT
CAST IRON

Fig. 14.101 Problem 41: Saddle support.

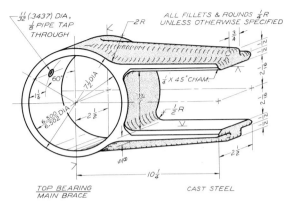

Fig. 14.102 Problem 42: Top bearing.

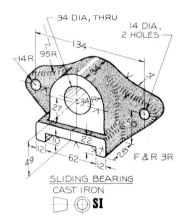

Fig. 14.105 Problem 45: Sliding bearing.

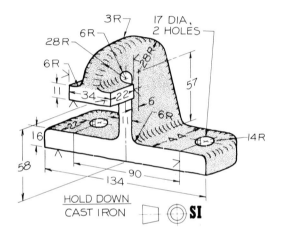

Fig. 14.103 Problem 43: Hold down.

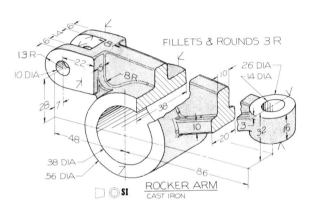

Fig. 14.106 Problem 46: Rocker arm.

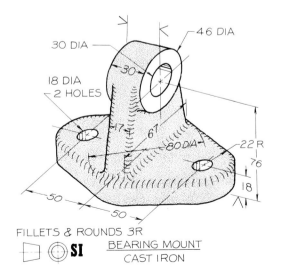

Fig. 14.104 Problem 44: Bearing mount.

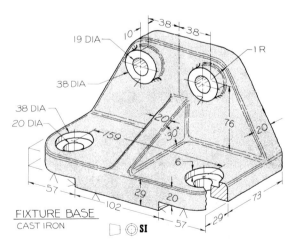

Fig. 14.107 Problem 47: Fixture base.

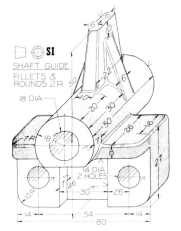

Fig. 14.108 Problem 48: Shaft guide.

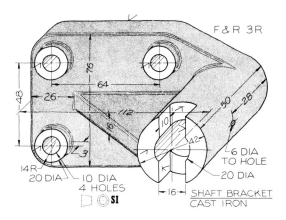

Fig. 14.111 Problem 51: Shaft bracket.

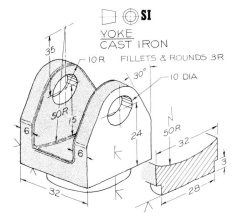

Fig. 14.109 Problem 49: Yoke.

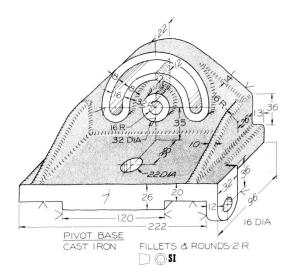

Fig. 14.112 Problem 52: Pivot base.

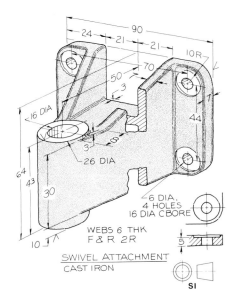

Fig. 14.110 Problem 50: Swivel attachment.

AUXILIARY VIEWS

15.1 INTRODUCTION

The auxiliary view is a type of orthographic view, but it is not a *principal* view as previously covered. The principal views, such as the top, front, and side views, are usually drawn so that their surfaces are parallel to the principal projection planes. In these views, the surfaces will appear true size.

However, if the surfaces of an object are not mutually perpendicular, the planes cannot be parallel to the projection planes. This means that a nonprincipal plane, an **auxiliary plane,** must be used on which the auxiliary view can be projected.

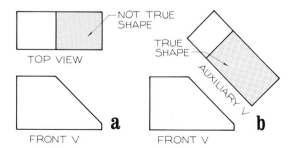

Fig. 15.1 When a surface appears as an inclined edge in a principal view, it can be found true size by an auxiliary view. The top view at a is foreshortened, but this plane is true size in an auxiliary view at b.

Such an example is shown in Fig. 15.1. The inclined plane is not true shape, but is foreshortened in the top view. The edge view of the surface is not horizontal in the front view; therefore, it cannot appear true shape in the top view. This surface is true shape in the auxiliary view that is projected from the front view.

This chapter will cover the auxiliary view and its usage in the preparation of engineering drawings.

15.2 THE FOLDING-LINE APPROACH

The three principal orthographic planes are the *frontal, horizontal,* and *profile* planes. These are the names of the planes of an imaginary glass box in which an object is placed for drawing its various views.

A primary auxiliary plane is a plane that is perpendicular to one of the principal planes, but is oblique to the other two planes. In other words, a *primary auxiliary* view is an auxiliary view that is projected from a *primary orthographic view.*

The auxiliary planes can be thought of as planes that fold from the principal planes as shown in Fig. 15.2. The plane at A folds from the

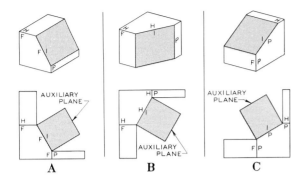

Fig. 15.2 A primary auxiliary plane can be folded from the frontal, horizontal, or profile planes. The fold lines are labeled F–1, H–1, and P–1, respectively.

frontal plane to make a 90° angle with it. The fold line between the two planes is labeled F-1. The F is an abbreviation for frontal and the numeral 1 represents *first* or *primary* auxiliary plane.

The examples at B and C illustrate the positions for auxiliary planes that fold from the horizontal and profile planes.

15.3 AUXILIARIES PROJECTED FROM THE TOP VIEW

The inclined plane in Fig. 15.3 is an edge in the top view, and it is perpendicular to the top projection plane, the horizontal plane. An auxiliary

plane can be drawn that is parallel to the inclined surface of the object, and the view projected onto it will be a true-size view of the inclined plane.

> It is necessary that a surface appear as an edge in a principal view before it can be found as true size in a primary auxiliary view.

Fold line H-1 is drawn parallel to the edge view of the inclined plane in step 1. The line of sight is drawn perpendicular to the edge view. Each corner of the inclined plane is projected perpendicularly to the auxiliary plane and is located by using the dimension of height (H) from the side or front view. This auxiliary view shows the true-size inclined surface.

A similar example is the object shown in the glass box in Fig. 15.4. Since this object has an inclined surface that appears as an edge in the top view, it can be found true size in a primary auxiliary view. The fold line between the horizontal and auxiliary plane is labeled H-1. The height dimension (H) is transferred from the front view to the auxiliary view, since both are measured from the same horizontal plane.

The auxiliary plane is rotated about the H-1 fold line into the plane of the top view in Fig. 15.5.

When drawn on a sheet of paper, the drawing of this part would appear as shown in Fig. 15.6.

Fig. 15.3 Auxiliary from the top—folding-line method

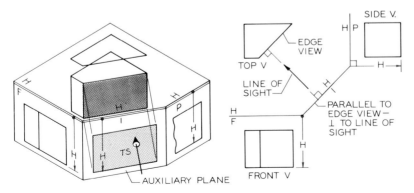
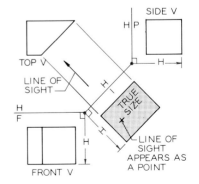

Given An object with an inclined surface shown pictorially.

Required Find the inclined surface true size by an auxiliary view.

Step 1 Construct a line of sight perpendicular to the edge view of the inclined surface. Draw the H–1 fold line parallel to the edge, and draw the H-F fold line between the top and front views.

Step 2 Project the four corners of the edge parallel to the line of sight. Locate the corners by measuring perpendicularly from the horizontal plane with height (H) dimensions.

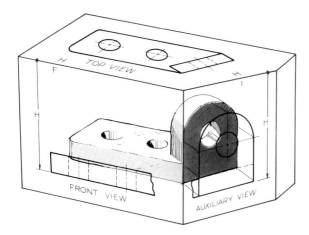

Fig. 15.4 A pictorial showing the relationship of the projection planes used to find the true-size view of the inclined surface.

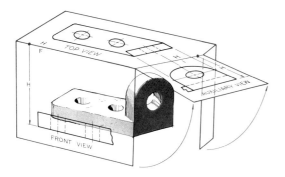

Fig. 15.5 The auxiliary plane is opened into the plane of the top view by revolving it about the H–1 fold line. This is how it would be positioned on your drawing surface.

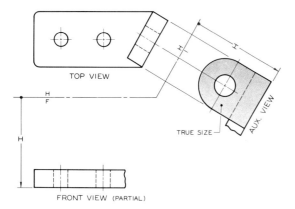

Fig. 15.6 When the object is drawn on a sheet of paper, it would be laid out in this manner. The front view is drawn as a partial view since the omitted part is shown true size in the auxiliary view.

The front view is shown as a partial view since the omitted portion would have been hard to draw, and it would not have been true size. The auxiliary view shows the true-size view of the inclined surface that is composed of circular features that can be drawn with a compass in this view.

15.4 CONSTRUCTING AN AUXILIARY FROM THE TOP VIEW—FOLDING-LINE METHOD

The steps of constructing an auxiliary view that is projected from the top view are shown in Fig. 15.7. The purpose of the auxiliary view will be to find the true-size view of the inclined surface.

This surface can be found true size in a primary auxiliary view since the inclined surface projects as an edge in the top view. The line of sight is drawn to be perpendicular to the edge view and the fold line is drawn parallel to the edge. The other dimension that must be transferred from the front view is height *(H)*. Height is perpendicular to the horizontal plane; consequently, it cannot be found in the top view. It must be taken from an adjacent view such as the front or side view.

The fold lines are drawn as thin, but black lines. They are labeled as H-1. It is also helpful to number or letter the points on the views. The projectors are construction lines, and they should be drawn with a hard pencil (3H–4H) just dark enough to be seen and used.

15.5 AUXILIARIES FROM THE TOP VIEW—REFERENCE-LINE METHOD

A similar method of locating an auxiliary view is a method that uses a **reference plane** instead of the folding-line technique. An example of this type of auxiliary view is shown in Fig. 15.8 where an auxiliary view is projected from the top view.

Instead of placing a fold line between the top and front views, a reference plane is passed through the bottom of the front view. This is shown pictorially in Fig. 15.8a. The height dimensions *(H)* are measured upward from the reference plane instead of downward from a fold line.

In Fig. 15.8b, the reference plane is shown as a horizontal edge in the front view where it is la-

Fig. 15.7 Construction of an auxiliary view

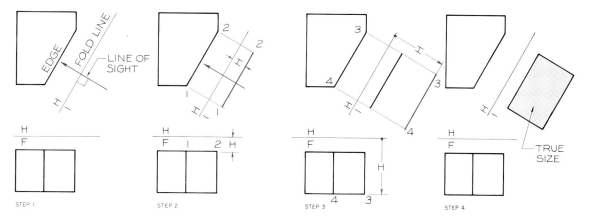

Step 1 The line of sight is drawn perpendicular to the edge views of the inclined surface. The H-1 fold line is drawn parallel to the edge view of the inclined surface. An H-F fold line is drawn between the given views.

Step 2 Points 1 and 2 are found by transferring the height (*H*) dimensions from the front view to the auxiliary view. This locates line 1–2.

Step 3 Points 3 and 4 are found in the same manner using the dimensions of height (*H*). This locates line 3–4.

Step 4 The corner points are connected to complete the true-size view of the inclined plane.

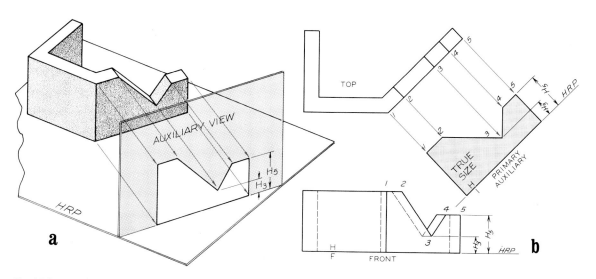

Fig. 15.8. A horizontal reference plane can be used instead of the folding-line technique to construct an auxiliary view. Instead of placing the reference plane between the top and auxiliary views, it is placed outside the auxiliary view (b).

Fig. 15.9 Auxiliary from the front—folding-line method

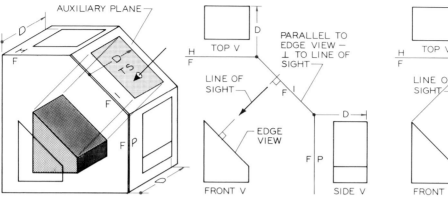

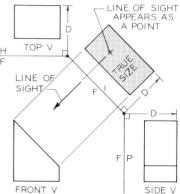

Given An object with an inclined surface shown pictorially.

Required Find the inclined surface true size by an auxiliary view.

Step 1 Draw a line of sight perpendicular to the edge view of the inclined surface. Draw the F–1 fold line parallel to the edge, and draw the H-F and/or F-P fold lines between the given views.

Step 2 Project the corners of the edge view parallel to the line of sight. Locate the corners by measuring perpendicularly from the frontal plane with depth (D) dimensions to find the true-size view.

beled HRP (horizontal reference plane). This plane will appear as an edge that is parallel to the edge view of the inclined surface from which the auxiliary view is projected. The auxiliary view will lie between the HRP and the top view.

When using the folding-line method, the fold line H-1 is located between the top view and the auxiliary view.

15.6 AUXILIARIES FROM THE FRONT VIEW—FOLDING-LINE METHOD

A plane that appears as an edge in the front view (Fig. 15.9) can be found true size in a primary auxiliary view projected from the front view. Fold line F-1 is drawn parallel to the edge view of the inclined plane in the front view at a convenient location.

The line of sight is drawn perpendicular to the edge view of the inclined plane in the front view. When an observer views the object in this direction, the frontal plane appears as an edge. Consequently, the measurements that are perpendicular to the frontal plane will be seen true length. These dimensions are depth (D) dimensions. Depth dimensions are transferred from the top view to the auxiliary view by using dividers.

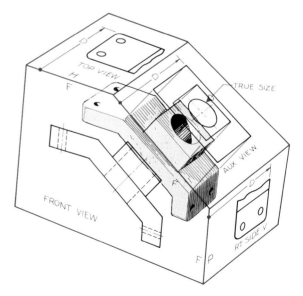

Fig. 15.10 A pictorial showing the relationship of the projection planes used to find the true-size view of the inclined surface of the object.

The inclined surface will be true size in Step 2. Note that the *line of sight* will appear as a *point* in this view.

A practical application of this type of auxiliary view is the part shown in Fig. 15.10. It is enclosed in a glass box where an auxiliary plane is constructed parallel to the inclined plane of the part.

When this arrangement is drawn on a sheet of paper, it will appear as shown in Fig. 15.11. The top and side views are drawn as partial views since the auxiliary view eliminates a need for more than is shown of them. The auxiliary view is located by using the depth dimension that is measured from the edge view of the frontal projection plane. The auxiliary view shows the surface's true size.

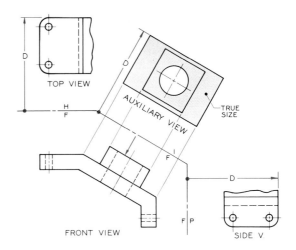

Fig. 15.11 The layout and construction of an auxiliary view of the object in Fig. 15.10 as it would appear on your drawing paper.

15.7 AUXILIARIES FROM THE FRONT VIEW— REFERENCE-PLANE METHOD

The object in Fig. 15.12 has an inclined surface that appears as an edge in the front view. Consequently, this plane can be found true size in a primary auxiliary view.

Since it is symmetrical, there is an advantage to using a reference plane that passes through the center of the symmetrical top view. The reference plane is a frontal plane; consequently this will be called a *frontal reference plane* (FRP) in the auxiliary view.

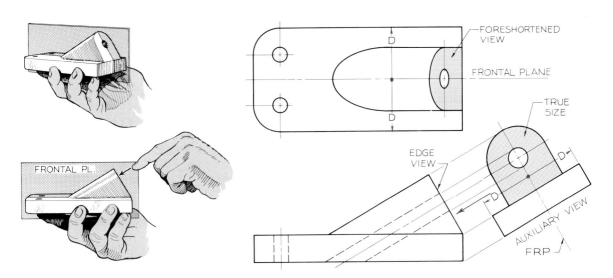

Fig. 15.12 Since the inclined surface of this part is symmetrical, it would be advantageous to use a frontal reference plane (FRP) that is passed through the object. The auxiliary view is projected perpendicularly from the edge view of the plane in the front view. The FRP appears as an edge in the auxiliary view, and depth (D) dimensions are made on each side of it to locate points on the true-size view of the inclined surface.

Fig. 15.13 Auxiliary from the side view—folding-line method

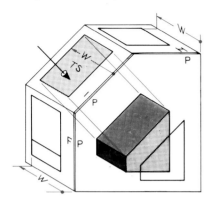

Given An object with an inclined surface shown pictorially.

Required Find the inclined surface true size by an auxiliary view.

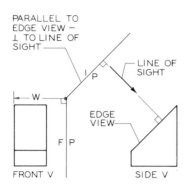

Step 1 Draw a line of sight perpendicular to the edge view of the inclined surface. Draw the P–1 fold line parallel to the edge, and draw the F-P fold line between the given views.

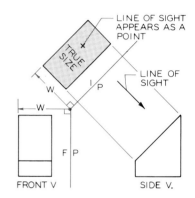

Step 2 Project the corners of the edge parallel to the line of sight. Locate the corners by measuring perpendicularly from the P-1 fold line with width (W) dimensions.

The FRP is located parallel to the edge view of the inclined plane. In the auxiliary view, the reference plane will pass through the center of the view (Fig. 15.13). The auxiliary fold line, P-1, is drawn parallel to the edge view of the inclined surface.

Whereas the folding line would be placed between the top and auxiliary view, the FRP is located through the center of the auxiliary view.

15.8 AUXILIARIES FROM THE PROFILE VIEW—FOLDING-LINE METHOD

Since the inclined surface appears as an edge in the profile plane, it can be found true size in a primary auxiliary view projected from the profile view (Fig. 15.13). The auxiliary fold line, P-1, is drawn parallel to the edge view of the inclined surface.

A line of sight that is perpendicular to the auxiliary plane will see the profile plane as an edge. Consequently, dimensions of width (W) will appear true length in the auxiliary view. Width dimensions are transferred from the front view to the auxiliary view to draw the auxiliary view.

15.9 AUXILIARIES FROM THE PROFILE—REFERENCE-PLANE METHOD

The object in Fig. 15.14 has two inclined surfaces that appear as edges in the right-side view, the profile view. These are found true size by using a *profile reference plane* (PRP) that is a vertical edge in the front view.

The two auxiliary views are drawn to find the inclined surface's true size. The profile reference planes are positioned at the far outsides of the auxiliary views instead of between the profile and auxiliary views as in the folding-line method. The views are found by transferring the width *(W)* dimensions from the edge view of the PRP in the front view to the auxiliary view.

15.10 AUXILIARIES OF CURVED SHAPES

When an auxiliary view is drawn to show a curve that is not a true arc, a series of points must be plotted using the principles previously covered. The cylinder in Fig. 15.15 has a beveled surface that appears as an edge in the front view. When

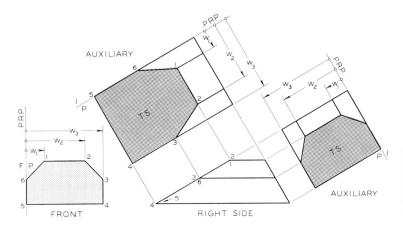

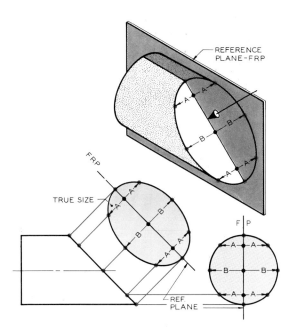

Fig. 15.14 Two auxiliary views are projected from the right-side view using a profile reference plane (PRP). The auxiliary views show the true-size views of the inclined surfaces.

Fig. 15.15 This auxiliary view requires that a series of points be plotted. Since the object is symmetrical, the reference plane is positioned through its center.

found true size, this surface will be elliptical in shape.

Since the cylinder is symmetrical, it is beneficial to use an FRP through the object so that dimensions can be measured on both sides of it. Points are located about the circular right-side view and these are projected to the edge view of the surface in the front view.

The FRP is located parallel to the edge view of the plane, and the points are projected perpendicularly from the edge view of the plane. Dimensions A and B are shown as examples for plotting points in the auxiliary view. More points are needed than are shown to achieve the best points for a smooth curve.

An irregular curve is found as an outline of a true-size surface in Fig. 15.16. Points are located on the curve in the top view and are projected to the front view. These are found in the auxiliary view, by plotting each point using the depth dimension (D) transferred from the top view.

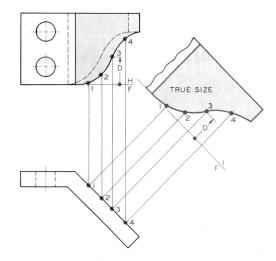

Fig. 15.16 The auxiliary view of this surface required that a series of points be located in the given view and then projected to the auxiliary view. The curve is drawn using an irregular curve through the plotted points.

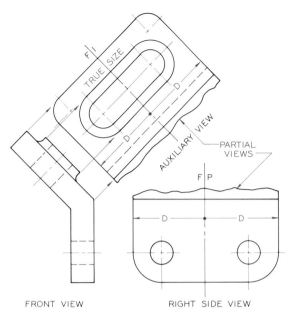

FRONT VIEW   RIGHT SIDE VIEW

Fig. 15.17 This auxiliary view was drawn as a partial view since it is supplemented by the true-size partial view shown in the right-side view. Note that the reference planes are labeled the same as a folding line, which is permissible.

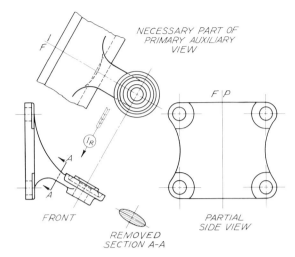

Fig. 15.18 This object is represented by a series of partial views. The hub, which would appear elliptical, is not shown in the side view at all.

15.11 PARTIAL VIEWS

Since an auxiliary view is a supplementary view, some views of an orthographic arrangement can be drawn as partial views. It is understood that the omitted features in one view will be shown in another view, perhaps in the auxiliary view.

The object in Fig. 15.17 is shown by a complete front view and partial auxiliary and side views. These views are easier to draw and they are more functional without sacrificing clarity.

A similar example is given in Fig. 15.18 where the front view is a complete view and the other two views are partial views. A frontal reference plane is passed through the center of the side view. Note that even though it is a reference plane, it is labeled F-1, which is a permissible practice. The guide bracket that was drawn in Fig. 15.18 is shown in Fig. 15.19.

You can see by comparison of the views in Fig. 15.20 that the partial views at A are correctly drawn but they are hard to read. The arrangement at B is much better since the views have been shown as partial views supplemented by an aux-

Fig. 15.19 A photograph of the guide bracket that was drawn orthographically in the previous figure.

iliary view. Note that a cross section is shown in part B to describe the part.

15.12 AUXILIARY SECTIONS

Auxiliary sections can be drawn to show the cross sections through a part to better describe it. A section through a part that is projected as an auxiliary view is shown in Fig. 15.21.

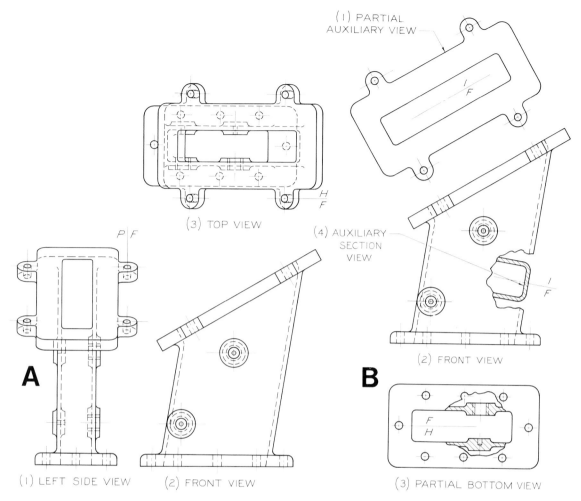

(1) PARTIAL AUXILIARY VIEW

(3) TOP VIEW

(4) AUXILIARY SECTION VIEW

(2) FRONT VIEW

A

(1) LEFT SIDE VIEW

(2) FRONT VIEW

B

(3) PARTIAL BOTTOM VIEW

Fig. 15.20 The three views in A are completely drawn orthographic views with all hidden lines shown. The better method of showing this part is the method in B where an auxiliary view is used to supplement the partial primary views. Hidden lines have been omitted to make the views easier to interpret.

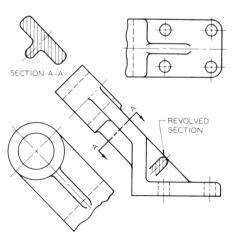

SECTION A-A

REVOLVED SECTION

Fig. 15.21 Auxiliary section A-A is projected from the cutting plane that is labeled A-A, to show the cross section of the object.

Fig. 15.22 Secondary auxiliary views

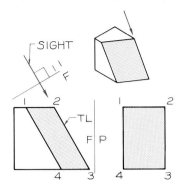

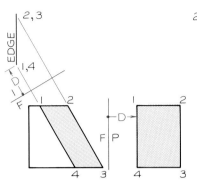

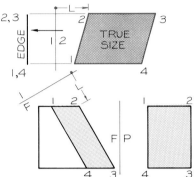

Step 1 A line of sight is drawn parallel to the true-length view of a line on the oblique surface. The folding line F–1 is drawn perpendicular to the line of sight.

Step 2 The primary auxiliary view of the oblique surface is an edge view. Dimension depth (*D*) is used to locate a point in the primary auxiliary view.

Step 3 A line of sight is drawn perpendicular to the edge view, and a secondary auxiliary view is projected in this direction. The dimension *L* is used to locate one of the points in the true-size view.

The section is labeled as Section A-A, and a cutting plane was passed through the object and labeled A-A. This was used to show from where the sectional view was projected and where the object was cut to show the section.

The auxiliary section gives a good description of the part that cannot be as readily understood from the given principal views. Sections will be covered in greater detail in Chapter 16.

15.13   SECONDARY AUXILIARY VIEWS

A **secondary auxiliary view** is an auxiliary view that has been projected from a primary auxiliary view. An example of a secondary auxiliary view is shown in Fig. 15.22.

The inclined plane is found as an edge view in the primary auxiliary view, and then a line of sight is established that is perpendicular to the edge. This second auxiliary view shows the inclined surface true size.

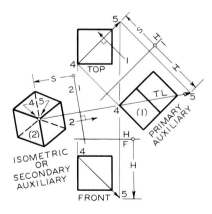

Fig. 15.23 A point view of a diagonal of a cube is found in the secondary auxiliary view. The diagonal is drawn from 4 to 5 in the given views. This line is found true length in the primary auxiliary view and then as a point in the secondary auxiliary view. The secondary auxiliary view is an isometric projection of the cube.

You will remember that it has been necessary in all cases to project from an edge view of a plane before it can be found true size.

The problem in Fig. 15.23 is a secondary auxiliary projection in which the point view of a diagonal of a cube is found. When this is found, the three surfaces of the cube are equally foreshortened. This view is an *isometric projection*, and it is the basis for isometric pictorial drawing. An *isometric drawing* is larger than a *projection*.

Since auxiliary views are supplementary views, the views can be partial views if all features are sufficiently shown. The object in Fig. 15.24 is shown as a series of orthographic views, some of which are partial views. This is an acceptable method of describing a complex part of this type.

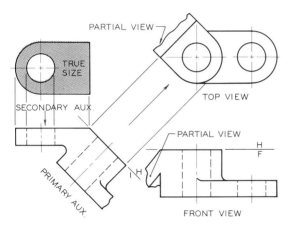

Fig. 15.24 An example of a secondary auxiliary view projected from a partial auxiliary view.

15.14 ELLIPTICAL FEATURES

There are occasions when circular shapes will project as ellipses, and these may need to be shown for clarity. Ellipses of this type can be drawn using any of the techniques introduced in Chapter 12 once the necessary points have been plotted.

The most convenient method of drawing ellipses is with the use of an ellipse guide (template). The angle of the ellipse guide is the angle

the line of sight makes with the edge view of the circular feature. In Fig. 15.25 the angle is found to be 45°. The right-side view is drawn as a 45° ellipse.

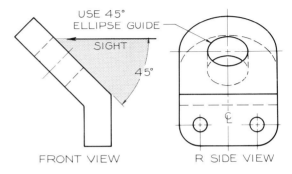

Fig. 15.25 The ellipse guide angle is the angle that the line of sight makes with the edge view of the circular feature. The ellipse guide for the right-side view is 45°.

A more complex problem that involves a secondary auxiliary view is shown in Fig. 15.26. The inclined surface is found true size in the secondary auxiliary view by projecting from the edge view of the surface found in the primary auxiliary view.

The edge view of the circular feature found in the primary auxiliary view can be used to complete the top view of the object. Since the line of sight makes a 30° angle with the edge, this is the angle of the ellipse guide that should be used for drawing the top view.

The elliptical features in the front view can be found by locating a box around the ellipses in the top view and projecting them to the front view. The front view of these points is located by transferring dimensions of height *(H)* from the primary auxiliary view to the front view. This gives the conjugate diameters that can be used for constructing an ellipse as covered in Chapter 12.

The auxiliary views in this problem could not have been drawn unless you knew what the dimensions and specifications of the object were before you started drawing the views. This can be thought of as an object that was drawn by a designer who was translating his or her ideas to the paper while working.

Fig. 15.26 Ellipses in secondary auxiliary views

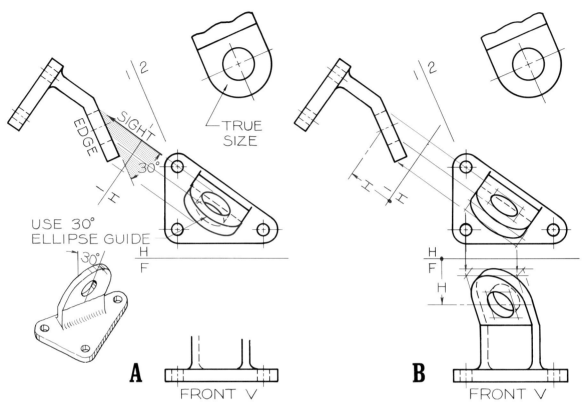

A. The inclined surface is found true size in a secondary auxiliary view. The edge view of the plane in the primary auxiliary view makes a 30° angle with the line of sight from the top view. This is the angle of the ellipse guide for drawing the circular features in the top view.

B. The elliptical features in the front view can be completed by constructing a box about the circular features as shown in the top view. When transferred to the front view, these boxes are used to find the conjugate diameters of the ellipses, and then the ellipses are constructed.

PROBLEMS

The following problems are to be solved on Size A or Size B sheets as assigned by your instructor.

1–10. (Fig. 15.27) Using the example layout, change the top and front views by substituting the top views given at the right in place of the one given in the example. The angle of inclination in the front view is 45° for all problems and the height is 1½ inches in the front view. Construct auxiliary views that show the inclined surface true size. Draw two problems per Size A sheet.

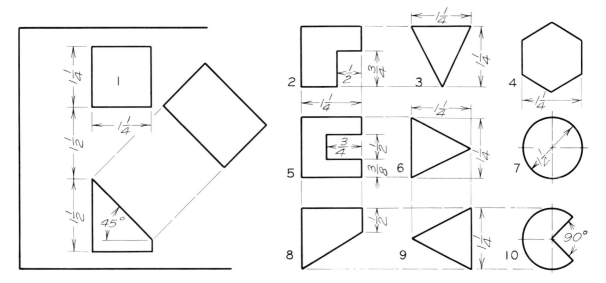

Fig. 15.27 Problems 1–10: Primary auxiliary views.

11–17. (Fig. 15.28) Using the example layout, change the top and right-side views using the side views given at the right of the example problem. Complete the front view and the auxiliary view. Draw two problems per Size AV sheet, showing each half size.

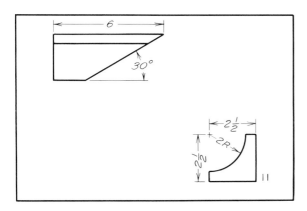

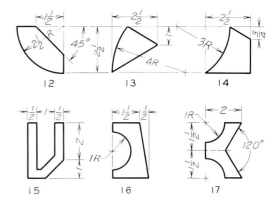

Fig. 15.28 Problems 11–17: Primary auxiliary views.

280

18–40. (Fig. 15.29–Fig. 15.51) Draw the necessary primary views and auxiliary views to describe the parts shown. Draw one per Size A or Size B sheet as assigned. Adjust the scale of each to fit the space on the sheet.

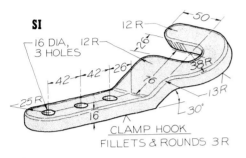

SI

CLAMP HOOK
FILLETS & ROUNDS 3 R

Fig. 15.29 Problem 18: Clamp hook.

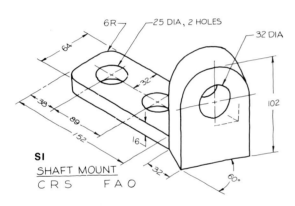

SI

SHAFT MOUNT
C R S   F A O

Fig. 15.30 Problem 19: Shaft mount.

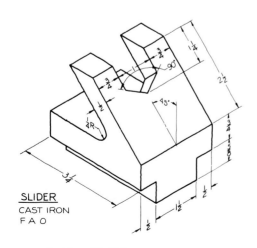

SLIDER
CAST IRON
F A O

Fig. 15.31 Problem 20: Slider.

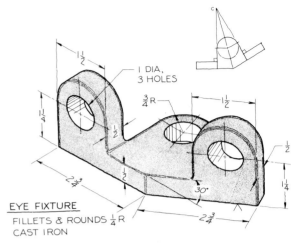

EYE FIXTURE
FILLETS & ROUNDS ¼ R
CAST IRON

Fig. 15.32 Problem 21: Eye fixture.

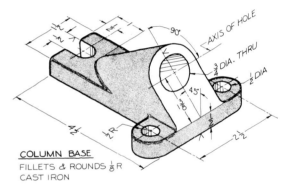

COLUMN BASE
FILLETS & ROUNDS ⅛ R
CAST IRON

Fig. 15.33 Problem 22: Column base.

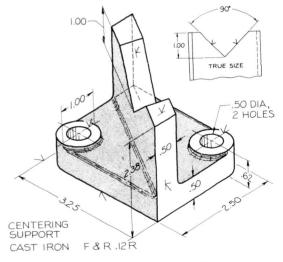

CENTERING
SUPPORT
CAST IRON   F & R .12 R

Fig. 15.34 Problem 23: Centering support.

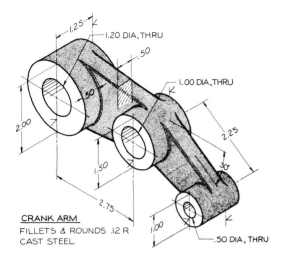

CRANK ARM
FILLETS & ROUNDS .12 R
CAST STEEL

Fig. 15.35 Problem 24: Crank arm.

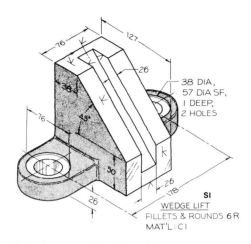

38 DIA,
57 DIA SF,
.1 DEEP,
2 HOLES

SI
WEDGE LIFT
FILLETS & ROUNDS 6 R
MAT'L: CI

Fig. 15.38 Problem 27: Wedge lift.

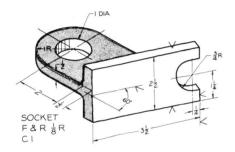

SOCKET
F & R ⅛R
CI

Fig. 15.36 Problem 25: Socket.

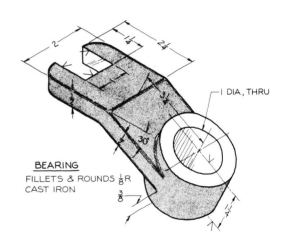

BEARING
FILLETS & ROUNDS ⅛R
CAST IRON

Fig. 15.39 Problem 28: Bearing.

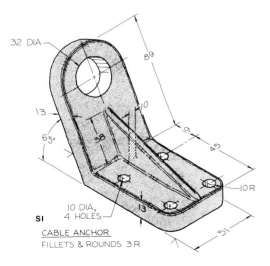

SI
CABLE ANCHOR
FILLETS & ROUNDS 3 R

Fig. 15.37 Problem 26: Cable anchor.

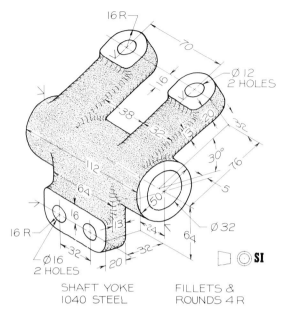

SHAFT YOKE
1040 STEEL

FILLETS &
ROUNDS 4 R

Fig. 15.40 Problem 29: Shaft yoke.

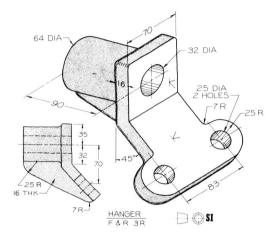

HANGER
F & R 3R

Fig. 15.43 Problem 32: Hanger.

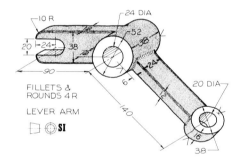

FILLETS &
ROUNDS 4 R

LEVER ARM

Fig. 15.41 Problem 30: Lever arm.

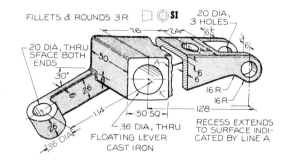

FLOATING LEVER
CAST IRON

RECESS EXTENDS
TO SURFACE INDI-
CATED BY LINE A

Fig. 15.44 Problem 33: Floating lever.

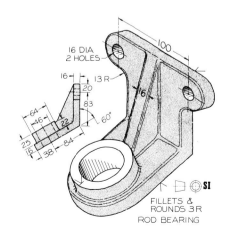

FILLETS &
ROUNDS 3 R
ROD BEARING

Fig. 15.42 Problem 31: Rod bearing.

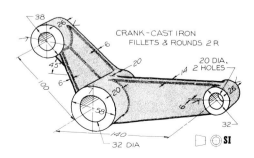

CRANK-CAST IRON
FILLETS & ROUNDS 2 R

Fig. 15.45 Problem 34: Crank.

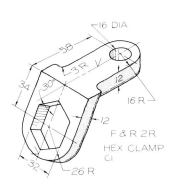

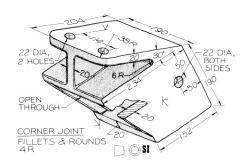

Fig. 15.49 Problem 38: Corner joint.

Fig. 15.46 Problem 35: Hexagon angle.

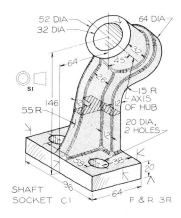

Fig. 15.50 Problem 39: Shaft socket.

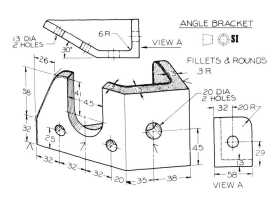

Fig. 15.47 Problem 36: Angle bracket.

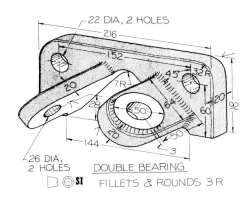

Fig. 15.48 Problem 37: Double bearing.

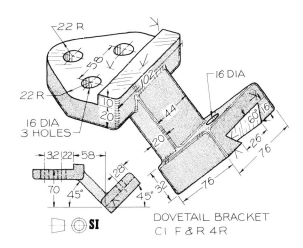

Fig. 15.51 Problem 40: Dovetail bracket.

41–42. (Fig. 15.52–Fig. 15.53) Lay out these orthographic views on Size B sheets and complete the auxiliary and primary views.

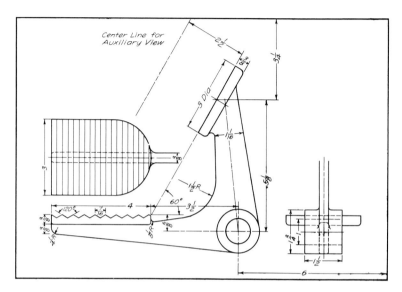

Fig. 15.52 Problem 41: Clutch pedal.

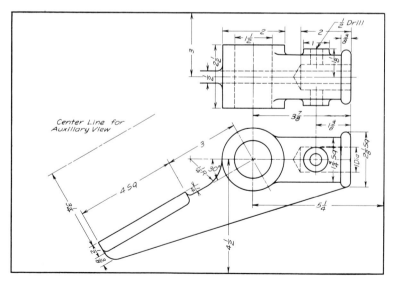

Fig. 15.53 Problem 42: Adjustment.

43–45. (Fig. 15.54–Fig. 15.56) Construct orthographic views of the given objects and draw auxiliary views that give the true-size views of the inclined surfaces using secondary auxiliary views. Draw one per Size B sheet.

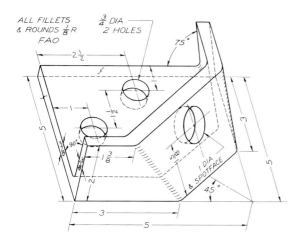

Fig. 15.54 Problem 43: Corner connector.

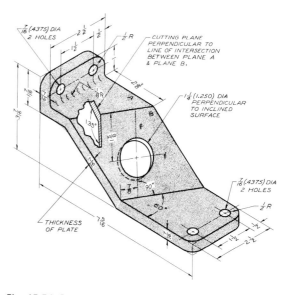

Fig. 15.56 Problem 45: Oblique support.

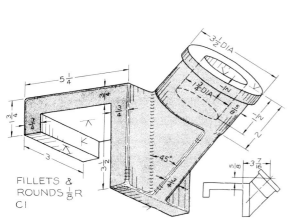

Fig. 15.55 Problem 44: Shaft bearing.

SECTIONS

16.1 SECTIONS

The standard orthographic views that show all hidden lines may not effectively reveal the true details of an object. This shortcoming can often be improved by using a technique of cutting away part of the object and looking at the cross section view. Such a cutaway view is called a **section.**

A section is shown pictorially in Fig. 16.1A, where an imaginary cutting plane is passed through the object in order to show its internal features. The standard top and front orthographic views are shown at B. The method of drawing a section is shown at C, where the front view has been converted to a *full section*, and the cut portion is cross-hatched or section-lined. Hidden lines have been omitted since they are not needed. The cutting plane is drawn as a heavy line with short dashes at intervals; this can be thought of as a knife-edge cutting through the object.

By referring to the top view and the front sectional view, you have no hidden lines to interpret, and you can understand the cross-sectional view of the object more easily.

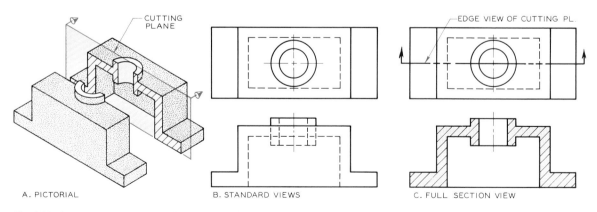

Fig. 16.1 A comparison of a regular orthographic view with a full-section view of the same object to show the internal features as well as the external features.

Two types of cutting planes are shown in Fig. 16.2. Either is acceptable, although the upper example is more often used. The spacing of the dashes depends upon the size of the drawing. The weight of the cutting plane is the same as that of a visible object line. Letters can be placed at each end of the cutting plane to label the sectional view, such as Section B-B, wherever it is drawn. Think of the cutting-plane line as a knife-edge cutting through a part.

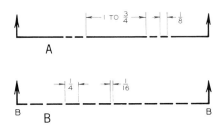

Fig. 16.2 Typical cutting plane lines used to represent sections. The cutting planes marked B-B will produce a section that will be labeled Section B-B.

The three basic views that may appear as sections are shown in Fig. 16.3 with their respective cutting planes. Each cutting plane has perpendicular arrows with the ends pointing in the direction of the line of sight for the section. For example, the cutting plane at A passes through the top view, the front of the top view is removed, and the line of sight is toward the remaining portion of the top view. The top view will appear as a section when the cutting plane passes through the front view and the line of sight is downward (B). When the cutting plane passes through the front view (C), the right side view will be a section.

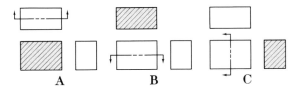

Fig. 16.3 The three standard positions of cutting planes that pass through given views to result in sectional views in the front, top, and side views. The arrows point in the direction of the line of sight for each section.

16.2 SECTIONING SYMBOLS

The symbols that are used to distinguish between different materials in section are shown in Fig. 16.4. Although the symbols can be used to indicate the materials within a section, it is good practice to provide supplementary notes specifying the materials to avoid misinterpretation.

> The *cast-iron* symbol of evenly spaced section lines can be used to represent *any material*, and is the most often used sectioning symbol.

Notes are even more essential to specify the type of material when this general symbol is used. Cast-iron symbols are usually drawn with a 2H pencil with lines that are slanted at 45°, or any other standard angle such as 30° or 60°, and spaced about $\frac{1}{16}''$ apart by eye.

The proper spacing of section lines for cast iron is shown at the top left of Fig. 16.5 where the lines are evenly spaced. Other common errors of section-lining are shown in the remainder of the figure.

Extremely thin parts such as sheet metal, washers, or gaskets (Fig. 16.6) are sectioned by blacking in the areas completely rather than using section lines. Large parts are sectioned with an *outline section* to save time and effort. The section lines are drawn closer together in small parts than in larger parts.

> Sectioned areas should be section-lined with line symbols that are neither parallel nor perpendicular to the outlines of the parts (Fig. 16.7), but are at some other angle with the part's outlines (A).

Parallel and perpendicular section lines may be confused as serrations or other machining treatments of the surface.

16.3 SECTIONING ASSEMBLIES

When an assembly of several parts is sectioned to give the relationship of the parts, it is important that the section lines be drawn at varying angles

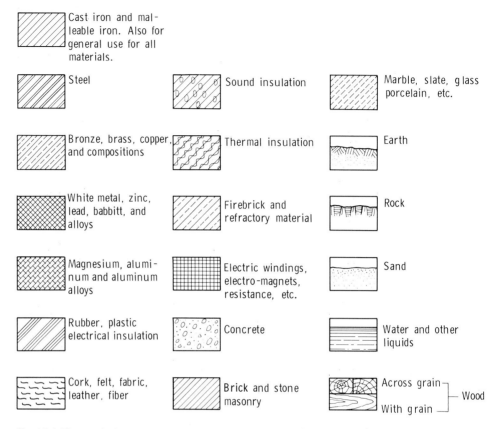

Fig. 16.4 The symbols used for section-lining parts in section. Note that the cast-iron symbol can be used for all materials.

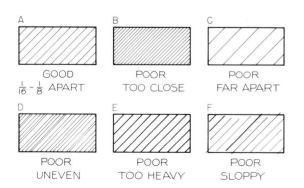

Fig. 16.5 Good section lines are drawn as thin lines and are spaced about $\frac{1}{16}$" to $\frac{1}{8}$" apart. Some typical section-lining errors are shown here.

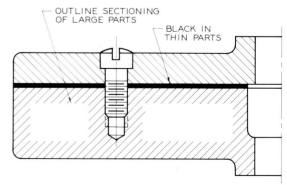

Fig. 16.6 Thin parts are blacked in and large areas are section-lined around their outlines to save time and effort.

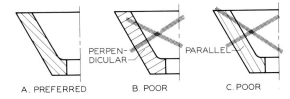

Fig. 16.7 Section lines should be drawn so that they are not parallel or perpendicular to the outline of a part.

to distinguish the parts (Fig. 16.8). The use of different material symbols, when the assembly is composed of parts made from different materials, is helpful in distinguishing the parts from one another. The same part is cross-hatched at the same angle and with the same symbol even though the part may be separated into different areas, as shown at Fig. 16.8B.

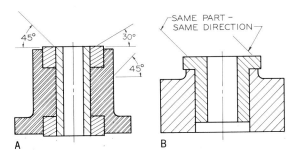

Fig. 16.8 The section lines of the same part should be drawn in the same direction. Section lines of different parts should be drawn at varying angles to separate the parts.

Two assemblies are illustrated in Fig. 16.9 where section lines are effectively used to identify the parts of the assembly.

16.4 FULL SECTIONS

A **full section** is a sectional view formed by passing a cutting plane fully through an object and removing half of it to give a view of its internal features.

The object shown at Fig. 16.10A is drawn as three orthographic views in which a number of hidden lines are shown. The front view can be drawn as a full section by passing a cutting plane fully through the top view (B), and removing the front portion as shown at C. The standard way of showing this section can be seen at D where the cutting plane is drawn through the top view with arrows indicating the direction of sight. The front view is then section-lined to give the full section.

A full section through a cylindrical part is shown in Fig. 16.11 where half of the object is removed. A common mistake in constructing sectional views is the omission of the visible lines behind the cutting plane (B); C shows the correctly drawn sectional view. Hidden lines are omitted in all sectional views unless they are believed to be necessary to provide a clear understanding of the view. Visible lines behind the sectional view are omitted also if they confuse the view.

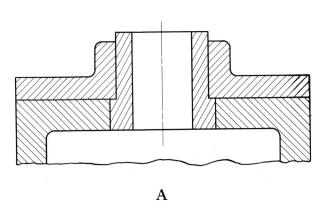

A

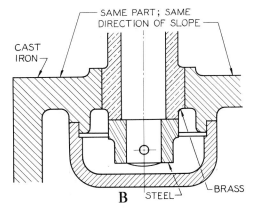

B

Fig. 16.9 A typical assembly in section with well-defined parts and correctly drawn section lines.

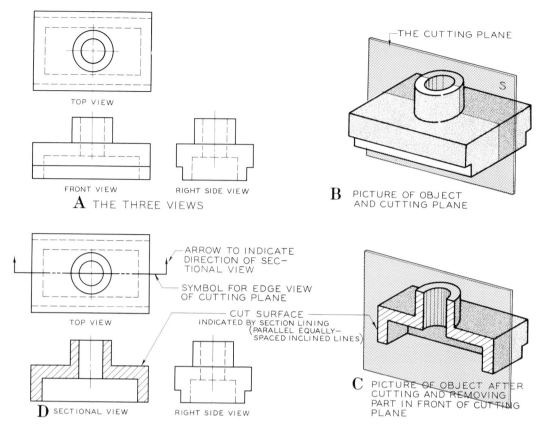

TOP VIEW

FRONT VIEW   **RIGHT SIDE VIEW**

A THE THREE VIEWS

THE CUTTING PLANE

B PICTURE OF OBJECT AND CUTTING PLANE

ARROW TO INDICATE DIRECTION OF SEC-TIONAL VIEW

SYMBOL FOR EDGE VIEW OF CUTTING PLANE

CUT SURFACE INDICATED BY SECTION LINING (PARALLEL EQUALLY-SPACED INCLINED LINES)

TOP VIEW

D SECTIONAL VIEW   RIGHT SIDE VIEW

C PICTURE OF OBJECT AFTER CUTTING AND REMOVING PART IN FRONT OF CUTTING PLANE

Fig. 16.10 A full section is a section formed by a cutting plane that passes fully through it. The cutting plane is shown passing through the top view, and the direction of sight is indicated by the arrows at each end. The front view is converted to a sectional view to give a clear understanding of the internal features.

Fig. 16.11 Full section—cylindrical part

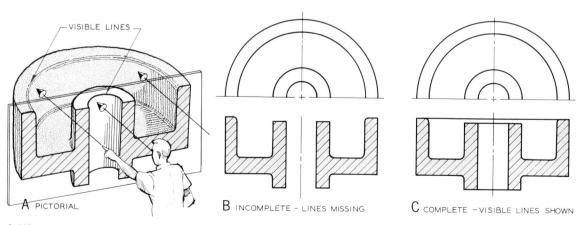

VISIBLE LINES

A PICTORIAL

B INCOMPLETE - LINES MISSING

C COMPLETE - VISIBLE LINES SHOWN

A. When a full section is passed through an object, you will see lines behind the sectioned area.

B. If only the sectioned area were shown, the view would be incomplete.

C. Visible lines behind the sectioned area must be shown also.

Figure 16.12 is an example of a part whose right-side view is shown as a full section. Likewise, the part in Fig. 16.13 illustrates a front view that appears as a full section. Note the lines behind the cutting plane are shown as visible lines.

16.5 PARTS NOT SECTION-LINED

Many standard parts are not section-lined even though the cutting plane passes through them. Examples of such parts in Fig. 16.14 are nuts and bolts, rivets, shafts, and set screws. These parts have no internal features, and sections through them would not be of value.

Other parts not section-lined are roller bearings, ball bearings, gear teeth, shafts, dowels, pins, set screws, and washers (Fig. 16.15).

16.6 RIBS IN SECTION

Ribs are not section-lined when the cutting plane passes flatwise through them as shown in Fig. 16.16a, since this would give a misleading impression of the rib. A similar example is the section shown in Fig. 16.16c. However, a rib is section-

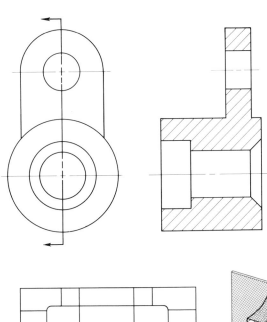

Fig. 16.12 A full section with the cutting plane shown.

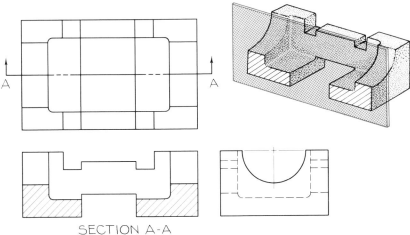

SECTION A-A

Fig. 16.13 A full section, Section A-A, is used to supplement the given views of the object.

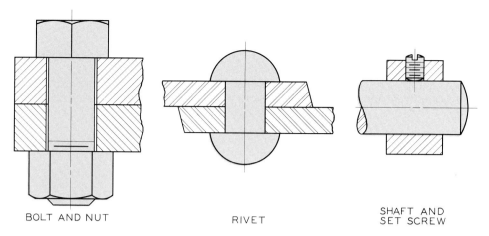

BOLT AND NUT

RIVET

SHAFT AND
SET SCREW

Fig. 16.14 These parts are not cross sectioned even though the cutting plane passes through them.

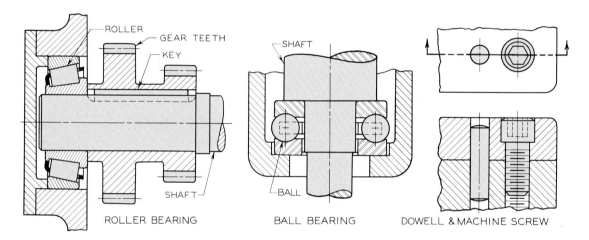

ROLLER
GEAR TEETH
KEY
SHAFT

ROLLER BEARING

SHAFT
BALL

BALL BEARING

DOWELL & MACHINE SCREW

Fig. 16.15 These parts are not section-lined even though cutting planes pass through them.

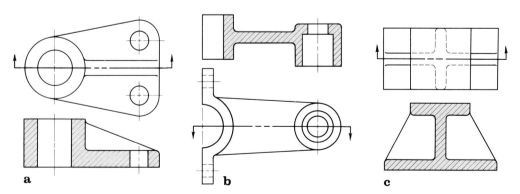

a

b

c

Fig. 16.16 A rib that is cut in a flatwise direction by a cutting plane is not section-lined. Ribs are section-lined when cutting planes pass perpendicularly through them as shown at (b) and (c).

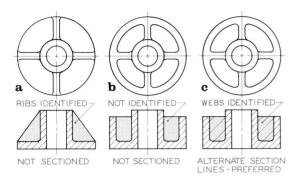

Fig. 16.17 Outside ribs in section are not section-lined. Webs that are poorly identified as at (b) should be identified by cross-hatching them with alternate section lines as shown at (c). This calls attention to the fact that webs are present.

lined when the cutting plane passes through it and shows its true thickness (Fig. 16.16b).

An alternate method of section-lining webs and ribs is shown in Fig. 16.17. At (a) the ribs are not section-lined since the cutting plane passes through them in a flatwise direction. The webs at (b) are symmetrically spaced about the hub. As a rule, webs are not cross-hatched, but this would leave the webs unidentified. Therefore, it is better to use the *alternate sectioning* technique and extend every other section line through the webs to call attention to them as shown at (c).

The ribs in Fig. 16.18 are not section-lined and thus afford a more descriptive view of the

part. If the ribs had been section-lined, the section would have given the impression that the part was solid and conical in shape (b). Note that the top views are partial views and the portion nearest the sectional view is the portion that has been omitted.

16.7 HALF-SECTIONS

A **half-section** is a view that results from passing a cutting plane halfway through an object, removing a quarter of it.

This type of section is most often used with symmetrical parts, cylinders in particular. A cylindrical part at Fig. 16.19A is shown as a pictorial half-section at B. The method of drawing the orthographic half-section is shown at C. In this view, both the internal and external features can be seen; the hidden lines are omitted in the sectional view.

A similar half-section is given in Fig. 16.20. Hidden lines are omitted unless they are believed to be essential for understanding or dimensioning the object. The half-section in Fig. 16.21 has been drawn without showing the cutting plane. This is permissible if it is obvious where the cutting plane was passed through the view to give the section.

Two half-sections are drawn in Fig. 16.22 to describe the object. The front and right-side views are constructed as half-sections as if the quarter cut away by the cutting plane had been removed. A few, but not all, of the hidden lines have been given in each view to clarify the sections. Note that, instead of using an object line, centerlines are given to separate the sectional half from the half that appears as an external view.

16.8 PARTIAL VIEWS

A conventional method of representing symmetrical views is shown in Fig. 16.23. A half-view is sufficient when it is drawn adjacent to the sectional view as at A. In full sections, the removed half is the portion nearest the section (part A). When drawing half-views that are associated with views (nonsectional views), the removed half of the partial view will be the half that is away from the adjacent view (Fig. 16.23B).

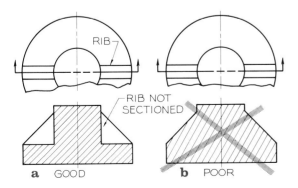

Fig. 16.18 It can be seen that, when ribs are not section-lined, the view is more descriptive of the part. Partial views are used in the top views to save space. The front part of the top view is removed when the front view is a section.

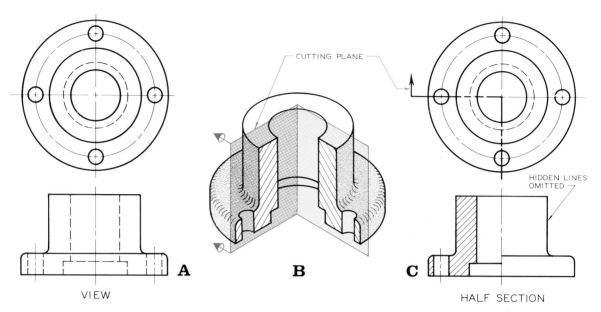

Fig. 16.19 The cutting plane of a half-section passes halfway through the object, which results in a sectional view that shows half of the outside and half of the inside of the object. Hidden lines are omitted unless they are necessary to clarify the view.

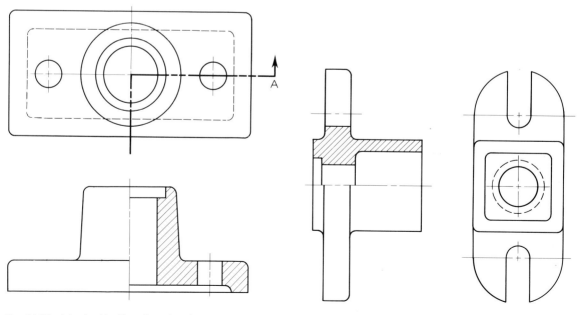

Fig. 16.20 A typical half-section drawing.

Fig. 16.21 When it is obvious where the cutting plane is located, it is unnecessary to show it as in this example.

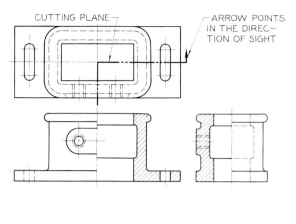

Fig. 16.22 Two separate half-sections are found in this example, using the same cutting plane for both.

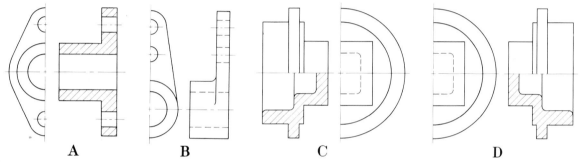

A   **B**   **C**   **D**

Fig. 16.23 Half-views can be used for symmetrical parts to conserve space and time. A. The omitted portion of the view is toward the full section. B. The omitted portion of the view is away from the adjacent view when it is not sectioned. C. and D. The omitted half can be toward or away from the section in the case of a half-section.

When the partial views are drawn in conjunction with half-sections, either the near or far halves of the partial views can be omitted, as shown in Fig. 16.23C and D. Partial views are faster to draw and require less space.

16.9 OFFSET SECTIONS

An **offset section** is a type of full section in which the cutting plane is offset to pass through important features that would be missed by the usual continuous full section formed by a flat plane.

An offset section is shown pictorially in Fig. 16.24A where the plane is offset to pass through the large hole and one of the small holes. The cut

object is shown at B. The method of drawing an offset section orthographically is given in part C. The cut formed by the offset is not shown in the section since this is an imaginary cut.

The part in Fig. 16.25 is one that lends itself to representation by an offset section. Another offset section is used to describe a part in Fig. 16.26. As previously discussed, the ribs are not section-lined in this case, whether the plane passes through them or not.

16.10 REVOLVED SECTIONS

A **revolved section** is used to describe a cross section of a part to eliminate the need for drawing an entirely separate view.

Fig. 16.24 Offset section

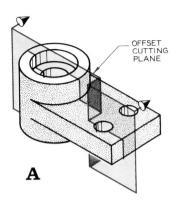

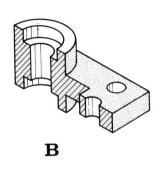

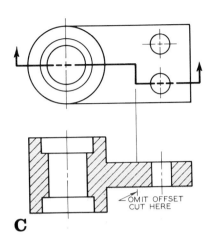

A

A. An offset section may be necessary to show all typical features of a sectioned part.

B

B. When the front portion is removed, the internal features can be seen. Note that the cutting plane has been offset.

C

C. In an offset section, the offset cut is not shown. The section is shown as if it were a typical full section.

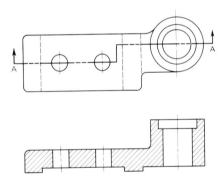

Fig. 16.25 An offset section.

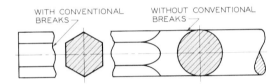

Fig. 16.27 Examples of revolved sections with and without conventional breaks.

For example, revolved sections are used to indicate cross sections of the parts in Fig. 16.27. One of the revolved sections is positioned within the view and conventional breaks are drawn on each side of the hexagonal section. The circular cross section is drawn superimposed on the cylindrical portion of the part without using conventional breaks. At the option of the drafter, either of these methods can be used when drawing revolved sections.

A more advanced type of revolved section is illustrated in Fig. 16.28 where a cutting plane is passed through the object (Step 1). The plane is imagined to be revolved in the top view to give a true-size revolved section in the front view (Step 2). Note that the object lines do not pass through the revolved section in the front view. It would have been permissible to use conventional breaks on each side of the revolved section as used in Fig. 16.27.

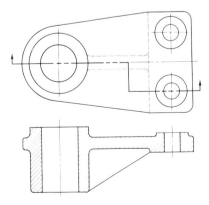

Fig. 16.26 An offset section.

Fig. 16.28 Revolved section

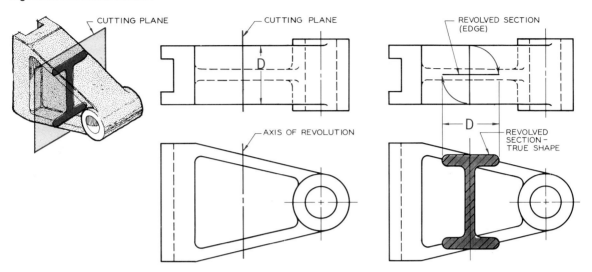

CUTTING PLANE   CUTTING PLANE   REVOLVED SECTION (EDGE)

AXIS OF REVOLUTION

REVOLVED SECTION – TRUE SHAPE

Step 1 An axis of revolution is shown in the front view. The cutting plane would appear as an edge in the top view if it were shown.

Step 2 The vertical section in the top view is revolved so that the section can be seen true size in the front view. Object lines are not drawn through the section.

Typical revolved sections are shown in Fig. 16.29 at A and B. These sections provide a direct method of giving a part's cross section without relying upon another complete orthographic view that would be time-consuming to draw and that might be less effective.

16.11 REMOVED SECTIONS

A **removed section** is a revolved section that has been removed from the view where it was revolved, as shown in Fig. 16.30.

Centerlines are used as axes of rotation to show from where the sections were taken. Removed sections may be necessary where room does not permit revolution on the given view (Fig. 16.31A); but instead, the cross section must be removed from the view as shown at B.

Removed sections do not have to be positioned directly along an axis of revolution adjacent to the view from where the sections were taken. Instead, cutting planes can be labeled at each end as shown in Fig. 16.32 to specify the sections. For example, the plane labeled with an A at each end is used to label Section A-A. Section B-B is found in a similar manner.

A series of removed sections is shown in Fig. 16.33. Removed sections may be put on different sheets when a set of drawings is composed of many pages. In this case, a cutting plane may be labeled as shown in Fig. 16.34. The A at each end identifies the section as Section A-A, and the numerals indicate on which page of the set of drawings the section is located.

Removed views can also be used to provide inaccessible orthographic views (nonsectional views) when these are needed, as shown in Fig. 16.35.

16.12 BROKEN-OUT SECTIONS

A **broken-out section** is a convenient method used to show interior features that would improve the understanding of a drawing without drawing a separate view as a section.

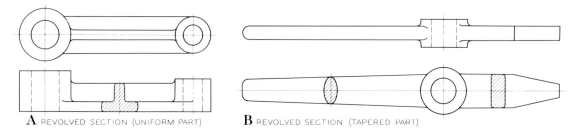

A REVOLVED SECTION (UNIFORM PART)   **B** REVOLVED SECTION (TAPERED PART)

Fig. 16.29 The revolved sections given here are helpful in describing the cross-sectional characteristics of the two parts. This is much more effective than using additional orthographic views.

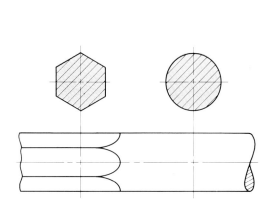

Fig. 16.30 Removed sections are similar to revolved sections, but they have been removed outside the object along an axis of revolution.

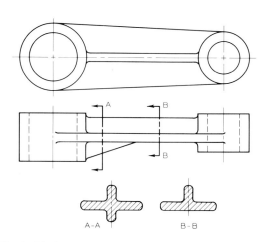

A-A   B-B

Fig. 16.32 Sections can be lettered at each end of a cutting plane such as A-A. This removed section can then be shown elsewhere on the drawing, where it is designated as Section A-A.

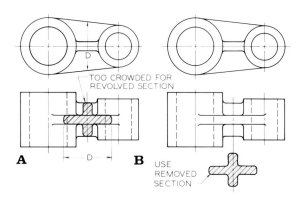

A   **B**

TOO CROWDED FOR REVOLVED SECTION

USE REMOVED SECTION

Fig. 16.31 Removed sections can be used to good advantage where space does not permit the use of a resolved section.

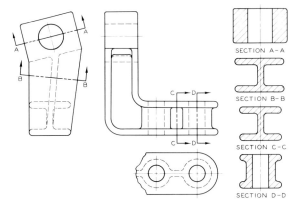

SECTION A-A

SECTION B-B

SECTION C-C

SECTION D-D

Fig. 16.33 A series of removed sections, each labeled according to the cutting plane that was used to find them.

CUTTING PLANE

Fig. 16.34 It may be necessary to remove a section to another page in a set of drawings. In this case, each end of the cutting plane can be labeled with a letter and a number. The letters refer to Section A-A and the numbers mean that this section is located on page 7.

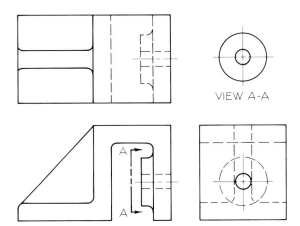

VIEW A-A

Fig. 16.35 A removed view (not a section) can also be used to view a part from an unconventional direction that would be more effective than an entire orthographic view.

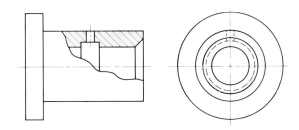

Fig. 16.36 A broken-out section is used to show an internal feature by using a conventional break.

A portion of the part in Fig. 16.36 is broken out to reveal details of the wall thickness to better explain the drawing. This reduces the need for hidden lines; therefore they may be omitted if it is desired to do so.

Figure 16.37 is a similar example of a broken-out section where hidden lines are given. The irregular lines that are used to represent the break

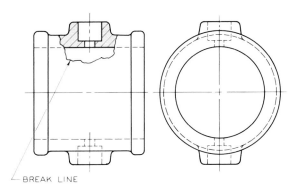

BREAK LINE

Fig. 16.37 A broken-out section that shows a section through the boss at the top of a collar.

are called *conventional breaks*; these are discussed in Section 16.14.

16.13 PHANTOM (GHOST) SECTIONS

A **phantom section** or **ghost section** is used occasionally to depict parts as if they were viewed by an x-ray.

An example is shown in Fig. 16.38. The cutting plane is drawn in the usual manner, but the section lines are drawn as dashed lines. You can see that if the object had been shown as regular full section, the circular hole through the front surface could not have been shown in the same view.

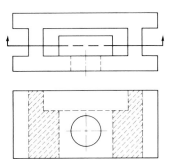

Fig. 16.38 Phantom sections give an "x-ray" view of an object. The section lines are shown as dashed lines. This makes it possible to show the section without removing the hole in the front of the part.

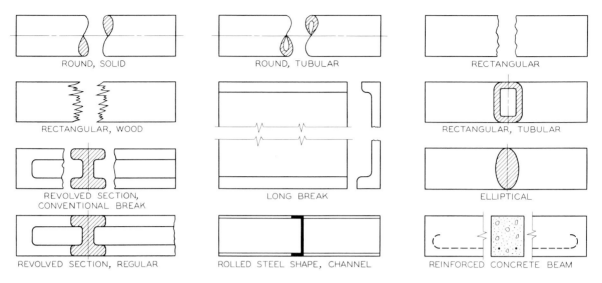

Fig. 16.39 These are often-used conventional breaks that are used to remove a portion of an object so it can be drawn at a larger scale.

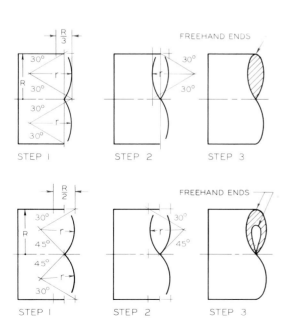

Fig. 16.40 The steps of constructing conventional breaks of solid and tubular shapes with instruments.

16.14 CONVENTIONAL BREAKS

Some of the previous examples have used conventional breaks to indicate the removal of parts of an object. Examples of conventional breaks are illustrated in Fig. 16.39. These should be reviewed and learned in order for these short-cut methods to be used in drawing views. The "figure 8" breaks that are used for cylindrical and tubular parts can be drawn freehand, or they can be drawn by using a compass as shown in Fig. 16.40 when they are drawn to a large scale.

One use of conventional breaks is to shorten a long piece that has a uniform cross section. The long part in Fig. 16.41 has been shortened and drawn at a larger scale for more clarity by using the conventional breaks shown at (b). The dimension specifies the true length of the part, and the breaks indicate that a portion of the length has been removed.

16.15 CONVENTIONAL REVOLUTIONS

Three conventional sections are shown in Fig. 16.42. The center hole is omitted at (a) since it does not really pass through the center of the circular plate. However, the hole at (b) does pass

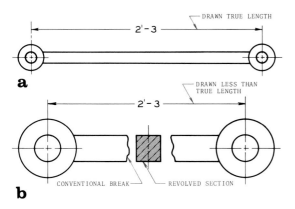

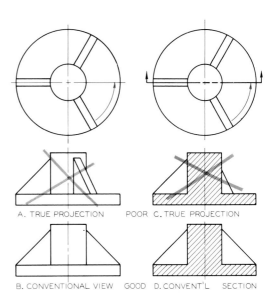

Fig. 16.41 By using conventional breaks and a revolved section, this part can be drawn on a larger scale that is easier to read.

Fig. 16.43 Symmetrically located ribs are shown revolved in both orthographic and sectional views as a conventional practice.

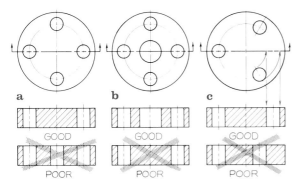

Fig. 16.42 Symmetrically spaced holes in a circular plate should be revolved in their sectional views to show them at their true radial distance from the center. At (a), no hole is shown at the center, but a hole is shown at the center in (b), since the hole is through the center of the plate. At (c), one of the holes is rotated to the cutting plane so that the sectional view will be symmetrical and more descriptive.

through the plate's center and is sectioned accordingly. At (c), the cutting plane does not pass through one of the symmetrically spaced holes in the top view, but the hole is revolved to the cutting plane in order to show the recommended full section.

When ribs are symmetrically spaced about a hub (Fig. 16.43) it is conventional practice to revolve them to where they will appear true size in either a view (B) or a section (D). A full section that shows both ribs and holes revolved to their true-size locations can be seen in Fig. 16.44.

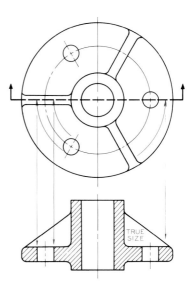

Fig. 16.44 A part with symmetrically located ribs and holes is shown in section with both ribs and holes rotated to the cutting plane.

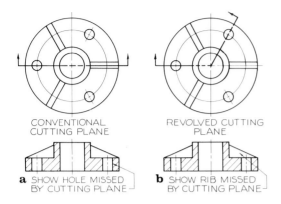

CONVENTIONAL   REVOLVED CUTTING
CUTTING PLANE   PLANE

a SHOW HOLE MISSED   **b** SHOW RIB MISSED
BY CUTTING PLANE   BY CUTTING PLANE

Fig. 16.45 Symmetrically located ribs are shown in section in revolved positions to show the ribs true size. Foreshortened ribs are omitted.

The cutting plane can be positioned as shown at Fig. 16.45. Even though the cutting plane does not pass through the ribs and holes in (a), the sectional view should be drawn as the view at (b), where the cutting plane is shown revolved through the hole. The cutting plane can be drawn in either position as preferred by the drafter. The revolved cutting plane can be used to show the path of the cutting plane in more complex parts.

Symmetrically spaced spokes are rotated and not section-lined, in the same manner as ribs in section. Figure 16.46 illustrates the preferred and poor-practice methods of representing spokes. Only the revolved, true-size spokes are drawn; the intermediate spokes are omitted.

The reason for not section-lining spokes can be seen in Fig. 16.47. If the spokes at B had been section-lined, the cross section of the part would be confused with the part at A where there are no spokes, but a continuous web.

The lugs that are symmetrically positioned about the central hub of the object in Fig. 16.48 are revolved to show them true size in both views and in sections. A more complex object that involves the same principle of rotation can be seen in Fig. 16.49. The oblique arm is drawn in the sec-

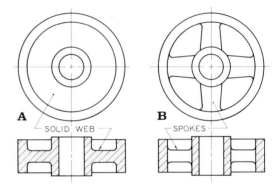

A SOLID WEB   **B** SPOKES

Fig. 16.47 Solid webs in sections of the type at A are section-lined. Spokes are not section-lined in section B when the cutting plane passes through them in this manner.

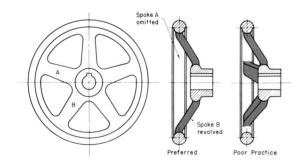

Spoke A omitted

Spoke B revolved

Preferred   Poor Practice

Fig. 16.46 Symmetrically positioned spokes are revolved to show the spokes true size in section. Spokes are not section-lined in section.

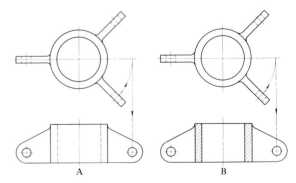

A   B

Fig. 16.48 Symmetrically spaced lugs (flanges) are revolved to show the front view and the sectional view as symmetrical.

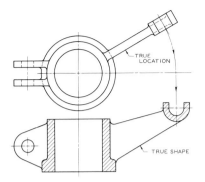

Fig. 16.49 A part with an oblique feature attached to the circular hub is revolved so it will appear true shape in the front view, which is a sectional view.

tion as if it had been revolved to the centerline in the top view and then projected to the sectional view.

16.16 AUXILIARY SECTIONS

Auxiliary sections can be used to supplement the principal views used in orthographic projections as shown in Fig. 16.50. Auxiliary cutting plane A-A is passed through the front view, and the auxil-

iary view is projected from the cutting plane as indicated by the sight arrows. The sectional view, A-A, gives the cross sectional description of the part.

Another auxiliary section is given in Fig. 16.51 to show a section through a threaded boss. The location of this threaded hole is located by a center line in the front view. In Fig. 16.52, Section A-A is drawn to clarify the front view where the cutting plane is positioned. The right-side view can be drawn as a partial view.

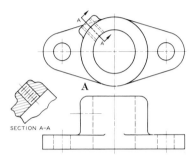

Fig. 16.51 A partial auxiliary section projected from the top view to show a threaded boss.

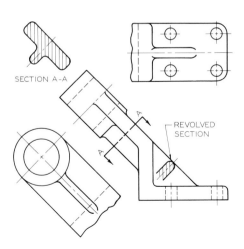

Fig. 16.50 Sectional views can be shown as auxiliary views for added clarity.

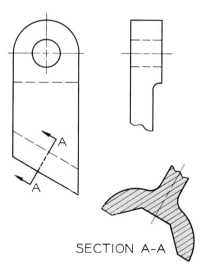

Fig. 16.52 An auxiliary section projected from the front view. The right-side view is a partial view since it is supplemented by the auxiliary section.

PROBLEMS

These problems can be solved on Size A or Size B sheets.

1–7. (Fig. 16.53) Full sections: Draw two of these problems per Size A sheet. Each grid is equal to ¼ inch or 5 mm. Complete the front views as full sections.

8–12. (Fig. 16.53) Half sections: Draw two of these problems per Size A sheet. Each grid is equal to ¼ inch or 5 mm. Complete the front views as half sections.

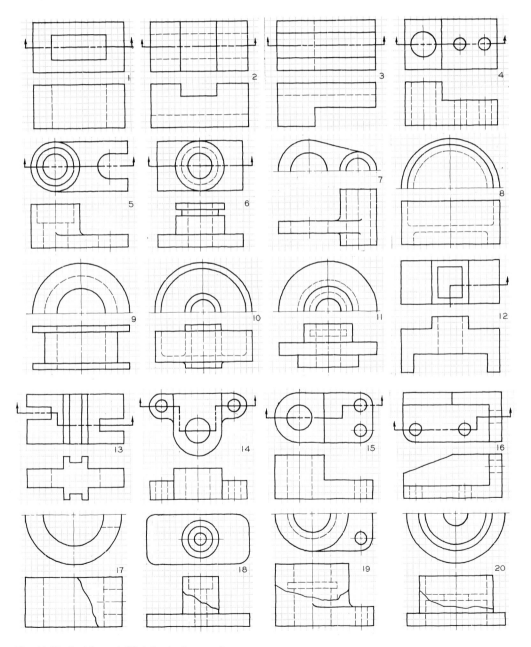

Fig. 16.53 Problems 1–20: Introductory sections.

13–16. (Fig. 16.53) Offset sections: Draw two of these problems per Size A sheet. Each grid is equal to ¼ inch or 5 mm. Complete the front views as offset sections.

17–20. (Fig. 16.53) Broken-out sections: Draw two of these problems per Size A sheet. Each grid is equal to ¼ inch or 5 mm. Complete the broken-out sections.

21–30. (Fig. 16.54, Fig. 16.55, Fig. 16.56) Half-sections: The views given are rectangular views of cylindrical parts. Draw one problem per Size A sheet. Show the circular view and draw the rectangular view as a half-section. Omit dimensions and show the cutting plane. Select appropriate scale.

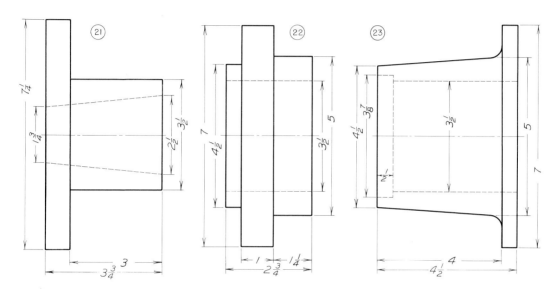

Fig. 16.54 Problems 21–23: Half sections.

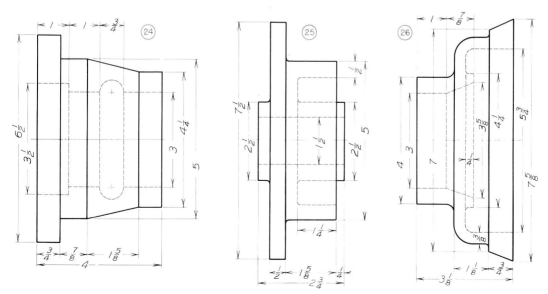

Fig. 16.55 Problems 24–26: Half sections.

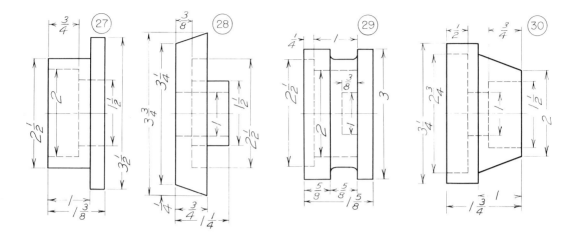

Fig. 16.56 Problems 27–30: Half sections.

31–36. (Fig. 16.57–Fig. 16.59) Full and offset sections: Draw the given views on Size A sheets, two per sheet. Complete the sections as indicated by the cutting planes. Problem 36 will require an offset section.

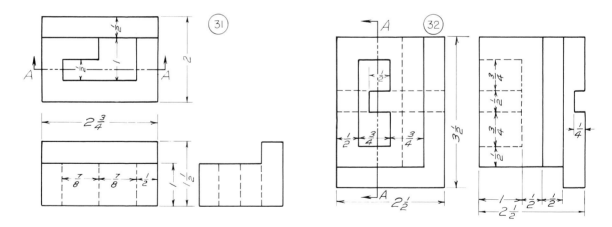

Fig. 16.57 Problems 31–32: Full sections.

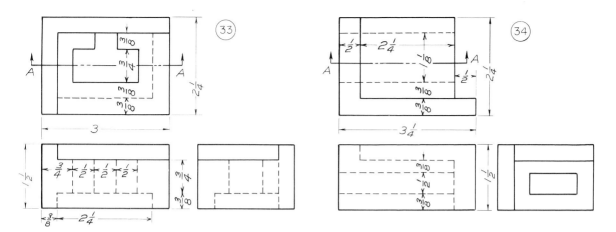

Fig. 16.58 Problems 33–34: Full sections.

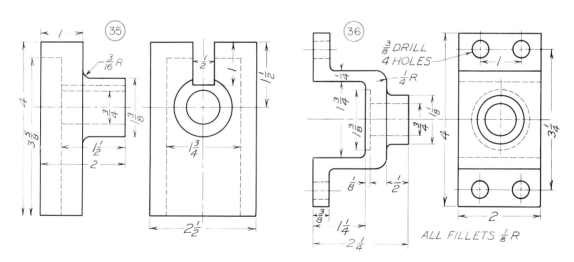

Fig. 16.59 Problems 35–36: Offset sections.

37–43. (Fig. 16.60–Fig. 16.64) Sections and conventions: Draw each problem on a Size B sheet. Complete the sectional views as indicated by the cutting planes or as good practice demands. Omit the dimensions.

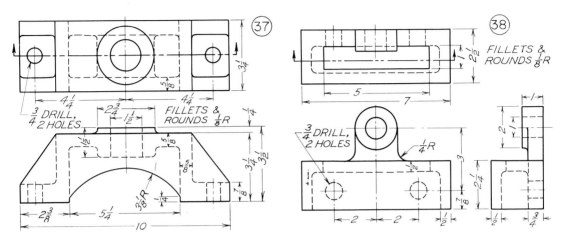

Fig. 16.60 Problems 37–38: Full sections.

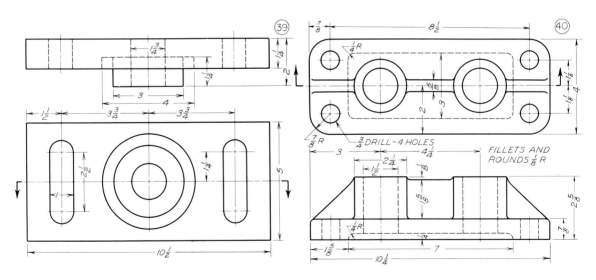

Fig. 16.61 Problems 39–40: Full sections.

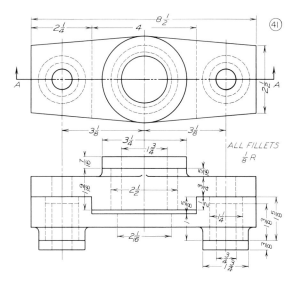

Fig. 16.62 Problem 41: Full section.

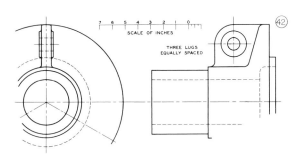

Fig. 16.63 Problem 42: Sections and conventions. (Show the lugs in all three positions, and show the entire view.)

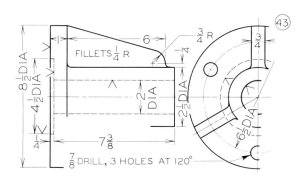

Fig. 16.64 Problem 43: Section and conventions. (Show a full section through this part and show the entire view.)

44–46. (Fig. 16.65–Fig. 16.67) Sections: Draw the necessary views to describe the parts using sections and conventional practices. Draw one problem per Size B sheet. Omit dimensions.

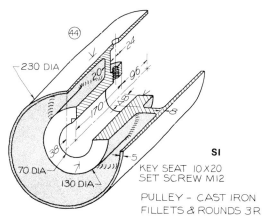

Fig. 16.65 Problem 44: Section.

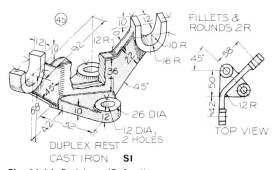

Fig. 16.66 Problem 45: Section.

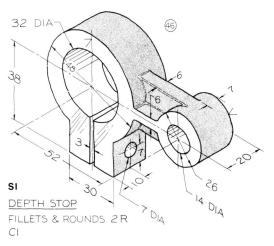

Fig. 16.67 Problem 46: Section.

17

SCREWS, FASTENERS, AND SPRINGS

17.1 THREADED FASTENERS

Screw threads provide a relatively fast and easy method of fastening two parts together and of exerting a force that can be used for adjustment of movable parts. For a screw thread to function, there must be two parts—an internal thread and an external thread. The internal threads may be tapped inside a part such as a motor block or, more commonly, they may be tapped inside a nut. Whenever possible, the nuts and bolts used in industrial projects should be stock parts that can be obtained from many sources. This reduces manufacturing expenses and improves the interchange-ability of parts, which is very important for repair or replacement of damaged fasteners.

Threaded fasteners made in different countries or by different manufacturers may have threads of different specifications that will not match. Progress has been made toward establishing standards that will unify threads both in this country and abroad by the introduction of metric standards. Other efforts have led to the adoption of the *Unified Screw Thread* by the United States, Britain, and Canada (ABC Standards), which is a modification of both the American Standard thread and the Whitworth thread.

17.2 DEFINITIONS OF THREAD TERMINOLOGY

Succeeding sections will discuss the uses and methods of representing screw threads. The terms used and defined below are illustrated in Fig. 17.1.

External Thread   This is a thread on the outside of a cylinder such as a bolt (Fig. 17.2).

Internal Thread   This is a thread cut on the inside of a part such as a nut (Fig. 17.2).

Major Diameter   This is the largest diameter on an internal or external thread.

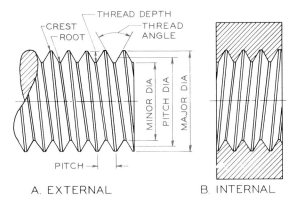

Fig. 17.1 Thread terminology.

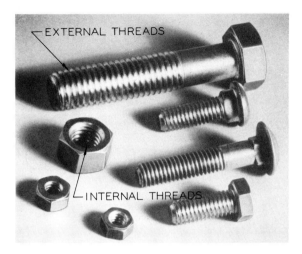

Fig. 17.2 Examples of external threads (bolts) and internal threads (nuts). (Courtesy of Russell, Burdsall & Ward Bolt and Nut Company.)

Minor Diameter This is the smallest diameter that can be measured on a screw thread.

Pitch Diameter This is the diameter of an imaginary cylinder passing through the threads at the points at which the thread width is equal to the space between the threads.

Pitch Pitch is the distance between crests of threads. Pitch is found mathematically by dividing one inch by the number of threads per inch of a particular thread.

Crest The crest is the peak edge of a screw thread.

Thread Angle This is the angle between threads cut by the cutting tool.

Root The root is the bottom of the thread cut into a cylinder.

Thread Form This is the shape of the thread cut into a threaded part.

Thread Series This is the number of threads per inch for a particular diameter, which results in three series: coarse, fine, and extra fine. Coarse se-

ries provides for rapid assembly, and extra-fine series provides for fine adjustment.

Thread Class This is a closeness of fit between two mating threaded parts. Class 1 represents a loose fit, and Class 3 a tight fit.

Right-Hand Thread This is a thread that will assemble when turned clockwise. A right-hand thread will slope downward to the right on an external thread when the axis is horizontal, and in the opposite direction on an internal thread.

Left-Hand Thread This is a thread that will assemble when turned counterclockwise. A left-hand thread slopes downward to the left on an external thread when the axis is horizontal, and in the opposite direction on an internal thread.

17.3 THREAD SPECIFICATIONS (ENGLISH SYSTEM)

Form

A thread form is the shape of the thread cut into a part as illustrated in Fig. 17.3. The Unified form, a combination of the American National and the British Whitworth, is most widely used, since it is a standard in several countries. It is referred to as UN in abbreviations and thread notes. The American National is signified by the letter N.

A new thread form, the UNR, was introduced into the 1974 ANSI standards. This designation is specified only for external threads—there is no UNR designation for internal threads. Figure 17.4 gives a comparison of the profiles of the external UN and UNR threads. The UN form has a flat root (rounded root is optional) in part A, whereas the UNR thread *must* have a rounded root, as shown in part B. The rounded root of the UNR thread is designed to reduce the wear of the threading tool and to improve the fatigue strength of the thread. UNR threads are usually made by rolling.

The transmission of power is achieved by the use of the *Acme, square, buttress,* and *worm* threads. These are commonly used in gearing and other pieces of machinery. The *sharp V* is used for set screws and in applications where friction in assembly is desired. The *knuckle* form is a fast-assembling thread used for light assemblies such as light bulbs and bottle caps.

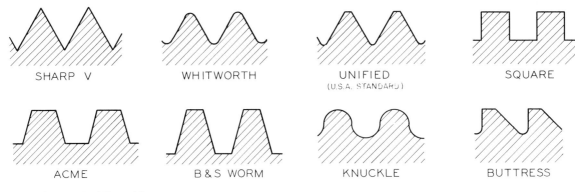

Fig. 17.3 Standard thread forms.

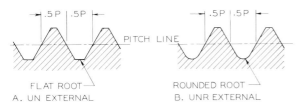

Fig. 17.4 The UN external thread (A) has a flat root (rounded root is optional); the UNR has a rounded root formed by rolling (B). The UNR form does not apply to internal threads.

Series

Thread series is closely related to thread form. It designates the type of thread specified for a given application and is abbreviated as C, F, and EF.

There are 11 standard series of threads listed under the American National form and the Unified National (UN/UNR) form. There are three series, with abbreviations coarse (C), fine (F), extra-fine (EF), and eight with constant pitches (4, 6, 8, 12, 16, 20, 28, and 32 threads per inch).

A Unified National form for a coarse-series thread is specified as UNC or UNRC, which is a combination of form and series in a single note. Similarly, an American National form for a coarse thread is written NC. The coarse-thread series (UNC/UNRC or NC) is suitable for bolts, screws, nuts, and general use with cast iron, soft metals, or plastics when rapid assembly is desired. The *fine* thread series (NF or UNF/UNRF) is suitable for bolts, nuts, or screws when a high degree of tightening is required. The *extra fine* series (UNEF/UNREF or NEF) is used for applications that will have to withstand high stresses. This series is suitable for sheet metal, thin nuts, ferrules, or couplings when length of engagement is limited.

The 8 thread series (8 UN), 12 thread series, (12 N or 12 UN/UNR), and 16 thread series (16 N or 16 UN/UNR) are threads with a uniform pitch for large diameters. The 8 N is used as a substitute for the coarse thread series on diameters larger than 1″ when a medium pitch thread is required. The 12 UN is used on diameters larger than 1½″, with a thread of a medium fine pitch as a continuation of the fine thread series. The 16 N series is used on diameters larger than 1¹¹⁄₁₆″, with threads of a fine pitch as a continuation of the extra-fine series.

Class of Fit

Thread classes are used to indicate the tightness of fit between a nut and a bolt or any two mating threaded parts. This fit is determined by the tolerances and allowances applied to threads. Classes of fit are indicated by the numbers 1, 2, or 3 followed by the letters A or B. For UN forms, the letter A represents an external thread, while the letter B represents an internal thread. These letters are omitted when the American National form is used.

Classes 1A and 1B These threads are used on parts that require assembly with a minimum of binding.

Classes 2A and 2B These are general-purpose threads for bolts, nuts, screws, and nominal appli-

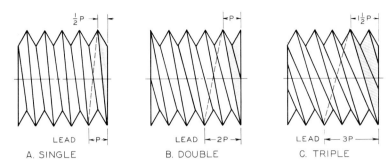

Fig. 17.5 Single and multiple threads.

cations in the mechanical field and are widely used in the mass production industries.

Classes 3A and 3B   These threads are used in precision assemblies where a close fit is desired to withstand stresses and vibration.

Single and Multiple Threads

A *single thread* (Fig. 17.5A) is a thread that will advance the distance of its pitch in one full revolution of 360°. In other words, its pitch is equal to its lead. In the drawing of a single thread, the crest line of the thread will slope ½P, since only 180° of the revolution is visible in a single view. A double thread is composed of two threads resulting in a lead equal to 2P, meaning that the threaded part will advance a distance of 2P in a single revolution of 360° (Fig. 17.5B). The crest line of a double thread will slope a distance equal to P in the view in which 180° can be seen. Similarly, a triple thread will advance 3P in 360° with a crest line slope of 1½P in the view in which 180° of the cylinder is visible (Fig. 17.5C). The lead of a double thread is 2P; that of a triple thread, 3P. Although power on multiple threads is somewhat limited, they are used wherever quick motion is required.

Thread Notes

Drawings of threads are only symbolic representations that are inadequate unless thread notes are applied to give the thread specifications (Fig. 17.6). The major diameter is given first, followed by the number of threads per inch, the form and series, the class of fit, and a letter denoting whether the thread is external or internal. This completes the note for a single, right-hand thread. However, if the thread is left-hand or double, the word DOUBLE or TRIPLE is included in the note along with the letters LH for left-hand thread.

   The UNR thread note is shown for the external thread in Fig. 17.7A (UNR does not apply to internal threads). When inches are used as the unit of measurement, fractions can be written as decimals or as common fractions. Decimal fractions are preferred, as shown in part B.

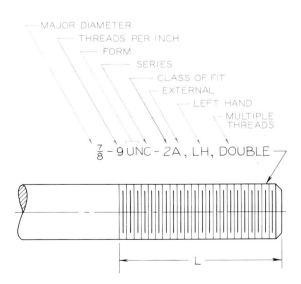

Fig. 17.6 Parts of a thread note for an external thread.

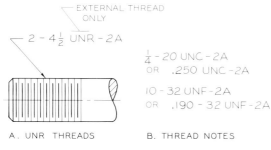

Fig. 17.7 The UNR thread notes apply to external threads only. Notes can be given as decimal fractions or as common fractions, as shown in part (B).

17.4 USING THREAD TABLES

The UN/UNR thread table is given in Appendix 15. A portion of this table is shown in Table 17.1. If an external thread (bolt) with a 1.500-inch diameter is to have a "fine" thread, it will have 12 threads per inch. Therefore, the thread note can be written.

1½-12 UNF-2A   or   1.500-12 UNF-2A.

If the thread had been an internal one (nut), the thread note would have been the same, but the letter B would have been used instead of the letter A.

A constant-pitch thread series can be selected for the larger diameters. The constant-pitch thread notes are written with the letter C, F, or EF omitted. For example, a 1¾-inch diameter bolt with a

Table 17.1 AMERICAN NATIONAL STANDARD UNIFIED INCH SCREW THREADS (UN AND UNR THREAD FORM)*

| Sizes | | Basic major diameter | Series with graded pitchers | | | Series with constant pitches | | | | | | | | Sizes |
Primary	Secondary		Course UNC	Fine UNF	Extra fine UNEF	UN	6 UN	8 UN	12 UN	16 UN	20 UN	28 UN	32 UN	
1		1.0000	8	12	20	—	—	UNC	UNF	16	UNEF	28	32	1
	1¹⁄₁₆	1.0625	—	—	18	—	—	8	12	16	20	28	—	1¹⁄₁₆
1⅛		1.1250	7	12	18	—	—	8	UNF	16	20	28	—	1⅛
	1³⁄₁₆	1.1875	—	—	18	—	—	8	12	16	20	28	—	1³⁄₁₆
1¼		1.2500	7	12	18	—	—	8	UNF	16	20	28	—	1¼
	1⁵⁄₁₆	1.3125	—	—	18	—	—	8	12	16	20	28	—	1⁵⁄₁₆
1⅜		1.3750	6	12	18	—	UNC	8	UNF	16	20	28	—	1⅜
	1⁷⁄₁₆	1.4375	—	—	18	—	6	8	12	16	20	28	—	1⁷⁄₁₆
1½		1.5000	6	12	18	—	UNC	8	UNF	16	20	28	—	1½
	1⁹⁄₁₆	1.5625	—	—	18	—	6	8	12	16	20	—	—	1⁹⁄₁₆
1⅝		1.6250	—	—	18	—	6	8	12	16	20	—	—	1⅝
	1¹¹⁄₁₆	1.6875	—	—	18	—	6	8	12	16	20	—	—	1¹¹⁄₁₆
1¾		1.7500	5	—	—	—	6	8	12	16	20	—	—	1¾
	1¹³⁄₁₆	1.8125	—	—	—	—	6	8	12	16	20	—	—	1¹³⁄₁₆
1⅞		1.8750	—	—	—	—	6	8	12	16	20	—	—	1⅞
	1¹⁵⁄₁₆	1.9375	—	—	—	—	6	8	12	16	20	—	—	1¹⁵⁄₁₆

*By using this table, a diameter of 1½ inches that is to be threaded with a fine thread would have the following thread note: 1½-12 UNF-2A. (Courtesy of ANSI; B1.1-1974.)

Table 17.2 BASIC THREAD DESIGNATIONS FOR COMMERCIAL SERIES OF ISO METRIC THREADS

Nominal size (mm)	Pitch, P (mm)	Basic thread designation*	Nominal size (mm)	Pitch, P (mm)	Basic thread designation*	Nominal size (mm)	Pitch, P (mm)	Basic thread designation*
1.6	0.35	M1.6	8	1.25	M8	22	2.5	M22
1.8	0.35	M1.8	8	1	M8 × 1	22	1.5	M22 × 1.5
2	0.4	M2	10	1.5	M10	24	3	M24
2.2	0.45	M2.2	10	1.25	M10 × 1.25	24	2	M24 × 2
2.5	0.45	M2.5	12	1.75	M12	27	3	M27
3	0.5	M3	12	1.25	M12 × 1.25	27	2	M27 × 2
3.5	0.6	M3.5	14	2	M14	30	3.5	M30
4	0.7	M4	14	1.5	M14 × 1.5	30	2	M30 × 2
4.5	0.75	M4.5	16	2	M16	33	3.5	M33
5	0.8	M5	16	1.5	M16 × 1.5	33	2	M33 × 2
6	1	M6	18	2.5	M18	36	4	M36
7	1	M7	18	1.5	M18 × 1.5	36	3	M36 × 3
			20	2.5	M20	39	4	M39
			20	1.5	M20 × 1.5	39	3	M39 × 3

*U.S. practice is to include the pitch symbol even for the coarse pitch series.
Basic descriptions shown are as specified in ISO Recommendations.
Source: Courtesy of Greenfield Tap and Die Corporation.

fine thread could be noted in constant-pitch series as

1¾-12 UN-2A   or   1.750-12 UN-2A.

This table can also be used for the UNR thread form (for external threads only) by substituting UNR in place of UN; for example, UNEF can be written as UNREF.

17.5 METRIC THREAD SPECIFICATIONS (ISO)

Metric thread specifications are recommended by the ISO (International Organization for Standard-ization). Thread specifications can be given using a *basic designation* that is suitable for general applications, or the *complete designation* can be used where detailed specifications are needed.

Basic Designation

Examples of metric screw thread notes are shown in Fig. 17.8. Each note begins with the letter M, which designates the note as a metric note, followed by the diameter in millimeters and the pitch in millimeters separated by the "×" sign. The pitch can be omitted in notes for coarse threads, but it is preferred by U.S. standards that it be shown. Table 17.2 shows the commercially available ISO threads recommended for general use. Additional ISO specifications are given in Appendix 18.

Complete Designation

For some applications it is necessary to show a complete thread designation, as shown in Fig. 17.9. The first part of this note is the same as the *basic designation*, but in addition it has a tolerance class designation separated by a dash. The 5g

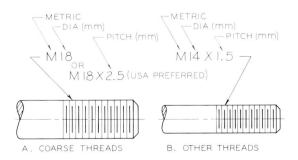

Fig. 17.8 Basic designations for metric threads.

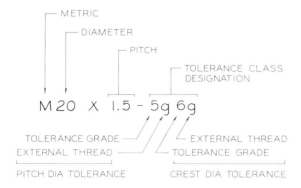

Fig. 17.9 A complete designation note for metric threads.

represents the pitch diameter tolerance and 6g represents the crest diameter tolerance.

The numbers 5 and 6 are *tolerance grades*. Grade 6 is commonly used for a medium, general-purpose thread that is nearly equal to the 2A and 2B classes of fit specified under the Unified system. Grades with numbers smaller than 6 are used for "fine" quality fits and short lengths of engagement. Grades represented by numbers greater than 6 are recommended for "coarse" quality fits and long lengths of engagement. Table 17.3 gives the tolerance grades for internal and external threads for the pitch diameter and the major and minor diameters.

The letters following the grade numbers designate *tolerance positions*. Lowercase letters represent external threads (bolts), as shown in Fig. 17.10. The lowercase letters e, g, and h represent large allowance, small allowance, and no allowance, respectively. Uppercase letters are used to designate internal threads (nuts); G designates small allowance and H designates no allowance. The letters are placed after the tolerance grade number. For example, 5g designates a medium tolerance with small allowance for the pitch diameter of an external thread, and 6H designates a medium tolerance with no allowance for the minor diameter of an internal thread.

TOLERANCE POSITIONS	
EXTERNAL THREADS LOWER-CASE LETTERS	**INTERNAL THREADS** UPPER-CASE LETTERS
e = LARGE ALLOWANCE	G = SMALL ALLOWANCE
g = SMALL ALLOWANCE	H = NO ALLOWANCE
h = NO ALLOWANCE	

LENGTH OF ENGAGEMENT		
S = SHORT	N = NORMAL	L = LONG

Fig. 17.10 Symbols used to represent tolerance grade, position, and class.

Tolerance classes are fine, medium, and coarse, as listed in Table 17.4. These classes of fit are combinations of tolerance grades, tolerance positions, and lengths of engagement—short (S), normal (N), and long (L). The length of engagement can be determined by referring to Appendix 17. Once it has been decided to use either a fine, medium, or coarse class of fit for a particular application, the specific designation should be selected first from the classes shown in large print in Table 17.4, second from the classes shown in medium-size print, and third from the classes

Table 17.3 TOLERANCE GRADES, ISO THREADS

External thread		Internal thread	
Major diameter (d_1)	Pitch diameter (d_2)	Minor diameter (D_1)	Pitch diameter (D_2)
—	3		
4	4	4	4
—	5	5	5
6	6	6	6
—	7	7	7
8	8	8	8
—	9	—	—

Grade 6 is medium; smaller numbers are finer and larger numbers are coarser. (Courtesy of ANSI B1-1972.)

Fig. 17.11 When both pitch and crest diameter tolerance grades are the same, the tolerance class symbol is shown only once (A). Letters S, N, and L are used to indicate the length of the thread engagement (B).

Table 17.4 PREFERRED TOLERANCE CLASSES, ISO THREADS*

Quality	External threads (bolts)									Internal threads (nuts)					
	Tolerance position e (large allowance)			Tolerance position g (small allowance)			Tolerance position h (no allowance)			Tolerance position G (small allowance)			Tolerance position H (no allowance)		
	Length of engagement			Length of engagement			Length of engagement			Length of engagement			Length of engagement		
	Group S	Group N	Group L	Group S	Group N	Group L	Group S	Group N	Group L	Group S	Group N	Group L	Group S	Group N	Group L
Fine Medium Coarse		**6e** 7e6e		5g6g	**6g** 8g	7g6g 9g8g	3h4h 5h6h	4h 6h	5h4h 7h6h	5G	6G 7G	7G 8G	4H **5H**	5H **6H** 7H	6H **7H** 8H

*In selecting tolerance class, select first from the large bold print, second from the medium-size print, and third from the small-size print. Classes shown in boxes are for commercial threads.

Fig. 17.12 A slash mark is used to separate the tolerance class designations of mating internal and external threads.

shown in small print. Classes shown in boxes are for commercial threads.

Variations in the complete designation thread notes are shown in Fig. 17.11. The tolerance class symbol is written as 6H if the crest and pitch diameters have identical grades (part A). Since an uppercase H is used, this is an internal thread. Where considered necessary, the length-of-engagement symbol may be added to the tolerance class designation, as shown in part B.

Designations for the desired fit between mating threads can be specified as shown in Fig. 17.12. A slash is used to separate the tolerance class designations of the internal and external threads.

Additional information pertaining to ISO threads may be obtained from *ISO Metric Screw Threads*, a booklet of standards published by ANSI in 1972. These standards were used as the basis for most of this section.

17.6 THREAD REPRESENTATION

The three major types of thread representations are (1) detailed, (2) schematic, and (3) simplified (Fig. 17.13). The detailed representation is the most realistic approximation of the true appearance of a thread, while the simplified representation is the most symbolic.

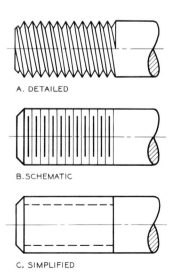

A. DETAILED

B. SCHEMATIC

C. SIMPLIFIED

Fig. 17.13 Three major types of thread representations.

17.7 DETAILED UN/UNR THREADS

Examples of detailed representations of internal and external threads are shown in Fig. 17.14. Instead of helical curves, straight lines are used to indicate crest and root lines. In this form of rep-

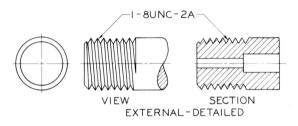

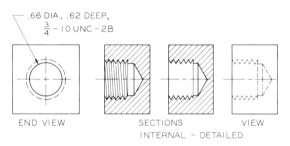

Fig. 17.14 Examples of detailed thread representations.

Fig. 17.15 Detailed thread representation

resentation, internal threads in section can be indicated in two ways. Thread notes are applied in all cases, regardless of the representation used.

The construction of a detailed representation is shown in Fig. 17.15. The pitch is found by dividing 1″ by the number of threads per inch. This can be done graphically as shown in Step 1. However, in most cases, this construction is unnecessary, since the pitch can be approximated by using a calibration close to the true pitch taken directly from an existing scale or by using dividers for spacing. Where threads are close, they should purposely be drawn at a larger spacing to facilitate the drawing process. Note in Step 4 that a 45° chamfer is used to indicate a bevel of the threaded end to improve ease of assembly of the threaded parts.

Metric threads would be drawn in the same manner; however, the pitch would not need to be computed. It is given in the metric thread table in millimeters.

17.8 DETAILED SQUARE THREADS

The method of drawing a detailed representation of a square thread is shown in four steps in Fig. 17.16. This method gives an approximation of the true projection of a square thread.

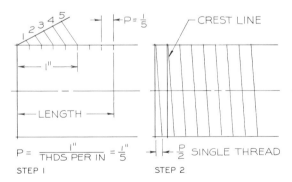

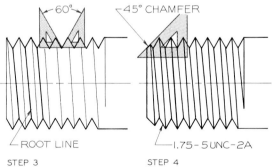

Step 1 To draw a detailed representation of a 1.75—5UNC—2A thread, the pitch is determined by dividing 1″ by the number of threads per inch, 5 in this case. The pitch is laid off the length of the thread.

Step 2 Since the thread is a right-hand thread, the crest lines slope downward to the right equal to ½P (for single threads). The crest lines will be final lines drawn with an H or F pencil.

Step 3 The root lines are found by constructing 60° vees between the crest lines. The root lines are drawn from the bottom of the vees. Notice that root lines are parallel to each other, but not to crest lines.

Step 4 A 45° chamfer is constructed at the end of the thread from the minor diameter. Strengthen all lines and add a thread note.

Fig. 17.16 Drawing the square thread

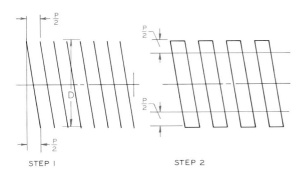

STEP 1   STEP 2   STEP 3   STEP 4

Step 1 Lay out the major diameter. Space the crest lines ½P apart. Slope them downward to the right for right-hand threads.

Step 2 Connect every other pair of crest lines. Find the minor diameter by measuring ½P inward from the major diameter.

Step 3 Connect the opposite crest lines with light construction lines. This will establish the profile of the thread form.

Step 4 Connect the inside crest lines with light construction lines to locate the points on the minor diameter where the thread wraps around the minor diameter.

In Step 1, the *major* diameter is laid off. The number of threads per inch is taken from the table in Appendix 19 for this size of thread. The pitch (*P*) is found by dividing 1″ by the number of threads per inch. Distances of *P*/2 are marked off with dividers.

Steps 2, 3, and 4 are completed and a thread note is added to complete the thread representation.

Square internal threads are drawn in the same manner, as shown in Fig. 17.17. Note that the threads in the section view are drawn in a slightly different way. The thread note for an internal thread is placed in the circular view whenever possible, with the leader pointing toward the center.

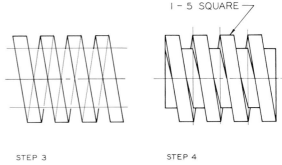

Fig. 17.18 Conventional method showing square threads without drawing each thread.

When a square thread is rather long, it need not be drawn continuously, but can be represented using the symbol shown in Fig. 17.18.

17.9 DETAILED ACME THREADS

The method involved in preparing detailed drawings of Acme threads is shown in Fig. 17.19 in four steps.

The length and the major diameter are laid off with light construction lines. From the table in Appendix 19, the pitch is found by dividing the number of threads per inch into 1″ to begin Step 1.

Steps 2, 3, and 4 complete the thread representation. The thread note is added in the last step.

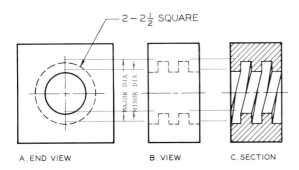

A. END VIEW   B. VIEW   C. SECTION

Fig. 17.17 Internal square threads.

Fig. 17.19 Drawing the acme thread

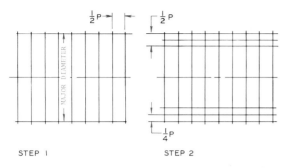

STEP 1   STEP 2

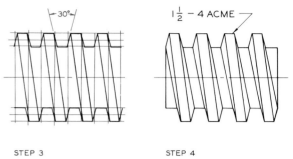

STEP 3   STEP 4

Step 1 Lay out the major diameter, the thread length, and divide the shaft into equal divisions ½P apart.

Step 2 Locate the minor diameter a distance ½P inside the major diameter. Locate the pitch diameter between the major and minor diameters.

Step 3 Draw lines at 15° with the vertical using the construction lines as shown to make a total angle of 30°. Draw the crest lines and the thread profile.

Step 4 Darken the lines, draw the root lines, and add the thread note to complete the drawing.

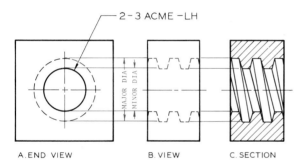

A. END VIEW   B. VIEW   C. SECTION

Fig. 17.20 Internal Acme threads.

Fig. 17.21 Cutting an Acme thread on a lathe. (Courtesy of Clausing Corporation.)

Internal Acme threads are shown in Fig. 17.20. Note that in the section view, left-hand internal threads are sloped so that they look the same as right-hand external threads.

Figure 17.21 shows a shaft that is being threaded on a lathe. These Acme threads are being cut as the tool travels the length of the shaft.

17.10 SCHEMATIC REPRESENTATION

Examples of schematic representations of internal and external threads are shown in Fig. 17.22. Note that the threads are indicated by parallel, nonsloping lines that do not show whether the threads are right-hand or left-hand. This information is given in the thread note. Since this representation is easy to construct and gives a good symbolic representation of threads, it is the most generally used thread symbol.

The method of constructing schematic threads is illustrated in Fig. 17.23 in four steps.

The method of drawing a schematic representation of metric threads is shown in Fig. 17.24. The pitch (in millimeters) can be taken directly from the metric thread tables as the distance used to separate the crest lines.

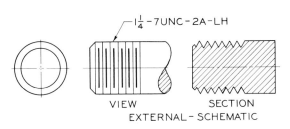

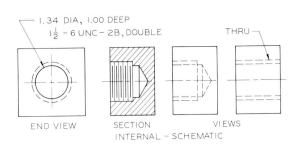

Fig. 17.22 Schematic representations of threads.

Fig. 17.23 Drawing schematic threads

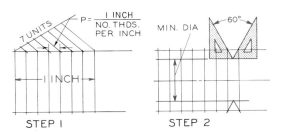

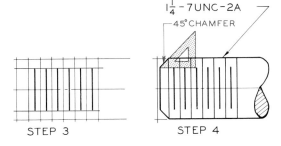

Step 1 Lay out the major diameter and divide the shaft into equal divisions a distance *P* apart. These crest lines should be drawn as thin lines.

Step 2 Find the minor diameter by drawing a 60° angle between two crest lines.

Step 3 Draw heavy root lines between the crest lines.

Step 4 Chamfer the end of the thread and give a thread note.

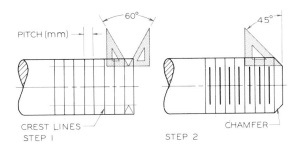

Fig. 17.24

Step 1 The pitch of metric threads can be taken directly from the metric tables, which can be used to find the minor diameter.

Step 2 The root lines are drawn in heavy between the crest lines. The end of the thread is chamfered.

17.11 SIMPLIFIED THREADS

Figure 17.25 illustrates the use of simplified representations with notes to specify thread details. Of the three types of thread representations covered, this is the easiest to draw. Hidden lines are positioned by eye to approximate the minor diameter.

The steps involved in constructing a simplified thread drawing are shown in Fig. 17.26.

17.12 DRAWING SMALL THREADS

Instead of using exact measurements to draw small threads, minor diameters can be drawn smaller by eye in order to separate the root and

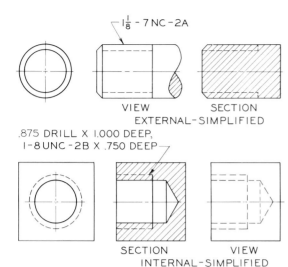

$1\frac{1}{8} - 7\,NC - 2A$

VIEW   SECTION
EXTERNAL—SIMPLIFIED

.875 DRILL X 1.000 DEEP,
1—8 UNC—2B X .750 DEEP

SECTION   VIEW
INTERNAL—SIMPLIFIED

Fig. 17.25 Simplified thread representations.

Fig. 17.26 Drawing simplified threads

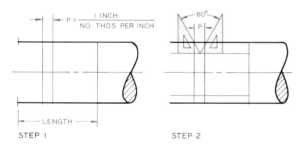

$P = \dfrac{1\ INCH}{NO.\ THDS.\ PER\ INCH}$

LENGTH

STEP 1   STEP 2

Step 1 Lay out the major diameter. Find the pitch (*P*) and lay out two lines a distance of *P* apart.

Step 2 Find the minor diameter by constructing a 60° angle between the two lines.

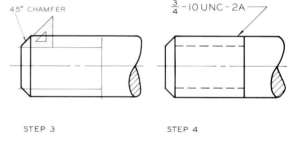

45° CHAMFER

$\frac{3}{4} - 10\,UNC - 2A$

STEP 3   STEP 4

Step 3 Draw a 45° chamfer from the minor diameter to the major diameter.

Step 4 Show the minor diameter as a dashed line. Add a thread note.

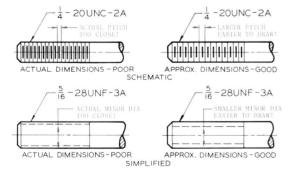

$\frac{1}{4} - 20\,UNC - 2A$

ACTUAL PITCH
TOO CLOSE!

$\frac{1}{4} - 20\,UNC - 2A$

LARGER PITCH
EASIER TO DRAW!

ACTUAL DIMENSIONS—POOR   APPROX. DIMENSIONS—GOOD
SCHEMATIC

$\frac{5}{16} - 28\,UNF - 3A$

ACTUAL MINOR DIA
TOO CLOSE!

$\frac{5}{16} - 28\,UNF - 3A$

SMALLER MINOR DIA
EASIER TO DRAW!

ACTUAL DIMENSIONS—POOR   APPROX. DIMENSIONS—GOOD
SIMPLIFIED

Fig. 17.27 Simplified and schematic threads should be drawn using approximate dimensions if the actual dimensions would result in lines drawn too close together.

crest lines, as illustrated in Fig. 17.27. This procedure makes the drawing more readable and easier to draw. Exactness is unnecessary, since the drawing is only a symbolic representation of a thread.

For both internal and external threads, a thread note is added to the symbolic drawing to give the necessary specifications and to complete the description of the threaded part.

17.13 NUTS AND BOLTS

Nuts and bolts come in many forms and sizes for different applications (Fig. 17.28). Drawings of the more common types of threaded fasteners are shown in Fig. 17.29. A *bolt* is a threaded cylinder with a head and a nut for holding two parts together (Fig. 17.29A). A *stud* does not have a head, but is screwed into one part with a nut attached to the other end (Fig. 17.29B). A *cap screw* is similar to a bolt, but it does not have a nut; instead it is screwed into a member with internal threads for greater strength (Fig. 17.29C). A *machine screw* is similar to a cap screw, but it is smaller. A *set screw* is used to adjust one member with respect to another, usually to prevent a rotational movement.

The types of heads used on standard bolts and nuts are illustrated in Fig. 17.30. These heads are used on both types of bolts: *regular* and *heavy*. The thickness of the head is the primary differ-

Fig. 17.28 Examples of nuts and bolts. (Courtesy of Russell, Burdsall & Ward Bolt and Nut Company.)

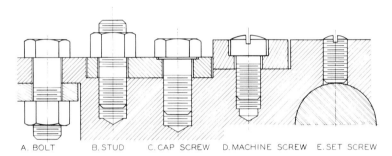

A. BOLT   B. STUD   C. CAP SCREW   D. MACHINE SCREW   E. SET SCREW

Fig. 17.29 Types of threaded bolts and screws.

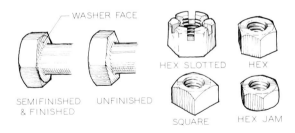

WASHER FACE

SEMIFINISHED & FINISHED   UNFINISHED

HEX SLOTTED   HEX

SQUARE   HEX JAM

Fig. 17.30 Types of finishes for bolt heads and types of nuts.

ence between the two series. Heavy-series bolts have the thicker heads and are used at points where bearing loads are heaviest. Bolts and nuts are classified as *finished* and *unfinished*. Figure 17.30 shows an unfinished head; that is, none of the surfaces of the head are machined. The finished head has a washer face that is ¹⁄₆₄″ thick to provide a circular boss on the bearing surface of the bolt head or the nut.

Other standard forms of bolt and screw heads are shown in Fig. 17.31. These heads are used primarily on cap screws and machine screws. Standard types of nuts are illustrated in Fig. 17.30. These can be machined to give a washer face for the finished series. A hexagon jam nut does not have a washer face, but it is chamfered on both sides.

Although ANSI tables are provided in Appendixes 21–25 to indicate the standard bolt lengths and their corresponding thread lengths; the following can be used as a general guide for square and hexagon head bolts.

Hexagon bolts lengths are available in increments of ¼″ up to 8″ in length, in ½″ increments from 8″ to 20″ of length, and in 1″ increments from 20″ to 30″ long.

Square head bolt lengths are available in increments of ⅛″ from lengths of ½″ to ¾″ long, ¼″ increments from ¾″ to 5″ long, ½″ increments from

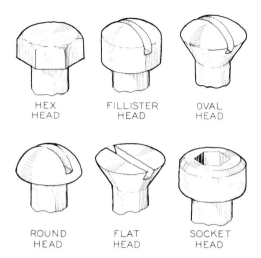

HEX HEAD   FILLISTER HEAD   OVAL HEAD

ROUND HEAD   FLAT HEAD   SOCKET HEAD

Fig. 17.31 Common types of bolt and screw heads.

5" to 12" long, and 1" increments from 12" to 30" long.

The lengths of the threads on both hexagon and square head bolts up to 6" in length can be found by the formula: Thread length = 2D + ¼", where D is the diameter of the bolt. The threaded length for bolts over 6" in length can be found by the formula: Thread length = 2D + ½".

The threads for bolts can be coarse, fine, or 8 pitch threads. It is understood that the class of fit for bolts and nuts will be 2A and 2B if no class of fit is specified.

Square and hexagon head bolts are designated by notes in the following form:

⅜ - 16 × 1½ SQUARE BOLT—STEEL, or

½ -13 × 3 HEX CAP SCREW, SAE GRADE 8 STEEL, or

0.75 × 5.00 HEX LAG SCREW—STEEL

The numbers represent bolt diameter, threads per inch (omit for lag screws), length, name of screw, and material. It is understood that these will have a class 2 fit.

Nuts are designated by notes in the following form:

½ = 13 SQUARE NUT—STEEL, or

¾ – 16 HEAVY HEX NUT, SAE GRADE 5—STEEL, or

1.00 – 8 HEX THICK SLOTTED NUT—CORROSION RESISTANT STEEL

When not noted as HEAVY, nuts are assumed to be REGULAR nuts. The class of fit is assumed to be 2B for nuts when not noted.

17.14 DRAWING THE SQUARE BOLT HEAD

Detailed tables are available in Appendix 20 and in published standards for various types of threaded parts. In most cases it is sufficient to draw nuts and bolts using only general proportions.

The first step in drawing a bolt head or a nut is to determine whether it is to be *across corners* or *across flats*. In other words, are the outlines at either side of the view going to represent corners, or are they going to be edge views of flat surfaces of the part? The head in Fig. 17.32 is drawn across corners. Nuts and bolts should be drawn across corners whenever possible; this type of drawing gives a better representation than drawing across flats.

17.15 DRAWING THE HEXAGON BOLT HEAD

It is desirable that nuts and bolts be drawn across corners since this gives a better impression of the parts. An example of constructing the head of a bolt is shown in Fig. 17.33

Note that the diameter of the bolt is D. The thickness of the head is drawn equal to ⅔D. The top view of the head is drawn as a circle with a radius of ¾D. For most applications, this proportionality based on D is sufficient for drawing bolt heads.

17.16 DRAWING NUTS

The construction of a drawing of a square nut or a hexagon nut across corners is exactly the same as the construction of a drawing of a bolt head across corners. The only variation is the thickness of the nut. The regular nut thickness is ⅞D, and for the heavy nut, the thickness is equal to the diameter (D).

Examples of square and hexagon nuts drawn across corners are shown in Fig. 17.34. Hidden lines are shown in the front view to indicate

Fig. 17.32 Drawing the square head

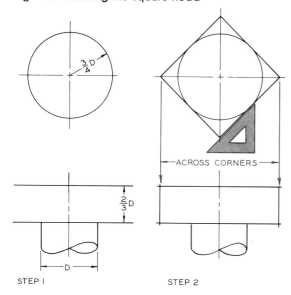

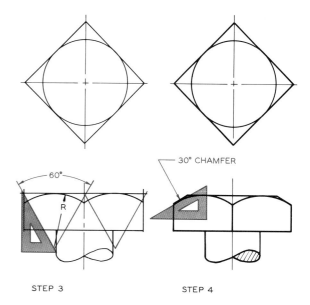

STEP I STEP 2 STEP 3 STEP 4

Step 1 Draw the diameter of the bolt. Use this to establish the head diameter and thickness.

Step 2 Draw the top view of the square head with a 45° triangle to give an across-corners view.

Step 3 Show the chamfer in the front view by using a 30°–60° triangle to find the centers for the radii.

Step 4 Show a 30° chamfer tangent to the arcs in the front view. Strengthen the lines.

Fig. 17.33 Drawing the hexagon head

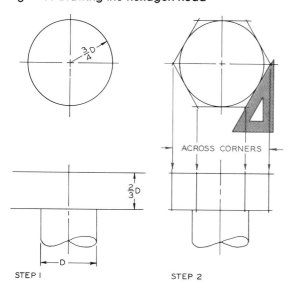

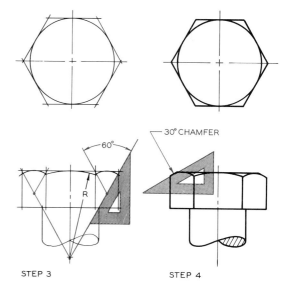

STEP I STEP 2 STEP 3 STEP 4

Step 1 Draw the diameter of the bolt. Use this to establish the head diameter and thickness.

Step 2 Construct a hexagon with a 30°–60° triangle to give an across-corners view.

Step 3 Find arcs in the front view to show the chamfer of the head.

Step 4 Draw a 30° chamfer that is tangent to the arcs in the front view. Strengthen the lines.

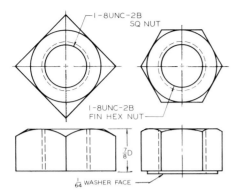

Fig. 17.34 Drawings of hexagon and square nuts are constructed in the same manner as drawings of bolt heads. Standard notes are added to give nut specifications.

threads. Since it is understood that nuts are threaded, these hidden lines may be omitted in general applications.

Note that a $\frac{1}{64}''$ washer face is shown on the hexagon nut. This is usually drawn thicker than $\frac{1}{64}''$ so that the face will be more noticeable in the drawing. Thread notes are placed in the circular views rather than the front views where possible. In the case of the square nut, the note tells us that the major diameter of the thread is 1″, that the nut

has 8 threads per inch, that the thread is of the Unified National form and coarse series, with a fit of 2, and that it is a regular square nut since it is not labeled as "HEAVY." The hexagon nut is similar except that it is a finished hexagon nut.

The leader from the note is directed toward the center of the circular view, but the arrow stops at the first visible circle it makes contact with.

Nuts can be drawn across flats in situations where doing so improves the drawing. Examples of nuts drawn across flats are shown in Fig. 17.35.

For regular nuts, the distance across flats is $1\frac{1}{2} \times D$ (D is the major diameter of the thread). For heavy nuts this distance is increased by $\frac{1}{8}''$. The top views are drawn in the same manner as in across-corners drawings except that they are positioned to give different front views.

In the case of the square nut (Fig. 17.35), the front view is a simple rectangle, with only the arc formed by the chamfer giving a hint that the object is a nut.

The hexagon nut drawn across flats looks more like a nut in the front view than does the square nut. Still, the hexagon nut drawn across corners is a better representation (Fig. 17.35). The centers for the arc used to show the chamfer are found with a 30°–60° triangle. Notes are added with leaders to complete the representation of the nuts. A washer face should be added to a nut if it is finished, except in the case of the square nut. Square nuts are always unfinished.

17.17 DRAWING NUTS AND BOLTS IN COMBINATION

The same rules followed in drawing nuts and bolts separately apply when drawing nuts and bolts in assembly. Examples are shown in Fig. 17.36.

The diameter of the bolt is used as the basis for other dimensions. The note is added to give the specifications of the nut and bolt. In the figure, the bolt heads are drawn across corners and the nuts across flats. The end views have been included to show how the front views were found by projection.

17.18 CAP SCREWS

Cap screws are used to hold two parts together without the use of a nut. One of these two parts

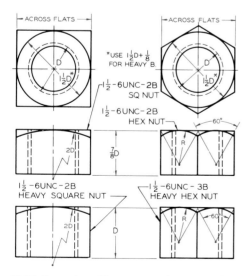

Fig. 17.35 Examples of hexagon and square nuts drawn across flats. Notes are added to give nut specifications.

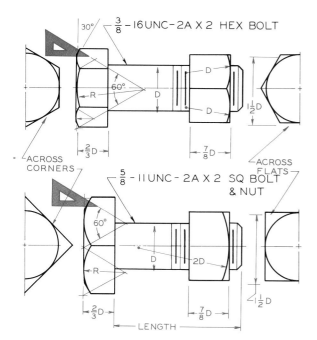

Fig. 17.36 Construction of nuts and bolts in assembly.

The proportions shown here can be used for drawing cap screws of all sizes. These types of cap screws range in diameter from No. 0 (0.060") to 1½".

17.19   MACHINE SCREWS

Machine screws are smaller than most cap screws, usually less than 1" in diameter. The machine screw is used to attach parts together; it is screwed either into another part or into a nut. Machine screws are threaded their full length when their length is 2" or shorter.

Drawings of common machine screws and their notes are given in Fig. 17.38. Many other types are available in addition to these types. The dimensions of round-head machine screws are given in Appendix 24.

The four types of machine screws in Fig. 17.38 are drawn on a grid to give the proportions of the head in relation to the major diameter of the screw. The proportions shown here can be used for drawing these screws regardless of their size or scale. Machine screws range in size from No. 0 (0.060" in diameter) to a diameter of ¾".

When slotted-head screws are drawn, it is conventional practice to show the slots positioned at a 45° angle in the circular view as illustrated in Fig. 17.39. Even though the slot is turned at this angle in the top view, the front view of the slot is drawn to show the width and depth of the slot, as in Fig. 17.39A. You can see that this gives a better representation of the screw head than does the example in Fig. 17.39C. This practice applies to all types of slotted fasteners.

has a threaded cylindrical hole and thus serves the same function as the nut. The other part is drilled with an oversize hole so that the cap screw will pass through it freely. When the cap screw is tightened, the two parts are held securely together.

The standard types of cap screws are illustrated in Fig. 17.37. Tables are available in Appendixes 22 and 23 to give the dimensions of several of these types of cap screws.

The cap screws in Fig. 17.37 are drawn on a grid in order to show the proportions of each type.

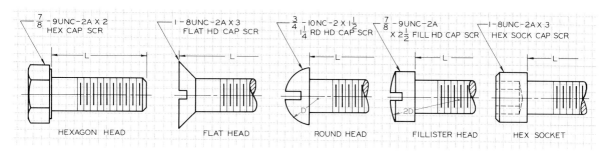

Fig. 17.37 The proportions of the standard types of cap screws are shown here for drawing cap screws of all sizes. Typical notes are given to provide typical specifications.

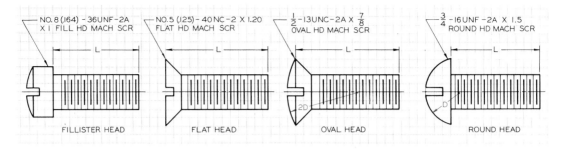

Fig. 17.38 Standard types of machine screws. The proportions shown here can be used for drawing machine screws of all sizes.

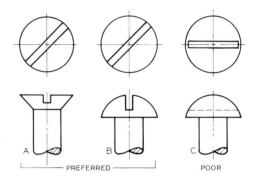

Fig. 17.39 Slotted-head screws should be drawn with the slot at 45° in the top view and with the notch shown in the front view.

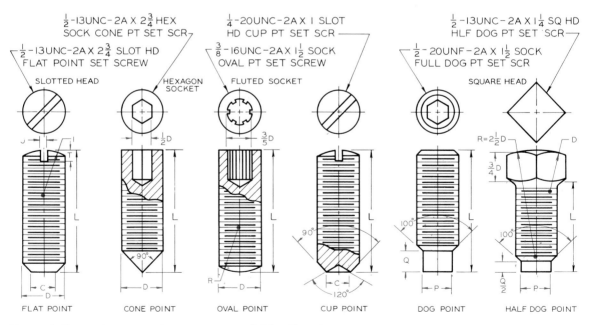

Fig. 17.40 Types of set screws. Set screws are available with various combinations of heads and points. Notes give their specifications. Dimensions are given in Table 17.5.

17.20 SET SCREWS

Parts such as wheels or pulleys are commonly attached to shafts. To attach these parts to a shaft, set screws or keys are used. Examples of various types of set screws are shown in Fig. 17.40.

Table 17.5 shows the dimensions of the various features of the set screws shown in Fig. 17.40. This table is useful in selecting the appropriate standard size of set screw for the application at hand. Drawings of set screws need not employ these dimensions precisely; like the other fasteners discussed in this chapter, set screw threads can be drawn as approximations.

Note that the points and the heads of these set screws are of different types. Set screws are available in any desired combination of point and head. The shaft against which the set screw is tightened may have a flat surface machined to give a good bearing surface for the set screw point. In this case, a dog point or a flat point would be most effective to press against the flat surface. The cup point gives good friction when applied to a round shaft where there is no flat surface.

Specifications for set screws are given in Appendixes 27, 28, and 29.

17.21 MISCELLANEOUS SCREWS

A few of the many types of specialty screws are shown in Fig. 17.41, each having its own special application. Others shown in Fig. 17.42 are: lag bolt, hanger bolt, Phillips head screw, and a drive screw.

The manufacture and design of specialized threaded fasteners is an essential career field within itself with an ever-growing need for additional fasteners of various types.

17.22 WOOD SCREWS

A wood screw is a pointed screw having a sharp thread of coarse pitch for insertion in wood. The three most common types of wood screws are shown in Fig. 17.43. These examples are drawn on a grid to show the proportions of the various heads in relation to the major diameter of the screw. The same proportions can be used for all sizes of wood screws. Detailed dimensions for wood screws are given in tables published by the American National Standards Institute.

Table 17.5 DIMENSIONS FOR THE SET SCREWS SHOWN IN FIG. 17.40 (ALL DIMENSIONS GIVEN IN INCHES)

D	I	J	T	R	C		P		Q	q
					Diameter of cup and flat points		Diameter of dog point		Length of dog point	
Nominal size	Radius of headless crown	Width of slot	Depth of slot	Oval point radius	Max	Min	Max	Min	Full	Half
5 0.125	0.125	0.023	0.031	0.094	0.067	0.057	0.083	0.078	0.060	0.030
6 0.138	0.138	0.025	0.035	0.109	0.074	0.064	0.092	0.087	0.070	0.035
8 0.164	0.164	0.029	0.041	0.125	0.087	0.076	0.109	0.103	0.080	0.040
10 0.190	0.190	0.032	0.048	0.141	0.102	0.088	0.127	0.120	0.090	0.045
12 0.216	0.216	0.036	0.054	0.156	0.115	0.101	0.144	0.137	0.110	0.055
¼ 0.250	0.250	0.045	0.063	0.188	0.132	0.118	0.156	0.149	0.125	0.063
⁵⁄₁₆ 0.3125	0.313	0.051	0.076	0.234	0.172	0.156	0.203	0.195	0.156	0.078
⅜ 0.375	0.375	0.064	0.094	0.281	0.212	0.194	0.250	0.241	0.188	0.094
⁷⁄₁₆ 0.4375	0.438	0.072	0.109	0.328	0.252	0.232	0.297	0.287	0.219	0.109
½ 0.500	0.500	0.081	0.125	0.375	0.291	0.270	0.344	0.344	0.250	0.125
⁹⁄₁₆ 0.5625	0.563	0.091	0.141	0.422	0.332	0.309	0.391	0.379	0.281	0.140
⅝ 0.625	0.625	0.102	0.156	0.469	0.371	0.347	0.469	0.456	0.313	0.156
¾ 0.750	0.750	0.129	0.188	0.563	0.450	0.425	0.563	0.549	0.375	0.188

Source: Courtesy of ANSI; B18.6.2–1956.

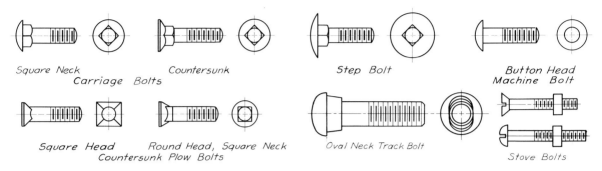

Fig. 17.41 Miscellaneous types of bolts.

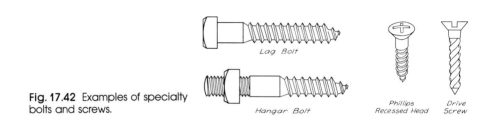

Fig. 17.42 Examples of specialty bolts and screws.

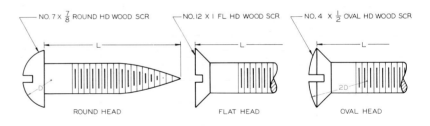

Fig. 17.43 Standard types of wood screws. The proportions shown here can be used for drawing wood screws of all sizes.

Sizes of wood screws are specified by single numbers such as 0, 6, or 16. From 0 to 10 each digit represents a different size. Beginning at 10, only even-numbered sizes are standard, i.e., 10, 12, 14, 16, 18, 20, 22, and 24. The following formula can be used to relate these numbered sizes to the actual diameter of the screws:

Actual DIA = 0.060 + screw number × 0.013.

For example, the actual diameter of a No. 5 screw is

$$0.060 + 5(0.013) = 0.125$$

17.23 USE OF TEMPLATES

Templates are available for drawing threads, nuts, and threaded fasteners. They are available for a range of sizes that is satisfactory for most applications, since thread representations are approximations at best.

Two typical templates are shown in Fig. 17.44. The black areas represent the holes cut into the thin plastic templates. The template is laid on the drawing and the threaded features are drawn using the template as a guide. Templates

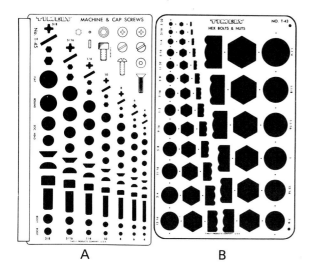

Fig. 17.44 Examples of templates that can be used for drawing threaded fasteners. (Courtesy of Timely Products Company.)

are also available for drawing nuts and bolts in pictorial. These are used primarily by the technical illustrator.

17.24 TAPPING A HOLE

A threaded hole is called a *tapped hole*, since the tool used to cut the threads is called a tap. The types of taps available for threading small holes by hand are shown in Fig. 17.45.

The taper, plug, and bottoming hand taps are identical in size, length, and measurements, their only difference being the chamfered portion of their ends. The taper tap has a long chamfer (8 to 10 threads), the plug tap has a chamfer of 3 to 5 threads, and the bottoming tap has a short chamfer of only 1 to 1 ½ threads.

When tapping by hand in open or "through" holes, the taper should be used for coarse threads, since it ensures straighter starting. The taper tap is also recommended for the harder metals. The plug tap can be used in soft metals or for fine-pitch threads. When it is desirable to tap a hole to the very bottom, all three taps—taper, plug, and bottoming—should be used in this order.

Notes are added to specify the depth of the drilled hole and the depth of the threads. For example, a note reading ⅞ DRILL, 3 DEEP, 1–8 UNC-2A, 2 DEEP means that the hole will be drilled deeper than it is threaded and the last usable thread will be 2″ deep in the hole. Note that the drill point has an angle of 120°.

17.25 WASHERS, LOCK WASHERS, AND PINS

Washers, called *plain washers*, are used with nuts and bolts to improve the assembly and the strength of the fastening. Plain washers are noted on a working drawing in the following manner:

0.938 × 1.750 × 0.134 TYPE A PLAIN WASHER

These numbers represent the inside diameter, outside diameter, and the thickness, in that order. These dimensions can be found in Appendix 36.

A lock washer is a washer that prevents a nut or cap screw from loosening as a result of vibration or movement. Two of the more common types shown in Fig. 17.46 are the external-tooth lock washer and the helical-spring lock washer. Other more varied types of locking washers and devices are shown in Fig. 17.47.

Tables for spring lock washers are given in Appendix 37 for regular and extra heavy-duty helical-spring lock washers. these are designated on drawings with notes in the following form:

HELICAL-SPRING LOCK WASHER–¼ REGULAR-
PHOSPHOR BRONZE

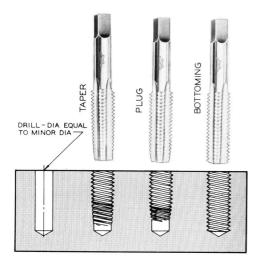

Fig. 17.45 Three types of taps for threading internal holes.

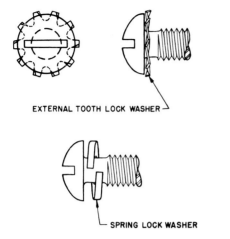

EXTERNAL TOOTH LOCK WASHER

SPRING LOCK WASHER

Fig. 17.46 Two types of lock washers for preventing a bolt from unscrewing.

(the ¼ is the washer's inside diameter).

Tooth lock washers are designated with notes in the following form:

INTERNAL-TOOTH LOCK WASHER–¼-TYPE A-STEEL,

or EXTERNAL-TOOTH LOCK WASHER–562-TYPE B-STEEL.

The specifications for tooth lock washers are given in Appendix 37.

Straight pins and taper pins are used to fix parts together in a specified alignment. Dimensions for these are given in Appendix 34.

Other locking devices are cotter pins, split taper pins, and straight pins (Fig. 17.48). Tables of specifications for cotter pins are given in the appendix.

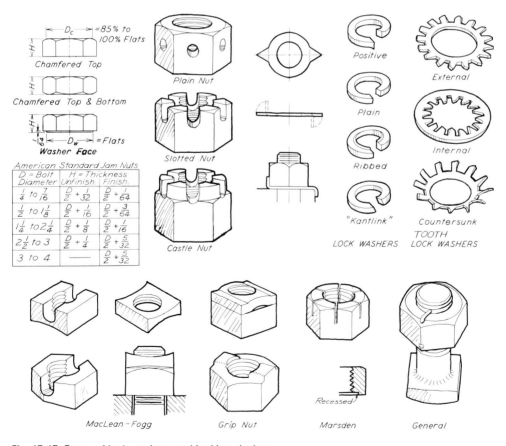

Fig. 17.47 Types of lock washers and locking devices.

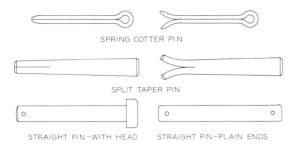

Fig. **17.48** Types of pins used to fix parts together.

17.26 PIPE THREADS

Pipe threads are used in connecting pipes, tubing, lubrication fittings, and other applications. The most commonly used pipe thread is tapered at a ratio of 1 to 16, but straight pipe threads are available (without a taper). Since pipe threads are usually tapered, the threads will only engage for an effective length determined by the formula below:

$$L = (0.80D + 6.8)P$$

where D is the outside diameter of the pipe and P is the pitch.

Methods of representing tapered threads are shown in Fig. 17.49. Notice that a taper of 1 to 16 is shown to call attention to the fact that the threads are tapered.

The abbreviations associated with pipe threads are:

N = National   G = Grease
P = Pipe   I = Internal

T = Taper   M = Mechanical
C = Coupling   L = Locknut
S = Straight   H = Hose coupling
F = Fuel and oil   R = Railing fittings

These abbreviations are used in combination for the following ANSI symbols.

NPT = National pipe taper
NPTF = National pipe thread (dryseal—for pressure-tight joints)
NPS = Straight pipe thread
NPSC = Straight pipe thread in couplings
NPSI = National pipe straight internal thread
NPSF = Straight pipe thread (dryseal)
NPSM = Straight pipe thread for mechanical joints
NPSL = Straight pipe thread for locknuts and locknut pipe threads
NPSH = Straight pipe thread for hose couplings and nipples
NPTR = Taper pipe thread for railing fittings

To specify a pipe thread in note form, the nominal pipe diameter (the internal diameter), the number of threads per inch, and the symbol which denotes the type of thread are given. For example:

1¼–11½ NPT   or   3–8 NPTR.

These specifications can be taken from Appendix 10. Examples of external and internal thread notes are shown in Fig. 17.50. The dryseal thread is used in applications where a pressure-tight joint is required without the use of a lubricant or sealer. Dryseal threads may be straight or tapered. No clearance between the mating parts of the joint is permitted, giving the highest quality fit.

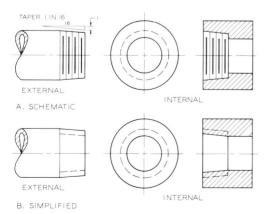

Fig. **17.49** Schematic and simplified techniques of representing external and internal pipe threads.

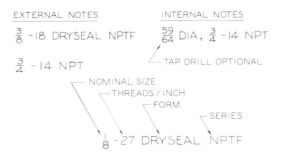

Fig. **17.50** Typical pipe-thread notes.

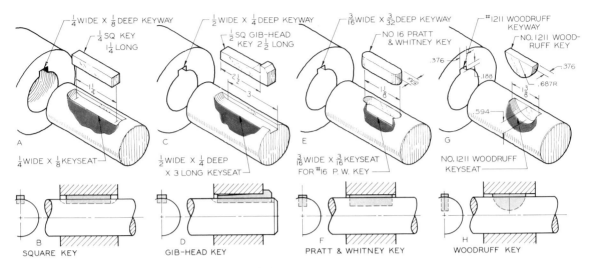

Fig. 17.51 Standard keys used to hold parts on a shaft.

The tap drill is sometimes given in the internal pipe thread note (Fig. 17.50), but this is optional.

17.27 KEYS

Keys are used to attach parts to shafts in order to transmit power to pulleys, gears, or cranks. Several types of keys are shown pictorially and orthographically in Fig. 17.51. The four types illustrated here are the most commonly used keys. To specify a key, notes must be given for the keyway, the key, and the keyseat, as shown in Fig. 17.51A,

C, E, and G. The notes given are typical of the notes used to give key specifications. These dimensions may be found in Appendixes 31 and 32 for various types of keys.

17.28 RIVETS

Rivets are fasteners used to join thin materials in a permanent joint. Rivets are designed to fit into holes that are slightly larger than the diameter of the rivet. The rivet is inserted in the hole and the headless end is formed into the specified shape by applying extreme pressure to the projecting end. This forming operation is done when the rivets are either hot or cold, depending on the application.

Typical shapes and proportions of small rivets are shown in Fig. 17.52. These rivets vary in diameter from $\frac{1}{16}''$ to $1\frac{3}{4}''$. Rivets are used extensively in pressure-vessel fabrication, in heavy structures such as bridges and buildings, and in construction with sheet metal.

The proportions for large rivets are shown in Fig. 17.53. The proportions of large and small rivets are based upon the diameters of the rivet bodies. Many ANSI tables of standard dimensions are available for sizing rivets.

Three types of lap joints are shown in Fig. 17.54 where the joints are held secure by one, two, or three rivets as shown in the sectional view.

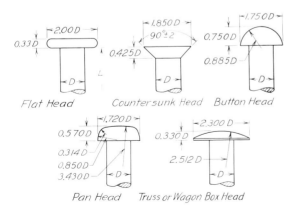

Fig. 17.52 Types and proportions of small rivets. Small rivets have shank diameters up to ½".

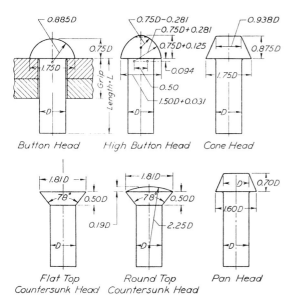

Button Head   High Button Head   Cone Head

Flat Top
Countersunk Head   Round Top
Countersunk Head   Pan Head

Fig. 17.53 Types and proportions of
large rivets. Large rivets have shank
diameters of ½" DIA to 1¾" DIA.

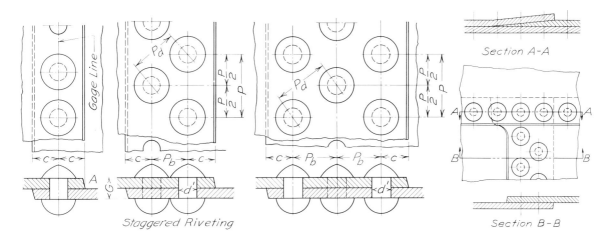

Gage Line

Staggered Riveting

Section A-A

Section B-B

Fig. 17.54 Examples of lap joints using single rivets, double rivets, and triple
rivets. The fourth example shows three plates that are fastened by double
rivets.

Note that the bodies of the rivets are drawn as hidden circles in the top views.

Single and double riveted butt joints are illustrated in Fig. 17.55.

The standard symbols recommended by ANSI for representing rivets are shown in Fig. 17.56. Rivets that are driven in the shop are called shop rivets, and those assembled on the job at the site are called field rivets.

17.29   SPRINGS

Of the many types of springs that are available, some of the more commonly used types are (1) compression, (2) torsion, (3) extension, (4) flat, and (5) constant force. Figure 17.57 shows the single-line, conventional representation of the first three of these. Also shown are the types of ends that can be used on compression springs and also

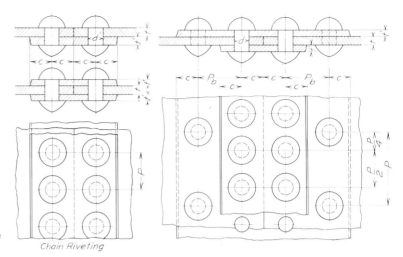

Fig. 17.55 Single and double riveted butt joints.

Chain Riveting

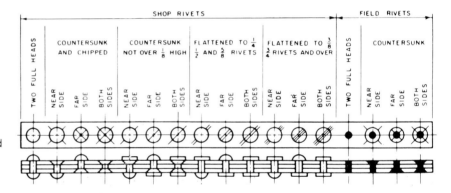

Fig. 17.56 The symbols used to represent rivets in a drawing (Courtesy of ANSI 14.14).

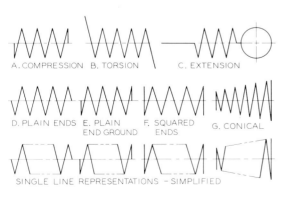

Fig. 17.57 Single-line representations of various types of springs.

the simplified, single-line representation of coil springs.

A typical working drawing of a compression spring is shown in Fig. 17.58. The ends of the spring are drawn using the double-line representation, and conventional lines are used to indicate the undrawn portion of the spring. Only the diameter and the free length of the spring are given on the drawing itself. The remaining specifications are given in tabular form near the drawing.

A working drawing of an extension spring (Fig. 17.59) is very similar to that of a compression spring. In a drawing of a helical torsion spring (Fig. 17.60), angular dimensions must be shown to specify the initial and final positions of the spring as torsion is applied to it. All types of springs require a table of specifications to describe their details.

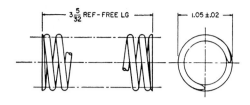

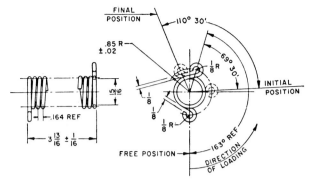

WIRE DIA .120
DIRECTION OF HELIX OPTIONAL
TOTAL COILS 12½ REF
LOAD AT COMPRESSED LG OF 2.05 IN.
= 39 LB ± 3.9 LB
LOAD AT COMPRESSED LG OF 1.69 IN.
= 51.5 LB ± 5.2 LB

Fig. 17.58 A conventional double-line drawing of a compressions spring and its specifications. (Courtesy of the U.S. Department of Defense.)

WIRE DIA .148
DIRECTION OF HELIX LEFT HAND
TOTAL COILS 20.55 REF
TORQUE 15 LB IN. ± 1.5 LB IN. AT INITIAL POSITION
TORQUE 33 LB IN. ± 3.3 LB IN. AT FINAL POSITION
MAXIMUM DEFLECTION WITHOUT SET BEYOND FINAL POSITION 56°
SPRING RATE .16 LB IN. / DEG REF

Fig. 17.60 A conventional double-line drawing of a helical torsion spring and its specifications. (Courtesy of the U.S. Department of Defense.)

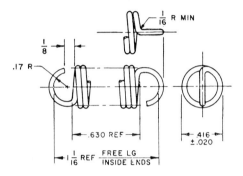

WIRE DIA .042
DIRECTION OF HELIX OPTIONAL
TOTAL COILS 14 REF
RELATIVE POSITION OF ENDS 180° ± 20°
EXTENDED LG INSIDE ENDS
WITHOUT PERMANENT SET 2.45 IN. (MAX)
INITIAL TENSION 1.0 LB ± .10 LB
LOAD 4 LB ± .4 LB AT 1.56 IN.
EXTENDED LG INSIDE ENDS
LOAD 6.3 LB ± .63 LB AT 1.95 IN.
EXTENDED LG INSIDE ENDS

Fig. 17.59 A conventional double-line drawing of an extension spring and its specifications. (Courtesy of the U.S. Department of Defense.)

17.30 DRAWING SPRINGS

Springs may be drawn as schematic representations using single lines to represent the springs. Examples of single-line drawings are shown in Fig. 17.61. Each is drawn by laying out the diameter of the coils and the lengths of the springs, and then the number of active coils are drawn by using the diagonal-line method.

In part B, the two end coils are "dead" coils and only four are active. An extension spring is drawn at C.

For applications where more realism is desired, a double-line drawing of a thread can be made as shown in Fig. 17.62. The end result is a good approximation of the spring.

A 4 COILS
COMPRESSION SPRING

B 4 ACTIVE COILS
COMPRESSION SPRING

C 5 ACTIVE COILS
EXTENSION SPRING

Fig. 17.61 In Part A, a schematic drawing of a spring with four active coils as shown. Once the length is laid out, the diagonal line method is used to divide it into four equally spaced coils. The spring at B has six coils, but only four of them are active cells. An extension spring with five active coils is shown at C.

Fig. 17.62 Detailed drawing of a spring

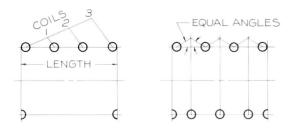 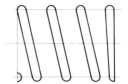 

Step 1 Lay out the diameter and the length of the spring, and locate the coils by the diagonal-line technique.

Step 2 Locate the coils on the lower side along the bisectors of the spaces between the coils on the upper side.

Step 3 Connect the coils on each side. This is a right-hand coil; a left-hand spring would slope in the opposite direction.

Step 4 Construct the back side of the spring and the end coils, to complete the detailed drawing of a compression spring.

PROBLEMS

These problems are to be completed on Size A sheets. Problems 1 through 16 are laid out one problem per sheet. Problems 17 through 30 are to be laid out two problems per sheet. The boxes drawn around the figures representing the problems are approximately 6″ wide × 5″ high, which equals about half of a Size A sheet.

Fasteners

1. The layout in Fig. 17.63 is to be used for constructing a detailed representation of an Acme thread with a major diameter of 3″. The thread note specifications are 3–1½ ACME. Show both external and internal thread representations. Show the thread note. Use inches or millimeters as instructed.

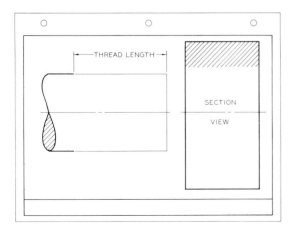

Fig. 17.63 Construction of thread symbols (Problems 1, 2, and 3).

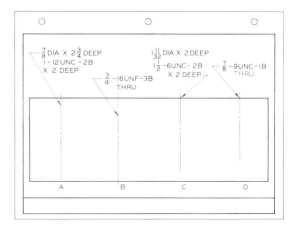

Fig. 17.64 Internal threads (Problems 4, 5, and 6).

2. Repeat Problem 1 but draw internal and external detailed representations of a square thread that is 3″ in diameter. The note specifications are 3–1½ SQUARE. Apply notes to both parts.

3. Repeat Problem 1, but draw internal and external detailed thread representations of an American National thread form. The major diameter of each part is 3″. The note specifications are 3–4 NC–2. Apply notes to both parts.

4. Notes are given in Fig. 17.64 to specify the depth of the holes that are to be drilled and the threads that are to be tapped in the holes. Following these notes, draw detailed representations of the threads as views according to specifications.

5. Repeat Problem 4, but use schematic representations.

6. Repeat Problem 4, but use simplified thread representations.

7. Figure 17.65 shows a layout of two external threaded parts and their end views. Also shown is a piece into which the external threads will be screwed. Complete all three views of each of the parts. Use detailed threads and apply notes to the internal and external threads. Use the table in Appendix 15 for thread specifications. Use UNC threads with a 2A fit.

8. Repeat Problem 7, but use schematic thread representations.

9. Repeat Problem 7, but use simplified thread representations.

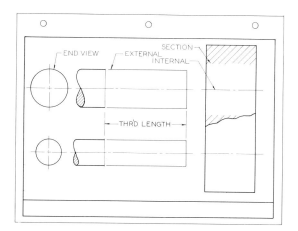

Fig. 17.65 Internal and external threads (Problems 7, 8, and 9).

10. Referring to Fig. 17.66, complete the drawing with instruments as a finished hexagon bolt and nut. The bolt head is a heavy nut drawn across corners. Use detailed thread representations. Show notes to specify the parts of the assembly. Thread specifications are 1½–6 UNC–3 or M36 × 4.

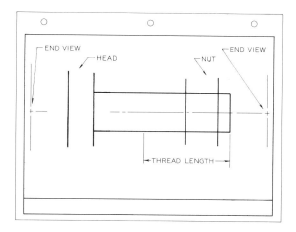

Fig. 17.66 Nuts and bolts in assembly (Problems 10, 11, and 12).

11. Referring to Fig. 17.66, complete the drawing with instruments as an unfinished squarehead bolt and nut. The bolt head is drawn across corners. The regular nut is to be drawn across corners. Use schematic thread representations. Show notes to specify the parts. Use the table in Appendix 15 or 18 for thread specifications (English or SI).

12. Referring to Fig. 17.66, complete the drawing with instruments as a finished hexagon nut and bolt. The regular bolt and nut are to be drawn across flats. Use simplified thread representations. Show notes to specify the parts. Use the table in Appendix 15 or 18 for specifications.

13. The notes in Fig. 17.67 apply to machine and cap screws that are to be drawn in the section view of the two parts. The holes in which the screws are to be drawn are through holes. Complete the drawings and show the notes as given. Show the remaining section lines. Use detailed thread symbols.

14. Repeat Problem 13, but use schematic thread symbols.

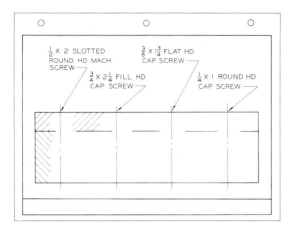

Fig. 17.67 Cap screws and machine screws (Problems 13, 14, and 15).

Fig. 17.68 Design involving threaded parts (Problem 16).

15. Repeat Problem 13, but use simplified thread symbols.

16. *Design.* The pencil pointer shown in Fig. 17.68 has a shaft of ¼" that fits into a bracket designed to clamp onto a desk top. A set screw holds the shaft in position. Make a drawing of the bracket, estimating its dimensions. Show the details and the method of using the set screw to hold the shaft. Give the specifications for the set screw.

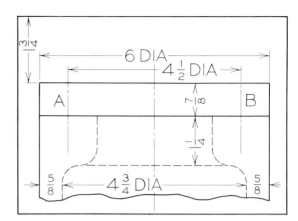

Fig. 17.69 Problem 17.

17. (Fig. 17.69). On axes A and B, construct ⅝" (16 mm) hexagonal head bolts across flats with a coarse thread, UNC. Convert the view to a half-section to show how the parts are assembled together.

18. (Fig. 17.70). On axes E and F, draw a stud with a hexagon head nut shown across corners. The stud is to have a diameter of ⅞" (22 mm) and should be a UNC form and series.

19. (Fig. 17.71). Draw a ⅞" (22 mm) DIA special stud with a collar that is 1¼" DIA. Show a plain washer and a regular nut at end B. On axes CD, draw a machine screw of your selection to hold the two parts together.

20. (Fig. 17.72). Draw a cap screw on axis AB to fasten the parts together. The following dimensions are bolt DIA 1¼" or 32 mm, C - 1.25", E - 1.12", H - 0.25", hexagon head across corners, G - 4.50", and D - 2.00".

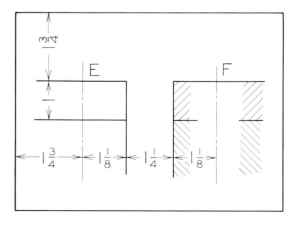

Fig. 17.70 Problem 18.

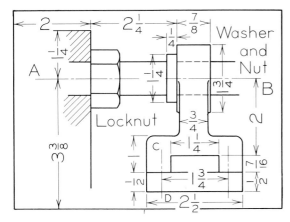

31

Fig. 17.71 Problem 19.

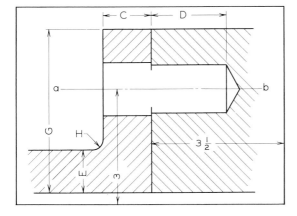

Fig. 17.72 Problem 20.

Keys

21. Figure 17.73 shows two parts assembled on a cylindrical shaft. These parts are to be held in position by a square key in part A and a gib-head key in part B. Show the necessary notes to specify the key, keyway, and keyseat.

22. Repeat Problem 21, but use a No. 16 Pratt & Whitney key in part A and a No. 1211 Woodruff key in part B. Show the necessary notes to specify the key, the keyway, and the keyseat.

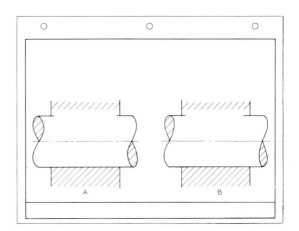

Fig. 17.73 Problems 21 and 22.

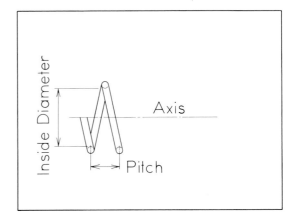

Fig. 17.74 Problems 23–30.

23-26. (Fig. 17.74 and Table 17.6). Using the double-line method, draw helical springs, two per Size A sheet. Use the specifications in the table for each.

27-30. (Fig. 17.74). Same as problems 23–26, except use single-line representations for the springs.

Table 17.6 PROBLEMS 23–26

Prob.	No. of Turns	Pitch	Size Wire	Inside Dia	Outside Dia	
23	4	1	No. 4 = 0.2253	3		RH
24	5	¾	No. 6 = 0.1920		2	LH
25	6	⅝	No. 10 = 0.1350	2		RH
26	7	¾	No. 7 = 0.1770		1¾	LH

18 GEARS AND CAMS

18.1 INTRODUCTION TO GEARS

Gears are toothed wheels that mesh together to transmit force and motion from one gear to the next. They are linked together by teeth cut into their circumferences that contact each other.

The most common types of gears (Fig. 18.1) are (A) spur gears, (B) bevel gears, and (C) worm gears. Each of these types of gears will be covered in this chapter. The many other more specialized types of gears deserve more coverage than is possible in this text.

Fig. 18.1 The three most basic types of gears are (A) spur gears, (B) bevel gears, and (C) worm gears. (Courtesy of the Process Gear Co.)

18.2 SPUR GEAR TERMINOLOGY

The spur gear is a circular gear with teeth cut around its circumference. Two mating spur gears can transmit power from a shaft to another parallel shaft.

> When the two meshing gears are unequal in diameter, the smaller gear is called the **pinion** and the larger one is called the **gear.**

The following terms are used to describe the parts of a spur gear. Many of these features are illustrated and labeled in Fig. 18.2 and 18.3. The corresponding formulas for each feature are given.

Pitch circle (PC) is the imaginary circle of a gear if it were a friction wheel without teeth that contacted the pitch circle of another friction wheel.

Pitch diameter (PD) is the diameter of the pitch circle; $PD = N / DP$ (where N = number of teeth and DP = diametral pitch).

Diametral pitch (DP) is the ratio between the number of teeth on a gear and its pitch diameter. For example, a gear with 20 teeth and a 4-inch pitch diameter will have a diametrical pitch of 5, which means that there are 5 teeth per inch of diameter; $DP = N / PD$ (where N = number of teeth).

Circular pitch (CP) is the circular measurement from one point on a tooth to the corresponding

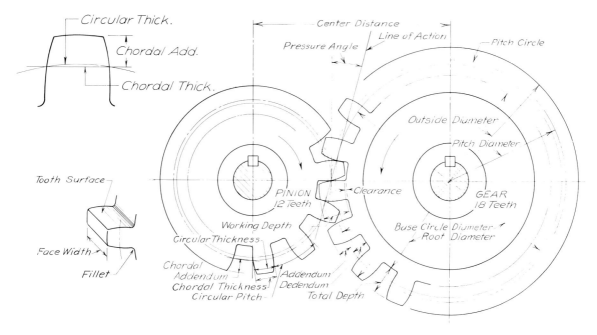

Fig. 18.2 Gear terminology for spur gears.

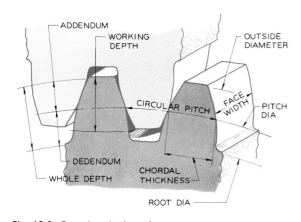

Fig. 18.3 Gear terminology for spur gears.

point on the next tooth measured along the pitch circle; $CP = 3.14 / DP$.

Center distance (CD) is the distance from the center of a gear to its mating gear's center; $CD = (N_p + N_s) / (2DP)$ (where N_p and N_s are the number of teeth in the pinion and spur, respectively).

Addendum (A) is the height of a gear above its pitch circle; $A = 1 / DP$.

Dedendum (D) is the depth of a gear below the pitch circle. $D = 1.157 / DP$.

Whole depth (WD) is the total depth of a gear tooth; $WD = A + D$.

Working depth (WKD) is the depth to which a tooth fits into a meshing gear; $WKD = 2 / DP$; or $WKD = 2A$.

Circular thickness (CRT) is the circular distance across a tooth measured along the pitch circle; $CRT = 1.57 / DP$.

Chordal thickness (CT) is the straight-line distance across a tooth at the pitch circle; $CT = PD$ $(\sin 90° / N)$ (where N = number of teeth).

Face width (FW) is the width across a gear tooth parallel to its axis. This is a variable dimension, but it is usually 3 to 4 times the circular pitch; $FW = 3$ to $4(CP)$.

Outside diameter (OD) is the maximum diameter of a gear across its teeth; $OD = PD + 2A$.

Root diameter (RD)   is the diameter of a gear measured from the bottom of its gear teeth. $RD = PD - (2D)$.

Pressure angle (PA)   is the angle between the line of action and a line perpendicular to the center line of two meshing gears. Angles of 14.5° and 20° are standard angles for involute gears.

Base circle (BC)   is the circle from which an involute tooth curve is generated or developed; $BC = PD (\cos PA)$.

18.3   TOOTH FORMS

The most common gear tooth is an *involute tooth* with a 14.5° pressure angle. The 14.5° angle is the angle of contact between two gears when the tangents of both gears pass through the point of contact. Gears with pressure angles of 20° and 25° are also used. Gear teeth with larger pressure angles are wider at the base and thus are stronger than the standard 14.5° teeth.

The standard gear face is an involute that keeps the meshing gears in contact as the gear teeth are revolved past one another. The principle of constructing an involute is illustrated in Fig. 18.4.

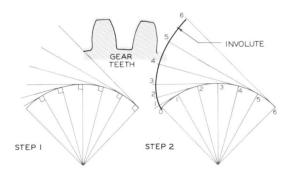

Fig. 18.4 Construction of an involute

Step 1   The base arc is divided into equal divisions with radial lines from the center. Tangents are drawn perpendicular to the radial lines on the arc.

Step 2   The chordal distance from 1 to 0 is used as the radius and 1 as the center to find point 1 on the involute curve. The distance from 2 to newly found 1 is revolved to the tangent line through 2 to locate a second point, and the process is continued.

An involute curve can be thought of as the path of a string that is kept taut as it is unwound from the base arc. It is unnecessary to use this procedure in drawing gear teeth since most detail drawings employ only approximations of gear teeth, if teeth are shown at all.

18.4   GEAR RATIOS

The diameters of two meshing spur gears establish ratios that are important to the function of the gears. Examples of these ratios are given in Fig. 18.5.

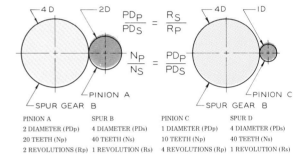

PINION A	SPUR B	PINION C	SPUR D
2 DIAMETER (PDp)	4 DIAMETER (PDs)	1 DIAMETER (PDp)	4 DIAMETER (PDs)
20 TEETH (Np)	40 TEETH (Ns)	10 TEETH (Np)	40 TEETH (Ns)
2 REVOLUTIONS (Rp)	1 REVOLUTION (Rs)	4 REVOLUTIONS (Rp)	1 REVOLUTION (Rs)

Fig. 18.5 Ratios between meshing spur gears.

If the radius of a gear is twice that of its pinion (the small gear), then the diameter is twice that of the pinion, and the gear has twice as many teeth as the pinion. In this case, the pinion must make twice as many turns as the larger gear. In other words, the revolutions per minute (RPM) of the pinion is twice that of the larger gear.

When the diameter of the gear is four times the diameter of the pinion, there must be four times as many teeth on the gear as on the pinion, and the number of revolutions of the pinion will be four times that of the larger gear.

The relationship between two meshing spur gears can be developed in formula form by finding the velocity of a point on the small gear that is equal to $\pi PD \times$ RPM of the pinion. The velocity of a point on the large gear is equal to: $\pi PD \times$ RPM of the spur. Since the velocity of points on each gear must be equal, the equation may be written as

$$\pi PD_p(\text{RPM}) = \pi PD_s(\text{RPM});$$

therefore,

$$\frac{PD_p}{PD_s} = \frac{RPM_s}{RPM_p}.$$

If the radius of the pinion is 1 inch, the radius of the spur is 4 inches, and the RPM of the pinion is 20 RPM, then the RPM of the spur can be found as follows:

$$\frac{2(1)}{2(4)} = \frac{RPM_s}{20 \ RPM_p}; \ RPM_s = \frac{2(20)}{2(4)} = 5 \ RPM$$

(one-fourth of the revolutions per minute of the pinion.)

The number of teeth on each gear is also proportional to the radii and diameters of a pair of meshing gears. This relationship can be written:

$$\frac{N_p}{N_s} = \frac{PD_p}{PD_s}$$

(where N_p and N_s are the numbers of teeth on the spur and pinion, respectively, and PD_p and PD_s are their pitch diameters.)

18.5 GEAR CALCULATIONS

Before a working drawing of a gear can be started, the drafter must perform a series of calculations to determine the dimensions of the gear using the definitions and formulas previously introduced. The following problem is given as an example.

PROBLEM 1: Calculate the dimensions for a spur gear that has a pitch diameter of 5 inches, a diametral pitch of 4, and a pressure angle of 14.5°. (The diametral pitch is the same for meshing gears.)

SOLUTION:
No. of teeth = PD × DP = 5 × 4 = 20
Addendum = 1 / 4 = 0.25
Dedendum = 1.157 / 4 = 0.2893
Circular thickness = 1.5708 / 4 = 0.3927
Outside diameter = (20 + 2) / 4 = 5.50
Root diameter = 5 − 2(0.2893) = 4.421
Chordal thickness = 5 (sine 90° / 20) = 5 (0.079) = 0.392
Chordal addendum = 0.25 + (0.3927²/(4 × 5)) = 0.2577

Face width = 3.5(0.79) = 2.75
Circular pitch = 3.14 / 4 = 0.785
Working depth = 0.6366 × (3.14 / 4) = 0.4997
Whole depth = 0.250 + 0.289 = 0.539

These dimensions can be used to draw the spur gear and to provide specifications necessary for its manufacture.

A second problem is given as an example of the method of determining the design information for two meshing gears when their working ratios are known.

PROBLEM 2: Find the number of teeth and other specifications for a pair of meshing gears with a driving gear that turns at 100 RPM and a driven gear that turns at 60 RPM. The dimetral pitch for each is 10. The center to center distance between the gears is 6 inches.

SOLUTION:

Step 1 Find the sum of the teeth on both gears:

Total teeth = 2 × (C to C dist.) × DP
= 2 × 6 × 10 = 120 teeth

Step 2 Find the number of teeth for the driving gear by two steps:

A. $\frac{Driver \ RPM}{Driven \ RPM} + 1 = \frac{100}{60} + 1 = 2.667$

B. $\frac{Total \ teeth}{\frac{100}{60} + 1} = \frac{120}{2.667} = 45 \ teeth*$

Step 3 Find the number of teeth for the driven gear.

Total teeth minus teeth on driver = teeth on driver 120 − 45 = 75' teeth

Step 4 The other specifications for the gears can be computed as shown in Problem 1 using the formulas in Section 18.2.

*The number of teeth must be a whole number since there cannot be fractional teeth on a gear. It may be necessary to adjust the center distance to yield a whole number of teeth.

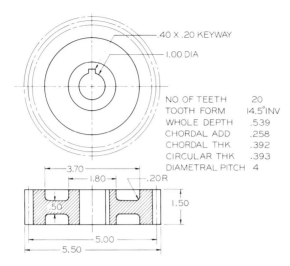

NO OF TEETH	20
TOOTH FORM	14.5°INV
WHOLE DEPTH	.539
CHORDAL ADD	.258
CHORDAL THK	.392
CIRCULAR THK	.393
DIAMETRAL PITCH	4

Fig. 18.6 A detail drawing of a spur gear with a table of values to supplement the dimensions shown on the drawing.

18.6 DRAWING SPUR GEARS

A conventional drawing of a spur gear is shown in Fig. 18.6. The teeth need not be drawn since this is time-consuming and unnecessary. It is possible to omit the circular view and show only a sectional view of the gear with a table of dimensions that have been calculated. Dimensions of this type are called "cutting data."

When the circular view is drawn, circular center lines are drawn to represent the root circle, pitch circle, and outside circle of the gear.

The table of dimensions is a necessary part of a gear drawing as shown in Fig. 18.7. This data can be calculated by formula or taken from tables of standards in gear handbooks such as *Machinery's Handbook*.

18.7 BEVEL GEAR TERMINOLOGY

Bevel gears are gears whose axes intersect at angles. Although the angle of intersection is usually 90°, other angles are used. The smaller of the

Fig. 18.7 A detail drawing of a spur gear that has been converted to metric units by multiplying dimensions in inches by 25.4 to yield millimeters.

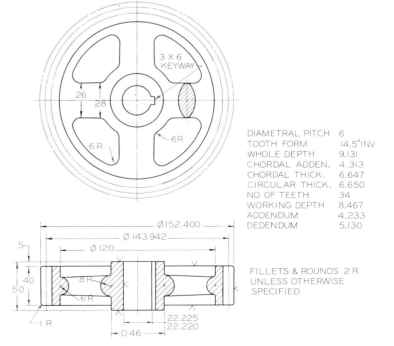

DIAMETRAL PITCH	6
TOOTH FORM	14.5°INV
WHOLE DEPTH	9.131
CHORDAL ADDEN.	4.313
CHORDAL THICK.	6.647
CIRCULAR THICK.	6.650
NO. OF TEETH	34
WORKING DEPTH	8.467
ADDENDUM	4.233
DEDENDUM	5.130

FILLETS & ROUNDS 2 R
UNLESS OTHERWISE
SPECIFIED

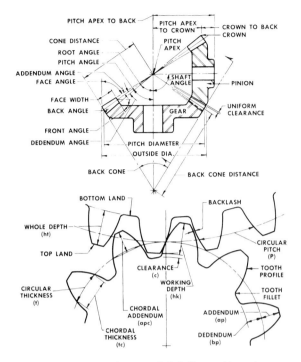

Fig. 18.8 The terminology and definitions of bevel gears. (Courtesy of Philadelphia Gear Corporation.)

two bevel gears is called the *pinion*, as in spur gearing.

The terminology of bevel gearing is illustrated in Fig. 18.8.

Pitch angle of pinion (small gear) (PA$_p$) is found by the following formula:

$$\text{Tan } PA_p = \frac{N_p}{N_g};$$

(N$_g$ and N$_p$ are the number of teeth on the gear and pinion respectively).

Pitch angle of gear (PA$_g$) is found by the following formula:

$$\text{Tan } PA_g = \frac{N_g}{N_p}.$$

Pitch diameter (PD) is the number of teeth (N) divided by the diametral pitch (DP).

$$PD = N / P$$

Addendum (A) is measured at the large end of the tooth: $A = 1 / DP$.

Dedendum (D) is measured at the large end of the tooth: $D = 1.157 / DP$.

Whole tooth depth (WD): $WD = 2.157 / DP$.

Thickness of tooth (TT) at pitch circle: $TT = 1.571 / DP$.

Dimetral pitch: $DP = N / PD$ (N = number of teeth).

Addendum angle (AA) is the angle formed by the addendum and the pitch cone distance.

$$\text{Tan} AA = \frac{A}{PCD}$$

(PCD = pitch cone distance).

Pitch cone distance: $PCD =$

$$PD / (2 \times \sin PA).$$

Dedendum angle (DA) is the angle formed by the dedendum and the pitch cone distance.

$$\text{Tan } DA = \frac{D}{PCD}$$

Face angle (FA) is the angle between the gear's center line and the top of its teeth: $FA = 90° - (PCD + AA)$.

Cutting angle (or root angle) (CA) is the angle between the gear's axis and the roots of the teeth: $CA = PCD - D$.

Outside diameter (OD) is the greatest diameter of a gear across its teeth: $OD = PD + 2A$.

Apex to crown distance (AC) is the distance from the crown of the gear to the apex of the cone measured parallel to the axis of the gear: $AC = OD / (2 \tan FA)$.

Chordal addendum (CA) is

$$A + \frac{TT^2 cosPA}{4(PD)}$$

Chordal thickness (CT) at the large end of the

tooth is found by the following formula:

$$CT = PD \times \sin \frac{90°}{N}$$

Face width (FW) can vary, but it is recommended that it be approximately equal to the pitch cone distance divided by 3: $FW = PCD / 3$.

Gear handbooks can be used for finding many of these dimensions from tables rather than using the formulas given above.

18.8 BEVEL GEAR CALCULATIONS

The following example of calculating the specifications for two bevel gears is given to demonstrate how the formulas in the previous section are used. You will note that some of the formulas result in the same specifications that apply to both the gear and pinion.

PROBLEM 3: Two gevel gears intersect at right angles. They have a dimetral pitch of 3, 60 teeth on the gear, 45 teeth on the pinion, and a face width of 4 inches. Find the dimensions of the gear.

SOLUTION:

Pitch cone angle of gear: Tan PCA = 60 / 45 = 1.33; PCA = 53° 7'.

Pitch cone angle of pinion: Tan PCA = 45 / 60; PCA − 36° 52'.

Pitch diameter of gear: 60/ 3 = 20.00".

Pitch diameter of pinion: 45 / 3 = 15.00".

The following formulas are the same for both the gear and pinion.

Addendum: 1 / 3 = 0.333".

Dedendum: 1.157 / 3 = 0.3857".

Whole depth: 2.157 / 3 = 0.719".

Tooth thickness on pitch circle: 1.571 / 3 = 0.5237.

Pitch cone distance: 20 / (2 sin 53° 7') = 12.5015.

Addendum angle: tan AA = 0.333 / 12.5015 = 1° 32'.

Dedendum angle: DA = 0.3857 / 12.5015 = 0.0308 = 1° 46'.

Face width: PCD / 3 = 4.00".

The remainder of the formulas must be applied separately to the gear and pinion.

Chordal addendum of gear:

$$0.333 + \frac{0.5237^2 \times \cos 53°7'}{4 \times 20} = 0.336".$$

Chordal addendum of pinion:

$$0.333 + \frac{0.5237^2 \times \cos 36°52'}{4 \times 15} = 0.338".$$

Chordal thickness of gear:

$$\sin \frac{90°}{60} \times 20 = 0.524".$$

Chordal thickness of pinion:

$$\sin \frac{90°}{45} \times 15 = 0.523".$$

Face angle of gear: 90° − (53° 7' + 1° 32') = 35° 21'.

Face angle of pinion: 90° − (36° 52' + 1° 32') = 51° 36'.

Cutting angle of gear: 53° 7' − 1° 46' = 51° 21'.

Cutting angle of pinion: 36° 52' − 1° 46' = 35° 6'.

Angular addendum of gear: 0.333 × cos 53° 7' = 0.1999".

Angular addendum of pinion: 0.333 × cos 36° 52' = 0.2667".

Outside diameter of gear: 20 + 2 (0.1999) = 20.4000".

Outside diameter of pinion: 15 + 2 (0.2667) = 15.533".

Apex to crown distance of gear: $\frac{20.400}{2}$ × tan 35° 7' = 7.173".

Apex to crown distance of pinion: $\frac{15.533}{2}$ × tan 51° 36' = 9.800".

18.9 DRAWING BEVEL GEARS

The dimensions calculated above are used to lay out the bevel gears in a detail drawing. Many of the calculated dimensions would be difficult to measure on a drawing within a high degree of accuracy; therefore, it is important to provide a table of "cutting data" for each gear.

Fig. 18.9 Construction of bevel gears.

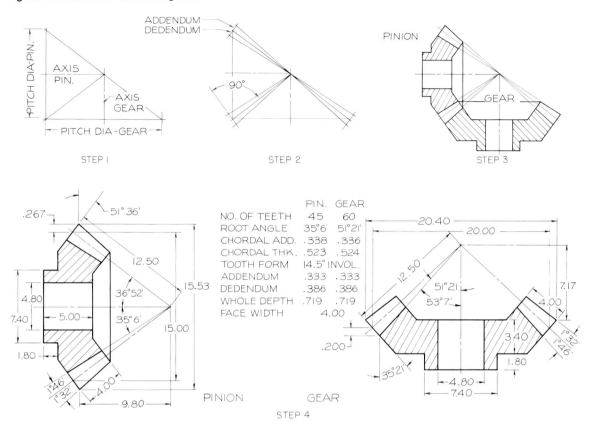

Step 1 Lay out the pitch diameters and axes of the two bevel gears.

Step 2 Draw construction lines to establish the limits of the teeth by using the addendum and dedendum dimensions.

Step 3 Draw the pinion and the gear using the specified dimensions or those that were calculated by formula.

Step 4 Complete the detail drawings of both gears and provide a table of cutting data.

The steps of drawing the bevel gears are shown in Fig. 18.9. The finished drawings are shown with a combination of dimensions and a table of dimensions, which is not shown.

18.10 WORM GEARS

A worm gear is composed of a thread shaft called a *worm* and a circular gear called a *spider*. (Fig. 18.10). The worm is revolved in a continuous motion, which causes the spider to revolve about its axis.

The following terminology is illustrated in Figs. 18.10 and 18.11. The following formulas can be used to calculate the dimensions associated with these terms.

Worm Specifications and Formulas

Linear pitch (P) is the distance from one thread to the next measured parallel to the worm's axis: $P = L/N$ (where N is number of threads: 1 if a single thread 2 if a double thread, etc.)

Lead (L) is the distance that a thread advances in a turn of 360°.

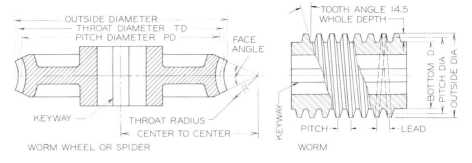

Fig. 18.10 The terminology and definitions of worm gears.

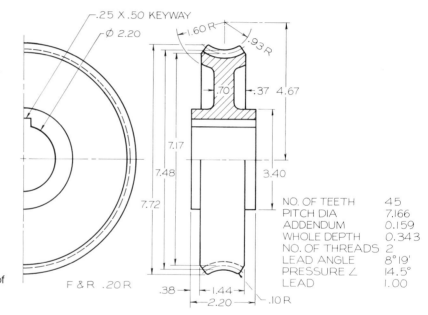

NO. OF TEETH	45
PITCH DIA	7.166
ADDENDUM	0.159
WHOLE DEPTH	0.343
NO. OF THREADS	2
LEAD ANGLE	8°19'
PRESSURE ∠	14.5°
LEAD	1.00

Fig. 18.11 A detail drawing of worm gear (spider) and the table of cutting data.

Addendum of Tooth: $AW = 0.3183\ P.$

Pitch Diameter: $PDW = OD - 2AW$ (OD is the outside diameter).

Whole Depth of Tooth: $WDT = 0.6866 \times P.$

Bottom Diameter of Worm: $BD = OD - 2\ WDT.$

Width of Thread at Root: $WT = 0.31P.$

Minimum Length of Worm: $MLW = \sqrt{8\ PDS \times AW}$; (PDS is the pitch diameter of the spider).

Helix angle of worm:

$$\mathrm{Cot}\ \beta = \frac{3.14\ PDW}{L}.$$

Outside Diameter: $OD = PD + 2A.$

Spider Specifications and Formulas

Pitch Diameter of Spider: $PDS = \dfrac{N(P)}{3.14}$; (N is the number of teeth on the spider).

Throat Diameter of Spider: $TD = PDS + 2A.$

Radius of Spider Throat: $RST =$

$$\frac{\text{OD of worm}}{2} - 2A.$$

Face angle (FA) may be selected to be between 60° and 80° for the average application.

Center to center distance (between the worm and spider): $CD = \dfrac{PDW + PDS}{2}.$

Outside Diameter of Spider: $ODS = TD + 0.4775\ P.$

Face width of gear: $FW = 2.38\ (P) + 0.25.$

18.11 WORM GEAR CALCULATIONS

The following is an example problem that has been solved for a worm gear using the formulas given above.

PROBLEM 4: Calculate the specifications for a worm and worm gear (spider). The gear has 45 teeth and the worm has an outside diameter of 2.50″. The worm has a double thread and a pitch of 0.5″.

SOLUTION:

Lead: $L = 0.5 \times 2 = 1''.$

Worm addendum: $AW = 0.3183P = 0.1592''.$

Pitch diameter of worm: $PDW = 2.50 - 2\ (0.1592) = 2.1818''.$

Pitch diameter of gear: $PDS = (45 \times 0.5)\ /\ 3.14 = 7.166''.$

Center distance between worm and gear: CD

$$= \frac{(2.182 + 7.166)}{2} = 4.674''.$$

Whole depth of worm tooth: $WDT = 0.687 \times 0.5 = 0.3433''.$

Bottom diameter of worm: $BD = 2.50 - 2(0.3433) = 1.813''.$

Helix angle of worm:

$$\operatorname{Cot} \beta = \frac{3.14\ (2.1816)}{1} = 8°\ 19'.$$

Width of thread at root: $WT = 0.31\ (1) = 0.155''.$

Minimum length of worm: $MLW = 8(0.1592)\ (7.1656) = 3.02''.$

Throat diameter of gear: $TD = 7.1656 + 2(0.1592) = 7.484''.$

Radius of gear throat: $RST = (2.5\ /\ 2) - (2 \times 0.1592) = 0.9318''.$

Face width: $FW = 2.38\ (0.5) + 0.25 = 1.44''.$

Outside diameter of gear: $ODS = 7.484 + 0.4775\ (0.5) = 7.723''.$

18.12 DRAWING WORM GEARS

The worm and worm wheel (spider) are drawn and dimensioned as shown in Figs. 18.11 and 18.12. Each gear must be dimensioned and supplemented with cutting data.

It is advantageous to use full or partial sections to show the details of the gears. In some instances, the circular views are omitted.

The specifications derived by the formulas in the previous section must be used for scaling and laying out the drawings.

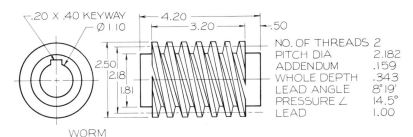

WORM

Fig. **18.12** A detail drawing of a worm using the dimensions that were calculated.

18.13 CAMS

Cams are irregularly shaped machine elements that produce motion in a single plane, usually up and down (Fig. 18.13). As the cam revolves about its center, the variation in the cam's shape produces a rise or fall in the follower that is in contact with it. The shape of the cam is determined graphically prior to the preparation of manufacturing specifications.

Only plate cams are covered in the brief review of this type of mechanism. Cams utilize the principle of the inclined wedge, with the surface of the cam causing a change in the slope of the plane, thereby producing the desired motion.

Fig. 18.13 Examples of machined cams. (Courtesy of Ferguson Machine Company.)

18.14 CAM MOTION

Cams are designed primarily to produce (1) uniform or linear motion, (2) harmonic motion, (3) gravity motion, or (4) combinations of these. Some cams are designed to serve special needs that do not fit these patterns, but are instead based on particular design requirements. Displacement diagrams are used to represent the travel of the follower relative to the rotation of the cam.

Uniform Motion

Uniform motion is shown in the displacement diagram in Fig. 18.14A. Displacement diagrams represent the motion of the cam follower as the cam rotates through 360°. The uniform-motion curve has sharp corners, indicating abrupt changes of velocity at two points, causing the follower to

bounce. Hence this motion is usually modified with arcs that smooth this change of velocity. The radius of the modifying arc is varied up to a radius of one-half the total displacement of the follower, depending on the speed of operation. Usually a radius of about one-third to one-fourth total displacement is best.

Fig. 18.14 DISPLACEMENT DIAGRAMS

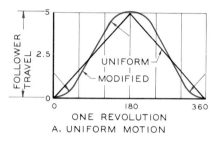

A. UNIFORM MOTION

Uniform-motion diagrams are modified with arcs of one-fourth to one-third of total displacement to smooth out the velocity of the follower at these points of abrupt change.

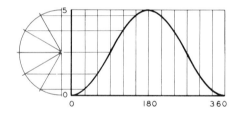

B. HARMONIC MOTION

Harmonic motion is plotted by projecting from a semicircle whose diameter is equal to the rise of the follower. The semicircle must be divided into the same number of sectors as the divisions on the x-axis of the graph to the point of maximum rise.

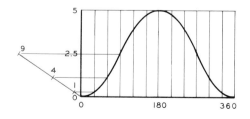

C. GRAVITY MOTION

Gravity-motion diagrams are constructed so that the rise of the follower is relative to the square of the units on the x-axis: 1^2, 2^2, 3^2, etc.

Harmonic Motion

Harmonic motion, plotted in Fig. 18.14B, is a smooth continuous motion based on the change of position of the points on the circumference of a circle. At moderate speeds, this displacement gives a smooth operation. A semicircle is drawn with its diameter equal to the total motion of the follower to locate the points on the curve.

Gravity Motion

Gravity motion (uniform acceleration), plotted in Fig. 18.14C, is used for high-speed operation. The variation of displacement is analogous to the force of gravity exerted on a falling body, with the difference in displacement being 1, 3, 5, 5, 3, 1, based on the square of the number. For instance, $1^2 = 1$; $2^2 = 4$; $3^2 = 9$. This same motion is repeated in reverse order for the remaining half of the motion of the follower. Intermediate points can be found by squaring fractional increments, such as $(2.5)^2$.

18.15 CONSTRUCTION OF A CAM

Plate Cam—Harmonic Motion

The steps of constructing a plate cam with harmonic motion are shown in Fig. 18.15. The drafter must know the following before designing a cam; the motion of the follower, the rise of the follower, size of the follower and type, the position of the follower, the diameter of the base circle, and the direction of rotation.

The specifications for the cam in Fig. 18.16

Fig. 18.15 Three basic types of cam followers—the flat surface, the roller, and the knife-edge.

Fig. 18.16 Construction of a plate cam with uniform acceleration

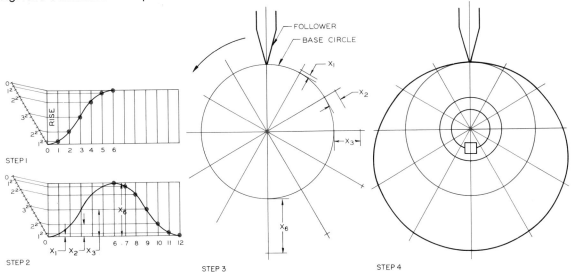

Step 1 Construct a displacement diagram to represent the rise of the follower. Divide the horizontal axis into angular increments of 30°. Draw a construction line through point 0; locate the 1^2, 2^2, and 3^2 divisions and project them to the vertical axis to represent half of the rise. The other half of the rise is found by laying off distances along the construction line with descending values.

Step 2 Use the same construction to find the right half of the symmetrical curve.

Step 3 Construct the base circle and draw the knife-edge follower. Divide the circle into the same number of sectors as there are divisions in the displacement diagram. Transfer distances from the displacement diagram to the respective radial lines of the base circle, measuring outward from the base circle.

Step 4 Connect the points found in step 3 with a smooth curve to complete the cam profile. Show also the cam hub and keyway.

Fig. 18.17 Construction of a plate cam with combination motions

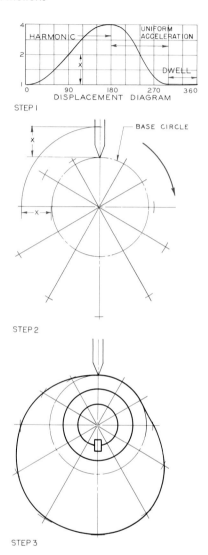

STEP 1

STEP 2

STEP 3

Step 1 The cam is to rise 4″ in 180° with harmonic motion, fall 4″ in 120° with uniform acceleration, and dwell for 60°. These motions are plotted on the displacement diagram.

Step 2 Construct the base circle and draw the knife-edge follower. Transfer distances from the displacement diagram to the respective radial lines of the base circle, measuring outward from the base circle.

Step 3 Draw a smooth curve through the points found in step 2 to complete the profile of the cam. Show also the cam hub and keyway.

are given graphically, The four steps of construction are followed as illustrated to construct the cam profile shown in Step 4.

Plate Cam—Uniform Acceleration

The steps of constructing a cam with uniform acceleration is done in the same manner as the previous example except for a different displacement diagram and a knife-edge follower.

The graphical layout of the problem is shown in Fig. 18.16. The profile of the cam is found by following the four steps of construction.

Plate Cam—Combination

In Fig. 18.17 a knife-edge follower is used with a plate cam to produce a 4″ rise with harmonic motion from 0° to 180°, a 4″ fall with a uniform acceleration from 180° to 300°, and dwell (no follower motion) from 300° to 360°. You are to draw the cam that will give this motion from the base circle.

The displacement diagram is drawn with a harmonic curve with a full rise of 4″. The curve is then drawn with a uniform acceleration drop of 4″. The values of rise must be known before beginning the problem. The dwell is a horizontal line to complete the diagram of the 360° rotation of the cam.

The profile of the plate cam is found by completing Steps 2 and 3 in the same manner as was done in the previous two examples.

18.16 CONSTRUCTION OF A CAM WITH AN OFFSET FOLLOWER

The cam in Fig. 18.18 is required to produce harmonic motion through 360°. This motion is plotted directly from the follower rather than using a displacement diagram, since there are no combinations of motion involved.

A semicircle is drawn with its diameter equal to the total motion of the follower. The base circle is drawn to pass through the center or roller of the follower. The centerline of the follower is extended downward, and a circle is drawn tangent to the extension with its center at the center of the base circle. The small circle is divided into 30° in-

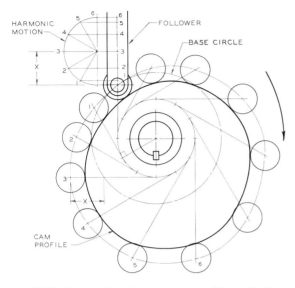

Fig. 18.18 Construction of a plate cam with an offset roller follower.

tervals to establish points through which construction lines will be drawn tangent to the circle.

The distances from the tangent points to the position points along the path of the follower are laid out along the tangent lines that were drawn at 30° intervals. Alternatively these points can be located in measuring from the base circle as shown in the example, where point 3 was located distance X from the base circle.

Draw the circular roller in all views and construct the profile of the cam tangent to the rollers at all positions.

PROBLEMS

Gears

Use Size A (8½″ × 11″) sheets for the following gear problems. Select the most appropriate scale in order for the drawings to utilize the available space.

1–5. Calculate the following dimensions for the following spur gears and make a detail drawing of each. Give the dimensions and cutting data for each gear.

Problem	Gear Teeth	Diametral pitch	14.5° Involute
1	20	5	″
2	30	3	″
3	40	4	″
4	60	6	″
5	80	4	″

Provide any other dimensions that are needed, using your judgement.

6–10. Calculate the gear sizes and number of teeth (similar to Problem 2 in Section 18.5), using the ratios and data below.

Problem	RPM pinion	RPM gear	Center to center	Diametral pitch
6	100 (driver)	60	6.0″	10
7	100 (driver)	50	8.0″	9
8	100 (driver)	40	10.0″	8
9	100 (driver)	35	12.0″	7
10	100 (driver)	25	14.0″	6

11–20 Make detail drawings of each of the gears for which calculations were made in Problems 6 through 10. Provide a table of cutting data and other dimensions that are needed to complete the specifications.

21–25. Calculate the specifications for the bevel gears that intersect at 90°, and make detail drawings of each with the necessary dimensions and cutting data.

Problem	Diametral pitch	No. of teeth on pinion	No. of teeth on gear
21	3	60	15
22	4	100	40
23	5	100	60
24	6	100	50
25	7	100	30

26–30. Calculate the specifications for worm gears, and make detail drawings of each, providing the necessary dimensions and cutting data.

Problem	No. of teeth in spider gear	Outside DIA of worm	Pitch of worm	Thread of worm
26	45	2.50	0.5	double
27	30	2.00	0.80	single
28	60	3.00	0.80	double
29	30	2.00	0.25	double
30	80	4.00	1.00	single

Cams

Draw the following on Size B sheets (11″ × 17″) with the following standard dimensions: base circle—3.50″; roller follower—0.60″ diameter; shaft-0.75″ diameter; hub—1.25″ diameter; direction of rotation—clockwise. The follower is positioned vertically over the center of the base circle except in Problems 40 and 41. Lay out the problems and displacement diagrams as shown in Fig. 18.19.

31. Draw a plate cam with a knife-edge follower for uniform motion and a rise of 1.00″.

32. Draw a displacement diagram and a cam that will give a modified uniform motion to a knife-edge follower with a rise of 1.7″. Modify the uniform motion with an arc of one-quarter of the rise in the displacement diagram.

33. Draw a displacement diagram and a cam that will give a harmonic motion to a roller follower with a rise of 1.60″.

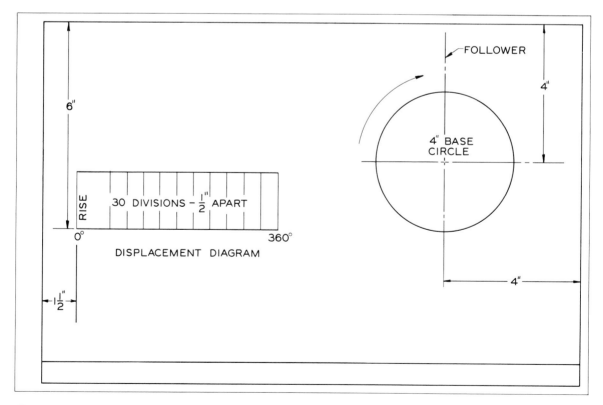

Fig. 18.19 Problem layout for the cam problems on Size B sheets.

34. Draw a displacement diagram and a cam that will give a harmonic motion to a knife-edge follower with a rise of 1.00″.

35. Draw a displacement diagram and a cam that will give uniform acceleration to a knife-edge follower with a rise of 1.70″.

36. Draw a displacement diagram and a cam that will give a uniform acceleration to a roller follower with a rise of 1.40″.

37. Draw a displacement diagram and a cam that will give the following motion to a knife-edge follower: Rise 1.25″ with harmonic motion in 120°, dwell for 120°, and fall 1.25″ with uniform acceleration.

38. Draw a displacement diagram and a cam that will give the following motion to a knife-edge follower: Dwell for 70°; rise 1″ with a modified uniform motion in 100°; fall 1″ with a harmonic motion in 100°; and dwell for 90°.

39. Draw a displacement diagram and a cam that will give the following motion to a roller follower: Rise 1.25″ with a harmonic motion in 120°; dwell for 120°; and fall 1.25″ with a uniform acceleration in 120°.

40. Repeat Problem 32, but offset the follower 0.60″ to the right of the vertical center line.

41. Repeat Problem 33, but offset the follower 0.60″ to the left of the vertical line.

19

MATERIALS AND PROCESSES

19.1 INTRODUCTION

This chapter presents an overview of the materials and processes used in manufacturing. The major emphasis is placed on the forming and fabrication of metals since these processes are fundamental to manufacturing (Fig. 19.1).

The study of metals, called metallurgy, is a highly complex area that is constantly changing as

Fig. 19.1 This furnace operator is pouring an aluminum alloy of manganese into ingots (shown at the right) that will be remelted and cast. (Courtesy of the Aluminum Company of America.)

new processes and alloys are being developed. The guidelines for designations of various types of metals have been standardized by three associations: the American Iron and Steel Institute (AISI), the Society of Automotive Engineers (SAE), and the American Society for Testing Materials (ASTM).

19.2 IRON

Metals that contain iron, even in small quantities, are called *ferrous metals*. The three types of iron are *gray iron*, *malleable iron*, and *nodular iron*. Iron is used in the manufacturing of machine parts by the casting process; consequently, iron is often called *cast iron*. Iron is cheaper than steel and it is easy to machine, but it does not have the ability to withstand the shock and forces that steel can withstand.

Iron is designated by a letter followed by four digits. The prefix letters for the types of iron are:

G—gray iron
M—malleable iron
D—nodular iron (ductile iron)

The first two digits represent the yield strength for malleable and nodular irons, and tensile strength for gray iron. The numbers represent 1000 pounds per square inch except for the number 10, which represents 100,000 psi. The second two digits of the designation are the percent of elongation for malleable and nodular iron. The last two digits for gray iron will always be 00.

An example designation for gray iron may be SAE G2500, which means it has a tensile strength of 25,000 psi. A designation of SAE D5506 specifies a nodular iron with a tensile strength of 55,000 psi and an elongation of 6 percent.

19.3 STEEL

Steel is an alloy of iron with the addition of other materials, but carbon is the ingredient that has the greatest effect on the grade of the steel. The three broad types of steel are *plain carbon steels, free-cutting carbon steels,* and *alloy steels.* The types of steels and their designations by four-digit numbers are shown in Table 19.1.

The number designations begin with a digit that indicates the type of steel: 1 is carbon steel, 2 is nickel steel, etc. The second digit gives the percentage content of the material represented by the first digit. The last two digits give the percentage of carbon in the alloy, where 100 is equal to 1% and 50 is equal to 0.50%.

19.4 COPPER

Copper is one of the first metals to be discovered. It is a soft metal that can be easily formed and bent

without breakage. Since it has a high resistance to corrosion and because of its high level of conductance, it is used in the manufacture of pipes, tubing, and electrical wiring. It is an excellent roofing and screening material since it withstands the weather well.

Copper has a number of alloys including brasses, tin bronzes, nickel silvers, and copper-nickel alloys. A few of the numbered designations of wrought copper are: C11000, C11100, C11300, C11400, C11500, C11600, C10200, C12000, and C12200. Brass is an alloy of copper and zinc, and bronze is an alloy of copper and tin.

Copper and copper alloys can be easily finished by buffing or plating. All of these can be joined by soldering, brazing, and welding, and can be easily machined and used for casting.

19.5 ALUMINUM

Aluminum is a corrosion-resistant, light-weight metal that has applications for many industrial products. Most of the materials that are called aluminum are actually alloys of aluminum. Alloys of aluminum possess greater strength than the pure metal and more applications are possible.

The types of wrought aluminum alloys are designated by four digits, as shown in Table 19.2. (Wrought aluminum has properties that permit it to be formed by hammering.) The first digit, from 2 through 9, indicates the alloying element that is combined with aluminum. The last two numbers identify the other alloying materials or indicate the aluminum purity. The second digit indicates

Table 19.1 NUMBERING AND APPLICATIONS OF TYPES OF STEEL

Type of Steel	Number	Application
Carbon steels		
Plain carbon	10XX	Tubing, wire, nails
Resulphurized	11 XX	Nuts, bolts (free-cut)
Resulphurized and rephosphorized	12XX	
Plain carbon 1.00%-1.65%	15XX	
Magnanese Steel	13XX	Gears and shafts
Nickel steel	23XX	Axles and shafts
	25XX	Structural shapes
Nickel-chromium	31XX	Crank shafts, gearing forgings
	32XX	
	33XX	Heavy duty axles
	34XX	
Molybdenum steel	40XX	Gearing shafts
	44XX	
Chromium-molybdenum	41XX	Machine parts
Nickel-chromium	43XX	Heavy duty

Source: Courtesy of the Society of Automotive Engineers.

Table 19.2 NUMBERING DESIGNATIONS FOR WROUGHT ALUMINUM AND ALUMINUM ALLOYS

Composition	Alloy Number
Aluminum (99% pure)	1XXX
Aluminum alloys	
Copper	2XXX
Manganese	3XXX
Silicon	4XXX
Magnesium	5XXX
Magnesium and silicon	6XXX
Zinc	7XXX
Other elements	8XXX
Unused series	9XXX

modifications of the original alloy or impurity limits. The second digit indicates modifications of the original alloy or impurity limits.

A four-digit numbering system with the last digit being to the right of a decimal point is used to designate types of cast aluminum and aluminum alloys. (When used for castings, the aluminum is melted and poured into a mold to form it.) The first digit indicates the alloy group as shown in Table 19.3. The next two digits identify the aluminum alloy or the aluminum purity. The numeral 1 to the right of the decimal point represents ingot aluminum, and 0 represent aluminum for casting. Ingots are blocks of cast metal that are to be remelted. Billets are castings of aluminum that are to be formed by forging.

Table 19.3 ALUMINUM CASTING AND INGOT DESIGNATIONS

Composition	Alloy Number
Aluminum, (99 percent pure)	1XX.X
Aluminum alloys	
Copper	2XX.X
Silicon with copper and/or magnesium	3XX.X
Silicon	4XX.X
Magnesium	5XX.X
Zinc	7XX.X
Tin	8XX.X
Other element	9XX.X

Source: Society of Automotive Engineers

19.6 MAGNESIUM

Magnesium is a light metal that is considered to be available in an inexhaustible supply since it is extracted from seawater and natural brines. It is approximately half the weight of aluminum, and therefore it is an excellent material for aircraft parts, clutch housing, crankcases for aircooled engines, and other applications where lightness of weight is desirable.

Magnesium and its alloys can be joined by bolting, riveting, and welding. Some numbered designations of magnesium alloys are: M10100, M11630, M11810, M11910, M11912, M12390, M13320, M16410, and M16620.

19.7 PROPERTIES OF MATERIALS

All materials have different properties that the designer must use to his or her best advantage. Consequently, it is important to be familiar with the terms used to describe these properties.

Ductility is a softness present in some materials, such as copper and aluminum, that permits them to be formed by stretching (drawing) or hammering without breaking. Wire is made of ductile materials that can be drawn through a die.

Brittleness is a characteristic of metals that will not stretch without breaking, such as cast irons and hardened steels.

Malleability is the ability of a metal to be rolled or hammered without breaking.

Hardness is a metal's ability to resist being dented when it receives a blow.

Toughness is the property of being shock-resistant while remaining malleable and resisting cracking or breaking.

Elasticity is the characteristic of a metal to return to its original shape after being bent or stretched.

19.8 HEAT TREATMENT OF METALS

The properties of different metals can be changed by various forms of heat treating. Steels are affected to a greater extent by heat treating than other materials.

Hardening of steel is performed by heating the material to a prescribed temperature depending upon its content and then quenching the hot steel in oil or water.

Quenching is the process of rapidly cooling heated metal by immersing it in liquids, gases, or solids (such as sand, limestone, or asbestos).

Tempering is the process of reheating previously hardened steel and then cooling it, usually by air. This increases the steel's toughness.

Annealing is the process of heating and cooling metals to soften them, to release their internal stresses, and to make them easier to machine and work with.

Normalizing is achieved by heating metals and letting them cool in air to room temperature, relieving their internal stresses.

Case hardening is the process of hardening a thin outside layer of a metal. In this process the outer layer is placed in contact with carbon or nitrogen compounds that become absorbed by the metal as it is heated. Afterwards, it is quenched to complete the case hardening.

Flame hardening is the method of heating a metal to a prescribed range with a flame and then quenching the metal.

Fig. 19.2 A large casting of a landing-gear mechanism of an aircraft is being removed from its mold. (Courtesy of Cameron Iron Works.)

19.9 CASTINGS

Two major methods of forming shapes are *casting* and *pressure forming*. Casting involves the preparation of a mold inside which is poured molten metal that cools and forms the part. The types of casting are *sand casting*, *permanent-mold casting*, *die casting*, and *investment casting*. These types of casting vary by the method in which the molds are made before pouring the metal.

Sand Casting

Sand casting is the most commonly used method. In the first step, a form or pattern is made that is representative of the final part that is to be cast. It is made of wood or metal. The pattern is placed in a metal box called a flask, and molding sand is packed around the pattern. When the pattern is withdrawn from the sand it leaves a void forming the mold. The molten metal is poured into the mold through sprues, or gates. After cooling, the casting is removed and cleaned (Fig. 19.2).

Cores are parts that are formed in sand and placed within a mold to leave holes or hollow portions within the finished casting (Fig. 19.3). Once the casting has been formed and set, the sand cores can be broken apart and removed, leaving behind the desired void within the casting,

Fig. 19.3 A typical sand mold. The opening that will be filled with molten metal is formed by a wood or metal pattern that is pressed into the sand.

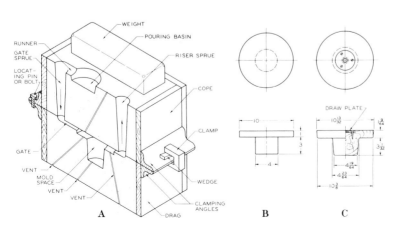

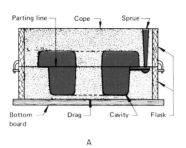
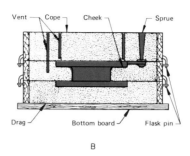
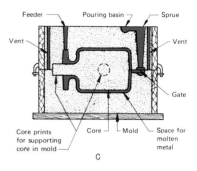

Fig. 19.4 Three types of sand molds: (A) two-section mold, (B) three-section mold, (C) two-section mold with a core. (Courtesy of General Motors Corporation.)

thereby reducing its weight. Cores add considerably to the cost of a casting, and they should not be used unless they are adequately offset by savings in materials.

Since the patterns must be placed in sand and then withdrawn before the metal is poured, it is necessary that the sides of the patterns be tapered for ease of withdrawal from the sand (Fig. 19.4). This taper is called *draft*. The amount of draft depends upon the depth of the pattern in the sand, and it usually varies from 2° to 8° for most applications. Also, patterns are made oversize to compensate for shrinkage that will occur when the metal cools.

The sand casting has a rough surface that is not desirable for contacting other moving parts or surfaces. Consequently, it is common practice to machine portions of a casting by drilling, grinding, shaping, or other machining operations (Fig. 19.5). When this is to be done, the pattern should be made larger in these areas to provide for removal of the metal caused by the machining operation.

Fillets and rounds at all intersections will increase the strength of a casting. Also, it is necessary to use fillets and rounds since it is difficult to form square corners by the sand casting process.

Permanent-Mold Castings

Permanent molds are often made for the mass production of parts. The molds are generally made of

Fig. 19.5 This casting of the outer cylinder of an aircraft's landing gear is being bored on a horizontal boring mill. (Courtesy of Cameron Iron Works.)

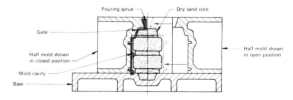

Fig. 19.6 Permanent molds are made of metal for repetitive usage when parts are mass produced. In this example a sand core is made from another mold and is placed in the permanent mold to give a hollow void within the casting. (Courtesy of General Motors Corporation.)

cast iron and are coated to prevent fusing with the molten metal that is poured into them. Permanent molds are often used with the manufacture of aluminum and magnesium parts. An example of a permanent mold is shown in Fig. 19.6.

Die Castings

Die castings are generally used for the mass production of parts made of aluminum, magnesium, zinc alloys, and copper; however, other materials are used also. Die castings are made by forcing molten metal into dies (or molds) under pressure. The dies are permanent molds that are used over and over again. The die casting can be produced at a low cost, a high rate of production, at close tolerances, and with good surface qualities.

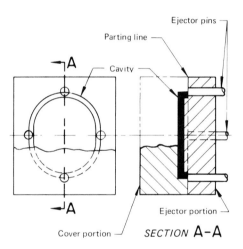

Fig. 19.7 A die for casting a simple part. Unlike the sand casting, the metal is forced into the die to form the die casting. (Courtesy of General Motors Corporation.)

The same general principles recommended for sand castings—using fillets and rounds, allowing for shrinkage, and specifying draft angles—apply to die castings also. An example of a die casting for a simple part is shown in Fig. 19.7.

A working drawing of a casting is shown in Fig. 19.8.

Investment Casting

Investment casting is a process used to produce complicated parts that would be difficult to form with uniformity by any other method. This technique is even used to form the complicated forms of artistic sculptures.

A mold or die is made to cast a master pattern that is made of wax since a new pattern must be used for each investment casting. The wax pattern will be identical to the finished casting. The wax pattern is placed inside a container and a plaster mixture or sand is poured (or invested) around the wax pattern.

The wax pattern is melted with heat once the investment has cured to remove the wax and leave a hollow cavity that will serve as the mold for the molten metal. When filled and set, the plaster or sand is broken away from the finished investment casting.

19.10 FORGINGS

Forging is the process of shaping or forming metal by hammering or squeezing the heated metal into a die. The resulting forging possesses high strength and a resistance to loads, vibrations, and impacts.

When preparing forging drawings, the following must be considered: (1) draft angles and parting lines, (2) fillets and rounds, (3) forging tolerances, (4) allowance of extra material for machining, and (5) heat treatment of the finished forging (Fig. 19.9). Some of the standard steels used for forging are steels designated by the following SAE numbers: 1015, 1020, 1025, 1045, 1137, 1151, 1335, 1340, 4620, 5120, and 5140. Other materials that can be forged are iron, copper, and aluminum.

Drop forges and press forges are used to hammer the metal (called billets) into the forging dies by multiple blows or forces. These repetitive blows sequentially form the metal into the desired shape.

Examples of dies are shown in Fig. 19.10. A single-impression die gives an impression on one side of the parting line between the mating dies; the double-impression die gives an impression on both sides of the parting line. The interlocking dies result in a forging whose impression may cross the parting line on either side.

An example of an object that is forged with auxiliary rams to hollow the forging is shown in Fig. 19.11. A drawing of a forged part is illustrated in Fig. 19.12.

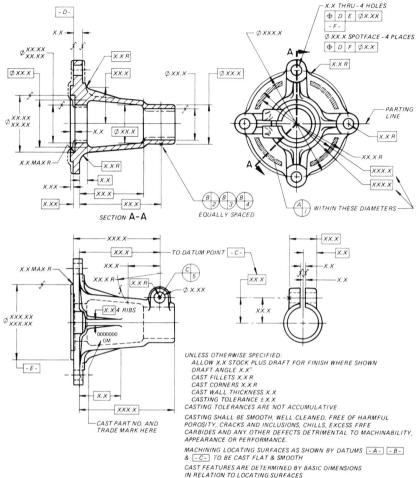

SECTION **A-A**

UNLESS OTHERWISE SPECIFIED:
 ALLOW X.X STOCK PLUS DRAFT FOR FINISH WHERE SHOWN
 DRAFT ANGLE X.X°
 CAST FILLETS X.X R
 CAST CORNERS X.X R
 CAST WALL THICKNESS X.X
 CASTING TOLERANCE ± X.X
CASTING TOLERANCES ARE NOT ACCUMULATIVE

CASTING SHALL BE SMOOTH, WELL CLEANED, FREE OF HARMFUL
POROSITY, CRACKS AND INCLUSIONS, CHILLS, EXCESS FREE
CARBIDES AND ANY OTHER DEFECTS DETRIMENTAL TO MACHINABILITY,
APPEARANCE OR PERFORMANCE.

MACHINING LOCATING SURFACES AS SHOWN BY DATUMS ⌐A⌐ ⌐B⌐
& ⌐C⌐ TO BE CAST FLAT & SMOOTH

CAST FEATURES ARE DETERMINED BY BASIC DIMENSIONS
IN RELATION TO LOCATING SURFACES

Fig. 19.8 A dimensioned drawing of a die casting. Note that the part has been cast oversize to allow for finished surfaces that must be machined. (Courtesy of General Motors Corporation.)

Fig. 19.9 Three stages of manufacturing a turbine fan are shown here. The blank is first formed by forging; it is then machined; and the fan blades are then attached in their machined slots. (Courtesy of Avco Lycoming.)

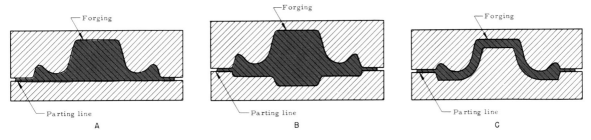

Fig. 19.10 Three types of forging dies: (A) single-impression die, (B) double-impression die, (C) interlocking die. (Courtesy of General Motors Corporation.)

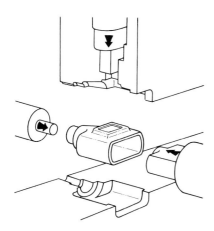

Fig. 19.11 Auxiliary rams can be used to form internal features on a part. (Courtesy of General Motors Corporation.)

Rolling

Rolling is a type of forging in which the stock is rolled between two rolls to give it a desired shape. Rolling can be done at right angles to the axis of the part or parallel to its axis (Fig. 19.13). The stock is usually heated before rolling if a high degree of shaping is required. If the forming requires only a slight change in configuration, the rolling can be performed when the metal is cold. This process is called cold rolling.

19.11 STAMPING

Stamping is a method of forming flat metal stock into three-dimensional shapes. In many applications, these can serve a design requirement more economically than a casting can. The first step of

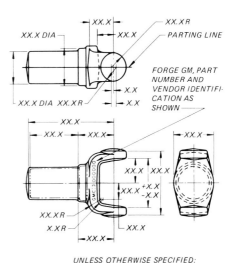

UNLESS OTHERWISE SPECIFIED:
DRAFT ANGLES X°.
ALL FILLETS X.X R, CORNERS X.X R.
+X.X - X.X TOLERANCES ON
FORGING DIM.

SNAG AND REMOVE SCALE.

SAMPLE FORGINGS ARE TO BE
APPROVED BY METALLURGICAL
AND ENGRG DEPTS FOR GRAIN
FLOW STRUCTURE.

FORGING DRAWING

Fig. 19.12 A drawing of a forging. The blank is forged oversize to allow for machining operations that will remove metal from it. (Courtesy of General Motors Corporation.)

stamping is to cut out the shapes, called *blanks*, that are to be bent.

Blanks are formed into shape by bending and pressing them against forms. Examples of box-shaped parts are shown in Fig. 19.14. An example of a flange stamping is illustrated in Fig. 19.15. Holes in stampings are made by punching, extruding, or piercing as shown in Fig. 19.16.

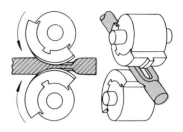

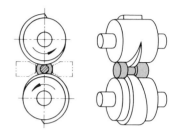

Fig. 19.13 Features on parts may be formed by rolling. In these examples, parts are being rolled parallel and perpendicular to their axes. (Courtesy of General Motors Corporation.)

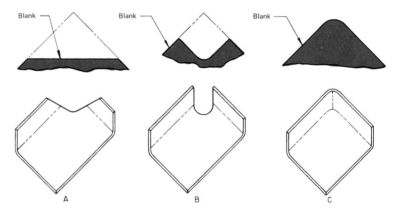

Fig. 19.14 Box-shaped parts formed by stamping. (A) A corner cut of 45° permits folding flanges and may require no further trim. (B) Notching has the same effect as the 45° cut, but it is often more attractive. (C) A continuous corner flange requires that the blank be developed so that it can be drawn into shape. (Courtesy of General Motors Corporation.)

Fig. 19.15 A sheet metal flange design, with notes calling attention to design details. (Courtesy of General Motors Corporation.)

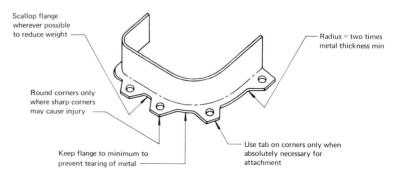

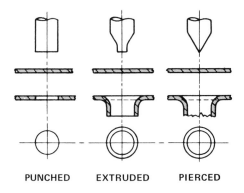

PUNCHED   **EXTRUDED**   **PIERCED**

Fig. 19.16 Three methods of forming holes in sheet metal. (Courtesy of General Motors Corporation.)

19.12 MACHINING OPERATIONS

After the metal has been formed into the shape of the final product, machining operations usually have to be performed to complete the part. The machining operations involve several basic types of equipment: the *lathe, drill press, milling machine, shaper,* and *planer.*

The Lathe

The lathe is a machine that shapes cylindrical parts or holes by rotating the work piece between the centers of the lathe (Fig. 19.17). The more fun-

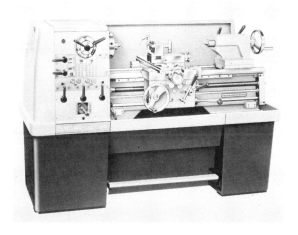

Fig. 19.17 A typical small-size metal lathe that holds the work pieces between centers as it is rotated about its axis. (Courtesy of the Clausing Corporation.)

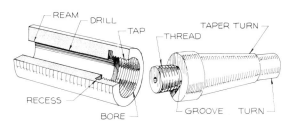

Fig. 19.18 The fundamental operations that are performed on a lathe are illustrated on the two parts shown above.

damental operations performed on the lathe are *turning, facing, drilling, boring, reaming, threading,* and *undercutting* (Fig. 19.18).

Turning is the process of forming a cylinder by a tool that advances against the cylinder being turned and moves parallel to its axis (Fig. 19.19).

Facing is the process of forming flat surfaces that are perpendicular to the axis of rotation of the part being rotated by the lathe. This is often done on the end of a cylinder.

Fig. 19.19 Turning is the most basic of all operations performed on the lathe. A continuous chip is removed by a cutting tool as the part is rotated.

Drilling is performed by mounting a drill in the tail stock of the lathe and rotating the work while the bit is advanced into the part (Fig. 19.20).

Boring is the process of making large holes that are too big to be drilled. Large holes are bored by enlarging smaller holes that have been drilled, or cored holes that have been formed by casting (Fig. 19.21).

Reaming is the removal of only a few thousandths of an inch of material inside a drilled hole

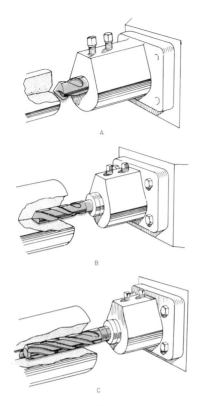

Fig. 19.20 Three steps of drilling a hole in the end of a cylinder involve (A) start drilling, (B) twist drilling, and (C) core drilling. These steps should be performed in sequence. Core drilling enlarges the previously drilled hole to the required size.

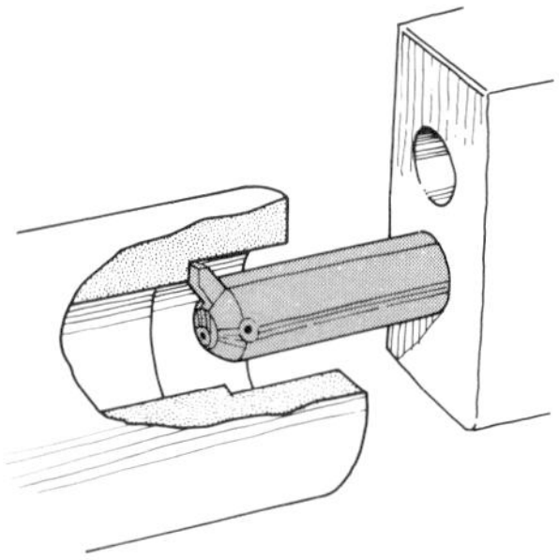

to bring it to its required level of tolerance. Conical as well as cylindrical reaming can be performed on the lathe (Fig. 19.22).

Threading of external and internal holes can be done on the lathe. The die used for cutting internal holes is called a *tap* (Fig. 19.23).

The turret lathe is a programmable lathe that can perform sequential operations on the same part, such as drilling a series of holes, boring them, and then reaming them in succession. The turret is mounted in such a way as to rotate each

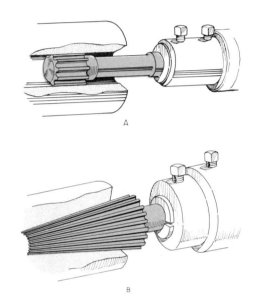

Fig. 19.22 Fluted reamers can be used to finish inside cylindrical and conical holes within a few thousandths of an inch.

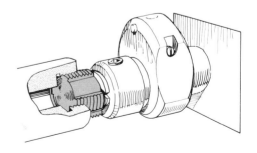

Fig. 19.21 Boring is the method of enlarging holes that are usually larger than available drill bits. The cutting tool is attached to the boring bar on the lathe.

Fig. 19.23 External and internal threads (shown here) can be cut on a lathe. The die being used to cut the threads is called a tap. Note that a recess has been formed at the end of the threaded hole prior to threading.

Fig. 19.24 A turret lathe that performs a sequence of operations as the turret head rotates in succession.

tool into position for its particular operation (Fig. 19.24).

Undercutting   is a method of cutting a recess inside of a cylindrical hole with a tool mounted on a boring bar. The groove is cut as the tool advances from the center of the axis of revolution into the part (Fig. 19.25).

The Drill Press

The drill press is used to drill small and medium size holes into stock that is held on the bed of the press. The work is held in position by a fixture or a clamp while the drilling is performed (Fig. 19.26).

In addition to drilling holes, the drill press can be used for *counterdrilling, countersinking, counterboring, spotfacing,* and *threading.* Examples of these operations are shown in Fig. 19.27.

Broaching   Cylindrical holes can be converted into square holes or hexagonal holes by using a tool called a *broach.* The broach has a similarity to a drill bit and is mounted in an overhead press.

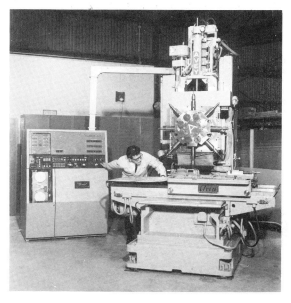

Fig. 19.26 A multiple-head drill press that can be programmed to perform a series of drill press operations in a desired sequence.

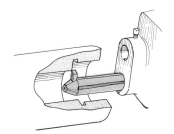

Fig. 19.25 A recess (undercut) can be formed by using the boring bar with the cutting tools attached as shown. As the boring bar is moved off center of the axis, the tool will form the recess.

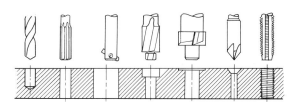

Fig. 19.27 The basic operations that can be performed on the drill press are (left to right): drilling, reaming, boring, counterboring, spotfacing, countersinking, and tapping (threading).

The broach has a series of teeth along its axis, beginning with teeth that are near the size of the hole to be broached, and tapering to the size of the finished hole that is to be broached. The broach is forced through the hole with each tooth cutting more from the hole as it passes through.

This mass-production process enables shops to form holes, other than cylindrical holes, in rapid succession.

Milling Machine

The milling machine uses a series of cutting tools that are rotated about a shaft (Fig. 19.28). The work piece is passed under the cutters and in contact with them to remove the metal.

The milling machine has a wide variety of cutters to form different grooved slots, threads, and gear teeth. Irregular grooves in cams can be cut by the milling machine. It can be used to finish a surface on a part within a high degree of tolerance.

The Shaper

The shaper is a machine that holds the work stationary while the cutter on the machine passes back and forth across the work to finish the surface or to cut a groove, one stroke at a time (Fig. 19.29). With each stroke of the cutting tool, the material is shifted slightly so as to align the part for the next overlapping stroke. The shaper can be used to cut a variety of slots and grooves in a part.

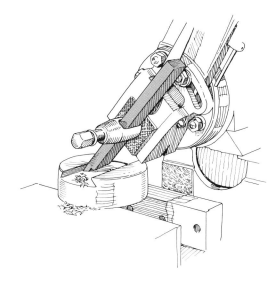

Fig. 19.29 The shaper moves back and forth across the part, removing metal as it advances. It can be used to finish surfaces, cut slots, and for many other operations.

The Planer

The planer is similar to the shaper except that the work is passed under the cutters by the planer rather than the work remaining stationary as in the case of the shaper (Fig. 19.30). Like the shaper, the planer can cut grooves or slots and finish surfaces that must meet tolerance specifications.

Fig. 19.28 This small milling machine is being used to cut a slot in the work piece. (Courtesy of the Clausing Corporation.)

Fig. 19.30 The planer has stationary cutters, and the work is fed past them to finish larger surfaces. This planer has a 30 foot bed. (Courtesy of Simmons Machine Tool Corporation.)

19.13 SURFACE FINISHING

The process of finishing a surface to the desired uniformity is called surface finishing. It may be accomplished by several methods including *grinding*, *polishing*, *lapping*, *buffing*, and *honing*. Each method removes a portion of the metal from the surface to produce a smoother finish than before the finishing.

Grinding is the finishing of a flat surface by holding it against a rotating abrasive wheel (Fig. 19.31). Grinding is used to smooth surfaces and to sharpen edges that are used for cutting, such as drill bits.

Polishing is performed in the same manner as grinding except the polishing wheel is flexible since it is made of felt, leather, canvas, or fabric.

Lapping is used to produce very smooth surfaces by holding the surfaces to be lapped against another large flat surface called a *lap*, which has been coated with a fine abrasive powder. As the lap rotates in contact with the work piece, the surface is finished to a high degree of smoothness. Lapping is done only after the surface has been previously finished by a less accurate technique such as grinding or polishing. Cylindrical parts can be lapped by using a lathe in conjunction with the lapping surface.

Buffing is a method of removing scratches from a surface with a rotating buffer wheel that is made of wool, cotton, or fabric. Sometimes the buffer is

Fig. 19.31 The upper surface of this part is being ground to a smooth finish by a grinding wheel. (Courtesy of the Clausing Corporation.)

a cloth or felt belt that is applied to the surface being buffed. An abrasive mixture is applied to the buffed surface from time to time to enhance the buffing. Polishing machines can be used for buffing by changing the wheels for each operation.

Honing is the process of finishing the outside or the inside of holes within a high degree of tolerance. The honing tool is rotated as it is passed through the holes to give the sort of finishes found in gun barrels, engine cylinders, and similar products where a high degree of smoothness is required.

DIMENSIONING

20.1 INTRODUCTION

Working drawings are the dimensioned drawings that are used to describe the details of a part or a project so the construction can be performed in accordance with the desired specifications.

Dimensions are needed, in addition to the drawings, to provide specifications and sizes of the various features. When properly applied, dimensions and notes will supplement the drawings so that they can be used as legal contracts for construction.

The techniques of dimensioning that are presented in this chapter are based primarily on the standards of the American National Standards Institute (ANSI) and especially their standards, Y14.5, *Dimensioning and Tolerancing for Engineering Drawings*. Various industrial standards from major corporations such as the General Motors Corporation have been utilized also.

Imperial (English) and metric (SI) units and the methods of using both are presented in keeping with current standards.

20.2 DIMENSIONING TERMINOLOGY

The Guide Slide in Fig. 20.1 has been properly dimensioned and is used as an example to identify some of the terms of dimensioning.

Dimension lines are thin lines (2H-4H pencil) with arrows at each end. Numbers placed near their midpoints specify a part's size.

Extension lines are thin lines (2H-4H pencil) that extend from a view of an object for dimensioning the part. The arrowheads of dimension lines end at these lines.

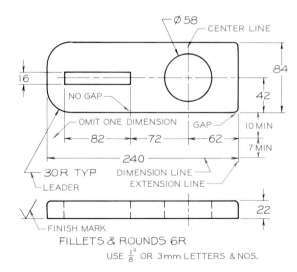

Fig. 20.1 This is a typical working drawing that is dimensioned in millimeters. The spacing of the dimension lines from the views are indicated in inches (0.40 and 0.25).

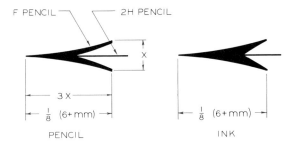

Fig. 20.2 Arrowheads are drawn as long as the height of the letters used on a drawing. They are one-third as wide as they are long.

Centerlines   are thin lines (2H-4H pencil) that are used to locate the centers of cylindrical parts, such as cylindrical holes.

Leaders   are thin lines (2H-4H pencil) drawn from a note to feature to which the note applies.

Arrowheads   are placed at the ends of dimension lines and leaders to indicate the endpoints of these lines. The arrowheads are drawn the same length as the height of the letters or numerals, ⅛″ in most cases. The form of the arrowhead is shown in Fig. 20.2. Use an F or HB pencil for arrowheads.

Dimension numbers   are placed near the middle of the dimension line, are usually ⅛″ in height, and units (″, IN, or mm) are omitted.

Placement techniques   of applying dimension and extension lines to parts are shown in Fig. 20.3. Di-

mension lines should not be closer than ⅜″ (10 mm) from a part.

20.3   UNITS OF MEASUREMENT

The two most commonly used units of measurement are the decimal inch, in the English (imperial) system, and the millimeter, in the metric (SI) system.

The inch in its common fraction form can be used, but it is preferable to give fractions in decimal form. Common fractions make addition, division, and general arithmetic very hard to perform. A comparison of dimensions in millimeters with those in inches is shown in Fig. 20.3.

A series of examples is shown in Fig. 20.4 where the units are given in millimeters, decimal inches, and fractional inches. Dimensions in millimeters are usually rounded off to whole numbers without decimal fractions. When a metric dimension is less than a millimeter, a zero precedes the decimal point.

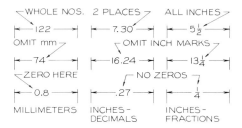

Fig. 20.4 When using SI units, dimensions are usually given to the nearest whole millimeter. When decimal inches are used, the fractions are carried to two decimal places. If fractions are given as common fractions, the fractions are twice as tall as whole numbers.

Regardless of the units of measurement, the units are omitted from the dimension numbers, and the units are normally understood to be in millimeters or inches (not feet and inches) as specified by the scale on the drawing. For example: 112, not 112 mm; and 67, not 67″ or 5′–7″.

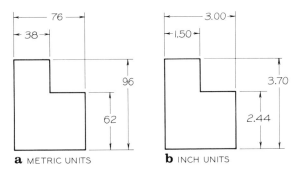

a METRIC UNITS   **b** INCH UNITS

Fig. 20.3 Dimension lines should be placed at least ⅜″ (10 mm) from an object. Other rows of dimensions should be located at least ¼″ (7 mm) apart.

Architects still use a combination of feet and inches, but the inch units can be omitted such as

7'–2 to represent seven feet and two inches. Feet and decimal fractions of feet are used by engineers to dimension large-scale projects such as road designs. These dimensions will be expressed as 252.7', for example.

When using decimal inches, it is common practice to show all dimensions with two-place decimal fractions even though the last numbers are zeros. For dimensions of less than an inch, no zero precedes the decimal point.

Common fractions, when used with inches, are lettered twice as tall as single numerals. Common fractions should not be used instead of decimal fractions unless necessary for a special application.

20.4 ENGLISH/METRIC CONVERSIONS

Dimensions in inches can be converted to millimeters by multiplying by 25.4. Similarly, dimensions in millimeters can be converted to inches by dividing by 25.4.

For most applications, the millimeter does not need more than a one-place decimal when it is found by conversion from inches. The last digit retained in a conversion of either mm or inches is unchanged if it is followed by a number less than 5. For example, 34.43 is rounded off to be 34.4 millimeters.

The last digit to be retained is increased by one if it is followed by a number greater than 5. The number 34.46 is rounded off to become 34.5.

The last digit to be retained is increased by one if it is odd and is followed by exactly 5. The number 34.75 is rounded off to become 34.8.

These same rules apply to decimal units that are taken to more than two decimal places, since decimal inches are customarily carried to two decimal places.

20.5 DUAL DIMENSIONING

Some drawings require that both metric and English units be shown on each dimension. This method of dimensioning is called *dual dimensioning*.

Dual dimensioning can be accomplished by placing the inch equivalent of millimeters either under or over the other units, as shown in Fig.

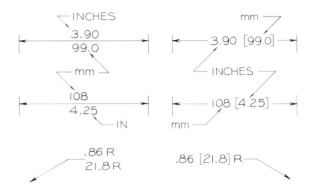

Fig. 20.5 Some dimensions are dual-dimensioned where both inches and millimeters are given on a single dimension line. If the drawing was originally made in inches, then the equivalent measurement in millimeters is placed under the inches or to the right in parentheses. If the drawing was originally made in millimeters, then the inch equivalents would be placed under or to the right. When inches are converted to millimeters, the millimeters may need to be written as decimal fractions.

20.5. If the drawing was originally dimensioned in inches, then the inch dimensions are placed on top of the millimeters. When it is converted to millimeters, inches are multiplied by 25.4 and the equivalent millimeters are given under the inch measurements.

If the drawing was originally dimensioned in millimeters and then converted to inches, the millimeters would be placed over the equivalent in inches. When the drawings are originally made in millimeters, these fractions are usually whole numbers with no fractions.

The other method of dual dimensioning involves the use of brackets placed around the converted dimensions. Examples of these types of dual dimensions are shown in Fig. 20.5.

It is important that you be consistent with whichever method you decide to use. Do not mix these two methods on the same drawing.

20.6 METRIC DESIGNATION

A comparison of the metric system with the English system is shown in Fig. 20.6. The metric system is the Système International d'Unités and is denoted by the letters SI. This system utilizes the European system of the first-angle of projection, which locates the front view over the top view and the right side view to the left of the front view.

When drawings are made for international circulation, it is customary to use one of the symbols at C or D in Fig. 20.6 to designate the angle of projection used. The large letters SI are used to indicate that the measurements are metric, or the word METRIC can be written prominently on the drawing or in the title block.

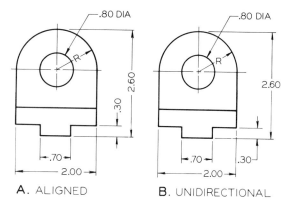

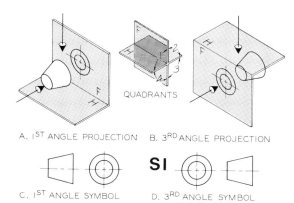

A. 1ST ANGLE PROJECTION   B. 3RD ANGLE PROJECTION

C. 1ST ANGLE SYMBOL

D. 3RD ANGLE SYMBOL & METRIC NOTE

Fig. 20.6 The European system of orthographic projection places the top view under the front view, the opposite of the American system. The system used on a drawing should be indicated by one of the symbols shown in parts C and D. If metric units are used, the large letters SI should be placed near the title block, or else the word METRIC should be used on the drawing.

A. ALIGNED   B. UNIDIRECTIONAL

Fig. 20.7 (A) Aligned dimensions are positioned to read from the bottom and right side of the sheet. (B) When the dimensions are positioned so all of them read from the bottom, they are unidirectional.

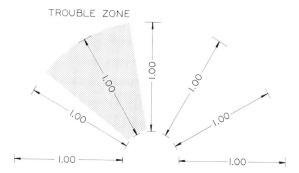

Fig. 20.8 Numbers on angular dimension lines should not be placed in the area called the trouble zone. This causes them to be read from the left side of the sheet rather than the right and bottom.

20.7 ALIGNED AND UNIDIRECTIONAL NUMBERS

The two methods of positioning dimension numbers on a dimension line are the *aligned* and *unidirectional* methods.

The unidirectional system is more widely accepted since it is easier to apply numerals that all read from the bottom of the sheet, as shown in Fig. 20.7B.

The aligned system places the numerals in an aligned direction with the dimension lines (Fig. 20.7A). The numbers must be readable from the bottom or the right side of the page.

Examples of aligned dimensions on angular dimension lines are shown in Fig. 20.8. Avoid placing aligned dimensions in the "trouble zone" since these numerals would read from the left instead of the right side or bottom of the sheet.

20.8 PLACEMENT OF DIMENSIONS

> In all dimensioning, it is good practice to dimension the views that are most descriptive.

You can see that the front view of Fig. 20.9 is more descriptive than the top view, and therefore this is the view that should be dimensioned.

The dimensions should be applied to the views in an organized manner such as those in

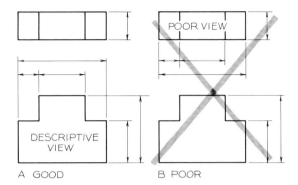

Fig. 20.9 It is a rule of dimensioning to place the dimensions on the most descriptive views where the true contour of the object can be seen.

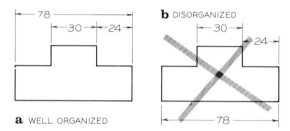

Fig. 20.10 Dimensions should be placed on the views in a well organized manner to make them as readable as possible.

Fig. 20.10. Place the dimension lines by beginning with the smaller ones to avoid crossing the dimension and extension lines.

> Always leave at least 0.4 inches between the object and the first row of dimensions (Fig. 20.11). The successive rows of dimensions should be at least 0.25 inches apart. If greater spaces are used, these same general proportions should be applied.

The Braddock-Rowe triangle can be used to space the dimension lines as well as be used as a lettering guide (Fig. 20.12).

When a row of dimensions is placed on a drawing, one of the dimensions is omitted as in Fig. 20.13A, since the overall dimension supplements the omitted dimension. If it is felt that the dimension needs to be given as a reference dimension, it is followed by the abbreviation REF for *ref-*

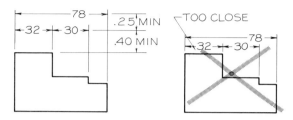

Fig. 20.11 The first row of dimensions should be placed at least 0.40 inches (⅜") from the view, and successive rows should be at least 0.25 inches (¼") from the first row. If greater spaces are used, these general proportions should be used.

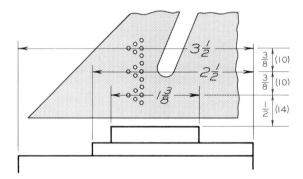

Fig. 20.12 When common fractions are used, the center holes of the triangular arrangements on the Braddock-Rowe triangle are aligned with the dimension lines to automatically space the lines as well as to draw the guide lines.

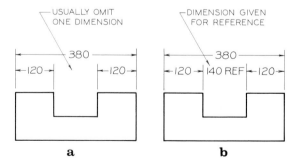

Fig. 20.13 One intermediate dimension is customarily omitted since the overall dimension provides this measurement. (A). If all the intermediate dimensions are given, one should be labeled REF to indicate that it has been given as a reference dimension.

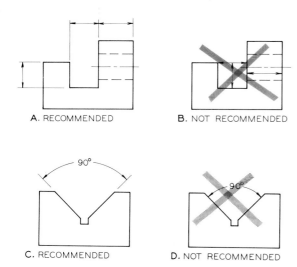

Fig. 20.14 DIMENSIONING RULES

It is preferred practice to place the dimensions outside of a part. Dimension lines should not be used as extension lines as at B.

(C and D) Angles are dimensioned with arcs and extension lines rather than showing the angle inside the angular cut as shown at D.

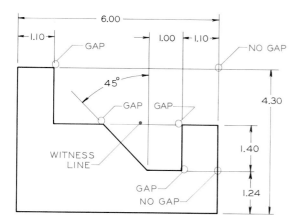

Fig. 20.15 Extension lines extend from the edge of an object leaving a small gap. They do not have gaps where they cross object lines or other extension lines. Note that a witness line is used to show that two planes of the part are aligned.

erence. This dimension is usually the least important of the dimensions.

The recommended techniques of dimensioning features are shown in Fig. 20.14.

Examples of the placement of extension lines are shown in Fig. 20.15. Extension lines may cross other extension lines or object lines with no gaps.

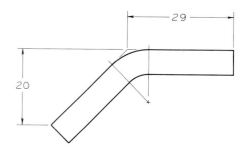

Fig. 20.16 A curved surface is dimensioned by locating the theoretical point of intersection with extension lines.

Extension lines are also used to locate theoretical points outside of curved surfaces (Fig. 20.16).

20.9 DIMENSIONING IN LIMITED SPACES

Several examples of dimensioning in limited spaces are shown in Fig. 20.17. Regardless of space limitations, the numerals should not be drawn smaller than they appear elsewhere on the drawing.

When numbers are crowded where dimension lines are closely grouped, they should be staggered to make them more readable (Fig. 20.18).

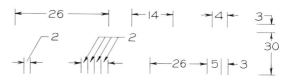

Fig. 20.17 Where room permits, the numerals and arrows should be placed inside the extension lines. Other placements are shown as the spacing becomes smaller.

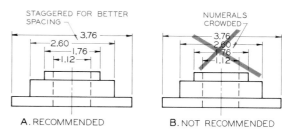

Fig. 20.18 When close spacing tends to crowd dimensioning numerals, they should be staggered for better spacing.

Fig. 20.19 DIMENSIONING PRISMS (METRIC)

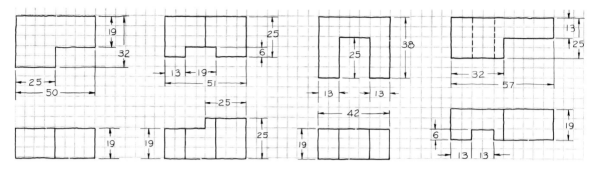

A. Dimensions should extend from the most descriptive view and be placed between the views to which they apply.

B. One intermediate dimension is not given. Extension lines may cross object lines.

C. It is permissible to dimension a notch inside the object if this improves clarity.

D. Whenever possible, dimensions should be placed on visible lines and not hidden lines.

20.10 DIMENSIONING PRISMS

The most basic element to be dimensioned is the prism, which is no more than a block when drawn in its simplest form. The following rules for dimensioning prisms are illustrated in Fig. 20.19.

1. Dimensions should extend from the most descriptive view (Fig. 20.19A).

2. Dimensions that apply to two views should be placed between these two views (Fig. 20.19A).

3. The first row of dimension lines should be placed a minimum of 0.40″ (10 mm) from the view. Successive rows are placed at least 0.25″ (5mm) apart.

4. Extension lines may cross, but dimension lines should not cross another line unless absolutely necessary.

5. In order to dimension each measurement in its most descriptive view, you may have to place dimensions in more than one view (Fig. 20.19B).

6. In the case of notches, clarity may be achieved more effectively by placing the dimension line inside the notch (Fig. 20.19C).

7. Whenever possible, dimensions should be applied to visible lines rather than hidden lines (Fig. 20.19D).

8. Dimensions should not be repeated nor should unnecessary information be given.

20.11 DIMENSIONING ANGLES

Angles can be dimensioned by using either coordinates to locate the ends of angular lines or planes, or angular measurements in degrees. Both methods are shown in Fig. 20.20. The two methods should not be mixed when dimensioning the same angle since these may not agree.

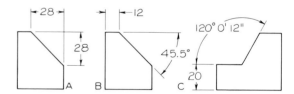

Fig. 20.20 Angular planes can be dimensioned by using coordinates (A), or an angle measured in degrees from the located vertex (B). The angles can be measured in decimal fractions (B) or in degrees, minutes, and seconds (C).

Units for angular measurements are degrees, minutes, and seconds, as shown in Fig. 20.20C. There are 60 minutes in a degree, and 60 seconds in a minute. Seldom will angular measurements need to be measured to the nearest second.

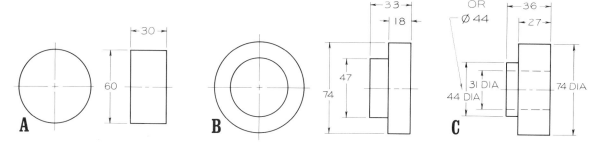

Fig. 20.21 (A) it is preferred that cylinders be dimensioned in their rectangular views using a diameter rather than a radius. (B) Dimensions should be placed between the views when possible. (C) The circular view can be omitted if DIA is written after the dimensions of the diameters.

20.12 DIMENSIONING CYLINDERS

The cylinders that are dimensioned may be either solid cylinders or cylindrical holes. Examples of solid cylinders that have been dimensioned are shown in Fig. 20.21.

> Solid cylinders are dimensioned in their rectangular views using diameters (not radii) since diameters are much easier to measure.

Parts composed of several concentric cylinders are dimensioned with diameters, beginning with the smallest cylinder (Fig. 20.21B). Parts of this type are assumed to be concentric unless otherwise noted.

A cylindrical part may be dimensioned with only one view if DIA is used with the diametrical dimension (Fig. 20.21C).

20.13 MEASURING CYLINDRICAL PARTS

Cylindrical parts are dimensioned with diameters rather than radii because diameters are easier to measure.

An internal cylindrical hole is measured with an internal micrometer caliper that has a built-in gauge permitting greater accuracy of measurement (Fig. 20.22).

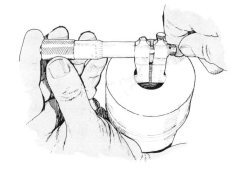

Fig. 20.22 A micrometer caliper is used to measure internal cylindrical diameters. This is why holes are dimensioned with radii instead of diameters.

Fig. 20.23 An outside micrometer caliper with a built-in gauge for measuring the diameter of a cylinder.

Likewise, an external micrometer caliper can be used for measuring the outside diameters of a part (Fig. 20.23). The choice of the diameter rather than a radius makes it possible to measure diameters during machining when the part is held between centers on a lathe.

20.14 CYLINDRICAL HOLES

Cylindrical holes may be dimensioned by one of the methods shown in Fig. 20.24. The diameter is used when a full circle is being dimensioned.

The preferred method of dimensioning cylindrical holes is to draw a leader from the circular view, and then add the dimension, followed by DIA, to indicate that the dimension is a diameter (Fig. 20.25). The metric system often uses the $\emptyset$ in front of the diametrical dimension.

Sometimes the note DRILL or BORE is added to specify the shop operation, but current standards prefer the use of DIA instead.

When dimension lines are drawn across the circular view so that it is obvious that the dimension is a diameter, we may omit the note DIA.

A part containing cylindrical features is dimensioned in Fig. 20.26 in both the circular and rectangular views to illustrate various methods of dimensioning.

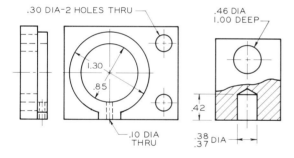

Fig. 20.24 Several acceptable methods of dimensioning cylindrical holes and shapes.

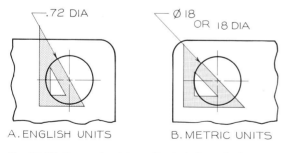

A. ENGLISH UNITS   B. METRIC UNITS

Fig. 20.25 The preferable method of dimensioning cylindrical holes is with a leader, a dimension, and the abbreviation DIA or the symbol $\emptyset$ to indicate that the dimension is a diameter. Note that the leader would pass through the center of the hole if it were extended.

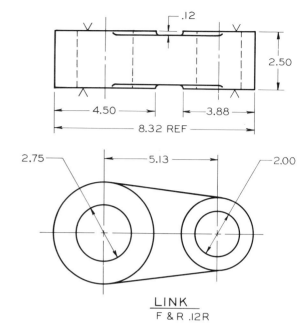

Fig. 20.26 This part composed of cylindrical features has been dimensioned using several approved methods. (F&R .12R means that fillets and rounds have a 0.12 radius).

20.15 PYRAMIDS, CONES, AND SPHERES

Three methods of dimensioning pyramids are shown in Fig. 20.27A, B, and C. The pyramids at B and C are truncated (portions removed). Note that in all three examples the apex of the pyramid is located in the rectangular view.

Two acceptable methods of dimensioning cones are shown in Fig. 20.27D and E.

A sphere is dimensioned using its diameter if it is a complete sphere (Fig. 20.27F); a radius is used if it is less than a full sphere (Fig. 20.27G). Only one view is necessary to describe a sphere.

20.16 LEADERS

Leaders are used to apply notes and dimensions to a feature that they describe. Leaders are drawn at a standard angle of a triangle, as previously illustrated in Fig. 20.25.

Examples of notes using leaders are shown in Fig. 20.28. The leader should be drawn from either

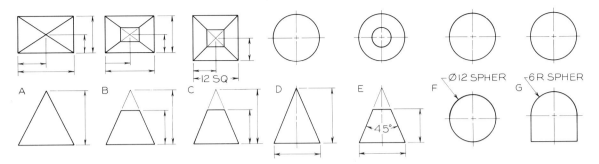

Fig. 20.27 Methods of dimensioning pyramids, cones, and spheres are shown here.

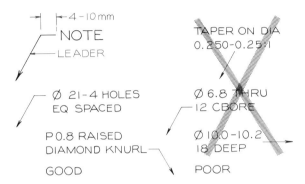

Fig. 20.28 Leaders from notes should extend with a horizontal bar from the first or last word of the note, not from the middle of the note.

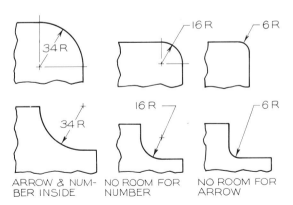

ARROW & NUM-BER INSIDE   NO ROOM FOR NUMBER   NO ROOM FOR ARROW

Fig. 20.29 When space permits, the dimension and arrow should be placed between the center and the arc. If room is not available for the number, the arrow is placed between the center and the arc with the number on the outside. If there is not room for the arrow, then both the dimension and the arrow are placed outside the arc with a leader.

the first word of the note or the last word of the note. Each leader should begin with a short horizontal line near the note.

20.17 DIMENSIONING ARCS

Cylindrical parts that are less than a full circle are dimensioned with radii as shown in Fig. 20.29. Where space permits, ths numeral should be placed between the center and the arc, and the letter R is used to indicate that it is a radius. When space does not allow this, the numeral can be placed in one of the other positions; however, in all cases, the arrowhead should touch the arc being dimensioned.

When the arc being dimensioned is a long arc, it may be dimensioned with a false radius as shown in Fig. 20.30A. A "zigzag" is placed in the radius to indicate that it is not a true radius. The

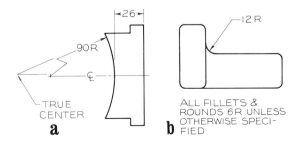

Fig. 20.30 When a radius is very long, it may be shown with a false radius with a "zigzag" to indicate that it is not true length. It should end on the center line of the true center. Fillets and rounds may be noted to reduce repetitive dimensions of small arcs.

center of the arc should lie on the centerline of the extension of the line on which the true center lies.

20.18   FILLETS AND ROUNDS

Fillets and *rounds* are rounded corners conventionally used on castings. A fillet is an internal rounding and a round is an external rounding.

When fillets and rounds are equal in radii, a note may be placed on the drawing to eliminate the need for repetitive dimensioning. The note may read as follows: ALL FILLETS AND ROUNDS 6 R. If most, but not all of the fillets and rounds have equal radii, the following note may be used: ALL FILLETS AND ROUNDS 6 R UNLESS OTH-

ERWISE SPECIFIED. In this case, only the fillets and rounds of different radii are dimensioned (Fig. 20.30B). The notes may be abbreviated such as F & R 10 R.

Fillets and rounds that are dimensioned with leaders are dimensioned as small arcs with separate leaders as shown in Fig. 20.31, rather than using long, confusing leaders as shown at B.

When repetitive features appear on a drawing, these may be noted as shown in Fig. 20.32. The note TYPICAL or TYP means that although only one of these features is dimensioned, the dimensions are typical of those that are undimensioned. The term PLACES is sometimes used to specify the number of places that a similar feature appears. Similarly, the number of holes sharing the same dimension may be indicated by including the number of holes in the note.

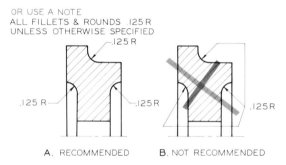

Fig. 20.31 When several arcs are dimensioned with the same radii, it is preferable that separate leaders be used rather than extending the leaders as shown in B.

20.19   CURVED SURFACES

An irregular shape composed of a number of tangent arcs of varying sizes (Fig. 20.33) can be dimensioned by using a series of radii.

When the curve is irregular rather than composed of arcs (Fig. 20.34), the coordinate method can be used to locate a series of points along the curve from two datum lines. The drafter must use judgment to determine the proper spacing for the points. Note that extension lines may be placed at an angle to provide additional space for showing dimensions.

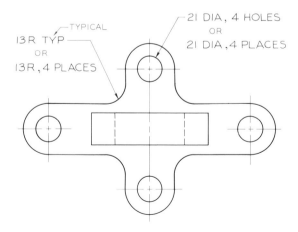

Fig. 20.32 Notes can be used to indicate that similar features and dimensions are repeated on drawings without having to show the dimensions repetitively.

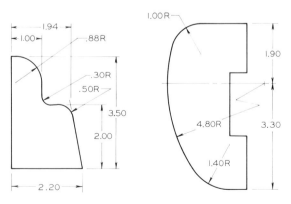

Fig. 20.33 Examples of dimensioned parts that are composed of a series of tangent arcs.

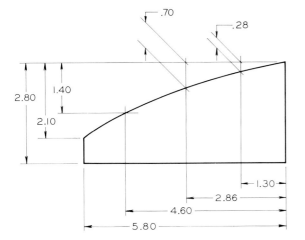

Fig. 20.34 This object with an irregular curve is dimensioned using coordinates.

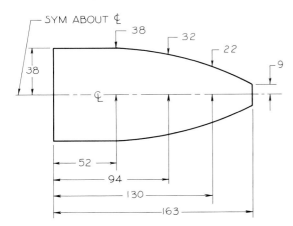

Fig. 20.35 This symmetrical part is dimensioned about its center line.

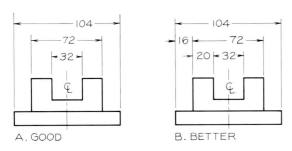

A. GOOD   B. BETTER

Fig. 20.36 Symmetrical parts may be dimensioned about their center lines as shown in A. It is better to dimension symmetrical parts of this type as shown at B.

A special case of an irregular curve is the symmetrical curve shown in Fig. 20.35. Note that dimension lines are used as extension lines in violation of a previously established rule.

20.20 SYMMETRICAL OBJECTS

When symmetrical objects are dimensioned, they may be dimensioned as shown in Fig. 20.36A where a centerline is used with the initials CL to identify it. In this case, it is assumed that the dimensions are each centered about the centerline.

The better method is the one shown in Fig. 20.36B where the need for an assumption is eliminated.

20.21 FINISHED SURFACES

Many parts are formed as castings in a mold that gives their exterior surfaces a rough finish. If the part is designed to come in contact with another surface, the rough finish must be machined by grinding, shaping, lapping, or a similar process.

To indicate that a surface is to be finished, finish marks are drawn on the surface where it appears as an edge (Fig. 20.37).

> Finish marks should be repeated in every view where the surface appears as an edge, even if it is a hidden line.

Three methods of drawing finish marks are shown in Fig. 20.37. The simple V mark is pre-

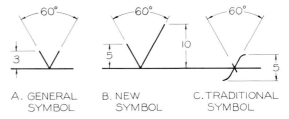

A. GENERAL SYMBOL   B. NEW SYMBOL   C. TRADITIONAL SYMBOL

Fig. 20.37 Finish marks of these types can be used to indicate that a surface has been machined to a smooth surface. The traditional V can be used for general applications, but the new finish mark at B is suggested as the new symbol. The f shown at C is also acceptable to indicate finished surfaces.

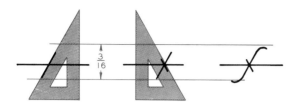

Fig. 20.38 The steps of drawing the f finish mark.

ferred in general cases. The uneven V (Fig. 20.37B is a newly recommended symbol that is related to surface texture and will be discussed in the next chapter. The steps of constructing the traditional f that is used as a finish mark are shown in Fig. 20.38.

When an object is finished on all surfaces, the note FINISHED ALL OVER is placed on the drawing. This can be abbreviated as FAO without periods after the letters. Cylindrical holes, although finished, are not specified with finish marks.

Fig. 20.39 LOCATION DIMENSIONS

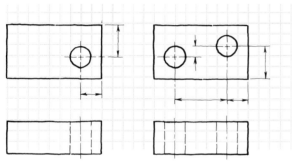 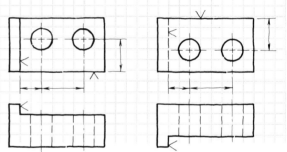

A. Cylindrical holes should be located in the circular view from two surfaces of the object.

B. When more than one hole is to be located, the other holes should be located in relation to the first from center to center.

C. A more accurate location is possible if holes are located from finished surfaces.

D. Holes should be located in the circular view and from finished surfaces even if the finished surfaces are hidden.

20.22 LOCATION DIMENSIONS

Location dimensions are used to locate the positions, not the sizes, of geometric elements, such as cylindrical holes. Figure 20.39 illustrates the basic rules for locating cylindrical parts. Since location dimensions do not involve the sizes of various features, the dimensions of the holes are omitted for clarity in Fig. 20.39. You will note that, in all of these examples of locating holes, the centers of the holes are located with coordinates in the circular view when possible.

When finished surfaces appear on the object, it is desirable that the holes be located from these surfaces since they can be located more accurately from a smooth, machined surface than from an unfinished surface. This rule is followed even if the finished surface is a hidden line as in Fig. 20.40.

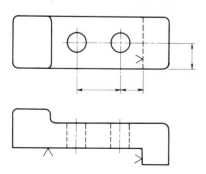

Fig. 20.40 These cylindrical holes are located from center to center in their circular view and from a finished surface, if one is available.

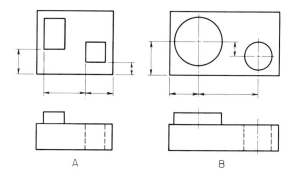

Fig. 20.41 Prisms and cylinders are located with coordinates in the view where both coordinates can be seen.

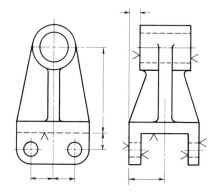

Fig. 20.42 An example of location dimensions applied to a part to locate its geometric features.

Prisms are located with respect to each other as shown in Fig. 20.41. Only a single corner of a prism need be located with respect to another when the sides of the prisms are parallel.

Location dimensions should be placed on views where both dimensions can be shown, the top view in Fig. 20.41. Cylinders are located in their circular views (Fig. 20.40B and Fig. 20.42).

20.23 LOCATION OF HOLES

When holes must be located very accurately, the dimensions should originate from a common reference plane on the part to reduce the accumulation of errors in measurement. Two examples of holes that are located from reference planes with coordinates are shown in Figs. 20.43A and B.

When several holes in a series are to be equally spaced, as in Fig. 20.43C, a note specifying that they are equally spaced can be used to locate the holes. The first and last holes of the series are determined by the usual location dimensions.

Holes on circular plates (Fig. 20.44) may be located by coordinates or by a note. When a note is used, the diameter of the circle passing through the centers of the hole must be given. This circle is sometimes referred to as the "bolt circle."

A similar method of locating holes is the polar system illustrated in Fig. 20.45. The radial dis-

Fig. 20.43 LOCATION OF HOLES

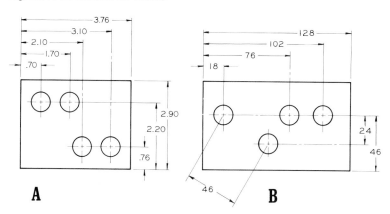

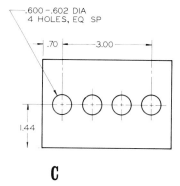

A. Holes can be more accurately located if common datum planes are used, from which all measurements are made.

B. A diagonal dimension can be used to locate a hole of this type from another hole's center.

C. A note can be used to specify the spacing between the centers of holes in an equally spaced series.

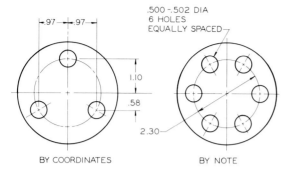

BY COORDINATES   BY NOTE

Fig. 20.44 Holes may be located in circular plates by coordinates (A) or by note (B).

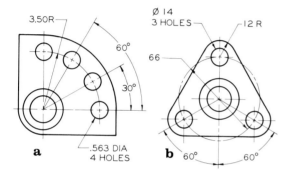

a   **b**

Fig. 20.45 Methods of locating cylindrical holes on concentric arcs.

the overall dimension. In this case the dimension of the radius must be given.

A part with partially rounded ends is dimensioned as shown in Fig. 20.47A. The overall dimension is given, and the dimensions of the radii are given in order that their centers may be located by measuring from the ends.

When an object has a rounded end that is less than a semicircle (Fig. 20.47B), location dimensions must be used to locate the center of the arc.

Slots with rounded ends are dimensioned in Fig. 20.48. Only one slot is dimensioned in part A, with a note indicating that there are two slots. The slot at B is dimensioned giving the overall dimension and the two arcs, which are understood to

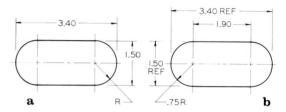

a   R   .75R   **b**

Fig. 20.46 The preferred method of dimensioning objects with rounded ends is shown at (a). A less desirable method is shown at (b).

tances from the point of concurrency and their angular measurements (in degrees) between the holes are used to locate the centers.

20.24 OBJECTS WITH ROUNDED ENDS

It is preferred that objects with rounded ends be dimensioned from end to end as shown in Fig. 20.46. This dimension provides the overall length without using arithmetic. The radius is shown with the letter R without a dimension to specify that the end is formed by an arc. Since the height is given, the radius is understood to be half of this measurement (1.50 in this case).

If the object is dimensioned from center to center of the rounded ends, the overall dimension should be given as a reference dimension (REF) to eliminate the calculations required to determine

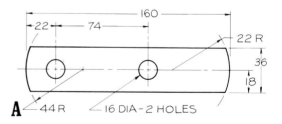

A

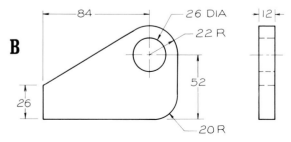

B

Fig. 20.47 Examples of dimensioned parts consisting of rounded ends and cylindrical features.

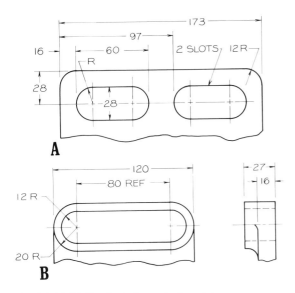

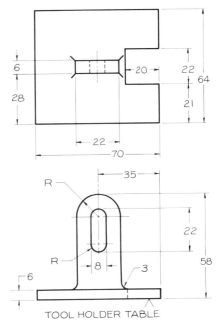

TOOL HOLDER TABLE

Fig. 20.48 Methods of dimensioning slots.

Fig. 20.49 This part is dimensioned using the previously presented principles.

apply to both ends. The distance between the centers is given as a reference dimension.

The dimensioned views of the tool holder table in Fig. 20.49 shows examples of arcs and slots. To prevent crossing of dimension lines, it is often necessary to place dimensions in a view that is not as descriptive as might be desired.

20.25 MACHINED HOLES

Machined holes are holes that are made or refined by a machine operation, such as drilling or boring.

These operations are specified by notes and leaders applied to the holes (Fig. 20.50).

It is preferred to give the diameter of the hole with the abbreviation DIA or to use the symbol ∅ with no reference to the machining method used. However, the note 32 DRILL may be used in some cases instead of 32 DIA.

Drilled Holes Several methods of specifying drilled holes are shown in Fig. 20.50. The depth

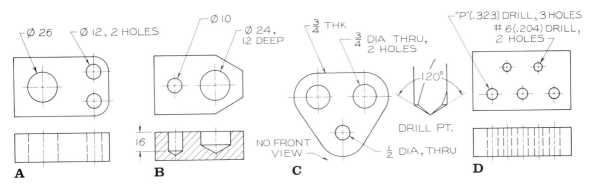

Fig. 20.50 Cylindrical holes may be dimensioned by either of the methods shown at A and B. Note that when one view is given at C, it is necessary to indicate in the note that the holes are "thru" holes, since this cannot be seen otherwise.

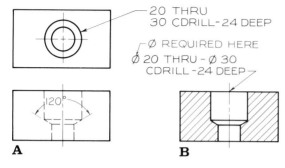

Fig. 20.51 Counterdrilling notes give the specifications for a larger hole drilled over a smaller hole. When the circular view is given, the abbreviation DIA or ∅ is sometimes omitted. When only the rectangular views are given, DIA or ∅ must be given. The 120° angle is not required as a dimension since this is understood to be the angle of a drill point.

of drilled holes can be specified in the note or it may be dimensioned in the rectangular view. Note that the depth of a drilled hole is dimensioned as the usable part of the hole; the conical point is disregarded, as shown in part B. The standard angle for the point of the drill is 120°.

Counterdrilled Holes A large hole drilled inside a smaller drilled hole to enlarge it is a counterdrilled hole, as illustrated in Fig. 20.51. When two views of counterdrilled holes are given at A, the reference to DIA or ∅ may be omitted from the note. (Also the 120° dimension is given to indicate the angle of the drill point and it need not be shown on the drawing.)

When only one view of the counterdrilled hole is shown as in the section at B, either ∅ or DIA needs to be used to specify that the dimensions are diameters.

Countersunk Holes Countersinking is the process of forming a conical head (Fig. 20.52). The diameter of the countersink hole (the maximum diameter on the surface) and the angle of the countersink are given in the notes.

Countersinking is also used to provide center holes in shafts, spindles, and other cylindrical parts that must be held between the centers of a lathe. Two methods of noting these countersinks are illustrated in Fig. 20.53.

Spotfacing This is a machining process used to finish the surface around the top of a hole in order to provide a level seat for a washer or a fastener head (Fig. 20.54a and b).

A spotfacing tool is shown in Fig. 20.55, where it has spotfaced a boss (a raised cylindrical element).

Counterboring Counterbored holes are made by enlarging the drilled hole to a larger diameter (Fig. 20.54). Note that the bottoms of the counterbored holes are flat with no taper as in counterdrilled and countersunk holes.

Bored Holes Boring is a machine operation that is usually performed on a lathe with a boring bar (Fig. 20.56) for making large holes.

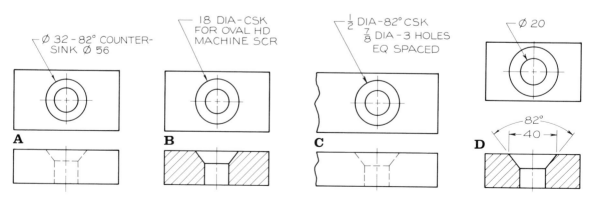

Fig. 20.52 Examples of notes specifying countersunk holes are shown here in various acceptable forms.

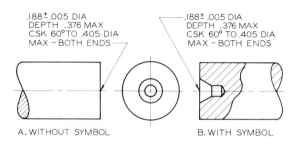

.188±.005 DIA
DEPTH .376 MAX
CSK 60°TO .405 DIA
MAX – BOTH ENDS

.188±.005 DIA
DEPTH .376 MAX
CSK 60° TO .405 DIA
MAX – BOTH ENDS

A. WITHOUT SYMBOL

B. WITH SYMBOL

Fig. 20.53 Methods of specifying countersinks in the ends of cylinders for mounting them on a lathe between centers.

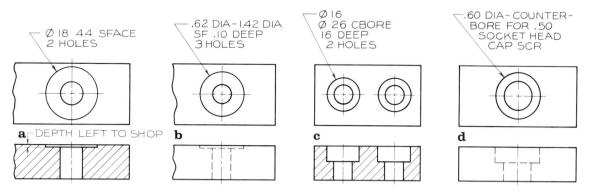

Ø 18 44 SFACE
2 HOLES

.62 DIA–1.42 DIA
SF .10 DEEP
3 HOLES

Ø 16
Ø 26 CBORE
16 DEEP
2 HOLES

.60 DIA–COUNTER-
BORE FOR .50
SOCKET HEAD
CAP SCR

a DEPTH LEFT TO SHOP

b

c

d

Fig. 20.54 Spotfaces can be specified as shown at (a) and (b). The depth of the spotface can be specified as shown if needed. Counterbores are dimensioned by giving the diameters of both holes and the depth of the larger hole (C and D).

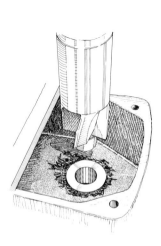

Fig. 20.55 This spotfacing tool is being used to spotface the cylindrical boss to provide a smooth bearing for a bolt.

Fig. 20.56 Boring a large hole on a lathe with a boring bar. (Courtesy of Clausing Corporation.)

Reamed Holes These are holes that have been finished or slightly enlarged after having been drilled or bored. This operation uses a ream, which is similar to a drill bit.

20.26 CHAMFERS

Chamfers are beveled edges that are used on cylindrical parts such as shafts and threaded fasteners. They eliminate rough edges and facilitate the assembly of parts.

When a chamfer angle is 45°, a note can be used in either of the forms shown in Fig. 20.57A. When the chamfer is at an angle other than 45°, the angle and the length are given as shown in part B.

Chamfers can also be specified at the openings of holes, as shown in Fig. 20.58.

20.27 KEYSEATS

A keyseat is a slot cut into a shaft for the purpose of aligning the shaft with a part mounted on it.

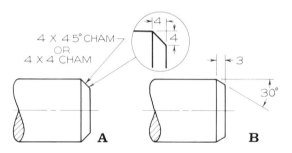

Fig. 20.57 Chamfers can be dimensioned by using either type of note shown at A when the angle is 45°. If the angle is other than 45°, it should be dimensioned as shown at B.

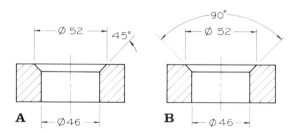

A   **B**

Fig. 20.58 Inside chamfers are dimensioned by using one of these methods.

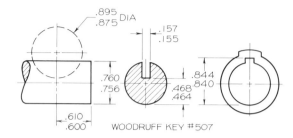

Fig. 20.59 Methods of dimensioning slots in a shaft and a slot for a Woodruff key that will hold the part on the shaft are shown here. The dimensions for these features are given in Appendix 32.

This part may be a pulley or a collar. The method of dimensioning a keyway and keyseat is shown in Fig. 20.59. The double dimensions are tolerances that will be discussed in more detail in the next chapter.

20.28 KNURLING

Knurling is the operation of cutting diamond-shaped or parallel patterns on cylindrical surfaces for gripping, for decoration, or for a press fit between two parts that will be permanently assembled.

A diamond knurl and a straight knurl are drawn and dimensioned in Fig. 20.60. Knurls should be dimensioned with specifications that give type, pitch, and diameter.

The "96 DP" means diametrical pitch, which is the ratio of the number of grooves on the circumference (N) to the length of the "pitch diame-

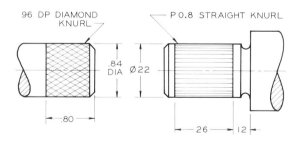

Fig. 20.60 A diamond knurl with a diametrical pitch of 96 is shown at A. The diametrical pitch is the number of teeth about the circumference divided by the diameter. A straight knurl is shown at B where the pitch (P) is 0.8 mm. Pitch is the distance between the grooves on the circumference.

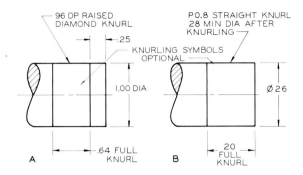

Fig. 20.61 Knurls need not be drawn if they are dimensioned as shown here.

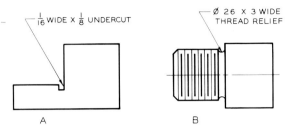

Fig. 20.63 An undercut can be dimensioned as shown at A. A thread relief, which is a type of neck, is dimensioned at B.

ter'' (D) which is found by the equation $DP = N/D$. The preferred diametrical pitches for knurling are 64DP, 96DP, 128DP, and 160DP. For diameters of 1 inch, knurling of 64DP, 96DP, 128DP, and 160DP will have 64, 96, 128, and 160 teeth respectively on the circumference. The note P 0.8 means that the knurling grooves are 0.8 mm apart. (Note: Calculations must be made using inches. Conversion to millimeters can be made afterwards.)

Knurls for press fits are specified with diameters before knurling and the minimum diameter after knurling. A simplified method of representing knurls is shown in Fig. 20.61 where notes are used and the knurls are not drawn.

20.29 NECKS AND UNDERCUTS

A neck is a recess cut into a cylindrical part. This recess is used where cylinders of different diameter join (Fig. 20.62). The neck ensures that a part

assembled on the smaller shaft will fit flush against the shoulder of the larger cylinder.

Undercuts are somewhat similar to necks and serve the same purpose (Fig. 20.63A). An undercut ensures that a part fitting in the corner of the surface will fit flush against both surfaces, and it permits space for trash to drop out of the way when entrapped in the corner. An undercut could also be a recessed neck on the inside of a cylindrical hole. A thread relief (Fig. 20.63B) is used to square the threads where they intersect a larger cylinder.

20.30 TAPERS

Tapers can be either conical surfaces or flat planes. Examples of conical features of tapers are dimensioned in Fig. 20.64. A taper can be dimensioned by giving the following information: (1) the diameter or width at each end of the taper, (2) the length of the tapered feature, (3) the rate of taper.

The rate of taper in Fig. 20.64 is 0.25 per inch of length and 3.00 inches per foot at part B. This taper can specified as a ratio such as 0.25:1.

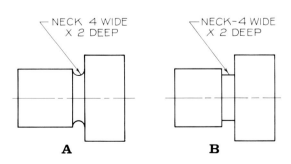

Fig. 20.62 Necks are recesses in cylinders that are used where cylinders of different sizes join together.

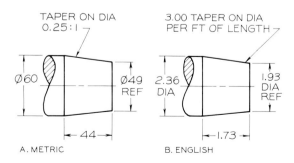

Fig. 20.64 Tapers can be dimensioned as shown here. The tapers are specified as the rate of taper per millimeter or per foot of length.

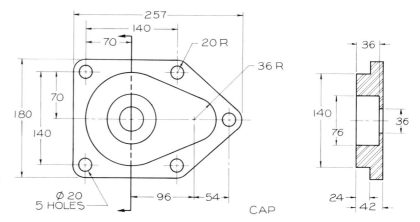

Fig. 20.65 An example of a part dimensioned with a variety of dimensioning principles.

20.31 DIMENSIONING SECTIONS

Sections are dimensioned in the same manner as regular views. A dimensioned section is shown in Fig. 20.65. Most of the principles of dimensioning covered in this chapter have been applied to this part.

20.32 MISCELLANEOUS NOTES

A variety of notes are used on detail drawings to provide information and specifications that would be difficult to represent by other means (Figs. 20.66 and 20.67). It is important that notes be placed on a drawing with adequate room for the notes to be lettered legibly without crowding or changing the height of the standard one-eighth inch lettering.

The notes are placed horizontally on the sheet with short dashes between the lines in the notes if they lie on the same horizontal line. The common abbreviations that are used in these notes can be used to save space. Refer to Appendix 1 for the standard approved abbreviations.

Note that the leaders for the notes originate at the first word of the note or the last word of the note.

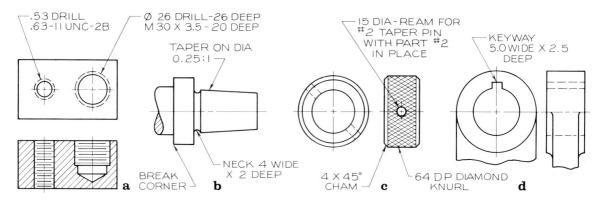

Fig. 20.66 (A and B) Treaded holes are sometimes dimensioned by giving the tap drill size in addition to the thread specifications. The tap drill size is not required, but it is permissible. The part at B is dimensioned to indicate a neck, a taper, and a break corner, which is a slight round to remove the sharpness from a corner. The collar at C has a knurl note, a chamfer note, and a note indicating the insertion of a #2 taper pin. The part at D gives a note for dimensioning a keyway.

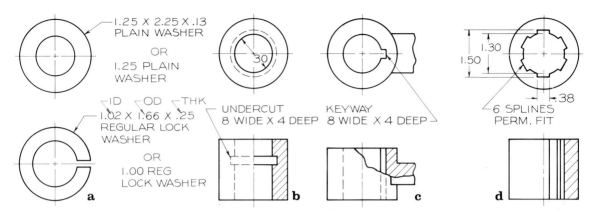

Fig. 20.67

(a and b) Methods of dimensioning washers and lock washers are shown. These dimensions can be found in the tables in Appendixes 36 and 37. An undercut is dimensioned at (b). (c and d) A keyway is dimensioned at (c) and a spline inside a hole is dimensioned at (d).

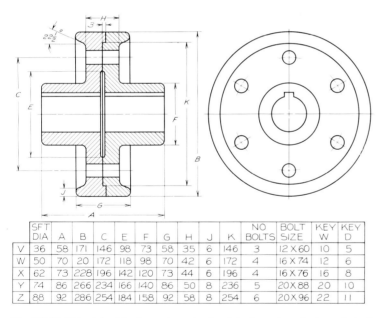

SFT DIA	A	B	C	E	F	G	H	J	K	NO BOLTS	BOLT SIZE	KEY W	KEY D
V	36	58	171	146	98	73	58	35	6	146	3	12 X 60	10	5
W	50	70	20	172	118	98	70	42	6	172	4	16 X 74	12	6
X	62	73	228	196	142	120	73	44	6	196	4	16 X 76	16	8
Y	74	86	266	234	166	140	86	50	8	236	5	20 X 88	20	10
Z	88	92	286	254	184	158	92	58	8	254	6	20 X 96	22	11

Fig. 20.68 This object is dimensioned using tabular values that apply to the same drawing. The letters in balloons on each dimension line are given in the table below the drawing.

20.33 TABULAR DRAWINGS

When parts have similar features except for variations in size, a table of values can be given to provide the variations in dimensions such as in Fig. 20.68. The part numbers are given in the tables with their respective dimensions A through Z that are enclosed in circles.

Tabular dimensions shorten drafting time while providing the essential information.

PROBLEMS

1–50. These problems shown in Figs. 20.69 and 20.70 are to be solved on Size A paper if they are drawn half-size. If drawn full size, Size B paper should be used. Note that the dimensions of the views are found by using your dividers to transfer drawing sizes to the scales beneath the problems. This will provide you with either metric or imperial units of measurement.

You will need to vary the spacing between the views to provide adequate room for the dimensions. It would be desirable to sketch the views and the dimensions in order to determine the spacing required before laying out the problems with instruments.

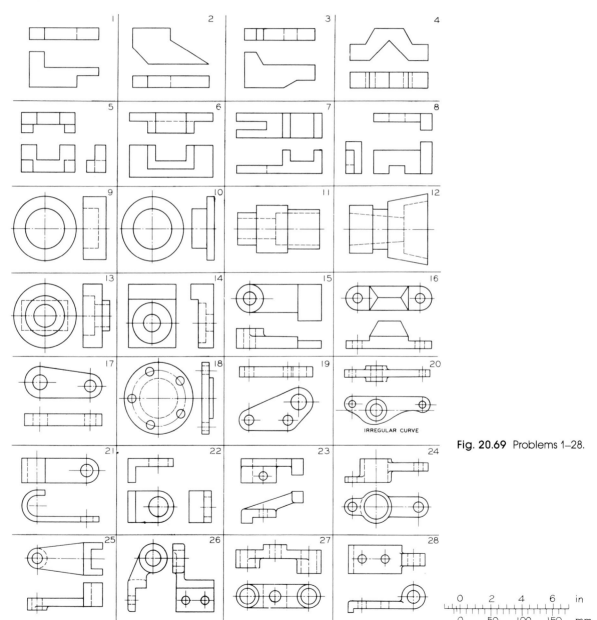

Fig. 20.69 Problems 1–28.

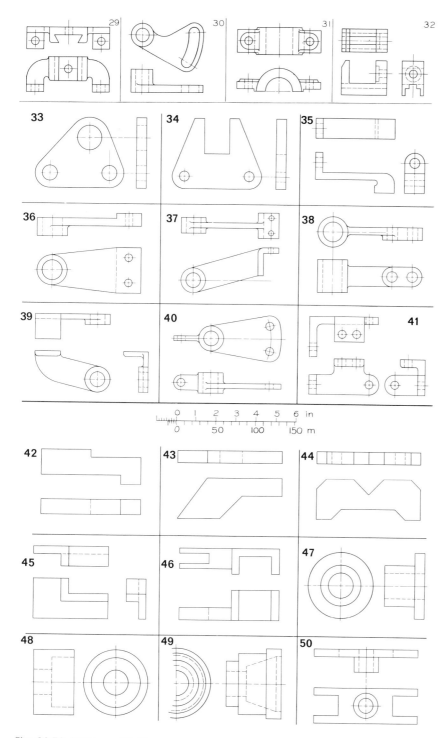

Fig. 20.70 Problems 29–50.

TOLERANCES

21.1 INTRODUCTION

Parts produced today require more accurate dimensions than did those produced in the past because many of today's parts are made by different companies in different locations. Therefore, these parts must be specified so they will be interchangeable.

The techniques of dimensioning parts to ensure interchangeability is called **tolerancing.** Each dimension is allowed a certain degree of variation within a specified zone called a tolerance. For example, a part's dimension might be expressed as 100 ± 0.50, which yields a tolerance of 1.00 mm.

> It is desirable that all dimensions be given as large a tolerance as possible without interfering with the function of a part. This reduces production costs since manufacturing to close tolerances is expensive.

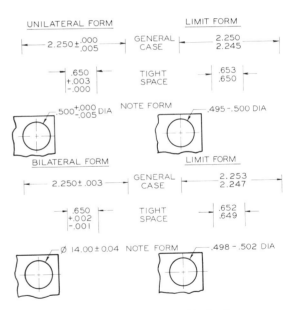

Fig. 21.1 Methods of positioning and indicating tolerances in unilateral and limit forms.

21.2 TOLERANCE DIMENSIONS

Several acceptable methods of specifying tolerances are shown in Fig. 21.1. When "plus-and-minus" tolerancing is used, tolerances are applied to a basic dimension. When plus-and-minus dimensions allow variation in only one direction, the tolerancing is *unilateral*. Tolerancing that permits variation in either direction from the basic dimension is *bilateral*.

The customary methods of indicating toleranced dimensions are shown in Fig. 21.2. The positioning and spacing of numerals of toleranced dimensions are shown in Fig. 21.3.

Tolerances may be given in the form of limits; that is, two dimensions are given that represent

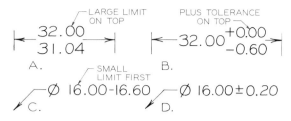

Fig. 21.2 When limit dimensions are given, the large limits are placed either above the small limits or to the right of the small limits. The plus limits are placed over the minus limits in plus-and-minus tolerancing.

Fig. 21.3 Positioning and spacing of numerals used to specify tolerances.

the largest and smallest size permitted for a feature of the part. When limits for dimensions are compared with their counterparts in plus-and-minus form, we see that both methods result in the same tolerance.

21.3 MATING PARTS

Mating parts are parts that fit together within a prescribed degree of accuracy (Fig. 21.4). The upper piece is dimensioned with two measurements that indicate the upper and lower limits of the

size. The notch is slightly larger allowing the parts to be assembled.

An example of mating cylindrical parts is shown in Fig. 21.5A. Part B of the figure illustrates the meaning of the tolerance dimensions. The size of the shaft can vary in diameter from 1.500″ (its maximum size) to 1.498″ (its minimum size). The difference between these limits on a single part is the tolerance, 0.002″ in this case. The dimensions of the hole in part A are given with limits of 1.503 and 1.505, for a tolerance of 0.002 (the difference between the limits as illustrated in part B).

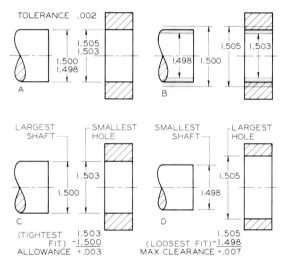

Fig. 21.5 The allowance (tightest fit) between these assembled parts is +0.003″. The maximum clearance is 0.007″.

21.4 TERMINOLOGY OF TOLERANCING

The meaning of most of the terms used in tolerancing can be seen by referring to Fig. 21.5.

Tolerance is the difference between the limits prescribed for a single part. The tolerance of the shaft in Fig. 21.5 is 0.002″.

Limits of tolerance are the extreme measurements permitted by the maximum and minimum sizes of a part. The limits of tolerance of the shaft in Fig. 21.5 are 1.500 and 1.498.

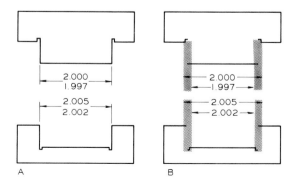

Fig. 21.4 Each of these mating parts has a tolerance of 0.003″ (variation in size). The allowance between the assembled parts (tightest fit) is 0.002″.

Allowance is the tightest fit between two mating parts. The allowance between the shaft and the hole in Fig. 21.5 is 0.003 (negative for an interference fit).

Nominal size is an approximate size that is usually expressed with common fractions. The nominal sizes of the shaft and hole in Fig. 21.5 are 1.50″ or 1½″.

Basic size is the exact theoretical size from which limits are derived by the application of plus-and-minus tolerances. There is no basic diameter if this is expressed with limits.

Actual size is the measured size of the finished part.

Fit signifies the type of fit between two mating parts when assembled. The types of fit are clearance, interference, transition, and line fits.

Clearance fit is a fit that gives a clearance between two assembled mating parts. The fit between the shaft and the hole in Fig. 21.5 is a clearance fit, since it provides a space between the two under the tightest condition.

Interference fit is a fit that results in an interference between the two assembled parts. The shaft in Fig. 21.6A is larger than the hole, which requires a force or press fit that has an effect similar to that of a weld between the two parts.

Transition fit can result in either an interference or a clearance. The shaft in Fig. 21.6B can be either smaller or larger than the hole and still be within the prescribed tolerances.

Line fit can result in a contact of surfaces or a clearance between them. The shaft in Fig. 21.6C can have contact or clearance when the limits are approached.

Selective assembly is a method of selecting and assembling parts by trial and error. Using this method, parts can be assembled that have greater tolerances, and consequently can be produced at a reduced cost. This process of hand assembly represents a compromise between a high degree of manufacturing accuracy and ease of assembly of interchangeable parts.

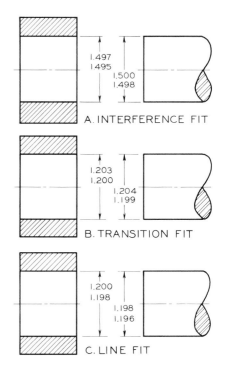

Fig. 21.6 Types of fits between mating parts. The clearance fit is not shown.

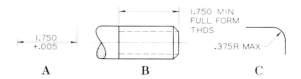

Fig. 21.7 Single tolerances can be given in some applications in MAX or MIN form.

Single limits are dimensions that are designated by MIN or MAX (minimum or maximum) instead of being labeled by both (Fig. 21.7). Depths of holes, lengths, threads, corner radii, chamfers, etc. are dimensioned in this manner in some cases. Caution should be exercised to prevent substantial deviations from the single limit.

21.5 BASIC HOLE SYSTEM

The basic hole system is a widely used system of dimensioning holes and shafts to give the required allowance between the two assembled parts. The smallest hole is taken as the basic diameter from which the limits of tolerance and allowance are applied.

The hole is used because many of the standard drills, reamers, and machine tools are designed to give standard hole sizes. Therefore, it is advantageous to use this diameter as the basic dimension.

If the smallest diameter of a hole is 1.500″, the allowance (0.003 in this example) can be subtracted from this diameter to find the diameter of the largest shaft (1.497″). The smallest limit for the shaft can then be found by subtracting the tolerance from 1.497″.

21.6 BASIC SHAFT SYSTEM

Some industries use the basic shaft system of applying tolerances to dimensions of shafts, since many come in standard sizes. In this system, the largest diameter of the shaft is used as the basic diameter from which the tolerances and allowances are applied.

For example, if the largest permissible shaft is 1.500″, the allowance can be added to this dimension to yield the smallest possible diameter of the hole into which the shaft must fit. Therefore, if the parts are to have an allowance of 0.004″, the smallest hole would have a diameter of 1.504″.

21.7 METRIC LIMITS AND FITS

The previous section dealt with the English system of specifying limits and fits between mating parts. This section will present the metric system as recommended by the International Standards Organization (ISO), which has been presented in ANSI B4.2-1978. These fits usually apply to cylinders—holes and shafts. However, these tables can also be used to determine the fits between any parallel surfaces, such as a key in a slot.

Metric Definitions of Limits and Fits

Most of the definitions given below are illustrated in Fig. 21.8.

Basic size   is the size from which the limits or deviations are assigned. Basic sizes, usually diameters, should be selected from Table 21.1 (pre-

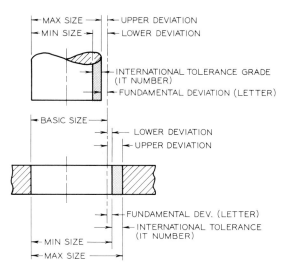

Fig. 21.8 Definition of terms related to metric fits.

sented in Section 21.8) under the heading of "first choice."

Deviation   is the difference in a hole or shaft size and the basic size.

Upper deviation   is the difference between the maximum permissible size of a part and its basic size.

Lower deviation   is the difference between the minimum permissible size of a part and its basic size.

Fundamental deviation   is the deviation that is closest to the basic size. In the note, 40H7, the letter H (an uppercase letter) represents the fundamental deviation for a hole. In the note, 40g6, the letter g (a lowercase letter) represents the fundamental deviation for a shaft.

Tolerance   is the difference between the maximum and minimum allowable sizes of a single part.

International tolerance grade (IT)   is a group of tolerances that vary in accordance with the basic size, but that provide a uniform level of accuracy within a given grade. In the note, 40H7, the number 7 represents the IT grade. There are 18 IT grades: IT01, IT0, IT1, . . . , IT16.

Tolerance zone   is the zone that represents the tolerance and its position in relation to the basic size. This is a combination of the fundamental deviation (represented by a letter) and the Interna-

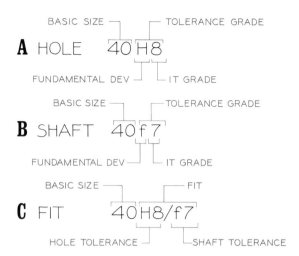

Fig. 21.9 Symbols and their definitions as applied to holes and shafts.

Fig. 21.10 Three methods of giving tolerance symbols. The numbers in parenthesis are for reference.

the tolerance zone. Capital letters are used to indicate the fundamental deviation for holes, and lowercase letters for shafts.

Three methods of specifying the tolerance information are shown in Fig. 21.10. The information in parentheses indicates that it is for reference only. The upper and lower limits are found in the tables in Appendix 44. This method will be covered in the next section.

21.8 PREFERRED SIZES

The preferred basic sizes for computing tolerances are shown in Table 21.1. Under the "first choice" heading the numbers increase by about 25% of the preceding numbers. Those in the "second choice" column increase at approximately 12% increments.

tional Tolerance grade (IT number). For example, in note 40H8, the H8 portion indicates the tolerance zone.

Hole basis   system is a system of fits based on the minimum hole size as the basic diameter. The fundamental deviation for a hole basis system is an uppercase letter, "H", for example (Fig. 21.9).

Shaft basis   system is a system of fits based on the maximum shaft size as the basic diameter. The fundamental deviation for a shaft basis system is a lowercase letter, "f", for example (Fig. 21.9).

Clearance fit   is a fit that results in a clearance between the two assembled parts under all tolerance conditions.

Interference fit   is a fit that results in an interference fit between the two assembled parts under all tolerance conditions.

Transition fit   is a fit that results in a clearance or an interference fit between two assembled parts.

Tolerance symbols   are notes that are used to communicate the specifications of tolerance and fit. Examples of these are given in Fig. 21.9. The basic size is the primary dimension from which the tolerances are determined; therefore, it is the first part of the symbol. It is followed by the fundamental position letter and the IT number to give

Table 21.1 PREFERRED SIZES

Basic Size, mm		Basic Size, mm		Basic Size, mm	
First Choice	Second Choice	First Choice	Second Choice	First Choice	Second Choice
1		10		100	
	1.1		11		110
1.2		12		120	
	1.4		14		140
1.6		16		160	
	1.8		18		180
2		20		200	
	2.2		22		220
2.5		25		250	
	2.8		28		280
3		30		300	
	3.5		35		350
4		40		400	
	4.5		45		450
5		50		500	
	5.5		55		550
6		60		600	
	7		70		700
8		80		800	
	9		90		900
				1000	

Table 21.2 DESCRIPTION OF PREFERRED FITS

ISO SYMBOL		DESCRIPTION
Hole Basis	Shaft Basis	
H11/c11	C11/h11	*Loose running* fit for wide commercial tolerances or allowances on external members.
H9/d9	D9/h9	*Free running* fit not for use where accuracy is essential, but good for large temperature variations, high running speeds, or heavy journal pressures.
H8/f7	F8/h7	*Close running* fit for running on accurate machines and for accurate location at moderate speeds and journal pressures.
H7/g6	G7/h6	*Sliding* fit not intended to run freely, but to move and turn freely and locate accurately.
H7/h6	H7/h6	*Locational clearance* fit provides snug fit for locating stationary parts; but can be freely assembled and disassembled.
H7/k6	K7/h6	*Locational transition* fit for accurate location, a compromise between clearance and interference.
H7/n6	N7/h6	*Locational transition* fit for more accurate location where greater interference is permissible.
H7/p6[1]	P7/h6	*Locational interference* fit for parts requiring rigidity and alignment with prime accuracy of location but without special bore pressure requirements.
H7/s6	S7/h6	*Medium drive* fit for ordinary steel parts or shrink fits on light sections, the tightest fit usable with cast iron.
H7/u6	U7/h6	*Force* fit suitable for parts which can be highly stressed or for shrink fits where the heavy pressing forces required are impractical.

On the left, from top to bottom: Clearance Fits, Transition Fits, Interference Fits.
On the right, from top to bottom: More Clearance, More Interference.

[1] Transition fit for basic sizes in range from 0 through 3 mm.

Where possible, you should select basic diameters from the first column since these correspond to standard sizes for round, square, and hexagonal metal products. This makes it possible to use stock sizes and reduce expenses.

Preferred fits are shown in Table 21.2 for clearance, transition, and interference fits. This table gives the tolerance symbols for hole basis and shaft basis fits. Where possible, fits should be taken from this table for mating parts. The tables in Appendixes 45–48 correspond to these fits. Many other fits are possible for second- and third-choice basic diameters, but these must be calculated in conjunction with tables not given in this textbook. The examples given are for preferred sizes and fits.

Preferred fits—Hole basis system Figure 21.11 illustrates the symbols that are used to show the combinations of fits that are possible using the hole basis system. There is a clearance between the two parts at A, a transitional fit at B, and an interference fit at C. This technique of representing fits is used in Fig. 21.12 to show a series of fits for a hole basis system. Note that the lower deviation of the hole is zero. In other words, the smallest size of the hole is the basic size.

The sizes of the shafts are varied to give a variety of fits from c11 to u6, where there is a maximum of interference. These fits correspond to those given in Table 21.1.

Preferred fits—Shaft basis system Figure 21.13 illustrates the preferred fits based on the shaft basis system where the largest shaft size is the basic diameter. The variation in the fit between the parts is caused by varying the size of the holes. This results in a range of fits from a clearance fit of C11/h11 to an interference fit of U7/h6.

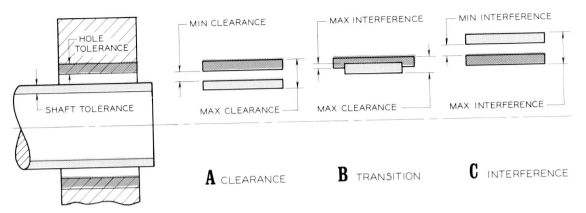

Fig. 21.11 Types of fits. (A) A clearance fit; (B) a transition fit where there can be an interference or a clearance; (C) an interference fit where the parts must be forced together.

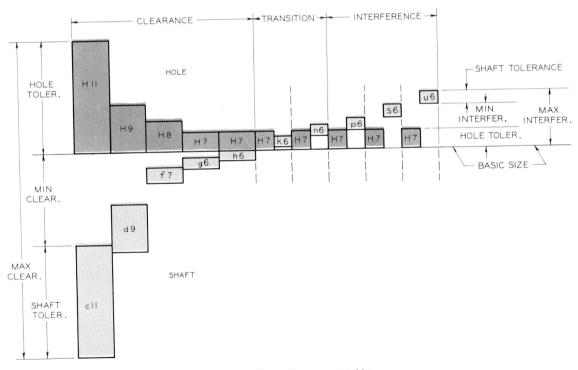

Fig. 21.12 The preferred fits for a hole basis system. These fits correspond to the fits given in Table 13.2.

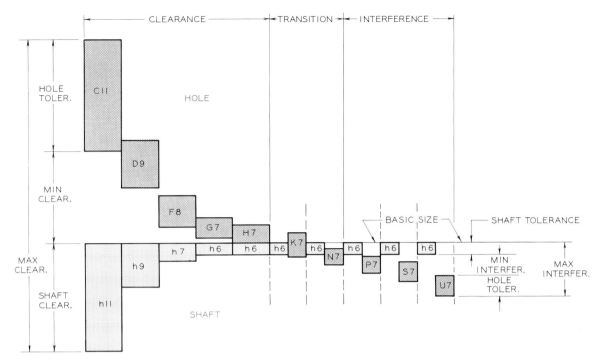

Fig. 21.13 The preferred fits for a shaft basis system. These fits correspond to the fits given in Table 13.2.

21.9 EXAMPLE PROBLEMS— METRIC SYSTEM

The following problems are given and solved as examples of determining the sizes and limits and the application of the proper symbols to mating parts. The solution of these problems requires the use of the tables in Appendix 45, the table of preferred sizes (Table 21.1), and the table of preferred fits (Table 21.2).

EXAMPLE 1: (Fig. 21.14) Given: Hole basis system, close running fit, basic diameter = 39 mm.

Solution: Use a basic diameter of 40 mm (Table 21.1) and fit of H8/f7 (Table 21.2).

Hole: Find the upper and lower limits of the hole in Appendix 45 under H8 and across from 40 mm. These limits are 40.000 and 40.039 mm.

Shaft: The upper and lower limits of the shaft are found under f7 and across from 40 mm in Appendix 45. These limits are 39.950 and 39.975 mm.

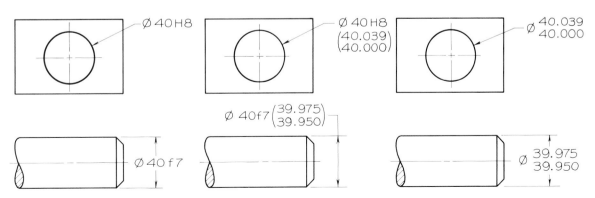

Fig. 21.14 Any of these three methods can be used to apply symbols to a detail drawing of two mating parts.

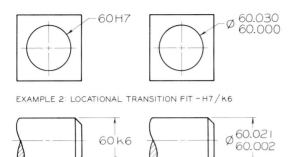

EXAMPLE 2: LOCATIONAL TRANSITION FIT –H7/k6

Fig. 21.15 Examples of tolerance symbols applied to a transitional fit. Both methods are acceptable.

Symbols: The methods of noting the drawings are shown in Fig. 21.14. Any of these methods is appropriate.

EXAMPLE 2: (Fig. 21.15) Given: Hole basis system, locational transition fit, basic diameter = 57 mm.

Solution: Use a basic diameter of 60 mm (Table 21.1) and a fit of H7/k6 (Table 21.2).

Hole: Find the upper and lower limits of the hole in Appendix 46 under H7 and across from 60 mm. These limits are 60.000 and 60.030 mm.

Shaft: The upper and lower limits of the shaft are found under k6 and across from 60 mm in Appendix 46. These limits are 60.021 and 60.002 mm.

Symbols: Two methods of applying the tolerance symbols to a drawing are shown in Fig. 21.15.

EXAMPLE 3: (Fig. 21.16) Given: Hole basis system, medium drive fit, basic diameter = 96 mm.

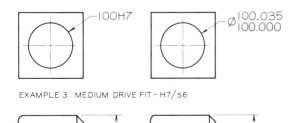

EXAMPLE 3: MEDIUM DRIVE FIT – H7/s6

Fig. 21.16 Examples of tolerance symbols applied to an interference fit. Both methods are acceptable.

Solution: Use a basic diameter of 100 mm (Table 21.1) and a fit of H7/s6 (Table 21.2).

Hole: Find the upper and lower limits of the hole in Appendix 46 under H7 and across from 100 mm. These limits are 100.035 and 100.000 mm.

Shaft: The upper and lower limits of the shaft are found under s6 and across from 100 mm in Appendix 46. These limits are found to be 100.093 and 100.071 mm. From the table, you can see that the tightest fit is an interference of 0.093 mm and the loosest fit is an interference of 0.036 mm. An interference is indicated by a minus sign in front of the numbers.

EXAMPLE 4: (Fig. 21.17) Given: Shaft basis system, loose running fit, basic diameter = 116 mm.

Solution: Use a basic diameter 120 mm (Table 21.1) and a fit of C11/h11 (Table 21.2).

Hole: Find the upper and lower limits of the hole in Appendix 47 under C11 and across from 120 mm. These limits are 120.400 and 120.180.

Shaft: The upper and lower limits of the shaft are found under h11 and across from 120 mm in Appendix 47. These limits are 119.780 and 120.000 mm.

EXAMPLE 4: LOOSE RUNNING FIT – C11/h11

Fig. 21.17 Examples of tolerance symbols applied to a clearance fit. Both methods are acceptable.

21.10 STANDARD FITS (ENGLISH UNITS)

The ANSI; B4.1-1955 standard specifies a series of fits between cylindrical parts that are based on the basic hole system in inches. The tolerances placed on the holes are unilateral and positive.

The types of fit covered in this standard are:

RC   Running and sliding fits
LC   Clearance locational fits

LT Transitional locational fits

LN Interference locational fits

FN Force and shrink fits

Each of these types of fit is listed in tables in Appendixes 39 to 43. There are several classes of fit under each of the types given above. The definition of each type of fit as given in the standards is listed below.

Running and Sliding Fits (RC) These are fits for which limits of clearance are specified to provide a similar running performance; with suitable lubrication allowance, throughout the range of sizes. The clearance for the first two classes (RC 1 and RC 2), used chiefly as slide fits, increases more slowly with diameter than that of the other classes, so that accurate location is maintained even at the expense of free relative motion.

Locational Fits (LC) These are fits intended to determine only the location of the mating parts; they may provide rigid or accurate location (interference fits), or some freedom of location (clearance fits). They are divided into three groups: clearance fits (LC), transition fits (LT), and interference fits (LN).

Force Fits (FN) These are special types of interference fits, typically characterized by maintenance of constant bore pressures throughout the range of sizes. The interference therefore varies almost directly with diameter, and the difference between its minimum and maximum value is small, to maintain the resulting pressures within reasonable limits.

Figure 21.18 illustrates how one uses the values from these tables in Appendix 39. The example is an RC 2 fit. The basic diameter for the hole and the shaft is 2.5000″, which is between the range of 1.97 and 3.15 given in the first column of the table. Since all limits are given in thousandths, the values can be converted by moving the decimal point three places to the left. For example, +0.7 is +0.0007″.

The upper and lower limits of the shaft (2.4996 and 2.4991) are found by subtracting the two limits (−0.0004 and −0.0009) from the basic diameter. The upper and lower limits of the hole (2.5007 and 2.5000) are found by adding the two limits (+0.007 and 0.000) to the basic diameter.

When the two parts are assembled, the tightest fit (+0.004) and the loosest fit (+0.0016) are

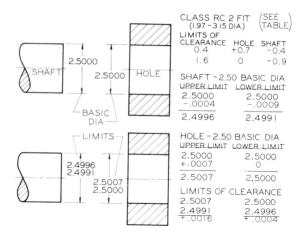

Fig. 21.18 The method of calculating the limits, allowances, and clearances for an RC 2 fit between a shaft and a hole. The basic diameter for both is 2.5000″. The limits are found in the appendix for the RC 2 fit.

found by subtracting the maximum and minimum sizes of the holes and shafts. Note that these values are provided in the second column of the table as a check on the limits.

The same method (but different tables) is used for calculating the limits for all types of fits. Plus values of clearance indicate that there is clearance, and minus values that there will be interference between the assembled parts.

These values should be converted to millimeters when using the metric system by multiplying inches by 25.4 or by using corresponding metric tables.

21.11 CHAIN DIMENSIONS

When parts are dimensioned to locate surfaces or geometric features by a chain of dimensions (Fig. 21.19A), variations may occur that exceed the tolerances specified. As successive measurements are made, with each based on the preceding one, the tolerances may accumulate as shown in Fig. 21.19A. For example, the tolerance between surface A and B is 0.002; between A and C, 0.004; between A and D, 0.006.

This accumulation of tolerances can be eliminated by measuring from a single plane called a datum plane. A datum plane is usually a plane on

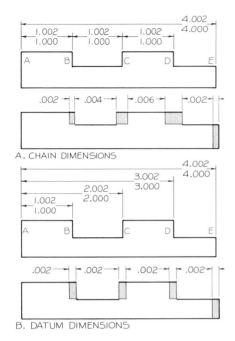

A. CHAIN DIMENSIONS

B. DATUM DIMENSIONS

Fig. 21.19 When dimensions are given as chain dimensions, the tolerances can accumulate to give a variation of 0.006″ at *D* instead of 0.002″. When dimensioned from a single datum plane, the variations of *X* and *Y* cannot deviate more than the specified 0.002″ from the datum.

the object, but it could be a plane on the machine used to make the part.

Note that the tolerances between the intermediate planes in Fig. 21.19B are uniform since each of the planes was located with respect to a single datum. In our example, this is 0.002, which represents the maximum tolerance.

21.12 TOLERANCE NOTES

All dimensions on a drawing are toleranced either by the previously covered rules or by a general note on a drawing placed in or near the title block. For example, the note, TOLERANCE ± 1/64 (or its decimal equivalent, 0.40 mm) might be given on a drawing for the less critical dimensions. These tolerances are for surfaces that do not come into contact with other surfaces.

Some industries may give dimensions in inches where decimals are carried out to two places, three places, and four places. A note might be given on the drawing in this manner: TOLER-

ANCES XX.XX ± 0.10; XX.XXX ± 0.005. Tolerances of four places would be given directly on the dimension lines, but those dimensions with two and three decimal places would have the tolerances indicated in the note.

The most common method of noting tolerances is to use as large a tolerance as feasible expressed in a note such as, TOLERANCES ± 0.05 (± 1 mm when using metrics), and give the tolerances for the mating dimensions that require smaller tolerances on the dimension lines.

Angular tolerances should be given in a general note in or near the title block such as: ANGULAR TOLERANCES ± 0.5° or ± 30′. When angular tolerances less than this are specified, they should be given on the drawing where these angles are dimensioned. Techniques of tolerancing angles are shown in Fig. 21.20.

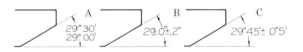

Fig. 21.20 Angles can be toleranced by any of these methods using limits or the plus-and-minus method.

21.13 GENERAL TOLERANCES— METRIC

All dimensions on a drawing are understood to have tolerances, and the amount of tolerance must be noted. This section is based on the metric system with the millimeter as the unit of measurement as outlined in ANSI B4.3, 1978, which is the standard governing the method of noting tolerances.

Tolerances may be specified by (1) applying tolerances directly to the dimensions, (2) giving tolerances in specification documents that are associated with the drawings, or (3) applying a general note on the drawing.

Linear dimensions may be toleranced by indicating ± one-half of an International Tolerance (IT) Grade as given in Appendix 44 (taken from ANSI B4.2-1978). The appropriate IT grade can be selected from the graph in Fig. 21.21 when the nature of the application is known.

You can see that IT grades for mass-produced items range from IT12 through IT16. When the

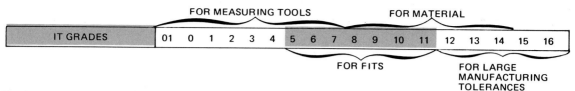

Fig. 21.21 The International Tolerance grades and their applications.

machining process is known, the IT grades can be selected from Table 21.3. Machining operations result in smaller tolerances since they are done with more precision than in mass production.

General tolerances using IT grades may be expressed in a note as follows:

UNLESS OTHERWISE SPECIFIED ALL

UNTOLERANCED DIMENSIONS ARE $\pm \dfrac{\text{IT14}}{2}$.

This means that tolerance of $\pm$ 0.700 mm is allowed for a dimension between 315 and 400 mm. The value of the tolerance is listed in Appendix 44 as 1.400.

Table 21.4 shows recommended tolerances for fine, medium, and coarse series for dimensions of graduated sizes. A medium tolerance, for example, can be specified by the following note:

GENERAL TOLERANCE SPECIFIED IN
ANSI B4.3 MEDIUM SERIES APPLY.

This same information can be given on the drawing in table form by selecting the grade—medium in this example—from Table 21.4 and presenting it on the drawing as shown in Fig. 21.22.

General tolerances can be expressed in a table of values that gives the tolerances for dimensions with one or no decimal places. An example of this table is shown in Fig. 21.23.

Table 21.3 IT GRADES AND THEIR RELATIONSHIP TO MACHINING PROCESSES

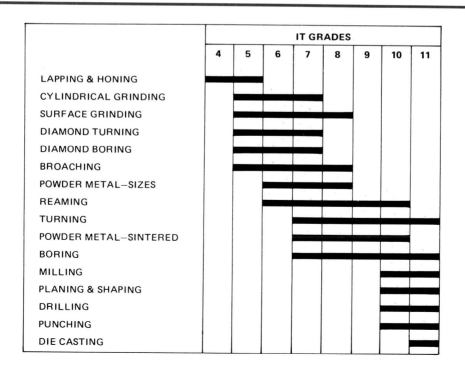

Table 21.4 FINE, MEDIUM, AND COARSE SERIES: GENERAL TOLERANCE—LINEAR DIMENSIONS

		Variations in mm						
Basic dimensions mm		0.5 to 3	over 3 to 6	over 6 to 30	over 30 to 120	over 120 to 315	over 315 to 1,000	over 1,000 to 2,000
Permissible variations	Fine series	±0.05	±0.05	±0.1	±0.15	±0.2	±0.3	±0.5
	Medium series	±0.1	±0.1	±0.2	±0.3	±0.5	±0.8	±1.2
	Coarse series		±0.2	±0.5	±0.8	±1.2	±2	±3

Dimensions in mm

General Tolerance
Unless Otherwise Specified The Following
Tolerances are Applicable

Linear	Over to	0.5 6	6 30	30 120	120 315	315 1000	1000 2000
Tol	±	0.1	0.2	0.3	0.5	0.8	1.2

Fig. 21.22 This table is a medium series of values taken from Table 21.4. It is placed on the working drawing to provide the tolerances for various ranges of sizes.

Dimensions in mm

General Tolerance
Unless Otherwise Specified The Following
Tolerances Are Applicable

Linear		Over To	— 120	120 315	315 1000	1000 —
Tol	One Decimal ±		0.3	0.5	0.8	1.2
	No Decimals ±		0.8	1.2	2	3

Fig. 21.23 This table of tolerances can be placed on a drawing to indicate the tolerances for dimensions with one decimal and those with no decimals.

Length of the shorter leg mm	Up to 10	Over 10 to 50	Over 50 to 120	Over 120 to 400
	±1°	±0° 30′	±0° 20′	±0° 10′

Fig. 21.24 This table can be placed on a drawing to indicate the general tolerances for angles. These values were extracted from Table 21.5.

Another method of giving general tolerances is a note of this form:

UNLESS OTHERWISE SPECIFIED ALL UNTOL-
ERANCED DIMENSIONS ARE ± 0.8 mm.

This method should be used only where the dimensions on a drawing have slight differences in magnitude.

Angular tolerances are expressed (1) as an angle in decimal degrees or in degrees and minutes, (2) as a taper expressed in percentage (number of millimeters per 100 mm), or (3) as milliradians. A milliradian is found by multiplying the angle in degrees by 17.45. The suggested tolerances for each of these units are shown in Table 21.5. Note that the angular tolerances are based on the length of the shorter leg of the angle.

General angular tolerances may be given on the drawing with a note in the following form:

UNLESS OTHERWISE SPECIFIED THE GENERAL
TOLERANCES IN ANSI B4.3 APPLY.

A second method shows the portion of Table 21.5 on the drawing using the units desired, as shown in Fig. 21.24.

A third method is a note with a single tolerance such as:

UNLESS OTHERWISE SPECIFIED THE GENERAL
ANGULAR TOLERANCES ARE ±0° 30′ (or ±0.5°).

21.14 TOLERANCES OF POSITION AND FORM

Two systems of improving the specification of tolerances of the position and form of geometric features are known as **true-position tolerancing** and

Table 21.5 GENERAL TOLERANCE—ANGLES AND TAPERS

	Length of the shorter leg mm	up to 10	Over 10 to 50	Over 50 to 120	Over 120 to 400
Permissible Variations	in degrees and minutes	±1°	±0°30'	±0°20'	±0°10'
	in millimeters per 100 mm	±1.8	±0.9	±0.6	±0.3
	in milliradians	±18	±9	±6	±3

geometric tolerancing. These systems are standardized by the ANSI Y14.5 Standards and the Military Standards (Mil-Std) of the U.S. Department of Defense.

True-position tolerancing is the method of locating geometric features (particularly holes and slots) with other features on a part. Geometric tolerancing deals with the form of the geometric characteristics of various features. For example, flatness, straightness, roundness, and parallelism of various features can be specified.

These systems have application in the area of computer-aided machining and manufacturing.

21.15 SYMBOLS OF POSITIONAL AND FORM TOLERANCES

The symbols of position and form are given in Fig. 21.25. These symbols are used as feature control symbols to give the desired specifications of a part (Fig. 21.26). The feature control symbols are boxes that are sized as shown. The lettering and numbers in the boxes are ⅛″ letters. The usage and meaning of these symbols will be explained in the following articles.

Other symbols not shown in Fig. 21.25 are M for "Maximum Material Condition," (MMC) and S for "Regardless of Feature Size" (RFS). A feature is at maximum material condition when it contains the maximum amount of material.

Thus a hole is at MMC when it is at its smallest size permitted by its tolerance. On the other hand, a shaft is at MMC when it is as large as permitted by its tolerance limits.

Two mating parts will have the tighest fit when both are at MMC.

In some cases, tolerances are specified to apply to parts at MMC, and in other cases RFS. Examples of these will be discussed later in this chapter.

In general the following rules will apply where the drawings are not noted otherwise:

1. Tolerances of form apply at RFS (regardless of feature size).
2. Tolerances of position apply at MMC (maximum material condition).

21.16 TOLERANCES OF POSITION

The traditional method of locating holes has been the coordinate method, which results in a square tolerance zone as shown in Fig. 21.27. The diagonal of this square is greater than the tolerances specified on each locational dimension.

In this example, each coordinate was given a tolerance of 0.01, which results in a diagonal variation of 1.4 × 0.01 or 0.014 instead of 0.01. This characteristic error can be overcome by using a circular tolerance zone of true-position tolerancing.

True-Position Tolerancing

True-position tolerancing is compared with coordinate dimensioning in Fig. 21.28. The coordinates are dimensioned with numerals inside a box. This is the symbolic method of indicating that the dimensions are *basic dimensions* without tolerances.

		CHARACTERISTIC	SYMBOL	NOTES
INDIVIDUAL FEATURES	FORM TOLERANCES	STRAIGHTNESS	—	1
		FLATNESS	▱	1
		ROUNDNESS (CIRCULARITY)	○	
		CYLINDRICITY	⌭	
INDIVIDUAL OR RELATED FEATURES		PROFILE OF A LINE	⌒	2
		PROFILE OF A SURFACE	⌓	2
RELATED FEATURES		ANGULARITY	∠	
		PERPENDICULARITY (SQUARENESS)	⊥	
		PARALLELISM	//	3
	LOCATION TOLERANCES	POSITION	⌖	
		CONCENTRICITY	◎	3,7
		SYMMETRY	≡	5
	RUNOUT TOLERANCES	CIRCULAR	↗	4
		TOTAL	↗	4,6

Note: 1) The symbol ⌒ formerly denoted flatness.

The symbol ⌒ or — formerly denoted flatness and straightness.

2) Considered "related" features where datums are specified.

3) The symbol || and ◉ formerly denoted parallelism and concentricity, respectively.

4) The symbol ↗ without the qualifier "CIRCULAR" formerly denoted total runout.

5) Where symmetry applies, it is preferred that the position symbol be used.

6) "TOTAL" must be specified under the feature control symbol.

7) Consider the use of position or runout.

Where existing drawings using the above former symbols are continued in use, each former symbol denotes that geometric characteristic which is applicable to the specific type of feature shown.

Fig. 21.25 These symbols of position and form are used in feature control symbols. (Courtesy of ANSI; Y14.5.)

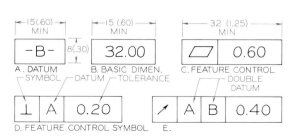

Fig. 21.26 Examples of symbols used to indicate datum planes, basic dimensions, and feature control symbols.

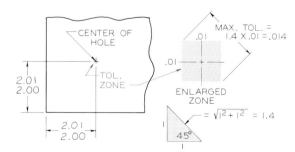

Fig. 21.27 The coordinate method results in a square tolerance zone. The diagonal of the zone exceeds the specified tolerance by a factor of 1.4.

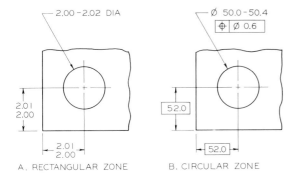

A. RECTANGULAR ZONE   B. CIRCULAR ZONE

Fig. 21.28 The method at A gives a rectangular tolerance zone for the center of the hole. The true-position method at B locates the center of the hole at true position and gives a circular tolerance zone of 0.6 mm DIA (shown in the feature control symbol).

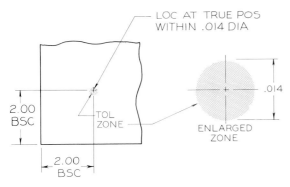

Fig. 21.29 The true-position method locates the center of the hole within a circular tolerance zone. When this method is used, the tolerance can be 1.4 larger than the same tolerance used in the coordinate method and still be as accurate.

The note specifies the tolerance in the diameter of the hole being located, and the feature control symbol indicates that the center is located within 0.014 DIA of the true position indicated by the basic dimensions. In other words, its center point must lie within a circular tolerance zone of 0.014. This ensures that the tolerance of position is the same in all directions from the theoretical center.

If the traditional coordinate method could accept a variation of 0.014 across the diagonal of the square tolerance zone, then the true-position tolerance should be acceptable with a circular zone of 0.014, which is a greater tolerance than the square zone permitted (Fig. 21.29).

True-position tolerances can be applied by notes as shown in Fig. 21.30A, or by symbol in part B. The symbol technique is preferred and is used throughout this chapter.

The circular tolerance zone specified in the circular view of a hole is assumed to extend the full depth of the hole. Therefore, the tolerance zone for the centerline of the hole is a cylindrical zone inside of which the axis must lie. The size of the hole and its position are both toleranced; consequently, these two tolerances are used to establish the diameter of a gage cylinder that can be used to check for the conformance of the dimensions to the specifications (Fig. 21.31).

The circle is found by subtracting the true position tolerance from the hole at maximum material condition (the smallest permissible hole). This zone represents the least favorable condition when the part is gaged or assembled with a mating part. When the hole is not at MMC, it is larger and permits a greater tolerance and easier assembly.

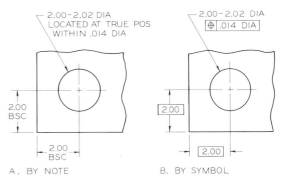

A. BY NOTE   B. BY SYMBOL

Fig. 21.30 True-position dimensioning can be shown by note (A) or by symbol (B). The symbol method is preferred.

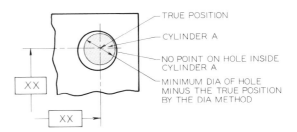

Fig. 21.31 When a hole is located at true position at MMC, no element of the hole shall be inside of the imaginary cylinder, cylinder A.

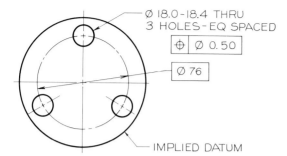

Fig. 21.32 Holes in a circular plate may be located at true position about the 76 mm DIA circle and within tolerance zones of 0.50 mm DIA.

An example of true-positioning applied to the location of equally spaced holes in a circular plate is shown in Fig. 21.32. The circle of centers (bolt circle) is the basic dimension with the centers of the holes located within 0.50 mm DIA.

Gaging a Two Hole Pattern

Gaging is a technique of checking dimensions to determine whether or not they have met the specifications of tolerance. In this case, the effects of the size and positional tolerances can be seen when two holes are dimensioned as shown in Fig. 21.33.

The two holes are positioned 26.00 mm apart with a basic dimension. The holes have limits of 12.70 and 12.84 for a tolerance of 0.14. They are located at true position within a diameter of 0.18.

The gage pin diameter is calculated to be 12.52 mm (the smallest hole's size minus the true-position tolerance) as illustrated in part B. This means that two pins with diameters of 12.52 mm that are spaced exactly 26.00 mm apart could be used to check the diameters and positions of the holes at MMC, the most critical size. If the pins assembled in the holes, then the holes are properly sized and located.

When the holes are not at MMC (when they are larger than their minimum size), these gage pins will permit a greater range of variation (Fig. 21.34). When the holes are at their maximum size of 12.84 mm, they can be located as close as 25.68 from center to center or as far apart as 26.32 center to center. When not specified, true-position tolerances are assumed to apply at MMC.

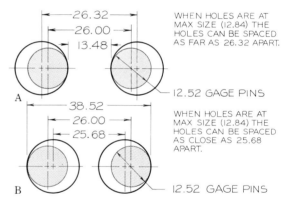

Fig. 21.34 When the two holes are at their maximum size, the centers of the holes can be spaced as far as 26.32 mm apart and still be acceptable. (A) The holes can be placed as close as 25.68 apart when the holes are at maximum size (B).

Concentricity

Surfaces of revolution are concentric when they have a common axis. Concentricity is specified in Fig. 21.35; it is used to position coaxial features about a common axis of rotation. Unless there is definite need for the control of axes, this should be specified as a runout instead of concentricity.

Note that the larger diameter is "flagged" as datum A. This means that this diameter is used as the datum for measuring the variation of the smaller cylinder's axis.

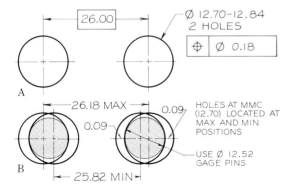

Fig. 21.33 When two holes are located at true position and they are at MMC, they may be gaged with pins 12.52 mm in diameter that are located 26.00 mm apart.

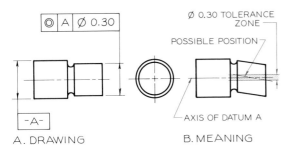

A. DRAWING   B. MEANING

Fig. 21.35 Concentricity is a tolerance of position. This relates the diameter of one cylinder with another within a 0.30 DIA.

Symmetry

A part or a feature is symmetrical when it has the same contour and size on opposite sides of a central plane. A symmetry tolerance positions features with respect to a datum plane.

The method of noting symmetry is shown in Fig. 21.36. Datum plane B is used to establish the position of the symmetry of the notch. The feature control symbol notes that the notch is symmetrical about datum B within a zone of 0.40 mm.

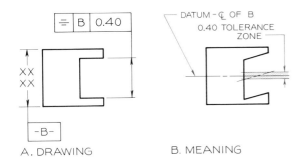

A. DRAWING   B. MEANING

Fig. 21.36 Symmetry is a tolerance of position that specifies that an object's feature is symmetrical about the center line of another feature on the same object.

21.17 TOLERANCES OF FORM

Flatness

A surface is flat when all its elements are in one plane. A feature control symbol is used to specify flatness within a 0.40 zone RFS in Fig. 21.37. No point on the surface may vary more than 0.40 from the highest to the lowest point on the surface.

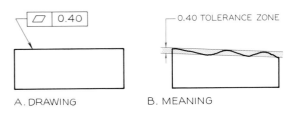

A. DRAWING   B. MEANING

Fig. 21.37 Flatness is a tolerance of form that specifies two parallel planes inside of which the object's surface must lie.

Straightness

A surface is straight if all its elements are straight lines. A feature control symbol is used to specify straightness of a cylinder in Fig. 21.38. A total of 0.12 mm RFS is permitted as the elements are gaged in a vertical plane parallel to the axis of the cylinder.

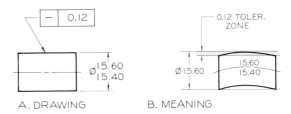

A. DRAWING   B. MEANING

Fig. 21.38 Straightness is a tolerance of form that indicates that the elements of a surface are straight lines. The symbol must be applied where the elements appear as straight lines.

Roundness

A surface of revolution (a cylinder, cone, or sphere) is round when all points on the surface intersected by a plane are equidistant from the axis. A feature control symbol is used to specify roundness of a cone and cylinder in Fig. 21.39. This symbol permits a tolerance of 0.34 mm RFS on the radius of each part.

The roundness of a sphere is specified in Fig. 21.40.

Cylindricity

A surface of revolution is cylindrical when all its elements form a cylinder. A cylindricity tolerance zone is specified in Fig. 21.41, where a tolerance of 0.54 mm RFS is permitted on the radius of the cylinder. Cylindricity is a combination of tolerances of roundness and straightness applied to a cylindrical object.

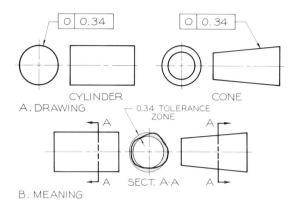

Fig. 21.39 Roundness is a tolerance of form that indicates that a cross section through a surface of revolution is round and lies within two concentric circles.

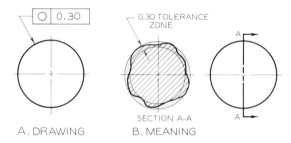

Fig. 21.40 Roundness of a sphere is indicated in this manner, which means that any cross section through it is round within the specified tolerance in the symbol.

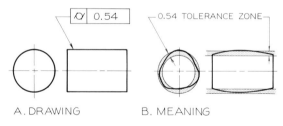

Fig. 21.41 Cylindricity is a tolerance of form that indicates that the surface of a cylinder lies within an envelope formed by two concentric cylinders.

Profile

Profile tolerancing is used to specify tolerances about a contoured shape formed by arcs or by irregular curves. Profile can apply to a single line or a surface.

The surface in Fig. 21.42 is given a profile tolerance that is unilateral (it can only be

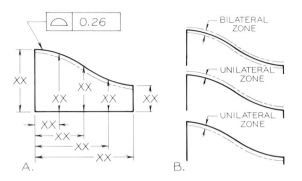

Fig. 21.42 Profile is a tolerance of form that is used to tolerance irregular curves of planes. The curving plane is located by coordinates and the tolerance is located by any of the methods shown at B.

smaller than the points located). Examples of specifying bilateral and unilateral tolerance zones are shown at B.

A profile tolerance for a single line can be specified as shown in Fig. 21.43. In this example, the curve is formed by tangent arcs whose radii are given as basic dimensions. The radii are permitted to vary by plus or minus 0.10 mm about the basic radii.

Parallelism

A surface or a line is parallel when all its points are equidistant from a datum plane or axis. Two

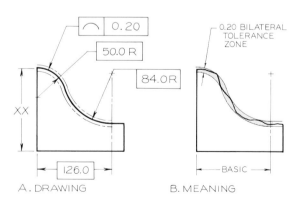

Fig. 21.43 Profile of a line is a tolerance of form that specifies the variation allowed from the path of a line. In this case the line is formed by tangent arcs. The tolerance zone may be either bilateral or unilateral as shown as Fig. 21.42B.

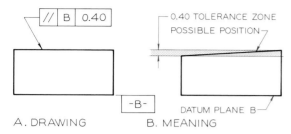

A. DRAWING   B. MEANING

Fig. 21.44 Parallelism is a tolerance of form that specifies that a plane is parallel to another within specified limits. Plane B is the datum plane in this case.

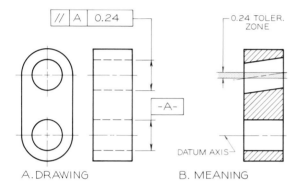

A. DRAWING   B. MEANING

Fig. 21.45 Parallelism of one center line to another can be specified by using the diameter of one of the holes as the datum.

types of parallelism are:

1. A tolerance zone between planes parallel to a datum plane within which the axis or surface of the feature must lie (Fig. 21.44). This tolerance also controls flatness.

2. A cylindrical tolerance zone parallel to a datum feature within which the axis of a feature must lie (Fig. 21.45).

The effect of specifying parallelism at MMC can be seen in Fig. 21.46 where the modifier M is given in the feature control symbol. Tolerances of form apply regardless of feature size when not specified. By specifying parallelism at MMC, this means that the axis of the cylindrical hole must vary no more than 0.20 mm when the holes are at the smallest permissible size.

As the hole approaches its upper limit of 30.30, this tolerance zone increases until it reaches 0.50 DIA. Therefore, a greater variation is given at MMC than at RFS.

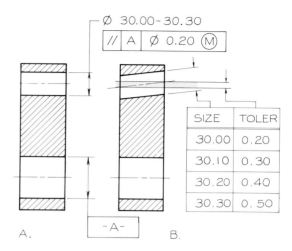

A.   B.

Fig. 21.46 The most critical tolerance will exist when features are at MMC. In this example, the upper hole must be parallel to the hole used as datum A within 0.20 DIA. As the hole approaches its maximum size of 30.30 mm, the tolerance zone approaches 0.50 mm.

Perpendicularity

Surfaces, axes, or lines that are at right angles to each other are perpendicular to each other. The perpendicularity of two planes is specified in Fig. 21.47. Note that datum plane C is "flagged," and the feature control symbol is applied to the perpendicular surface.

A hole is specified as perpendicular to a surface in Fig. 21.48, where surface A is indicated as the datum plane.

Angularity

A surface or line is angular when it is at a specified angle (other than 90°) from a datum or axis.

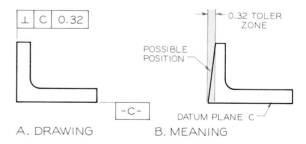

A. DRAWING   B. MEANING

Fig. 21.47 Perpendicularity is a tolerance of form that gives a tolerance zone for a plane that is perpendicular to a specified datum plane.

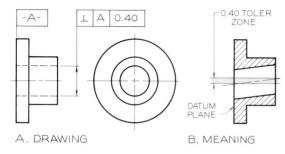

Fig. 21.48 Perpendicularity can apply to the axis of a feature, such as the center line of a cylinder.

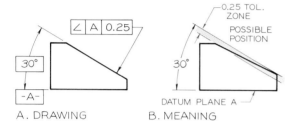

Fig. 21.49 Angularity is a tolerance of form that specifies the tolerance for an angular surface with respect to a datum plane. The 30° angle is a true angle, a basic angle. The tolerance of 0.25 mm is from this basic angle.

The angularity of a surface is specified in Fig. 21.49 where the angle is given a basic dimension of 30°. The angle is permitted to vary within a tolerance zone of 0.25 mm about the basic angle.

Runout

Runout tolerance is a means of controlling the functional relationship of two or more features of a part within the allowable errors of concentricity, perpendicularity, and alignment features. It also takes into account variations in roundness, straightness, flatness, and parallelism of individual surfaces. In essence, it establishes a composite form of control of those features having a common axis.

An example of this tolerance applied to a part is shown in Fig. 21.50. This shows the method of noting the large cylinder and the conical feature, and the meaning of these feature control symbols. The indicator is applied to the part's feature and the part is revolved about its axis.

Two surfaces on the part in Fig. 21.51 are noted with feature control symbols that indicate a

runout tolerance. Two datum surfaces, C and D, are indicated for this part. This means that the part is to be mounted on surface C and rotated about datum axis D to gage the tolerances.

The word TOTAL is placed beneath the feature control symbol to indicate that the variation of runout cannot exceed the tolerance specified in the feature control symbol. This gives a composite control of all surface elements such as roundness, straightness, concentricity, angularity, taper, and surface profile. The tolerance is applied simulta-

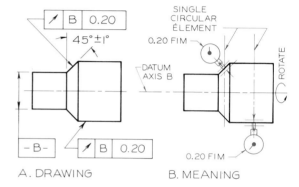

Fig. 21.50 Runout tolerance of a surface is a tolerance of form that is a composite of several form characteristics. It is used to specify concentric cylindrical parts. The part is mounted on one of the axes, the datum axis, and the part is gaged as it is rotated about the datum axis.

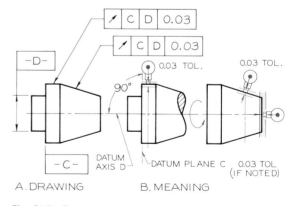

Fig. 21.51 The runout tolerance in this example is measured by mounting the object on the primary datum plane, surface C, and the secondary datum plane, cylinder D. The cylinder and conical surface is gaged to determine if it conforms to a tolerance zone of 0.03 mm. The end of the cone could have been noted to specify its runout (perpendicularity to the axis).

neously to all circular and profile measuring positions as the part is revolved 360° about its axis.

When the word TOTAL is not used with the note, this implies that the tolerance applies only to single circular elements as the cylindrical part is rotated.

21.18 THE THREE DATUM PLANE CONCEPT

In the previous sections, we have discussed form and position tolerances with respect to a single plane, but more accuracy is achieved if the part is positioned with respect to three mutually perpendicular datum planes.

The planes that are used as datums are usually not on the part but are on the manufacturing or inspection equipment. The priority of the three planes is selected by the designer based on knowledge of the desired function of the part.

For example, the part in Fig. 21.52 has its horizontal base specified as the primary datum. This means that the plane on the base is established by the three highest points on it that contact the datum plane. The part is further related to the secondary datum plane by the two highest points on the second datum feature. The third datum feature is brought into contact with the third datum plane, which contacts the highest point on the datum feature.

The priority of these datum planes is noted on the drawing of the part by feature control symbols as shown in Fig. 21.53. Feature control symbol ex-

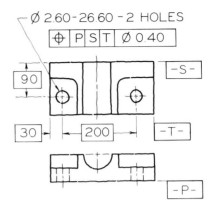

Fig. 21.53 The sequence of the three-plane reference system (shown in Fig. 21.52) is labeled where the planes appear as edges. Note that the primary datum plane, P, is listed first in the feature control symbol, the secondary plane, S, next, and the third plane, T, last.

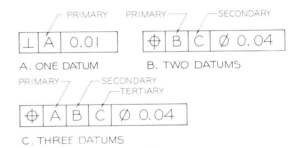

Fig. 21.54 Feature control symbols may have from one to three datum planes indicated. These are listed in order of priority.

amples are given in Fig. 21.54, where the primary, secondary, and tertiary (third) datum planes are listed in order in the symbol.

Cylindrical datum features vary from flat datum planes since they must be located with respect to their axes. The axis of a cylinder is established by two datum planes that intersect at the center of the cylinder at right angles. This is usually represented by the centerlines in the circular view. Thus, one cylindrical datum feature is associated with two datum planes (Fig. 21.55). Consequently, in the feature control symbol, only two datum features would be specified: (1) primary datum, K, a flat surface, and (2) secondary datum feature, M, a cylindrical feature associated with the second and third datum planes.

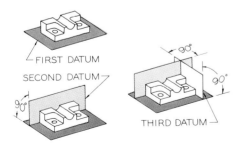

Fig. 21.52 When a part is referenced to a primary datum plane, the object contacts the plane with its highest three points. The vertical surface contacts the secondary vertical datum plane with two points. The third datum plane is contacted by one point of the plane on the object. The datum planes are listed in this order in the feature control symbol.

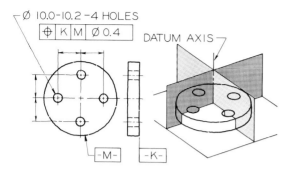

Fig. 21.55 These true-position holes are located with respect to datum *K* and datum *M*. Since datum *M* is a circle, this implies that the holes are located about two intersecting datum planes formed by the crossing centerlines in the circular view. This satisfies the three-plane concept.

The sequence of the datum reference in the feature control symbol is significant to the manufacturing and inspection process. The part in Fig. 21.56 is dimensioned with an incomplete feature control symbol; it does not specify the primary and secondary datum planes. The schematic drawing at (b) illustrates the effect of specifying diameter A as the primary and surface B as the secondary datum plane. This means that the part is centered about cylinder A by mounting the part in a chuck, mandrel, or centering device on the processing equipment, which centers the part at RFS (regardless of feature size). Surface B is assembled to contact at least one point of the third datum plane.

If surface B were specified as the primary datum feature, it would be assembled to contact da-

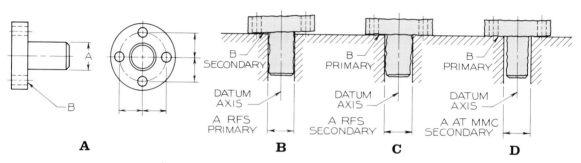

Fig. 21.56 Three examples are shown to illustrate the effects of selection of the datum planes in order of priority and the effect of RFS and MMC.

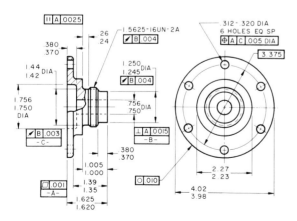

Fig. 21.57 A fully dimensioned part utilizing geometric tolerancing.

tum plane B in at least three points (Fig. 21.57). The axis of datum feature A will be gaged by the smallest true cylinder that is perpendicular to the first datum that will contact surface A and RFS.

In Fig. 21.56(d), plane B is specified as the primary datum feature, and cylinder A is specified as the secondary datum feature at MMC (maximum material condition). The part is mounted on the processing equipment where at least three points on feature B are in contact with datum B. The second and third planes intersect at the datum axis to complete the three-plane relationship. This utilization of the modifier to specify MMC gives a more liberal tolerance zone than would be otherwise acceptable when RFS was specified.

The dimensional part in Fig. 21.57 is an example of a part that makes use of the three-plane system.

21.19 SURFACE TEXTURE

The surface texture of a part will have a bearing on its function; consequently, this must be more precisely specified than by the general V that does not elaborate on the finish desired. The following terms of surface texture are defined below and in Fig. 21.58.

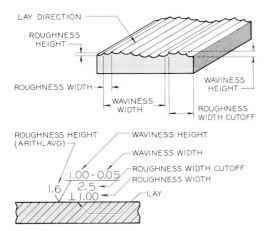

Fig. 21.58 The definition of terms identified with surface texture.

Surface texture   is the variation in the surface and it includes roughness, waviness, lay, and flaws.

Roughness   describes the finest of the irregularities in the surface. These are usually caused by the manufacturing process used to smooth the surface.

Roughness height   is the average deviation from the mean plane of the surface. It is measured in microinches (μin.) or micrometers (μm) which are millionths of an inch or meter.

Roughness width   is the width between successive peaks and valleys that form the roughness. This is measured in microinches or micrometers.

Roughness width cutoff   is the largest spacing of repetitive irregularities that includes average roughness height (measured in inches or millmeters). When not specified, a value of 0.8 mm (0.030 in.) is assumed.

Waviness   is a widely spaced variation that exceeds the roughness width cutoff. Roughness may

be considered as superimposed on a wavy surface. Waviness is measured in inches or millimeters.

Waviness height   is the peak-to-valley distance between waves. It is measured in millimeters or inches.

Waviness width   is the spacing between peaks or wave valleys measured in inches or millimeters.

Lay   is the direction of the surface pattern. This is determined by the production method that is used.

Flaws   are irregularities or defects that occur infrequently or at widely varying intervals on a surface. These include cracks, blow holes, checks, ridges, scratches, etc. Unless otherwise specified, the effect of flaws shall not be included in roughness height measurements.

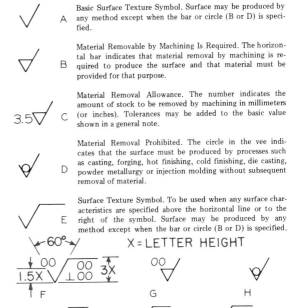

Fig. 21.59 Surface control symbols for specifying surface finish. General notes (I, J, and K) can be used to specify the recommended machining operation for finishing a surface.

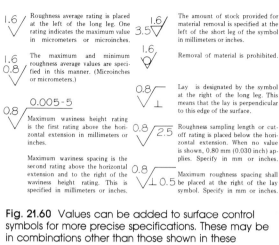

Fig. 21.60 Values can be added to surface control symbols for more precise specifications. These may be in combinations other than those shown in these examples.

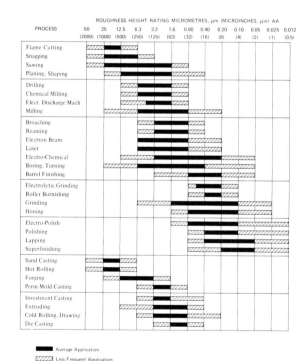

■ Average Application

▨ Less Frequent Application

The ranges shown above are typical of the processes listed.
Higher or lower values may be obtained under special conditions.

Fig. 21.61 The surface roughness heights produced by various types of production methods are shown here in micrometers (microinches). (Courtesy of the General Motors Corporation.)

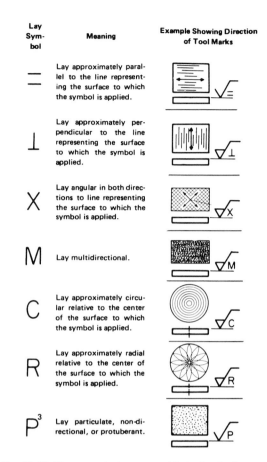

Fig. 21.62 These symbols are used to indicate the direction of lay with respect to the surface where the control symbol is placed. (Courtesy of ANSI; B46.1.)

Contact area is the surface that will make contact with its mating surface.

The symbols used to specify surface texture are given in Fig. 21.59. The point of the V must contact the edge view of the surface being specified, an extension line from the surface, or a leader that points to the surface. The top horizontal bar can be extended as far to the right as necessary.

In Fig. 21.60, values of surface texture that can be applied to surface texture symbols, individually or in combination, are given. The roughness height values are related to manufacturing processes used to finish the surface (Fig. 21.61).

Lay symbols that indicate the direction of texture of a surface are given in Fig. 21.62. These symbols can be incorporated into surface texture symbols as shown in Fig. 21.63. An example of a part with a variety of surface texture symbols applied to it is shown in Fig. 21.64.

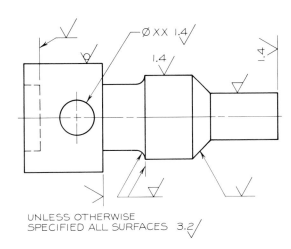

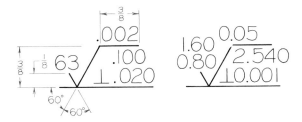

Fig. 21.63 Examples of fully specified surface control symbols.

UNLESS OTHERWISE
SPECIFIED ALL SURFACES 3.2

Fig. 21.64 This drawing illustrates the techniques of applying surface texture symbols to a part.

PROBLEMS

These problems can be solved on Size A sheets. The problems are laid out on a grid of 0.20 inches (5 mm).

Cylindrical Fits

1. Construct the drawing of the shaft and hole as shown in Fig. 21.65 (it need not be drawn to scale), give the limits for each diameter, and complete the table of values. Use a basic diameter of 1.00 inches (25 mm) and a class RC 1 fit, or a corresponding metric fit.

2. Same as Problem 1, but use a basic diameter of 1.75 inches (45 mm) and a class RC 9 fit, or a corresponding metric fit.

3. Same as Problem 1, but use a basic diameter of 2.00 inches (51 mm) and a class RC 5 fit, or a corresponding metric fit.

4. Same as Problem 1, but use a basic diameter of 12.00 inches (305 mm) and a class LC 11 fit, or a corresponding metric fit.

5. Same as Problem 1, but use a basic diameter of 3.00 inches (76 mm) and a class LC 1 fit, or a corresponding metric fit.

6. Same as Problem 1, but use a basic diameter of 8.00 inches (203 mm) and a class LC 1 fit, or a corresponding metric fit.

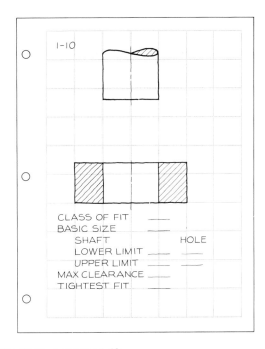

Fig. 21.65 Problems 1–10.

7. Same as Problem 1, but use a basic diameter of 102 inches (2 591 mm) and a class LN 3 fit, or a corresponding metric fit.

8. Same as Problem 1, but use a basic diameter of 11.00 inches (279 mm) and a class LN 2 fit, or a corresponding metric fit.

9. Same as Problem 1, but use a basic diameter of 6.00 inches (152 mm) and a class FN 5 fit, or a corresponding metric fit.

10. Same as Problem 1, but use a basic diameter of 2.60 inches (66 mm) and a class FN 1 fit, or a corresponding metric fit.

Tolerances of Position

11. Make an instrument drawing on Size A paper of the part shown in Fig. 21.66. Locate the two holes with a size tolerance of 1.00 mm and a true-position tolerance of 0.50 DIA. Show the proper symbols and dimensions for this arrangement.

12. Same as Problem 11, except locate three holes using the same tolerances for size and position.

13. Give the specifications for a two-pin gage that can be used to gage the correctness of the two holes specified in Problem 11. Make a sketch of the gage and show the proper dimensions on it.

14. Using true positioning, locate the holes and properly note them to provide a size tolerance of 1.50 mm and a locational tolerance of 0.60 DIA (Fig. 21.67).

15. Same as Problem 14, except locate six equally spaced holes of the same size using the same tolerances of position.

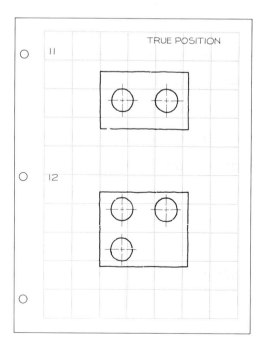

Fig. 21.66 Problems 11–13.

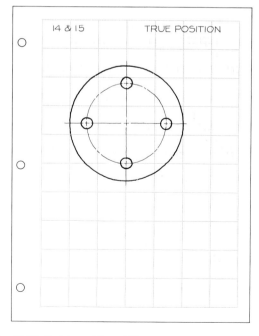

Fig. 21.67 Problems 14 and 15.

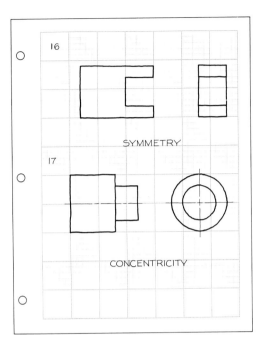

Fig. 21.68 Problems 16 and 17.

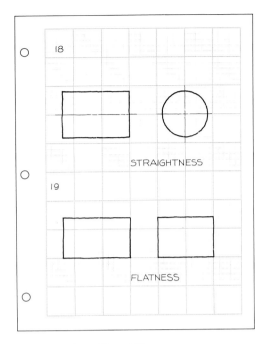

Fig. 21.69 Problems 18 and 19.

16. Using a feature control symbol and the necessary dimensions, indicate that the notch is symmetrical to the left-hand end of the part in Fig. 21.68 within 0.60 mm.

17. Using a feature control symbol and the necessary dimensions, indicate that the small cylinder is concentric with the large one (the datum cylinder) within a tolerance of 0.80 (Fig. 21.68).

18. Using a feature control symbol and the necessary dimensions, indicate that the elements of the cylinder are straight within a tolerance of 0.20 mm (Fig. 21.69).

19. Using a feature control symbol and the necessary dimensions, indicate that the upper surface of the object is flat within a tolerance of 0.08 mm (Fig. 21.69).

20–22. Using feature control symbols and the necessary dimensions, indicate that the cross sections of the cylinder, cone, and sphere are round within a tolerance of 0.40 mm (Fig. 21.70).

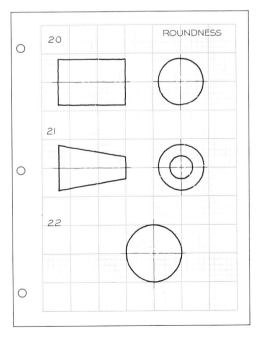

Fig. 21.70 Problems 20–22.

23. Using a feature control symbol and the necessary dimensions, indicate that the profile of the irregular surface of the object lies within a tolerance zone 0.40 mm, either bilateral or unilateral (Fig. 21.71).

24. Using a feature control symbol and the necessary dimensions, indicate that the profile of the line formed by tangent arcs has a profile that lies within a tolerance zone of 0.40 mm, either bilateral or unilateral (Fig. 21.71).

25. Using a feature control symbol and the necessary dimensions, indicate that the cylindricity of the cylinder is 0.90 mm (Fig. 21.72).

26. Using a feature control symbol and the necessary dimensions, indicate that the angularity tolerance of the inclined plane is 0.7 mm from the bottom of the object, the datum plane (Fig. 21.72).

27. Using a feature control symbol and the necessary dimensions, indicate that the upper surface of the object is parallel to the lower surface, the datum, within 0.30 mm (Fig. 21.73).

28. Using a feature control symbol and the necessary dimensions, indicate that the small hole is parallel to the large hole, the datum, within a tolerance of 0.80 mm (Fig. 21.73).

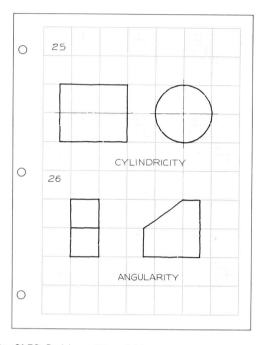

Fig. 21.72 Problems 25 and 26.

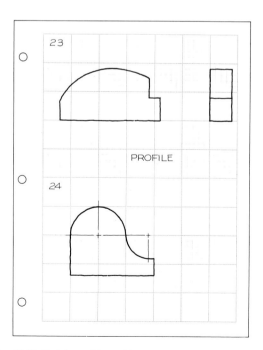

Fig. 21.71 Problems 23 and 24.

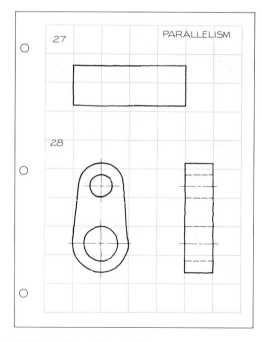

Fig. 21.73 Problems 27 and 28.

29. Using a feature control symbol and the necessary dimensions, indicate that the vertical surface is perpendicular to the bottom of the object, the datum, within a tolerance of 0.20 mm (Fig. 21.74).

30. Using a feature control symbol and the necessary dimensions, indicate that the hole is perpendicular to datum A within a tolerance of 0.08 mm (Fig. 21.74).

31. Using a feature control symbol and cylinder A as the datum, indicate that the conical feature has a runout of 0.80 mm (Fig. 21.75).

32. Using a feature control symbol with cylinder B as the primary datum, and surface C as the secondary datum, indicate that surfaces D, E, and F have a runout of 0.60 mm (Fig. 21.75).

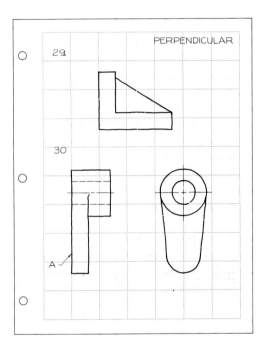

Fig. 21.74 Problems 29 and 30.

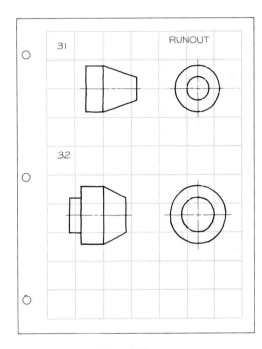

Fig. 21.75 Problems 31 and 32.

WELDING

22.1 INTRODUCTION

Welding is the process of joining metal by heating a joint to a suitable temperature with or without the application of pressure, and with or without the use of filler material. Welding is used to join assemblies permanently when it will be unnecessary to disassemble them for maintenance or other purposes.

The welding practices presented in this chapter are in compliance with the standards developed by the American Welding Society and the American National Standards Institute (ANSI). Reference is also made to the drafting standards used by General Motors Corporation.

Some of the advantages of welding over other methods of fastening are: (1) simplified fabrication, (2) economy, (3) increased strength and rigidity, (4) ease of repair, (5) creation of gas- and liquid-tight joints, and (6) reduction in weight and/or size.

The various welding processes are given in Fig. 22.1. The three main types of welding are *gas welding*, *arc welding*, and *resistance welding*.

Gas welding is a process in which gas flames are used to melt and fuse metal joints. Gases such as acetylene or hydrogen are mixed within a torch

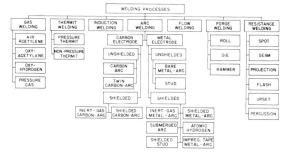

Fig. 22.1 The types of welding processes available to the designer. The three major types are gas welding, arc welding, and resistance welding. (Courtesy of General Motors Corporation.)

and are burned with air or oxygen (Fig. 22.2). The oxyacetylene method is the best known process of gas welding, and it is widely used for repair work and field construction.

Most oxyacetylene welding is done manually with a minimum of equipment. Filler material in the form of welding rods is used to deposit metal at the joint as it is heated. Most metals, except for low- and medium-carbon steels, will require fluxes that aid in the process of melting and fusing the metals together.

Arc welding is a process that uses an electric arc to heat and fuse the joints (Fig. 22.3). In some cases pressure may be applied in addition to the

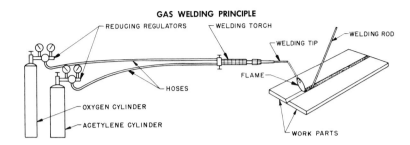

Fig. 22.2 The gas welding process burns gases, such as oxygen and acetylene, in a torch to apply heat to a joint while the welding rod supplies the filler material. (Courtesy of General Motors Corporation.)

heat, and in others, pressure will not be required. The filler material is supplied by a consumable electrode through which the electric arc is transmitted or through a nonconsumable electrode to fuse the metals. Metals that are well-suited to arc welding are wrought iron, low- and medium-carbon steels, stainless steel, copper, brass, bronze, aluminum, and some nickel alloys.

Flash welding is a form of arc welding that is similar to resistance welding since both pressure and an electric current are used to join two pieces (Fig. 22.4). The two parts are brought together causing a heat buildup between them. As the metal burns, the current is turned off and the pressure between the parts is increased to fuse the parts together.

Resistance welding is a group of processes where metals are fused together by heat produced from the resistance of the parts to the flow of electric

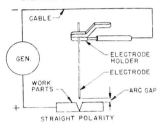

Fig. 22.3 The arc welding process can use either DC or AC current that is passed through an electrode to heat the joint to be welded. (Courtesy of General Motors Corporation.)

current, and by the application of pressure. Fluxes and filler materials are normally not used. All resistance welds are either lap- or butt-type welds.

Figure 22.5 illustrates how resistance spot welding is performed on a lap joint. The two parts are lapped, pressed together, and an electric cur-

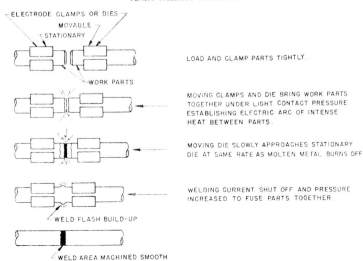

Fig. 22.4 Flash welding is a type of arc welding that uses a combination of electric current and pressure to fuse two parts together. (Courtesy of General Motors Corporation.)

SPOT WELDING PRINCIPLE

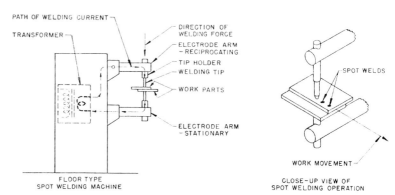

Fig. 22.5 Resistance welding can be used to join lap and butt joints to fuse metals by heat generated by passing an electrical current through them. (Courtesy of General Motors Corporation.)

rent fuses the parts together where they join. A series of spots spaced at intervals, called *spot welds*, are used to secure the parts. Other welds besides spot welds can be produced by resistance welding. Table 22.1 suggests the processes of welding that can be used for various types of materials.

22.2 WELD JOINTS

The five standard weld joints are illustrated in Fig. 22.6. The butt joint can be joined with the following types of welds: square groove, V-groove, bevel groove, U-groove, and J-groove. The corner joint can be joined with the same welds as the butt joint, but with the addition of the fillet weld as well.

The tee joint can be joined with the following welds: bevel groove, J-groove, and fillet welds.

Table 22.1 RECOMMENDED RESISTANCE WELDING PROCESSES

Material	Spot Welding	Flash Welding
Low Carbon Mild Steel:		
SAE 1010	R	R
SAE 1020	R	R
Medium Carbon Steel:		
SAE 1030	R	R
SAE 1050	R	R
Wrought Alloy Steel:		
SAE 4130	R	R
SAE 4340	R	R
High Alloy Austenitic Stainless Steel:		
SAE 30301–30302	R	R
SAE 30309–30316	R	R
Ferritic and Martensitic Stainless Steel:		
SAE 51410–51430	S	S
Wrought Heat Resisting Alloys:*		
10–9–DL	S	S
16–25–6	S	S
Cast Iron	NA	NR
Gray Iron	NA	NR
Aluminum & Aluminum Alloys	R	S
Nickel & Nickel Alloys	R	S

S–Satisfactory   NA–Not applicable
R–Recommended   NR–Not recommended
*–For composition see American Society of Metals Handbook.
Source: Courtesy of General Motors Corporation.

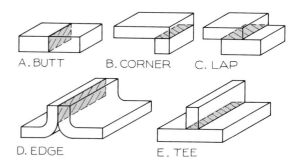

Fig. 22.6 The standard types of joints encountered in the welding process.

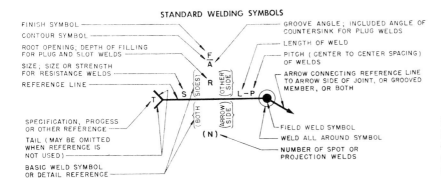

STANDARD WELDING SYMBOLS

FINISH SYMBOL
CONTOUR SYMBOL
ROOT OPENING; DEPTH OF FILLING FOR PLUG AND SLOT WELDS
SIZE; SIZE OR STRENGTH FOR RESISTANCE WELDS
REFERENCE LINE

GROOVE ANGLE; INCLUDED ANGLE OF COUNTERSINK FOR PLUG WELDS
LENGTH OF WELD
PITCH (CENTER TO CENTER SPACING) OF WELDS
ARROW CONNECTING REFERENCE LINE TO ARROW SIDE OF JOINT, OR GROOVED MEMBER, OR BOTH

SPECIFICATION, PROCESS OR OTHER REFERENCE
TAIL (MAY BE OMITTED WHEN REFERENCE IS NOT USED)
BASIC WELD SYMBOL OR DETAIL REFERENCE

FIELD WELD SYMBOL
WELD ALL AROUND SYMBOL
NUMBER OF SPOT OR PROJECTION WELDS

Fig. 22.7 The welding symbol. It is not necessary to show the entire symbol in all applications. It may be used in modified form.

Welds used to join lap joints are fillet, bevel groove, J-groove, slot, plug, spot, projection, and seam welds. The edge joint uses the same welds that are used for lap joints with the addition of the square groove, V-groove, U-groove, and seam welds.

22.3 WELDING SYMBOLS

The specification of welds on a working drawing is done by the application of symbols. If a drawing has a general note such as ALL JOINTS WELDED or WELDED THROUGHOUT, the designer has transferred the design responsibility to the welder. Welding is too important to be left to chance; it must be completely specified.

The method of providing specifications on a drawing is by the use of a welding symbol as shown in Fig. 22.7. This example gives the symbol in its entirety, which is seldom needed in its complete form. Instead, the symbol is usually modified to a simpler form when not all of the specifications are necessary for a particular application.

The scale of the welding symbol is shown in Fig. 22.8 where it is drawn on a 3 mm (⅛″) grid. Its size can be scaled down using these same proportions where space is limited. The lettering used is the standard height of ⅛″ or 3 mm.

The *ideograph* is the symbol used to denote the type of weld desired. In general, the ideograph depicts the cross section of the type of weld used.

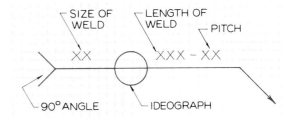

SIZE OF WELD
LENGTH OF WELD
PITCH
XX
XXX – XX
90° ANGLE
IDEOGRAPH

Fig. 22.8 When the grid is drawn as a full-size ⅛″ (3 mm) grid, the size of the welding symbol can be determined. It can be drawn smaller at these same proportions when necessary.

The more often used ideographs are drawn to scale on the ⅛″ (3 mm) grid to represent their full size when added to the welding symbol (Fig. 22.9).

22.4 TYPES OF WELDS

The more commonly used welds are shown in Fig. 22.10 along with their corresponding ideographs. The fillet weld is a built-up weld at the angular intersection between two surfaces. The square, bevel, V-groove, J-groove, and U-groove welds involve grooves inside of which the weld is placed. Slot and plug welds have intermittent holes or openings where the parts are welded. Holes are unnecessary when resistance welding is used. Note that the symbols (ideographs) are symbolic of the cross sections of the welds and grooves.

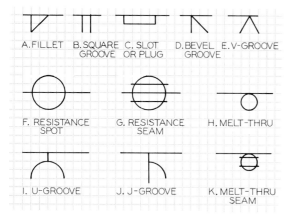

Fig. 22.9 The sizes of the ideographs are shown on the ⅛″ (3 mm) grid. These sizes should be used to conjuction with the symbol shown in the previous figure.

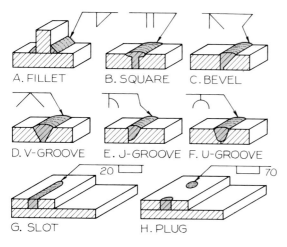

Fig. 22.10 The standard types of welds and their corresponding ideographs.

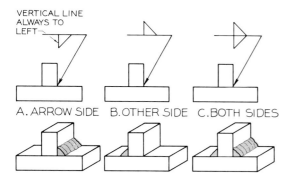

Fig. 22.11 Fillet welds are indicated by abbreviated symbols. When the ideograph is on the lower side it refers to the arrow side, and to the other side when it is placed above the horizontal line.

22.5 APPLICATION OF SYMBOLS

Fillet welds are applied to two parts in Fig. 22.11 to indicate three methods of welding. At A, the fillet ideograph is placed on the lower side of the horizontal line of the symbol, which indicates that the weld is to be at the joint on the *arrow side*, the side of the arrow.

> The vertical leg of the ideograph is *always* on the left side.

By placing the ideograph on the upper side of the horizontal, the weld is specified to be on the *other side*, the joint on the other side of the part away from the arrow. Again, the vertical leg is on the left side.

When the part is to be welded on both sides of the vertical part, the ideograph at C is used. Note that the tail has been omitted from the symbol along with other written specifications. This is permitted when detailed specifications are given in another form to specify the details of the welds used.

A single arrow can be used to specify a weld that is to be all around two joining parts (Fig. 22.12). A circle of 6 mm diameter placed at the bend in the leader of the symbol gives this specification. If this process is to be done *in the field*, a black circle of 3 mm diameter can be used to denote this joint. This means that the parts will be assembled on the site rather than in a shop as part of the manufacturing process. Both symbols can be used separately as well as together.

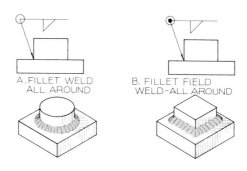

Fig. 22.12 Symbols for indicating fillet welds all around two types of parts.

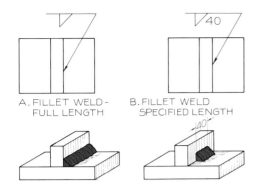

Fig. 22.13 Symbol for indicating full-length fillet welds and fillet welds less than full length.

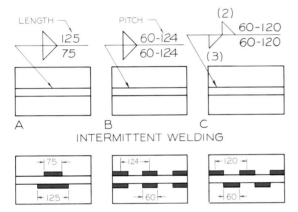

Fig. 22.14 Symbols for specifying varying and intermittent welds.

When a fillet weld is to be the full length of the two parts, it may be specified as shown in Fig. 22.13. Since the ideograph is on the lower side, the weld will be on the arrow side. If the weld is to be less than full length, it can be specified as shown at B where the number, 40, represents its length in millimeters, and it is centered about the approximate location of the arrow.

When fillet welds are of different lengths and are positioned on both sides, they may be specified as shown in Fig. 22.14A. The dimensions on the lower side of the horizontal give the length of the weld on the arrow side and the number on the upper side gives the length on the other side.

Intermittent welds are welds of a given length that are spaced uniformly apart from center to center by a distance called the *pitch*. Since the welds are on both sides, 60 mm long, and have pitches of 124 mm, this can be indicated by a symbol as shown on Fig. 22.14B. If the intermittent welds are to be staggered to alternate positions on both sides, this can be specified by the symbols shown in part C.

22.6 GROOVE WELDS

The more standard groove welds are illustrated in Fig. 22.15 with their respective symbols. When the depth of the grooves, the angle of the chamfer, and

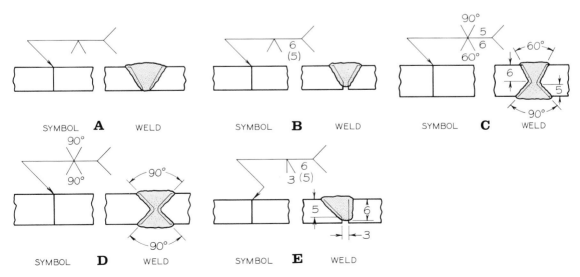

Fig. 22.15 Types of groove welds and their general specifications.

the root openings are not given on a symbol, this information needs to be specified elsewhere on the drawings or in supporting documents. At B and E, the depth of the chamfer of the prepared joint is given in parentheses above the size dimension of the weld, which takes into account the penetration of the weld beyond this chamfer. The size of the joint is equal to the depth of the prepared joint when only one number is given.

When the chamfer is different on each side of the joint it can be noted with a symbol as shown at Fig. 22.15C. If the spacing between the two parts, the *root opening*, is to be specified, this is done by placing its dimensions in millimeters between the groove angle number and the weld ideograph, part E.

A bevel weld is groove beveled from one of the parts being joined; consequently the symbol must indicate which is to be beveled as shown in Fig. 22.15E. To call attention to this operation, the leader from the symbol is bent and aimed toward the piece to be beveled. This practice also applies to J-welds where one side is grooved and the other is not (Fig. 22.16).

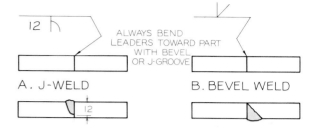

Fig. 22.16 J-welds and bevel welds are specified by bent arrows pointing to the side of the joint that is to be grooved.

22.7   EXAMPLE WELDS

A series of joints and welds is shown in Fig. 22.17 that incorporates the previously covered principles. Letters are used instead of numerals to represent specified dimensions.

22.8   SURFACE CONTOURED WELDS

Contour symbols are used to indicate which of the three types of contours is desired on the surface of

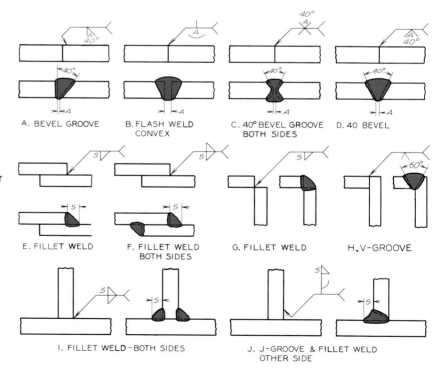

Fig. 22.17 An assortment of welds and grooves with their accompanying symbols. Letters are used instead of numbers to give the weld specification.

A. BEVEL GROOVE

B. FLASH WELD CONVEX

C. 40° BEVEL GROOVE BOTH SIDES

D. 40 BEVEL

E. FILLET WELD

F. FILLET WELD BOTH SIDES

G. FILLET WELD

H. V-GROOVE

I. FILLET WELD-BOTH SIDES

J. J-GROOVE & FILLET WELD OTHER SIDE

the weld: flush, concave, or convex. Flush welds are those that are smooth with the surface or flat across the hypotenuse of a fillet weld. A concave contour is a weld that bulges inward with a curve, and a convex contour is one that bulges outward with a curve (Fig. 22.18).

In many cases, it is necessary to finish the weld by a supplementary process to bring it to the desired contour. These processes may be added to the contour symbols to specify their operations in finishing the welds. These processes and their letter specifications are: chipping (C), grinding (G), hammering (H), machining (M), rolling (R), and peening (P). Examples of these contour symbols are given in Fig. 22.19.

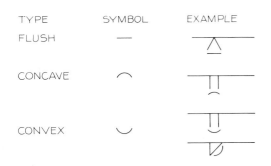

Fig. 22.18 The contour symbols that are used to specifiy the surface finish of a weld.

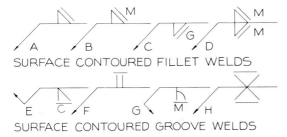

Fig. 22.19 Examples of contoured surface symbols and letters of finishing applied to them.

22.9 SEAM WELDS

A seam weld is a weld that joins two lapping parts with a continuous weld or a series of closely spaced spot welds. The process used for seam welds must be given by abbreviations placed in the tail of the weld symbol. The ideograph for a

resistance weld is about 12 mm in diameter and it is placed with the horizontal line of the symbol through its center. The weld's width, length, and pitch are indicated by numbers, as shown in Fig. 22.20.

When the seam weld is made by arc welding, the diameter of the ideograph is about 6 mm and it is placed on the upper or lower side of the horizontal bar of the symbol to indicate whether the seam will be on the arrow or the other side when applied (part B). When the numeral that represents the length of the weld is omitted from the symbol, it is understood that the seam weld extends between abrupt changes in the seam or as it is dimensioned.

Spot welds are specified in a similar manner with ideographs, and specifications, as shown in Fig. 22.21, are given by diameter, number of welds, and pitch between the welds. The process,

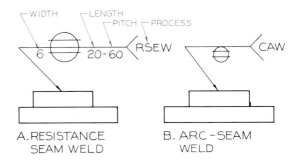

Fig. 22.20 Resistance and arc seam welds with the processes are indicated in the tail of the symbol for each. The arc weld must specify arrow side or other side in the symbol.

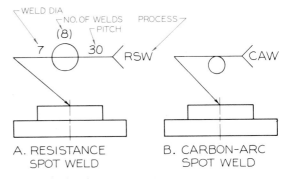

Fig. 22.21 Resistance and arc spot welds with the process indicated in tail of the symbol. The arc weld must specify arrow or other side location in the symbol.

Table 22.2 WELDING PROCESS SYMBOLS

CAW	Carbon arc welding	FRW	Friction welding	PGW	Pressure gas welding
CW	Cold welding	FW	Flash welding	RB	Resistance brazing
DB	Dip brazing	GMAW	Gas metal arc welding	RPW	Projection welding
DFW	Diffusion welding	GTAW	Gas tungsten welding	RSEW	Resistance seam welding
EBW	Electron beam welding	IB	Induction brazing	RSW	Resistance spot welding
ESW	Electroslag welding	IRB	Infrared brazing	RW	Resistance welding
EXW	Explosion welding	OAW	Oxyacetylene welding	TB	Torch brazing
FB	Furnace brazing	OHW	Oxyhydrogen welding	UW	Upset welding
FOW	Forge welding				

resistance spot welding (RSW), is noted in the tail of the symbol. For arc welding, the arrow side or other side must be indicated by a symbol as shown in part B.

(The abbreviations of various welding processes can be seen in Table 22.2.)

22.10 BUILT-UP WELDS

When the surface of a part is to be enlarged by welding, called *building up*, this can be indicated by the symbol shown in Fig. 22.22. The width of the built-up weld is dimensioned in the view, and the height of the weld above the surface is specified in the symbol to the left of the ideograph. The radius of the circular segment is 6 mm.

22.11 WELDING STANDARDS

Figure 22.23 (pages 436 and 437) gives an overview of the welding symbols and specifications that have been covered in the previous sections. The chart was prepared by the American Welding Society of Miami, Florida. It can be used as a single reference for most general types of welding and their associated symbols.

22.12 BRAZING

Brazing is a method of joining pieces of metal that is similar to welding. It is a process in which the joints are heated above 800 degrees Fahrenheit and a nonferrous filler material with a melting point below the base materials is distributed by capillary action between the closely fit parts.

Prior to brazing, the parts to be brazed must be cleansed, and flux is added to the joints. The brazing filler is also added prior to or just as the joints are heated beyond the melting point of the filler. There are two basic joints for brazing: lap and butt joints as shown in Fig. 22.24. The filler material is allowed to flow between the parts to form the joint after it has melted.

Brazing is used to hold parts together, to provide gas- and liquid-tight joints, to assure electrical conductivity, and to aid in repair and salvage. Brazed joints will withstand more stress, higher temperature, and more vibration than will soft-soldered joints.

The method of noting a brazed joint in a drawing is illustrated in Fig. 22.25. A heavy dark line is used to indicate the brazed joint.

22.13 SOFT SOLDERING

Soldering is the process of joining two metal parts with another metal that melts below the temperature of the metals being joined. Solders are alloys of nonferrous metals that melt below 800° F. This method of joining is widely used in the automotive and electrical industries.

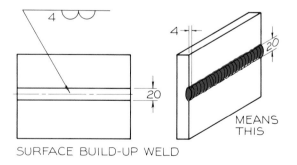

SURFACE BUILD-UP WELD

Fig. 22.22 The method of applying a symbol to a built-up weld on a surface.

LAP JOINTS   **BUTT JOINTS**

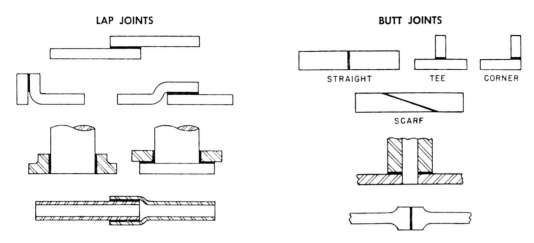

STRAIGHT   TEE   CORNER

SCARF

Fig. 22.24 Examples of the two basic types of brazing joints: lap joints and butt joints. (Courtesy of General Motors Corporation.)

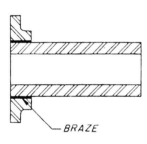

— BRAZE

Fig. 22.25 The method of indicating a brazed joint in a drawing. (Courtesy of General Motors Corporation.)

Soldering is one of the basic techniques of welding and is often done by hand with a soldering iron of the type shown in Fig. 22.26. The iron is placed on the joint to heat it and to melt the solder that fuses the joint. The method of indicating a soldered joint is shown in Fig. 22.26 where a heavy dark line is used with a note.

SOLDERING COPPER

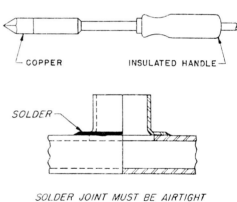

— COPPER   INSULATED HANDLE —

SOLDER —

SOLDER JOINT MUST BE AIRTIGHT UNDER _____ P. S. I. PRESSURE

Fig. 22.26 A typical hand-held soldering iron used to soft-solder two parts together, and the method of indicating a soldered joint on a drawing. (Courtesy of General Motors Corporation.)

PROBLEMS

Make working drawings of the following problems given in other chapters. Wherever feasible, change the joints of the features of the parts to be welded instead of joined by one-piece casting. These drawings should be made on Size B sheets.

1. Use Fig. 23.36.
2. Use Fig. 23.40.
3. Use Fig. 23.45.
4. Use Fig. 23.47.
5. Use Fig. 23.51.
6. Use Fig. 23.53
7. Use Fig. 23.65.
8. Use Fig. 23.67.
9. Use Fig. 23.72, Part 1.
10. Use Fig. 23.78, Part 1.

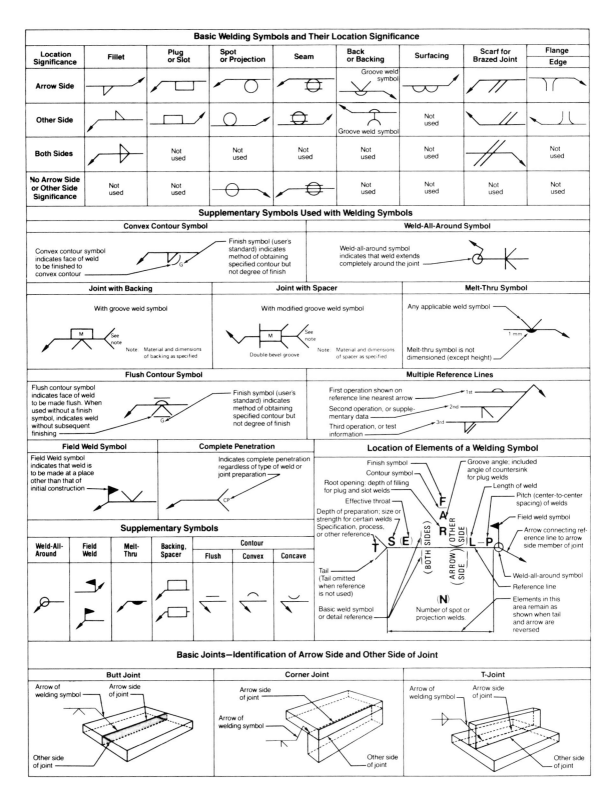

Fig. 22.23 The American Welding Society Standard Welding Symbols. This chart serves as a review of the principles covered in this chapter.

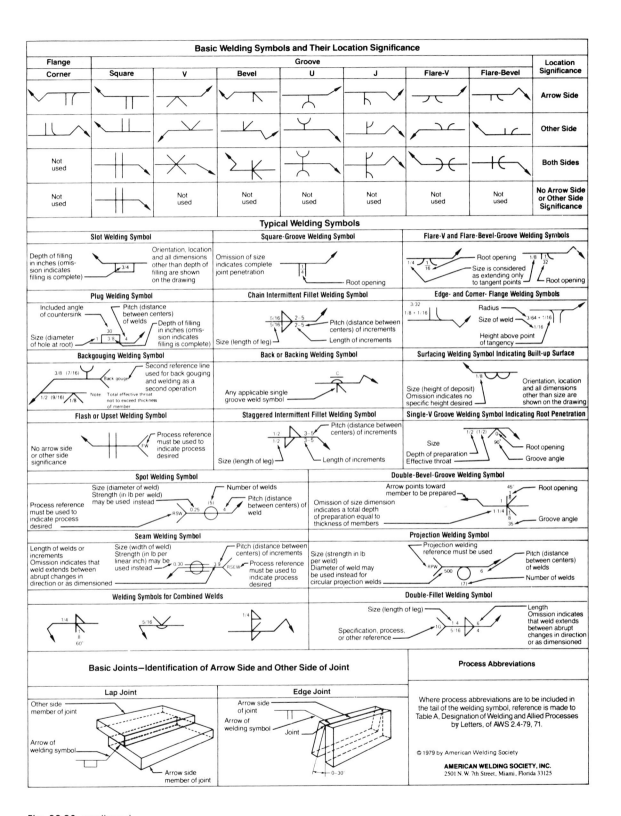

Fig. 22.23 continued

WORKING DRAWINGS

23.1 INTRODUCTION

A set of working drawings are drawings from which a design is constructed. The set may contain any number of sheets from one to over one hundred, depending upon the complexity of the project. The written instructions that accompany working drawings are called *specifications*. When the project can be represented on several drawing sheets, the written specifications are very often written on the drawings to consolidate the information into a single format. Although much of the work in preparing working drawings is done by the drafter, the designer, who is most often an engineer, is responsible for their correctness.

> A working drawing is often called a detail drawing because it describes and dimensions the details of the parts being presented.

All of the principles of orthographic projection, sections, conventions, dimensioning, tolerancing, pictorials, and practically all other types of graphical techniques are utilized to communicate the details as effectively as possible.

23.2 WORKING DRAWINGS— INCH SYSTEM

An example of a working drawing is shown in Fig. 23.1, where the base-plate mount is detailed in three orthographic views. Dimensions and notes are used to give the necessary information to construct the piece without misinterpretation. This particular drawing is dimensioned using decimal inches.

Decimal inches are preferable to common fractions, although both of these systems are still in widespread use. The English system, which is based on the inch, is giving way to the metric system, which is based on the millimeter. Decimal inches make it possible to handle arithmetic involving the dimensions with much greater ease than is possible with fractions that must be first converted to decimals. Inch marks such as X.XX″ are omitted from dimensions on a working drawing since it is understood that these dimensions are in inches.

An example of a part represented in a set of working drawings dimensioned with common fractions using the inch as the unit of measurement is the revolving clamp assembly shown in Fig. 23.2. The detail drawings of the parts of the assembly are shown in Figs. 23.3 through 23.5. Several dimensioned parts are shown on each

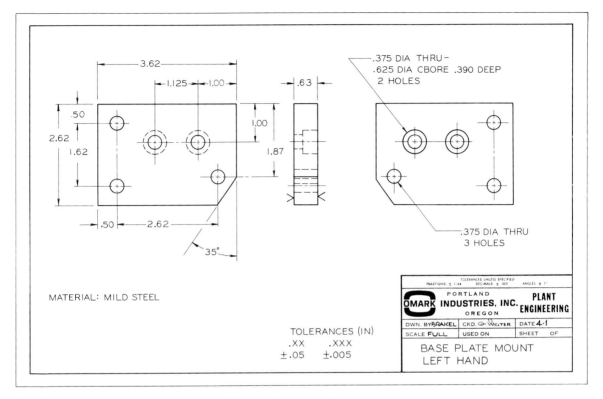

Fig. 23.1 A working drawing of a single part that is dimensioned in inches. (Courtesy of Omark Industries, Inc.)

sheet as orthographic views. The arrangement of these parts on the sheet has no relationship to how the parts fit together. They are simply positioned to take advantage of the available space. Each part is given a number and a part name for identification purposes. The material that each part is made of is indicated along with the necessary notes to fully explain any manufacturing procedures that are necessary.

Each sheet is numbered in the title block and the other title block information is completed.

An orthographic assembly is given on sheet 3 (Fig. 23.5) that explains how the parts fit together. The parts are numbered to correspond to the part numbers in the parts list, which serves as a bill of materials.

Fig. 23.2 A revolving clamp assembly manufactured to hold parts stationary while they are being machined. (Courtesy of Jergens, Inc.)

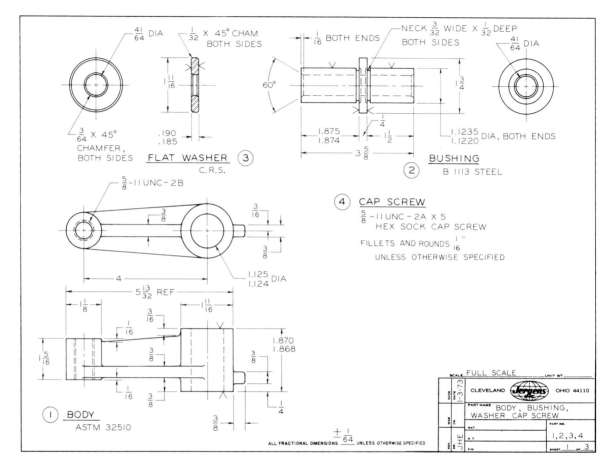

Fig. 23.3 Detail drawings showing parts of the clamp assembly. (Courtesy of Jergens, Inc.)

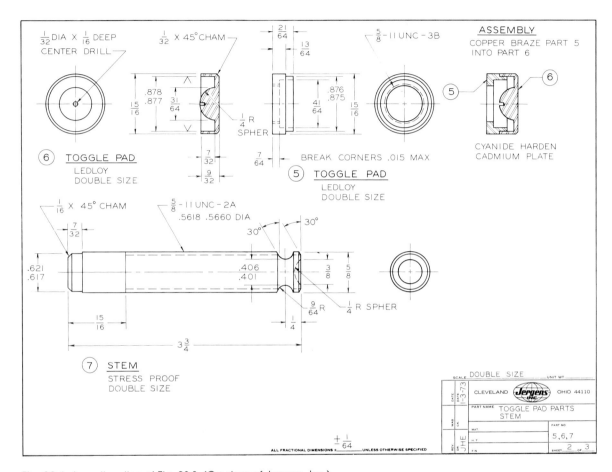

Fig. 23.4 A continuation of Fig. 23.3. (Courtesy of Jergens, Inc.)

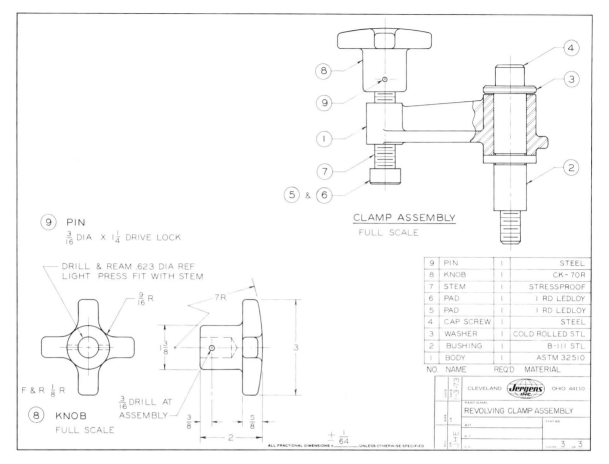

Fig. 23.5 A detail drawing and an orthographic assembly of the clamp assembly. (Courtesy of Jergens, Inc.)

23.3 WORKING DRAWINGS— METRIC SYSTEM

The millimeter is the basic unit of the metric system, which is the recommended system. A single part, a back tool post, is detailed in Fig. 23.6. In this case, all dimensions are measured to the nearest whole millimeter with no decimal fractions except where tolerances are shown. Metric units such as XX mm are omitted from dimensions on a working drawing since it is understood from the title block that the units are in millimeters. Remember that the width of the fingernail of your index finger is about 10 millimeters wide. A single part is dimensioned using the metric system in

Fig. 23.7. The left-side view is shown as a full section.

The lifting device that is used to level heavy equipment (Fig. 23.8) is detailed in working drawings shown in Fig. 23.9 and 23.10. All dimensions are given in millimeters; this is indicated by the SI symbol near the title block along with the symbol for the third angle of projection.

In Fig. 23.10 the parts are shown in an assembly that is shown as a full section. This explains how the parts are assembled after they have been constructed. Assemblies are not dimensioned since they have been dimensioned and noted elsewhere in the drawings. A parts list is given on the same page with the assembly, above the title block for easy reference to the assembly.

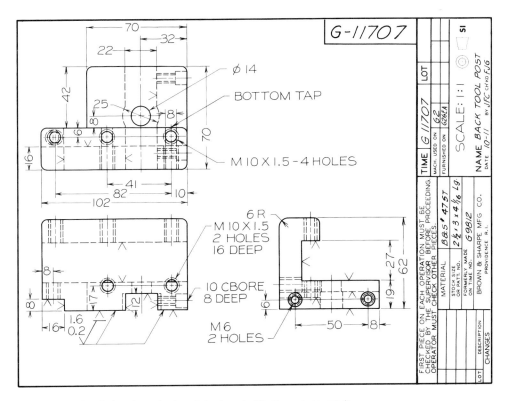

Fig. 23.6 A detail drawing of a back tool post. All dimensions are in millimeters.

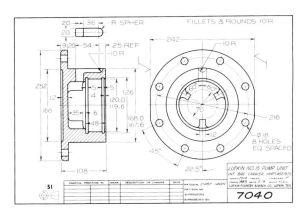

Fig. 23.7 A single part dimensioned using the metric system (SI units).

Fig. 23.8 A Lev-L-ine lifting device used to level heavy machinery. This product is the basis of the working drawings in Fig. 23.9 and 23.10. (Courtesy of Unisorb Machinery Installation Systems.)

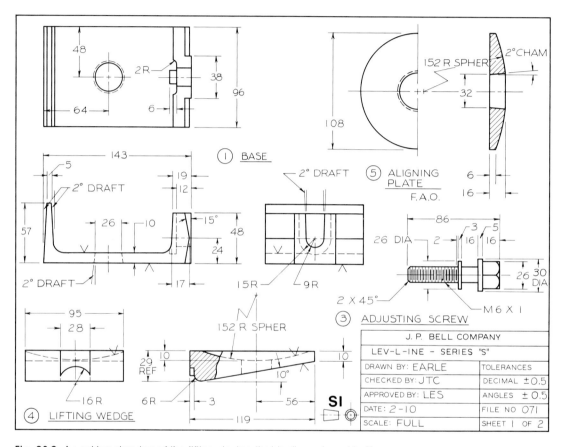

Fig. 23.9 A working drawing of the lifting device that is dimensioned in SI units. (Courtesy of Unisorb Machinery Installation Systems.)

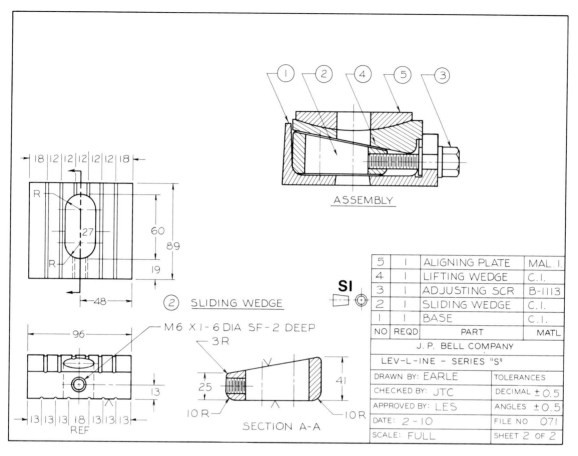

Fig. 23.10 A working drawing and assembly of the lifting device that is dimensioned using SI units. (Courtesy of Unisorb Machinery Installation Systems.)

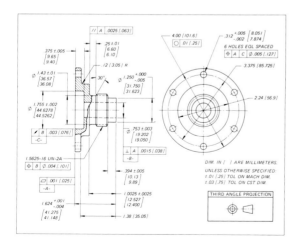

Fig. 23.11 A dual-dimensioned drawing. The dimensions are given in inches with their equivalents in millimeters given in parentheses. (Courtesy of General Motors Corporation.)

23.4 WORKING DRAWINGS— DUAL DIMENSIONS

Some working drawings are dimensioned with both inches and millimeters for those who may work in both systems. A typical example is shown in Fig. 23.11, where the dimensions in parentheses are millimeters and the dimensions above the parentheses are inches. Converting from one unit to the other will result in fractional units that must be rounded off.

The units may first be made in millimeters and then converted to inches equally as well. Usually, the converted units are shown in parentheses. An explanation of the system used should be noted in the title block.

23.5 LAYING OUT A WORKING DRAWING

The working drawing is laid out by beginning with the border, if a printed border is not provided. At least 0.25" (7 mm) should be allowed for the border at the edge of the sheet (Fig. 23.12, Step 1). The title block is drawn to size at the lower right corner of the sheet and views and dimensions are drawn lightly to take advantage of the

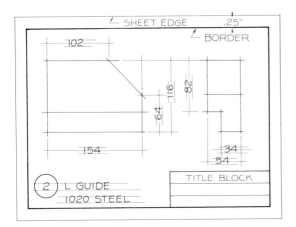

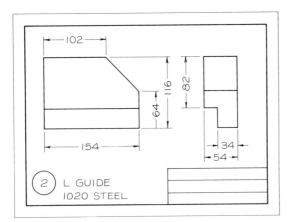

Fig. 23.12 Laying out a detail drawing

Step 1 The border and title strip are drawn on a preliminary sheet that will be traced. The views are positioned to allow adequate room for their dimensions. Guidelines are constructed for all dimensions and notes.

Step 2 The layout is overlaid with vellum or film and the lines are drawn to their proper weights, the dimensions and notes are lettered, and the title block is completed to finish the detail drawing.

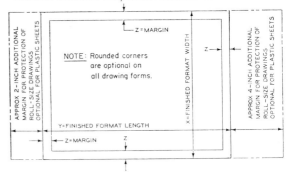

FLAT SIZES				ROLL SIZES				
SIZE DES LTR	X WIDTH	Y LENGTH	Z MARGIN	SIZE DES LTR	X WIDTH	Y MIN LENGTH	Y MAX LENGTH	Z MARGIN
A(HORIZ)	8.50	11	.25 & .38*	G	11	42	144	.38
A(VERT)	11	8.50	.25 & .38*	H	28	48	144	.50
B	11	17	.38	J	34	48	144	.50
C	17	22	.50	K	40	48	144	.50
D	22	34	.50					
E	34	44	.50					
F	28	40	.50					

*HORIZONTAL MARGINS .38-INCH; VERTICAL MARGIN .25-INCH

Fig. 23.13 The standard sheet sizes for working drawings dimensioned in inches. (Courtesy of the U.S. Department of Defense.)

available space. When using tracing paper or film, it is more efficient to lay out the views and dimensions on a different sheet of paper and then overlay this drawing with vellum or film for tracing the final working drawing. Guidelines for lettering are drawn for each dimension line on this preliminary layout.

The lines are darkened to their proper weight and the drawing is completed as shown in Step 2. Properly positioning the views to provide adequate space is one of the major concerns of the drafter in laying out a drawing.

The standard sheet sizes of working drawings are shown in Fig. 23.13. Papers, films, cloths, and reproduction materials are available in these modular sizes. Most of your student assignments will be completed on Size A and Size B sheets

23.6 TITLE BLOCKS AND PARTS LISTS

A typical parts list and title block for student assignments are shown in Fig. 23.14. These are usually located in the lower right-hand corner of the

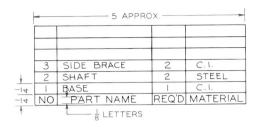

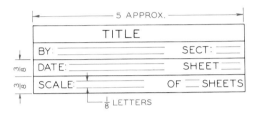

Fig. 23.14 A typical parts list and title strip that is suitable for most student assignments.

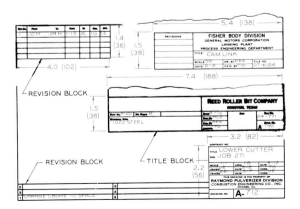

Fig. 23.15 A title block that is typical of those used in industry. (Courtesy of General Motors Corporation.)

Fig. 23.16 Several examples of title blocks and revision blocks that are used in industry.

drawing sheet against the borders. You will notice from observation that practically all title blocks contain the following information: title or part name, drafter, date, scale, company, and sheet number. Other information such as tolerances, checkers, and materials may be shown in more elaborate title blocks.

The standard parts list is shown in Fig. 23.14 with the usual elements.

> The parts list should be placed directly over the title block in the lower right corner of the drawing.

A title block is used by General Motors Corporation is shown in Fig. 23.15. You will notice a note to the left of the title block that lists John F. Brown as an inventor. This is used to establish the ownership of patent rights when drawings have been made of devices that might be patentable. One or two associates are asked to date and sign the drawings as witnesses of the work of the inventor. This establishes the ownership of the ideas and dates the time of their development, in case this becomes an issue in obtaining a patent at a later date.

Other examples of title blocks and revision blocks are shown in Fig. 23.16. These are typical of those that are printed on sheets by various industries. Revision blocks are used to indicate modifications of parts at later dates to improve the design.

It is important to give the number of sheets in the set on each sheet. For example: sheet 2 of 6, sheet 3 of 6, etc.

23.7 SCALE SPECIFICATION

The scale of a working drawing should be indicated in or near the title block when all drawings are the same scale. If several drawings are made at different scales, the scales used should be indicated in close association to the drawings to specify the scales of each.

Several methods of indicating scales are shown in Fig. 23.17. When the colon is used, such as 1:2, the metric system is implied, whereas the English system is implied when the equal sign is

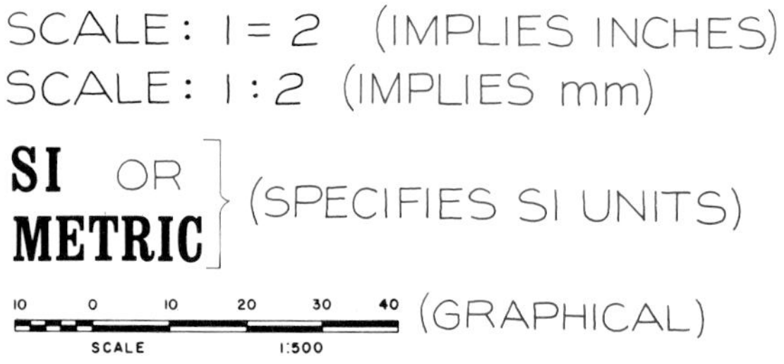

SCALE: 1 = 2   (IMPLIES INCHES)
SCALE: 1 : 2   (IMPLIES mm)

SI OR
METRIC } (SPECIFIES SI UNITS)

(GRAPHICAL)

Fig. 23.17 Methods of specifying scales and metric units on working drawings.

used: 1 = 2. The symbol SI or the word METRIC on a drawing clearly specifies that the units of measurement are millimeters.

In some cases a graphical scale is given with calibrations that permit the interpretation of linear units by transferring dimensions from the drawing with your dividers to the scale. This is commonly used for scaling distances on maps.

23.8   TOLERANCES

You should recall from Chapter 21 that general tolerance notes can be given to working drawings to specify the tolerances of dimensions that do not have tolerances applied in the drawing. Such a table of values is shown in Fig. 23.18 where a check can be made to indicate whether the units will be in inches or millimeters.

Plus-or-minus tolerances are given in the blanks under the number of digits under each decimal fraction. For example, this table specifies that each dimension with two-place decimals will have a tolerance of ±0.01″.

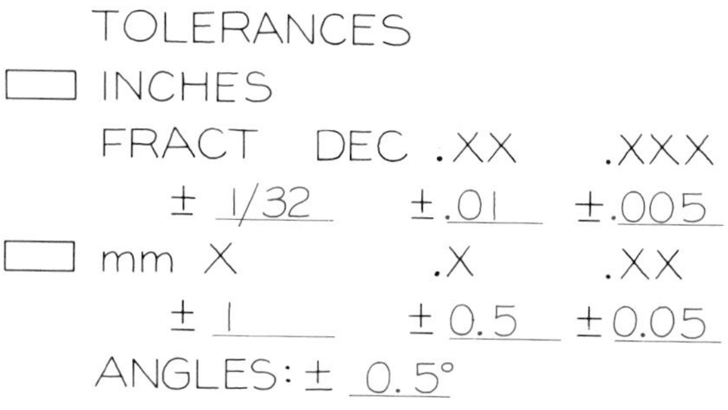

TOLERANCES
☐ INCHES
  FRACT   DEC .XX   .XXX
    ± 1/32   ±.01   ±.005
☐ mm X   .X   .XX
    ± 1   ± 0.5   ± 0.05
  ANGLES: ± 0.5°

Fig. 23.18 General tolerance notes that are given on working drawings to specify the tolerance permitted on dimensions.

Angular tolerances can be given in general notes also. Refer to Chapter 21 for more detailed examples of using general tolerance notes.

23.9   PART NAMES

Each part should be given a name and a number, as shown in Fig. 23.19. The letters should be ⅛″ (3 mm) high. The part numbers are placed inside circles, which are called *balloons*. Balloons are drawn approximately four times the height of the numbers.

The part numbers should be placed close to the parts on the working drawing so that is clear which part they are associated with. Balloons are especially important on assembly drawings since the numbers of the parts refer to the same numbers in the parts list that serves as a cross reference.

Fig. 23.19 Each part of a working drawing should be named and numbered for listing in the parts list.

23.10   CHECKING A DRAWING

All drawings must be checked before they are released for production, since a slight mistake could prove very expensive when many parts are made. The people who check drawings have special qualifications that enable them to suggest revisions and modifications that will result in a better product at less cost. The checker may be a chief drafter who is experienced in this type of work, or the engineer or designer who originated the project. In larger companies the drawings are reviewed by the various shops involved to determine whether the most efficient methods of production are specified for each particular part.

The checker never checks an original drawing, but instead checks a diazo print (a blue-line print). He or she marks the print with a colored pencil, making notes and corrections that he or she feels are desirable. The print is returned to the

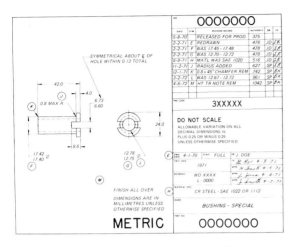

Fig. 23.20 The revisions of this working drawing are noted near the revisions with letters in balloons that are cross-referenced in a table of revisions. (Courtesy of General Motors Corporation.)

	Max Value	Points Earned
TITLE BLOCK		
Student's name	1	_____
Checker	1	_____
Date	1	_____
Scale	1	_____
Sheet number	1	_____
REPRESENTATION OF DETAILS		
Selection of views	5	_____
Assembly drawings	10	_____
Positioning of views	5	_____
DRAFTING PRACTICES		
Line quality	8	_____
Lettering	8	_____
Proper dimensioning	10	_____
Proper use of sections	5	_____
Proper use of auxiliary views	5	_____
DESIGN INFORMATION		
Indication of tolerances	5	_____
General tolerance notes	5	_____
Fillets and rounds notes	5	_____
Finish marks	5	_____
Parts list	8	_____
Thread notes and symbols	5	_____
PRESENTATION		
Properly trimmed	2	_____
Properly folded	2	_____
Properly stapled	2	_____
	100	

grade [___]

Fig. 23.21 A checklist for evaluating a working drawing assignment.

drafter for revision of the original drawing and another print is made for approval.

In Fig. 23.20, a detail drawing of a special bushing is shown. In this drawing the various modifications made by checkers are labeled with letters that are circled and placed near the revisions. The changes are listed and dated in the revision record by the drafter. Note that the change numbers are placed in a row approximately below the revisions. This procedure serves as a check on the various revisions to prevent one from being overlooked.

Note that several drafters and checkers were involved in the approval and preparation of the drawing. Tolerances and general information are printed in the title block to ensure uniformity in the production of similar parts.

Checkers are responsible for checking the soundness of the design and its functional characteristics. They are also responsible for the completeness of the drawing, the quality of the drawing, its readability, lettering, drafting techniques, and clarity. A poorly drawn view must be redrawn to meet company requirements so that it will reproduce well and be clearly understood by those using it. Quality of lettering is very important. Since working drawings should not be scaled, the craftsmen in the shop must rely on lettered notes and dimensions for their information. The best method for students to check their drawings is to make a scale drawing of the part from the working drawings when they are complete. It is often easier to find another's mistakes rather than one's own. It is good exercise to exchange drawings with a classmate and check each other's drawing.

A grading scale for checking working drawings prepared by students is given in Fig. 23.21. This general list can be used as an outline for reviewing working drawings to ensure that the major requirements have been met.

23.11 DRAFTER'S LOG

The drafter will find that many changes and revisions must be made before a final drawing is approved. Drafters should keep a record called a *log* to show all changes and modifications and decisions that were made during the project. Changes, dates, and the people involved should be recorded for references as the project progresses, and as a review of the finished project.

DRAFTSMAN'S DESIGN LOG	Sheet No. /
	of / Sheet

Detailed Description: *Layout and details of new transmission low speed gear and mainshaft combination. Low speed gear to have 10 of splines accurately ground with respect to gear teeth, and mainshaft to have three ground lands for mounting low speed gear on these surfaces.*

Job no.	*9344-97*
Job name	*Trans. Mainshaft*
	First and Reverse Gear
Models	*1950*
Engineer	*Poe*
Job started	*3-25-48*
Job finished	*4-7-48*
Layout numbers	*L-56042*

Job Objective: *To eliminate selective fit on mating parts.*

References: *L-33827*

Progress, Decisions and Authority:
3-30-48 - Messrs. Poe and Poe decided to change from a 22 tooth basic spline with 3 unevenly spaced lands to both a 24 tooth basic spline with 6 evenly spaced lands and a 24 tooth basic spline with 8 evenly spaced lands. Engineers also requested study of a longer hub for 1st reverse gear to reduce runout.
4-6-48 After preliminary investigation Messrs. Poe and Poe decided to cancel the 8 lands construction and the elongated hub.
4-7-48 Mr. Poe decided to have an additional layout made of a 22 tooth basic spline with alternating teeth and lands.

Calculations and sketches to be dated and attached.

R. Poe
Signature

Fig. 23.22 A drafter's log should be kept as a record of the project to explain the actions taken that might otherwise be forgotten. (Courtesy of General Motors Corporation.)

An example of a drafter's log is shown in Fig. 23.22. The description of the project and its objectives are given first. Each change and the reason for it are tabulated under "Progress, Decisions and Authority." The people responsible for the changes are mentioned by name.

These notes serve to refresh the memory of anyone who wishes to review the project. Calculations are often made during the process of preparing a drawing. If they are lost or if they are poorly done, it may be necessary to make them again. Consequently, they should be made a permanent part of the log and attached to the log. All notes should be complete enough to be understood by anyone who may read the log.

23.12 ASSEMBLY DRAWINGS

When the parts have been made according to the specifications of the working drawings, they will be assembled (Fig. 23.23). This requires a drawing called an **assembly drawing.**

> Two general types of assembly drawings are drawn using (1) pictorial techniques, or (2) orthographic projection.

Both can be used to good advantage.

An assembly drawing in pictorial is shown in Fig. 23.24. The parts are numbered and cross-referenced to the parts list where more information

Fig. 23.23 An assembly drawing is used to explain how the parts of a product such as this Ford tractor are assembled. (Courtesy of Ford Motor Company.)

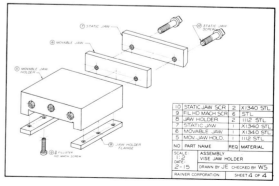

10	STATIC JAW SCR	2	X1340 STL
9	FIL HD MACH SCR	6	STL
8	JAW HOLDER	2	1112 STL
7	STATIC JAW	1	X1340 STL
6	MOVABLE JAW	1	X1340 STL
5	MOV. JAW HOLD.	1	1112 STL
NO	PART NAME	REQ	MATERIAL

SCALE: 1:2	ASSEMBLY	
DATE: 2-15	VISE JAW HOLDER	
	DRAWN BY JE	CHECKED BY WS
RAINER CORPORATION		SHEET 4 OF 4

Fig. 23.24 An exploded pictorial assembly. Each part is listed by number in the parts list.

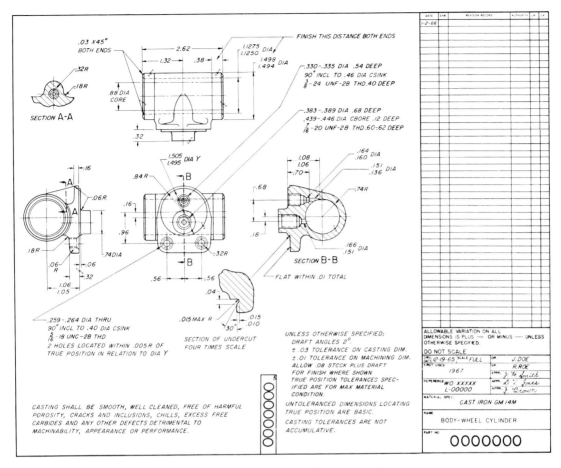

Fig. 23.34 A detail drawing of a casting that shows the machining operations and specifications also. (Courtesy of General Motors Corporation.)

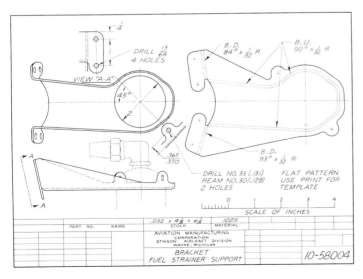

Fig. 23.35 A detail drawing of a sheet metal part. It is shown as a flat pattern and as orthographic views when bent into shape.

PROBLEMS

The following problems (Figs. 23.36–23.82) are to be drawn on sheet sizes assigned or those suggested. Each problem should be drawn with the appropriate dimensions and notes to fully describe the parts and assemblies being drawn.

Working drawings may be made on film or tracing vellum in ink or in pencil. Select a suitable title block and complete it using good lettering practices. Some problems will require more than one sheet to show all of the parts properly.

Assemblies should be prepared with a parts list where there are several parts. These may be orthographic or pictorial assemblies, either exploded, assembled, or partially exploded.

The dimensions given in the problems do not always represent good dimensioning practices because of space limitations; but the dimensions given are usually adequate for you to complete the detail drawings. In some cases, there may be omitted dimensions that you must approximate using your own judgment. When making the detail drawings, strive to provide all of the necessary information, notes, and dimensions to describe the views completely. Utilize any of the previously covered principles, conventions, and techniques to present the views with the maximum of clarity and simplicity.

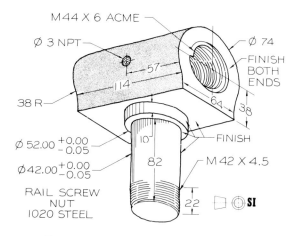

Fig. 23.36 Make a detail drawing on a Size B sheet.

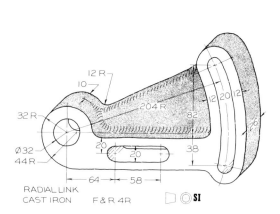

Fig. 23.37 Make a detail drawing on a Size B sheet.

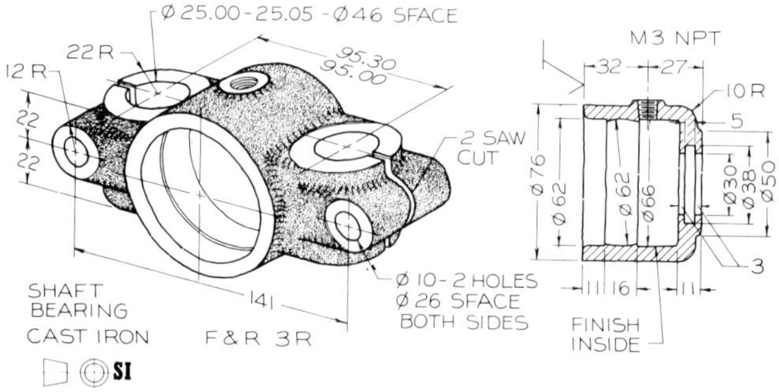

Fig. 23.38 Make a detail drawing on a Size C sheet.

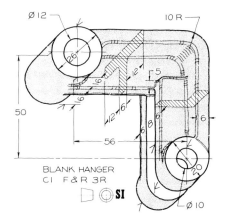

Fig. 23.39 Make a detail drawing on a Size B sheet.

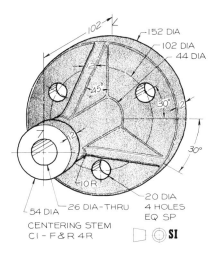

152 DIA
102 DIA
44 DIA
102
45°
30°
30°
10R
54 DIA
26 DIA-THRU
20 DIA
4 HOLES
EQ SP

CENTERING STEM
CI - F & R 4R

Fig. 23.40 Make a detail drawing on a Size B sheet.

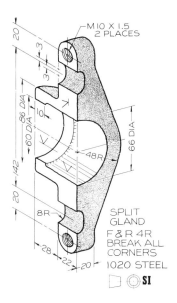

M10 X 1.5
2 PLACES
20
3
3
86 DIA
60 DIA
142
10
48R
66 DIA
20
8R
SPLIT
GLAND
F & R 4R
BREAK ALL
CORNERS
1020 STEEL
28
22
20

Fig. 23.43 Make a detail drawing on a Size B sheet.

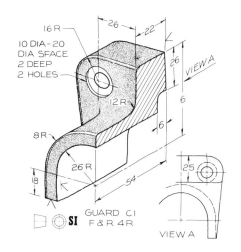

16 R
26
22
10 DIA-20
DIA SFACE
2 DEEP
2 HOLES
26
VIEW A
6
12 R
6
8R
26 R
18
25
54
GUARD CI
F & R 4R
VIEW A

Fig. 23.41 Make a detail drawing on a Size B sheet.

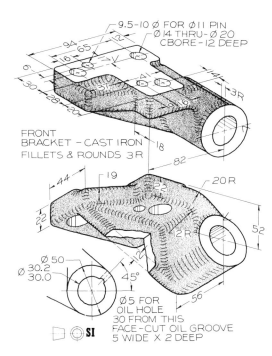

9.5-10 Ø FOR Ø11 PIN
Ø 14 THRU-Ø 20
CBORE-12 DEEP
94
65
16
9
30
28
20
14
3R
18
82
FRONT
BRACKET - CAST IRON
FILLETS & ROUNDS 3R

44
19
22
20 R
22
12 R
52
Ø 50
Ø 30.2
30.0
45°
Ø 5 FOR
OIL HOLE
30 FROM THIS
FACE-CUT OIL GROOVE
5 WIDE X 2 DEEP
56

Fig. 23.44 Make a detail drawing on a Size B sheet.

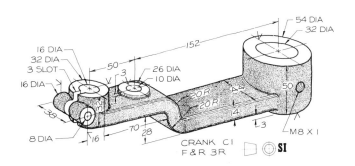

54 DIA
32 DIA
16 DIA
32 DIA
3 SLOT
50
26 DIA
10 DIA
16 DIA
3
152
20R
20R
50
38
4
3
8 DIA
16
70
28
M8 X 1
CRANK CI
F & R 3R

Fig. 23.42 Make a detail drawing on a Size B sheet.

458

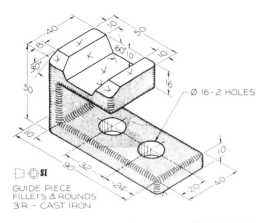

GUIDE PIECE
FILLETS & ROUNDS
3 R – CAST IRON

Fig. 23.45 Make a detail drawing on a Size B sheet.

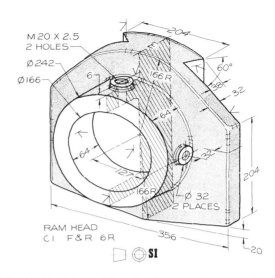

RAM HEAD
CI   F & R   6R

Fig. 23.48 Make a detail drawing on a Size B sheet.

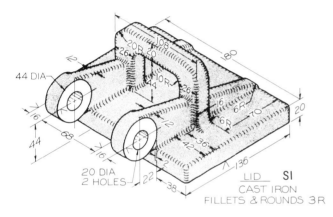

44 DIA

20 DIA
2 HOLES

LID   SI
CAST IRON
FILLETS & ROUNDS 3R

Fig. 23.46 Make a detail drawing on a Size B sheet.

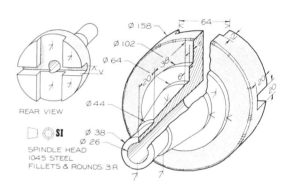

REAR VIEW

SPINDLE HEAD
1045 STEEL
FILLETS & ROUNDS 3R

Fig. 23.49 Make a detail drawing on a Size B sheet.

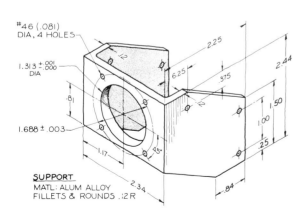

#46 (.081)
DIA, 4 HOLES

1.313 +.001
      –.000
DIA

1.688 ± .003

SUPPORT
MATL: ALUM ALLOY
FILLETS & ROUNDS .12 R

Fig. 23.47 Make a detail drawing on a Size B sheet using (A) decimal inches, or (B) convert the inches to millimeters.

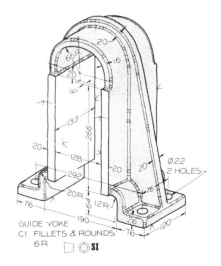

GUIDE YOKE
CI   FILLETS & ROUNDS
6R

Fig. 23.50 Make a detail drawing on a Size B sheet.

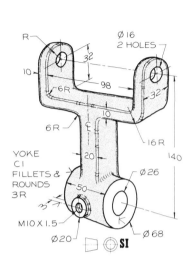

Fig. 23.51 Make a detail drawing on a Size B sheet.

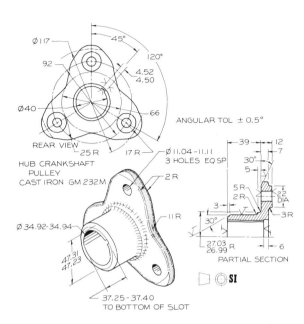

Fig. 23.53 Make a detail drawing on a Size B sheet.

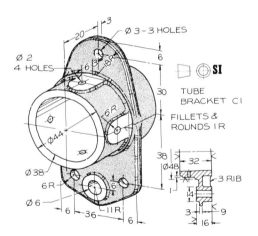

Fig. 23.52 Make a detail drawing on a Size B sheet.

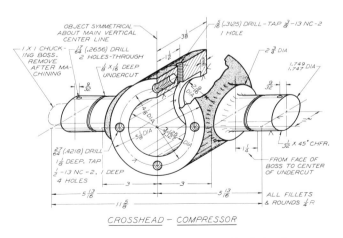

CROSSHEAD − COMPRESSOR

Fig. 23.54 Make a detail drawing on a Size B sheet using (A) decimal inches, or (B) convert the inches to millimeters.

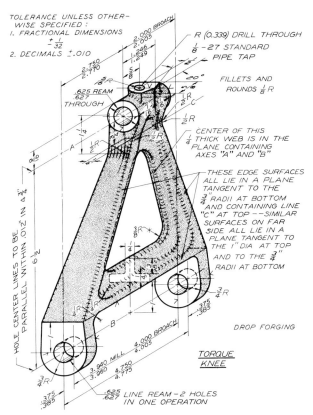

Fig. 23.55 Make a detail drawing on a Size B sheet using (A) decimal inches, or (B) convert the inches to millimeters.

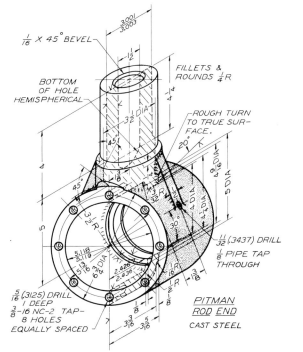

Fig. 23.56 Make a detail drawing on a Size B sheet using (A) decimal inches, or (B) convert the inches to millimeters.

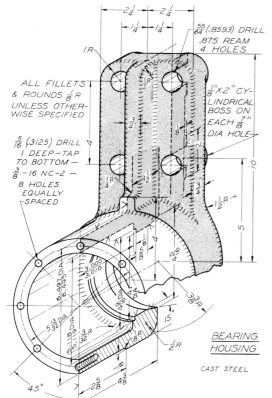

461

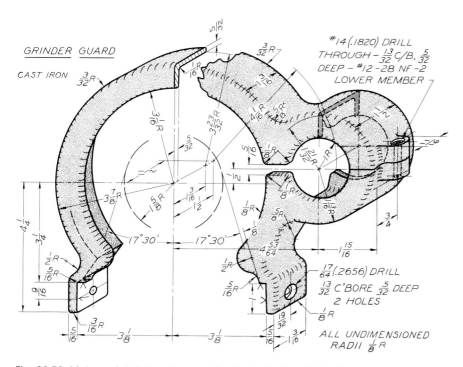

Fig. 23.57 Make a detail drawing on a Size B sheet using (A) decimal inches, or (B) convert the inches to millimeters.

Fig. 23.58 Make a detail drawing on a Size B sheet using (A) decimal inches, or (B) convert the inches to millimeters.

462

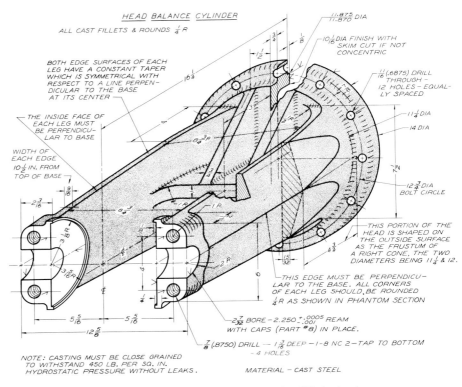

HEAD BALANCE CYLINDER
ALL CAST FILLETS & ROUNDS ¼ R

NOTE: CASTING MUST BE CLOSE GRAINED
TO WITHSTAND 450 LB. PER SQ. IN.
HYDROSTATIC PRESSURE WITHOUT LEAKS.

MATERIAL — CAST STEEL

Fig. 23.59 Make a detail drawing on a Size B sheet using (A) decimal inches, or (B) convert the inches to millimeters.

Fig. 23.60 Make a detail drawing on a Size C sheet using (A) decimal inches, or (B) convert the inches to millimeters.

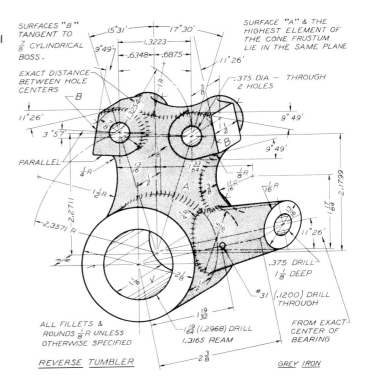

REVERSE TUMBLER

GREY IRON

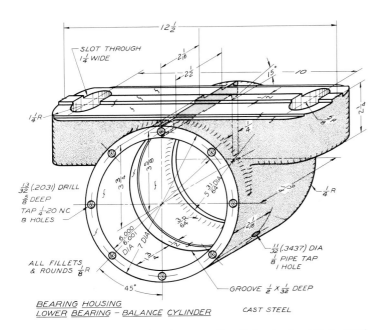

SLOT THROUGH
1¼ WIDE

12½

2⅙

2½

15°

10

2¼

1¼R

5⅝

13/32 (.2031) DRILL
⅝ DEEP
TAP ¼-20 NC
8 HOLES

3¾

3⅝

5 31/64 DIA

¼R

2⅙

¼R

6.000
6.001
DIA 7 DIA

63/64

2⅙

11/32 (.3437) DIA
⅛ PIPE TAP
1 HOLE

ALL FILLETS
& ROUNDS ⅛R

3¾

45°

GROOVE ½ × 1/32 DEEP

BEARING HOUSING
LOWER BEARING – BALANCE CYLINDER   CAST STEEL

Fig. 23.61 Make a detail drawing on a Size C sheet using (A) decimal inches, or (B) convert the inches to millimeters.

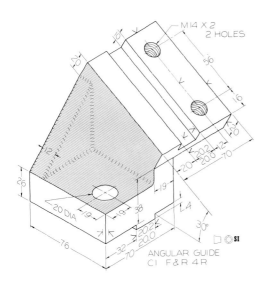

M14 × 2
2 HOLES

20

50

16

12

20.2
20.0
20.2

20

26

2R

20 DIA

19

19

19

38

4

30°

76

32

20.2
20.0

70

□ ◎ **SI**

ANGULAR GUIDE
C1 F & R 4R

Fig. 23.62 Make a detail drawing on a Size B sheet using (A) decimal inches, or (B) millimeters.

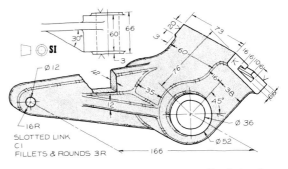

30°

60

66

20

3

73

3

60

16 61/64

□ ◎ **SI**

Ø 12

12

35

16

38

45°

16

12

Ø 36

16R

Ø 52

SLOTTED LINK
C1
FILLETS & ROUNDS 3R

166

Fig. 23.63 Make a detail drawing on a Size B sheet using (A) decimal inches, or (B) millimeters.

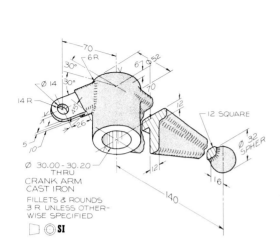

Fig. 23.64 Make a detail drawing on a Size B sheet.

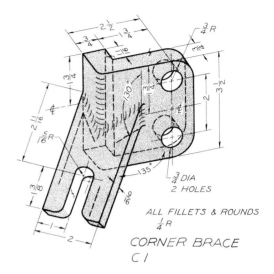

Fig. 23.66 Make a detail drawing on a Size B sheet. Convert the fractional inches to (A) decimal inches, or (B) millimeters.

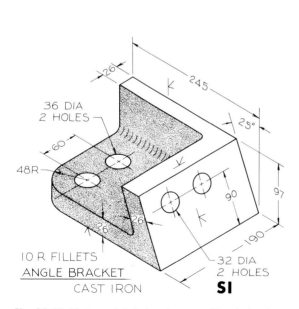

Fig. 23.65 Make a detail drawing on a Size A sheet.

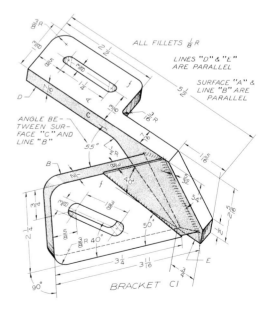

Fig. 23.67 Make a detail drawing on a Size C sheet. Convert the fractional inches to (A) decimal inches, or (B) millimeters.

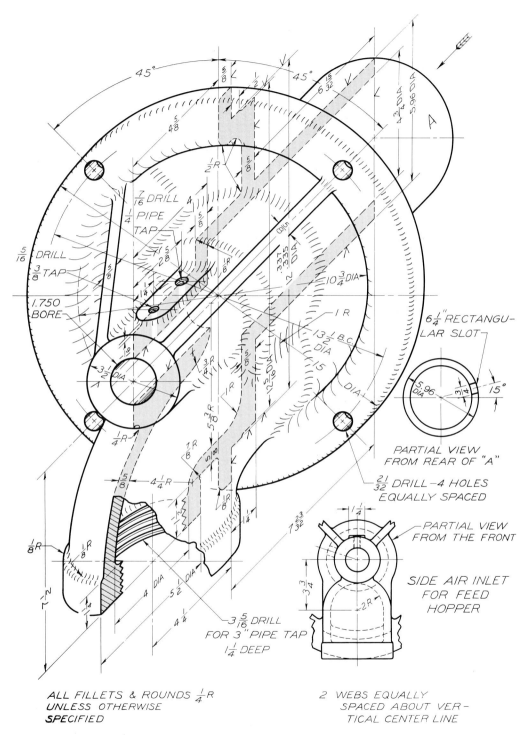

Fig. 23.68 Make a detail drawing on a Size C sheet. Convert the fractional inches to (A) decimal inches, or (B) millimeters.

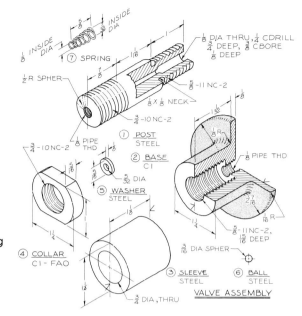

Fig. 23.69 Make a detail drawing of the parts of the valve assembly on Size B sheets. Convert the fractional inches to (A) decimal inches, or (B) millimeters. Draw an assembly and provide a parts list.

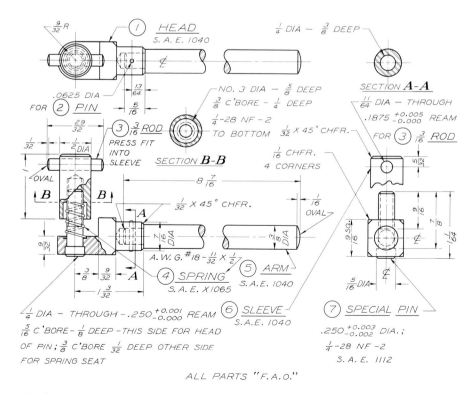

Fig. 23.70 Make a detail drawing of the parts of the cut-off crank on size B sheets. Convert the fractional dimensions to (A) decimal inches, or (B) millimeters. Draw an assembly and provide a parts list.

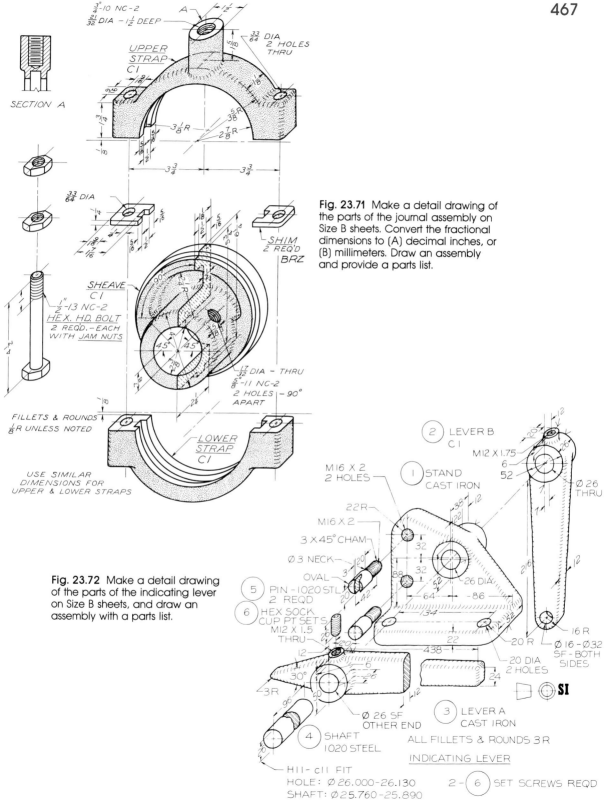

3/4"-10 NC-2
21/32 DIA – 1 1/2 DEEP

A

1/2

33/64 DIA
2 HOLES
THRU

5/8

UPPER
STRAP
C1

9/16

OVAL

3/4

5/16

5/8

3 1/8 R

2 7/8 R

5/8 R
– 3/8

1/8

1/2

3 3/4

3 3/4

SECTION A

33/64 DIA

1/4

5/16

9/16

1/16

1/8

5/16

2

1/2

SHIM
2 REQ'D
BRZ

SHEAVE
C1

1/2"-13 NC-2
HEX. HD. BOLT
2 REQD. – EACH
WITH JAM NUTS

45° 45°

17/32 DIA – THRU
5/8"-11 NC-2
2 HOLES – 90°
APART

2 1/4

7 1/4

1/8

FILLETS & ROUNDS
1/8 R UNLESS NOTED

LOWER
STRAP
C1

USE SIMILAR
DIMENSIONS FOR
UPPER & LOWER STRAPS

Fig. 23.71 Make a detail drawing of
the parts of the journal assembly on
Size B sheets. Convert the fractional
dimensions to (A) decimal inches, or
(B) millimeters. Draw an assembly
and provide a parts list.

Fig. 23.72 Make a detail drawing
of the parts of the indicating lever
on Size B sheets, and draw an
assembly with a parts list.

② LEVER B
C1

M12 X 1.75

6

52

① STAND
CAST IRON

Ø 26
THRU

M16 X 2
2 HOLES

22R

M16 X 2

38

32

3 X 45° CHAM

Ø 3 NECK

88

32

OVAL

⑤ PIN –1020 STL
2 REQD

⑥ HEX SOCK
CUP PT SET S
M12 X 1.5
THRU

26 DIA

64   86

34

22

20 R

438

Ø 16 – Ø 32
SF – BOTH
SIDES

16 R

20 DIA
2 HOLES

24

12

30°

6

26

3R

96

Ø 26 SF
OTHER END

③ LEVER A
CAST IRON

④ SHAFT
1020 STEEL

ALL FILLETS & ROUNDS 3 R

INDICATING LEVER

H11– c11 FIT
HOLE: Ø 26.000 – 26.130
SHAFT: Ø 25.760 – 25.890

2 – ⑥ SET SCREWS REQD

SI

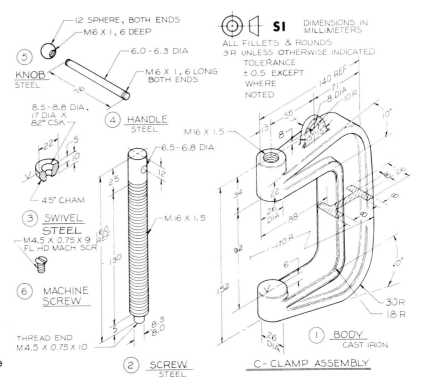

Fig. 23.73 Make a detail drawing of the parts of the C-clamp assembly on Size B sheets; show an assembly of the parts, and provide a parts list.

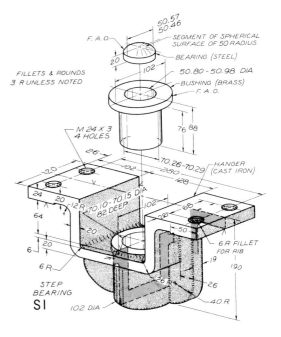

Fig. 23.74 Make a detail drawing of the parts of the step bearing on Size B sheets, and draw an assembly with a parts list.

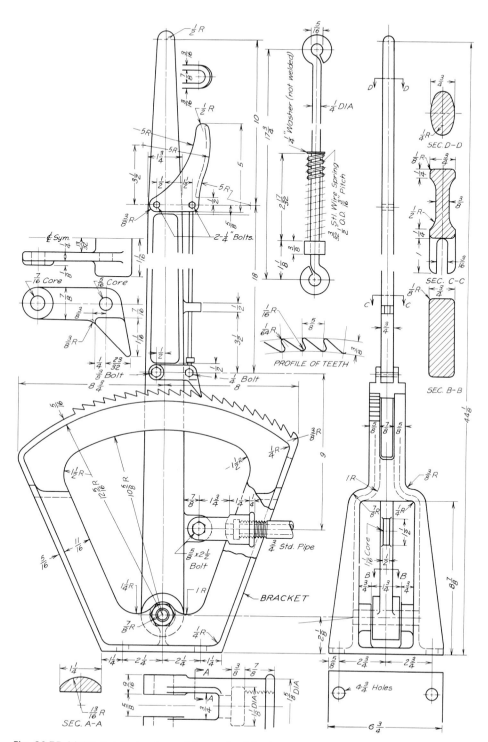

Fig. 23.75 Make detail drawings of the parts of the brake lever on Size B sheets. Convert the fractional inches to (A) decimal inches, or (B) millimeters. Draw an assembly of the parts and provide a parts list.

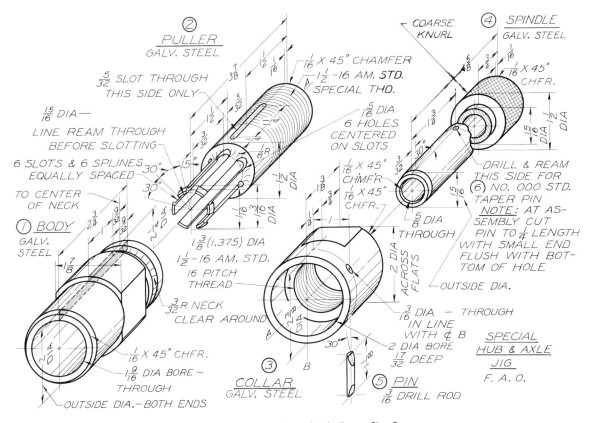

② PULLER
GALV. STEEL

$\frac{5}{32}$ SLOT THROUGH
THIS SIDE ONLY

$\frac{15}{16}$ DIA—
LINE REAM THROUGH
BEFORE SLOTTING

6 SLOTS & 6 SPLINES
EQUALLY SPACED
TO CENTER
OF NECK

① BODY
GALV.
STEEL

$\frac{1}{16}$ X 45° CHFR.

$1\frac{9}{16}$ DIA BORE —
THROUGH

OUTSIDE DIA. – BOTH ENDS

$\frac{1}{38}$ $\frac{1}{2}$ $\frac{1}{16}$

$\frac{5}{32}$

$\frac{1}{16}$ X 45° CHAMFER
$1\frac{1}{2}$ -16 AM. STD.
SPECIAL THD.

$\frac{5}{16}$ DIA
6 HOLES
CENTERED
ON SLOTS

COARSE
KNURL

④ SPINDLE
GALV. STEEL

$\frac{5}{8}$ $\frac{3}{4}$ $\frac{1}{16}$

$\frac{1}{16}$ X 45°
CHFR.

$\frac{3}{4}$

30°

$\frac{15}{16}$ DIA $\frac{1}{2}$ DIA

DRILL & REAM
THIS SIDE FOR
⑥ NO. 000 STD.
TAPER PIN
NOTE: AT AS-
SEMBLY CUT
PIN TO $\frac{1}{4}$ LENGTH
WITH SMALL END
FLUSH WITH BOT-
TOM OF HOLE

OUTSIDE DIA.

$1\frac{3}{8}$ (1.375) DIA
$1\frac{1}{2}$ -16 AM. STD.
16 PITCH
THREAD

$\frac{3}{32}$ R NECK
CLEAR AROUND

30° 30°

$\frac{1}{16}$ X 45°
CHMFR
$\frac{1}{16}$ X 45°
CHFR.

$\frac{5}{8}$ DIA
THROUGH

2 DIA
ACROSS
FLATS

$\frac{3}{16}$ DIA — THROUGH
IN LINE
WITH ₵ B
2 DIA BORE
$\frac{17}{32}$ DEEP

③ COLLAR
GALV. STEEL

30°

⑤ PIN
$\frac{3}{16}$ DRILL ROD

SPECIAL
HUB & AXLE
JIG
F. A. O.

Fig. 23.76 Make detail drawings of the parts of the hub and axle jig on Size B
sheets. Convert the fractional inches to (A) decimal inches, or (B) millimeters.
Draw an assembly of the parts and provide a parts list.

Fig. 23.77 Make a detail drawing
of the parts of the special puller on
Size B sheets. Convert the fractional
inches to (A) decimal inches, or (B)
millimeters. Draw a pictorial
assembly of the parts and give a
parts list.

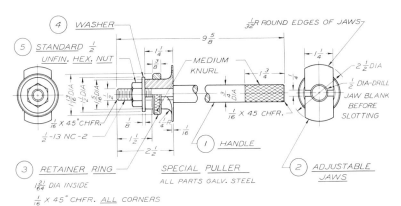

④ WASHER

⑤ STANDARD $\frac{1}{2}$
UNFIN. HEX. NUT

$\frac{1}{16}$ X 45° CHFR.

$\frac{1}{2}$ -13 NC-2

③ RETAINER RING
$1\frac{21}{64}$ DIA INSIDE

$\frac{1}{16}$ X 45° CHFR. ALL CORNERS

$9\frac{5}{8}$

MEDIUM
KNURL

$\frac{3}{4}$ DIA

$\frac{1}{16}$ X 45 CHFR.

① HANDLE

SPECIAL PULLER
ALL PARTS GALV. STEEL

$\frac{1}{32}$ R ROUND EDGES OF JAWS

$2\frac{1}{2}$ DIA
$\frac{1}{2}$ DIA-DRILL
JAW BLANK
BEFORE
SLOTTING

② ADJUSTABLE
JAWS

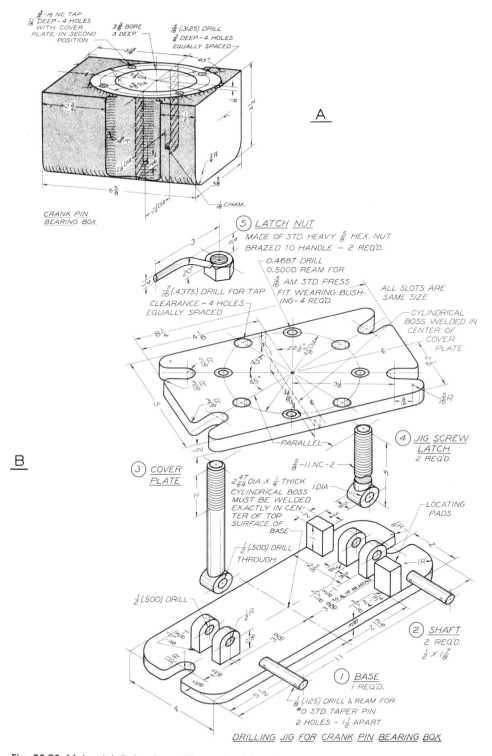

A

CRANK PIN
BEARING BOX

B

5 LATCH NUT
MADE OF STD. HEAVY ⅝ HEX. NUT
BRAZED TO HANDLE – 2 REQ'D.

ALL SLOTS ARE
SAME SIZE

CYLINDRICAL
BOSS WELDED IN
CENTER OF
COVER
PLATE

4 JIG SCREW
LATCH
2 REQ'D.

3 COVER
PLATE

LOCATING
PADS

2 SHAFT
2 REQ'D.
½ × ⅞

1 BASE
1 REQ'D.

DRILLING JIG FOR CRANK PIN BEARING BOX

Fig. 23.78 Make detail drawings of the parts of the drilling jig and crank pin bearing box. Convert the fractional inches to (A) decimal inches, or (B) millimeters. Add finish marks to the surfaces that should be finished, but are not indicated as being finished. Draw an assembly of the parts and provide a parts list.

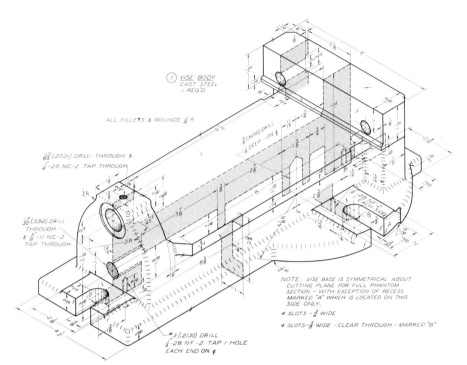

Fig. 23.79 Make a detail drawing of the vise body on Size C sheets. Convert the fractional inches to (A) decimal inches, or (B) millimeters.

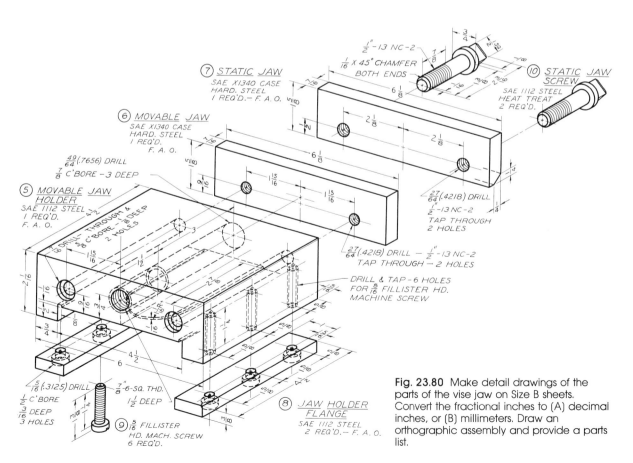

Fig. 23.80 Make detail drawings of the parts of the vise jaw on Size B sheets. Convert the fractional inches to (A) decimal inches, or (B) millimeters. Draw an orthographic assembly and provide a parts list.

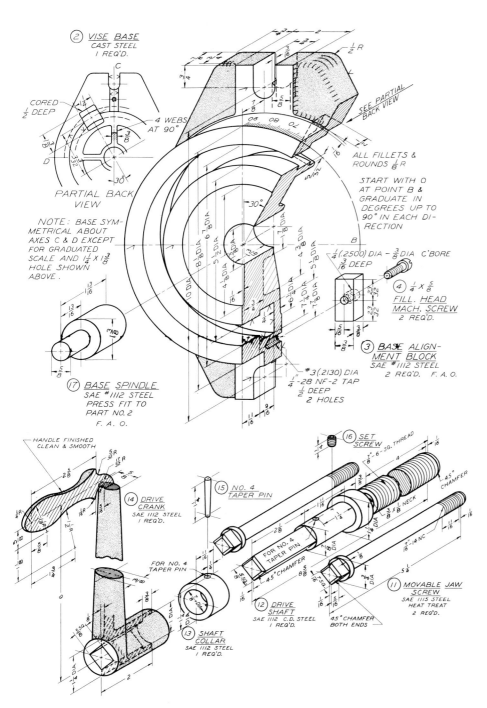

Fig. 23.81 Make detail drawings of the parts of the vise body and crank assembly on Size B sheets. Convert the fractional inches to (A) decimal inches, or (B) millimeters. Draw an assembly of the parts and provide a parts list.

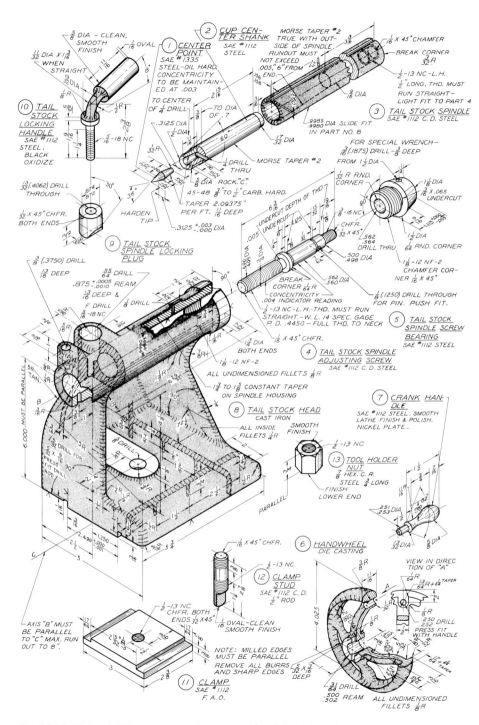

Fig. 23.82 Make detail drawings of the parts of the tail stock assembly on Size C sheets. Convert the fractional inches to (A) decimal inches, or (B) millimeters. Draw an assembly of the parts and provide a parts list.

REPRODUCTION METHODS AND DRAWING SHORTCUTS

24.1 INTRODUCTION

So far this text has dealt with the processes of preparing drawings and specifications to communicate ideas in a technical manner. This has progressed through the working-drawing stage where a detailed drawing is completed on tracing film or paper. Now the drawing must be reproduced, folded, and prepared for filing or transmittal to the users of the drawings. These steps will be discussed in this chapter.

Shortcuts that aid in the completion of a set of working drawings will be briefly covered to familiarize you with time-saving methods.

24.2 REPRODUCTION OF WORKING DRAWINGS

A drawing made by a drafter is of little use in its original form. It would be impractical for the original to be handled by checkers and, even more so, by workers in the field or in the shop. The drawing would quickly be damaged or soiled and no copy would be available as a permanent record of the job. Consequently, reproduction of drawings is necessary so that copies can be available for use by the various people concerned. A checker can

mark corrections on a work copy without damaging the original drawing. The drafter in turn can make the corrections on the original from the work copy.

Several methods of reproduction are used for making the copies that have traditionally been called "blueprints." This term comes from the original reproduction process that gave a blue background with white lines. The term blueprint is still used, although incorrectly, to describe almost all reproduced working drawings regardless of the process. However, you should become familiar with the various processes so that you can refer to them properly.

The processes discussed here are (1) diazo printing, (2) blueprinting, (3) microfilming, (4) xerography, and (5) photostatting. These are the most often used processes of reproducing engineering drawings.

Diazo Printing

The diazo print is more correctly called a "white-print" or a "blue-line print" than a blueprint, since it has a white background and blue lines. Other colors of lines are available depending on the type of paper used. The white background makes notes and corrections drawn on the draw-

Fig. 24.1 A typical whiteprinter that operates on the diazo process. (Courtesy of Blu-Ray, Inc.)

ing more clearly visible than does the blue background of blueprint.

Both blueprinting and diazo printing require that the original drawing be made on semitransparent tracing paper, cloth, or film that will allow light to pass through the drawing. The paper on which the copy is made, the diazo paper, is chemically treated so that it has a yellow tint on one side. This paper must be stored away from heat and light to prevent spoilage.

The tracing paper or film drawing is placed face up on the yellow side of the diazo paper and is run through the diazo-process machine, which exposes the drawing to a built-in light. The light passes through the tracing paper and burns out the yellow chemical on the diazo paper except where the drawing lines have shielded the paper from the light. After exposure to light, the diazo paper is a duplicate of the original drawing except that the lines are light yellow and are not permanent. The diazo paper is then passed through the developing unit of the diazo machine where the yellow lines are developed into permanent blue lines by exposure to ammonia fumes. Diazo printing is a completely dry process.

A typical diazo printer-developer, sometimes called a whiteprinter, is shown in Fig. 24.1. This machine will take sheets up to 42″ wide.

The speed at which the drawing passes under the light determines the darkness of the copy. A slow speed burns out more of the yellow and produces a clear white background; however, some of the lighter lines of the drawing may be lost. Most diazo copies are made at a somewhat faster speed to give a light tint of blue in the background and

stronger lines in the copy. Ink drawings give the best reproductions since the lines are uniform in quality.

It is important to remember that the quality of the diazo print is determined by the quality of the original drawing. A print will not be clear and readable unless the lines of the drawing are dark and dense. Light will pass through gray lines and the result will be a fuzzy print that will not be satisfactory.

Blueprinting

Blueprints are made with paper that is chemically treated on one side. As in the diazo process the tracing-paper drawing is placed in contact with the chemically treated side of the paper and exposed to light. The exposed blueprint paper is washed in clear water for a few seconds and is coated with a solution of potassium dichromate. The print is washed again and dried. The wet sheets can be hung on a line to dry or they can be dried by special equipment made for this purpose.

This process is still used but to a lesser degree than in the past. Since it is a wet process, it requires more time than does the diazo process.

Microfilming

Microfilming is a photographic process that converts large drawings into film copies—either aperture cards or roll film. Drawings must be photographed on either 16 mm or 35 mm film. A camera and copy table are shown in Fig. 24.2.

The roll film or aperture cards can be placed in a microfilm enlarger-printer (Fig. 24.3), where the individual drawings can be viewed on a built-in screen. The selected drawings can then be printed from the film to give standard-size drawings. The range of enlargement varies with the equipment used. Microfilm copies are usually smaller than the original drawings; this saves paper and makes the drawings easier to use.

Microfilming makes it possible to eliminate large, bulky files of drawings, since hundreds of drawings can be stored in miniature size on a small amount of film. The aperture cards shown in Fig. 24.3 are data processing cards that can be cataloged and recalled by a computer to make them accessible with a minimum of effort.

Fig. 24.2 The Micro-Master 35 mm camera and copy table are used for microfilming engineering drawings. (Courtesy of Keuffel & Esser Company, Morristown, N.J.)

Fig. 24.3 The Bruning 1200 microfilm enlarger-printer makes drawings up to 18″ × 24″ from aperture cards and roll film. (Courtesy of Bruning Company.)

Xerographic Reproduction

Xerography is an electrostatic process of duplicating drawings on ordinary, unsensitized paper. This process was developed originally for office duplication uses, but has recently been used for the reproduction of engineering drawings.

An advantage of the xerographic process is the possibility of making copies of drawings at a reduced size (Fig. 24.4). The new Xerox 2080 reduces drawings as large as 24″ × 36″ directly from the original to paper sizes ranging from 8″ × 10″ to 14″ × 18″.

Photostatic Reproduction

A method of enlarging or reducing drawings using a camera is called the **photostatic process.** The combination camera and processor shown in Fig. 24.5 is used for photographing drawings and producing high-contrast photographic copies.

Fig. 24.4 This Xerox printer permits the reduction of drawings to less than half size and enlargements to nearly fifty percent larger. (Courtesy of Xerox.)

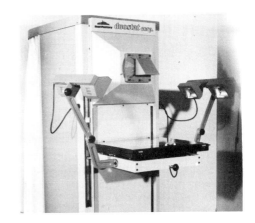

Fig. 24.5 A combination camera-processor for enlarging and reducing drawings to be reproduced as photostats. (Courtesy of the Duostat Corporation.)

Fig. 24.6 The steps in making a photostat with the Duostate camera-processor.

Step 1 The drawing or artwork is placed under the glass on the copyboard.

Step 2 The image is projected inside the darkroom compartment onto photographically sensitive negative paper.

Step 3 The exposed negative paper is placed in contact with the receiver paper and the two are fed through the developing chemicals in the processor.

The drawing or artwork is placed under the glass of the exposure table (Fig. 24.6), which is lit by built-in lamps. The image can be seen on the glass inside the darkroom where it is exposed on photographically sensitive paper. The negative paper that has been exposed to the image is placed in contact with receiver paper and the two are fed through the developing solution to obtain a photostatic copy.

These high-contrast reproductions are often used to prepare artwork that is to be printed by offset printing presses. Additionally, this process can be used to make reproductions on transparent films, and for the reproduction of photographs with tones of gray, called halftones.

24.3 FOLDING THE DRAWING

Once the prints have been finished, the original drawings should be stored in a flat file for future use and updating. The original drawings should not be folded, and handling of any kind should be kept to a minimum.

The printed drawings, on the other hand, are usually folded for transmittal from office to office, often through the mail. The methods of folding Size B, C, D, and E sheets are shown in Fig. 24.7. In each case, the final size after folding is 8½″ × 11″ (or 9″ × 12″). This is the standard modular size the fits most mailing envelopes and file cabinets.

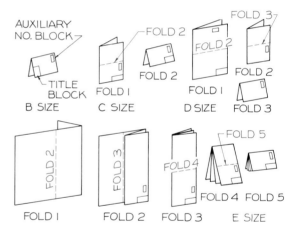

Fig. 24.7 Standard folds for engineering drawing sheets. The final size in each case is 8½″ × 11″.

Note that the drawings are folded with the title blocks positioned to be visible at the top and the lower right of the drawing. This is essential in order for a drawing to be retrieved from a file cabinet with the greatest ease.

24.4 OVERLAY DRAFTING TECHNIQUES

Valuable drafting time can be saved by taking advantage of current processes and materials that utilize a series of overlays to separate parts of a single drawing. For example, engineers and architects often work from a single site plan or floor plan on which a variety of drawings will be made that all utilize the same base plan. The floor plan of a building will be used for the electrical plan, furniture arrangement plan, air-conditioning plan, floor materials plan, etc. You can see that it would be expensive to retrace the plan for each application.

A series of overlays can be used in a system referred to as **pin drafting,** where accurately spaced holes are punched in the polyester drafting film at the top edge of the sheets. These holes are aligned on pins attached to a metal strip that match the holes that were punched in the film (Fig. 24.8). This method ensures accurate alignment or registration of a series of sheets, and the polyester film ensures stability of the material since it does not stretch or sag with changes in humidity.

The steps in using the pin drafting system are shown in Fig. 24.9. The title block can be printed on all drawing film sheets along with the border lines. The overlay sheets need not have borders or a title block. The base plan is the sheet that will be common to several drawings.

The composite of the various overlays is shown in the third part of Fig. 24.9. Note that the base plan is printed in gray, which means that it has been screened to the desired percentage of black by a photographic process. This makes the additional information provided by the subcontractor or consultant on the plan more noticeable and easier to read from the base plan.

The set of overlays could be attached by the alignment pins or taped together and run through a diazo machine for full-size prints. Another option of reproduction is the use of a flat-bed process camera (Fig. 24.10) to photograph and reduce the drawings to a standard $8\frac{1}{2}'' \times 11''$ size.

You can see that the process of using overlays makes it possible for base plans to be sent to consultants who will overlay them (using the pin registration system) for making their particular additions as overlays. This ensures that the consultants are each working with uniform plans and that they will not accidentally erase or modify any part of the design that was sent to them.

When a large number of prints are needed, reproductions may be printed by offset lithography. This permits them to be printed in multicolors to highlight certain features on drawings. In these ex-

Fig. 24.8 In the pin system, separate overlays are aligned by seven pins mounted on metal strips. (Courtesy of Keuffel & Esser Company, Morristown, N.J.)

amples, the primary objective has been to economize the preparation of the drawings and increase their clarity.

24.5 PASTE-ON PHOTOS

There are many occasions when repetitive details must be shown on a drawing. The engineer who is laying out a manufacturing plant will need to represent many identical machines on a drawing along with other features that are repeated. The architect encounters the same situation when locating furniture or common details such as offices, bathrooms, window details, etc.

When a number of repetitive drawings are necessary, it is often more economical to use the photographic process to make a number of reproductions of the drawings on transparent film. These features can be "pasted" into position on the master drawing. (The term "pasted" is commonly used to describe this attachment, but in reality, tape is used when the drawings are reproduced on transparent film. If the reproductions are made as opaque photostats, then rubber cement can be used.)

The office arrangements shown in Fig. 24.11 are transparencies that have been photographi-

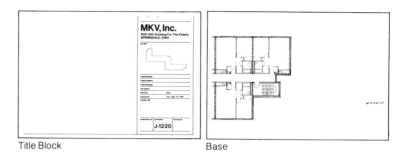

Title Block   Base

Fig. 24.9 Examples of overlays that are overlayed and reproduced to give a combination of several sheets in the final reproduced drawing. (Courtesy of Keuffel & Esser Company, Morristown, N.J.)

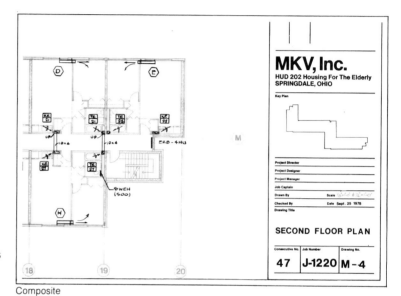

Composite

Fig. 24.10 The process camera that is used to reduce and enlarge engineering drawings is the heart of the pin system. (Courtesy of Keuffel & Esser Company, Morristown, N.J.)

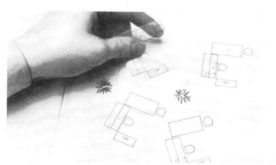

Fig. 24.11 When drawings of parts or arrangements are to be used repetitively on a set of drawings, it may be more economical to photographically reproduce them than draw them. (Courtesy of Eastman Kodak Company.)

Fig. 24.12 The photographically reproduced drawing features are taped in position to complete the overall drawing. (Courtesy of Eastman Kodak Company.)

cally duplicated from a single drawing. These can be efficiently attached to the master drawing to save drawing time. The architect in Fig. 24.12 is composing an entire drawing sheet with paste-on images of repetitive features that were previously drawn.

The steps in preparing final drawings from paste-on photos are schematically shown in Fig. 24.13. Photographic prints of the repetitive details

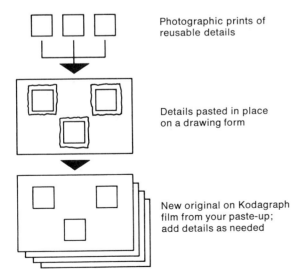

Photographic prints of reusable details

Details pasted in place on a drawing form

New original on Kodagraph film from your paste-up; add details as needed

Fig. 24.13 The steps in photographically revising an engineering drawing. (Courtesy of Eastman Kodak Company.)

are made, and then are pasted into position on the new drawing. In the third step, the drawings are reproduced on film or paper. If reproduced on film, successive prints can be made by the diazo process.

24.6 PHOTO REVISIONS

When a previously made drawing is in need of revision, unnecessary drawing time can be saved by photographically modifying the drawing. If you wish to change the old drawing in Fig. 24.14 to look like the example shown, this can be done by making a clear film reproduction of the parts, cutting them out and taping them in position on a new form. The new drawing is then photographed onto a new film on which additional notes and lines can be provided to complete the drawing. This new drawing can be used as a master for making diazo prints.

Opaquing is another method of revising a drawing. A photographic negative is made from the original drawing. The area to be removed is opaqued out by using a brush and an ink-like opaquing solution (Fig. 24.15). You can see that it is advantageous to reduce the negative to a smaller size to lessen the area and effort of opaquing.

The negative is then used to make a positive on polyester drafting film, on which the revision can be drawn and noted in the conventional manner. You now have a new master from which photographic or diazo prints can be made.

24.7 STICK-ON MATERIALS

A number of companies market stick-on symbols, screens, and lettering that can be applied to drawings to economize on time and improve the appearance of drawings. The three standard types of materials are stick-ons, burnish-ons, and tape-ons.

The stick-on symbols or letters are printed on thin plastic sheets that are cut out with a razor-sharp blade and are transferred to the drawing where the cutout adheres to the drawing. It is burnished permanently to it (Fig. 24.16). The plastic cutout remains as a part of the drawing. This material is available in glossy and matte finishes.

The burnish-on symbols are applied by placing the entire sheet over the drawing and burnish-

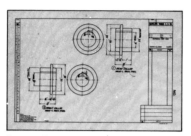

Say you have an existing drawing:

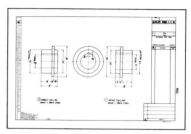

and you want to revise it like this.

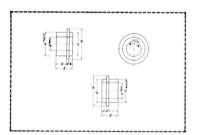

Step 1: First you make a *clear* film reproduction of the original and cut out the elements.

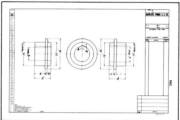

Step 2: Then tape the elements in their new positions on a new form.

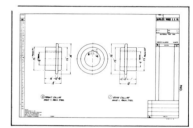

Step 3: Photograph it on film with a matte finish (the tapes and film edges will disappear). Draw in whatever extra detail you want—and you have a new original drawing.

Fig. 24.14 The steps of photographically revising an engineering drawing. (Courtesy of Eastman Kodak Company.)

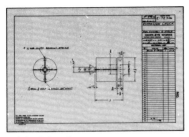

Original drawing

Step 1: Make a reduced-size negative and opaque the area to be revised. (The small size makes it quicker.)

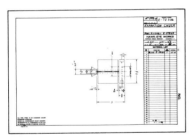

Step 2: Then make a positive on matte film, enlarged to original size.

Fig. 24.16 Stick-on lettering can be applied to a drawing by cutting the letters from the plastic sheet, applying them to the drawing surface in alignment with a guideline, and burnishing the letters to the sheet. (Courtesy of Graphic Products Corporation.)

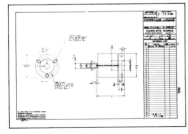

Step 3: Draw in the new detail, and you have a new second original.

Fig. 24.15 The method of modifying a drawing by opaquing the negative. (Courtesy of Eastman Kodak Company.)

ing the desired symbol into place with a rounded-end object such as the end of a pencil cap. The symbol thus is transferred from the plastic sheet to the drawing surface.

Sheets of symbols can be custom-printed for users who have repetitive needs for trademarks and other often-used symbols (Fig. 24.17). Title blocks are sometimes printed in this manner for application to drawings, to reduce drawing time.

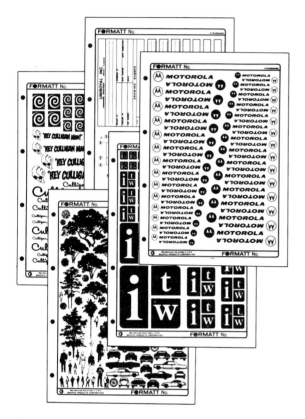

Fig. 24.17 Stick-on symbols are available in a wide range of design and they can be custom-printed to suit the needs of the client. (Courtesy of Graphic Products Corporation.)

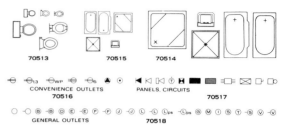

Fig. 24.18 Examples of architectural symbols that can be used on drawings by sticking them to the drawing surface. (Courtesy of Zip-a-Tone Incorporated.)

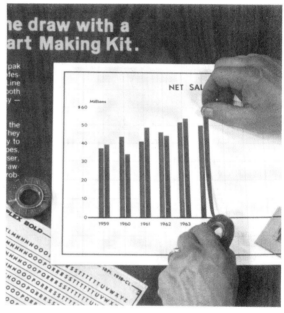

Fig. 24.19 Colored tapes in various widths can be used to speed the process of preparing charts and graphs. (Courtesy of Chartpak Rotex.)

A number of symbols that are used on architectural plans are shown in Fig. 24.18. Plumbing and electrical symbols can be rapidly applied and thereby shorten the drafting time required to complete a drawing where these are required.

Colored adhesive tapes in varying widths are available for the preparation of charts and graphs (Fig. 24.19). Tapes are also used to represent wide lines on large drawings. Although tapes can be burnished on tightly, they can be removed for modification when this is desired.

A matte finished sheet is available from several sources that can be typed on with a standard typewriter and then transferred to the drawing where it is attached with its adhesive backing to become a permanent part of the drawing.

A specially designed typewriter can be used that allows a secretary to assist the drafter by typing notes directly on the surface of the drawing to shorten the time that would be required to hand-letter the notes (Fig. 24.20). It is suggested that a carbon ribbon be used in the typewriter to obtain dense, opaque letters that reproduce well.

24.8 PHOTO DRAFTING

Photography can be used as an effective means of economizing when preparing certain types of drawings. This is especially true when existing

Fig. 24.20 This especially-made typewriter can be used by a secretary to type notes on a drawing, thereby relieving the drafter of this chore. (Courtesy of Vari-Typer.)

built as models, then photographed, noted, and reproduced as photodrawings.

The steps of preparing a photodrawing are shown in Fig. 24.22, where it is desired to specify certain parts of a sprocket-and-chain assembly. In Step 1, a halftone print of the photograph is made. (This is the process of screening the photograph, or representing it by a series of dots that give varying tones of gray.) In Step 2 the halftone is taped to a white drawing sheet, which is then photographed to give a negative. The negative is used to produce a positive on polyester drafting film. The notes can be lettered on this drawing film to complete the master drawing (Step 4). The master drawing can then be used to make diazo prints, or it can be microfilmed or reproduced photographically to the desired size.

Aerial photographs can be used to aid the drafter and technologist in preparing topographical drawings (Fig. 24.23). The photographs are overlaid with polyester film and the details added to include streets, structures, contours, pipelines, utilities, and similar features.

The size of the photograph can be reduced or enlarged to the desired scale when a single dimension on the photograph is known between two points. The photographic process enables the drafter to reproduce the final drawing at the desired size for ease of filing and transmittal.

designs are being modified, and where the existing parts can be photographed.

An example of a photodrawing is shown in Fig. 24.21, where an assembly of parts has been noted. In some cases it would be worthwhile to even build a model for photographing in order to clarify assembly details. This is especially true in the piping industry where complex refineries are

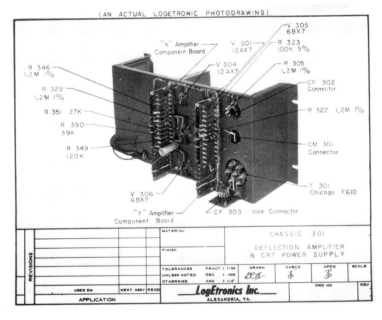

Fig. 24.21 This assembly is an example of a photodrawing that has saved a considerable amount of drafting time. (Courtesy of LogEtronics, Inc., and Eastman Kodak Company.)

Step 1: Make a halftone print of the photograph.

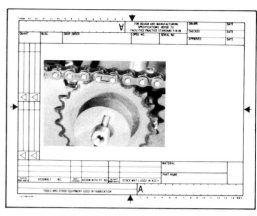

Step 3: Make a positive reproduction on matte film.

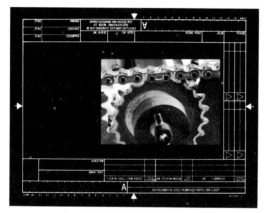

Step 2: Tape the halftone print to a drawing form, and photograph it to produce a negative.

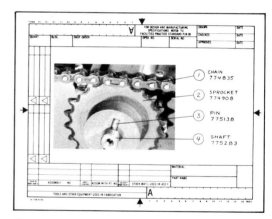

Step 4: Now draw in your callouts—and the job is done.

Fig. 24.22 The steps in making a photodrawing. (Courtesy of Eastman Kodak Company.)

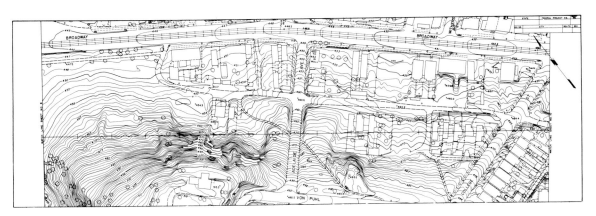

Fig. 24.23 This topographic drawing was made from the aerial photograph to save drafting time. (Courtesy of Jack Amman, Inc., and Eastman Kodak Company.)

PICTORIALS

25.1 INTRODUCTION

A pictorial is an effective means of communicating an idea. This is especially true if a design is unique or if the person to whom it is being explained has difficulty with the interpretation of multiview drawings.

Pictorials are also helpful in developing a design since they enable the designer or drafter to work in three dimensions.

Pictorials are sometimes called technical illustrations. They are widely used in catalogs, parts manuals, and maintenance publications to describe various products. Practically everyone has used technical illustrations and instructions to assemble a product that was purchased unassembled. Pictorials are used in industry for the same purpose of putting parts together properly.

This chapter will cover the basic types of pictorial methods: (1) oblique drawing, (2) isometric drawing, (3) axonometric projection, and (4) perspective projection. Techniques of rendering and commercial materials will be covered as an introduction to technical illustration.

25.2 TYPES OF PICTORIALS

The three commonly used forms of pictorials are (1) obliques, (2) axonometrics/isometrics, and (3) perspectives. Examples of these are shown in Fig. 25.1.

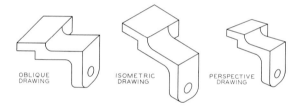

Fig. 25.1 The three standard pictorial systems: oblique, isometric, and perspective.

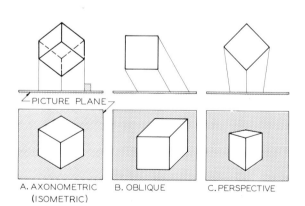

Fig. 25.2 Types of projection systems for pictorials. (A) Axonometric pictorials are formed by parallel projectors that are perpendicular to the picture plane. (B) Obliques are formed by parallel projectors that are oblique to the picture plane. (C) Perspectives are formed by converging projectors that make varying angles with the picture plane.

Oblique pictorials are three-dimensional pictorials made on a plane of paper by projecting from the object with parallel projectors that are *oblique* to the picture plane (Fig. 25.2B).

Axonometric (isometric) projection is a three-dimensional pictorial on a plane of paper drawn by projecting from the object to the picture plane as illustrated in Fig. 25.2A. The parallel projectors are perpendicular to the picture plane.

Perspective pictorials are drawn with projectors that converge at the viewer's eye and make varying angles with the picture plane (Fig. 25.2C).

25.3 OBLIQUE PROJECTIONS

The system of oblique *projection* is used as the basis of oblique *drawings*; however, oblique projections are seldom used. In Fig. 25.3a, a number of lines of sight are drawn through point 2 of line 1–2. Each line of sight makes a 45° angle with the picture plane, which creates a cone with its apex at 2, and each element on the cone makes a 45° angle with the plane. A variety of projections of line 1–2′ on the picture plane can be seen in the front view (b). Each of these projections of 1–2′ is equal in length to the true length of 1–2.

This is called a **cavalier oblique** projection because the projectors make 45° angles with the picture plane, and measurements along the receding axis can be made true length and in any direction.

Examples of cavalier obliques are shown in Fig. 25.4. The front surface is usually positioned parallel to the picture plane; therefore it will appear true size and as an orthographic view. The measurements along the receding axes are made true length and at any angle.

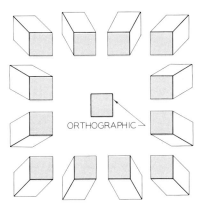

Fig. 25.4 A cavalier oblique is drawn with one surface as a true-size orthographic view. The dimensions along the receding axis are true length and the axes are drawn at any angle.

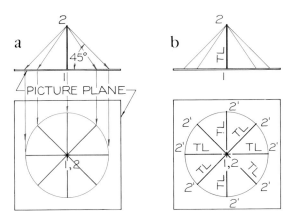

Fig. 25.3 The underlying principle of the cavalier oblique can be seen here where a series of projectors form a cone. Each element makes a 45° angle with the picture plane. Consequently, the projected lengths of 1–2′ are true length and are equal in length to line 1–2 that is perpendicular to the picture plane.

The top and side views are given as orthographic views, and the front view is drawn as a projection by using the two views of a selected line of sight (Fig. 25.5). Projectors are drawn from the object parallel to the lines of sight to locate their respective points in the oblique view. Surfaces that are parallel to the picture plane will appear true size and shape.

The dimensions along the receding axes are less than true length and greater than half-length; therefore this is called a **general oblique.**

If the angle between the line of sight and the picture plane is less than 45°, the measurements along the receding axes will be greater than true length. This is objectionable since the oblique will be distorted.

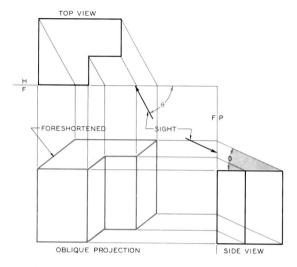

Fig. 25.5 An oblique projection can be drawn at any angle of sight to obtain an oblique pictorial. However, the line of sight should not make an angle less than 45° with the picture plane. This would result in a receding axis longer than true length, thereby distorting the pictorial.

25.4 OBLIQUE DRAWINGS

Oblique projections are seldom used in the manner just illustrated.

Instead, three basic types of *oblique drawings* are used that are based on these principles. The three types are (1) cavalier, (2) cabinet, and (3) general (Fig. 25.6).

In each case, the angle of the receding axis can be at any angle between 0° and 90° (Fig. 25.6). Measurements along the receding axes of the *cavalier oblique* are true length (full scale). The *cabinet oblique* has measurements along the receding axes reduced to half-length. The *general oblique* has measurements along the receding axes reduced to between half and full length.

Three examples of cavalier obliques of a cube are shown in Fig. 25.7. Each has a different angle for the receding axes, but the measurements along the receding axes are true length.

A comparison of cavalier and cabinet obliques is given in Fig. 25.8. The cabinet oblique reduces the distortion of an object with a long depth, thereby giving a more pleasing appearance.

Fig. 25.6 Types of obliques

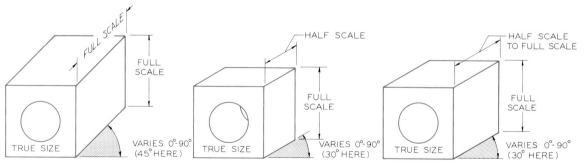

A. The *cavalier oblique* can be drawn with a receding axis at any angle, but measurements along this axis are true length.

B. The *cabinet oblique* can be drawn with a receding axis at any angle, but the measurements along this axis are half size.

C. The *general oblique* can be drawn with a receding axis at any angle, but the measurements along this axis can vary from half to full size.

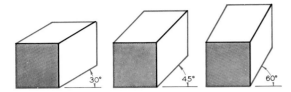

Fig. 25.7 The cavalier oblique is usually drawn with the receding axis at the standard angles of the drafting triangles. Each gives a different view of a cube.

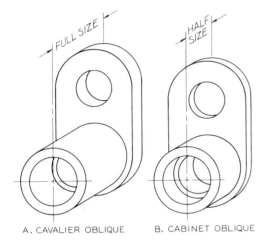

A. CAVALIER OBLIQUE   B. CABINET OBLIQUE

Fig. 25.8 Measurements along the receding axis of a cavalier oblique are full size, and those in a cabinet oblique are half-size.

25.5 CONSTRUCTION OF AN OBLIQUE

An oblique should be drawn by constructing a box using the overall dimensions of height, width, and depth with light construction lines. In Fig. 25.9, the front view is drawn true size in Step 1. In Step 2, the receding axis is drawn at 30° and the depth dimensions are measured true length. This will be a cavalier oblique. True measurements can be made parallel to the three axes.

The notches are removed from the blocked-in construction box to complete the oblique. These measurements are transferred from the given orthographic views with your dividers.

25.6 ANGLES IN OBLIQUE

Angular measurements can be made on the true-size plane of an oblique that is parallel to the picture plane. However, angular measurements will not be true size on the other two planes of the oblique.

To construct an angle in an oblique, coordinates must be used as shown in Fig. 25.10. The sloping surface of 30° must be found by locating the vertex of the angle H distance from the bottom. The inclination is found by measuring the distance of D along the receding axis to establish the slope. This angle is not equal to the 30° angle that was given in the orthographic view.

Fig. 25.9 Oblique construction

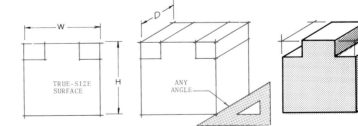

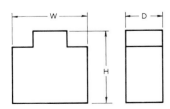

Step 1 The front surface of the oblique is drawn as a true-size plane. The corners are removed.

Step 2 The receding axis is selected and the true dimensions are measured along this axis.

Step 3 The finished cavalier oblique is strengthened to complete the drawing.

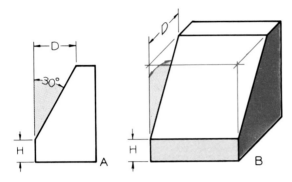

Fig. 25.10 Angles in oblique must be located by using coordinates. They cannot be measured true size except on a true-size plane.

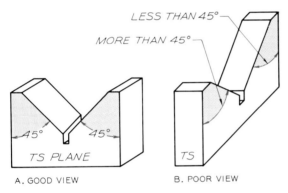

A. GOOD VIEW   B. POOR VIEW

Fig. 25.11 The best view is the view that takes advantage of the ease of construction offered by oblique drawings. The view in (B) is less descriptive and more difficult to construct than the view in (A).

You can see in Fig. 25.11 that a true angle can be measured on a true-size surface. At B, angles along the receding planes are either smaller or larger than their true angles.

It requires less effort and gives a better appearance when obliques are drawn where angles will appear true size as shown in Fig. 25.11A, rather than as shown at B.

25.7 CYLINDERS IN OBLIQUE

The major advantage of obliques is that circular features can be drawn as true circles when they are parallel to the picture plane. This is illustrated in Fig. 25.12 where an oblique of a cylinder is drawn.

The centerlines of the circular end at A are drawn and the receding axis is drawn at any desired angle. The end at B is located by measuring along the axis. Circles are drawn at each end using centers A and B. The circles are connected with tangent lines parallel to the axis. Hidden lines are omitted and the lines are darkened to complete the oblique.

These same principles are used to construct an object with semicircular features (Fig. 25.13). The oblique is positioned to take advantage of the option of drawing circular features as true circles. Centers C_2 and C_3 are located for drawing two semicircles (Step 3).

Fig. 25.12 A cylinder in oblique

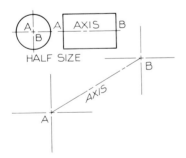

Step 1 The centerlines of the circular ends are located at A and B. The axis is drawn at any angle and since it is true length, this is a cavalier oblique.

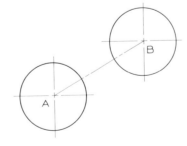

Step 2 Circles are drawn using centers A and B. These are true circles, since they lie in the orthographic planes of the oblique.

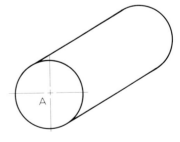

Step 3 Lines are drawn parallel to the axis and tangent to the two end circles. Hidden lines are omitted. Centerlines may also be omitted if they are not used for dimensioning purposes.

Fig. 25.13 Construction of an oblique   493

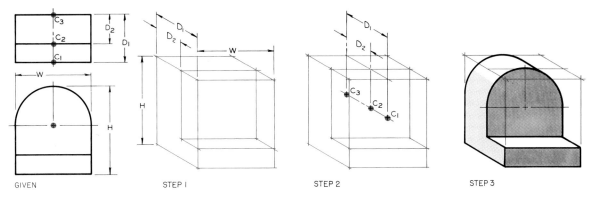

GIVEN   STEP I   STEP 2   STEP 3

Step 1 The overall dimensions are used to block in the oblique pictorial. The notch is removed.

Step 2 The three centers, C_1 C_2, and C_3, are located on each of the planes.

Step 3 The three centers found in step 2 are used to draw the semicircular features of the oblique. Lines are strengthened.

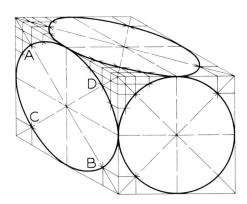

Fig. 25.14 Circular features on the faces of a cavalier oblique of a cube appear as two ellipses and one true circle.

25.8 CIRCLES IN OBLIQUE

Although circular features will be true size on a true-size plane of a cavalier oblique pictorial, circular features on the other two planes will appear as ellipses (Fig. 25.14). The ellipses can be found by using coordinates to locate a series of points along the elliptical curves.

A technique of drawing elliptical views in oblique that is used more often is the four-center ellipse technique, which is shown in Fig. 25.15. A rhombus is drawn that would be tangent to the circle at four points. Perpendicular construction lines are drawn from where the centerlines cross the sides of the rhombus. This construction is performed as shown in Steps 2 and 3 to locate four

Fig. 25.15 Four-center ellipse in oblique

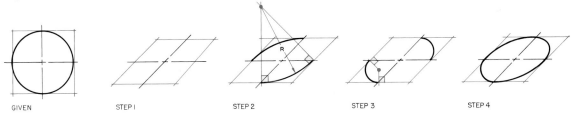

GIVEN   STEP I   STEP 2   STEP 3   STEP 4

Step 1 The circle that is to be drawn in oblique is blocked in with a square which is tangent to the circle at four points. This square will appear as a rhombus on the oblique plane.

Step 2 Construction lines are drawn perpendicularly from the points of tangency to locate the centers for arcs for drawing two of the four segments of the ellipse.

Step 3 The centers for the two remaining arcs are located.

Step 4 When the four arcs have been drawn, the final result is an approximate ellipse.

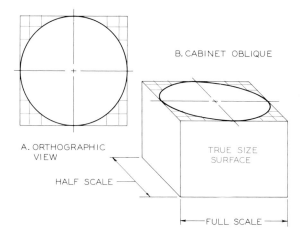

Fig. 25.16 The four-center ellipse technique cannot be used to locate circular shapes on the foreshortened surface of a cabinet oblique. These ellipses must be plotted with coordinates.

centers that are used to draw four arcs that join together to form the ellipse in oblique.

The four-center ellipse method will not work for the cabinet or the general oblique. In these cases, coordinates must be used to locate a series of points on the curve. Coordinates in Fig. 25.16 are shown in orthographic (part A) and then in the cabinet oblique (B). The coordinates that are parallel to the receding axis are reduced to half-size in the oblique. Those in a horizontal direction are drawn full size. The ellipse can be drawn with an irregular curve, or with an ellipse template that approximates the plotted points.

Whenever possible, oblique drawings of objects with circular features should be positioned with circles on true-size planes so they can be drawn as true circles.

In Fig. 25.17, the view at A is better than the one at B since it gives a more descriptive view of

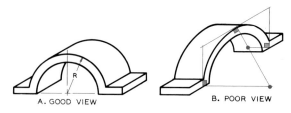

Fig. 25.17 An oblique should be positioned to enable circular features to be drawn with the greatest of ease.

the part and it was easier to draw. The view at B required much more construction since the ellipses were drawn using the four-center ellipse method.

25.9 CURVES IN OBLIQUE

Irregular curves in oblique must be plotted point by point using coordinates (Fig. 25.18). The coordinates are transferred from the orthographic view to the oblique view and the curve is drawn through these points with an irregular curve.

If the object has a uniform thickness, the lower curve can be found by projecting vertically downward from the upper points a distance equal to the height of the object.

In Fig. 25.19 the elliptical feature on the inclined surface was found by using a series of coordinates to locate the points along its curve. These points are then connected with an irregular curve or by using an ellipse template of approximately the same size.

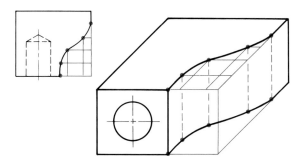

Fig. 25.18 Coordinates are used to establish irregular curves in oblique. The lower curve is found by projecting the points downward a distance equal to the height of the oblique.

25.10 OBLIQUE SKETCHING

The ability to make rapid freehand sketches is a valuable asset to the technical person. An understanding of the mechanical principles of oblique construction is essential for sketching obliques.

As shown in Fig. 25.20, the use of light guidelines is helpful in developing a sketch. These guidelines should be drawn lightly so they will

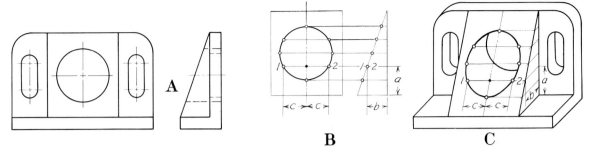

Fig. 25.19 The construction of a circular feature on an inclined surface must be found by plotting points using three coordinates, *a*, *b*, and *c*, to locate points 1 and 2 in this example. The plotted points are connected to complete the elliptical feature.

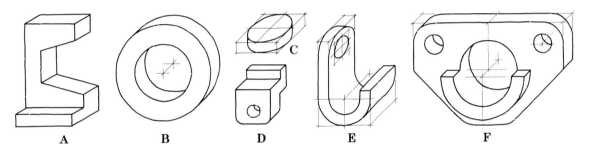

Fig. 25.20 Obliques may be drawn as freehand sketches by utilizing the same principles that were used for instrument pictorials. It is beneficial to use light construction lines to locate the more complex features.

not need to be erased when the finished lines of the sketch are darkened.

When sketching on tracing vellum, it is helpful if a printed grid is placed under the vellum to provide guidelines. Grids are available commercially that have a combination of rectangular and angular grids that can be used for drawing oblique sketches.

25.11 DIMENSIONED OBLIQUES

A dimensioned oblique full section is given in Fig. 25.21 where the interior features and the dimensions of the part are shown. This drawing is sufficient for completely describing the part.

When dimensions and notes are given in oblique, the numerals and lettering should be ap-

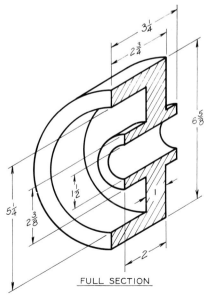

Fig. 25.21 Oblique pictorials can be drawn as sections and dimensioned to serve as working drawings.

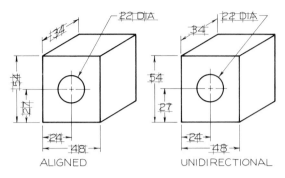

Fig. 25.22 Oblique pictorials can be dimensioned by using either of these methods of applying numerals to the dimension lines.

plied using one of the methods shown in Fig. 25.22. In the aligned method, the numerals are aligned with the dimensioned lines. In the unidirectional method, the numerals are all positioned in a single direction regardless of the direction of the dimension lines. Notes that are connected with leaders are positioned horizontally in both methods.

25.12 ISOMETRIC PICTORIALS

An *isometric projection* is a type of axonometric projection in which parallel projectors are perpendicular to the picture plane and the diagonal of a cube is seen as a point (Fig. 25.23). The three axes are spaced 120° apart and the sides are foreshortened to 82% of their true length.

> The term isometric, which means "equal measurement," is used to describe this type of pictorial since the planes are equally foreshortened.

An *isometric drawing* is similar to an *isometric projection* except that it is not a true axonometric projection, but an approximate method of drawing a pictorial. Instead of reducing the measurement along the axes to 82%, they are drawn true length (Fig. 25.23B).

A comparison between an isometric projection and an isometric drawing is shown in Fig. 25.24. By using the isometric drawing instead of the isometric projection, pictorials can be mea-

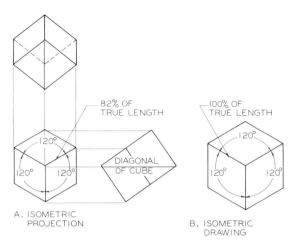

Fig. 25.23 A true isometric projection is found by constructing a view that shows the diagonal of a cube as a point. An isometric drawing is not a true projection since the dimensions are drawn true size rather than reduced in size as in projection.

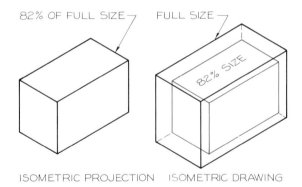

Fig. 25.24 The isometric projection is foreshortened to 82% of full size. The isometric drawing is drawn full size for convenience.

sured using standard scales, the only difference being the 18% increase in size. This is the most commonly used method of drawing isometrics.

The axes of isometric drawings are separated by 120° (Fig. 25.25). Although one of the axes is usually drawn vertically, this is not necessary.

Construction of an Isometric Drawing

An isometric drawing is begun by drawing three axes 120° apart. Lines that are parallel to these

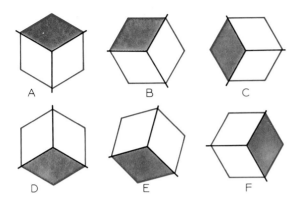

Fig. 25.25 Isometric axes are spaced 120° apart, but they can be revolved into any position.

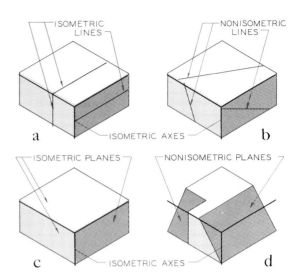

Fig. 25.26 (a) True measurements can be made along isometric lines (parallel to the three axes). (b) Nonisometric lines cannot be measured true length. (c) The three isometric planes are indicated; there are no nonisometric planes in this drawing. (d) Nonisometric planes are planes that are inclined to any of the three isometric planes of a cube.

axes are called *isometric lines* (Fig. 25.26a). True measurements can be made along isometric lines. Measurements cannot be made along nonisometric lines (part b).

The three surfaces of a cube in isometric are called *isometric planes* (Fig. 25.26c). Planes that are parallel to these planes are isometric planes, and planes that are not parallel to them are called nonisometric planes.

To draw an isometric, you will need a scale and a 30°–60° triangle (Fig. 25.27). Begin by constructing a plane of the isometric using the dimensions of height (*H*) and depth (*D*). In Step 3, the

Fig. 25.27 Construction of an isometric drawing

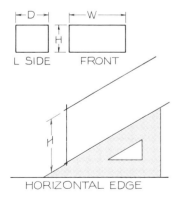

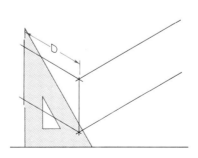

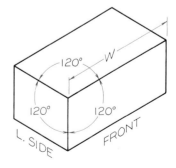

A 30°–60° triangle is used in combination with a horizontal straightedge to begin the isometric. The dimensions of *W, D,* and *H* from the orthographic views were doubled in this case.

The depth dimension, *D,* in the left side view is laid off in isometric. Parallel lines appear parallel in isometric drawings, as well as in orthographic views.

The final sides of the isometric are drawn and the lines are strengthened. The 30°–60° triangle automatically separates the axes by 120°.

third dimension, width (W), is used to complete the isometric drawing.

It is recommended that all isometric drawings be blocked in using light guidelines as shown in Fig. 25.28, and overall dimensions of W, D, and H. Other dimensions can be taken from the given views and measured along the isometric axes to locate notches and portions that are removed from the "blocked-in" drawing.

A more complex isometric is shown in Fig. 25.29. Again, the object is blocked in using H, W, and D, and portions of the block are removed to complete the isometric drawing.

25.13 ANGLES IN ISOMETRIC

Angles cannot be measured true size in an isometric drawing since the surfaces of an isometric are not true size. Angles must be located by using co-ordinates measured along isometric lines as shown in Fig. 25.30. Lines AB and BC are equal in length in the orthographic view, but they are shorter and longer than true length in the isometric drawing.

A similar example can be seen in Fig. 25.31 where two angles are drawn in isometric. The equal size angles in orthographic are less and

Fig. 25.28 Layout of an isometric drawing (simple)

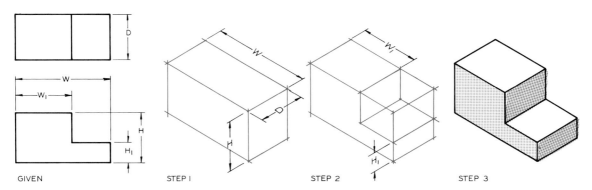

GIVEN   STEP 1   STEP 2   STEP 3

Step 1 The object is blocked in using the overall dimensions. The notch is removed.

Step 2 The notch is located by establishing its end points.

Step 3 The lines are strengthened to complete the drawing.

Fig. 25.29 Layout of an isometric drawing (complex)

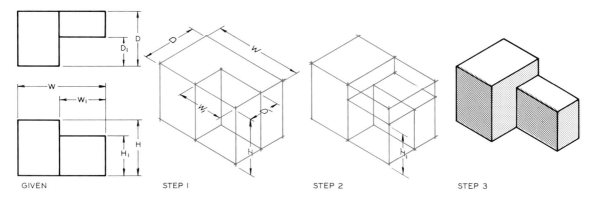

GIVEN   STEP 1   STEP 2   STEP 3

Step 1 The overall dimensions of the object are used to lightly block in the object. One notch is removed by using dimensions taken from the given views.

Step 2 The second notch is removed using dimension H_1.

Step 3 The final lines of the isometric are strengthened.

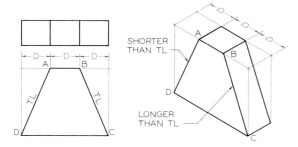

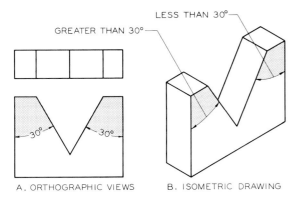

Fig. 25.30 Inclined surfaces must be found by using coordinates measured along the isometric axes. The lengths of angular lines will not be true length in isometric.

Fig. 25.31 Angles in isometric must be found by using coordinates. Angles will not appear true size in isometric.

greater than true size in the isometric drawing.

An isometric drawing of an object with an inclined surface is drawn in three steps in Fig. 25.32. The object is blocked in pictorially and portions are removed. The extreme ends of the inclined plane are located along the isometric axes, and these points are connected to complete the pictorial.

25.14 CIRCLES IN ISOMETRIC

The drawing of circular features in isometric is the most difficult construction problem in this type of pictorial.

Three methods of constructing circles in isometric drawings are (1) point plotting, (2) four-center ellipse construction, and (3) ellipse templates.

Circles: Point Plotting

A series of points located on a circle can be located in an isometric drawing by using two dimensions that are parallel to the isometric axes. These two dimensions are called *coordinates* (Fig. 25.33).

The cylinder is blocked in and drawn pictorially, with the centerlines added in Step 1. Coordinates, *A, B, C,* and *D* are used in Step 2 to locate points along the ellipse and are connected using an irregular curve.

Fig. 25.32 Construction of an isometric drawing with an inclined plane

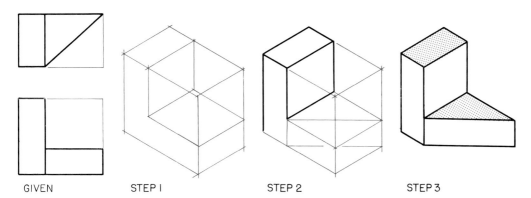

GIVEN STEP I STEP 2 STEP 3

Step 1 The object is blocked in using the overall dimensions. The notch is removed.

Step 2 The inclined plane is located by establishing its end points.

Step 3 The lines are strengthened to complete the drawing.

Fig. 25.33 Plotting circles

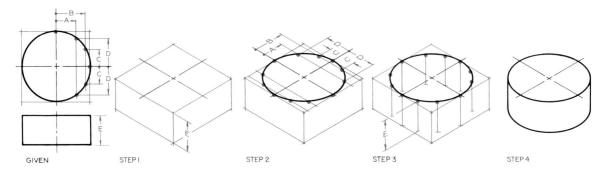

GIVEN STEP 1 STEP 2 STEP 3 STEP 4

Step 1 The cylinder is blocked in using the overall dimensions. The center lines locate the points of tangency of the ellipse.

Step 2 Coordinates are used to locate points on the circumference of the circle.

Step 3 The lower ellipse is found by dropping each point a distance equal to the height of the cylinder, *E.*

Step 4 The two ellipses can be drawn with an irregular curve and connected with tangent lines to complete the cylinder.

The lower ellipse located on the bottom plane of the cylinder can be found by using a second set of coordinates. The most efficient method is by measuring the distance, *E,* vertically beneath each point that was located on the upper ellipse (Step 3). A plotted ellipse is a true ellipse and it is equivalent to a 35° ellipse on an isometric plane.

An example of a design that is composed of circular features that were drawn in isometric is the handwheel shown in Fig. 25.34.

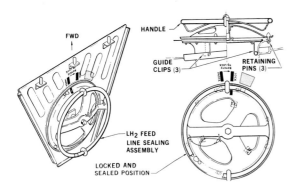

Fig. 25.34 An example of parts that have been drawn using ellipses in isometric to represent circles. This is a handwheel that was proposed for use in an orbital workshop to be launched into space in the future. (Courtesy of NASA.)

Circles: Four-Center Ellipse Construction

The four-center ellipse method can be used to construct an approximate ellipse in isometric by using four arcs that are drawn with a compass (Fig. 25.35).

The four-center ellipse is drawn by blocking in the orthographic view of the circle with a square that is tangent to the circle at four points. This square is drawn in isometric as a rhombus (Step 1). The four centers are found by constructing perpendiculars to the sides of the rhombus at the midpoints of the sides (Step 2). The four arcs are drawn to give the completed four-center ellipse (Step 3). This method can be used to draw ellipses on any of the three isometric planes since each is equally foreshortened as illustrated in Fig. 25.36.

You can see in Fig. 25.37 that the four-center ellipse is only an approximate ellipse when it is compared with a true ellipse.

Circles: Ellipse Templates

A specially designed ellipse template can be purchased for drawing ellipses in isometric. A typical example is shown in Fig. 25.38.

The diameters of the ellipses on the template are measured along the direction of the isometric

Fig. 25.35 The four-center ellipse

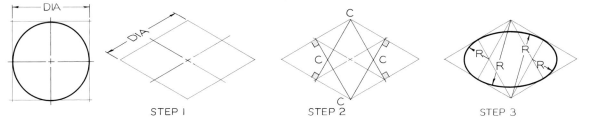

STEP 1   STEP 2   STEP 3

Step 1 The diameter of the circle is used to construct a rhombus that is tangent to the ellipse.

Step 2 Perpendicular lines are drawn from the midpoints of the sides of the rhombus to locate four centers.

Step 3 Using the four centers and two radii, the four-center ellipse is drawn tangent to the rhombus.

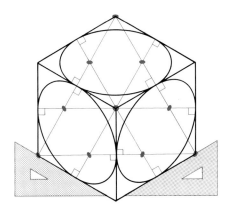

Fig. 25.36 Four-center ellipses can be drawn on all three surfaces of an isometric drawing.

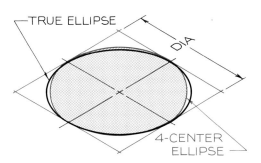

Fig. 25.37 The four-center ellipse is not a true ellipse, but an approximate ellipse.

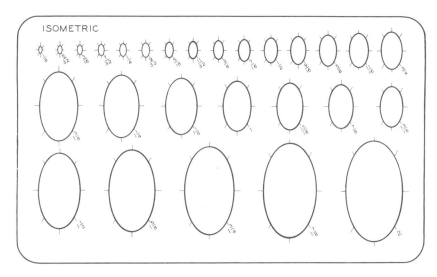

ISOMETRIC

Fig. 25.38 The diameter of a circle in isometric is measured along the direction of the isometric axes. Therefore, the major diameter of an isometric ellipse is greater than the measured diameter. The minor diameter is perpendicular to the major diameter.

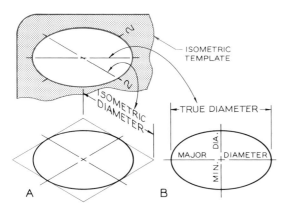

Fig. 25.39 The isometric ellipse template is a special template designed to reduce drafting time. Note that the true diameters of the circles are not the major diameters of the ellipses, but the diameters that are parallel to the isometric axes.

lines, since this is how diameters are measured in an isometric drawing (Fig. 25.39). The maximum diameter that can be measured across the ellipse is the *major diameter,* which is a true diameter. Consequently, the size of the diameter marked on the template is less than the ellipses' major diameter, the true diameter.

The isometric ellipse template can be used to draw an ellipse by constructing the centerlines of the ellipse in isometric and aligning the ellipse template with these isometric lines (Fig. 25.39A).

25.15 CYLINDERS IN ISOMETRIC

A cylinder can be drawn in isometric by using the four-center ellipse method as shown in Fig. 25.40. A rhombus is drawn at each end of the cylinder's axis with the centerlines drawn as isometric lines (Step 1). The ellipses are drawn using the four-center ellipse method at each end (Step 2). The ellipses are connected with tangent lines and the lines are darkened in Step 3.

A cylinder can be drawn using the ellipse template as illustrated in Fig. 25.41. The axis of the cylinder is drawn and perpendiculars are constructed at each end (Step 1). Since the axis of a right cylinder is perpendicular to the major diameter of its elliptical end, the ellipse template is positioned as shown in Step 2. The size of the ellipse is marked near the elliptical hole on the template. The ellipses are drawn at each end and are connected with tangent lines (Step 3).

Internal cylinders (holes) are drawn using the principles used for cylinders that were just covered. To construct a cylindrical hole in the block (Fig. 25.42), begin by locating the center of the hole on the isometric plane. The axis of the cylinder is drawn parallel to the isometric axis that is perpendicular to this plane through its center

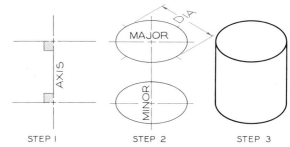

Fig. 25.41 Cylinders: ellipse template

Step 1 The axis of the cylinder is drawn its proper length, and perpendiculars are drawn at each end.

Step 2 The elliptical ends are drawn by aligning the major diameter with the perpendiculars at the ends of the axis. The isometric diameter of the isometric ellipse template is given along the isometric axis.

Step 3 The ellipses are connected with tangent lines to complete the isometric drawing. Hidden lines are omitted.

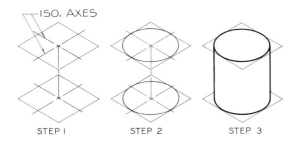

Fig. 25.40 Cylinder: four-center method

Step 1 A rhombus is drawn in isometric at each end of the cylinder's axis.

Step 2 A four-center ellipse is drawn within each rhombus.

Step 3 Lines are drawn tangent to each rhombus to complete the isometric drawing.

Fig. 25.42 Cylinders in isometric

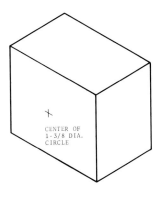

CENTER OF
1-3/8 DIA.
CIRCLE

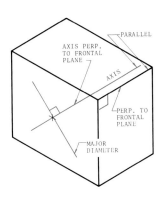

PARALLEL

AXIS PERP.
TO FRONTAL
PLANE

AXIS

PERP. TO
FRONTAL
PLANE

MAJOR
DIAMETER

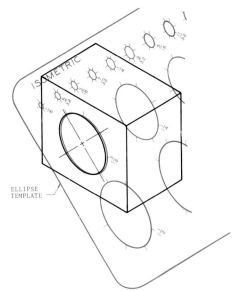

ISOMETRIC

ELLIPSE
TEMPLATE

Step 1 The center of the hole with a given diameter is located on a face of the isometric drawing.

Step 2 The axis of the cylinder is drawn from the center parallel to that isometric axis which is perpendicular to the plane of the circle. The major diameter is drawn perpendicular to the axis.

Step 3 The 1⅜″ ellipse template is used to draw the ellipse by aligning the major and minor diameters with the guidelines on the template.

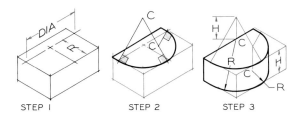

| STEP 1 | STEP 2 | STEP 3 |

DIA   R   C   C   H   C   R   C   H   R

Fig. 25.43 Semicircular features

Step 1 Objects with semicircular features can be drawn by blocking in the objects as if they had square ends. The centerlines are drawn to locate the centers and tangent points.

Step 2 Perpendiculars are drawn from each point of tangency to locate two centers. These are used to draw half of a four-center ellipse.

Step 3 The lower surface can be drawn lowering the centers by the distance of H, the thickness of the part. The same radii are used with these new centers to complete the isometric.

(Step 2). The ellipse template is aligned with the major and minor diameters to complete the elliptical view of the cylindrical hole (Step 3).

25.16 PARTIAL CIRCULAR FEATURES

When an object has a semicircular end as in Fig. 25.43, the four-center ellipse method can be used with only two centers to draw half of the circle (Step 2). To draw the lower ellipse at the bottom of the object, the centers are projected downward a distance of H, which is equal to the height of the object. These centers are used with the same radii that were used on the upper surface to draw the arcs.

In Fig. 25.44, an object with rounded corners is blocked in and the centerlines are located at each rounded corner (Step 1). An ellipse template is used to construct the rounded corners at Step 2. The rounded corners could have been constructed by the four-center ellipse method or by plotting points on the arcs.

Fig. 25.44 To construct rounded corners on an object, the centerlines of the ellipse are drawn at A. The elliptical corners are drawn with an ellipse template at B.

A

B

Fig. 25.45 Construction of a cone in isometric

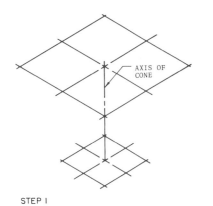

STEP 1

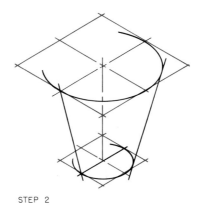

STEP 2

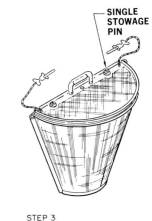

SINGLE STOWAGE PIN

STEP 3

Step 1 The axis of the cone is constructed. Each circular end of the cone is blocked in.

Step 2 An ellipse guide is used for constructing the circles in isometric at each end. These ends are connected to give the outline of the object.

Step 3 The remaining details of the screen storage provisions are added to complete the isometric. (Courtesy of the National Aeronautics and Space Administration.)

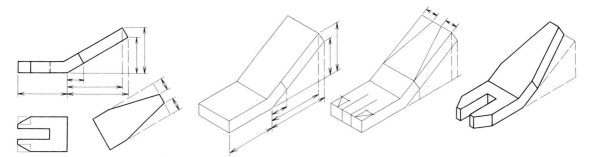

Fig. 25.46 Inclined surfaces in isometric must be located by using coordinates laid off parallel to the isometric axes. True angles cannot be measured in isometric drawings.

A similar drawing involving the construction of ellipses is the conical shape in Fig. 25.45. The ellipses are blocked in at the top and bottom surfaces (Step 1), and the half ellipses are drawn in Step 2 by using a template or the four-center method. The oblique drawing is shown in Step 3.

25.17 MEASURING ANGLES

Angles in isometric may be located by coordinates, as shown in Fig. 25.46, since angles will not appear true in an isometric.

A second method of measuring and locating angles is the ellipse template method shown in Fig. 25.47. Since the hinge line is perpendicular to the path of revolution, an ellipse is drawn in Step 1 with the major diameter perpendicular to the hinge line. A true circle is drawn with a diameter that is equal to the major diameter of the ellipse.

In Step 2, point A is located on the ellipse and is projected to the circle. This locates the direction of a horizontal line. From this line, the angle of revolution of the hinged part can be measured true size, 120° in this example (Step 3). Point B is pro-

jected to the ellipse to locate a line at 120° in isometric.

The thickness of the revolved part is found by drawing line C perpendicular to line A and projecting back to the ellipse. A smaller ellipse through point D is drawn to locate the thickness of the revolved part in the isometric. The remaining lines are drawn parallel to these key lines.

25.18 CURVES IN ISOMETRIC

Irregular curves must be plotted point by point using coordinates to locate each point. Points A through F are located in the orthographic view with coordinates of width and depth (Fig. 25.48). These coordinates are transferred to the isometric view of the blocked-in part (Step 1). The plotted points are connected with an irregular curve.

Each point on the upper curve is projected downward for a distance of H, which is equal to the height of the part, to locate points on the lower curve (Step 2). The points are connected with an irregular curve to complete the isometric.

Fig. 25.47 Measuring angles with an ellipse template

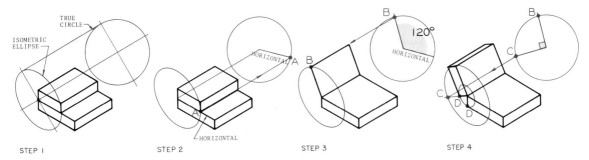

Step 1   An ellipse is drawn with the major diameter perpendicular to the hinge line of the two parts. Any size of ellipse could be used. A true circle is drawn with its center on the projection of the hinge line and with a diameter equal to the major diameter of the ellipse.

Step 2   Point A is projected to the circle to locate the direction of the horizontal in the circular view.

Step 3   The position of rotation is measured 120° from the horizontal to locate point B, which is then projected to the ellipse to locate the position of the revolved surface.

Step 4   To locate the perpendicular to the surface, a 90° angle is drawn in the circular view and point C is projected to the ellipse; a line is drawn from point C on the ellipse to the center of the ellipse. A smaller ellipse is drawn to pass through point D on the lower part of the object. The point where this ellipse intersects the line from C to the center of the ellipse establishes the thickness of the revolved part.

Fig. 25.48 Plotting irregular curves

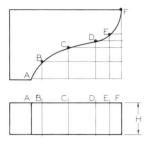
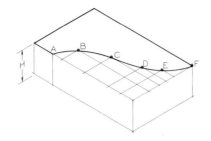
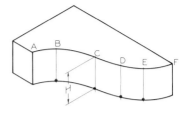

Step 1 Coordinates are established in the orthographic views.

Step 2 One set of coordinates is transferred to the isometric view.

Step 3 The second set of coordinates is transferred to the isometric to establish points on the ellipse.

Step 4 The plotted points are connected with an elliptical curve. An ellipse template can usually serve as a guide for connecting the points.

Fig. 25.49 Construction of ellipses on an inclined plane

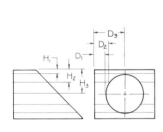
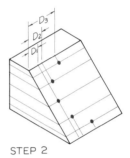

STEP 1   STEP 2   STEP 3

Step 1 Draw two coordinates to locate a series of points on the irregular curve. These coordinates must be parallel to the standard W, D, and H dimensions.

Step 2 Block in the shape using overall dimensions. Locate points on the irregular curve using the coordinates from the orthographic views.

Step 3 Since the object has a uniform thickness, the lower curve can be found by projecting downward the distance H from the upper points. Connect the points with an irregular curve.

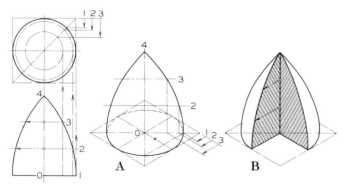
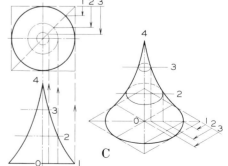

Fig. 25.50 Surfaces of revolution must be constructed by drawing a series of cross sections that are connected to give their overall shapes. The cross sections in these examples are circles that are drawn as ellipses in isometric.

25.19 ELLIPSES ON NONISOMETRIC PLANES

When an ellipse lies on a nonisometric plane such as the one shown in Fig. 25.49, points on the ellipse can be plotted to locate the ellipse.

Coordinates are located in the orthographic views and then transferred to the isometric as shown in Steps 1 and 2. The plotted points can be connected with an irregular curve, or an ellipse template can be selected that will approximate the plotted points (Step 3). The isometric ellipse template cannot be used for this purpose.

25.20 SURFACES OF REVOLUTION

A surface of revolution is a solid form made by revolving a plane about an axis of revolution. Two examples of surfaces of revolution are shown in Fig. 25.50. To draw these in isometric, the axes were drawn through point 0 to point 4. Circular cross sections were located along this axis using the correct elliptical diameters taken from the orthographic views for each.

The elliptical cross sections are connected with tangent lines to find the outlines of the objects in isometrics. The more cross sections that are used, the more accurate will be the location of the tangent lines.

25.21 MACHINE PARTS IN ISOMETRIC

Orthographic views of a spotface, countersink, and boss are shown in Fig. 25.51. The isometric pictorials of each are shown also. The isometric draw-

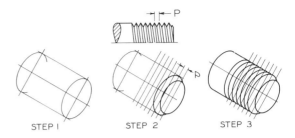

Fig. 25.52 Threads in isometric

Step 1 Draw the cylinder that is to be threaded by using an ellipse template.

Step 2 Lay off perpendiculars that are spaced by a distance equal to the pitch of the thread, *P*.

Step 3 Draw a series of ellipses to represent the threads. The chamfered end is drawn using an ellipse whose major diameter is equal to the root diameter of the threads.

ings of these features could be drawn by point-plotting the circular features, by the four-center method, or by using an ellipse template. The template is by far the easiest method.

A threaded shaft can be drawn in isometric as shown in Fig. 25.52, by first drawing the cylinder that is to be threaded in isometric (Step 1). In Step 2, the major diameters of the crest lines of the thread are drawn separated at a distance of *P* the pitch of the thread. In Step 3, ellipses are drawn by aligning the major diameter of the ellipse template with the perpendiculars to the cylinder's axis. Note that the 45° chamfered end is drawn using a smaller ellipse at the end.

A hexagon head nut (Fig. 25.53) is drawn in three steps using an ellipse template. The nut is blocked in and an ellipse drawn tangent to the rhombus. The hexagon is constructed by locating the distance across a flat, *W*, parallel to the isometric axes. The other sides of the hexagon are found in step 2 by drawing lines tangent to the ellipse. The distance *H* is laid off at each corner to establish the amount of chamfer at each corner (Step 3).

A hexagon head bolt is drawn in two positions in Fig. 25.54. The washer face can be seen on the lower side of the head and the chamfer on the upper side of the bolt head.

Many machine parts are composed of spherical shapes. The sphere in isometric is constructed

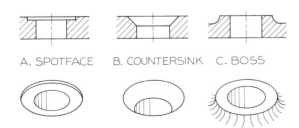

A. SPOTFACE   B. COUNTERSINK   C. BOSS

Fig. 25.51 Examples of circular features drawn in isometric. These can be drawn by using ellipse templates.

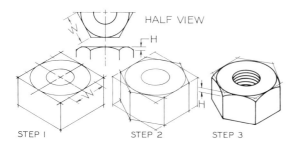

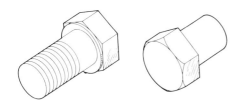

Fig. 25.53 Construction of nut

Step 1 The overall dimensions of the nut are used to block in the nut.

Step 2 The hexagonal sides are constructed at the top and bottom.

Step 3 The chamfer is drawn with an irregular curve. Threads are drawn to complete the isometric.

Fig. 25.54 Isometric drawings of the upper and lower sides of a hexagon-head bolt.

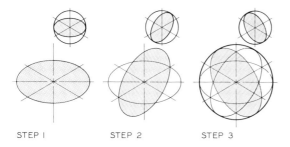

Fig. 25.55 An isometric sphere

Step 1 The three intersecting isometric axes are drawn. The ellipse template is used to draw the horizontal elliptical section.

Step 2 The isometric ellipse template is used to draw one of the vertical elliptical sections.

Step 3 The third vertical elliptical section is drawn and the center is used to draw a circle tangent to the three ellipses.

as shown in Fig. 25.55. Three ellipses are drawn as isometric planes with a common center. The center is used to construct a circle that will be tangent to each ellipse as shown in Step 3. The sphere is larger in isometric than in a true axonometric projection.

A portion of a sphere is used to draw a round-head screw in Fig. 25.56. A hemisphere is constructed in Step 1. The centerline of the slot is located along one of the isometric planes. The thickness of the head is measured as distance of E from the highest point on the sphere. The slot is drawn in Step 3 to complete the isometric drawing.

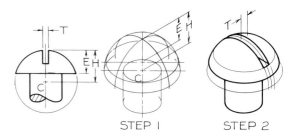

Fig. 25.56 Spherical features

Step 1 An isometric ellipse template is used to draw the elliptical features of a round-head screw.

Step 2 The slot in the head is drawn and the lines are darkened to complete the isometric of the head.

25.22 ISOMETRIC SECTIONS

A full section can be drawn in isometric to clarify internal details that might otherwise by over-looked (Fig. 25.57). Half-sections can also be used as illustrated in Fig. 25.58. The same part shown as a full section and a half-section is given in Fig. 25.59.

25.23 DIMENSIONED ISOMETRICS

When it is advantageous to dimension and note a part shown in isometric, either of the aligned or unidirectional methods illustrated in Fig. 25.60 can be used to apply the notes. In both cases, notes connected with leaders are positioned horizontally. Always use guidelines for your lettering and numerals.

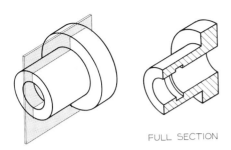

Fig. 25.57 Parts can be shown in isometric sections to clarify internal features, such as this full section.

FULL SECTION

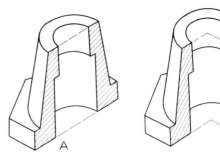

Fig. 25.59 A comparison of isometric full and half-sections of the same part.

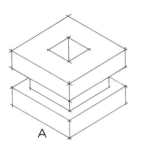

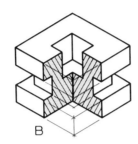

Fig. 25.58. An isometric drawing of a half-section can be constructed.

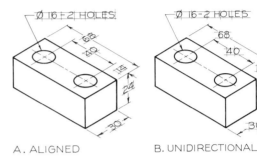

A. ALIGNED   B. UNIDIRECTIONAL

Ø 16 – 2 HOLES

68
40
14
24
30

Ø 16 – 2 HOLES

68
40
14
24
30

Fig. 25.60 Dimensions can be placed on isometric drawings using either of the techniques shown here. Guidelines should always be used for the lettering.

Fig. 25.61 Representation of fillets and rounds

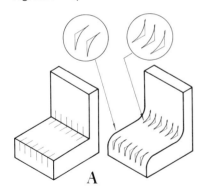

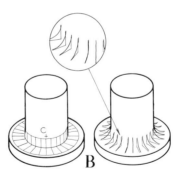

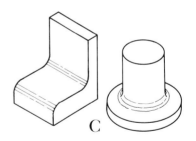

A. Fillets and rounds can be represented by segments of an isometric ellipse if guidelines are constructed at intervals.

B. Fillets and rounds can be represented by elliptical arcs by constructing radial guidelines.

C. Fillets and rounds can be represented by lines that run parallel to the fillets and rounds.

25.24 FILLETS AND ROUNDS

Fillets and rounds in isometric can be represented by either of the methods shown in Fig. 25.61 to give added realism to a pictorial drawing. The examples at *A* and *B* show how intersecting guidelines are drawn equal in length to the radii of the fillets and rounds, and arcs are drawn tangent to these lines. These arcs can be drawn freehand or with an ellipse template. The method in part C utilizes freehand lines drawn parallel or concentric with the directions of the fillets and rounds.

An example of these two methods is shown in Fig. 25.62. The stipple shading was applied by us-

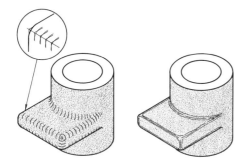

Fig. 25.62 Two methods of representing fillets and rounds on a part.

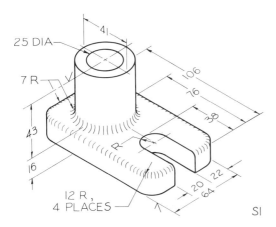

Fig. 25.63 An isometric drawing with complete dimensions and fillets and rounds represented.

ing an adhesive overlay film that can be purchased.

When fillets and rounds are illustrated as shown in Fig. 25.63 and dimensions are applied, it is much easier to understand the features of the part than when it is represented by orthographic views. The dimensions in this example are shown using the aligned method.

25.25 ISOMETRIC ASSEMBLIES

Assemblies are used to explain how a series of parts is assembled. The more common mistakes in applying leaders to an assembly are shown in Fig. 25.64A. The more acceptable techniques are shown in part B. The numbers in the circles ("balloons") refer to the number given to each part that appears in the parts list.

An assembly used in a parts manual is shown in Fig. 25.65, along with its parts list. This assembly is "exploded" apart so the parts are separated, but it is clear how they would be assembled. The

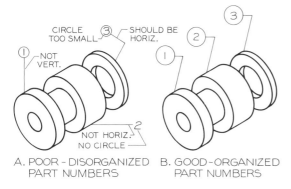

A. POOR – DISORGANIZED PART NUMBERS

B. GOOD – ORGANIZED PART NUMBERS

Fig. 25.64 Part numbers should be applied to assemblies as shown at B and by avoiding the errors made at A.

ELECTRIC OPERATORS for Jenkins Ball Valves
PARTS LIST

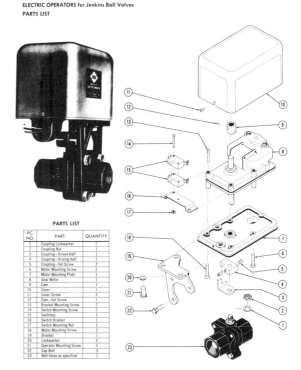

PC. NO.	PART	QUANTITY
1	Coupling Lockwasher	1
2	Coupling Nut	1
3	Coupling—Driven Half	1
4	Coupling—Driving Half	1
5	Coupling—Set Screw	2
6	Motor Mounting Screw	2
7	Motor Mounting Plate	1
8	Gear Motor	1
9	Cam	1
10	Cover	1
11	Cover Screw	2
12	Cam—Set Screw	1
13	Bracket Mounting Screw	2
14	Switch Mounting Screw	2
15	Switches	2
16	Switch Bracket	1
17	Switch Mounting Nut	2
18	Motor Mounting Screw	2
19	Bracket	1
20	Lockwasher	3
21	Operator Mounting Screw	2
22	Cap Bolt	3
23	Ball Valve as specified	1

Fig. 25.65 An industrial example of an assembly.

THREAD CUTTER

*25 Draper thread cutter

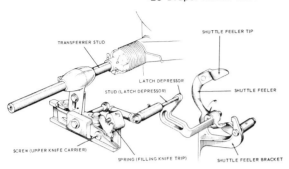

Fig. 25.66 An assembly drawing with parts assembled and named. (Courtesy of Draper Corporation.)

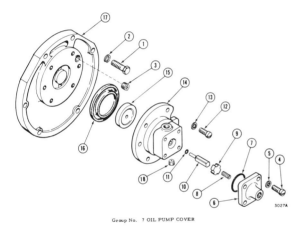

Group No. 7 OIL PUMP COVER

Fig. 25.67 An exploded isometric assembly.

assembly in Fig. 25.66 is totally assembled, and each part is named by note. An exploded isometric assembly is given in Fig. 25.67.

25.26 PIPING SYSTEMS IN ISOMETRIC

Piping layouts are often drawn as isometrics to clarify complex systems that are hard to interpret when shown in orthographic views. Some of the more often used piping connectors are shown in Fig. 25.68. The components are shown both in orthographic and isometric views. Ellipse templates are used to represent circular arcs in isometric. These are single-line representations since the pipes are represented by single, heavy lines.

A portion of a pipe system is shown in Fig. 25.69. Additonal pipe symbols are given in Appendix 9. All symbols are drawn in isometric using the same principles presented in this article.

25.27 AXONOMETRIC PROJECTION

An *axonometric projection* is a form of orthographic projection in which the pictorial view is projected perpendicularly onto the picture plane with parallel projectors. The object is positioned in an angular position with the picture plane so that its pictorial projection will be a three-dimensional view rather than a two-dimensional view as an orthographic view.

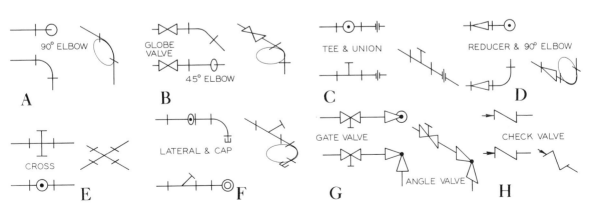

Fig. 25.68 A comparison of orthographic views and isometric pictorials of single-line representations of piping symbols. Note that the isometric template is used for constructing rounded corners in the isometric drawings.

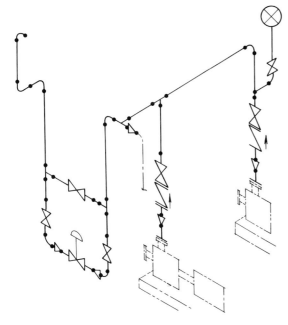

Fig. 25.69 A typical piping system drawn in isometric using the appropriate symbols.

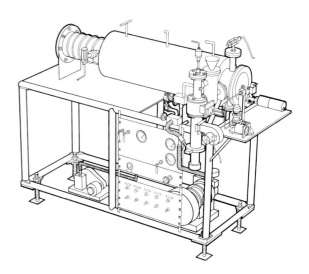

Fig. 25.71 A trimetric projection of a cold diffusion pump. (Courtesy of Aro, Inc.)

Three types of axonometric pictorials are possible: (1) isometric, (2) dimetric, or (3) trimetric. The **isometric projection** is the view where the diagonal of a cube is viewed as point. The planes will be equally foreshortened and axes equally spaced 120° apart (Fig. 25.70). The measurements along the three axes will be equal, but less than true length since this is true projection.

A **dimetric projection** is an axonometric projection in which two planes are equally foreshortened and two of the axes are separated by equal angles (part B). The measurements along two of the axes are equal, but less than true length.

A **trimetric projection** is an axonometric projection where all three planes are unequally foreshortened and the angles between the three axes are different (Fig. 25.70c).

Figure 25.71 is a trimetric projection of a cold diffusion pump. Note that the angles between the three axes are unequal; this identifies it as a trimetric.

25.28 AXONOMETRIC CONSTRUCTION

All axonometric constructions can be made in the same manner as the trimetric constructed in Fig. 25.72.

A cube should always be used instead of the object to be drawn as an axonometric. By using a cube, you can construct axonometric scales that can be used to draw trimetrics of many objects, not just a single part.

The trimetric scales found in Step 3 can be lengthened to accommodate any size part. Using axonometric scales eliminates the need for duplicating the construction when trimetrics of similar angles are drawn.

Trimetric scales are not complete unless the ellipse angles are found for each plane so that an ellipse template can be used. In a trimetric, a different ellipse angle will be used for each plane since each plane is foreshortened differently.

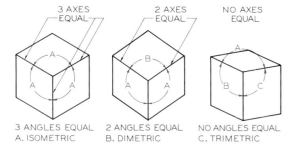

Fig. 25.70 The three types of axonometric projection.

Fig. 25.72 Trimetric-scale construction

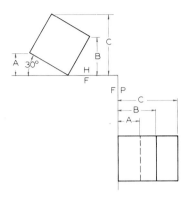

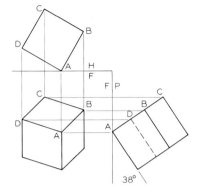

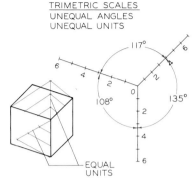

TRIMETRIC SCALES
UNEQUAL ANGLES
UNEQUAL UNITS

Step 1 Revolve the top view 30° clockwise. Find the side view by transferring dimensions A, B, and C from the top view. If the revolution had been 45° in the top view, the resulting projection would be either a dimetric or an isometric.

Step 2 Tilt the side view 34°. This will change the projection of the top view but not the width dimension; consequently it is unnecessary to change the top view. Determine the trimetric projection of the cube by projecting from the top and side views as in orthographic projection.

Step 3 All sides of a cube are equal; therefore divide each axis (or edge) into an equal number of units by proportional divisions as shown, even though the three axes in a trimetric have different lengths. The three axes can be extended and scaled into as many units as desired.

Fig. 25.73 Ellipse template angles for a trimetric scale

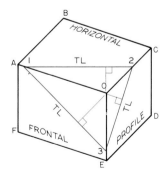

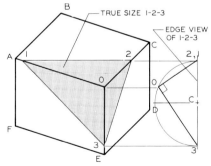

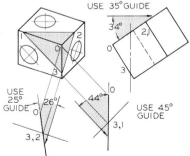

Step 1 The planes of a cube are mutually perpendicular; therefore line OA is perpendicular to plane OCDE and line OE is perpendicular to plane ABCO. On each plane, construct true-length lines, which are located perpendicular to the axis lines AO, CO, and EO. The true-length lines are 1–2, 2–3, and 3–1. These lines will intersect at points on the axes forming triangle 1–2–3.

Step 2 Since plane 1–2–3 is composed of true-length lines, it is true size in the trimetric projection. Determine the side view of the plane, which is a frontal plane by projection. Find the 90° angle of the cube in the side view by constructing a semicircle, using the edge view of plane 1–2–3 as the diameter. Project point O to the semicircle where the 90° angle is inscribed.

Step 3 Two views permit auxiliary views to be found. Determine the edge view of each principal plane by locating the point views of lines 1–2, 2–3, and 3–1. The ellipse guide angle is the angle between the edge views of 1–0–2, 2–0–3, and 1–0–3 and the line of sight. Position the ellipse guides on each plane so that the major diameter is parallel to the true-length lines on that plane.

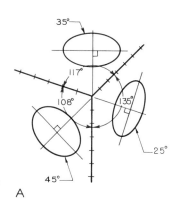

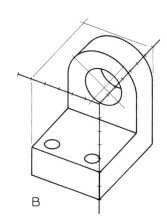

Fig. 25.74 The complete trimetric scales showing the units along the axes and the ellipse template angles. The object at B was drawn using the axonometric scales

A

B

When a trimetric is given (Fig. 25.73) a true-size plane can be found by drawing lines 1–2, 2–3, and 3–4 perpendicular to the three axes (Step 1). The side view of this true-size plane is found as a vertical edge in Step 2. Using these two views, the angles between the lines of sight and each plane can be found in Step 3. These angles are the ellipse angles for each surface.

The ellipses are positioned on each plane with their major diameters perpendicular to the axes intersecting each plane. This gives you a set of trimetric scales with calibrations, and the ellipse angles for each plane, Fig. 25.74A. These

scales can be used to construct a trimetric by overlaying the scales with tracing vellum as shown in Fig. 25.74B.

A second technique of constructing an axonometric is shown in Fig. 25.75. The three axes are located in a convenient position and plane 1–2–3 is constructed with three true-length lines in Step 1. Semicircles are drawn with diameters equal to these true-length lines. Angles are inscribed inside of each to give a 90° angle at point 0 for each semicircle (Step 2). The top, front, and side views of the object are located at point 0 for each view, and each view is projected back where the three pro-

Fig. 25.75 Axonometric projection

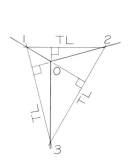

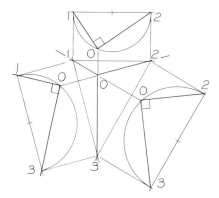

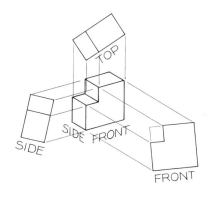

Step 1 The three axes of an axonometric projection can be selected and drawn at the angle of your choice. True-length perpendiculars are drawn so they intersect at 1, 2, and 3.

Step 2 Semicircles are projected from each true-length line. Right angles are inscribed in the semicircle, 2–0–3 for example.

Step 3 The three orthographic views of the object are located with the same corner placed at the right angles found in step 2. These views are projected back to intersect and form a trimetric in this case.

jectors converge at common points. This locates the points of the axonometric projection to form the pictorial.

25.29 PERSPECTIVES

A perspective is a view that is normally seen by the eye or camera, and is the most realistic form of pictorial. All parallel lines converge at infinite vanishing points as they recede from the observer.

The three basic types of perspectives are (1) one-point, (2) two-point, and (3) three-point, depending on the number of vanishing points used in their construction.

Examples of each of these types are shown in Fig. 25.76.

The **one-point perspective** has one surface of the object that is parallel to the picture plane;

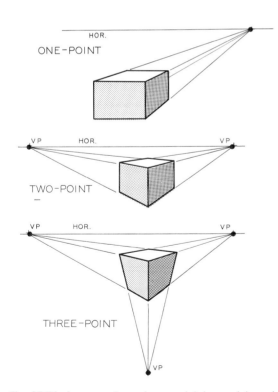

Fig. 25.76 A comparison of one-point, two-point, and three-point perspectives.

therefore it is true shape. The other sides vanish to a single point on the horizon called a vanishing point.

A **two-point perspective** is a pictorial that is positioned with two sides at an angle to the picture plane; this requires two vanishing points (Fig. 25.76). All horizontal lines converge at the vanishing points, but vertical lines remain vertical and have no vanishing point.

The **three-point perspective** utilizes three vanishing points since the object is positioned so that all sides of it are at an angle with the picture plane (Fig. 25.76). The three-point perspective is used in drawing larger objects such as buildings.

25.30 ONE-POINT PERSPECTIVES

The steps of drawing a one-point perspective are shown in Fig. 25.77. Here are given the top and side view of the object, the picture plane, station point (S.P.), the horizon, and the ground line.

The picture plane is the plane on which the perspective is projected. It appears as an edge in the top view.

The station point is the location of the observer's eye in the plan view. The front view of the station point will always lie on the horizon.

The horizon is a horizontal line in the front view that represents an infinite horizontal, such as the surface of the ocean.

The ground line is an infinite horizontal line in the front view that passes through the base of the object being drawn.

The center of vision (C.V.) is a point that lies on the picture plane in the top view and on the horizon in the front view. In both cases, it is on the line from the station point that is perpendicular to the picture plane.

When drawing any perspective, the station point (S.P.) should be located far enough away from the object so that the perspective can be contained in a cone of vision that is not more than 30°, Fig. 25.78. If a larger cone of vision is required, the perspective will be distorted.

Fig. 25.77 Construction of a one-point perspective

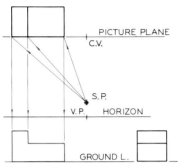 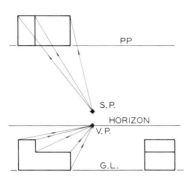 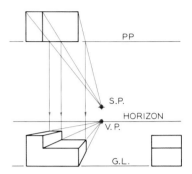

Step 1 Since the object is parallel to the picture plane, there will be only one vanishing point, which will be located on the horizon below the station point. Projections from the top and side views establish the front plane. This surface is true size, since it lies in the picture plane.

Step 2 Draw projectors from the station point to the rear points of the object in the top view and from the front view to the vanishing point on the horizon. In a one-point perspective, the vanishing point is the front view of the station point.

Step 3 Construct vertical projectors from the top view to the front view from the points where the projectors cross the picture plane. These projectors intersect the lines leading to the vanishing point. This is called a one-point perspective, since the lines converge at a single point.

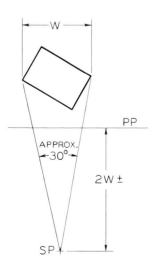

Fig. 25.78 The station point (SP) should be placed far enough away from the object to permit the cone of vision to be less than 30° to reduce distortion.

Measuring Points for One-Point Perspective

A measuring point is an additional vanishing point that is used to locate measurements along the receding lines that vanish to the horizon. In Fig. 25.79, the measuring point of a one-point perspective is found by revolving line 0–2 into the picture plane to 0–2'. The measuring point is found by drawing a construction line from the station point to the picture plane parallel to 2–2'. The measuring point is located on the horizon by projection from the picture plane (Step 1).

Since the distance 0–2 is equal to 0–2', depth dimensions can be laid off along the ground line and then projected to the measuring point. This locates the rear corner of the one-point perspective, Step 2. The depth from the picture plane to the front of the object is found in the same manner.

The use of the measuring-point method eliminates the need for placing the top view in the customary top-view position. Instead the dimensions can be transferred to the ground line in a more convenient manner.

Fig. 25.79 One-point perspective—measuring points

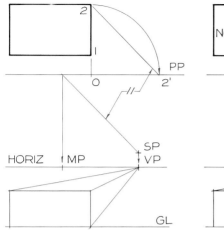

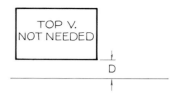

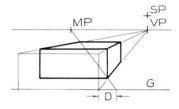

Step 1 Line 0–2 is revolved into the picture plane to locate point 2'. A line is drawn parallel to 2'–2 through SP to the PP. The measuring point is located on the horizon.

Step 2 Distance D is laid off along the ground line from the front corner of the perspective. This distance is projected to the measuring point to locate the rear corner of the perspective.

Step 3 The front of the object is located by laying off distance D from the corner of the perspective and projecting to the measuring point. This locates the front surface of the object behind the picture plane.

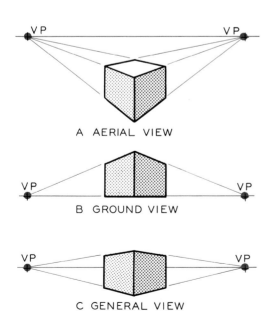

A AERIAL VIEW

B GROUND VIEW

C GENERAL VIEW

Fig. 25.80 Different perspectives can be obtained by locating the horizontal over, under, and through the object.

25.31 TWO-POINT PERSPECTIVES

If two surfaces of an object are positioned at an angle to the picture plane, two vanishing points will be required to draw it as a perspective. Different views can be obtained by changing the relationship between the horizontal and the ground line, as illustrated in Fig. 25.80.

An **aerial view** will be obtained when the horizon is placed above the ground line and the top of the object in the front view. When the ground line and the horizon coincide in the front view, a **ground-level view** will be obtained. This would give the view that would be seen if your eye was looking from the ground. A **general view** is one where the horizon is placed above the ground line and through the object, usually at a height equal to the height of a person (part C).

The steps of constructing a two-point perspective are shown in Fig. 25.81. The horizon has been placed above the object to give a slight aerial view. The vanishing points are found by drawing construction lines parallel to the sides of the object from the station point in the top view.

Fig. 25.81 Construction of a two-point perspective

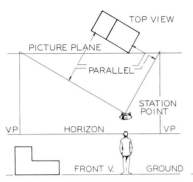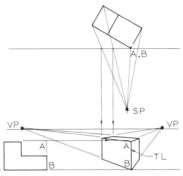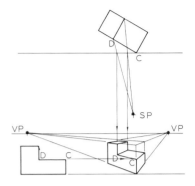

Step 1 Construct projectors which extend from the top view of the station point to the picture plane parallel to the forward edges of the object. Project these points vertically to a horizontal line in the front view to locate vanishing points. Locate the horizon in a convenient position. Draw the ground line below the horizon and construct the side view on the ground line.

Step 2 Since all lines in the picture plane are true length, line *AB* is true length. Consequently, line *AB* is projected from the side view to determine its height. Then project each end of *AB* to the vanishing points to determine two perspective planes. Draw projectors from the station point to the exterior edges of the top view. Project the intersections of these projectors with the picture plane to the front view to determine the limits of the object.

Step 3 The box obtained in step 2 must have a notch removed. Determine point *C* in the front view by projecting from the side view to the true-length line *AB*. Draw a projector from point *C* to the left vanishing point. Point *D* will lie on this projector beneath the point where a projector from the station point to the top view of point *D* crosses the picture plane. Complete the notch by projecting to the respective vanishing points.

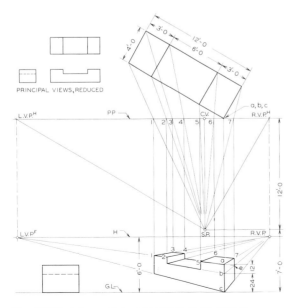

Fig. 25.82 A two-point perspective of an object.

Since line *AB* lies in the picture plane, it will be true length in the perspective. All height dimensions must originate at this vertical line because it is the only true-length line in the perspective. Points *C* and *D* are found by projecting to *AB*, and then projecting toward the vanishing points.

A typical two-point perspective is shown in Fig. 25.82 where the construction has not been separated into steps. By referring to Fig. 25.80, you will be able to understand the development of the construction used.

The object in Fig. 25.83 does not make contact with the picture plane in the top view as in the previous examples. To draw a perspective of this object, the lines of the object must be measured where the plane intersects the picture plane. The height is measured, and the infinite plane is drawn to the vanishing point.

The corner of the object can be located on this infinite plane by projecting the corner to the picture plane in the top view with a projector from the station point. This point does not contact the picture plane; none of the height dimensions will be true length.

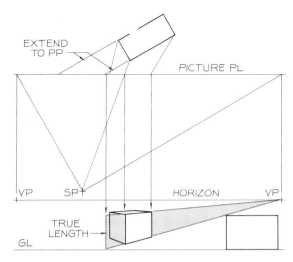

Fig. 25.83 A two-point perspective of an object that is not in contact with the picture plane.

Two-Point Perspectives: Measuring Points

Measuring points are found in Fig. 25.84 to aid in the construction of two-point perspectives. The use of measuring points eliminates the need to have the top view in the top-view position, after the vanishing points have been found.

Two vanishing points are found for two-point perspectives. These are used to locate dimensions along the receding planes that vanish to the horizon as shown in Step 2. Depth dimensions are laid off along the ground line from point B, and secondary planes are passed from these points to their measuring points (Step 3).

Measuring points are used in Fig. 25.85 to construct a two-point perspective. The same principles were used in this construction as in the previous example. No top view is needed since the measuring point system is used to locate measurements along the receding lines.

Arcs in Perspective

Arcs in two-point perspectives must be found by using coordinates to locate points along the curves in perspective (Fig. 25.86). Points 1 through 7 are located along the semicircular arc in the orthographic view. These same points are found in the perspective by projecting coordinates from the top and side views.

The points do not form a true ellipse, but a slightly egg-shaped oval. An irregular curve is used to connect the points.

A two-point perspective of a cylindrical object is constructed in Fig. 25.87. Again, it is necessary to locate points along the circle in perspective.

Fig. 25.84 Construction of measuring points

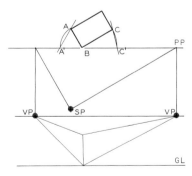

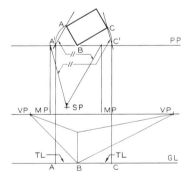

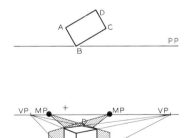

Step 1 In the top view, revolve lines AB and BC into the picture plane using point B as the center of revolution. Draw construction lines through point A–A' and points C–C'. These lines represent edge views of vertical planes passing through the corner points A and C.

Step 2 Draw lines from the station point parallel to lines A–A' and C–C' to the picture plane. Project these points of intersection to the horizon to locate two measuring points. Distances AB and BC can be laid off true length on the ground line (GL), since they have been revolved into the picture plane in the top view.

Step 3 Extend imaginary planes from points A and C on the ground line to their respective measuring points. These planes intersect the infinite planes, which are extended to the vanishing points, to locate the two corners of the block. True measurements can to be laid off on the ground line and projected to measuring points to find corner points.

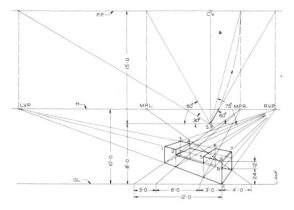

Fig. 25.85 A two-point perspective drawn with the use of measuring points.

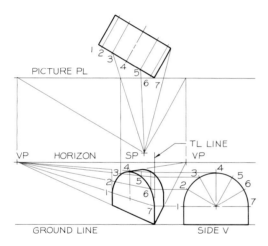

Fig. 25.86 A two-point perspective of an object with semicircular features.

Eight points are located in this example to approximate the perspective of the circular ends.

Sloping Planes

A sloping plane that is not horizontal will not have its vanishing point on the horizon, but below or above the horizon. In Fig. 25.88, a roof slopes 25° with the horizontal in two directions. The vanishing points of these two planes are found in Step 2 by revolving AB into the picture plane to AB', and B' is projected to the horizon. From this point, two lines are drawn at 25° with the horizon to the vertical projector through the vanishing point

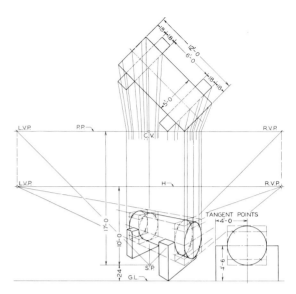

Fig. 25.87 A two-point perspective of an object with circular features.

found in Step 1. This locates the vanishing points of the two planes of the roof.

The planes of the roof are located in the perspective and are projected to their respective vanishing points (Step 3). Only horizontal planes vanish to the horizon.

25.32 THREE-POINT PERSPECTIVES

A three-point perspective can be constructed by positioning the object so that all of its sides make angles with the picture plane. In Fig. 25.89, the top and side views are orthographic views that have been tilted with respect to the picture plane. The station point is the same distance from the picture plane in the side view.

Three vanishing points are located by constructing lines from the two views of the station point that are parallel to the sides of object. The corner points in perspective are found by projecting to the picture plane and then to the perspective view where the lines vanish to their respective vanishing points.

In laying out a three-point perspective, you must begin by positioning the top and side views. The vanishing point for vertical lines will lie

Fig. 25.88 Perspectives of sloping surfaces

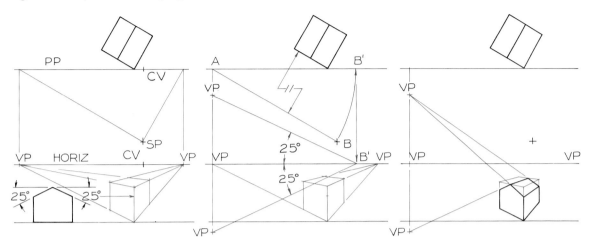

Step 1 The vanishing points are located and the perspective of the object is drawn as if it had no sloping planes. The point where the sloping plane begins its slope is found on the true-length vertical line by projecting from the side view.

Step 2 Line *AB* is rotated in the top view to locate *B'* on PP which is then projected to the horizon. Since both planes slope 25°, their vanishing points are drawn at 25° with the horizon from *B'* to a vertical line that passes through the left VP.

Step 3 Each sloping plane is found by using the two vanishing points found in step 2 and projecting from points found on the vertical lines where the slopes begin.

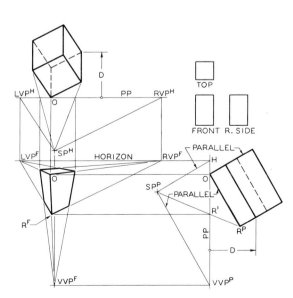

Fig. 25.89 The construction of a three-point perspective.

along a vertical construction line through the station point (S.P.).

A short-cut method of drawing a three-point perspective is shown in Fig. 25.90. An equilateral triangle with 60° angles is drawn and a convenient point *O* is located within the triangle. The vertexes of the triangle are used as the three vanishing points.

Width and depth are laid off on either side of Point *O*. These dimensions converge at the two horizontal vanishing points to locate the top view of the box. Height is laid off along a construction line through *O* that is parallel to one on the other sides of the triangle. Height is projected back to the vertical line from *O* to the lower vanishing point. Now that the three dimensions have been located, the three-point perspective can be completed.

25.33 PERSPECTIVE CHARTS

Perspective charts are time-saving grids that eliminate the need for constructing vanishing points and other tedious constructions (Fig. 25.91). The grid is drawn to scale so that it can be overlaid by

Fig. 25.90 Three-point perspective

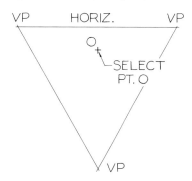 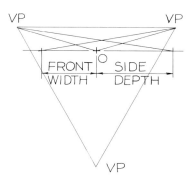 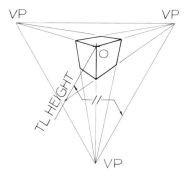

Step 1 Draw an equilateral triangle. The vanishing points lie at each vertex. Locate point O anywhere within this triangle and it will be the beginning point of the perspective.

Step 2 Draw a horizontal through point O. The width and depth can be laid off from point O and projectors are drawn to the two upper vanishing points. This locates the top surface.

Step 3 The height is laid off along a line that is parallel to one of the triangle's sides. The height is projected back to where it intersects a line from point O to the lower VP. The object three-point perspective is completed.

Fig. 25.91 Use of a perspective grid. (Courtesy of Graphicraft.)

tracing vellum and perspectives drawn by using the printed grid. Scales can be assigned to the grid for varying sizes of perspectives.

Perspective charts come in a variety of angles and views to suit most of your perspective needs. Perspective grids are excellent for preliminary sketches, since they give a general idea of the appearance of the finished perspective in the shortest time.

25.34 SHADES AND SHADOWS IN ORTHOGRAHIC

Shades and shadows are added to drawings to increase their realistic appearance. A surface is in shade when the light does not strike it. For example, if a block were facing the sunlight, its backside would not be in the direct rays of light and would be in *shade*.

When an object shields another surface from the sunlight, the dark area that is cast on this surface is called *shadow*.

The triangular prism in Fig. 25.92A is drawn in orthographic projection. The rays of light are chosen to be 45° backward and downward in the two given views. The shadow of the object on the frontal plane is found by projecting to the vertical plane in the top view and downward to the projectors from the front view. One surface is in shade since it is away from the light rays. The shadow of the object helps you instinctively visualize the shape of the object.

The shades and shadows of a cylinder are found in Fig. 25.92B. Points are located around the circular ends of the cylinder. These are projected to the horizontal plane, and upward to the projectors from the top view. The shadow will be an edge in the front view, but it can be seen true size in the top view. Since the circular ends of the

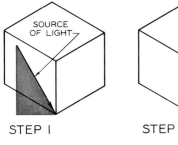

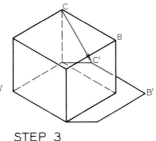

Fig. 25.92 A. The construction of the shadow of a triangular prism in orthographic on a vertical plane. B. The construction of the shadow of the cylinder in orthographic on a horizontal plane.

cylinder are parallel to the horizontal, their shadows will be true circles that are connected with tangent lines.

Since the shadow does not begin at the edge of the cylinder in the top view, you have the impression that the object is suspended above the horizontal surface on which the shadow is cast.

25.35 SHADES AND SHADOWS—A BLOCK IN PERSPECTIVE

A light ray is established in Fig. 25.93 that is parallel to the picture plane since the vertical and horizontal legs of the triangle are perpendicular and true length. The shadows of two vertical lines are found in Step 2 when the rays are passed through points A and B. The shadows are found at A' and B' where horizontal lines from the vertical lines intersect the light rays. The process is continued to find the shadows A', B', and C', which are connected to complete the shadow and the shaded area is indicated.

These same principles are used to find the shades and shadows of a more complex object as shown in Fig. 25.94

25.36 SHADES AND SHADOWS—CYLINDERS IN ISOMETRIC

The shadow of a cylinder is found by locating the shadows of a number of its elements as if they were individual lines (Fig. 25.95).

The light source is parallel to the picture plane in this example. The shadows of the point along the upper circle are connected to complete the shadow. The shaded area is located where the

Fig. 25.93 Shades and shadows of a cube

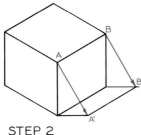

STEP 1   STEP 2   STEP 3   STEP 4

Step 1 The light-source triangle is constructed for a ray of light that is parallel to the drawing paper.

Step 2 The shadows cast from vertical corners passing through the points A and B are found. Note that the projectors from each end of the vertical corners are parallel to the sides of the light-source triangle.

Step 3 The shadow of the vertical line through point C is found even though it is hidden. Line $C'B'$ is drawn to complete the outline of the shadow.

Step 4 The object is shaded to indicate the surfaces in shade and the shadow.

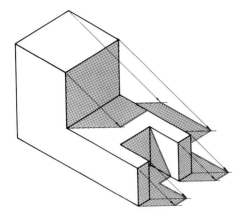

Fig. 25.94 Shades and shadows constructed for an object in isometric.

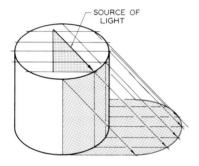

Fig. 25.95 Construction of the shades and shadows of a vertical cylinder. The shadows of each element are found one at a time. The light is parallel to the picture plane.

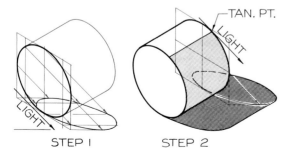

Fig. 25.96 The construction of shades and shadows of a horizontal cylinder. The light is parallel to the picture plane.

light source is tangent to the upper surface of the cylinder.

When a cylinder is in a horizontal position, the circular ends are blocked in and coordinates are used to project points to the horizontal surface, Fig. 25.96. This gives an elliptical shadow for each end. The elliptical shadows of both ends are found in the same manner to complete the shadow. The area of shade is located where the ray of light is tangent to the circular end of the cylinder.

25.37 SHADES AND SHADOWS—OBLIQUE LIGHT

It is unnecessary for the ray of light to be parallel to the picture plane. It can be in any position. A triangle and source of light is established in Fig. 25.97, Step 1. The ray of light passes through the end of a vertical line at A and intersects with the projector that passes through the bottom of the line, parallel to the triangle (Step 2). This locates the shadow, A'.

Note that horizontal lines and their shadows are parallel and equal in length.

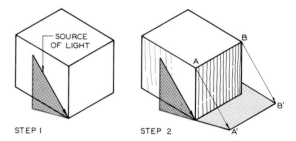

Fig. 25.97 The construction of the shadow of an isometric with a source of light that is oblique to the picture plane.

25.38 SHADES AND SHADOWS—OBLIQUES

An oblique drawing can have shades and shadows applied in the same manner as has been used for isometrics. A triangle is formed to establish the ray of light in Step 1 of Fig. 25.98. The shadows of the vertical corner lines are found. Points C', B', and E' are connected to complete the outline of the shadow. Both the vertical surfaces of the block that are visible are in shade.

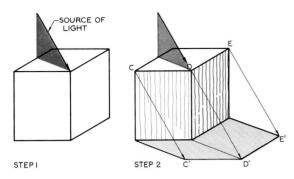

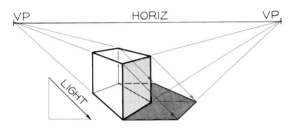

Fig. 25.98 The construction of the shadow of an oblique drawing of a cube with an angular light source.

Fig. 25.99 The construction of shades and shadows of a perspective. The light source is parallel to the picture plane.

25.39 SHADES AND SHADOWS—PERSPECTIVES

To find shades and shadows of perspectives, begin by establishing the ray of light. Since the coordinates of the ray of light in Fig. 25.99 intersect at 90°, the light is parallel to the picture plane.

The shadows of the vertical corners of the object are found by passing projectors through the upper corners that are parallel to the light source. They intersect on the horizontal plane where the projectors from the bottom of vertical lines are extended parallel to the horizontal leg of the light triangle.

The shadows of each upper point are found in this manner and are connected to form the outline of the shadow. Note that horizontal lines and their shadows converge to the same vanishing points since they are parallel in reality.

The shades and shadows of an object are shown in Fig. 25.100 where the source of light is at an angle with the picture plane. The light ray is drawn in the top and front views at selected angles. The vanishing points of horizontal and an-

Fig. 25.100 Shades and shadows—oblique light source

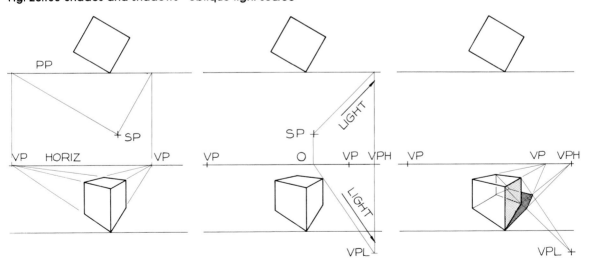

Step 1 The perspective of the object is drawn in the conventional manner.

Step 2 The top and front views of the ray of light are established as desired. The SP is projected to the horizon to locate point O. The vanishing point of the light ray (VPL) is found by drawing a construction line parallel to it in the front view from point O.

Step 3 A shadow of a vertical line is found by locating the intersection between the projectors to VPL and VPH.

gular shadows are found on the horizontal and below the ground line.

The shadows of the vertical line project the VPL (vanishing point, light) where they intersect the shadows that converge at VPH (vanishing point, horizontal). This is repeated for each corner to complete the outline of the shadow.

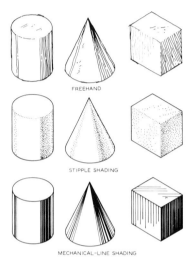

Fig. 25.101 Line and dot shading add realism to a pictorial.

25.40 RENDERING TECHNIQUES

Rendering or shading an object is a technique of adding realism to a pictorial. Examples of freehand and instrument rendering are shown on several basic shapes in Fig. 25.101. These illustrations were made using India ink, but a similar effect can be achieved by using black pencil lines.

A stipple shading technique was used to render the components of a jet turbo engine in Fig. 25.102. This rendered drawing is more realistic than it would be without surface shading.

25.41 OVERLAY FILM

Overlay film is an acetate film on which is printed a pattern that can be applied to a drawing to shade an illustration. It is applied as shown in Fig. 25.103. These films have adhesive backings and can be burnished permanently to the surface by a firm rubbing pressure. It is trimmed to size by using a pointed stylus or razor blade.

Patterns, symbols, letters, numbers, arrowheads, etc., are available on film. Arrowheads are applied to an assembly in Fig. 25.104. Overlay films are available with both dull and glossy surfaces. Both have excellent reproduction qualities.

Fig. 25.103 The steps of applying overlay film to shade an area. (Courtesy of Artype Incorporated.)

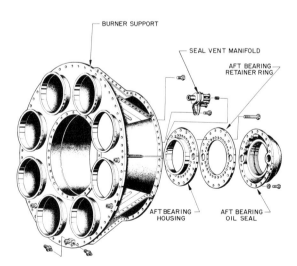

Fig. 25.102 A pictorial assembly that was drawn using a stippling technique with ink. (Courtesy of General Motors Corporation.)

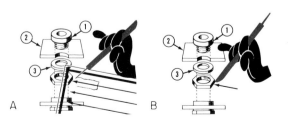

Fig. 25.104 The steps of applying leaders and numbers to an assembly drawing. (Courtesy of Artype Incorporated.)

Examples of various patterns and symbols are shown in Fig. 25.105. Film is also available in percentage screens that are labeled in percent and lines per inch (Fig. 25.106). The percentage indicates the percentage of solid black; the lines per inch represent the number of rows of dots per inch. A 27.5-line screen is an open screen, whereas a 55-line screen is composed of smaller dots that are twice as close together.

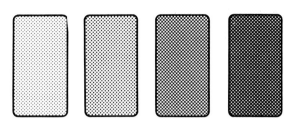

Fig. 25.106 Overlay film is available in screens at 10-percent increments. A 10-percent screen is 10 percent solid black, and a 70-percent screen is 70 percent solid black.

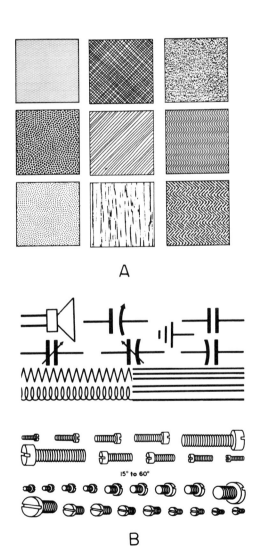

A

B

Fig. 25.105 Examples of a few of the (A) patterns and (B) symbols that are available in 9″ × 12″ sheets of overlay film. (Courtesy of Artype Incorporated.)

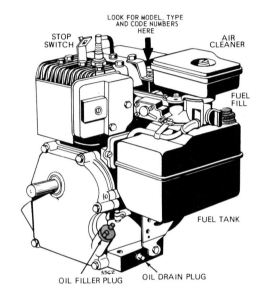

LOOK FOR MODEL, TYPE AND CODE NUMBERS HERE

STOP SWITCH

AIR CLEANER

FUEL FILL

FUEL TANK

OIL FILLER PLUG   OIL DRAIN PLUG

Fig. 25.107 An example of a line drawing made from photograph to improve its reproduction when reduced to a small size. (Courtesy of Briggs & Stratton Corporation.)

25.42 PHOTOGRAPHIC ILLUSTRATIONS

Technical illustrations can be made from photographs as shown in Fig. 25.107. Drawings can emphasize certain aspects of the photograph that might not be otherwise clear. The photograph can be converted to a drawing by projecting it onto the drawing surface by an opaque projector where it can be outlined and then rendered.

Photographs of a large size can be overlaid with tracing vellum or polyester film and the drawing made by tracing from the photograph. Considerable skill is required to convert a photograph into a drawing using bold simple techniques that properly portray an art being illustrated. A bold illustration is required if it is to be reduced to a much smaller size for publication.

Photographs can be used to illustrate parts such as the assembly in Fig. 25.108. These parts were positioned and photographed and the background was masked (whited out) to eliminate the props that were used to support the parts while it was being photographed.

Photographs are usually retouched with an airbrush to improve their quality. An airbrush is a small spray gun that applies ink or paint to the drawing surface to give soft tones of gray. The airbrush has adjustable controls that allow the variation of the application of the spray.

An airbrush is shown in Fig. 25.109 as it is used to shade a drawing. The illustrator who uses an airbrush must use numerous masks to shield the portion of the drawing surface where shading

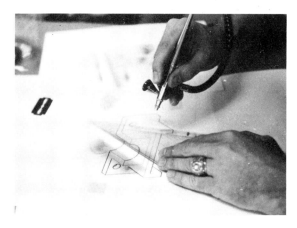

Fig. 25.109 The airbrush is a device that allows the illustrator to apply a light spray of ink or paint on an illustration to obtain gradual tones.

is unwanted. The masks are called **friskets.** They are applied to the surface and windows are cut from them to open the areas to be sprayed.

25.43 MULTIMEDIA ILLUSTRATIONS

The technical illustrator may use any media to achieve the desired effect when making a technical illustration. These can include pencil, ink, charcoal, overlay film, airbrush, and liquid-tipped markers.

The freehand sketch of the automobile in Fig. 25.110 was made with a combination of pencil and liquid-tipped markers. The result is a pleasing sketch that enables the designer to develop and communicate styling ideas while working.

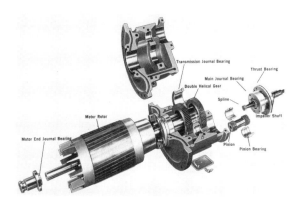

Fig. 25.108 A photographic assembly that has been retouched with an airbrush. (Courtesy of Carrier Air Conditioning Company.)

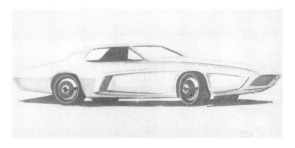

Fig. 25.110 This drawing was rendered using a multimedia method with pencil, ink, and liquid tipped markers. (Courtesy of Ford Motor Company.)

PROBLEMS

The following problems are to be solved on Size A or B sheets as assigned by your teacher. Select the appropriate scale that will best take advantage of the space available on each sheet.

Obliques

1–31. Construct either cavalier, cabinet, or general obliques of the objects assigned.

Isometrics

1–31. Construct isometric drawings of the objects assigned.

Axonometrics

1–31. Construct axonometric scales and find the ellipse template angles for each surface by using a cube rotated into positions of your choice. Calibrate the scales to represent ½″ intervals. Overlay the scales and draw axonometrics of the objects assigned.

Perspectives

32–37. Lay out these perspective problems on Size B sheets and complete the perspectives as assigned.

38–42. Construct three-point perspectives of parts 1 through 5 as assigned. Select the most appropriate scale to take best advantage of the space available.

Shades and Shadows

43–55. Construct pictorials (any type as assigned) of parts 1–13 and find the shades and shadows of each. Establish the source of light that will best enhance the pictorial.

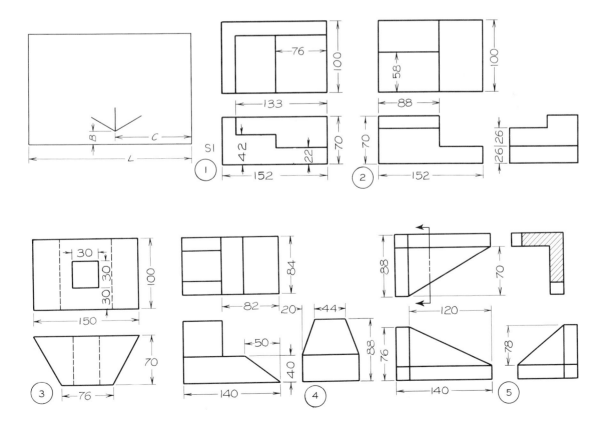

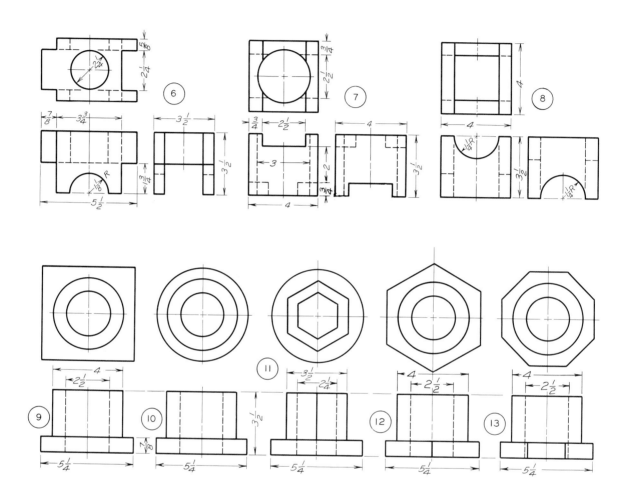

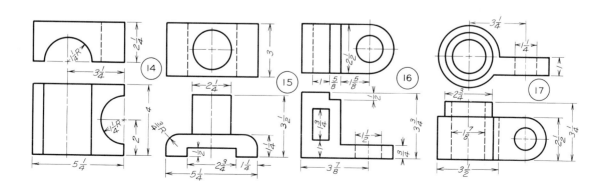

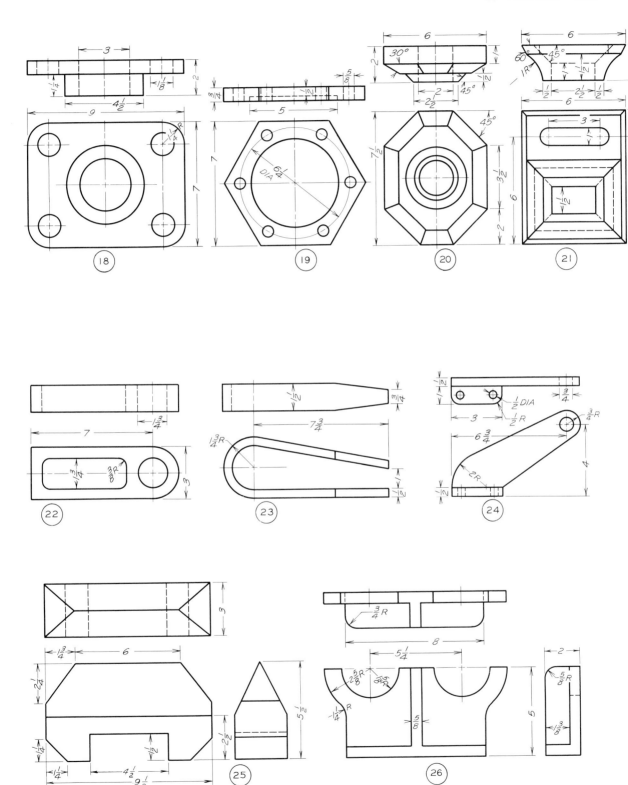

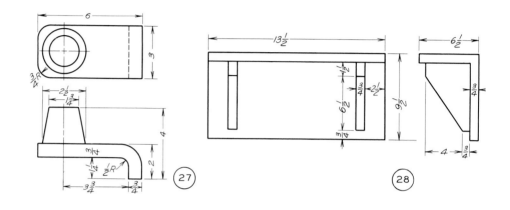

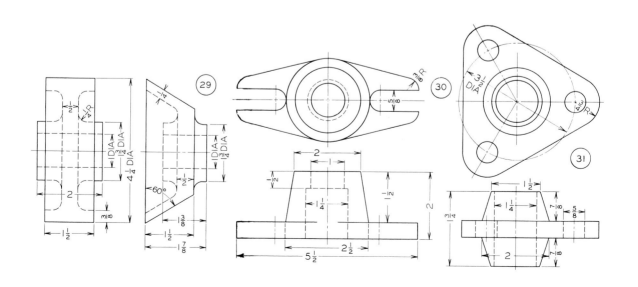

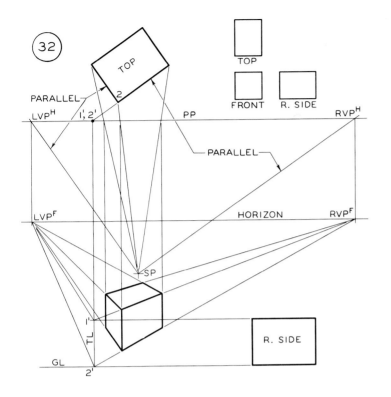

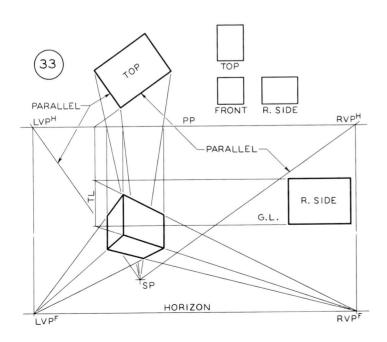

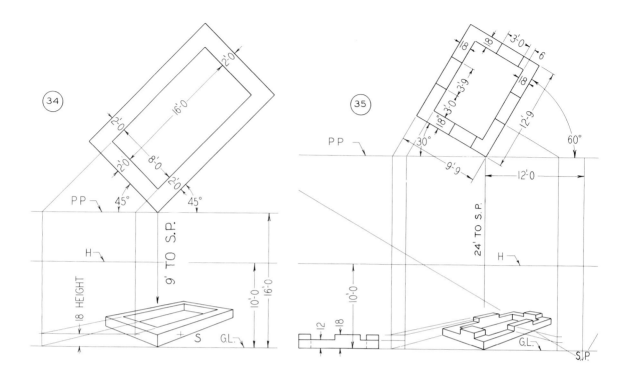

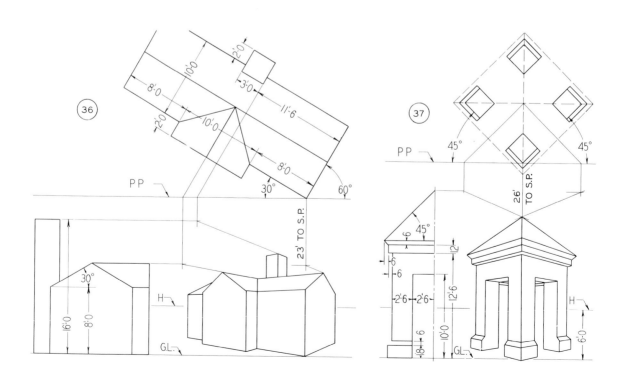

POINTS, LINES, AND PLANES

26.1 INTRODUCTION

Points, lines, and planes are the basic geometric elements that comprise the physical world in which we live. These elements are used extensively in descriptive geometry, which is the discipline that deals with graphically describing the three-dimensional geometry of our technological environment.

The method of presenting descriptive geometry is through the application of orthographic projection.

The following rules of solving and labeling descriptive problems are illustrated in Fig. 26.1.

Lettering: All point, lines, and planes should be labeled using ⅛″ letters with guidelines. Lines should be labeled at each end and planes at each corner. Either letters or numbers can be used.

Points in space should be indicated by two short dashes that are perpendicular to form a cross, *not* a dot. Each dash should be approximately ⅛″ long.

Points on a line should be indicated with a short perpendicular dash on the line, *not* a dot. Label the point with a letter or numeral.

Reference lines are thin black lines that should be labeled in accordance with the text in Chapter 13.

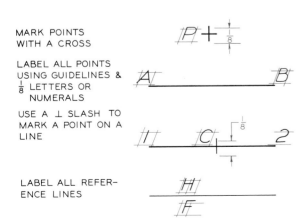

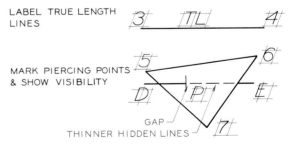

Fig. 26.1 Standard practices for labeling points, lines, and planes.

Object lines used to represent points, lines, and planes should be drawn heavier than reference lines, with an H or F pencil. Hidden lines are drawn thinner than visible lines.

True-length lines should be labeled by the full note, TRUE LENGTH, or by the abbreviation, TL.

True-size planes should be labeled by a note, TRUE SIZE, or by the abbreviation, TS.

Projection lines that are used in constructing the solution to a problem should be precisely drawn with a 4H pencil. These should be thin gray lines, just dark enough to be visible. They need not be erased after the problem is completed.

26.2 ORTHOGRAPHIC PROJECTION OF A POINT

A point is a theoretical location in space and it has no dimension. However, a series of points can establish areas, volumes, and lengths, which are the basis of our physical world.

A point must be projected perpendicularly onto at least two principal planes to establish its true position (Fig. 26.2). Note that when the planes of the projection box at (a) are opened into the plane of the drawing surface in (b), the projectors from each view of point 2 are perpendicular to the reference lines between the planes. The let-

ters H, F, and P are used to represent the horizontal, frontal, and profile planes, respectively, the three principal projection planes.

A point can be located from verbal descriptions with respect to the principal planes. For example, point 2 in Fig. 26.2 can be described as being (1) 4 units left of the profile plane, (2) 3 units below the horizontal plane, and (3) 2 units behind the frontal plane. Each of these measurements must be made in the view where the principal plane being used appears as an edge.

When looking at the front view, the horizontal and profile planes appear as edges. The frontal and profile planes appear as edges in the top view, and the frontal and horizontal planes appear as edges in the side view.

26.3 LINES

A line is a straight path between two points in space. A line can appear in three forms: (1) as a foreshortened line, (2) as a true-length line, or (3) as a point (Fig. 26.3).

Oblique lines are lines that are neither parallel or perpendicular to a principal projection plane, as shown in Fig. 26.4. When line 1–2 is projected onto the horizontal, frontal, and profile planes, it appears foreshortened in each view. This is the general case of a line.

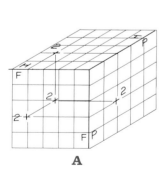

A

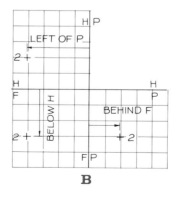

B

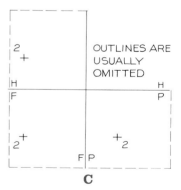

C

Fig. 26.2. The three projections of point 2 are shown pictorially at A and orthographically at B where the projection planes are opened into the plane of the drawing paper. Point 2 is 4 units to the left of the profile, 3 below the horizontal, and 2 behind the frontal. The outlines of the projection planes are usually omitted in orthographic projection as shown at C.

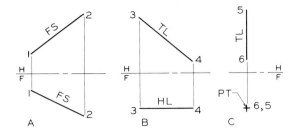

Fig. 26.3 A line in orthographic projection can appear as a point (PT), foreshortened (FS), or true length (TL).

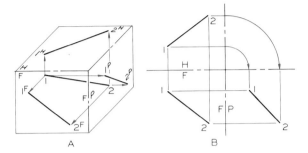

Fig. 26.4 A pictorial of the orthographic projection of a line is shown at A, and as a standard orthographic projection at B.

Principal lines are lines that are parallel to at least one of the principal projection planes.

> The three principal lines are (1) horizontal, (2) frontal, and (3) profile lines, since these are the three principal projection planes.

A principal line is true length in the view where the principal plane to which it is parallel appears true size.

A horizontal line is shown in Fig. 26.5A where it appears true length in the horizontal view, the top view. It may be shown in an infinite number of positions in the top view and still appear true length provided it is parallel to the horizontal plane.

An observer cannot tell whether the line is horizontal or not when looking at the top view. This must be determined from looking at the front or side views. A horizontal line will be parallel to the edge view of the horizontal in the front and

Fig. 26.5 Principal lines

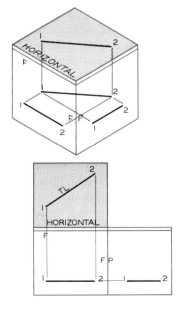

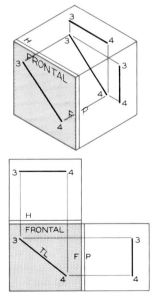

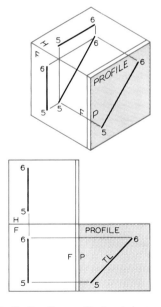

A. Horizontal line The horizontal line is true length in the horizontal view (the top view). The horizontal line will appear parallel to the edge view of the horizontal projection plane in the front and side views.

B. Frontal line The frontal line is true length in the front view. It will appear parallel to the edge view of the frontal plane in the top and side views.

C. Profile line The profile line is true length in the profile view (the side view). It will appear parallel to the edge view of the profile plane in the top and front views.

side views which is the *H-F* fold line. A line that projects as a point in the front view is a combination horizontal and profile line.

A frontal line is parallel to the frontal projection plane and it appears true length in the front view since the observer's line of sight is perpendicular to it in this view. You cannot see that a line is a frontal line by looking at the front view, but you must observe one of the adjacent views, the top or side views. Line 3–4 in Fig. 26.5B is determined to be a frontal line by observing its top and side view of the frontal plane.

Profile lines are parallel to the profile projection planes and they appear true length in the side views, the profile views. It is necessary to look at a view adjacent to the profile view to tell whether or not a line is a profile line. In Fig. 26.5C line 5–6 is parallel to the edge view of the profile plane in the top and side views.

26.4 LOCATION OF A POINT ON A LINE

The top and front views of line 1–2 are shown in Fig. 26.6. Point *O* is located on the line in the top view and it is required that the front view of the point be found.

Since the projectors between the views are perpendicular to the *H-F* fold line between the views in orthographic projection, point *O* is found by projecting in this same direction from the top view to the front view of the line. Point *O* is located on the line to complete the solution.

If a point is to be located at the midpoint of a line, it will be at the line's midpoint whether the line appears true length or foreshortened.

26.5 INTERSECTING AND NONINTERSECTING LINES

Lines that intersect have a point of intersection that lies on both lines and is common to both. Point *O* is Fig. 26.7a is a point of intersection since it projects to a common crossing point in the three views given.

On the other hand, the crossing point of the lines in Fig. 26.7b in the front view is not a point of intersection. Point *O* does not project to a common crossing point in the top view; point *O* is not aligned with the projector. Therefore the lines do not intersect although they do cross. This is verified in the side view where it can be clearly seen that they do not cross.

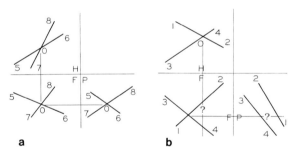

Fig. 26.7 The lines at (b) cross in the top and front views, but they do not intersect. The common point of intersection does not project from view to view. The lines at (a) do intersect because the point of intersection *O* projects as a point of intersection in all views.

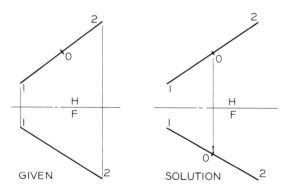

GIVEN   SOLUTION

Fig. 26.6 A point on a line that is shown orthographically can be found on the front view by projection. The direction of the projection is perpendicular to the reference line between the two views.

26.6 VISIBILITY OF CROSSING LINES

Lines *AB* and *CD* in Fig. 26.8 do not intersect, however; it is necessary to determine the visibility of the lines by analysis.

Select a crossing point in one of the views, the front view in Step 1, and project it to the top view to determine which line is in front of the other.

Since *AB* is encountered before *CD*, then *AB* is in front of *CD* and is visible in the front view.

In Step 2 the crossing point in the top view is projected to the front view to find that line *CD* is higher than *AB*; therefore *CD* is visible in the top view.

This process of determining visibility is done by analysis rather than visualization. Two views must be utilized since it would be impossible if only one view were available.

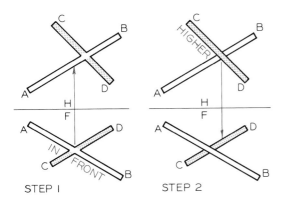

STEP I STEP 2

Fig. 26.8 Visibility of lines

Required Find the visibility of the lines in both views.

Step 1 Project the point of crossing from the front to the top view. This projector strikes *AB* before *CD*; therefore, line *AB* is in front and is visible in the front view.

Step 2 Project the point of crossing from the top view to the front view. This projector strikes *CD* before *AB*; therefore, line *CD* is above *AB* and is visible in the top view.

26.7 VISIBILITY OF A LINE AND A PLANE

The principle of visibility of intersecting lines is used in determining the visibility for a line and a plane. In Step 1 of Fig. 26.9, the intersections of *AB* and lines 1–3 and 2–3 are projected to the top view to determine that the lines of the plane are in front of *AB* in the front view. Consequently, the line is shown as a dashed line in the front.

Similarly, the two intersections on *AB* in the top view are projected to the front view where line

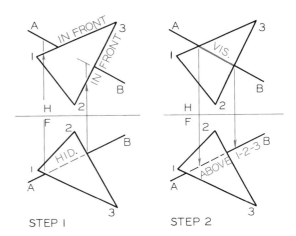

STEP I STEP 2

Fig. 26.9 Visibility of line and a plane

Required Find the visibility of the plane and the line in both views.

Step 1 Project the points where *AB* crosses the plane in the front view to the top view. These projectors encounter lines 1–3 and 2–3 of the plane first; therefore, the plane is in front of the line, making the line invisible in the front view.

Step 2 Project the points where *AB* crosses the plane in the top view to the front view. These projectors encounter line *AB* first; therefore, the line is higher than the plane, and the line is visible in the top view.

AB is found to be above the two lines of the plane, 2–3 and 1–3. Therefore *AB* is drawn as visible in the top view since it is over the plane.

26.8 PLANES

Planes may be considered as infinite in some problems, but it is necessary to determine planes by establishing their limits. A plane can be represented by any of the four methods shown in Fig. 26.10.

Planes in orthographic projection can appear in one of the forms shown in Fig. 26.11: (1) as an edge, (2) as true size, (3) as foreshortened.

Oblique planes are planes that are not parallel or perpendicular to principal projection planes in any view, as shown in Fig. 26.12. This is a general case of a plane.

Fig. 26.10 Representations of a plane

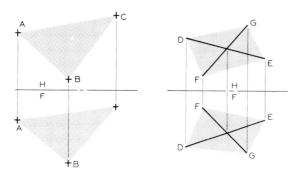

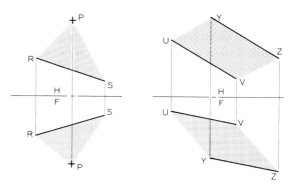

A. Three points not in a straight line can be used to represent a plane.

B. Two intersecting lines can be used to represent a plane.

C. A line and a point not on the line or its extension can be used to represent a plane.

D. Two parallel lines can be used to represent a plane.

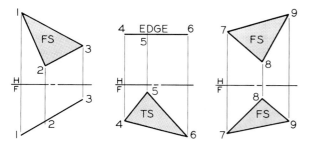

Fig. 26.11 A plane in orthographic projection can appear as an edge, true size (TS), or foreshortened (FS). If a plane is foreshortened in all principal views, it is an oblique plane.

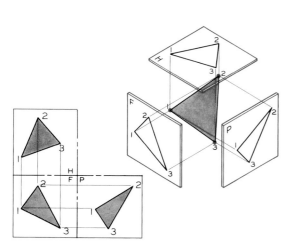

Fig. 26.12 An oblique plane is one that is not parallel or perpendicular to a projection plane; it can be called a general-case plane. The orthographic projection of an oblique plane is shown here.

Principal planes are planes that are parallel to the projection plane as shown in Fig. 26.13, where the three types of principal planes are given: frontal, horizontal, and profile planes.

A frontal plane is parallel to the frontal projection plane and it appears true size in the front view. To tell that the plane is frontal, you must observe the top or side views where its parallelism to the edge view of the frontal plane can be seen.

A horizontal plane is parallel to the horizontal projection plane and it is true size in the top view. To tell that the plane is horizontal, you must observe the front or side views where its parallelism to the edge view of the horizontal plane can be seen.

A profile plane is parallel to the profile projection plane and it is true size in the side view. To tell that the plane is a profile plane, you must observe the top or front view where its parallelism to the edge view of the profile plane can be seen.

26.9 A LINE ON A PLANE

Line AB is given on the front view of the plane in Fig. 26.14. It is required to find the top view of the line. Points A and B that lie on lines 1–4 and 2–3 of the plane can be projected to the top view to the same lines of the plane.

Points A and B are found in the top view and are connected to complete the top view of line AB. This is an application of the principle covered in Section 26.4.

Fig. 26.13 Principal planes

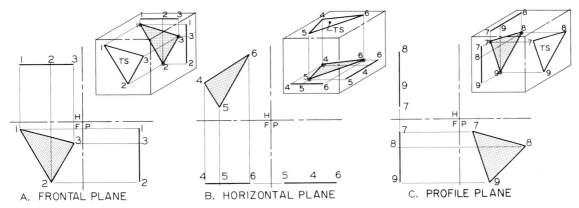

A. FRONTAL PLANE   B. HORIZONTAL PLANE   C. PROFILE PLANE

A. Frontal planes are true size and shape (TS) in the front view. They will appear as edges parallel to the frontal plane in the top and side views.

B. Horizontal planes are true size in the horizontal views (top views). They appear as edges parallel to the horizontal plane in the front and side views.

C. Profile planes are true size in the profile views (side views). They appear as edges parallel to the profile plane in the top and front views.

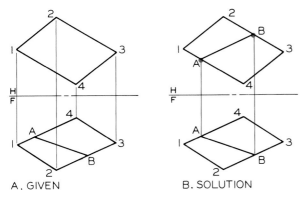

A. GIVEN   B. SOLUTION

Fig. 26.14 If line *AB* lying on the plane is given, the top view of the line can be found. Points *A* and *B* are projected to lines 1–4 and 2–3, respectively, and are connected to form line *AB*.

26.10 A POINT ON A PLANE

Point *O* is given on the front view of plane 4–5–6 in Fig. 26.15. It is required to locate the point on the plane in the top view.

In Step 1, a line in any direction other than vertical is drawn through the point to establish a line on the plane. The line is projected to the top view in Step 2, and the point is projected from the front view to the top view of the line.

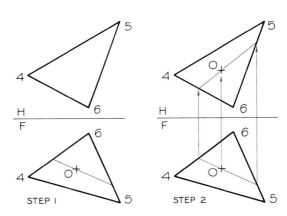

STEP 1   STEP 2

Fig. 26.15 Location of a point on a plane.

Find the top view of point *O* that lies on the plane.

Step 1 Draw a line through the given view of point *O* in any convenient direction except vertical.

Step 2 Project the ends of the line to the top view and draw the line. Point *O* is projected to the line.

Fig. 26.16 Principal lines on a plane.

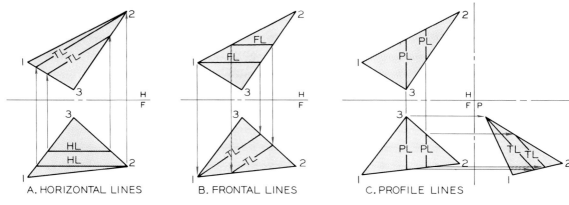

A. HORIZONTAL LINES

B. FRONTAL LINES

C. PROFILE LINES

A. Horizontal lines are drawn first in the front view parallel to the edge view of the horizontal plane. These lines are found true length when they are projected to the top view.

B. Frontal lines are drawn first in the top view parallel to the edge view of the frontal plane. These lines are found true length when they are projected to the front view.

C. Profile lines are drawn first in the front view parallel to the edge view of the profile plane. These lines are found true length when they are projected to the profile view (side view)

26.11   PRINCIPAL LINES ON A PLANE

Principal lines—horizontal, frontal, and profile—may be found in any view of a plane when at least two views of the plane are given.

Horizontal lines are drawn in the front view of the plane in Fig. 26.16A that are parallel to the edge view of the horizontal projection plane. These horizontal lines are projected to the top view of the plane where they will be horizontal and true length.

Frontal lines are drawn parallel to the frontal projection plane in the top view in Fig. 26.16B. When projected to the front view, the lines are true length since the frontal plane appears true size in this view.

Profile lines are drawn parallel to the profile projection plane in the top and front views in Fig. 26.16C. When projected to the side view, the lines will appear true length.

An infinite number of principal lines can be drawn on a single plane. Only two have been shown in each case as examples.

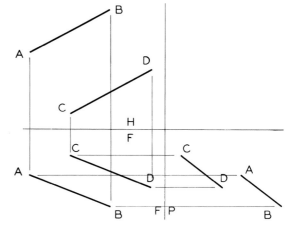

Fig. 26.17 When two lines are parallel, they will project as parallel in all orthographic views.

26.12   PARALLELISM OF LINES

If two lines are parallel, they will appear parallel in all views in which they are seen, except where both appear as points. Lines *AB* and *CD* appear parallel in three views in Fig. 26.17. Parallelism of lines in space cannot be determined if only one view is given; two or more views are required.

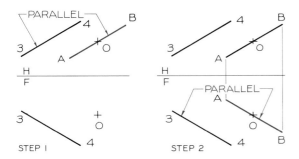

Fig. 26.18 A line parallel to a line.

Draw a line through O that is parallel to the given line.

Step 1 Draw line AB parallel to the top view of line 3–4 with its midpoint at O.

Step 2 Draw the front view of line AB parallel to the front view of 3–4 through point O.

Using this principle, a line can be drawn parallel to a given line through a specified point as shown in Fig. 26.18. In Step 1, line AB is drawn parallel to line 3–4 and through point O, the midpoint of the line.

In Step 2, the front view of AB is drawn parallel to the front view of 3–4 and through point O. Line AB is parallel to 3–4 since it is parallel to 3–4 in both views.

26.13 PARALLELISM OF A LINE AND A PLANE

> A line is parallel to a plane if it is parallel to any line in the plane.

In Fig. 26.19 it is required that a line with its midpoint at point O be drawn that is parallel to plane 1–2–3. This is done by drawing line AB parallel to a line in the plane, line 1–3 in this case, in the top and front views.

The line could have been drawn parallel to *any* line in the plane; therefore, there are an infinite number of positions for lines that are parallel to a given plane.

A similar example is shown in Fig. 26.20 where it is required to draw a line with its midpoint at O that is parallel to the plane formed by two intersecting lines. In Step 1, AB is drawn parallel to line 1–2 on the plane in the top view. In Step 2, AB is drawn parallel to the front view of 1–2, which completes the solution of the problem.

26.14 PARALLELISM OF PLANES

> Two planes are parallel when intersecting lines in one plane are parallel to intersecting lines in the other, as shown in Fig. 26.21.

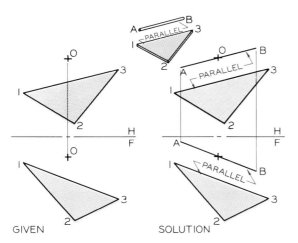

GIVEN

SOLUTION

Fig. 26.19 A line can be drawn through point O that is parallel to the given plane if the line is parallel to any line in the plane. Line AB is drawn parallel to line 1–3 in the front and top views making it parallel to the plane.

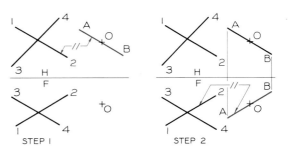

STEP 1

STEP 2

Fig. 26.20 A line parallel to plane

Draw a line parallel to the plane represented by two intersecting lines.

Step 1 Line AB is drawn parallel to line 1–2 through point O.

Step 2 Line AB is drawn parallel to the same line, line 1–2, in the front view which makes AB parallel to the plane.

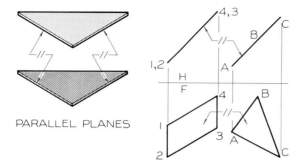

PARALLEL PLANES

Fig. 26.21 Two planes are parallel when intersecting lines in one are parallel to intersecting lines in the other. When parallel planes appear as edges, the edges will be parallel.

It is easy to determine that planes are parallel when both appear as parallel edges in a view.

It is required that a line be drawn through point O and parallel to plane 1–2–3 in Fig. 26.22. In Step 1, EF is drawn through point O and parallel to line 1–2 in both the top and front views. In Step 2, a second line is drawn through point O parallel to line 2–3 of the plane in the front and top views. These two intersecting lines form a plane that is parallel to plane 1–2–3.

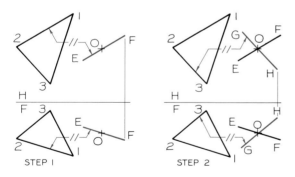

Fig. 26.22 A plane through a point parallel to a plane

Draw a plane through point O that is parallel to the given plane.

Step 1 Draw line EF parallel to any line in the plane, line 1–2 in this case. Show the line in both views.

Step 2 Draw a second line parallel to line 2–3 in the top and front views. These two intersecting lines represent a plane parallel to 1–2–3.

26.15 PERPENDICULARITY OF LINES

> When two lines are perpendicular, they will project at true 90° angles when one or both are true length (Fig. 26.23).

It can be seen that the axis is true length in the front view; therefore, any spoke will be shown perpendicular to the axis in the front view. Spokes OA and OB are examples where one is true length and the other is foreshortened.

When two lines are perpendicular but neither is true length, they will not project with a true 90° angle. This angle will be either less than or greater than 90°.

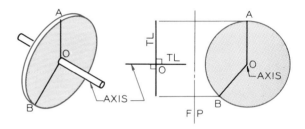

Fig. 26.23 Perpendicular lines will intersect at 90° angles in a view where one or both of the lines appear true length.

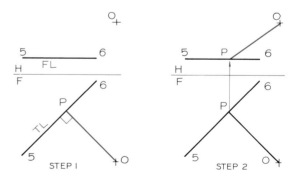

Fig. 26.24 A line perpendicular to a frontal line

Draw a line from point O perpendicular to line 5–6.

Step 1 Since line 5–6 is a frontal line and is true length in the front view, a perpendicular from point O will make a true 90° angle with it.

Step 2 Project point P to the top view and connect it to point O. Since neither of the lines is true length they will not intersect at 90° in the top view.

26.16 A LINE PERPENDICULAR TO A PRINCIPAL LINE

It is required in Fig. 26.24 to construct a line through point O that is perpendicular to frontal line 5–6 that is true length in the front view. In Step 1, OP is drawn perpendicular to 5–6 since it is true length. In Step 2, point P is projected to the top view of 5–6. Line OP in the top view cannot be drawn as a true 90° angle since neither of the lines is true length in this view.

26.17 A LINE PERPENDICULAR TO AN OBLIQUE LINE

It is required in Fig. 26.25 to construct a line from point O that is perpendicular to oblique line 1–2.

In Step 1, horizontal line OE is drawn at a convenient length. In Step 2, OE can be drawn perpendicular to line 1–2 since OE is true length in the top view.

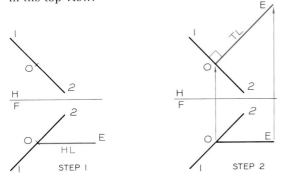

Fig. 26.25 A line perpendicular to an oblique line

Draw a line from point O that is perpendicular to the given line, 1–2.

Step 1 Draw a horizontal line from point O in the front view.

Step 2 Horizontal line OE will be true length in the top view; therefore, it can be drawn perpendicular to line 1–2 in this view.

26.18 PERPENDICULARITY INVOLVING PLANES

A line can be drawn perpendicular to a plane if it is drawn perpendicular to any two intersecting lines in the plane, as shown in Fig. 26.26A. Also,

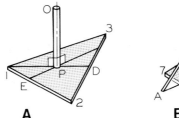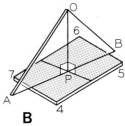

Fig. 26.26 (A) A line is perpendicular to a plane if it is perpendicular to two intersecting lines on the plane. (B) A plane is perpendicular to another plane if the plane contains a line that is perpendicular to the other plane.

a plane is perpendicular to another plane if a line in one is perpendicular to the other. This is illustrated in Fig. 26.26B.

26.19 A LINE PERPENDICULAR TO A PLANE

It is required in Fig. 26.27 that a line be drawn from point O on the plane perpendicular to the plane.

In Step 1, a frontal line is drawn on the plane in the top view and is projected to the front view

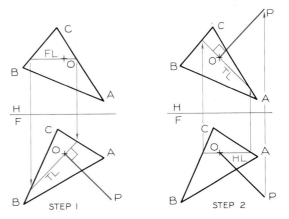

Fig. 26.27 A line perpendicular to a plane

Draw a line from point O that is perpendicular to the plane.

Step 1 Construct a frontal line on the plane through O in the top view. This line is true length in the front view; therefore, line OP can be drawn perpendicular to the true length line.

Step 2 Construct a horizontal line through point O in the front view. This line is true length in the top view; therefore, line OP can be drawn perpendicular to it.

where the line is true length. Line *OP* is drawn perpendicular to the true-length line.

In Step 2, a horizontal line is drawn in the front view and then in the top view of the plane through point *O*. The top view of line *OP* is drawn perpendicular to the true-length line in the top view. This results in a line perpendicular to the plane since the line is perpendicular to two intersecting lines in the plane, a horizontal and a frontal line.

26.20 A PLANE PERPENDICULAR TO AN OBLIQUE LINE

It is required in Fig. 26.28 to construct a plane through point *O* that is perpendicular to line 1–2.

In Step 1, *AB* is drawn as a frontal line in the top view. Since it will be true length in the front view, *AB* is drawn perpendicular to 1–2 in the front view.

In Step 2, *CD* is drawn as a horizontal line in the front view. Line *CD* is drawn perpendicular to 1–2 in the top view since *CD* is true length in this

view. These two intersecting lines, *AB* and *CD*, form a plane that is perpendicular to line 1–2.

26.21 PERPENDICULARITY OF PLANES

In Fig. 26.29 it is required that a plane be constructed through line *AB* that is perpendicular to plane 1–2–3.

In Step 1, a true-length frontal line is found on the plane. Line *CD* is drawn through *AB* and perpendicular to the frontal line in the plane.

In Step 2, the intersection point between *AB* and *CD* is projected to the top view. A true-length horizontal line is found on the plane. The top view of *CD* is drawn through the point of intersection and perpendicular to the true-length line in the plane. These two intersecting lines, *AB* and *CD*, form a plane that is perpendicular to the plane since a single line in the plane, *CD*, is perpendicular to the plane.

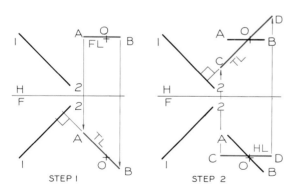

Fig. 26.28 A plane through a point perpendicular to a line

Draw a plane through *O* that is perpendicular to line 1–2.

Step 1 Draw line *AB* as a frontal line in the top view. Since it will be true length in the front view, it can be drawn perpendicular to line 1–2.

Step 2 Draw line *CD* as a horizontal line in the front view. Since it will be true length in the top view, it can be drawn perpendicular to line 1–2. The intersecting lines form a plane that is perpendicular to the line.

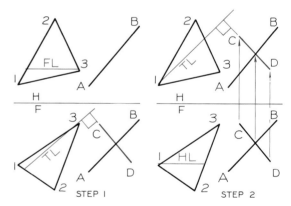

Fig. 26.29 A plane through a line perpendicular to a plane

Draw a plane through the line that is perpendicular to the plane.

Step 1 Draw a frontal line on the plane in the top view and find its true-length view in the front view. Line *CD* is drawn through line *AB* and perpendicular to the true length line.

Step 2 Draw a horizontal line on the plane in the front view and find its true-length view in the top view. Draw line *CD* perpendicular to the true-length line and through line *AB* at the point of intersection projected from the front view.

PROBLEMS

Use Size A (8½" × 11") sheets for the following problems, and lay out the problems using instruments. Each square on the grid is equal to 0.20" or about 5 mm. The problems can be laid out on grid paper or plain paper. Label all reference planes and points in each problem with ⅛" letters or numbers, using guidelines.

1. (Fig. 26.30) Find the missing third views of the given points in Problems 1A through 1D. Find the missing third views of the lines between the given points in Problems 1E through 1F.

2. (Fig. 26.31) (2A) Find the front and top views of the profile line. (2B) Find the front and side views of the horizontal line. (2C) Find the side view of 5–6. (2D) Find the front and side views of 7–8. (2E and 2F) Find the top and side views of the given lines.

3. (Fig. 26.32) (3A) Find the side view of the line. (3B) Find the front and side views of the horizontal line. (3C) Draw three views of a line from point 2 to point O on the line. (3D) Find the side view of the line and locate the midpoint of the line in each view. (3E and 3F) Find the piercing

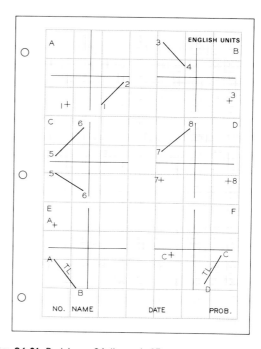

Fig. 26.31 Problems 2A through 2F.

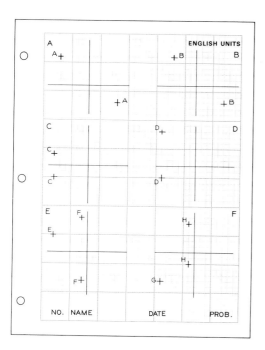

Fig. 26.30 Problems 1A through 1F.

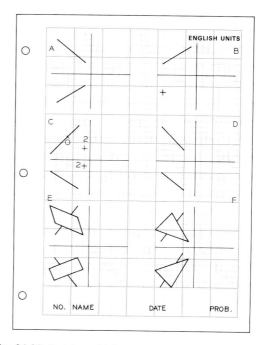

Fig. 26.32 Problems 3A through 3F.

points between the lines and planes, show visibility, and complete the missing third view.

4. (Fig. 26.33) (4A) Find the front and side views of the horizontal plane. (4B) Draw the top and side views of the frontal plane. (4C) Draw the front and top views of the profile plane. (4D) Draw the top view of the plane. (4E) Find the side view of the plane and locate the line upon it in each view. (4F) Find the missing view of the plane and locate point O on each view of the plane.

5. (Fig. 26.34) (5A) Draw a line with its midpoint at O that is parallel to the line. (5B) Construct the front view of a second plane that is parallel to the given plane. (5C and 5D) In each problem, draw lines through point O that are parallel to the given planes.

6. (Fig. 26.35) In each problem, draw lines through point O that are perpendicular to the given lines.

7. (Fig. 26.36) (7A and 7B) Draw a line through each point O that is perpendicular to the given planes. (7C) Draw a plane through point O that is perpendicular to the given line. (7D) Draw a plane through the given line that is perpendicular to the plane.

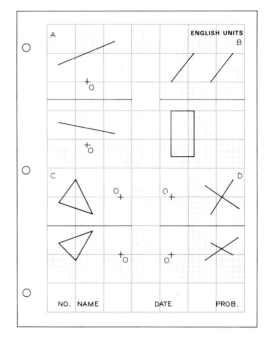

Fig. 26.34 Problems 5A through 5D.

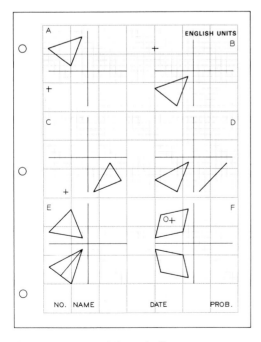

Fig. 26.33 Problems 4A through 4F.

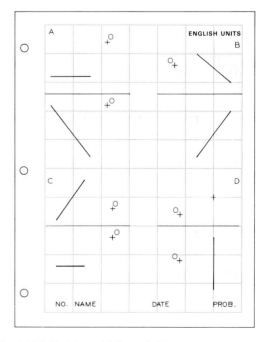

Fig. 26.35 Problems 6A through 6D.

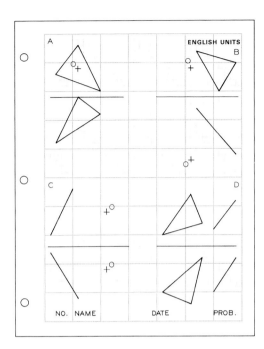

Fig. 26.36 Problems 7A through 7D.

PRIMARY AUXILIARY VIEWS IN DESCRIPTIVE GEOMETRY

27.1 INTRODUCTION

Descriptive geometry can be defined as the projection of three-dimensional figures onto a two-dimensional plane of paper in such a manner as to allow geometric manipulations to determine lengths, angles, shapes, and other geometric information by means of graphics. Orthographic projection is the system used for laying out descriptive geometry problems.

The primary auxiliary view is a powerful tool of descriptive geometry that permits the analysis of three-dimensional geometry that would be difficult by other means. This area of study permits the measurement and determination of distances, lengths, angles, sizes, and areas that are essential to the development of solutions to technical problems.

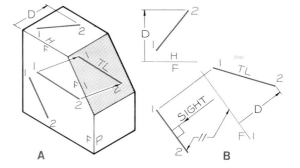

Fig. 27.1 A pictorial of line 1–2 is shown inside a projection box where a primary auxiliary plane is established that is perpendicular to the frontal plane and parallel to the line. The orthographic arrangement of this auxiliary view is shown at B where the auxiliary view is projected from the front view to find 1–2 true length.

27.2 PRIMARY AUXILIARY VIEW OF A LINE

The top and front views of line 1–2 are shown pictorially and orthographically in Fig. 27.1. Since the line is not a principal line, it is not true length in the principal views. To find its true-length view, a primary auxiliary view must be used.

At B the line of sight is drawn perpendicular to the front view of the line and reference line F–1 is drawn parallel to the frontal view. You can see in the pictorial that the auxiliary plane is par-

allel to the line and perpendicular to the frontal plane, which accounts for it being labeled as F–1.

The auxiliary view is found by projecting parallel to the line of sight and perpendicular to the F–1 reference line. Point 2 is found by transferring distance D with your dividers to the auxiliary view, since the frontal plane appears as an edge in both the top and auxiliary views. Point 1 is located in the same manner and the points are connected to find the true-length view of the line. It is labeled TL in this view. The reference lines are labeled as shown in the figure.

Fig. 27.2 True length of a line by a primary auxiliary view

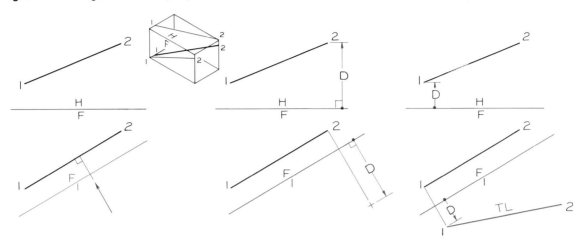

Step 1 To find 1–2 true length, construct the line of sight perpendicular to the line. Draw reference line F–1 parallel to the front view of the line.

Step 2 Project point 2 perpendicularly from the front view. Dimension D from the top view locates point 2.

Step 3 Point 1 is located by transferring dimension D from the top view. Line 1–2 is true length in the auxiliary view.

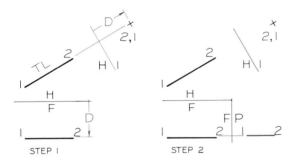

Fig. 27.3 Point view of a line

Step 1 The point view of a line can be found in a primary auxiliary view that is projected from the true-length view of the line.

Step 2 An auxiliary view projected from a foreshortened view of a line will result in a foreshortened view of the line; not a point view.

Figure 27.2 separates the sequential steps required to find the true length of an oblique line. It is beneficial to letter all reference planes using the notation suggested in the various steps with the exception of the dimensions (such as D) that are transferred from one view to another with your dividers to locate desired points and lines.

A primary auxiliary view can result in a point view of the line if projected from a true-length view of the line in a principal view (Fig. 27.3). In Step 1, the point view is found by projecting from the horizontal line that is true length in the top view. The auxiliary view that is projected from the front view of the line does not give a point view since the line is foreshortened in the front view.

27.3 TRUE LENGTH BY ANALYTICAL GEOMETRY

You can see in Fig. 27.4 that the length of a frontal line can be found in the front view by the application of analytical geometry (mathematics) and

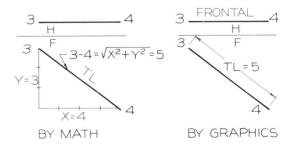

Fig. 27.4 A line that appears true length in a view (the front view in this case) can have its length calculated by application of the Pythagorean theorem and mathematics. Since the line is a frontal line and is true length in the front view, its length can be measured graphically.

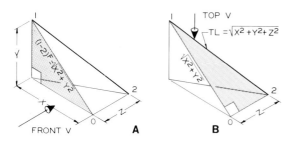

Fig. 27.5 A three-dimensional line that is not true length in a principal view can be found true length by the Pythagorean theorem in two steps. The frontal projection, 1–0, is found using the x- and y-coordinates. At B, the hypotenuse of right triangle 1–0–2 is found using x-, y-, and z-coordinates.

the Pythagorean theorem. The Pythagorean theorem states that the hypotenuse of a right triangle is equal to the square root of the sum of the squares of the other two sides. The x- and y-coordinates of 4 and 3 result in a length of 5 units for line 3–4.

The length of the line can be graphically measured in the front view since it is true length in this view. The accuracy of the measurement will depend upon the accuracy of your drafting of the views.

The line shown pictorially in Fig. 27.5 can be found true length by analytical geometry by determining the length of the front view where the x- and y-coordinates form a right triangle at A. A second right triangle at B, (1–0–2), is solved to find its hypotenuse, which is the true length of the line in question, 1–2. You can see that the true length of an oblique line is the square root of the sum of the squares of the x-, y- and z-coordinates that correspond to width, height, and depth.

The steps in determining the true length of line 1–2 by analytical geometry are shown in Fig. 27.6.

27.4 THE TRUE-LENGTH DIAGRAM

A true-length diagram is constructed with two perpendicular lines to find a line of true length as shown in Fig. 27.7. This method does not give a direction for the line, but merely its true length.

The two measurements that are laid out on the true-length diagram can be transferred from any two adjacent orthographic views. One measurement is the distance between the endpoints in one of the views. The other measurement, taken

Fig. 27.6 True length of a line—analytical method

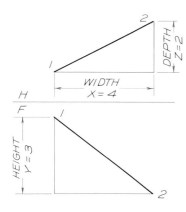

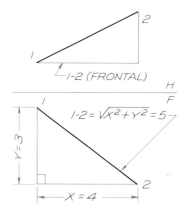

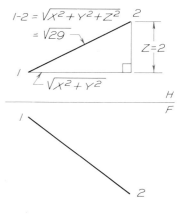

Step 1 Right triangles are drawn with line 1–2 as the hypotenuse in the top and front views. The coordinates, or legs, of the right triangles are drawn parallel and perpendicular to the H-F reference line.

Step 2 The true length of the front view of 1–2 is found by the Pythagorean theorem. The resulting length is found to be 5 units.

Step 3 The true length of the line is found by combining the length of the front projection with the true length of the z-coordinate in the top view. The length is $\sqrt{29}$.

Fig. 27.7 True-length diagram

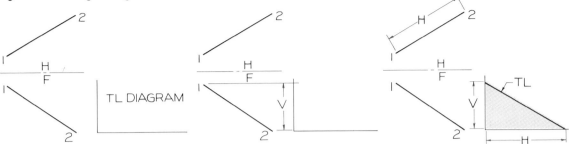

Required Find line 1–2 true length in a TL diagram.

Step 1 Transfer the vertical distance between the ends of 1–2 to the vertical leg of the TL diagram.

Step 2 Transfer the horizontal length of the line in the top view to the horizontal leg of the TL diagram.

Fig. 27.8 Angles between lines and principal planes

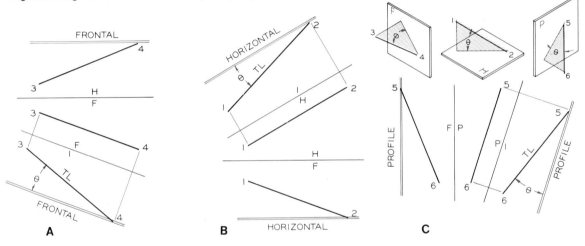

A

B

C

A. When an auxiliary view is projected from the front view, the frontal plane appears as an edge and the line is true length. The angle with the frontal plane can be measured in this view.

B. When an auxiliary view is projected from the top view, the horizontal plane appears as an edge and the line is true length. The angle with the horizontal plane can be measured in this view.

C. When an auxiliary view is projected from the side view, the profile plane appears as an edge and the line is true length. The angle with the profile plane can be measured in this view.

from the adjacent view, is measured between the endpoints in a direction perpendicular to the reference line between the two views.

27.5 ANGLES BETWEEN LINES AND PRINCIPAL PLANES

To measure the angle between a line and a plane, the line must appear true length in the view where the plane appears as an edge. Since a principal plane will appear as an edge in a primary auxil-

iary view, the angle a line makes with this plane can be measured if the line is found true length in this view.

Line 3–4 in Fig. 27.8A is found true length and the frontal plane as an edge when the auxiliary view is projected from the front view. The angle between 3–4 and the frontal plane can be measured in this view.

Similarly, the angles between 1–2 and the horizontal and the angle between the profile and 5–6 are found in primary auxiliary views where the lines are true length and the planes are edges.

27.6 SLOPE OF A LINE

> **Slope** is defined as the angle a line makes with the horizontal plane.

It may be specified by either of the three methods in Fig. 27.9; it can be indicated as *slope angle*, *percent grade*, or *slope ratio*.

Slope Angle

The slope of a line can be measured in a view where the line is true length and the horizontal plane appears as an edge. Consequently, the slope of *AB* in Fig. 27.10 can be measured in the front view where Θ is found to be 31°. This angle can

also be found by converting its tangent of 0.60 to 31° by using the trigonometric tables.

Percent Grade

The percent grade of a line is found in the view where the line is true length and the horizontal plane appears as an edge. Grade is the ratio of the vertical (rise) divided by the horizontal (run) between the ends of a line expressed as a percentage.

The percent grade of *AB* is found in Fig. 27.10 by using a combination of mathematics and graphics. Ten units are laid off parallel to the horizontal from *A* using a convenient scale with decimal units in Step 1. In Step 2 the vertical drop of the line after 10 units along the horizontal is measured as 6 units. Since 10 units were used along the horizontal, your arithmetic is simplified in finding the tangent of the angle to be 0.60, which is easily converted into −60% grade. The grade is negative from *A* to *B* since this is downhill. It would be positive from *B* to *A*, which is uphill. Line *CD* has a positive fifty-percent grade (+50%) from *C* to *D* in Fig. 27.11A.

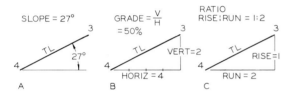

Fig. 27.9 The inclination of a line with the horizontal can be measured and expressed by (A) slope angle, (B) percent grade, or (C) slope ratio.

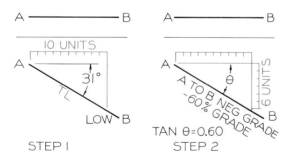

Fig. 27.10 Percent grade of a line

Step 1 The percent grade of a line can be measured in the view where the horizontal appears as an edge and the line is true length (the front view here). Ten units are laid off parallel to the horizontal from the end of the line.

Step 2 A vertical distance from *A* to the line is measured to be 6 units. The percent grade is 6 divided by 10 or 60%. This is negative when the direction is from *A* to *B*. The tangent of this slope angle is 6/10 or 0.60.

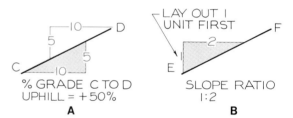

Fig. 27.11 The percent grade of a line is positive if uphill; negative if downhill (part A). The slope ratio is expressed as 1: *xx* where 1 is the rise and *xx* is the horizontal distance. The 1 unit must be drawn first and then the horizontal distance can be drawn (part B).

Slope Ratio

Slope ratio is the same as the percent grade except for the method of expressing the relationship between vertical and horizontal distances. The first number of the ratio is always one, such as 1:10, 1:200, and so on. The first number is the rise (always one) and the second number is horizontal run (see Fig. 27.9).

The graphical method of finding the slope ratio is shown in Fig. 27.11B where the rise of one is laid off on the true-length view of *EF*. The cor-

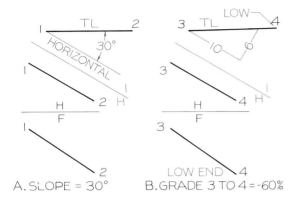

A. SLOPE = 30° B. GRADE 3 TO 4 = -60%

Fig. 27.12 Slope of an oblique line

A. The slope angle can be measured in a view where the horizontal appears as an edge and the line is true length. The slope of 30° is found in an auxiliary view projected from the top view.

B. The percent grade can be measured in a true length view of the line projected from the top view. Line 3–4 has a –60% grade from 3 to the low end at 4.

Compass bearings always begin with the north or south directions and the angles with north and south are measured toward east or west.

The line in part A that makes 30° with north has a bearing of N 30° W. A line making 60° with south toward the east has a compass bearing of South 60° East, or S 60° E. Since a compass can be read only when held horizontally, the compass bearings of a line can be determined only in the top view, the horizontal view.

An azimuth bearing is measured from north in clockwise direction to 360° (Fig. 27.13B). Azimuth bearings are used to avoid confusion that might be caused by reference to the four points of a compass. Azimuth bearings of a line are written as N 120°, N 210°, etc., with this notation indicating that the measurements are made from north.

responding horizontal is found to be 2, which results in a slope ratio of 1:2. The tangent of the slope angle in this case is ½ or 0.50; therefore this line has a percent grade of 50%

Oblique Lines When a line is oblique and does not appear true length in the front view, it must be found true length by an auxiliary view projected from the top view. This auxiliary view shows the horizontal as an edge and the line true length making it possible to measure the slope angle (Fig. 27.12A).

Similarly, an auxiliary view projected from the top view must be used to find the percent grade of a line of an oblique line (Fig. 27.12B). Ten units are laid off horizontally, parallel to the H–1 reference line and the vertical distance is found to be 6. This gives a –60% grade of the line downhill from 3 to 4.

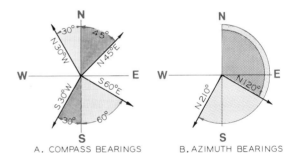

A. COMPASS BEARINGS B. AZIMUTH BEARINGS

Fig. 27.13 A. Compass bearings are measured with respect to north and south directions on the compass. B. Azimuth bearings are measured with respect to north in a clockwise direction up to 360°.

The compass bearing (direction) of a line is assumed to be toward the low end of the line unless otherwise specified.

For example, line 2–3 in Fig. 27.14 has a bearing of N 45° E since the line's low end is point 3. It can be seen in the front view that point 3 is the lower end.

The compass bearing and slope of a line are found in Fig. 27.15. In Step 1, the bearing of the line is found in the top view toward point 6, the low end of the line. The line is found true length by projecting from the top view in Step 2 where

27.7 COMPASS BEARING OF A LINE

Two types of bearings of a line's direction are (A) compass bearings, and (B) azimuth bearings (Fig. 27.13).

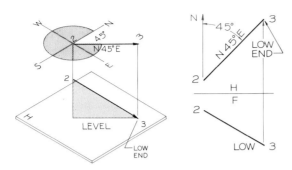

Fig. 27.14 The compass bearing of a line is measured in the top view toward its low end (unless specified toward the high end). Line 2–3 has a bearing of N 45° E toward the low end at 3.

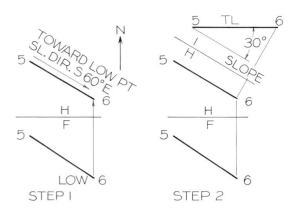

Fig. 27.15 Slope and bearing of a line

Step 1 Slope bearing can be found in top view toward its low end. Direction of slope is S 60° E.

Step 2 The slope angle of 30° is found in an auxiliary view projected from the top view where the line is found true length.

Fig. 27.16 A line from slope specifications

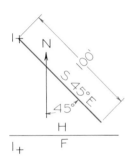

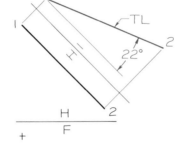

Step 1 It is required to draw a line through point 1 that bears S 45° E for 100' horizontally and slopes 22°. The bearing and horizontal are drawn in the top view.

Step 2 An auxiliary view is projected from the top view where the 22° slope angle is measured.

Step 3 The front view of 1–2 is found by locating point 2 in the front view.

the slope angle of 30° is found. This information can be used to describe verbally the line as having a compass bearing of S 60° E and a slope of 30° from 5 to 6.

When given verbal information and one point in the top and front views, a line can be drawn as shown in Fig. 27.16. If it is known that the line has a bearing of S 45° from point 1 for a horizontal distance of 100', the top view can be drawn (Step 1). If it is known that the line has a 22° slope, the true length auxiliary view can be constructed

(Step 2). The front view can be completed by locating point 2 in Step 3.

27.8 CONTOUR MAPS AND PROFILES

The contour map is the method of representing irregular surfaces of the earth. A pictorial view of a sectional plane and a portion of the earth are shown in Fig. 27.17, along with the conventional

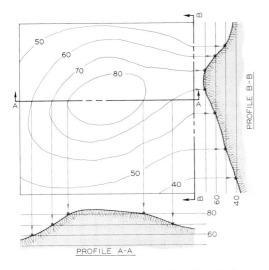

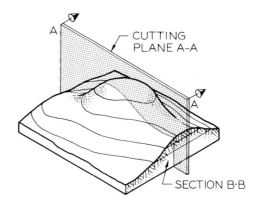

Fig. 27.17 A contour map uses contour lines to show variations in elevation on an irregular surface. Vertical sections taken through a contour map are called profiles.

orthographic views of the contour map and profiles. The following definitions must be understood prior to further discussion.

Contour lines   are horizontal lines that represent constant elevations from a horizontal datum such as sea level. Contour lines can be thought of as the intersections of horizontal planes with the surface of the earth. The vertical interval of spacing between the contours in Fig. 27.17 is 10′.

Contour maps   are maps on which contour lines are drawn to represent irregularities of the surface (Fig. 27.17). The closer the contour lines are to each other on the map, the steeper the terrain.

Profiles   are vertical sections through a contour map that are used to show the earth's surface at any desired location. Two profiles are shown in Fig. 27.17. When applied to topography, a vertical section is called a profile regardless of the direction of the cutting plane in the top view. Contour lines appear as edge views of equally spaced horizontal planes in profiles. The true representation of a profile is drawn with the vertical scale equal to the scale of the contour map; however, this scale is often increased to emphasize changes in elevation that would not otherwise be apparent.

Contoured surfaces   such as airfoils, automobile bodies, ship hulls, and household appliances must also be depicted on the drawing board by using contour lines. When applied to objects other than the earth's surface, this technique is representing contours is called **lofting.**

27.9   VERTICAL SECTIONS

In Fig. 27.18, vertical sections (called profiles) are passed through the top view of an underground pipe system that begins at 1 and ends at 3. Auxiliary views are projected from the top view to find the surface of the earth and the pipe under the ground in Steps 1 and 2. The pipe is known to be located 15′ under the ground at points 1, 2, and 3. The percent grade of the pipes is found in Step 3 to complete the problem.

The same scale was used to draw the profiles as was used to construct the contour map. This makes it possible to measure the true lengths and angles of slope in the profiles. The percent grade and the compass bearing of each line is labeled on the contour map.

27.10   PLAN-PROFILES

A plan-profile is a combination drawing that includes a plan with contours and a vertical section called a profile. A plan-profile is used to show an underground drainage system from manhole 1 to manhole 3 in Fig. 27.19 and Fig. 27.20.

Fig. 27.18 Vertical sections

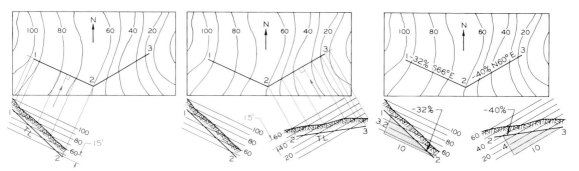

Step 1 To locate pipes 15 feet under the surface from point 1 to 2 to 3, vertical sections are passed through the pipes in the plan view. A profile is taken perpendicularly from pipe 1–2 to find the surface of the ground, the ends of the pipe 15 feet under the surface, and pipe 1–2 true length.

Step 2 A second profile is projected perpendicularly from the plan view of pipe 2–3 to find the surface and pipe 2–3 true length. Notice that the cutting plane is extended beyond point 3 in the top view to provide a more descriptive profile section.

Step 3 The percent grades of pipes 1–2 and 2–3 are found by construction in the profiles. These are negative since the pipes run downhill from 1 to 3. The percent grades and bearings are used to label the top view of the pipes. The elevations of each point can be measured in the profile section.

Fig. 27.19 Plan-profile

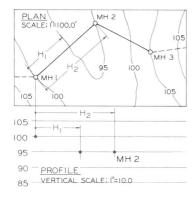

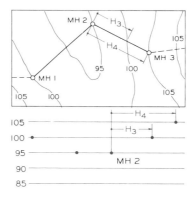

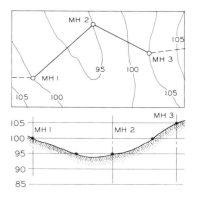

Required Find the profile of the earth over the underground drainage system.

Step 1 Distances H_1 and H_2 from manhole 1 are transferred to their respective elevations in the profile. This is not an orthographic projection.

Step 2 Distances H_3 and H_4 are measured from manhole 2 in the plan and are transferred to their respective elevations in the profile. These points represent elevations of points on the earth above the pipe.

Step 3 The five points are connected with a freehand line and the drawing is crosshatched using the symbol given in Chapter 15 to represent the earth's surface. Center lines are drawn to show the locations of the three manholes that will be located in Fig. 27.20

Fig. 27.20 Plan-profile, manhole location

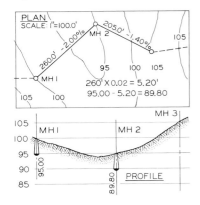

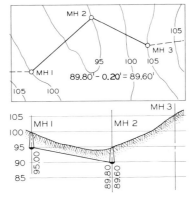

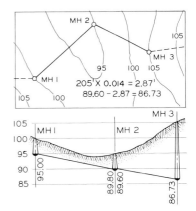

Step 1 The horizontal distance from the MH1 to MH2 is multiplied by the percent grade. The elevation of the bottom of manhole 2 is calculated by subtracting from the elevation of manhole 1.

Step 2 The lower side of manhole 2 is 0.20' lower than the inlet side to compensate for loss of head (pressure) due to the turn in the pipeline. The lower side is found to be 89.60' and is labeled.

Step 3 The elevation of manhole 3 is calculated to be 86.73' since the grade is 1.40% from manhole 2 to manhole 3. The flow line of the pipeline is drawn from manhole to manhole and the elevations are labeled.

The profile is drawn with an exaggerated vertical scale to emphasize the variations in the earth's surface and the grade of the pipe. Although the vertical scale is usually increased, it can be drawn at the same scale as is used in the plan if desired.

Manhole 1 is projected to the profile using orthographic projection, but the other points are not orthographic projections (Fig. 27.19). The points where the contour lines cross the top view of the pipe are transferred to their respective elevations in the profile with your dividers. These points are connected to show the surface of the earth over the pipe and the location of manhole center lines.

The vertical drop from one end of a pipe to the other is calculated by multiplying the horizontal distances by the percent grade (Fig. 27.20). The drop from manhole 1 to manhole 2 is found to be 5.20' by multiplying the horizontal distance of 260.00' by a −2.00% grade. This drop is subtracted from the depth of manhole 1 to find the elevation of manhole 2; this turns out to be 89.80'.

Since the pipes intersect at manhole 2 at an angle, the flow of the drainage is disrupted at the turn; consequently, a drop of 0.20' is given from the inlet across the floor of the manhole to compensate for the loss of pressure (head) through the manhole. The elevations on both sides of the manhole are specified as shown in Step 2 of Fig. 27.20.

The true lengths of the pipes cannot be accurately measured in the profile when the vertical scale is different from the horizontal scale. Trigonometry must be used for this computation.

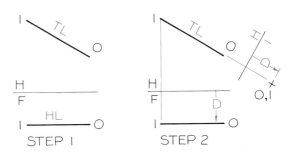

Fig. 27.21 Point view of a line

Step 1 Line 1–0 is horizontal in the front view and is therefore true length in the top view.

Step 2 The point view of 1–0 is found by projecting parallel from the top view to the auxiliary view.

Fig. 27.22 Edge view of a plane

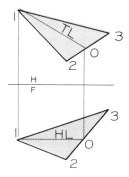 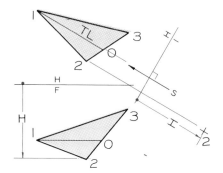 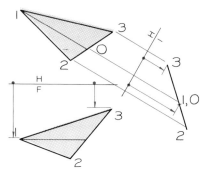

Step 1 To find plane 1–2–3 as an edge, horizontal line 1–0 is drawn on the plane in the front view. Line 1–0 is projected to the top view where it is true length.

Step 2 A line of sight is constructed parallel to the true-length line 1–0. Reference line H-1 is drawn perpendicular to the line of sight. Point 2 is found in the auxiliary view by transferring dimension H from the front view.

Step 3 Points 1 and 3 are found by transferring their height dimensions from the front view. These points will lie in a straight line which is the edge of the plane. Line 1–0 will appear as a point in this view.

27.11 EDGE VIEW OF A PLANE

The edge view of a plane can be found in any view where a line on the plane appears as a point. A line can be found as a point by projecting from a view where the line is true length (Fig. 27.21).

A true-length line can be drawn on any plane by drawing the line parallel to one of the principal planes and projecting it to the adjacent view, as shown in Step 1 of Fig. 27.22. Since line 1–0 is true length in the top view, its point view may be found in Steps 2 and 3. The remainder of the plane appears as an edge in this auxiliary view.

27.12 DIHEDRAL ANGLES

The angle between two planes is called a *dihedral angle*. This angle can be found in the view where the line of intersection between two planes appears as a point.

The line of intersection, 1–2, between the two planes in Fig. 27.23 is true length in the top view. This makes it possible to find the point view of line 1–2 and the edge view of both planes in a primary auxiliary view that is projected from the true-length view of 1–2.

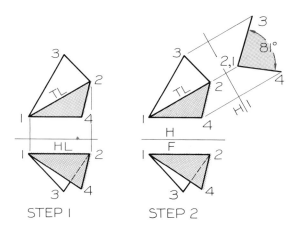

Fig. 27.23 Dihedral angle

Step 1 The line of intersection between the planes, 1–2, is true length in the top view.

Step 2 The angle between the planes (the dihedral angle) can be found in the auxiliary view where the line of intersection appears as a point.

27.13 PIERCING POINTS BY PROJECTION

Figure 27.24 gives the sequential steps necessary to find the piercing point of line 1–2 that passes through the plane. A cutting plane is passed through the line and plane in the top view. The trace of this cutting plane, line *DE*, is then projected to the front view where the piercing point *P* is found in Step 1. The top view of *P* is located in Step 2, and the visibility of the line is found in Step 3 by applying the principles covered in the previous chapter.

27.14 PIERCING POINTS BY AUXILIARY VIEWS

The piercing point of a line and a plane can be found by an auxiliary view as shown in Fig. 27.25. Piercing point *P* can be seen in Step 2 where the plane is found as an edge. Point *P* is projected back to the line in the top and front views from the auxiliary view in Step 3. The location of *P* in the front view is checked by transferring dimension *H* from the auxiliary view with your dividers.

Visibility is easily determined in the top view since it can be seen that *AP* is higher than the plane in the auxiliary view, and is therefore visible in the top view. Analysis of the top view shows that endpoint *A* is the most forward point; therefore this end of the line is visible in the front view.

27.15 PERPENDICULAR TO A PLANE

In Fig. 27.26 it is required that line be drawn from point *O* that is perpendicular to the plane.

> This perpendicular line will appear as a true-length perpendicular in the view where the plane appears as an edge.

The true-length perpendicular is drawn in Step 2 to locate piercing point *P*. Point *P* is found in the top view by drawing line *OP* parallel to the H–1 reference line. It must lie in this direction since *OP* is true length in the auxiliary view. It

Fig. 27.24 Piercing points by projection

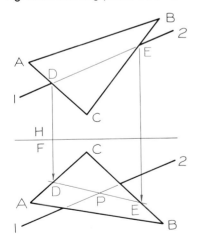

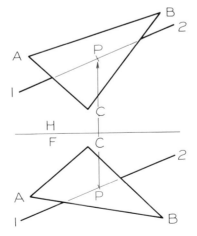

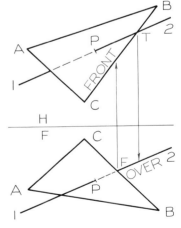

Step 1 Assuming that a vertical cutting plane is passed through the top view of 1–2, the plane intersects *AC* and *BC* at *D* and *E*. The intersection of *DE* with 1–2 in the front view locates the piercing point *P*.

Step 2 Point *P* is projected to the top view of line 1–2 from the front view where it was first located.

Step 3 Lines *CB* and *P2* cross at point *T* in the top view. By projecting downward from *T*, *P2* is found to be over *CB*; therefore, *PT* is visible in the top view. Intersection *F* can be used to find that line *CB* is in front of *P2*; therefore, *PF* is hidden in the front view.

Fig. 27.25 Piercing points by auxiliary view

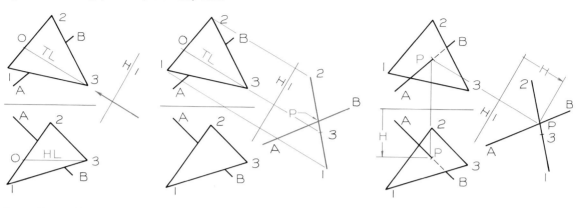

Step 1 Draw a horizontal line on the plane in the front view, project it to the top view to find *TL* line 0–3 on the plane. The line of sight is drawn parallel to the *TL* line.

Step 2 Find the edge view of the plane and project the line *AB* to this view. Point *P* is the piercing point in the auxiliary view.

Step 3 Point *P* is projected to the top and front views to locate the piercing point. Point *A* of the auxiliary view is the highest point and *AP* will be visible in the top view. Point *B* in the top view is the farthest back and it is therefore hidden in the front view.

Fig. 27.26 Line perpendicular to a plane

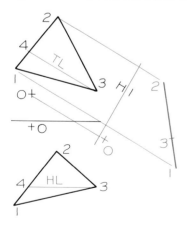
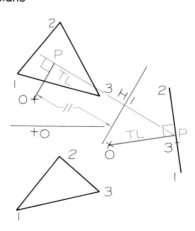
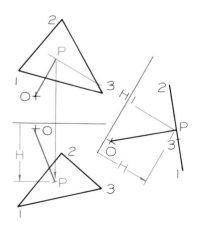

Step 1 Find the edge view of the plane by finding the point view of a line on it. Project from either view. Project point *O*.

Step 2 Line *OP* is drawn perpendicular to the edge view of the plane. Since *OP* is *TL* in the auxiliary view, it will be parallel to the H-1 reference line in the top view and perpendicular to a *TL* line on the plane.

Step 3 Piercing point *P* is found in the front view by projecting from the top view. Point *P* is accurately located by transferring dimension *H* from the auxiliary view to the front view.

will also be perpendicular to a true-length line in the top view of the plane. The front view of point P is found by projection and measurement in Step 3 along with its visibility.

27.16 INTERSECTIONS BY PROJECTION

The line of interection between two planes can be found by locating the piercing points of two lines on one plane with the other. In Fig. 27.27 a cutting plane is used in Step 1 to find piercing point P by projection. In Step 2, piercing point T is found by the same method. Line PT is the line of intersection, which will always be visible.

The two lines that are selected to be analyzed for their piercing points should be lines of a plane that cross the other plane. Lines AB and 1–2 would be poor selections since they lie outside the other plane. Each line is then analyzed to find its piercing point as if it were a single line.

The visibility is determined by analyzing the points where lines of each plane cross, as covered in the previous chapter.

27.17 INTERSECTIONS BY AUXILIARY VIEW

The intersection between planes can be found by finding the edge view of one of the planes as shown in Fig. 27.28, Step 1. Piercing points L and M are projected from the auxiliary view to their respective lines, 5–6 and 4–6, in the top view in Step 2.

The visibility of plane 4–5–6 in the top view is apparent by inspection of the auxiliary view, where sight line S_1 has an unobstructed view of the 4–L–M portion of the plane. Plane 4–5–L–M is visible in the front view, since sight line S_2 has an unobstructed view of the top view of this portion of the plane.

27.18 SLOPE OF A PLANE

Planes can be established by using verbal specifications of *slope* and *direction of slope* of a plane as defined below.

Slope   of a plane is the angle its edge view makes with the edge of the horizontal plane.

Fig. 27.27 Intersection by projection

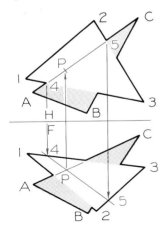
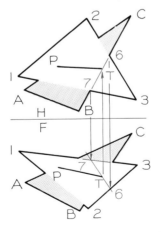
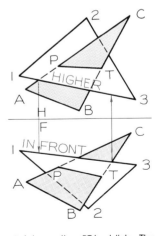

Step 1 Pass a cutting plane through line AC in the top view to locate points 4 and 5 that are projected to the front view. Point P is found on 4–5 in the front view and point P is projected to the top view.

Step 2 Pass a cutting plane through the top view of BC to locate points 6 and 7 that are projected to the front view. Piercing point T is located on 6–7 in the front view and then project point T to the top view.

Step 3 Intersection PT is visible. The intersection between AC and 1–3 is projected to the front view to determine that 1–3 is higher in the top view. Line 1–3 is found to be in front of BC in the front view to establish the visibility of the plane.

Fig. 27.28 Intersection by auxiliary view

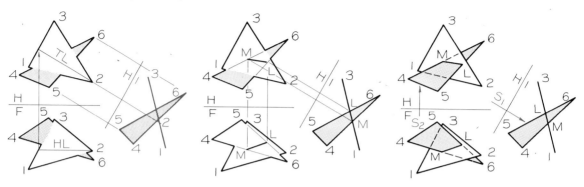

Step 1 Find the edge view of one of the planes and project the other plane to this view also.

Step 2 Piercing points *L* and *M* can be seen on the edge view of the plane. *LM* is projected to the top and front views.

Step 3 The line of sight from the top view strikes 1–5 first in the auxiliary view, which makes this portion of the plane visible in the top view. Line 4–5 is farthest forward in the top view and is visible in the front view.

Direction of slope is the compass bearing of a line that is perpendicular to a true-length line in the top view of a plane toward its low side. This is the direction in which a ball would roll on the plane.

It can be seen in Fig. 27.29A that a ball would roll perpendicular to all horizontal lines of the roof toward the low side. This is the direction of slope of the roof. The slope is seen when the roof and the horizontal are edges in a single view.

The direction of slope of a plane is indicated as a compass bearing measured in the top view (Fig. 27.29B).

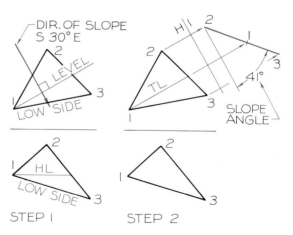

Fig. 27.30 Slope and direction of slope of a plane

Step 1 Slope direction can be found as perpendicular to a true-length level line in the top view toward the low side of the plane, S 30° E in this case.

Step 2 Slope is measured in an auxiliary view where the horizontal is an edge and the plane is an edge, 41° in this case.

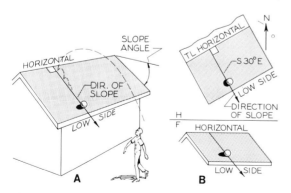

Fig. 27.29 The direction of slope of a plane is the compass bearing of a line on the plane. This is measured in the top view toward the low side of the plane.

Figure 27.30 gives the steps of finding the direction of slope and the slope angle of the plane. An understanding of these terms enables you to verbally describe a sloping plane.

27.19 CUT AND FILL

A level roadway routed through irregular terrain or the embankment of a fill used to build a dam involve the principles of cut-and-fill (Fig. 27.31). This is the process of cutting away equal amounts of the high ground to fill the lower areas to form a nearly level roadway where possible.

In Fig. 27.32, it is required that a level roadway be a given width and an elevation of 60' be constructed about the given centerline in the contour map using the specified angles of cut and fill.

In Step 1 the roadway is drawn in the top view, and the contour lines in the profile view are

Fig. 27.31 The road across the top of this dam was built by applying the principles of cut and fill. (Courtesy of the Bureau of Reclamation, U.S. Department of the Interior.)

Fig. 27.32 Cut and fill of a level roadway

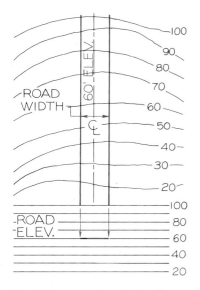

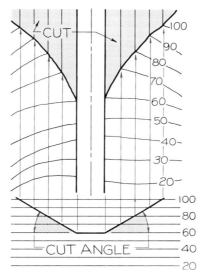

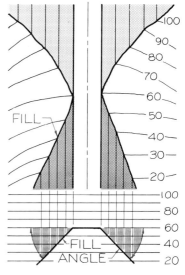

Step 1 Draw a series of elevation planes in the front view at the same scale as the map and label them to correspond to the contours on the map. Draw the width of the roadway in the top view and in the front view at the given elevation, 60' in this case.

Step 2 Draw the cut angles on the upper sides of the road in the front view according to the given specifications. The points of intersection between the cut angles and the contour planes in the front view are projected to their respective contour lines in the top view to determine the limits of cut.

Step 3 Draw the fill angles on the lower sides of the road in the front view. The points in the front view where the fill angles cross the elevation planes are projected to their respective contour lines in the top view to give the limits of the fill. Note that the countour lines have been changed in the cut-and-fill areas to indicate the new contours parallel to the roadway.

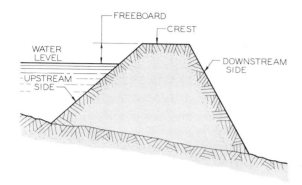

Fig. 27.33 The terms and symbols used in the construction drawing of a dam.

drawn 10' apart since the contours in the top view are this far apart. The cut angles are measured and drawn on both sides of the roadway on the upper side, Step 2. The points, on the various elevation lines crossed by the cut embankments are projected to the top view to find the limits of cut in this view.

The fill angles are laid off in the profile in Step 3 from given specifications. The crossing points on the profile view of the elevation lines are projected to the top view to find the limits of fill. New contour lines are drawn in Step 3 inside the areas of cut and fill to indicate that they have been changed by the process of cut and fill.

27.20 DESIGN OF A DAM

The design of a dam is a type of cut-and-fill problem. The basic definitions of terms associated with dams are shown in Fig. 27.33. These are: (1) **crest,** the top of the dam, (2) **water level,** the level of water held by the dam, and (3) **freeboard,** the height of the crest above the water level.

An earthen dam is located on the contour map in Fig. 27.34. It makes an arc with its center at point C. The top of the dam is specified to be level to provide a roadway. This method of drawing the top view of the dam and indicating the level of the water held by the dam is shown in the three steps of Fig. 27.34.

These same principles were used in the design and construction of the 726-foot-high Hoover Dam that was built in the 1930s. Since this dam was made of concrete instead of earth, the dam was built in the shape of an arch that is bowed toward the water to take advantage of the compressive strength of concrete (Fig. 27.35).

27.21 STRIKE AND DIP

Strike and **dip** are terms used in geological and mining engineering to describe strata of ore under the surface of the earth. These terms are closely related to slope and direction of slope of a plane.

Fig. 27.34 Graphical design of a dam

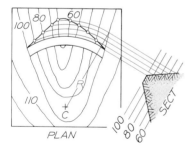

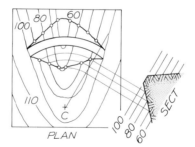

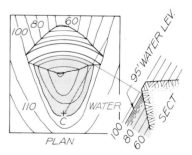

Step 1 A dam in the shape of an arc with its center at C has an elevation of 100'. Draw radius R from C and project perpendicularly from this line and draw a section through the dam from specifications. The downstream side of the dam is projected to radial line, R. Using the radii from C, use your compass to locate points on their respective contour lines.

Step 2 The elevations of the dam on the upstream side of the section are projected to the radial line, R. Using the center C and your compass, locate points on their respective contour lines in the plan view as they are projected from the section.

Step 3 The elevation of the water level is 95' in this case and is drawn in the section. The point where the water intersects the dam is projected to the radial line in the plan view, and is drawn as an arc using center C. The limit of the water is drawn between the 90' and 100' contour lines in the top view.

Strike is defined as the compass bearing of a level line in the top view of a plane. Strike has two possible compass bearings since it is the direction of a level line.

Dip is the angle the edge view of a plane makes with the horizontal plus its general compass direction, such as NW or SW. The dip angle is found in the primary auxiliary view projected from the top view, and its general direction measured in the top view. Dip direction is measured perpendicular to a level line in the top view of the plane toward the low side.

The steps of finding the strike and dip of a plane are given in Fig. 27.36. Strike can be measured in the top view by finding a true-length line on the plane in this view. The dip angle requires an auxiliary view that must be projected from the top view in order for the horizontal to appear as an edge and the plane as an edge.

27.22 DISTANCES FROM A POINT TO A PLANE

Descriptive geometry principles can be used to find various distances from a point to a plane. An example of this is shown in Fig. 27.37 where the distance from point O on the ground to an ore vein under the ground is found.

Fig. 27.36 Strike and dip of a plane

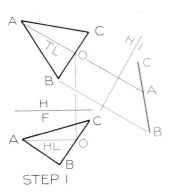

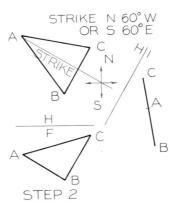

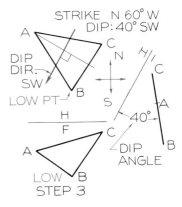

Step 1 Find the edge view of the plane by projecting from the top view.

Step 2 Strike is the compass direction of a level line on the plane and is measured in the top view. The strike of the plane is either N 60° W or S 60° E.

Step 3 The dip of a plane is the angle it makes with the horizontal (40° in the auxiliary view) plus the general compass direction toward the low side in the top view (SW). Dip direction is perpendicular to a *TL* line in the top view. Dip is written 40° SW.

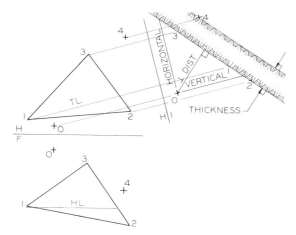

Fig. 27.37 The vertical, horizontal, and perpendicular distances from a point to an ore vein can be found in an auxiliary view projected from the top view. The thickness of an ore vein is perpendicular to the upper and lower planes of the vein.

Three points are located on the top plane of an ore vein. Point O is the point on the earth from which the tunnels are to be drilled to the vein for mining. Point 4 is a point on the lower plane of the vein.

The edge view of plane 1–2–3 is found by projecting from the top view. The lower plane is drawn parallel to the upper plane through point 4. The horizontal distance from point O to the plane is drawn parallel to the H–1 reference line. The vertical distance is perpendicular to the H–1 line. The shortest distance to the plane is perpendicular to the plane. Each of these lines is true length in this view where the ore vein appears as an edge.

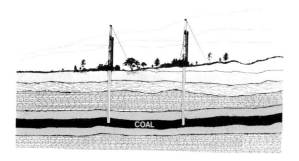

Fig. 27.38 Test wells are drilled into coal zones to determine which coal seams will contribute significantly to the production of gas. (Courtesy of Texas Eastern News.)

The process of finding the distance from a point to a plane or a vein is a technique often used in solving mining and geological problems. Figure 27.38 illustrates test wells that are drilled into coal zones to learn more about them.

27.23 OUTCROP

Strata of ore or rock formations usually approximate planes of a uniform thickness. This assumption is employed in analyzing data concerning the orientation of ore veins that are underground. A vein of ore may be inclined to the surface of the earth and may actually outcrop on its surface. Outcrops permit open-surface mining operations at a minimum of mining expense.

The steps of finding the outcrop of an ore vein are given in Fig. 27.39. The locations of sample drillings, A, B, and C are shown on the contour map and their elevations are located on the contours of the profile. These points are known to lie on the upper surface of the vein. Point D is known to lie on the lower plane of the vein.

The edge view of the ore vein can be found in an auxiliary view projected from the top view (Step 1). The points on the upper surface are projected back to their respective contour lines in the top view in Step 2. The points on the lower surface of the vein are projected to the top view in Step 3. These two lines are drawn to show the limits of the outcrop in the top view. If the ore vein does continue uniformly at its angle of inclination to the surface, the space between these two lines will be the edge of the vein on the surface of the earth.

27.24 INTERSECTIONS BETWEEN PLANES—CUTTING PLANE METHOD

The top and front views of two planes are given in Fig. 27.40 where it is required to find the line of intersection between them if the planes are infinite in size. Cutting planes are passed through either view at any angle and projected to the adjacent view. The two points L and M that are found in the top view are connected to form the top view of the line of intersection. The compass direction of this line can be used to describe its direction of slope toward its low end.

Fig. 27.39 Ore vein outcrop

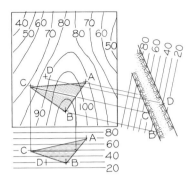

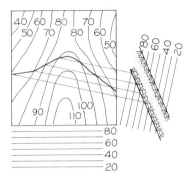

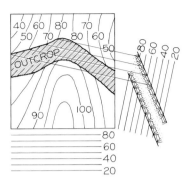

Step 1 Using points A, B, and C on the upper surface of the plane, find its edge view by projecting an auxiliary off the top view. The lower surface of the plane is found by drawing it parallel to the upper surface through point D, a known point on the lower surface.

Step 2 Points of intersection between the upper surface and the contour lines in the auxiliary view are projected to their respective contour lines in the top view to find one line of the outcrop.

Step 3 Points from the lower surface in the auxiliary view are projected to their respective contour lines in the top view to find the second line of outcrop. Cross-hatch this area to indicate the outcrop area.

Fig. 27.40 Intersection of planes by cutting-plane method

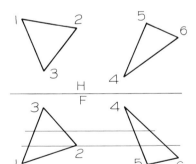

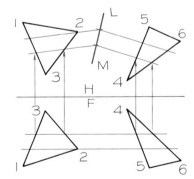

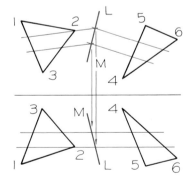

Step 1 Pass cutting planes through the front view of the planes. These planes can be drawn in any direction.

Step 2 Project the intersections of the cutting planes in the front view to the top view of the planes. Line of intersection, LM, is found in the top view.

Step 3 Points L and M are projected to their respective cutting planes in the front view to complete the solution.

Fig. 27.41 Intersection between ore veins by auxiliary view

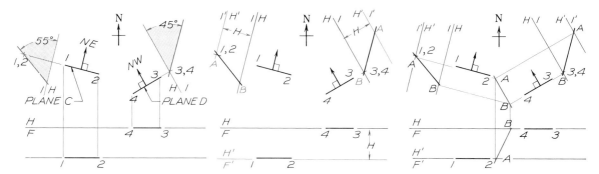

Step 1 Lines 1–2 and 3–4 are strike lines and are true length in the horizontal view. The point view of each strike line is found by auxiliary views, using a common reference plane. The edge views of the ore veins can be found by constructing the dip angles with the H-1 line through the point views. The low side is the side of the dip arrow.

Step 2 A supplementary horizontal plane, H'-F', is constructed at a convenient location in the front view. This plane is shown in both auxiliary views located H distance from the H-1 reference line. The H'-1' plane cuts through each ore vein edge in the auxiliary views to locate point A on each plane.

Step 3 Points A, which were established on each auxiliary view by the H'-1' plane, are projected to the top view, and they intersect at point A. Points B on the H-1 plane are projected to their intersection in the top view at point B. Points A and B are projected to their respective planes in the front view. Line AB is the line of intersection between the two planes.

The front view of the line of intersection is found by projecting the points from the top view to their respective planes in the front view. This is the line on which all lines on the planes would intersect.

This technique of geometry has application in the area of geology when the intersections of underground veins are being analyzed.

27.25 INTERSECTION BETWEEN PLANES—AUXILIARY METHOD

In Fig. 27.41 two planes have been located and specified using strike and dip. Since the given strike lines are true-length level lines in the top view, the edge view of the planes can be found in the view where the strike appears as a point, Step 1. The edge views are drawn using the given dip angles and directions.

Horizontal datum planes H-F and H'-F', are used to find lines on each plane that will intersect when projected from the auxiliary views to the top view. Points A and B are connected to determine the line of intersection between the two planes in the top view. These points are projected to the front view, where line AB is found.

27.26 SOLUTION OF DESCRIPTIVE GEOMETRY PROBLEMS

Figure 27.42 illustrates the techniques of labeling and solving a descriptive geometry problem. Note that some of the lettering and numbering is aligned with inclined lines and reference lines to which the labeling applies, and other lettering is not aligned but is parallel to the edge of the paper. You may use either technique or a combination of the two.

Always use guidelines and ⅛" lettering for best results. Observe the difference in line qualities that are used in the problem solution. Guidelines and projection lines need be only dark enough to be seen and used as guides.

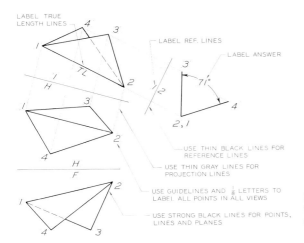

LABEL TRUE
LENGTH LINES

LABEL REF. LINES

LABEL ANSWER

71°

USE THIN BLACK LINES FOR
REFERENCE LINES

USE THIN GRAY LINES FOR
PROJECTION LINES

USE GUIDELINES AND $\frac{1}{8}$ LETTERS TO
LABEL ALL POINTS IN ALL VIEWS

USE STRONG BLACK LINES FOR POINTS,
LINES AND PLANES

Fig. 27.42 Rules that should be followed in solving descriptive geometry problems.

PROBLEMS

Use Size A (8½″ × 11″) sheets for the following problems, and lay out the problems using instruments. Each square on the grid is equal to 0.20″ or about 5 mm. The problems can be laid out on grid or plain paper. Label all reference planes and points in each problem with ⅛″ letters and/or numbers, using guidelines.

1. Fig. 27.43. (1A–1D) Find the true length views of the lines as indicated by the given lines of sight by an auxiliary view.

2. Fig. 27.44. (2A–2D) Find the angles that these lines make with the respective principal planes indicated by the given auxiliary reference lines.

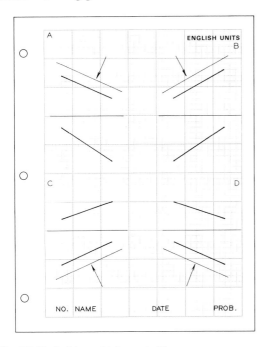

Fig. 27.43 Problems 1A through 1D.

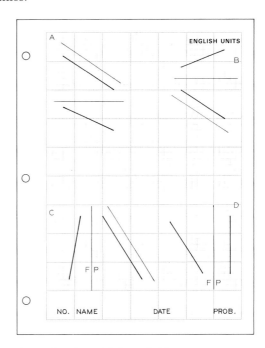

Fig. 27.44 Problems 2A through 2D.

3. Fig. 27.45. (3A and 3B) Find the lines' true length by the true-length diagram method, using the same diagram for both lines. (3C and 3D) Find the point views of the lines.

4. Fig. 27.46. (4A and 4B) Find the slope angle, tangent of the slope angle, and the percent grade of the four lines.

5. Fig. 27.47. (5A and 5B) Find the edge views of the two planes.

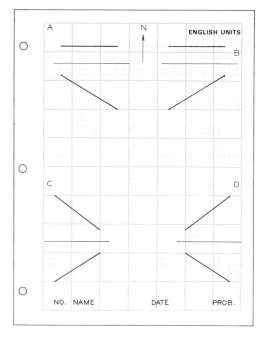

Fig. 27.46 Problems 4A through 4D.

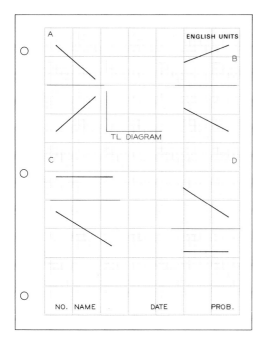

Fig. 27.45 Problems 3A through 3D.

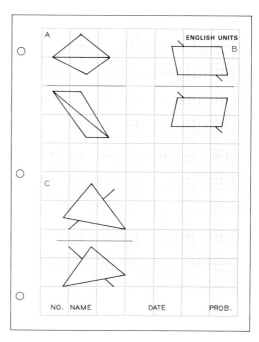

Fig. 27.47 Problems 5A and 5B.

6. Fig. 27.48. (6A) Find the angle between the planes. (6B) Find the piercing point by projection. (6C) Find the piercing point by an auxiliary view.

7. Fig. 27.49. (7A) Construct a line perpendicular to the plane and through point O on the plane by an auxiliary view. (7B) Construct a line perpendicular to the plane from point O by an auxiliary view.

8. Fig. 27.50. Find the line of intersection between the planes by projection in part A and by an auxiliary view in part B. Show visibility.

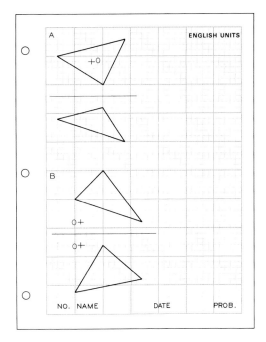

Fig. 27.49 Problems 7A and 7B.

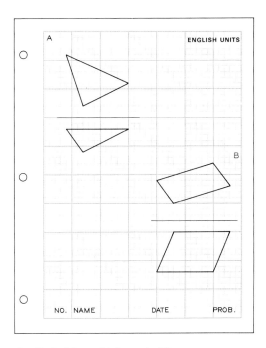

Fig. 27.48 Problems 6A through 6C.

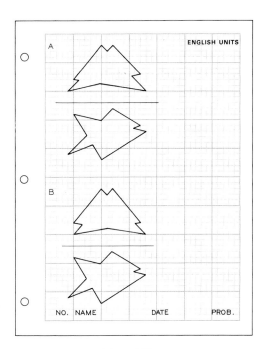

Fig. 27.50 Problems 8A and 8B.

9. Fig. 27.51. (9A and 9B) Find the direction of slope and slope of angle of the planes.

10. Fig. 27.51. (10A and 10B) Find the strike and dip of the planes.

11. Fig. 27.52. Find the shortest distance, the horizontal distance, and the vertical distance from point *O* on the ground to the underground ore vein represented by the triangle. Point *B* is on the lower plane of the vein. Find the thickness of the vein.

12. Fig. 27.53. (12A) Find the line of intersection between the two planes by the cutting-plane method. (12B) Find the line of intersection between the two planes indicated by strike lines 1–2 and 3–4. The plane with strike 1–2 has a dip of 30°, and the one with strike 3–4 has a dip of 60°.

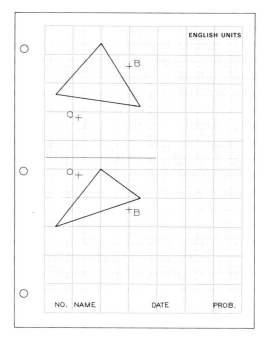

Fig. 27.52 Problem 11.

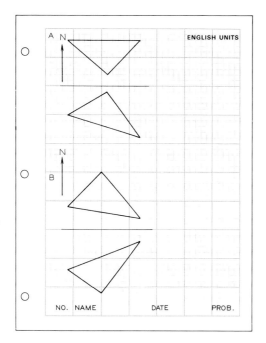

Fig. 27.51 Problems 9A, 9B, 10A, and 10B.

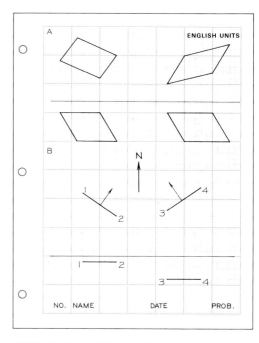

Fig. 27.53 Problems 12A and 12B.

13. Fig. 27.54. Find the limits of cut and fill in the plan view of the roadway. The roadway has a cut angle of 35° and a fill angle of 40°.

14. Fig. 27.55. Find the outcrop of the ore vein represented by the triangle (upper surface). Point *B* is on the lower surface.

15. Fig. 27.56. Complete the plan-profile drawing of the drainage system from manhole 1, through manhole 2, to manhole 3, using the grades indicated. Allow a drop of 0.20' across each manhole to compensate for loss of pressure.

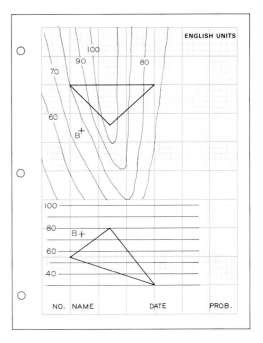

Fig. 27.55 Problem 14.

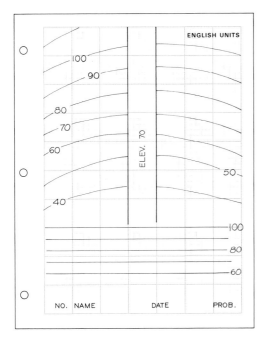

Fig. 27.54 Problem 13.

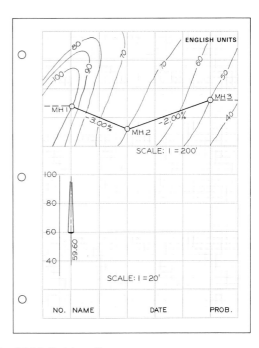

Fig. 27.56 Problem 15.

SUCCESSIVE AUXILIARY VIEWS

28.1 INTRODUCTION

A design cannot be detailed with complete specifications necessary for construction unless its complete geometry has been determined. This usually requires the application of descriptive geometry principles. Typical details that are needed are true shapes of planes, angles between planes, distances from points to lines, and angles between lines and planes. The Comsat satellite (Fig. 28.1) is an example of a design where various problems of geometry were solved by the use of successive auxiliary views.

You will recall that a primary auxiliary view is a supplementary view that is found by projecting orthographically from a primary view—a horizontal, frontal, or profile view.

> A **secondary auxiliary view** is an auxiliary view that is projected from a primary auxiliary view. A **successive auxiliary view** is an auxiliary view of a secondary auxiliary view.

Usually three auxiliary views projected from the primary view are adequate to solve the more complex problems of descriptive geometry.

28.2 POINT VIEW OF A LINE

When a line appears true length, its point view can be found by projecting an auxiliary view with

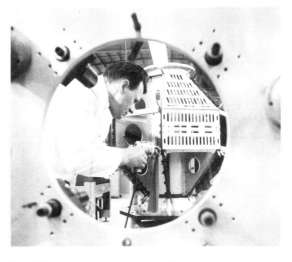

Fig. 28.1 Applications of descriptive geometry requiring successive auxiliary views are seen in the structure of this Comsat satellite. The angles between the planes and the true-size views of the surfaces were determined by applying the principles of descriptive geometry. (Courtesy of TRW Systems.)

parallel projectors from it. In Fig. 28.2, line 3–4 is true length in the top view since it is horizontal in the front view. The point view is found in the primary auxiliary view by constructing reference line H–1 perpendicular to the true-length line. The height dimension, H, is transferred to the auxiliary view to locate the point view of 4–3.

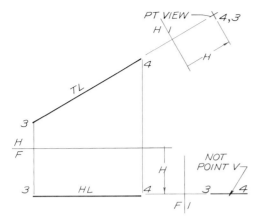

Fig. 28.2 The point view of a line can be found by projecting an auxiliary view from the true-length view of the line.

The line in Fig. 28.3 is not true length in either view, which requires that the line be found true length by a primary auxiliary view. In this example, the line is found true length by projecting from the front view, but this view could have been projected from the top as well. The point view of the line is found by projecting from the true-length line to a secondary auxiliary view. This point is labeled 4, 3 since point 4 is seen first.

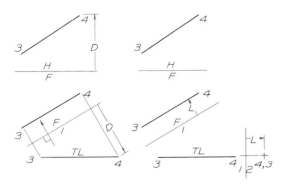

Fig. 28.3 Point view of an oblique line

Step 1 A line of sight is drawn perpendicular to one of the views, the front view in this example. Line 3–4 is found true length by projecting perpendicularly from the front view.

Step 2 A secondary reference line 1–2 is drawn perpendicular to the true-length view of 3–4. The point view is found by transferring dimension *L* from the front view to the secondary auxiliary view.

28.3 ANGLE BETWEEN PLANES

> The angle between two planes is called a **dihedral angle.** This angle can be found in the view where the line of intersection appears as a point.

Since this results in the point view of a line that lies on both planes, the planes will appear as edges in this view.

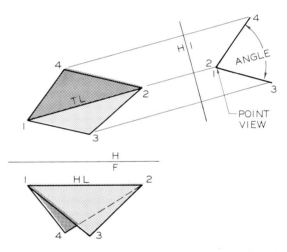

Fig. 28.4 The angle between two planes can be found in the view where the line of intersection between them projects as a point. Since the line of intersection, 1–2, is true length in the top view, it can be found as a point in a view that is projected from the top view.

The two planes in Fig. 28.4 represent a special case since the line of intersection, 1–2, is true length in the top view. This permits you to find its point view in a primary auxiliary view where the true angle can be measured.

A general case is given in Fig. 28.5 where the line of intersection between the two planes is not true length in either of the principal views.

The line of intersection, 1–2, is found true length in a primary auxiliary view, and the point view of the line is then found in the secondary auxiliary view. The dihedral angle is measured in the secondary auxiliary view.

It is apparent that this principle must be used to determine the angles between intersecting planes such as those shown in Fig. 28.6, where the

Fig. 28.5 Angle between two planes

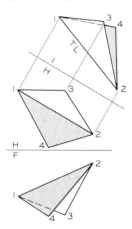
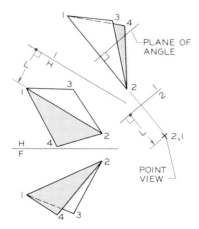
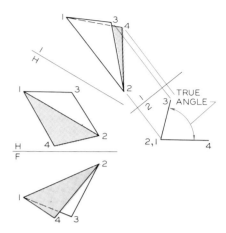

Step 1 The angle between two planes can be measured in a view where the line of intersection appears as a point. The line of intersection is first found true length by projecting a primary auxiliary view perpendicularly from the top view, in this case.

Step 2 The point view of the line of intersection is found in the secondary auxiliary view by projecting parallel to the true length view of 1–2. Note that the plane of the dihedral angle is an edge and perpendicular to the true-length line of intersection.

Step 3 The edge views of the planes are completed in the secondary auxiliary view by locating points 3 and 4. The angle between the planes, the dihedral angle, can be measured in this view.

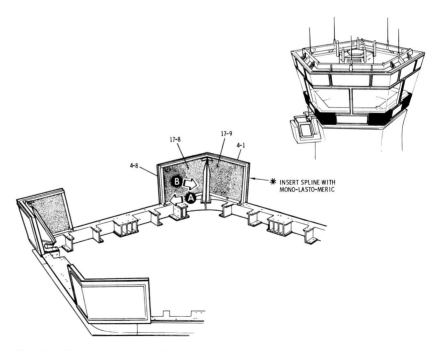

Fig. 28.6 The determination of the angle between the planes of the corner panels of the control tower utilized principles of descriptive geometry. (Courtesy of the Federal Aviation Agency.)

Fig. 28.7 True size of a plane

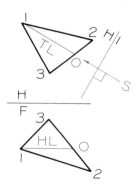
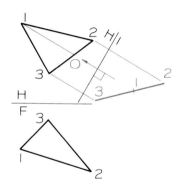
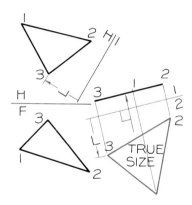

Step 1 The edge view of a plane can be found by finding the point view of any line that lies on it. Horizontal line 1-O is drawn on the front view and is then projected to the top view. A line of sight is drawn parallel to 1-O

Step 2 Line 1-O is found as a point in the primary auxiliary view and the plane appears as an edge.

Step 3 The true-size view of a plane can be found by locating a line of sight that is perpendicular to an edge view of the plane. A secondary auxiliary view is projected perpendicularly from the edge of plane 1–2–3 to find its true-size view.

side panels of a control tower join. This is necessary in order to design and fabricate corner braces to hold the structure together.

28.4 TRUE SIZE OF A PLANE

A plane can be found true size in a view that is projected perpendicularly from an edge view of a plane.

In Fig. 28.7, the true size of a plane 1–2–3 is found by first finding the edge view of the plane in Steps 1 and 2. In Step 3, the secondary auxiliary view is projected perpendicularly from the edge view. The result is a true-size view of the plane where each angle is true size.

This principle can be used to find the angle between lines such as bends in a fuel line of an aircraft engine (Fig. 28.8). A problem of this type is shown in Fig. 28.9 where the top and front views of intersecting lines are given. It is required

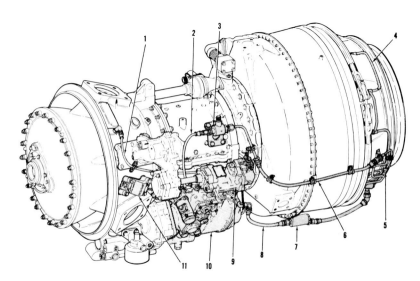

Fig. 28.8 The angles of bend in the fuel line were found by the application of the principle of finding the angle between two lines. (Courtesy of Avco Lycoming.)

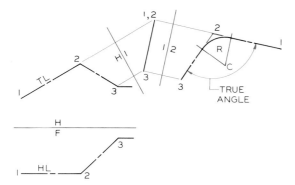

Fig. 28.9 The angle between two lines can be found by finding the plane of the lines' true size.

that the angles be determined at each bend, and that a given radius of curvature be shown.

The angle 1–2–3 is found as an edge in the primary auxiliary view and as a true-size angle in the secondary view. The angle can be measured in this view and the radius of curvature drawn using principles of geometric construction.

28.5 SHORTEST DISTANCE FROM A POINT TO A LINE

> The shortest distance from a point to a line can be measured in the view where the line appears as a point.

This distance will appear perpendicular to the line whenever the line appears true length.

This type of problem is solved in Fig. 28.10 by finding the line 1–2 true length in a primary auxiliary view along with point 3. The line is found as a point in the secondary auxiliary view, where the distance from point 3 is true length. Since line O–3 is true length in this view, it must be parallel to the 1–2 reference line in the preceding view, the primary auxiliary view. It is also perpendicular to the true length view of line 1–2 in the primary auxiliary view. Point O is projected back to the principal views to complete the solution.

Fig. 28.10 Shortest distance from a point to a line

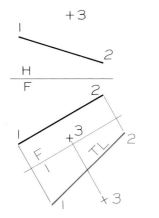

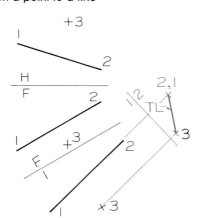

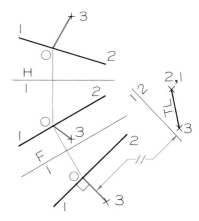

Step 1 The shortest distance from a point to a line can be found in the view where the line appears as a point. Line 1–2 is found true length by projecting from the front view.

Step 2 Line 1–2 is found as a point in a secondary auxiliary view projected from the true-length view of 1–2. The shortest distance appears true length in this view.

Step 3 Since 3-O is true length in the secondary auxiliary view, it must be parallel to the 1–2 reference line in the primary auxiliary view and perpendicular to the line. The front and top views of 3-O are found by projecting from the primary auxiliary view in sequence.

28.6 SHORTEST DISTANCE BETWEEN SKEWED LINES—LINE METHOD

The shortest distance between two skewed lines (randomly positioned lines) can be measured in the view where one of the lines appears as a point.

The shortest distance between two lines is perpendicular to both lines. The location of the shortest distance is both functional and economical, as demonstrated by the connector between two pipes in Fig. 28.11, since a standard connector is a 90° Tee.

A problem of this type is solved by the line method in Fig. 28.12. Line 3–4 is found as a point in the secondary auxiliary view where the shortest distance is drawn perpendicular to line 1–2. Since the distance is true length in the secondary auxiliary view, it must be parallel to the 1–2 reference line in the primary auxiliary view. Point O is found by projection and OP is drawn perpendicular to line 3–4. The line is projected back to the given principal views.

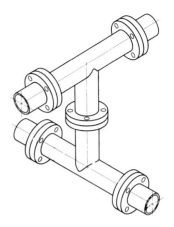

Fig. 28.11 The shortest distance between two lines, or pipes, is a line that is perpendicular to both. This is the most economical connection and the most functional since perpendicular connectors are standard.

28.7 SHORTEST DISTANCE BETWEEN SKEWED LINES—PLANE METHOD

The distance between skewed lines can be solved using the alternative plane method. This involves the construction of a plane through one of the

Fig. 28.12 Shortest distance between skewed lines—line method

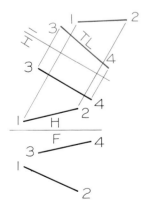
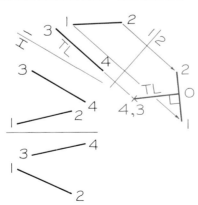
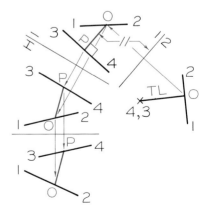

Step 1 The shortest distance between two skewed lines can be found in the view where one of the lines appears as a point. Line 3–4 is found true length by projecting from the top view along with line 1–2.

Step 2 The point view of line 3–4 is found in a secondary auxiliary view projected from the true-length view of 3–4. The shortest distance between the lines is drawn perpendicular to line 1–2.

Step 3 Since the shortest distance is true length in the secondary auxiliary view, it must be parallel to the reference line in the preceding view. Points O and P are projected back to the given view to show the shortest distance in all views.

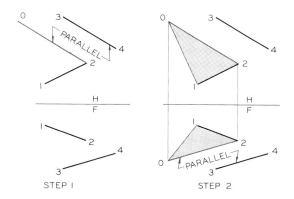

Fig. 28.13 A plane can be constructed through a line and parallel to another line by construction.

Step 1 Line O-2 is drawn parallel to line 3–4 to a convenient length.

Step 2 The front view of line O-2 is drawn parallel to the front view of line 3–4. The length of the front view of O-2 is found by projecting from the top view of O. Plane 1–2–O is parallel to line 3–4.

lines parallel to the other, as shown in Fig. 28.13. The top and front views of 0–2 are drawn parallel to their respective views of line 3–4. The resulting plane, 1–2–0, is parallel to line 3–4. Both lines will appear parallel in a view where the plane appears as an edge.

In Fig. 28.14, plane 3–4–0 is constructed and its edge view is found and both lines appear parallel (Step 1). A secondary auxiliary view is projected perpendicularly from these parallel lines to find the view where the lines cross (Step 2). This crossing point represents the point view of the shortest distance between the two lines. It will appear true length and perpendicular to the two lines when it is projected to the primary auxiliary view, where it is labeled as line *LM*. It is projected back to the given views to complete the solution of locating the shortest distance between two lines (Step 3).

This principle was applied to the design of the separation of power lines shown in Fig. 28.15 where the clearance is critical.

28.8 SHORTEST LEVEL DISTANCE BETWEEN SKEWED LINES

When it is required to find the shortest level (horizontal) distance between two skewed lines, the plane method must be used instead of the line method. Also, the primary auxiliary view must be projected from the top view in order to find a view where the horizontal plane appears as an edge.

Fig. 28.14 Shortest distance between skewed lines—plane method

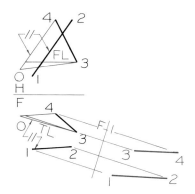

Step 1 Construct a plane through line 3–4 that is parallel to line 1–2. When this plane is found as an edge by projecting from the front view, the two lines will appear parallel in this auxiliary view.

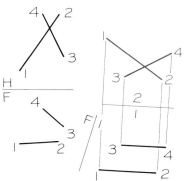

Step 2 The shortest distance will appear true length in the primary auxiliary view, where it will be perpendicular to both lines. Draw a secondary auxiliary view by projecting perpendicularly from the lines in the primary auxiliary view.

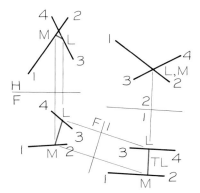

Step 3 The crossing point of the two lines is the point view of the perpendicular distance, *LM*, between the lines. This distance is projected to the primary auxiliary view, where it is true length, and back to the given views.

Fig. 28.15 The determination of clearance between power lines is a critical problem for the electrical engineer. This is an application of the shortest distance between two lines. (Courtesy of the Tennessee Valley Authority.)

In Fig. 28.16, plane 3–4–O is drawn parallel to line 1–2 and an edge view of the plane is found in the primary auxiliary view. The lines appear parallel in this view and the horizontal (H–1) appears as an edge. A line of sight is drawn parallel to the H–1, and the secondary reference line, 1–2,

is drawn perpendicular to the H–1 (Step 2). The crossing point of the lines found in the secondary auxiliary views locates the point view of the shortest horizontal distance between the lines (Step 3). This line, LM, is true length in the primary auxiliary view and parallel to the H–1 plane.

Line LM is projected back to the given views. As a check, this line must be parallel to the H-line in the front view since it is a level or horizontal line.

28.9 SHORTEST GRADE DISTANCE BETWEEN SKEWED LINES

Many lines representing highways, power lines, or conveyors are connected by lines at a specified grade other than horizontal or perpendicular. Conveyors, such as the one shown in Fig. 28.17, are used to transport aggregates or grain for mixing. The slopes of such a system cannot exceed specified grades for efficient operation.

If a 50% grade connector between two lines must be found between the two lines in Fig. 28.18, the plane method is used as in the two previous examples. A view of the lines where they appear

Fig. 28.16 Shortest level distance between skewed lines—plane method

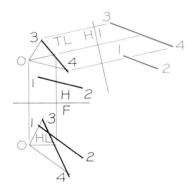

Step 1 Construct plane O–3–4 parallel to line 1–2 by drawing line O–4 parallel to 1–2. Find the edge view of O–3–4 by projecting off the top view. The lines will appear parallel in this view. *Note:* The auxiliary view *must* be projected from the *top view* to find the horizontal plane as an edge.

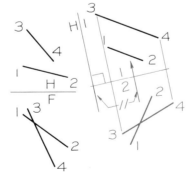

Step 2 An infinite number of horizontal (level) lines can be drawn parallel to H-1 between the lines in the auxiliary view, but only the shortest level line will appear true length in the primary auxiliary view. Construct the secondary auxiliary view by projecting parallel to the horizontal (H-1) to find the point view of the shortest level line.

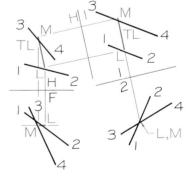

Step 3 The crossing point of the two lines in the secondary auxiliary view establishes the point view of the level connector, *LM*. Project *LM* back to the given views. *LM* is parallel to the H-plane in the front view, which verifies that it is a level line.

Fig. 28.17 These conveyors represent the application of skewed-line problems that must be solved using descriptive geometry principles.

parallel is constructed and a 50% grade line is constructed from the edge view of the horizontal (H-1). It should be noted that the auxiliary views must be projected from the top view in order to have an edge view of the horizontal from which the 50% grade is constructed.

The grade line can be constructed in two directions from the H-1 line, but the shortest distance will be the most nearly perpendicular to both lines, which is the one shown in Step 2. The secondary auxiliary view is projected parallel with this 50% grade line to find the crossing point of the lines to locate the shortest connector, *LM*, that is at a 50% grade.

Line *LM* is projected back to all views. This line appears true length in the primary auxiliary view where the lines appear parallel.

By now, it should be apparent that any connector between skewed lines will appear true length in the view where the lines appear parallel. Perpendicular lines, horizontal lines, and grade lines are true length in this view.

28.10 ANGULAR DISTANCE TO A LINE

Standard connectors used to connect pipes and structural members are available in two standard angles—90° and 45°. Consequently, it is economical to incorporate these into a design rather than having to design specially made connectors.

In Fig. 28.19 it is required to locate the point of intersection on line 1–2 of a line drawn from

Fig. 28.18 Grade distance between skewed lines

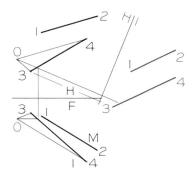

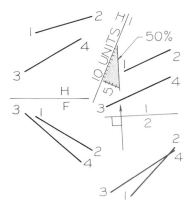

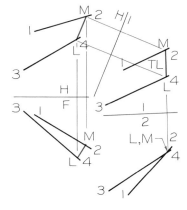

Step 1 To find a level line or a line on a grade between two skewed lines, the primary auxiliary must be projected from the top view. Plane 3–4–O is constructed parallel to line 1–2. The edge view of the plane is found where the lines appear parallel.

Step 2 Construct a 50-percent grade line with respect to the edge view of the H-1 line in the primary auxiliary view that is as nearly perpendicular to the lines as possible. Project the secondary auxiliary view parallel to the grade line. The shortest grade distance will appear true length in the primary auxiliary view.

Step 3 The point of crossing of the two lines in the secondary auxiliary view establishes the point view of the 50-percent grade line, *LM*. This line is projected back to the previous views in sequence.

Fig. 28.19 Line through a point with a given angle to a line

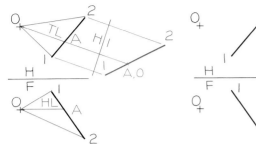
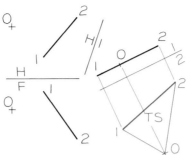
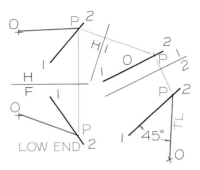

Step 1 Connect point O to each end of the line to form a plane 1–2–O in both views. Draw a horizontal line in the front view of the plane and project it to the top view, where it is true length. Find the edge view of the plane by obtaining the point view of line A-O.

Step 2 Find the true size of plane 1–2–O projecting perpendicularly from the edge view of the plane in the primary auxiliary view. The plane can be omitted in this view and only line 1–2 and point O are shown.

Step 3 Line OP is constructed at the specific angle with line 1–2, 45° in this case. Note that, if the angle is toward point 2 the line slopes downhill, and if toward point 1 it slopes uphill. Point P is projected back to the other views in sequence to show the line, OP, in all views.

point O at a 45° angle to the line. The plane of the line and point, 1–2–O, is found as an edge in the primary auxiliary view. The angle can be measured in this view where the plane of the line and point is true size.

The 45° connector is drawn from O to the line toward point 2 to slope downhill, or toward point 1 if it is to slope uphill. This can be determined by referring to the front view where height can be easily seen. Line OP is projected back to the given views.

28.11 ANGLE BETWEEN A LINE AND A PLANE—PLANE METHOD

> The angle between a line and plane can be measured in the view where the plane appears as an edge and the line true length.

In Fig. 28.20, the edge view of the plane is found in a primary auxiliary view projected from any primary view. The plane is then found true size in Step 2 where the line is foreshortened. The line, AB, can be found true length in a third auxiliary

view projected perpendicularly from the secondary auxiliary view of AB. The line appears true length and the plane as an edge in the third successive auxiliary view.

The piercing point is projected back to the views in sequence and the visibility is determined for each view.

28.12 ANGLE BETWEEN A LINE AND A PLANE—LINE METHOD

To find the angle between a line and a plane by the line method, the line is first found as a point, and then true length as shown in Fig. 28.21. The plane is foreshortened in Step 2 where the line AB appears as a point.

The plane is found as an edge in a third auxiliary view by finding the point view of a line on the plane in this view. Since the view is projected from the point view of line AB, the line will appear true length. This view satisfies the condition that the line be true length and the plane an edge. The angle is measured in the third view. The piercing point is projected back to previous views and the visibility is determined to complete the solution.

Fig. 28.20 Angle between a line and a plane—plane method

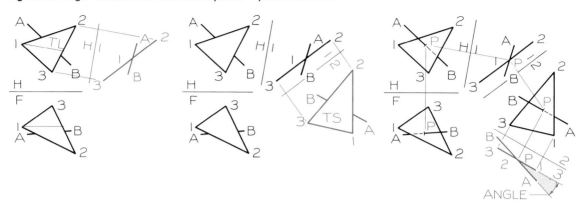

Step 1 The angle between a line and a plane can be measured in the view where the plane is an edge and the line is true length. The plane is found as an edge by projecting off the top view. The line is not true length in this view.

Step 2 The plane is found true size by projecting perpendicularly from the edge view of the plane.

Step 3 A view projected in any direction from a true-size view of a plane will show the plane as an edge. A third successive auxiliary view is projected perpendicularly from line *AB*. The line appears true length and the plane as an edge in this view. The angle is measured in this view. The piercing points and visibility are shown in the views by projecting back in sequence.

Fig. 28.21 Angle between a line and a plane—line method

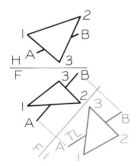
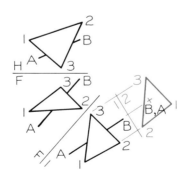
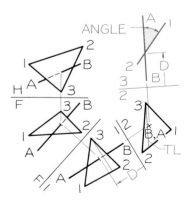

Step 1 The angle between a line and a plane can be measured in the view where the plane appears as an edge and the line is true length. Find a view where line *AB* is true length and project the plane to this view also.

Step 2 Construct a point view of line *AB* in the secondary auxiliary view. Plane 1–2–3 does not appear true size in this view unless the line is perpendicular to the plane. The point view of the line in this view is the piercing point on the plane.

Step 3 An edge view of the plane is found by finding the point view of a line on the plane. Line *AB* will appear true length in this view since it was a point in the secondary auxiliary view. Measure the angle, project back to each view to locate the piercing point, and determine the visibility of the line.

28.13 ELLIPTICAL VIEWS OF A CIRCLE

Circles appear as circles only when your line of sight is perpendicular to the plane of the circle. When this line of sight is oblique to the plane, the circle will appear as an ellipse. The following definitions are given to explain the terminology of ellipses (Fig. 28.22).

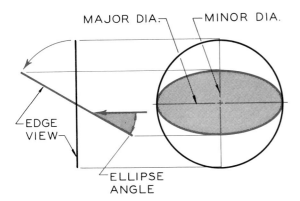

Fig. 28.22 The ellipse angle is the angle the line of sight makes with the edge view of a circle. An ellipse template can be used to construct the ellipse by aligning it with the major and minor axes.

Major diameter is the largest diameter measured across an ellipse. It is always true length.

Minor diameter is the shortest diameter across an ellipse and it is perpendicular to the major diameter at its midpoint.

Ellipse angle is the angle between the line of sight and the edge view of the plane of the circle.

Cylindrical axis is the centerline of a cylinder that connects the centers of the circular ends of a right cylinder.

Ellipse template is a template of various sizes of ellipses used to draw the ellipses by aligning the major and minor diameters. The templates are available in 5° intervals (Fig. 28.23).

It is required in Fig. 28.24 to draw a circle through the three vertexes of the triangular plane, and to show the circle in all views. This is done by finding the true size of the triangle and then

Fig. 28.23 Typical ellipse templates used for ellipse representation. (Courtesy of the A. Lietz Company.)

finding the center of the circle where the perpendicular bisectors of the sides intersect. The circle is drawn in this view.

The circle is projected to the primary auxiliary view where it appears as an edge. The major and minor diameters are projected back to the top view where they are parallel and perpendicular to the H-1 reference line. The ellipse guide angle is found to be 45° since this is the angle the line of sight makes with the edge view of the circle.

The elliptical view of the circle is found in the front view projecting an edge view of the plane from the front view. The major and minor diameters are drawn parallel and perpendicular to the F-1 reference line.

> The angle of the ellipse template is the angle the line of sight makes with the edge view of the circle, 40°.

The elliptical ends of right cylinders will be perpendicular to the cylinder's axis (Fig. 28.25A). Consequently, the major diameters will be perpendicular to the axis. When the major diameter is not drawn perpendicular to the axis (Fig. 28.25B), it is apparent to even the untrained eye that the cylinder is not a right cylinder. That is, the ends of the cylinder are not perpendicular to the cylinder's axis.

Fig. 28.24 Elliptical views of a circle

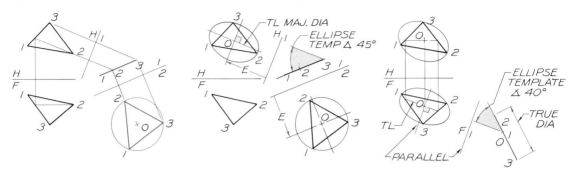

Step 1 To construct a circle that will pass through each vertex of a triangle, find plane 1–2–3 true size. The center of the circle is found at the intersection of the perpendicular bisectors of each side of the triangle.

Step 2 Draw the diameters, *AB* and *CD,* parallel and perpendicular to the 1–2 line, respectively, in the secondary auxiliary view. Project these lines to the primary auxiliary and top views, where they will represent the major and minor diameters of an ellipse. Select the ellipse template for drawing the top view by measuring the angle between the line of sight and the edge view of the plane.

Step 3 Determine the ellipse template for drawing the ellipse in the front view by finding the edge view of the plane in an auxiliary view projected from the front view. The ellipse angle is measured in the auxiliary view as shown. The major diameter is true length and is parallel to a true-length line on the plane in the front view. The minor diameter is perpendicular to it.

Fig. 28.25 The axis of the cylinder (the centerline) is drawn perpendicular to the major diameter of its elliptical end, if it is a right cylinder. It should be apparent to you by inspection that the cylinder at B is not a right cylinder.

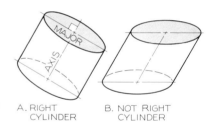

PROBLEMS

Use Size A (8½″ × 11″) sheets for the following problems and lay out the problems using instruments. Each square on the grid is equal to 0.20″ or about 5 mm. The problems can be laid out on grid or plain paper. Label all reference planes and points in each problem with ⅛″ letters and/or numbers, using guidelines.

The crosses marked "1" and "2" are to be used for placing the primary and secondary reference lines. The primary reference line should pass through "1" and the secondary through "2."

1 and 2. Find the point views of the line in Fig. 28.26.

3 and 4. Find the angles between the planes in Fig. 28.26.

5 and 6. Find the true size views of the planes in Fig. 28.27. Project from the front view in Problem 5 and from the top view in Problem 6.

7 and 8. Find the angles between the lines in Fig. 28.28. Project from the top views of both problems.

9. Find the shortest distance from the point to the line in Fig. 28.29 and show the distance in all views. Use the plane method and project from the left side view. Scale: full size.

10. Find the shortest distance from point *O* to the line in Fig. 28.29 and show the distance in all

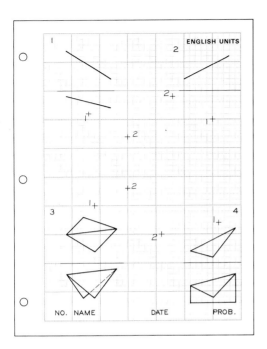

Fig. 28.26 Problems 1 through 4.

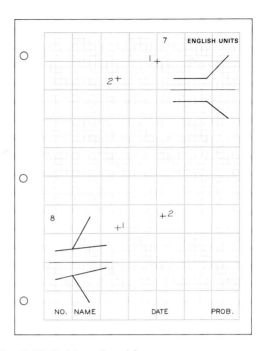

Fig. 28.28 Problems 7 and 8.

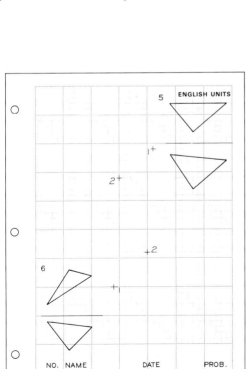

Fig. 28.27 Problems 5 and 6.

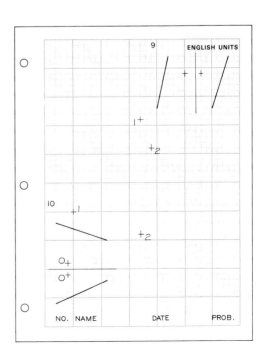

Fig. 28.29 Problems 9 and 10.

views. Use the line method and project from the top view. Scale: full size.

11 and 12. Find the shortest distances between the two skewed lines in Fig. 28.30 using the line method, and show the distances in all views. Begin each problem by finding line 3–4 true length, using the cross marks given. Scale: full size.

13. Find the shortest horizontal distance between the two lines in Fig. 28.31 by the plane method. Project from the top view. Scale: full size.

14. On a separate sheet of paper, redraw Problem 11 (Fig. 28.31) and find the shortest 20 percent grade between two lines. Project the first view from the top view. Scale: full size.

15. Find the shortest 25-percent grade distance between the two lines in Fig. 28.32. Show the distance in all views. Scale: full size.

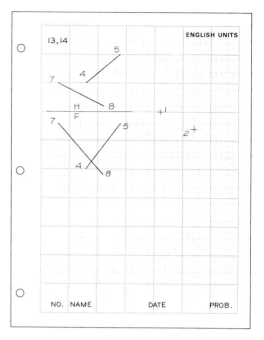

Fig. 28.31 Problems 13 and 14.

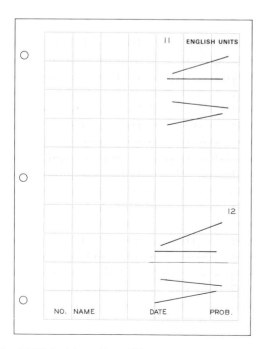

Fig. 28.30 Problems 11 and 12.

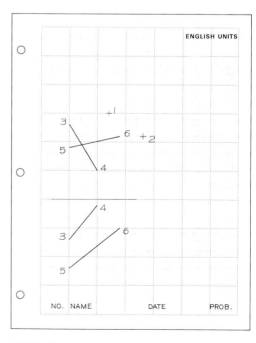

Fig. 28.32 Problem 15.

16. Find the connector from point O that will intersect line 1–2 at a 60° angle (Fig. 28.33). Show this line in all views. Project from the top view. Scale: full size.

17. Find the angle between the line and the plane in Fig. 28.34 by using the plane method. Project from the front view and show the visibility in all views. Scale: full size.

18. Same as Problem 17, except use the line method.

19. Construct a circle that will pass through each vertex of the triangle in Fig. 28.35. Project from the top view and show the elliptical views of the circle in all views.

20. Find the front view of the elliptical path of a circular section through the sphere in Fig. 28.35. The edge view of the section is shown in the top view.

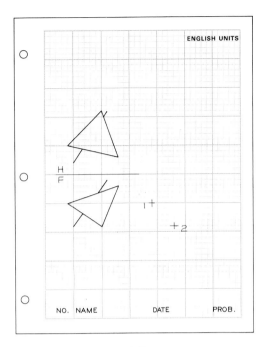

Fig. 28.34 Problems 17 and 18.

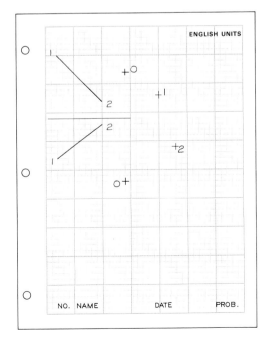

Fig. 28.33 Problem 16.

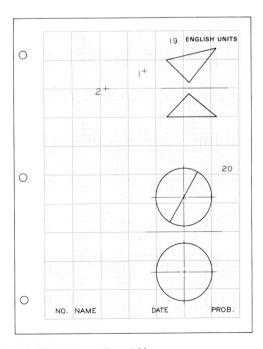

Fig. 28.35 Problems 19 and 20.

21. The line in Fig. 28.36 represents the centerline of a right cylinder that has each circular end perpendicular to the axis. Complete the views of the cylinder and show the ends, which will appear as ellipses.

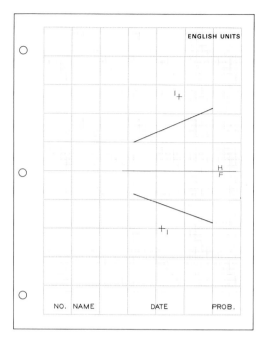

Fig. 28.36 Problem 21.

REVOLUTION

29.1 INTRODUCTION

The orthicon camera in Fig. 29.1 was designed to permit the camera to be revolved about three axes; thus it is possible to aim it in any direction for tracking space vehicles. This is just one of many

Fig. 29.1 This orthicon camera is an advanced example of a design that utilizes principles of revolution. Its cradle was designed to permit the camera to be revolved into any position by revolving it about three axes. (Courtesy of ITT Industrial Laboratories.)

designs that was based on the principles of revolution.

Revolution is another method of solving problems that can be solved by auxiliary views. In fact, revolution techniques were developed and used before auxiliary views came into use. The understanding of revolution will reinforce your understanding of auxiliary-view principles, which is necessary for the solution of spatial problems.

29.2 TRUE LENGTH OF A LINE IN THE FRONT VIEW

The simple object in Fig. 29.2 is used to demonstrate how an inclined surface can be found true size by auxiliary view and by revolution. When the auxiliary view method is used, the observer changes position to an auxiliary vantage point where he or she can look perpendicularly at the inclined surface.

When the revolution method is used, the top view of the object is revolved about an axis until the edge view of the inclined plane is perpendicular to the standard line of sight from the front view. In other words, the observer's line of sight does not change, and the conventional lines of sight between adjacent orthographic views are used. Note that the height dimension, H, in Fig. 29.2B does not change when the object is revolved.

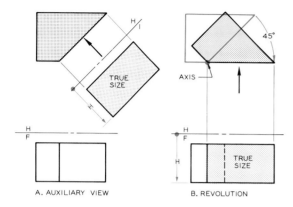

A. AUXILIARY VIEW   B. REVOLUTION

Fig. 29.2 The surface is found true size by an auxiliary view in Part A. At B, the surface is found true size by revolving the top view.

A single line can be found true length in the front view by revolution as shown in Fig. 29.3. By establishing the point view of an axis in the top view, line *AB* is revolved into a position that is parallel to the frontal plane. Therefore, the line will appear true length in the front view.

Note that the top view represents the circular base of a right cone and the front view is the triangular view of a cone. Line *AB′* is the outside element of the cone and is therefore true length.

Figure 28.4 illustrates the technique of finding line 1–2 true length in the front view. When in its first position, the observer's line of sight is not

perpendicular to the triangular plane or line 1–2. But when it is revolved to be perpendicular to the line of sight, the triangle appears true size and line 1–2′ is true length.

29.3 TRUE LENGTH OF A LINE IN THE TOP VIEW

A surface that appears as an edge in the front view can be found true size in the top view by a primary auxiliary view or by a single revolution (Fig. 29.5). When revolution is used, you need not change your position, but the standard line of sight that gives the top view can be used.

The axis of revolution is located as a point in the front view, and is true length in the top view. The edge view of the plane is revolved until it is a horizontal edge in the front view (Fig. 29.5A). It is projected to the top view to find the surface true size. As in the auxiliary-view method, the depth dimension *(D)* does not change.

Line *CD* is found true length in the top view by revolving the line into a horizontal position in Step 2 of Fig. 29.6. The arc of revolution in the front view represents the base of a cone of revolution. Line *CD′* is true length in the top view since it is an outside element of the cone. Note that the depth dimension in the top view does not change.

Fig. 29.3 True length in the front view

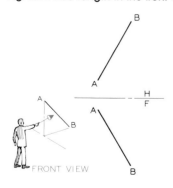

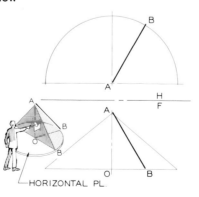

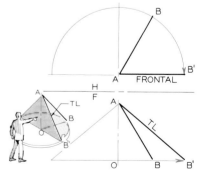

Given The top and front views of line *AB*.

Required Find the true-length view of line *AB* in the front view by revolution.

Step 1 The top view of line *AB* is used as a radius to draw the base of a cone with point *A* as the apex. The front view of the cone is drawn with a horizontal base through point *B*. Line *AO* is the axis of the cone.

Step 2 The top view of line *AB* is revolved to be parallel to the frontal plane *AB′*. When projected to the front view, frontal line *AB′* is the outside element of the cone and is true length.

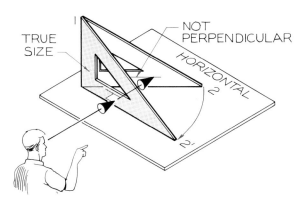

TRUE
SIZE

NOT
PERPENDICULAR

Fig. 29.4 Line 1–2 of the triangle does not appear true length in the front view because your line of sight is not perpendicular to it. When the triangle is revolved to a position where your line of sight is perpendicular to it, line 1–2' can be seen true length.

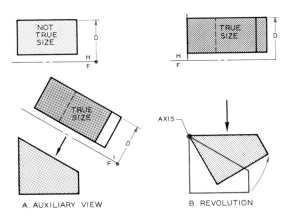

A. AUXILIARY VIEW

B. REVOLUTION

Fig. 29.5 At A, the inclined plane is found true size by an auxiliary with a line of sight that is perpendicular to the surface. At B, the surface is found true size by revolving the front view until it is perpendicular to the standard line of sight from the top view.

Fig. 29.6 True length of a line in the top view

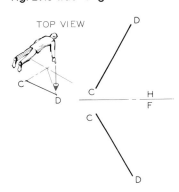

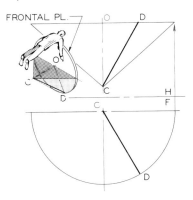

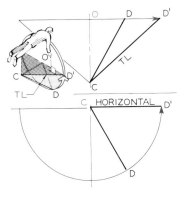

Given The top and front views of line CD.

Required Find the true-length view of line CD in the top view by revolution.

Step 1 The front view of line CD is used as a radius to draw the base of a cone with point C as the apex. The top view of the cone is drawn with the base shown as a frontal plane. The axis, CO, is perpendicular to the frontal base.

Step 2 The front view of line CD is revolved into position CD' where it is horizontal. When projected to the top view, CD' is the outside element of the cone and is true length.

Fig. 29.7 True length of a line in the side view

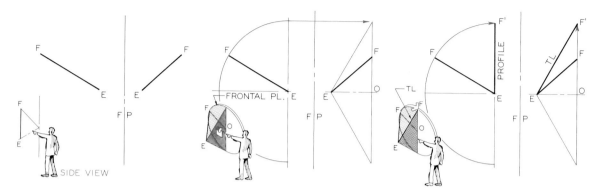

Given The front and side views of line *EF*.

Required Find the true-length view of line *EF* in the profile review by revolution.

Step 1 The front view of line *EF* is used as a radius to draw the circular view of the base of a cone. The side view of the cone is drawn with a base through point *F* that is a frontal edge.

Step 2 Line *EF* in the frontal view is revolved to position *EF'* where it is a profile line. Line *EF'* in the profile view is true length, since it is a profile line and the outside element of the cone.

29.4 TRUE LENGTH OF A LINE IN THE PROFILE VIEW

Line *EF* in Fig. 29.7 is found true length by revolving it in the front view until it is parallel to the edge view of the profile plane, Step 1. The circular view of the cone is projected to the side view, where the triangular shape of the cone is seen. Since *EF'* is a profile in line in Step 2, *EF'* is true length in the side view, where it is the outside element of the cone.

In the previously covered examples, each line has been revolved about one of its ends. However, a line can be revolved about any point on its length. Line 5–6 in Fig. 29.8 is an example of a line that is found true length by revolving it about point *O* near its midpoint.

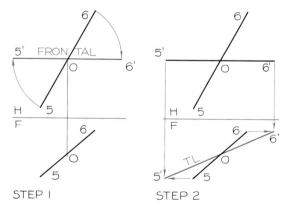

Fig. 29.8 In the preceding examples, the lines have been revolved about their ends, but they can be found true length by revolving them about any point on them. Line 5–6 is revolved into a frontal position in the top view and is found true length in the frontal view.

29.5 ANGLES WITH A LINE AND PRINCIPAL PLANES

The angle between a line and a plane will appear true size in the view where the plane is an edge and the line is true length. Two principal planes appear as edges in all principal views. Consequently, when a line appears true length in a principal view the angle between the line and two principal planes can be measured.

Two examples of finding lines true length by revolution are shown in Fig. 29.9. The angles with the principal planes can be measured in these views where the lines are true length.

The angle between the horizontal and the profile planes can be measured in part A in the front view. The angle with frontal and profile planes can be measured in the top view, part B.

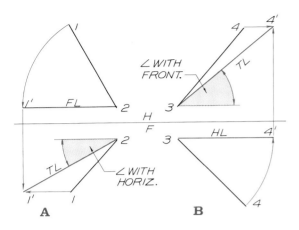

Fig. 29.9 Angles with principal planes

A. The angle with the horizontal plane can be measured in the front view if the line appears true length.

B. The angle with the frontal plane can be measured in the top view if the line appears true length.

29.6 TRUE SIZE OF A PLANE

A plane that appears as an edge in an orthographic view can be found true size by revolving the edge until it is parallel to a reference line between it and the adjacent view. In Fig. 29.10A, the edge view in the top view is revolved until it is parallel to the frontal plane, Step 1.

In Step 2, the true-size view is found by projecting down from the revolved view and horizontally across the given front view.

A plane that does not appear as an edge in a given view is found true size by an auxiliary view and one revolution in Fig. 29.10B. The plane is found as an edge (Step 1) and the edge is revolved to be parallel to the F-1 reference line (Step 2). The revolved view is found true size in Step 3.

Fig. 29.10A True size of a plane

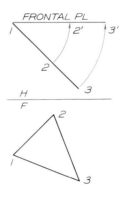 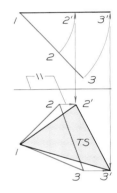

Step 1 The edge view of the plane is revolved to be parallel to the frontal plane.

Step 2 Points 2' and 3' are projected to the horizontal projectors from 2 and 3 in the front view.

Fig. 29.10B True size of a plane

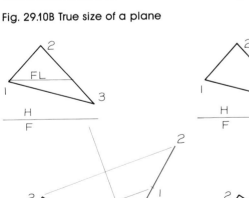 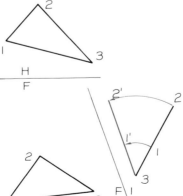 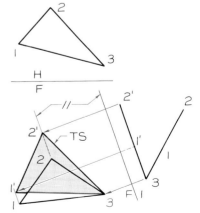

Step 1 An edge view of the plane is found by projecting from the front view.

Step 2 The plane is revolved to a position parallel to the F-1 reference line.

Step 3 Points 1' and 2' are found in the front view by projecting from the auxiliary view to find the true size of the plane.

Fig. 29.11 Edge view of a plane

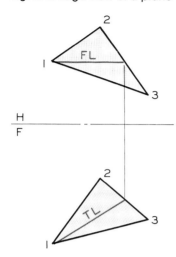
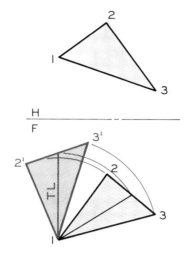
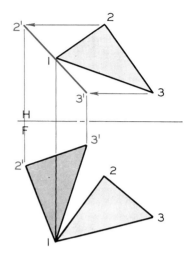

Step 1 It is required that we find the edge view of plane 1–2–3. A frontal line is found true length on the front view of the plane.

Step 2 The front view of the plane is revolved until the true-length line is vertical.

Step 3 Since the true-length line is vertical, it will appear as a point in the top view and the plane will appear as an edge, 1–2'–3'.

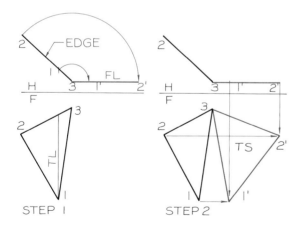

Fig. 29.12 True size of a plane

Step 1 When a plane appears as an edge in a principal view, it can be revolved to a position parallel to a reference line, the frontal plane in this case.

Step 2 Points 1' and 2' are projected to the front view to intersect with the horizontal projectors from the original points 1 and 2. The plane is true size in this view.

29.7   TRUE SIZE OF A PLANE BY DOUBLE REVOLUTION

The edge view of a plane can be found by revolution without the use of auxiliary views (Fig. 29.11). A frontal line is drawn on plane 1–2–3 and the line appears true length in the front view. The plane is revolved until the true-length line becomes vertical in the front view (Step 2). The true-length line will project as a point in the top view, and therefore, the plane will appear as an edge in this view (Step 3). Note that the projectors from the top view of points 2 and 3 are parallel to the H-F reference line.

A second revolution, called a **double revolution,** can be made to revolve this edge view of the plane until it is parallel to the frontal plane as shown in Fig. 29.12. The top views of the points 1' and 2' are projected to the front view where the plane 1'–2'–3 is true size.

This second revolution could have been performed in Fig. 29.11, but this would have resulted in an overlapping of view that would have made it difficult to observe the separate steps.

Double revolution is used in Fig. 29.13 to find the oblique plane of the object, plane 1–2–3, true

Fig. 29.13 True size by double revolution

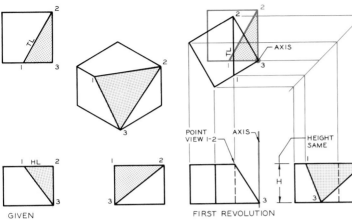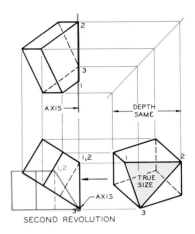

GIVEN   FIRST REVOLUTION   SECOND REVOLUTION

Given Three views of a block with an oblique plane across one corner.

Required Find the plane true size by revolution.

Step 1 Since line 1–2 is horizontal in the frontal view, it is true length in the top view. The top view is revolved into a position where line 1–2 can be seen as a point in the front view.

Step 2 Since plane 1–2–3 was found as an edge in step 1, this plane can be revolved into a vertical position in the front view, so that it will appear true size in the side view. The depth dimension does not change, since it is parallel to the axis of revolution.

Fig. 29.14 Double revolution

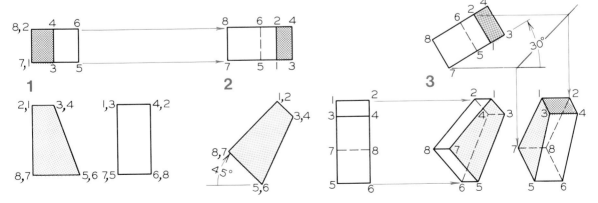

Given Three views of a surface with an inclined plane.

Step 1 The front view is revolved 45°, and the new width is projected to the top view where the depth is unchanged.

Step 2 The top view is revolved 30° to change the width and depth, but the height remains the same. The resulting front view is an axonometric pictorial.

size. In Step 1, the true-length line 1–2 on the plane is revolved in the top view until it is perpendicular to the frontal plane. Consequently, line 1–2 appears as a point in the front view, and the plane appears as an edge. This changes the width and depth dimensions, but the height dimension does not change.

In Step 2, the edge view of the plane is revolved into a vertical position parallel to the profile plane. The plane is found true size by projecting to the profile view where the depth dimension is unchanged and the height dimension has been increased.

A second example of double revolution of a solid is shown in Fig. 29.14 with dimensions transferred from view to view. The front view is revolved clockwise 45° and new top and side views are drawn with the depth dimension remaining constant. The top view is revolved 30° counterclockwise in Step 2. New front and side views are constructed by using projectors from the side view in Step 1 and the top view in Step 2. The resulting views are axonometric pictorials and none of the surfaces is true size.

29.8 ANGLE BETWEEN PLANES

The engine mount frame of a helicopter is an application where the angle between two intersect-

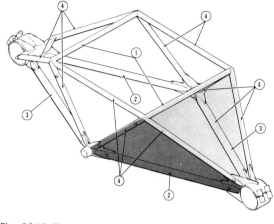

Fig. 29.15 The angle between any two planes of the helicopter engine mount can be found by using revolution principles. (Courtesy of Bell Helicopter Corporation.)

ing planes must be found, in order to provide its design specifications (Fig. 29.15). If orthographic views of this frame were given, the angles could be found by revolution.

In Fig. 29.16, the angle between two planes is found by drawing the edge view of the dihedral angle (the angle between the planes) perpendicular to the line of intersection and the plane of the angle is projected to front view (Step 1). The edge

Fig. 29.16 Angle between planes

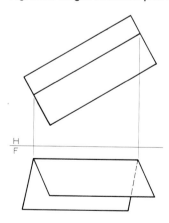

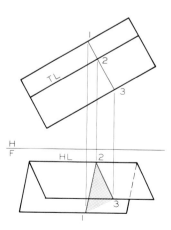

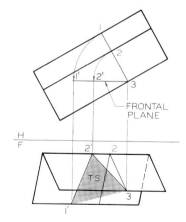

Given The top and front views of two intersecting planes.

Required Find the angle between the planes by revolution

Step 1 A right section is drawn perpendicular to the true-length line of intersection between the planes in the top view and is projected to the front view. The section is not true size in the front view.

Step 2 The edge view of the right section is revolved to position 1'–2'–3 in the top view to be parallel to the frontal plane. This section is projected to the front view, where it is true size since it is a frontal plane.

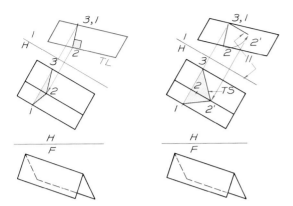

Fig. 29.17 Angle between oblique planes

Step 1 A true-length view of the line of intersection is found in an auxiliary view projected from the top view. The right section is constructed perpendicular to the true length of the line of intersection and is projected to the top view.

Step 2 The edge view of the right section is revolved to be parallel to the H-1 reference line so the plane will appear true size in the top view after being revolved. The angle between the planes can be found by measuring angle 1–2′–3.

view of the plane in the top view is revolved until it is a frontal plane in the top view, and it is then projected to the front view where its true-size view is found (Step 2).

A similar problem is solved in Fig. 29.17. In this example the line of intersection does not appear true length in one of the given views; consequently an auxiliary view is used in Step 1 to find the line of intersection true length. The plane of the angle between the planes can be drawn as an edge perpendicular to the true-length line of intersection (Step 1). The foreshortened view of plane 1–2–3 is projected to the top view in Step 1. The edge view of plane 1–2–3 is then revolved in the primary auxiliary view until it is parallel to the H-1 line (Step 2). When it is projected back to the top view, angle 1–2′–3 is true size.

29.9 LOCATION OF DIRECTIONS

It is necessary that you be able to locate the basic directions of up, down, forward, and back in any view that you are given in order to solve more ad-

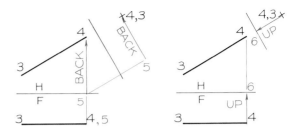

Fig. 29.18 To find the direction of back, forward, up, and down in an auxiliary view, construct an arrow pointing in the desired direction in the given principal views and project this arrow to the auxiliary view. The directions of back and up are shown here.

vanced problems of revolution. In Fig. 29.18, the directions of back and up are located by first drawing directional arrows in the given top and front views.

Line 4–5 is drawn pointing back in the top view, and its front view appears as a point. This directional arrow, 4–5, is projected to the auxiliary view as any other line would be drawn to locate the direction of back. By drawing the arrow on the other end of the line, you would find the direction of forward.

The direction of up is located in Fig. 29.18 by drawing line 4–6 in the direction of up in the front view and in the top view as a point. The arrow is found in the primary auxiliary by the usual pro-

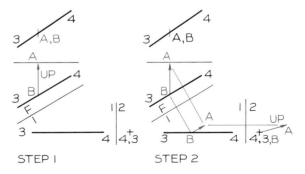

Fig. 29.19 Direction in a secondary auxiliary view

Step 1 To find the direction of up in the secondary auxiliary view, arrow AB is drawn pointing upward in the front view. It appears as a point in the top view.

Step 2 Arrow AB is projected to the primary and secondary auxiliary views like any other line. The direction of up is located in the secondary auxiliary view.

jection method, and the direction of up is located. The direction of down would be in the opposite direction.

The location of directions in secondary auxiliary views is found in the same manner. The direction of up is found in Fig. 29.19 by beginning with an arrow pointing upward in the front view and appearing as a point in the top view (Step 1). The arrow, AB, is projected from the front view to the primary and then to a secondary auxiliary view to give the direction of up in all views. The other directions can be found in the same manner by beginning with the two given principal views of a known directional arrow.

29.10 REVOLUTION OF A POINT ABOUT AN AXIS

In Fig. 29.20 it is required that point O be revolved about axis 3–4 into its most forward position. The circular path of revolution is drawn in the primary auxiliary view where the axis is a point (Step 1). The direction of forward is drawn in Step 2 and the new location of point O is found at O'. By projecting back through the successive views, point O' is found in each view. Note that O' lies on the

line in the front view; this verifies that O' is in its most forward position.

The problem in Fig. 29.21 requires an additional auxiliary view since the axis 3–4 is not true length in the given views. Therefore the line must be found true length before it can be found as a point where the path of revolution can be drawn as a circle. Point O is revolved into its highest position, O', where the "up" arrow, 3–5, is found in the secondary auxiliary view.

By projecting back to the given views, O' is located in each view. Its position in the top view is over the axis, which verifies that the point is located at its highest position.

The paths of revolution will appear as edges when the axis is true length, and as ellipses when the axis is not true length. The angle of the ellipse guide for drawing the ellipse in the front view is the angle the projectors from the front view make with the edge view of the revolution in the primary auxiliary view. To find the ellipse in the top view, an auxiliary view must be used to find the path of revolution as an edge perpendicular to the true-length axis projected from the top view.

The handcrank of a casement window (Fig. 29.22) is an example of a problem that must be solved using this principle to determine the clearances between the sill and the window frame.

Fig. 29.20 Revolution about an axis

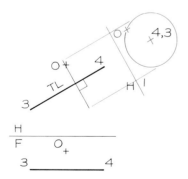

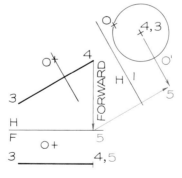

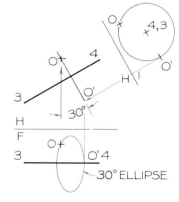

Step 1 To rotate point O about axis 3–4, it is necessary to find the point view of the axis in a primary auxiliary view. The circular path is drawn and the path of revolution is shown in the top view as an edge that is perpendicular to the axis.

Step 2 If it is required to rotate point O to its most forward position, draw an arrow pointing forward in the top view. It will appear as a point in the front view. The arrow, 4–5, is found in the auxiliary view to locate point O'.

Step 3 Point O' is projected back to the given views. The path of revolution appears as an ellipse in the front view since the axis is not true length in this view. A 30° ellipse is drawn since this is the angle your line of sight makes with the circular path in the front view.

Fig. 29.21 Revolution of a point about an axis

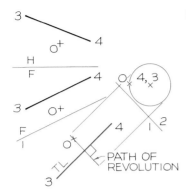

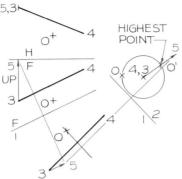

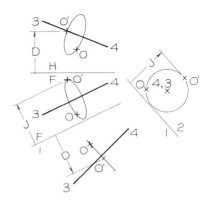

Step 1 To rotate O about axis 3–4, the axis is found as a point in a secondary auxiliary view where the circular path is drawn. The path appears as an edge in the primary auxiliary view where the axis is true length.

Step 2 If it is required to rotate O to its highest position, construct an arrow 3–5 in the front and top views that points upward. The direction of this arrow in the secondary auxiliary view locates the highest position, O'.

Step 3 Point O' is projected back to the given views by transferring the dimensions J and D using your dividers. The highest point lies over the line in the top view to verify its position. The path of revolution is elliptical wherever the axis is not true length.

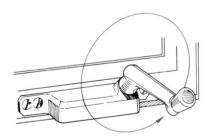

Fig. 29.22 This handcrank that is used on casement windows is an example of a problem that must be solved by using revolution principles. The handle must be properly positoned so to not interfere with the window sill or wall.

29.11 REVOLUTION OF A LINE ABOUT AN AXIS

Line 3–4 is revolved about axis 1–2 in Fig. 29.23. The point view of the axis 1–2 is found as a point in Step 1. A circle is drawn tangent to line 3–4 with its center at the point view of 1–2, and arcs are drawn through each end of the line as well. The perpendicular is revolved into the desired position and the new endpoints are found, 3′ and 4′.

The top vew of line 3′–4′ is found by projecting parallel to the H-1 line from the original points of 3 and 4 in the top view, as shown in Step 2. These projectors intersect those from the auxili-

ary view. The front view is obtained by projecting from the top view and transferring the height dimensions from the primary auxiliary view (Step 2).

29.12 REVOLUTION OF A RIGHT PRISM ABOUT ITS AXIS

Conveyors, ducts, and stairways are used to connect various parts of industrial installations. Such an example is the coal chute between two buildings shown in Fig. 29.24, which is used to convey coal at a continuous rate. You can easily understand why it is necessary to have the sides of the enclosed chute vertical and the bottom of the chute's right section horizontal in order for the coal to be transported efficiently.

In Fig. 29.25, it is required that the right section be positioned about centerline AB so that two of its sides will be vertical. This is done by finding the point view of the axis in Step 1, and the direction of up is projected to this view. In Step 2, the right section is found in the other views in Step 2. In Step 3, the sides of the chute are constructed parallel to the axis.

The bottom of this chute will be horizontal, and will be properly positioned for conveying materials such as coal.

Fig. 29.23 Revolution of a line about an axis

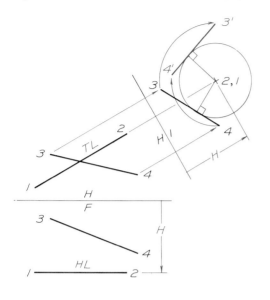

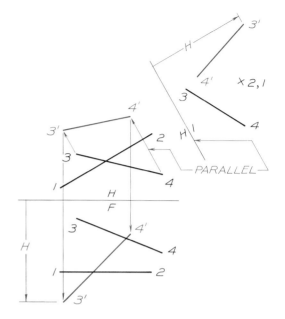

Step 1 The axis, 1–2, is found as a point in the auxiliary view. Line 3–4 is revolved to its specified position in this view.

Step 2 The new position of line 3–4 is projected back to the top view where projectors parallel to the top view of the H-1 reference line intersect those from the auxiliary view to find line 3'–4'. Line 3'–4' is located in the front view.

Fig. 29.24 A conveyor chute must be properly installed so that two edges of its right section are vertical so the conveyors will function properly. This requires the application of the revolution of a prism about its axis. (Courtesy of Stephens-Adamson Manufacturing Company.)

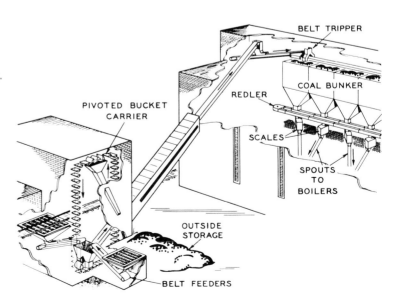

Fig. 29.25 Revolution of a prism about its axis

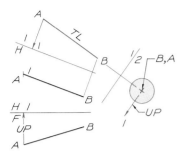 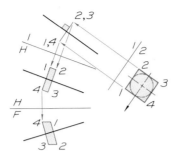 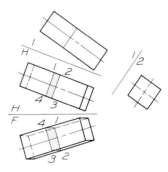

Step 1 Locate the point view of centerline AB in the secondary auxiliary view by drawing a circle about the axis with a diameter equal to one side of the square right section. Draw a vertical arrow in the front and top views and project it to the secondary auxiliary view to indicate the direction of vertical in this view.

Step 2 Draw the right section, 1–2–3–4, in the secondary auxiliary view with two sides parallel to the vertical directional arrow. Project this section back to the successive views by transferring measurements with dividers. The edge view of the section could have been located in any position along centerline AB in the primary auxiliary view, so long as it was perpendicular to the centerline.

Step 3 Draw the lateral edges of the prism through the corners of the right section so that they are parallel to the centerline in all views. Terminate the ends of the prism in the primary auxiliary view where they appear as edges that are perpendicular to the centerline. Project the corner points of the ends to the top and front views to establish the ends in these views.

29.13 ANGLE BETWEEN A LINE AND PLANE

The angle between the line and plane is found by a combination of auxiliary views and revolution in Fig. 29.26. The plane is found true size in a secondary auxiliary view in Step 1.

The line is revolved in Step 2 until it is parallel to the 1–2 reference line. The line can then be found true length in the primary auxiliary view in Step 3. Since the line appears true length and the plane as an edge in this view, the true angle can be measured here.

29.14 A LINE AT A SPECIFIED ANGLE WITH TWO PRINCIPAL PLANES

In Fig. 29.27 it is required that a line be drawn through point O that will make angles of 35° with the frontal plane and 44° with the horizontal plane, and slopes forward and downward.

The cone containing elements making 35° with the frontal plane is drawn in Step 1. In Step 2, the cone with elements making 44° with the horizontal plane is drawn. In Step 2 the length of the elements of the cone must be equal to elements of the first cone so that the cones will intersect with equal elements.

Lines 0–1 and 0–2 are found in Step 3 where these lines are elements that lie on each cone and make the specified angle with the principal planes.

These principles can be applied to the determination of intersections between piping systems that are joined by standard connectors.

29.15 REVOLUTION OF PARTS ON DETAIL DRAWINGS

It is standard practice to revolve features such as those shown in Fig. 29.28 to make the views more descriptive. The front view of the part at A is true size because the top view has been revolved. Similarly, the front view of the part at B is more descriptive with the revolution in the top view.

These are called conventional practices, and they should be used to conserve effort while gaining additional clarity in the preparation of working drawings.

Fig. 29.26 Angle between a line and plane

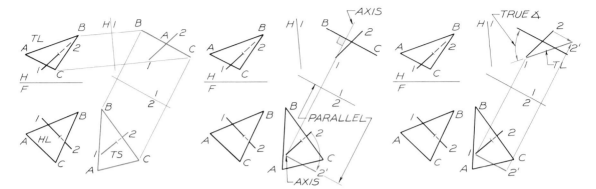

Step 1 Construct plane *ABC* as an edge in a primary auxiliary view, which can be projected from either view. Determine the true size of the plane in the secondary auxiliary view, and project line 1–2 to each view.

Step 2 Revolve the secondary auxiliary view of the line until it is parallel to the 1–2 reference line. The axis of revolution appears as a point through point 1 in the secondary auxiliary view. The axis appears true length and is perpendicular to the 1–2 line and plane *ABC* in the primary auxiliary view.

Step 3 Point 2′ is projected to the primary auxiliary view where the true length of line 1–2′ is found by projecting the primary auxiliary view of point 2 parallel to the 1–2 line as shown. Since the plane appears as an edge and the line appears true length in this view, the true angle between the line and the plane is found.

Fig. 29.27 A line at specified angles

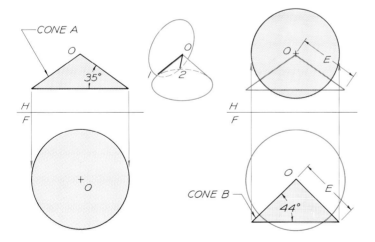
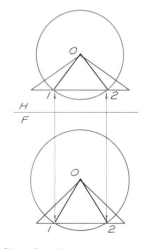

Step 1 Draw a triangular view of a cone in the top view such that the extreme elements make an angle of 35° with the frontal plane. Construct the circular view of the cone in the front view, using point *O* as the apex. All elements of this cone make an angle of 35° with the frontal plane.

Step 2 Draw a triangular view of a cone in the front view such that the elements make an angle of 44° with the horizontal plane. Draw the elements of this cone equal in length to element *E* of cone *A*. All elements of cone *B* make an angle of 44° with the horizontal plane.

Step 3 Since the elements of cones *A* and *B* are equal in length, there will be two common elements that lie on the surface of each cone, elements *O*–1 and *O*–2. Locate points 1 and 2 at the point where the bases of the cone intersect in both views. Either of these lines will satisfy the problem requirements.

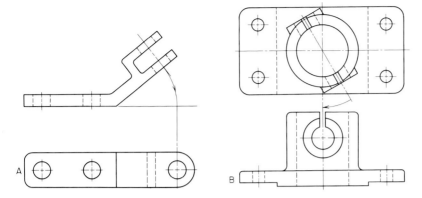

Fig. 29.28 It is conventional practice to revolve features of parts in order to show the features true size in the adjacent orthographic views. When this is done, it is unnecessary to show the arrows of rotation since this is understood as standard practice.

PROBLEMS

Use Size A (8⅛" × 11") sheets for the following problems and lay out the problems using instruments. Each square on the grid is equal to 0.20" or about 5 mm. The problems can be laid out on grid or plain paper. Label all reference planes and points in each problem with ⅛" letters and/or numbers, using guidelines.

The crosses marked "1" and "2" are to be used for placing primary and secondary reference lines. The primary reference line should pass through "1" and the secondary through "2".

1–4. Find the true-length views of the lines by revolution in Fig. 29.29.

5. Find the true-size view of the plane by an auxiliary view and a single revolution in Fig. 29.30.

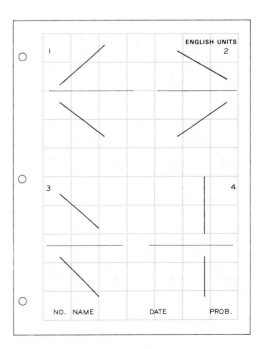

Fig. 29.29 Problems 1 through 4.

6. Find the true-size view of the plane by revolution in Fig. 29.30.

7 and 8. Find the true-size views of the planes by double revolution in Fig. 29.31.

9 and 10. Find the dihedral angles between the planes in Fig. 29.32.

11 and 12. Revolve the points about the given axes in Fig. 29.33 and show the points in all views. In Problem 11, revolve the point into its most forward position, and in Problem 12 into its highest position.

13. The centerline of a conveyor chute is given in the top and front views of Fig. 29.34. The chute has a 10-foot-square cross section. Construct the necessary views to revolve the 10-foot-square into a position where two sides of the right section will be vertical planes. Show the chute in all views. Scale: $1'' = 10'$.

14–19. Lay out these problems on a size B sheet ($11'' \times 17''$) using the horizontal format. Position the crossing division lines at the center of the sheet. The grid is spaced at 0.25″ or approximately 6 mm apart. Problem 14 requires that you lay out the prism in area 1 (Fig. 29.35) and rotate it as specified about its corner point 0 in the remaining areas of the sheet.

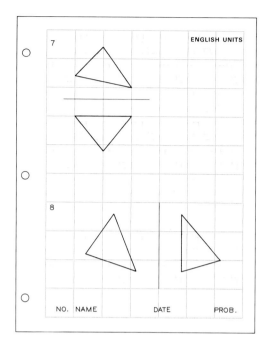

Fig. 29.31 Problems 7 and 8.

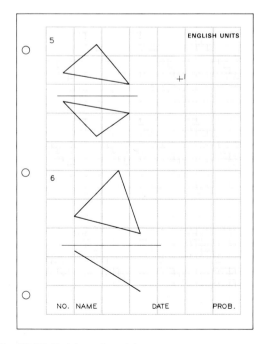

Fig. 29.30 Problems 5 and 6.

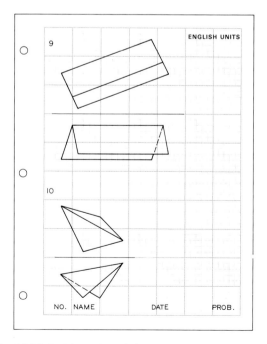

Fig. 29.32 Problems 9 and 10.

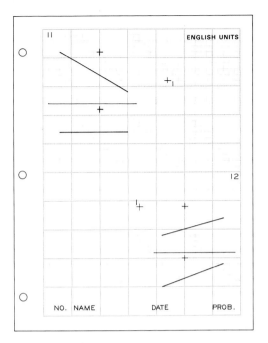

Fig. 29.33 Problems 11 and 12.

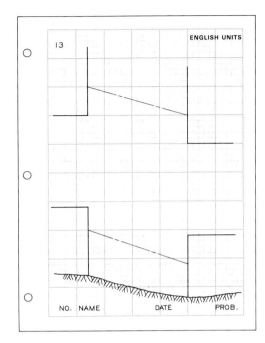

Fig. 29.34 Problem 13.

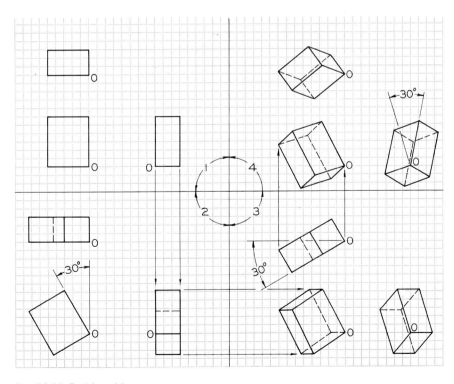

Fig. 29.35 Problem 14.

Problems 15 through 19 require that you replace the object in Problem 14 with one of those given in Fig. 29.36, and rotate these objects through the angles specified in each step.

20. Draw the object in Fig. 29.37, but complete the top view by showing the inclined surface revolved into a true-size position. This will eliminate the need for an auxiliary view as presently shown.

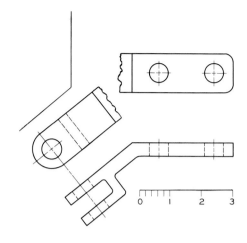

Fig. 29.37 Problem 20.

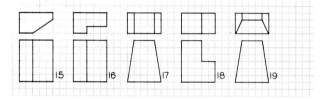

Fig. 29.36 Problems 15 through 19.

VECTOR GRAPHICS

30.1 INTRODUCTION

When analyzing a system for strength, it is necessary to consider the forces of tension and compression within the system. These forces are represented by vectors. Vectors may also be used to represent other quantities such as distance, velocity, and electrical properties.

Graphical methods are useful in the solution of vector problems, which are often very complicated to solve by conventional trigonometric and algebraic methods. Each method can serve as an effective check on the solutions found by the other methods.

30.2 BASIC DEFINITIONS

A knowledge of the terminology of graphical vectors is necessary to understand the techniques of problem-solving with vectors. The following definitions will be used throughout this chapter.

Force   A push or a pull that tends to produce motion. All forces have (1) magnitude, (2) direction, (3) a point of application, and (4) sense. A force is represented by the rope being pulled in Fig. 30.1A.

Vector   A graphical representation of a quantity of force that is drawn to scale to indicate magni-

tude, direction, sense, and point of application. The vector shown in Fig. 30.1B represents the force of the rope pulling the weight, W.

Magnitude   The amount of push or pull. In drawings, this is represented by the length of the vector line. Magnitude is usually measured in pounds or kilograms of force.

Direction   The inclination of a force (with respect to a reference coordinate system).

Point of application   The point through which the force is applied on the object or member, point A in Fig. 30.1A.

Sense   Either of the two opposite ways in which a force may be directed, i.e., toward or away from the point of application. The sense is shown by an

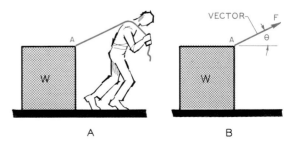

Fig. 30.1 Representation of a force by a vector.

arrowhead attached to one end of the vector line. It is shown in part B of Fig. 30.1 by the arrowhead at *F*.

Compression The state created in a member by subjecting it to opposite pushing forces. A member tends to be shortened by compression (Fig. 30.2A). Compression is represented by a plus sign (+) or the letter C.

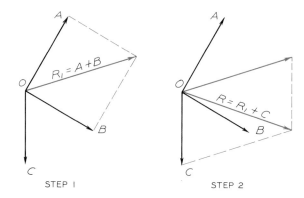

STEP I   STEP 2

Fig. 30.3 Resultant by the parallelogram method

Step 1 Draw a parallelogram with its sides parallel to vectors *A* and *B*. The diagonal R_1, drawn from point *P* to point *O*, is the resultant of forces *A* and *B*.

Step 2 Draw a parallelogram using vectors R_1 and *C* to find diagonal *R* from *P* to *Q*. This is the resultant that can replace forces *A*, *B*, and *C*.

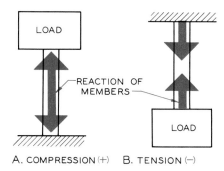

A. COMPRESSION (+)   B. TENSION (−)

Fig. 30.2 Comparison of tension and compression in a member.

Tension The state created in a member by subjecting it to opposite pulling forces. A member tends to be stretched by tension, as shown in Fig. 30.2B. Tension is represented by a minus sign (−) or the letter T.

Force system The combination of all forces acting on a given object. Figure 30.3 shows a force system.

Resultant A single force that can replace all the forces of a force system and have the same effect as the combined forces.

Equilibrant The opposite of a resultant; it is the single force that can be used to counterbalance all forces of a force system.

Components Any individual forces which, if combined, would result in a given single force. For example, Forces *A* and *B* are components of resultant R_1 in Step 1 of Fig. 30.3

Space diagram A diagram depicting the physical relationship between structural members. The force system in Fig. 30.3 is given as a space diagram.

Vector diagram A diagram composed of vectors that are scaled to their appropriate lengths to represent the forces within a given system. The vector diagram is used to solve for unknowns.

Statics The study of forces and force systems that are in equilibrium.

Metric units The kilogram (kg) is the standard unit for indicating mass (loads). A comparison of kilograms with pounds is shown in Fig. 30.4. The metric ton is 1000 kilograms. One kilogram = 2.2 pounds.

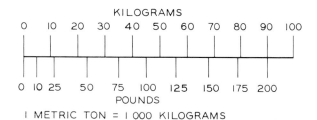

Fig. 30.4 The kilogram (kg) is the standard metric unit for measuring forces, which are represented by pounds in the English system: 1 kilogram = 2.2 pounds.

30.3 COPLANAR, CONCURRENT FORCE SYSTEMS

When several forces, represented by vectors, act through a common point of application, the system is said to be **concurrent.** Vectors A, B, and C act through a single point in Fig. 30.3; therefore this is a concurrent system. When only one view is necessary to show the true length of all vectors, as in Fig. 30.3, the system is **coplanar.**

Engineering designs are analyzed to determine the total effect of the forces applied in a system. Such an analysis requires that the known forces be resolved into a single force—the **resultant**—that will represent the composite effect of all forces on the point of application. The resultant is found graphically by two methods: (1) the parallelogram method and (2) the polygon method.

30.4 RESULTANT OF A COPLANAR, CONCURRENT SYSTEM—PARALLELOGRAM METHOD

In the system of vectors shown in Fig. 30.3, all the vectors lie in the same plane and act through a common point. The vectors are scaled to a known magnitude.

The vectors for a force system must be known and drawn to scale in order to apply the parallelogram method to determine the resultant. Two vectors are used to find a parallelogram; the diagonal of the parallelogram is the resultant of these two vectors and has its point of origin at point P (Fig. 30.3). Resultant R_1 can be called the *vector sum* of vectors A and B.

Since vectors A and B have been replaced by R_1, they can be disregarded in the next step of the solution. Again, resultant R_1 and vector C are resolved by completing a parallelogram, i.e., by drawing a line parallel to each vector. The diagonal of this parallelogram, PQ, is the resultant of the entire system and is the vector sum of R_1 and C. This resultant, R, can be analyzed as though it were the only force acting on the point; therefore the analysis of a particular point-of-force application is simplified by finding the resultant.

30.5 RESULTANT OF A COPLANAR, CONCURRENT SYSTEM—POLYGON METHOD

The system of forces shown in Fig. 30.3 is shown again in Fig. 30.5, but in this case the resultant is found by the polygon method. The forces are drawn to scale and in their true directions, with each force being drawn head-to-tail to form the polygon. In this example, the vectors are drawn in a counterclockwise sequence, beginning with vector A. Note that the polygon does not close; this means that the system is not in *equilibrium.* In other words, it would tend to be in motion, since the forces are not balanced in all directions. The resultant R is drawn from the tail of vector A to the head of vector C to close the polygon. The resultant is equal in length, direction, and sense to the resultant found by the parallelogram method of the previous article.

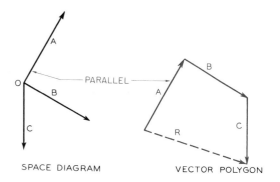

Fig. 30.5 Resultant of a coplanar, concurrent system as determined by the polygon method, in which the vectors are drawn head-to-tail.

30.6 RESULTANT OF A COPLANAR, CONCURRENT SYSTEM—ANALYTICAL METHOD

Vectors can be solved analytically by application of algebra and trigonometry. The analytical example in Fig. 30.6 is given to afford a comparison between the graphical and analytical methods.

In Step 1, the vertical components, which are parallel to the Y-axis, are drawn from the ends of both vectors to form right triangles, and their lengths are found through the use of the trigonometric functions of the angles the vectors make with the X-axis.

Fig. 30.6 Resultant by the analytical method

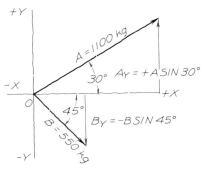

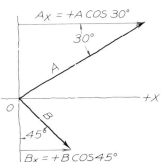

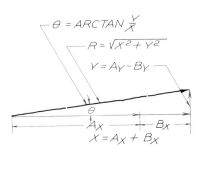

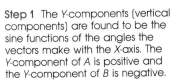

Step 1 The Y-components (vertical components) are found to be the sine functions of the angles the vectors make with the X-axis. The Y-component of A is positive and the Y-component of B is negative.

Step 2 The X-components (horizontal components) are the cosine functions of 30° and 45° in this case, both in the positive direction.

Step 3 The Y-components and X-components are summed to find the components of the resultant, X and Y. The Pythagorean theorem is applied to find the magnitude of the resultant. Its angle with the X-axis is the arctangent of Y/X.

The horizontal component of each vector is drawn parallel to the X-axis through the end of each vector. The lengths of these components are found to be the cosine functions of the given vectors in Step 2.

The Y-components of each vector, A_y and B_y, can be added, since each lies in the same direction (Step 3). The resulting value is $Y = A_y - B_y$, since the components have opposite senses. The horizontal component is $X = A_x + B_x$, since both components have equal directions and senses.

A right triangle is sketched using the X- and Y-components that were found trigonometrically. The vertical and horizontal components are laid off head-to-tail and the head of the horizontal component is connected to the tail of the vertical to form a right triangle of forces. The magnitude of the resultant is found by the Pythagorean theorem,

$$R = \sqrt{X^2 + Y^2}$$

The direction of the resultant is

$$\text{angle } \theta = \arctan Y/X$$

and it is measured from the horizontal X-axis.

Law of Sines

The law of sines is illustrated in Fig. 30.7A. When any three values are known, the remaining un-

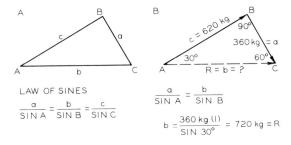

Fig. 30.7 The law of sines is illustrated in part A. This principle is used in part B to solve for resultant R (b) when two angles and vectors are known.

knowns of a triangle can be computed. An example is given (Fig. 30.7B) where two sides of a triangle are vectors of known magnitude and direction. This enables you to find the resultant mathematically, as shown.

An *equilibrant* has the same magnitude, direction, and point of application as the *resultant* in any system of forces.

The difference is the sense. Note that the resultant of the system of forces shown in Fig. 30.8 is solved for through the parallelogram method. The equili-

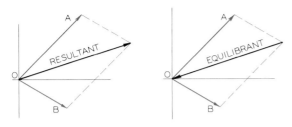

Fig. 30.8 The resultant and equilibriant are equal in all respects except in sense (position of arrowhead).

brant can be applied at point O to balance the forces A and B and thereby cause the system to be in a state of equilibrium.

30.7 RESULTANT OF NONCOPLANAR, CONCURRENT FORCES—PARALLELOGRAM METHOD

When vectors lie in more than one plane of projection, they are said to be **noncoplanar;** therefore more than one view is necessary to analyze their

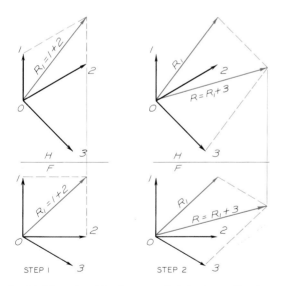

Fig. 30.9 Resultant by the parallelogram method

Step 1 Vectors 1 and 2 are used to construct a parallelogram in the top and front views. The diagonal, R_1, is the resultant of these two vectors.

Step 2 Vectors 3 and R_1 are used to construct a second parallelogram to find the views of the overall resultant, R_2.

spatial relationships. The resultant of a system of noncoplanar forces can be found by the parallelogram method, regardless of their number, if their true projections are given in two adjacent orthographic views.

Vectors 1 and 2 in Fig. 30.9 are used to construct the top and front views of a parallelogram. The diagonal of the parallelogram, R_1, is found in both views. As a check, the front view of R_1 must be an orthographic projection of its top view; if it is not, there is an error in construction.

In Step 2, resultant R_1 and vector 3 are resolved to form resultant R_2 in both views. The top and front views of R_2 must project orthographically if there is no error in construction. Resultant R_2 can be used to replace vectors 1, 2, and 3. Since R_2 is an oblique line, its true length can be found by auxiliary view, as shown in Fig. 30.10 or by revolution.

30.8 RESULTANT OF NONCOPLANAR, CONCURRENT FORCES—POLYGON METHOD

The same system of forces that was given in Fig. 30.9 is solved in Fig. 30.10 for the resultant of the system by the polygon method.

In Step 1, each vector is laid head-to-tail in a clockwise direction, beginning with vector 1. The vectors are drawn in each view to be orthographic projections at all times (Step 2). Since the vector polygon did not close, the system is not in equilibrium. The resultant R is constructed from the tail of vector 1 to the head of vector 3 in both views.

Resultant R is an oblique line and requires an auxiliary view to find its true length. The magnitude of the resultant can be measured in the true-length auxiliary view by using the same scale as was used to draw the original views.

30.9 RESULTANT OF NONCOPLANAR, CONCURRENT FORCES—ANALYTICAL METHOD

We are required to solve for the resultant of the system in Fig. 30.11 by the analytical method, using trigonometry and algebraic equations. The projected lengths of the vectors are known in both views.

In Step 1, the summation of the forces in the X-direction is found in the front view. Since this

Fig. 30.10 Resultant by the polygon method

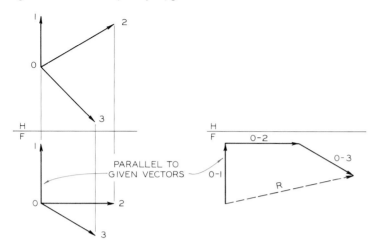 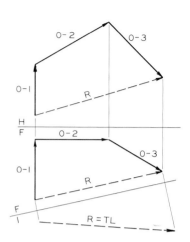

Required Find the resultant of this system of concurrent, noncoplanar forces by the polygon method.

Step 1 Each vector is laid off head-to-tail in the front view. The front view of the resultant is the vector found.

Step 2 The same vectors are laid off head-to-tail in the top view to complete the three-dimensional polygon. The resultant is found true length by an auxiliary view.

30.11 Resultant by the analytical method

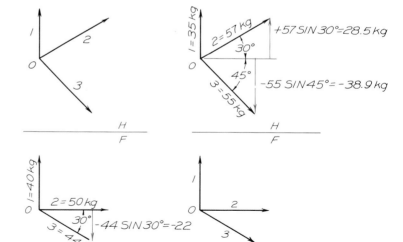 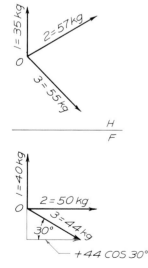

Step 1 The X-component is found in the front or top view. The X-components are found to be: force 1 = 0 kg; force 2 = 50 kg; force 3 = 44 kg cos 30°. These values are positive.

Step 2 The Y-component must be found in the front view. The Y-components are found to be: force 1 = 40 kg; force 2 = 0 kg; force 3 = 44 kg sin 30°.

Step 3 The Z-component must be found in the top view. The Z-components are found to be: force 1 = 35 kg; force 2 = 57 kg sin 30°; force 3 = 55 kg sin 45°. The resultant is found in Fig. 30.12

left and right direction can be seen in both the top and front views, either view can be used for finding the X-component of the system. The summation in the X-direction is expressed in the following equation:

$$\sum F_x = (2) + (3) \cos 30°$$
$$= 50 + 44 \cos 30° = 88.2 \text{ kg}(+).$$

The X-component is found to be 88.2 lb in the positive direction, which is toward the right. Vector 1 is vertical and consequently has no component in the X-direction.

Summation of forces in the Y-direction is found in the front view. This summation is expressed in the following equation:

$$\sum F_y = (1) - (3) \sin 30°$$
$$= 40 - 44 \sin 30° = 18 \text{ kg}(+).$$

Vector 2 is horizontal and has no vertical component.

The summation of forces in the Z-direction is found in the top view. Positive direction is considered to be backward, and negative to be forward. This summation is expressed in the following equation:

$$\sum F_z = (1) + (2) \sin 30° - (3) \sin 45°$$
$$= 35 + 57 \sin 30° - 55 \sin 45°$$
$$= 24.16 \text{ lb}(+).$$

The resultant that can be used to replace vectors 1, 2, and 3 can be found from these three components by the following equation:

$$R = \sqrt{X^2 + Y^2 + Z^2}.$$

By substitution of the X-, Y-, and Z-components found in the three previous summations, the equation can be solved as follows:

$$R = \sqrt{88.2^2 + 18^2 + 24.6^2} = 93.3 \text{ kg}.$$

The resultant force of 93.3 kg is of no value unless its direction and sense are known. To find this information, we must refer to the two orthographic views of the force system, as shown in Fig. 30.12. The X- and Z-components, 88.2 kg and 24.6 kg, are drawn to form a right triangle in the top view. The hypotenuse of this triangle depicts the direction and sense of the resultant in the top view. The angular direction of the top view of re-

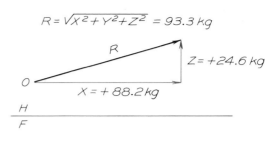

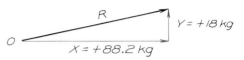

Fig. 30.12 The three components X, Y, and Z found in Fig. 30.11 are used to find the resultant $R = X^2 + Y^2 + Z^2 = 93.3$ kg.

sultant is found in the following equation:

$$\tan \theta = \frac{24.6}{88.2} = 0.279; \quad \theta = 15.6°.$$

The angular direction of the resultant is found in the front view by constructing a triangle with the X- and Y-components, 88.2 kg and 18 kg. The hypotenuse of this right triangle is the direction of the resultant. The direction of the resultant in the front view is expressed in the following equation:

$$\tan \phi = \frac{18}{88.2} = 0.204; \quad \phi = 11.5°.$$

These two angles, found in the top and front views, establish the direction of the resultant vector, whose sense can be described as upward, to the right, and back.

30.10 FORCES IN EQUILIBRIUM

In the previous examples, the vectors were drawn from given or known magnitudes and directions. The same principles can be applied to structural systems in which the magnitudes and senses are not given.

An example of a coplanar, concurrent structure in equilibrium can be seen in the loading cranes in Fig. 30.13, which are used for the handling of cargo on board ship.

The coplanar, concurrent structure given in Fig. 30.14 is designed to support a load of $W = 1000$ kg. The maximum loading in each is used to

Fig. 30.13 The cargo cranes on the cruise ship *Santa Rosa* are examples of coplanar, concurrent force systems that are designed to remain in equilibrium. (Courtesy of Exxon Corporation.)

diagram. A vector that acts toward a point of application is in compression. Vector *A* points away from point *A* when transferred to the structural diagram and is, therefore, in tension.

The length versus the cross section of a member will be considered when selecting a member, but the determination of force in the member is found in the same manner in the vector polygon regardless of member length.

A similar example of a force system involving a pulley is solved in Fig. 30.15 to determine the loads in the structural members caused by the weight of 100 lb. The only difference between this solution and the previous one is the construction of two equal vectors to represent the loads in the cable on both sides of the pulley.

30.11 TRUSS ANALYSIS

Vector polygons can be used to analyze structural trusses to determine the loads in each member by two graphical methods: (1) joint-by-joint analysis, and (2) Maxwell diagrams.

Joint-by-Joint Analysis

The truss shown in Fig. 30.16 is called a Fink truss, and is loaded with forces of 3000 lb that are concentrated at joints of the structural members.

determine the type and size of structural members used in the structural design.

In Step 1, the only known force, $W = 1000$ kg, is laid off parallel to the given direction. Unknown forces *A* and *B* are drawn as vectors to close the force polygon. Each vector must be drawn head-to-tail.

In Step 2, vectors *A* and *B* can be analyzed to determine whether they are in tension or compression. Vector *B* points upward to the left, which is toward point *O* when transferred to the structural

Fig. 30.14 Coplanar forces in equilibrium

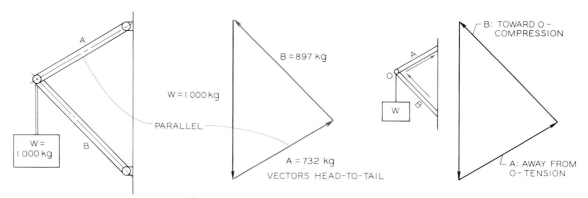

Required Find the forces in the two structural members caused by the load of 1000 kg.

Step 1 Draw the known load of 1000 kg as a vector. Draw the vectors *A* and *B* to the same scale and parallel to their directions. Arrowheads are drawn head-to-tail.

Step 2 Vector *A* points away from the point *O* when transferred to the structural diagram. Therefore, vector *A* is in tension. Vector *B* points toward point *O* and is in compression.

Fig. 30.15 Determination of forces in equilibrium   619

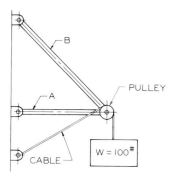

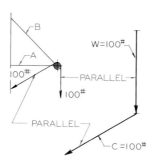

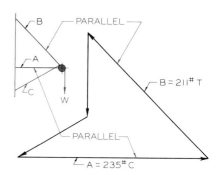

Required Find the forces in the members caused by the load of 100 lb supported by the pulley.

Step 1 The force in the cable is equal to 100 lb on both sides of the pulley. These two forces are drawn as vectors head-to-tail parallel to their directions in the space diagram.

Step 2 A and B are drawn to close the polygon, and arrowheads are placed to give a head-to-tail arrangement. The sense of A is toward the point of application, and in compression; B is away from the point and is in tension.

Fig. 30.16 Joint analysis of a truss

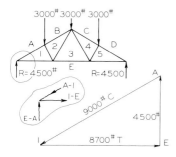

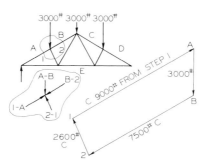

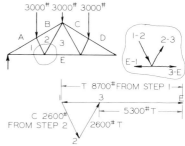

Step 1 The truss is labeled using Bow's notation, with letters between the exterior loads and numbers between interior members. The lower left joint can be analyzed, since it has only two unknowns, A–1 and 1–E. These vectors are found by drawing them parallel to their directions from both ends of the reaction of 4500 lb. The vectors are laid off in a head-to-tail order.

Step 2 Using the vector 1–A found in step 1 and load AB, the two unknowns B–2 and 2–1 can be found. The known vectors are laid out beginning with vector 1–A and moving clockwise about the joint. Vectors B–2 and 2–1 close the polygon. If the sense of a vector is toward the point of application, it is in compression; if away from the point, it is in tension.

Step 3 The third joint can be analyzed by laying out the vectors E–1 and 1–2 from the previous steps. Vectors 2–3 and 3–E close the polygon and are parallel to their directions in the space diagram. The senses of 2–3 and 3–E are away from the point of application; these vectors are in tension.

A special method of designating forces, called **Bow's notation,** is used. The exterior forces applied to the truss are labeled with letters placed between the forces. Numerals are placed between the interior members.

Each vector used to represent the load in each member is referred to by the number on each of its sides by reading in a clockwise direction. For example, the first vertical load at the left is called AB, with A at the tail and B at the head of the vector.

We first analyze the joint at the left where the reaction of 4500 pounds (denoted by #) is known. This force, reading in a clockwise direction about the joint, is called EA with an upward sense. The tail is labeled E and the head A. Continuing in a clockwise direction, the next force is A-1 and the next 1–E, which closes the polygon and ends with the beginning letter, E. The arrows are placed, beginning with the known vector EA, in a head-to-tail arrangement.

Tension and compression can be determined

Fig. 30.17 Truss analysis

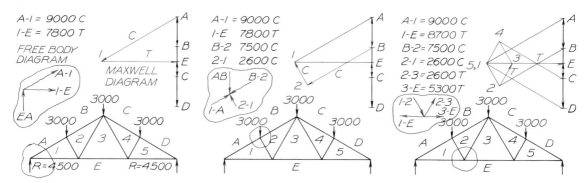

Step 1 Label the spaces between the outer forces of the truss with letters and the internal spaces with numbers, using Bow's notation. Add the given load vectors in a Maxwell diagram, and sketch a free-body diagram of the first joint. Using vectors *EA*, *A–1*, and *1–E* drawn head-to-tail, draw a vector diagram to find their magnitudes. Vector *A-1* is in compression (+) because its sense is toward the joint, and *1–E* is in tension (−) because its sense is away from the joint.

Step 2 Draw a sketch of the next joint to be analyzed. Since *AB* and *A-1* are known, we have to determine only two unknowns, *2–1* and *B–2*. Draw these parallel to their direction, head-to-tail, in the Maxwell diagram using the existing vectors found in step 1. Vectors *B–2* and *2–1* are in compression since each has a sense toward the joint. Note that vector *A–1* becomes *1–A* when read in a clockwise direction.

Step 3 Sketch a free-body diagram of the next joint to be analyzed. The unknowns in this case are *2–3* and *3–E*. Determine the true length of these members in the Maxwell diagram by drawing vectors parallel to given members to find point 3. Vectors *2–3* and *3–E* are in tension because they act away from the joint. This same process is repeated to find the loads of the members on the opposite side.

by relating the sense of each vector to the original joint. For example, *A–1* has a sense toward the joint and is in compression, while *1–E* is away and in tension.

Since the truss is symmetrical and equally loaded, the loads in the members on the right will be equal to those on the left.

The other joints are analyzed in the same manner in Steps 2 and 3; the procedure is to begin with known vectors found in the previous polygons and then solve for the unknowns. Note that the sense of the vectors is opposite at each end. Vector *A–1* has a sense toward the left in Step 1, and toward the right in Step 2.

Maxwell Diagrams

The Maxwell diagram is exactly the same as the joint-by-joint analysis except that the polygons are positioned to overlap, with some vectors common to more than one polygon. Again, Bow's notation is used to good advantage.

The first step (Fig. 30.17) is to lay out the exterior loads beginning clockwise about the truss—*AB, BC, CD, DE,* and *EA*—head-to-tail. A letter is placed at each end of the vectors. Since they are parallel, this polygon will be a straight line.

The structural analysis begins at the joint through which reaction *EA* acts. A free-body diagram is drawn to isolate this joint for easier analysis. The two unknowns are members *A–1* and *1–E*. These vectors are drawn parallel to their direction in the truss in Step 1 of Fig. 30.17 with *A–1* beginning at point *A* and *1–E* beginning at point *E*. These directions are extended to a point intersection, which locates point 1. Since this joint is in equilibrium, as are all joints of a system in equilibrium, the vectors must be drawn head-to-tail. Because resultant *EA* has an upward sense, vector *A–1* must have its tail at *A*, giving it a sense toward point 1. By relating this sense to the free-body diagram, we can see that the sense is toward the point of application, which means that *A–1* is a compression member. Vector *1–E* has a sense away from the joint, which means that it is a tension member. The vectors are coplanar and can be scaled to determine their loads.

In Step 2, vectors 1–*A* and *AB* are known, while vectors *B*–2 and 2–1 are unknown. Since there are only two unknowns it is possible to solve for them. A free-body diagram showing the joint to be analyzed is sketched. Vector *B*–2 is drawn parallel to the structural member through point *B* in the Maxwell diagram and the line of vector 2–1 is extended from point 1 until it intersects with *B*–2, where point 2 is located. The sense of each vector is found by laying off each vector head-to-tail. Both vectors *B*–2 and 2–1 have a sense toward the joint in the free-body diagram; therefore, they are in compression.

The next joint is analyzed in sequence to find the stresses in 2–3 and 3–*E* (Step 3). The truss will have equal forces on each side, since it is symmetrical and is loaded symmetrically. The total Maxwell diagram is drawn to illustrate the completed work in Step 3.

If all the polygons in the series do not close at every point with perfect symmetry, there is an error in construction. If the error of closure is very slight, it can be disregarded, since safety factors are generally applied in derivation of working stresses of structural systems to assure safe construction. Arrowheads are usually omitted on Maxwell diagrams, since each vector will have opposite senses when applied to different joints.

Fig. 30.18 The structural members of this tripod support for a moon vehicle can be analyzed graphically to determine design loads. (Courtesy of NASA.)

30.12 NONCOPLANAR STRUCTURAL ANALYSIS—SPECIAL CASE

Structural systems that are three-dimensional require the use of descriptive geometry, since it is necessary to analyze the system in more than one plane. The manned flying system (MFS) in Fig. 30.18 can be analyzed to determine the forces in the support members (Fig. 30.19). Weight on the moon can be found by multiplying earth weight by a factor of 0.165. A tripod that must support 182 lb on earth has to support only 30 lb on the moon. This is a special case; since members *B* and *C* lie in the same plane and appear as an edge in the front view, we need to determine only two unknowns.

A vector polygon is constructed in the front view in Step 1 of Fig. 30.19 by drawing force *F* as a vector and using the other vectors as the other sides of the polygon. One of these vectors is actually a summation of vectors *B* and *C*. The top view is drawn using the vectors *B* and *C* to close the polygon from each end of vector *A*. In Step 2, the front view of vectors *B* and *C* is found.

The true lengths of the vector are found in a true-length diagram in Step 3. The vectors are measured to determine their loads. Vector *A* is found to be in compression because its sense is toward the point of concurrency. Vectors *B* and *C* are in tension.

30.13 NONCOPLANAR STRUCTURAL ANALYSIS— GENERAL CASE

The structural frame shown in Fig. 30.20 is attached to a vertical wall to support a load of $W = 600$ lb. Since there are three unknowns in each of the views, we are required to construct an auxiliary view that will give the edge view of a plane containing two of the vectors, thereby reducing the number of unknowns to two (Step 1). We no longer need to refer to the front view.

A vector polygon is drawn by constructing vectors parallel to the members in the auxiliary view (Step 1). An adjacent orthographic view of the vector polygon is also drawn by constructing its vectors parallel to the members in the top view (Step 2). A true-length diagram is used in Step 3 to find the true length of the vectors to determine their magnitudes.

Fig. 30.19 Noncoplanar structural analysis—a special case

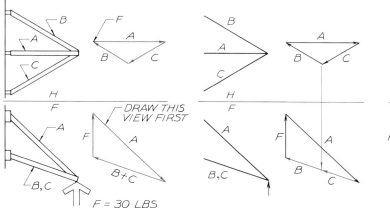

A = 67.5 LBS
B = 29.0 LBS
C = 29.0 LBS

DRAW THIS VIEW FIRST

F = 30 LBS

TL DIAGRAM

Step 1 Two forces, *B* and *C*, coincide in the front view, resulting in only two unknowns in this view. Vector *F* (30 lb) is drawn, and the other two unknowns are drawn parallel to their front view to complete the front view of the vector polygon. The top view of *A* can be found by projection, from which vectors *B* and *C* can be found.

Step 2 The point of intersection of vectors *B* and *C* in the top view is projected to the front view to separate these vectors. All vectors are drawn head-to-tail. Vectors *B* and *C* are in tension because their vectors are acting away from the point in the space diagram, while *A* is in compression.

Step 3 The completed top and front views found in step 2 do not give the true lengths of vectors *B* and *C*, since they are oblique. The true lengths of these lines are determined by a true-length diagram where they are scaled to find the forces in each member.

Fig. 30.20 Noncoplanar structural analysis—general case

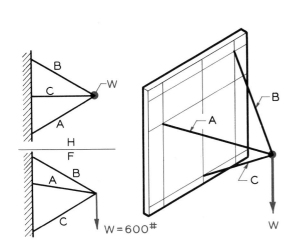

Given The top and front views of a three-member frame which is attached to a vertical wall and supports a weight of 600 lb.

Required Find the loads in the structural members.

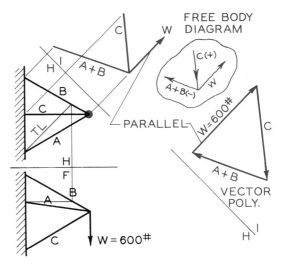

Step 1 To limit the unknowns to two, construct an auxiliary view to find two vectors lying in the edge of a plane. Use the auxiliary view and top view in the remainder of the problem. Draw a vector polygon parallel to the members in the auxiliary view in which $W = 600$ lb is the only known vector. Sketch a free-body diagram for preliminary analysis.

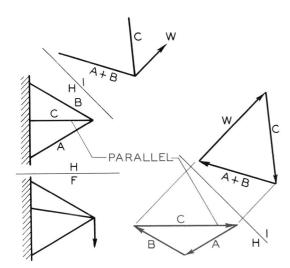

Step 2 Construct an orthographic projection of the view of the vector polygon found in step 1 so that its vectors are parallel to the members in the top view. The reference plane between the two views is parallel to the H–1 plane. This portion of the problem is closely related to the problem in Fig. 30.19.

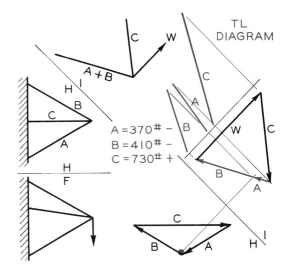

Step 3 Project the intersection of vectors A and B in the horizontal view of the vector polygon to the auxiliary view polygon to establish the lengths of vectors A and B. Determine the true lengths of all vectors in a true-length diagram to determine their magnitudes. Analyze for tension or compression, as covered in Section 30.11.

Fig. 30.21 Tractor sidebooms represent noncoplanar, concurrent systems of forces that can be solved graphically. (Courtesy of Trunkline Gas Company.)

A three-dimensional system is the side-boom tractors used for lowering pipe into a ditch during pipeline construction (Fig. 30.21).

30.14 NONCONCURRENT, COPLANAR VECTORS

Forces may be applied in such a manner that they are not concurrent, as illustrated in Fig. 30.22. Bow's notation can be used to locate the resultant of this type of nonconcurrent system.

In Step 1, the vectors are laid off to form a vector diagram in which the closing vector is the resultant, $R = 68$ lb. Each vector is resolved into two components by randomly locating point O on the interior or exterior of the polygon and connecting point O with the end of each vector. The components, or strings, from point O are equal and opposite components of adjacent vectors. For example, component o-b is common to vectors AB and BC. Since the strings from point O are equal and opposite, the system has not changed statically.

In Step 2, each string is transferred to the space diagram of the vectors where it is drawn between the respective vectors to which it applies. (The figure thus produced is called a *funicular diagram*.) For instance, string o-b is drawn in the

Fig. 30.22 Resultant of nonconcurrent forces

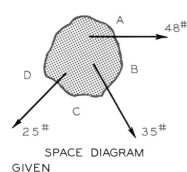

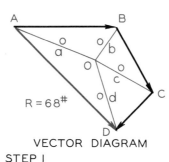

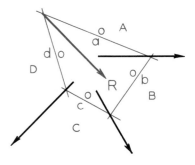

SPACE DIAGRAM	VECTOR DIAGRAM	FUNICULAR DIAGRAM
GIVEN	STEP I	STEP 2

Required Find the resultant of the known forces applied to the above object. The forces are nonconcurrent.

Step 1 The vectors are drawn head-to-tail to find resultant R. Point O is conveniently located for the construction of strings to the ends of each vector.

Step 2 Each string is drawn between the two vectors to which it applies in Step 1. *Example:* o–c between BC and CD. These strings are connected in sequence until the strings o–a and o–d establish the position of R, found in step 1.

area between the vectors *AB* and *BC*. String *o-c* is drawn in the *C*-area to connect at the intersection of *o-b* and vector *BC*. The point of intersection of the last strings, *o-a* and *o-d*, locates a point through which the resultant *R* will pass. The resultant has now been determined with respect to magnitude, sense, direction, and point of application.

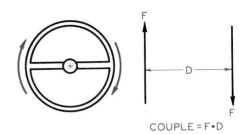

COUPLE = F•D

Fig. 30.23 Representation of a couple or moment.

30.15 NONCONCURRENT SYSTEMS RESULTING IN COUPLES

A *couple* is the descriptive name given to two parallel, equal, and opposite forces that are separated by a distance and are applied to a member in such a manner that they cause the member to rotate.

An important quantity associated with a couple is its *moment.*The moment of any force is a measure of its rotational effect. An example is shown in Fig. 30.23, in which two equal and opposite forces are applied to a wheel. The forces are separated by the distance *D*. The moment of the couple is found by multiplying one of the forces by the perpendicular distance between it and a point on the line of action of the other: $F \times D$. If the force is 20 lb and the distance is 3 ft, the moment of the couple would be given as 60 ft-lb.

A series of parallel forces is applied to a beam in Fig. 30.24. The spaces between the vectors are labeled with letters that follow Bow's notation. We are required to determine the resultant.

After constructing a vector diagram, we have a straight line that is parallel to the direction of the forces and that closes at point *A*. We then locate pole point *O* and draw the strings of a funicular diagram.

The strings are transferred to the space diagram and are drawn in their respective spaces. For example, *o–c* is drawn in the *C*-space between vectors *BC* and *CD*. The last two strings, *o–d* and *o–a*, do not close at a common point, but are found to be parallel; the result is therefore a cou-

Fig. 30.24 Couple resultants

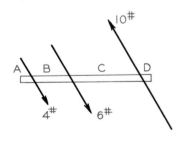

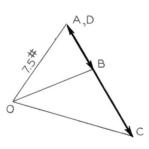

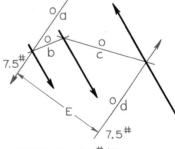

COUPLE = 7.5#(E)

SPACE DIAGRAM	VECTOR DIAGRAM	FUNICULAR DIAGRAM

Required Find the resultant of these nonconcurrent forces applied to this beam.

Step 1 The spaces between each force are labeled in Bow's notation.

Step 2 The vectors are laid out head-to-tail; they will lie in a straight line since they are parallel. Pole point *O* is located in a convenient location and the ends of each vector are connected with point *O*.

Step 3 Strings *o–a, o–c,* and *o–d* are successively drawn between the vectors to which they apply. Since strings *o–a* and *o–d* are parallel, the resultant will be a couple equal to 7.5 lb × *E*, where *E* is the distance between *o–a* and *o–d*.

Fig. 30.25 The boom of this crane can be analyzed for its resultant as a parallel, nonconcurrent system of forces when the cables have been disregarded.

ple. The distance between the forces of the couple is the perpendicular distance, E, between strings o–a and o–d in the space diagram, using the scale of the space diagram. The magnitude of the force is the scaled distance from point O to A and D in the vector diagram, using the scale of the vector diagram. The moment of the couple is equal to 7.5 lb $\times$ E in a counterclockwise direction.

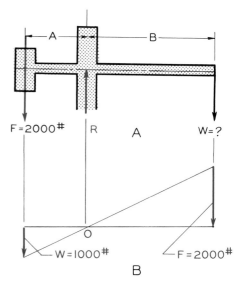

Fig. 30.26 Determining the resultant of parallel, noncurrent forces.

Fig. 30.27 Beam analysis with parallel loads

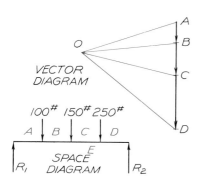

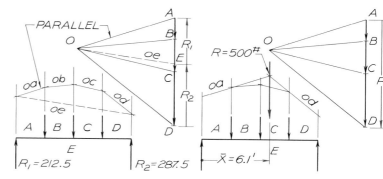

Step 1 Letter the spaces between the loads with Bow's notation. Find the graphical summation of the vectors by drawing them head-to-tail in a vector diagram at a convenient scale. Locate pole point O at a convenient location and draw strings from point O to each end of the vectors.

Step 2 Extend the lines of force in the space diagram, and draw a funicular diagram with string o–a in the A-space, o–b in the B-space, o–c in the C-space, etc. The last string, which is drawn to close the diagram, is o–e. Transfer this string to the vector polygon and use it to locate point E, thus establishing the lengths of R_1 and R_2, which are EA and DE, respectively.

Step 3 The resultant of the three downward forces will be equal to their graphical summation, line AD. Locate the resultant by extending strings o–a and o–d in the funicular diagram to a point of intersection. The resultant $R = 500$ lb will act through this point in a downward direction. $\bar{X}$ is a locating dimension.

30.16 RESULTANT OF PARALLEL, NONCONCURRENT FORCES

Forces applied to beams, such as those shown in Fig. 30.25, are parallel and nonconcurrent in many instances, and they may have the effect of a couple, tending to cause a rotational motion.

The beam in Fig. 30.26 is on a rotational crane that is used to move building materials in a limited area. The magnitude of the weight W is unknown, but the counterbalance weight is known to be 2000 lb; column R supports the beam as shown. Assuming that the support cables have been omitted, we desire to find the weight W that would balance the beam.

This problem can be solved by the application of the law of moments, i.e., the force is multiplied by the perpendicular distance to its line of action from a given point, or $F \times A$. If the beam is to be in balance, the total effect of the moments must be equal to zero, or $F \times A = W \times B$.

The graphical solution (Fig. 30.26B) is found by constructing a line to represent the total distance between the forces F and W. Point O is projected from the space diagram to this line. Point O is the point of balance where the summation of the moments will be equal to zero. Vectors F and W are drawn to scale at each end of the line by transposing them to the opposite ends of the beam. A line is drawn from the end of vector F through point O and extended to intersect the direction of vector W. This point represents the end of vector W, which can be scaled, resulting in a magnitude of 1000 lb.

30.17 RESULTANT OF PARALLEL, NONCONCURRENT FORCES ON A BEAM

The beam given in Fig. 30.27 is supported at each end and must in turn support three given loads. We are required to determine the magnitude of each support, R_1 and R_2, along with the resultant of the loads and its location. The spaces between all vectors are labeled in a clockwise direction with Bow's notation in Step 1, and a force diagram is drawn.

In Step 2 the lines of force in the space diagram are extended and the strings from the vector diagram are drawn in their respective spaces, parallel to their original direction. *Example:* String oa is drawn parallel to string OA in space A between forces EA and AB, and string ob is drawn in space B beginning at the intersection of oa with vector AB. The last string, oe, is drawn to close the funicular diagram. The direction of string oe is transferred to the force diagram, where it is laid off through point O to intersect the load line at point E. Vector DE represents support R_2 (refer to Bow's notation as it was applied in Step 1). Vector EA represents support R_1.

The magnitude of the resultant of the loads (Step 3) is the summation of the vertical downward forces, or the distance from A to D, or 500 lb. The location of the resultant is found by extending the extreme outside strings in the funicular diagram, oa and od, to their point of intersection. The resultant is found to have a magnitude of 500 lb, a vertical direction, a downward sense, and a point of application established by $\overline{X}$.

PROBLEMS

Problems should be presented in instrument drawings on Size A ($8\frac{1}{2}'' \times 11''$) paper, grid or plain. Each grid square represents .20″. All notes, sketches, drawings, and graphical work should be neatly prepared in keeping with good practices. Written matter should be legibly lettered using $\frac{1}{8}''$ guidelines.

1. In Fig. 30.28A, determine the resultant of the force system by the parallelogram method at the left of the sheet. Solve the same system using the vector polygon method at the right of the sheet. Scale: $1'' = 100$ lb (note that each grid square equals .20″). (B) In part B of the figure, determine the resultant of the concurrent, coplanar force system shown at the left of the sheet by the parallelogram method. Solve the same system using the polygon method at the right of the sheet. Scale $1'' = 100$ lb.

2. (A and B) In Fig. 30.29 solve for the resultant of each of the concurrent, noncoplanar force systems by the parallelogram method at the left of

the sheet. Solve for the resultant of the same systems by the vector polygon method at the right of the sheet. Find the true length of the resultant in both problems. Letter all construction. Scale: 1" = 600 lb.

3. (A and B) In Fig. 30.30, the concurrent, coplanar force systems are in equilibrium. Find the loads in each structural member. Use a scale of 1" = 300 lb in part A and a scale of 1" = 200 lb in part B. Show and label all construction.

4. In Fig. 30.31, solve for the loads in the structural members of the truss. Vector polygon scale: 1" = 2000 lb. Label all construction.

5. In Fig. 30.32, solve for the loads in the structural members of the concurrent, noncoplanar force system. Find the true length of all vectors. Scale: 1" = 300 lb.

6. In Fig. 30.33, solve for the loads in the structural members of the concurrent, noncoplanar force system. Find the true length of all vectors. Scale: 1" = 400 lb.

7. (A) In Fig. 30.34, find the resultant of the coplanar, nonconcurrent force system. The vectors are drawn to a scale of 1" = 100 lb. (B) In part B of the figure, solve for the resultant of the co-

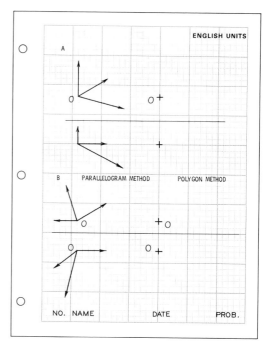

Fig. 30.29 Resultant of concurrent, noncoplanar vectors.

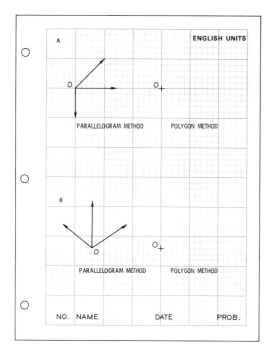

Fig. 30.28 Resultant of concurrent, coplanar vectors.

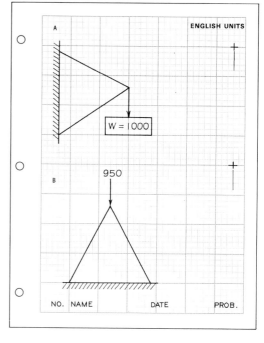

Fig. 30.30 Coplanar, concurrent forces in equilibrium.

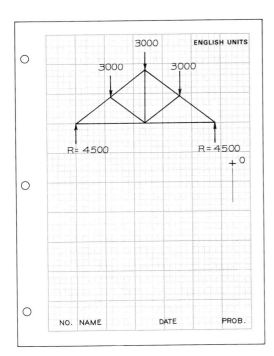

Fig. 30.31 Truss analysis.

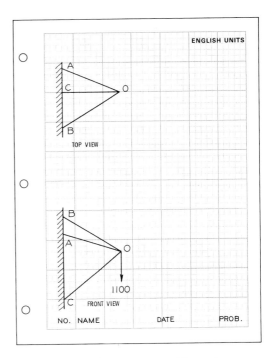

Fig. 30.33 Noncoplanar, concurrent forces in equilibrium.

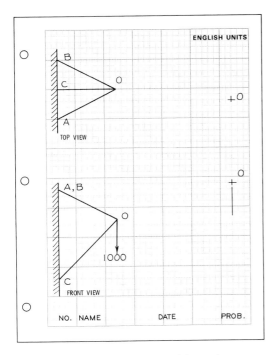

Fig. 30.32 Noncoplanar, concurrent forces in equilibrium.

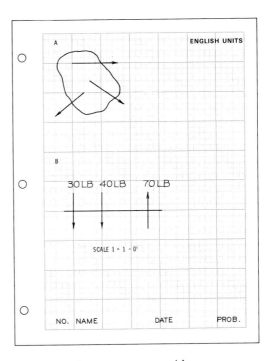

Fig. 30.34 Coplanar, nonconcurrent forces.

planar, nonconcurrent force system. The vectors are given in their true positions and at the true distances from each other. The space diagram is drawn to a scale of 1″ = 1.0′. Draw the vectors to a scale of 1″ = 30 lb. Show all construction.

8. (A) In Fig. 30.35, determine the force that must be applied at A to balance the horizontal member supported at B. Scale 1″ = 100 lb. (B) In part B of the figure, find the resultants at each end of the horizontal beam. Find the resultant of the downward loads and determine where it would be positioned. Scale: 1″ = 600 lb.

9. Determine the forces in the three members of the tripod in Fig. 30.36. The tripod supports a load of W = 250 lb. Find the true lengths of all vectors.

10. The vectors in Fig. 30.37 each make an angle of 60° with the structural member on which they are applied. Find the resultant of this force system.

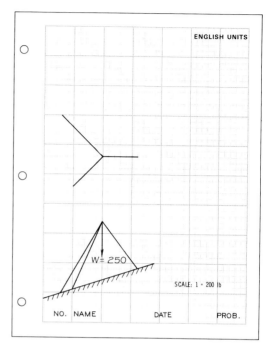

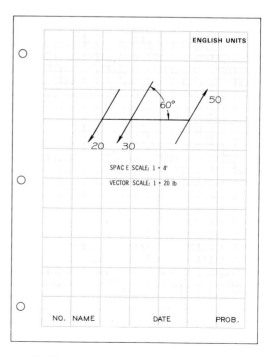

Fig. 30.36 Beam analysis.

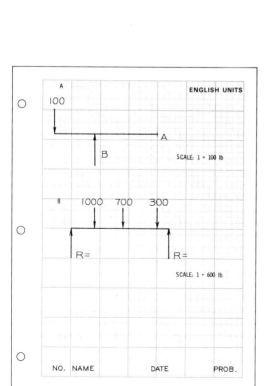

Fig. 30.35 Beam analysis.

Fig. 30.37 Noncoplanar, concurrent forces in equilibrium.

INTERSECTIONS AND DEVELOPMENTS

31.1 INTRODUCTION

This chapter deals with the methods of finding lines of **intersections** between parts that join together. Usually these parts are made of sheet metal or plywood used for forming concrete to a desired shape.

Once the intersections have been found, flat patterns called **developments** can be found graphically. The patterns can then be laid out on the sheet metal and cut to conform to the desired shape. Consequently, intersection and development problems are closely related. Developments cannot be found until after the intersections are determined.

You can see many examples of intersections and developments in Fig. 31.1 where a refinery is under construction. Although this is a massive project, the principles covered in this chapter must be used in solving problems of this type.

31.2 INTERSECTIONS OF LINES AND PLANES

The basic steps of finding the intersection between a line and a plane are illustrated in Fig. 31.2. This is a special case since the plane appears as an edge, and the point of intersection can be easily seen in this view, Step 1. It is projected to the

Fig. 31.1 This refinery installation illustrates many examples of the application of principles of intersections and developments.

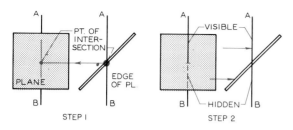

Fig. 31.2 Intersection of a line and a plane

Step 1 The point of intersection can be found in the view where the plane appears as an edge, the side view in this example.

Step 2 Visibility in the front view is determined by looking from the front view to the right side view.

front view, and the visibility of the line is found in Step 2.

This principle is used in Fig. 31.3 to find the line of intersection between two planes. Since *EFGH* appears as an edge in the side view, points of intersection 1 and 2 can be found and projected to the front view and the visibility determined in Step 2. Note that the intersection was found by finding the piercing points of lines *AB* and *DC* and connecting these points.

The intersection of a plane at a corner of a prism results in a line of intersection that bends around the corner (Fig. 31.4). Piercing points 2′ and 1′ are found in Step 1.

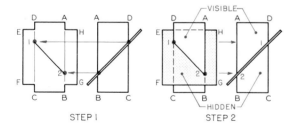

Fig. 31.3 Intersection between planes

Step 1 The points where plane *EFGH* intersects lines *AB* and *DC* are found in the view where the plane appears as an edge. These points are projected to the front view.

Step 2 Line 1–2 is the line of intersection. Visibility is determined by looking from the front view to the right side view.

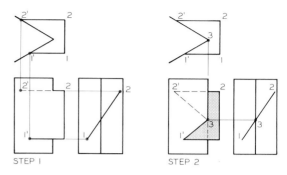

Fig. 31.4 Intersection of a plane at a corner

Step 1 The intersecting plane appears as an edge in the side view. Intersection points 1′ and 2′ are projected from the top and side views to the front view.

Step 2 The line of intersection from 1′ to 2′ must bend around the vertical corner at point 3 in the top and side views. This point is projected to the front view to locate line 1′–2′–3′.

Corner point 3 is seen in the side view of Step 2 where the vertical corner pierces the plane. Point 3 is projected to the corner in the front view. Point 2′ is hidden in the front view since it is on the back side of the assembly.

The intersection of a plane and a prism is found in Fig. 31.5, where the plane appears as an edge. The points of intersection are found for each corner line and are connected; visibility is shown to complete the line of intersection.

An intersection between a plane and a prism is shown in Fig. 31.6 where the vertical corners of the prism are true length in the front view and the plane appears foreshortened in both views. Imaginary cutting planes are passed vertically through the planes of the prism in the top view to find the piercing points of the corners in the front view. The points are connected and the visibility is determined to complete the solution.

The intersection between a foreshortened plane and an oblique prism is found in Fig. 31.7. The plane is found as an edge in a primary auxiliary view. The piercing points of the corners of the prism are located in the auxiliary view and are projected back to the given views.

Points 1, 2, and 3 are projected from the auxiliary view to the given views as examples. Visibility is determined by analysis of crossing lines, as previously covered.

Fig 31.5 Intersection of a plane and a prism

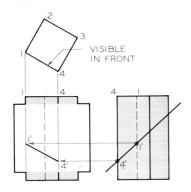

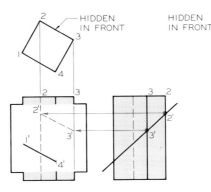

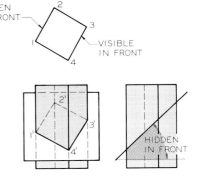

Step 1 Vertical corners 1 and 4 intersect the edge view of the plane in the side view at 1' and 4'. These points are projected to the front view and are connected with a visible line.

Step 2 Vertical corners 2 and 3 intersect the edge of the inclined plane at 2' and 3' in the side view. 2' and 3' are connected in the front view with a hidden line. Inspection of the top view tells us that this line is hidden.

Step 3 Lines 1'–2' and 3'–4' are drawn as hidden and visible lines, respectively. Visibility is determined by inspection of the top and side views and by projection to the front view.

Fig. 31.6 Intersection of an oblique plane and a prism

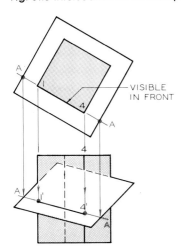

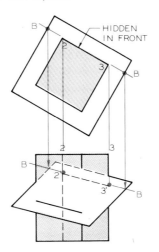

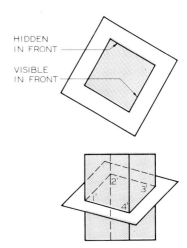

Step 1 Vertical cutting plane A–A is passed through the vertical plane, 1–4, in the top view and is projected to the front view. Piercing points 1' and 4' are found in this view.

Step 2 Vertical plane B–B is passed through the top view of plane 2–3 and is projected to the front view where piercing points 2' and 3' are found. Line 2'–3' is a hidden line.

Step 3 The line of intersection is completed by connecting the four points in the front view. Visibility in the front view is found by inspection of the top view.

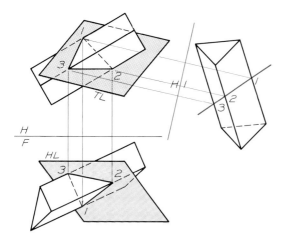

Fig. 31.7 The intersection between a plane and a prism can be found by constructing a view in which the plane appears as an edge.

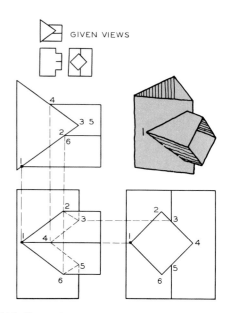

GIVEN VIEWS

Fig. 31.8 Three views of intersecting prisms. The points of intersection can be seen where intersecting planes appear as edges.

31.3 INTERSECTIONS BETWEEN PRISMS

The same principles used to find the intersection between a plane and a line are used to find the intersection between two prisms in Fig. 31.8. Piercing points 1, 2, 4, and 6 are found in the front view by projecting from the side and top views. Points 3 and 5 are located in the side view where the lines of intersection 2–4 and 4–6 bend around the vertical corner of the other prism. These points

are connected in sequence and visibility is determined.

In Fig. 31.9, an inclined prism intersects an inclined prism. In Step 1, the end view of the inclined prism is found by an auxiliary view. In the auxiliary view, you can see where plane 2–3 bends around corner AB at point X in Step 2.

Fig. 31.9 Intersection between prisms

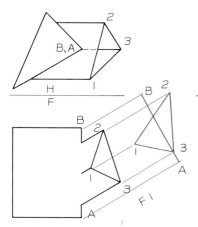

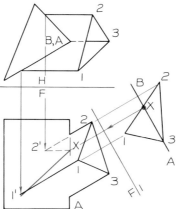

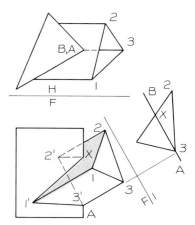

Step 1 Construct the end view of the inclined prism by projecting an auxiliary view from the front view. Show only line AB of the vertical prism in the auxiliary view.

Step 2 Locate piercing points 1′ and 2′ in the top and front views. Intersection line 1′–2′ will not be a straight line, but it will bend around corner AB at point X, which is projected from the auxiliary view.

Step 3 Intersection lines from 2′ and 1′ to 3′ do not bend around the corner. Therefore, these are drawn as straight lines. Line 1′–3′ is visible and line 2′–3′ is invisible.

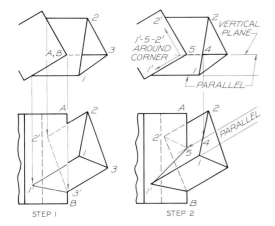

STEP I   STEP 2

Fig. 31.10 Cutting planes can be passed through one prism parallel to its edges to locate point 2 where the plane wraps around corner 3–4.

Points of intersection 2′ and 3′ are projected from the top to the front view. The line of intersection 2′–X–3′ can be drawn to complete this portion of the line of intersection. The remaining lines, 1′–3′ and 2′–1′, are connected to complete the solution (Step 3).

An alternative method of solving a problem of this type is shown in Fig. 31.10 where vertical cutting planes are used to locate point 2 where the intersection bends around corner 3–4. The plane is passed through the corner and parallel to the

sides of the inclined prism in the top view. It is projected to the front view where its projection intersects line 3–4 at point 2, where the intersection bends around the corner. The other piercing points are found in the top view and are projected to the front view.

The conduit connector shown in Fig. 31.11 is an example of the application of intersecting planes and prisms.

31.4 INTERSECTION OF A PLANE AND CYLINDER

The intersections of the components of this gas transmission system in Fig. 31.12 offer numerous examples of problems that were solved using the principles of intersections.

The intersection between a plane and a cylinder is found in Fig. 31.13 where the plane appears as an edge in one of the given views. Cutting planes are passed vertically through the top view

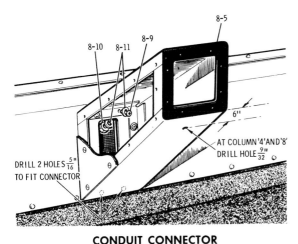

CONDUIT CONNECTOR

Fig. 31.11 This conduit connector was designed through the use of the principles of intersection of a plane and a prism. (Courtesy of the Federal Aviation Administration.)

Fig. 31.12 This complex of pipes and vessels contains many applications of intersections. (Courtesy of Trunkline Gas Company.)

Fig. 31.13 Intersection between a cylinder and a plane

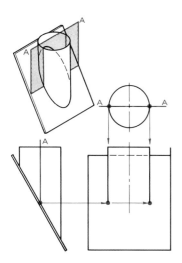

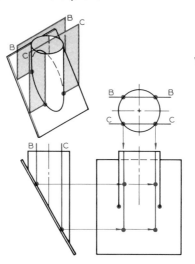

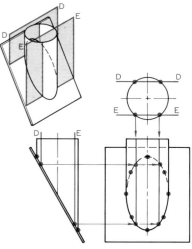

Step 1 A vertical cutting plane, *A–A*, is passed through the cylinder parallel to its axis to find two points of intersection.

Step 2 Two more cutting planes, *B–B* and *C–C*, are used to find four additional points in the top and left-side views; these points are projected to the front view.

Step 3 Additional cutting planes are used to find more points. These points are connected to give an elliptical line of intersection.

Fig. 31.14 Intersection of a cylinder and an oblique plane

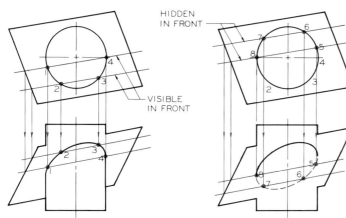

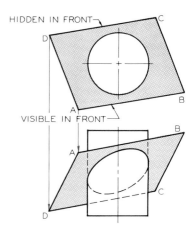

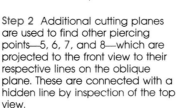

Step 1 Vertical cutting planes are passed through the cylinder in the top view to establish elements on its surface and lines on the oblique plane. Piercing points 1, 2, 3, and 4 are projected to the front view to their respective lines and are connected with a visible line.

Step 2 Additional cutting planes are used to find other piercing points—5, 6, 7, and 8—which are projected to the front view to their respective lines on the oblique plane. These are connected with a hidden line by inspection of the top view.

Step 3 Visibility of the plane and cylinder is completed in the front view. Line *AB* is found to be visible by inspection of the top view, and line *CD* is found to be hidden.

of the cylinder to establish elements on the cylinder and their piercing points. The piercing points are projected to each view to find the line of intersection, which is an ellipse.

A more general problem is solved in Fig. 31.14 where the cylinder is vertical, but the plane is oblique. Vertical cutting planes are passed through the cylinder and the plane in the top view to find piercing points of the cylinder's elements on the plane. These points are projected to the front view to complete the line of intersection, an ellipse. The more cutting planes that are used, the more accurate will be the plotted line of intersection.

The general case of the intersection between a plane and cylinder is solved in Fig. 31.15 where both are oblique in the given views. The edge view of the plane is found in an auxiliary view. Cutting planes are passed through the cylinder parallel to the cylinder's axis in the auxiliary view to find the piercing points.

The piercing points of the elements are connected to give elliptical lines of intersection in the given views. Visibility is determined by analysis.

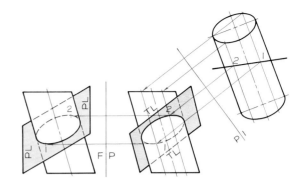

Fig. 31.15 The intersection between an oblique cylinder and an oblique plane can be found by constructing a view that shows the plane as an edge.

31.5 INTERSECTIONS BETWEEN CYLINDERS AND PRISMS

A series of vertical cutting planes are used in Fig. 31.16 to establish lines that lie on the surfaces of the cylinder and prism. A primary auxiliary view is drawn to show the end view of the inclined prism. The vertical cutting planes are shown in this view also, spaced the same distance apart as in the top view (Step 1).

Fig. 31.16 Intersection between a cylinder and a prism

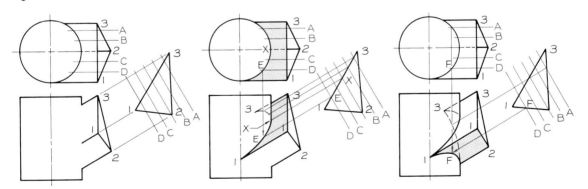

Step 1 Project an auxiliary view of the triangular prism from the front view to show three of its surfaces as edges. Pass frontal cutting planes through the top view of the cylinder and project them to the auxiliary view. The spacing between the planes is equal in both views.

Step 2 Locate points along the line of intersection of plane 1–3 in the top view and project them to the front view. *Example:* Point E on cutting plane D is found in the top and primary auxiliary views and projected to the front view where the projectors intersect. Point X on the centerline is the point where visibility changes from visible to hidden in the front view.

Step 3 Determine the remaining points of intersection by using the same cutting planes. Point F is shown in the top and primary auxiliary views and is projected to the front view on line of intersection 1–2. Connect the points and determine visibility. Judgment should be used in spacing the cutting planes so that they will produce an accurate plot of the line of intersection.

Fig. 31.17 Intersection between two cylinders

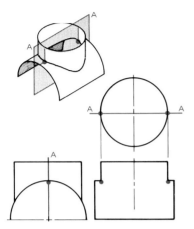

Step 1 A cutting plane, *A–A*, is passed through the cylinders parallel to the axes of both. Two points of intersection are found.

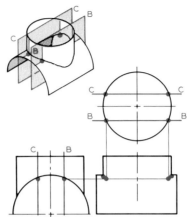

Step 2 Cutting planes *C–C* and *B–B* are used to find four additional points of intersection.

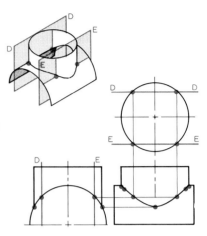

Step 3 Cutting planes *D–D* and *E–E* locate four more points. Points found in this manner are connected to give the line of intersection.

The line of intersection from 1 to 3 is projected from the auxiliary view to the front view, Step 2, where the intersection is an elliptical curve. The change of visibility of this line is found at point X in the top and auxiliary views, and is projected to the front view. In Step 3, the process is continued to find the lines of intersection of the other two planes of the prism.

31.6 INTERSECTIONS BETWEEN TWO CYLINDERS

The method of finding the line of intersection between two perpendicular cylinders is to pass a cutting plane through the cylinders parallel to the centerlines of each (Fig. 31.17). Each cutting plane locates the piercing points of two elements of one cylinder on an element of the other cylinder. The points are connected and visibility is determined to complete the solution.

The intersection between nonperpendicular cylinders is found in Fig. 31.18 by a series of vertical cutting planes. Each cutting plane is passed through the cylinders parallel to the centerline of each. Points 1 and 2 are labeled on cutting plane *D* as examples of points on the line of intersection. Other points are found in the same manner. The auxiliary view is an optional view that is not required for the solution of this problem, but it as-

Fig. 31.18 The intersection between these cylinders is found by finding the end view of the inclined cylinder in an auxiliary view. Vertical cutting planes are used to find the piercing points of the elements of the cylinder and the line of intersection.

sists you in visualizing the problem. Points 1 and 2 are shown on cutting plane *D* in the auxiliary view, where they can be projected to the front view as a check on the solution found when projecting from the top view.

31.7 INTERSECTIONS BETWEEN PLANES AND CONES

To find points of intersection on a cone, cutting planes can be used that are (1) perpendicular to the cone's axis, or (2) parallel to the cone's axis. Horizontal cutting planes are shown in Fig. 31.19A where they are labeled as H_1 and H_2. The horizontal planes cut circular sections that appear true size in the top view.

The cutting planes in Fig. 31.19B are passed radially through the top view to establish elements on the surface of the cone that are projected to the front view. Points 1 and 2 are found on these elements in both views by projection. These two types of cutting planes are used to solve intersections involving cones.

A series of radial cutting planes is used to find elements on the cone in Fig. 31.20. These ele-

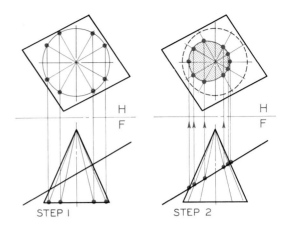

Fig. 31.20 Intersection of a plane and a cone

Step 1 Divide the base into even divisions in the top view and connect these points with the apex to establish elements on the cone. Project these to the front view.

Step 2 The piercing point of each element on the edge view of the plane is projected to the top view to the same elements, where they are connected to form the line of intersection. Visibility is shown to complete the drawing.

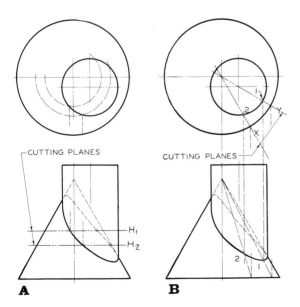

Fig. 31.19 Intersections on conical surfaces can be found by cutting planes that pass through the cone parallel to its base (part A). A second method at B shows radial cutting planes that pass through the cone's centerline and perpendicular to its base.

ments cross the edge view of the plane in the front view to locate piercing points that are projected to the top view of the same elements. The points are connected to form the line of intersection.

A cone and an oblique plane intersect in Fig. 31.21, and the line of intersection is found by using a series of horizontal cutting planes. The sections cut by these imaginary planes will be true circles in the top view. Also, the cutting planes locate lines on the oblique plane that intersect the same circular sections cut by each respective cutting plane. The points of intersection are found in the top view and are projected to the front view.

The horizontal cutting-plane method could have been used to solve the example in Fig. 31.20 as an alternative method.

31.8 INTERSECTIONS BETWEEN CONES AND PRISMS

A primary auxiliary view is used to find the end view of the inclined prism that intersects the cone in Fig. 31.22, Step 1. Cutting planes that radiate from the apex of the cone in the top view are

Fig. 31.21 Intersection of an oblique plane and a cone

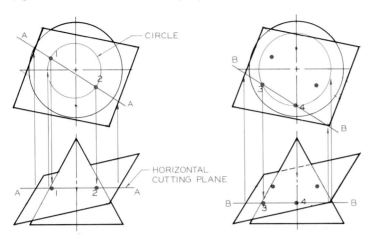

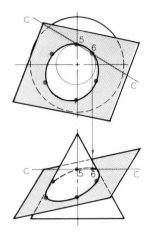

Step 1 A horizontal cutting plane is passed through the front view to establish a circular section on the cone and a line on the oblique plane in the top view. The piercing point of this line must lie on the circular section. Piercing points 1 and 2 are projected to the front view.

Step 2 Horizontal cutting plane B–B is passed through the front view in the same manner to locate piercing points 3 and 4 in the top view. These points are projected to the horizontal plane in the front view from the top view.

Step 3 Additional horizontal planes are used to find sufficient points to complete the line of intersection. Determination of the visibility completes the solution.

Fig. 31.22 Intersection between a cone and a prism

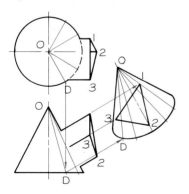

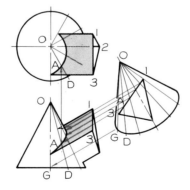

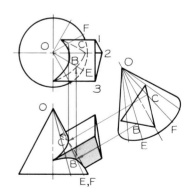

Step 1 Construct a primary auxiliary view to obtain the edge views of the lateral surfaces of the prism. In the auxiliary view, pass cutting planes through the apex to establish elements on the cone. Project the elements to the front and auxiliary views.

Step 2 Locate the piercing points of the cone's elements with the edge view of plane 1–3 in the primary view and project them to the front and top views. *Example:* point A lies on element OD in the auxiliary view, so it is projected to the front and top views of element OD. Locate other points in this manner.

Step 3 Locate the piercing points where the conical elements intersect the edge views of the other planes of the prism in the auxiliary view. *Example:* Point B is found on OE in the primary auxiliary view and is projected to the front and top views of OE. Show visibility of the lines of intersection in each view.

drawn in the auxiliary view to locate elements on the cone's surface that intersect the prism. These elements are drawn in the front view by projection.

Wherever the edge view of plane 1–3 intersects an element in the auxiliary view, the piercing points are projected to the same element in the front and top views, Step 2. An extra cutting plane is passed through point 3 in the auxiliary view to locate an element that is projected to the front and top views. Piercing point 3 is projected to this element in sequence from the auxiliary view to the top view.

This same procedure is used to find the piercing points of the other two planes of the prism in Step 3. All projections of points of intersection originate in the auxiliary view, where the planes of the prism appear as edges.

A similar problem is solved in Fig. 31.23 where a cylinder intersects a cone. The circular view of the cylinder is found in a primary auxiliary view. Points 2 and 2′ are found in the auxiliary view on element O–X where a radial cutting

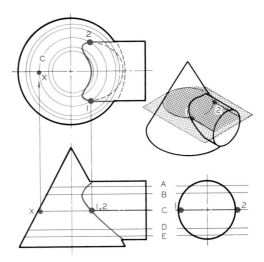

Fig. 31.24 Horizontal cutting planes are used to find the intersection between the cone and the cylinder. The cutting planes form circles in the top view.

plane is passed through apex O. Element O–X is found in the top and front views by projecting from the auxiliary view to locate points 2 and 2′.

In Fig. 31.24, horizontal cutting planes are passed through the cone and the intersecting perpendicular cylinder to locate the line of intersection. A series of circular sections are found in the top view. Points 1 and 2 are found on cutting plane C in the top view as examples, and are projected to the front view. Other points are found in this same manner.

This method is feasible only when the centerline of the cylinder is perpendicular to the axis of the cone, so that circular sections can be found in the top view, rather than elliptical sections that would be difficult to draw.

The distributor housing in Fig. 31.25 is an example of an intersection between cylinders and a cone.

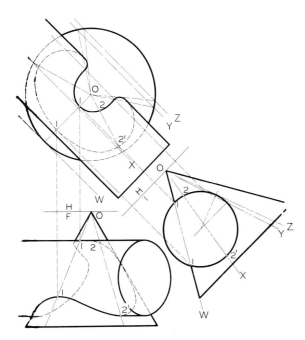

Fig. 31.23 The intersection between these two cylinders is found by projecting an auxiliary view from the top view of the cone to find the circular view of the cylinder. Radial cutting planes are passed through the cone and the cylinder in the auxiliary view to locate piercing points of the cylinder's elements.

31.9 INTERSECTIONS BETWEEN PYRAMIDS AND PRISMS

The intersection between an inclined prism and a pyramid is solved in Fig. 31.26. The end view of the inclined prism is found in the primary auxiliary view and the pyramid is shown in this view also (Step 1). Radial lines OB and OA are passed through corners 1 and 3 in the auxiliary view

Fig. 31.25 This electrically operated distributor illustrates intersections between a cone and a series of cylinders. (Courtesy of GATX.)

(Step 2). The radial lines are projected from the auxiliary view to the front and top views. Intersecting points 1 and 3 are located on OB and OA in each of these views by projection. Point 2 is the point where line 1–3 bends around corner OC. Lines of intersection 1–4 and 4–3 are found in Step 3; the visibility is determined; and the solution is completed.

A prism that is parallel to the base of a pyramid is shown in Fig. 31.27. Its lines of intersection are found by using a series of horizontal cutting planes that pass through the pyramid parallel to its base to form triangular sections in the top view.

The same cutting planes are passed through the corner lines of the prism in the front and auxiliary views. Each corner edge is extended in the top view to intersect the triangular section formed by the cutting plane passing through it. Point 1 is given as an example.

Corner point X is found by passing cutting plane B through it in the auxiliary view where it crosses the corner line. This is where the line of intersection of this plane bends around the corner. The radial cutting plane method could have been used as an alternative method to solve this problem.

Fig. 31.26 Intersection between a prism and a pyramid

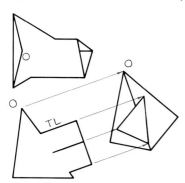

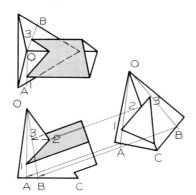

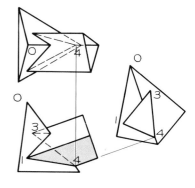

Step 1 Find the edge view of the surfaces of the prism by projecting an auxiliary view from the front view. Project the pyramid into this view also. Only the visible surfaces need be shown in this view.

Step 2 Pass planes A and B through apex O and points 1 and 3 in the auxiliary view. Project the intersections of the planes OA and OB to the front and top views. Project points 1 and 3 to OA and OB in the principal views. Point 2 lies on line OC. Connect points 1, 2, and 3 to give one line of intersection.

Step 3 Point 4 lies on line OC in the auxiliary view. Project this point to the principal views. Connect points 1, 4, and 3 to complete the intersection. Visibility is indicated. Note that these geometric shapes are assumed to be hollow as though constructed of sheet metal.

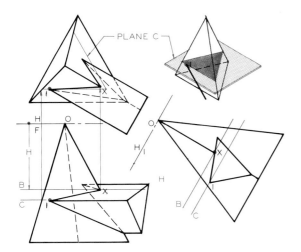

Fig. 31.27 The intersection between this pyramid and prism is found by finding the end view of the prism in an auxiliary view. Horizontal cutting planes are passed through the fold lines of the prism to find the piercing points and the line of intersection.

31.10 INTERSECTIONS BETWEEN SPHERES AND PLANES

The line of intersection between a sphere and a plane is found in Fig. 31.28 where the plane appears as an edge in the front view. Horizontal cutting planes are passed through the front view to form circular sections in the top view. Two points are located on each cutting plane in the top view by projecting from the front view where the cutting plane crosses the edge view of the intersecting plane. The points are connected and the visibility is shown in the top view.

The elliptical intersection could have been drawn with an ellipse template that was selected by measuring the angle between the edge view of the plane in the front view and the projectors from the top view. The major diameter of the ellipse would be equal to the true diameter of the sphere, since the plane passes through the center of the sphere.

Fig. 31.28 Intersection of a sphere and a plane

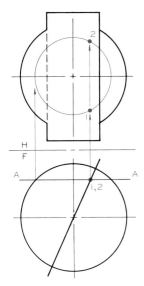

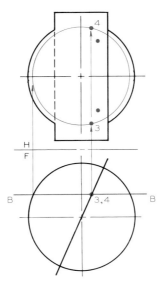

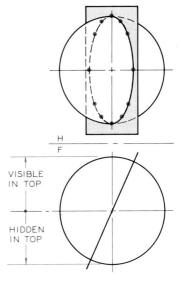

Step 1 Horizontal cutting plane A–A is passed through the front view of the sphere to establish a circular section in the top view. Piercing points 1 and 2 are projected from the front view to the top view, where they lie on the circular section.

Step 2 Horizontal cutting plane B–B is used to locate piercing points 3 and 4 in the top view by projecting to the circular section cut by the plane in the top view. Additional horizontal planes are used to find sufficient points in this manner.

Step 3 Visibility of the top view is found by inspection of the front view. The upper portion of the sphere will be visible in the top view and the lower portion will be hidden.

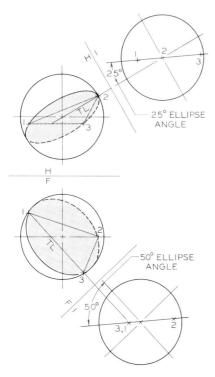

Fig. 31.29 Three points are given on the surface of the sphere through which a circle passes. This plane is found as an edge by projecting from the top and front views. Ellipse angles of 25° and 50° are found for drawing the top and front views of the elliptical intersections.

A general case of the intersection between a plane and a sphere is given in Fig. 31.29 where three points, 1, 2, and 3, are located on the sphere's surface. A circle is to be drawn through these three points and lie on the surface of the sphere.

The edge view of the plane is found by projecting from the top view to a primary auxiliary view. The circle on the sphere through 1, 2, and 3 will have an elliptical line of intersection in the top and front views. The ellipse template angle for the top view is the angle between the edge of the plane and the projector from the top view, 25°. The major diameter is drawn parallel to the true-length lines on plane 1–2–3 in the top view.

The ellipse for the front view of the intersection is found in the same manner by finding the edge view of the plane in an auxiliary view projected from the front view. The ellipse template angle is found to be 50°.

31.11 INTERSECTIONS BETWEEN SPHERES AND PRISMS

The intersection between a sphere and a prism is found in Fig. 31.30 by drawing a series of vertical cutting planes in the top and side views. The planes form circular sections in the front view.

The intersections of the edges with the cutting planes in the side view are projected to their respective circles in the front view. Points 1 and 2 are located on cutting plane A in the side view and on the circular path of A in the front view. Point X in the side view locates the point where the visibility changes in the front view.

Point Y in the side view is the point where the visibility of the intersection changes in the top view. Both of these points lie on the centerlines of the sphere on the side view.

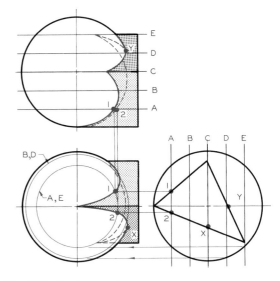

Fig. 31.30 Vertical cutting planes are used to find the intersection between the prism and sphere.

31.12 MISCELLANEOUS INTERSECTIONS

A series of intersections of cutting planes passed through different figures is shown in Fig. 31.31. Radial lines are used to find a hyperbolic section on a cone at A. Horizontal cutting planes are used to locate a section in the top view of a torus (a donut-shaped object) at B.

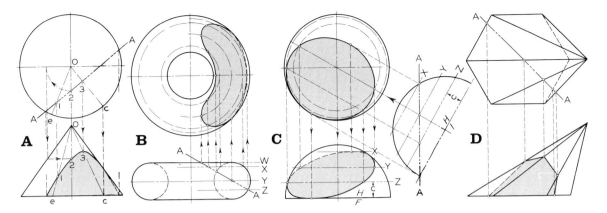

Fig. 31.31 Examples of cutting-plane intersections through four types of figures are shown here. (A) Radial lines are used to find a hyperbolic section through a cone. (B) An inclined plane is shown intersecting a torus. (C) The section cut by a plane through a hemisphere is shown. (D) A section cut by a vertical plane through a pyramid.

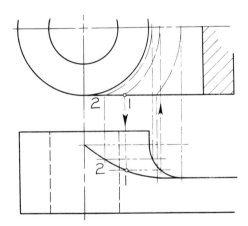

Fig. 31.32 Horizontal cutting planes are used to find the runout (intersection) where surfaces intersect the cylindrical ends of these parts.

Horizontal cutting planes are drawn to locate an elliptical section through a hemisphere at C with a supplementary auxiliary view. The section of a cutting plane through an oblique pyramid is shown at D.

The construction of runouts formed by fillets that intersect cylinders is shown in Fig. 31.32. Horizontal cutting planes are passed through the fillets in the front view and are projected to the top view to locate arcs formed by the cutting planes.

Points along the runout in the front view are found by projecting from the top view. Points 1 and 2 are shown as examples.

31.13 PRINCIPLES OF DEVELOPMENTS

The processing plant shown in Fig. 31.33 illustrates numerous examples of sheet metal shapes that were designed using the principles of developments. In other words, their patterns were laid out on flat stock and then formed to the proper shape by bending and seaming the joints.

Fig. 31.33 Most of the surfaces shown in this refinery were made from flat stock that was fabricated to form these irregular shapes. These flat patterns are called developments.

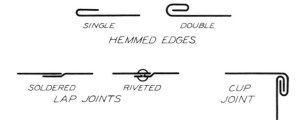

SINGLE   DOUBLE

HEMMED EDGES

SOLDERED   RIVETED   CUP
LAP JOINTS   JOINT

Fig. 31.34 Examples of the types of seams that are used to join developments.

Examples of standard hemmed edges and joints are shown in Fig. 31.34. The application of the sheet metal design will determine the best method of connecting the seams.

The developments of the surfaces of three typical shapes into a flat pattern are shown in Fig. 31.35. The sides of a box are imagined to be unfolded into a common plane. The cylinder is rolled out for a distance that is equal to its circum-

ference. The pattern of a right cone is developed using the length of the element as the radius.

Patterns of shapes with parallel elements such as the prisms and cylinders shown in Fig. 31.36(a) and (b) are begun by constructing stretch-out lines that are parallel to the edge view of the right section of the parts. The distance around the right section is laid off along the stretch-out line. The prism and cylinder at (c) and (d) are inclined; consequently the right sections must be drawn perpendicular to their sides, not parallel to their bases. The distances around the right sections are laid out along these stretch-out lines.

An inside pattern (development) is more desirable than an outside pattern because most bending machines are designed to fold metal so that markings are folded inward, and because markings and scribings will be hidden when the pattern is assembled in final form. The method of denoting

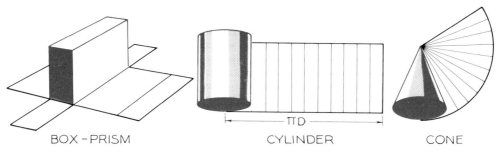

BOX – PRISM   CYLINDER   CONE

Fig. 31.35 Three standard types of developments: the box, cylinder, and cone.

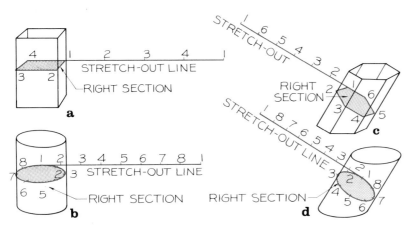

Fig. 31.36 The developments of right prisms and cylinders are found by rolling out the right sections along a stretch-out line (parts A and B). When these figures are oblique, the right sections are found to be perpendicular to the sides of the prism and cylinder. The development is laid out along the stretch-out line that is parallel to the edge view of the right section (parts C and D).

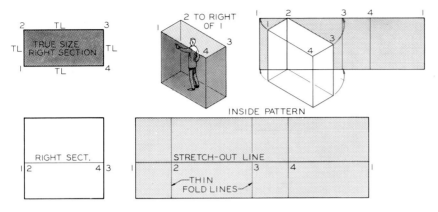

Fig. 31.37 The development of a rectangular prism to give an inside pattern. The stretch-out line is parallel to the edge view of the right section.

a pattern is by a series of lettered or numbered points about its layout. All lines on a development must be true length.

It is economical if seam lines (lines where the pattern is joined) are selected as the shortest lines. This results in the least expense of riveting or welding the pattern together to form the final shape.

31.14 DEVELOPMENT OF PRISMS

A flat pattern for a prism is developed in Fig. 31.37. Since the edges of the prism are vertical in the front view, its right section is perpendicular to these sides. The top view shows the right section true size. The stretch-out line is drawn parallel to the edge of the right section, beginning with point 1.

If an inside pattern is drawn and it is to be laid out to the right, point 2 will be to the right of point 1. This is determined by looking from the inside of the top view where 2 is seen to the right of 1.

Lines 2–3, 3–4, and 4–1 are transferred from the right section in the top view with your dividers to the stretch-out line to locate the fold lines on the pattern. The length of each fold line is found by projecting its true length from the front view. The ends of the fold lines are connected to form the limits of the developed surface. Fold lines are drawn as thin lines and the outside lines of a development are drawn as visible object lines.

Fig. 31.38 Numerous examples of developments fabricated from sheet metal are shown in this plant. (Courtesy of Heat Engineering.)

The complex installation in Fig. 31.38 is composed of many developments ranging from simple prisms to more complicated shapes.

The development of the prism in Fig. 31.39 is similar to the example in Fig. 31.37 except that one end is beveled rather than square. The stretch-out line is drawn parallel to the edge view of the right section in the front view. The true-length distances around the right section are laid off

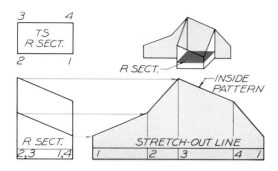

Fig. 31.39 The development of a rectangular prism with a beveled end to give an inside pattern. The stretch-out line is parallel to the right section.

along the stretch-out line and the fold lines are located. The lengths of the fold lines are found by projecting from the front view of these lines.

31.15 DEVELOPMENT OF OBLIQUE PRISMS

The prism in Fig. 31.40 is inclined to the horizontal plane, but its fold lines are true length in the front view. The right section is drawn as an edge perpendicular to these fold lines and the stretch-out line is drawn as shown in Step 1. A true-size view of the right section is constructed in the auxiliary view.

In Step 2, the distances between the fold lines are transferred from the true-size right section to the stretch-out line. The lengths of the fold lines are found by projecting from the front view.

In Step 3, the ends of the prism are found and are attached to the pattern so that they can be folded into position. All lines that are laid out in a flat pattern must be true length.

Fig. 31.40 Development of an oblique prism

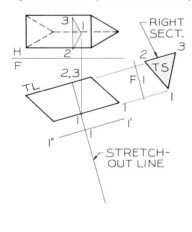

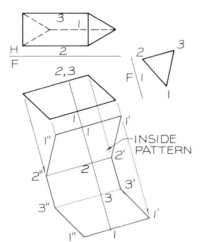

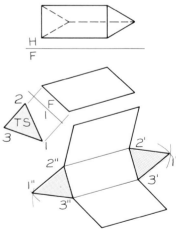

Step 1 The edge view of the right section will appear perpendicular to the true-length axis of the prism in the front view. Determine the true-size view of the right section by constructing an auxiliary view. Draw the stretch-out line parallel to the edge view of the right section. Project bend line 1'–1" as the first line of the development.

Step 2 Since the pattern is developed toward the right, beginning with line 1'–1", the next point is found to be line 2'–2" by referring to the auxiliary view. Transfer true-length lines 1–2, 2–3, and 3–1 from the right section to the stretch-out line to locate the elements. Determine the lengths of the bend lines by projection.

Step 3 Find the true-size views of the end pieces by projecting auxiliary views from the front view. Connect these surfaces to the development of the lateral sides to form the completed pattern. Fold lines are drawn with thin lines, while outside lines are drawn as regular object lines.

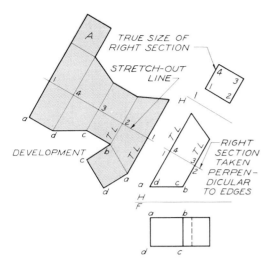

Fig. 31.41 The development of this oblique chute is found by finding the right section true-size in the auxiliary view. The stretch-out line is drawn parallel to the right section.

A similar example is solved in Fig. 31.41. The fold lines are true length in the top view; this enables you to draw the edge view of the right section perpendicular to the fold lines in the top view. The stretch-out line is drawn parallel to the edge view of the section, and the true size of the right section is found in an auxiliary view projected from the top view. The distances about the

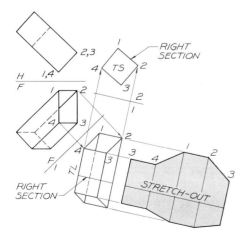

Fig. 31.42 The development of an oblique cylinder is found by finding an auxiliary view in which the fold lines are true length, and a secondary auxiliary view in which the right section appears true size. The stretch-out line is drawn parallel to the right section.

right section are transferred to the stretch-out line to locate the fold lines. The lengths of the fold lines are found by projecting from the top view. The end portions of the pattern are attached to the pattern to complete the construction.

A prism that does not project true length in either view can be developed as illustrated in Fig. 31.42. The fold lines are found true length in an auxiliary view projected from the front view. The right section will appear as an edge perpendicular to the fold lines in the auxiliary view. The true size of the right section is found in a secondary auxiliary view.

The stretch-out line is drawn parallel to the edge view on the right section. The fold lines are located on the stretch-out line by measuring around the right section in the secondary auxiliary view. The lengths of the fold lines are then projected to the development from the primary auxiliary view.

31.16 DEVELOPMENT OF CYLINDERS

The development of a cylinder is found in Fig. 31.43. Since the elements of the cylinder are true length in the front view, the right section will appear as an edge in this view, and true size in the top view. The stretch-out line is drawn parallel to the edge view of the right section, and point 1 is chosen as the beginning point since it is the shortest element. To draw an inside pattern, assume that you are standing on the inside looking at point 1 and you will see that point 2 is to the right of 1. Therefore, the pattern is laid out with point

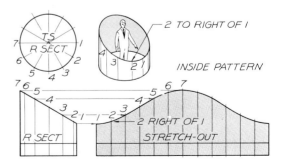

Fig. 31.43 The development of a right cylinder's inside pattern. The stretch-out line is parallel to the right section. Point 2 is to the right of point 1 for an inside pattern.

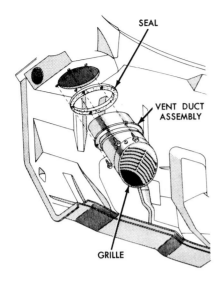

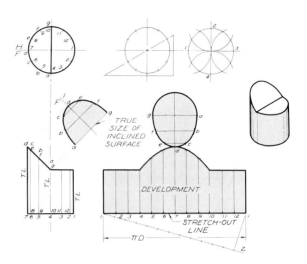

Fig. 31.44 This ventilator air duct was designed through the use of development principles. (Courtesy of Ford Motor Company.)

Fig. 31.45 The development of a right cylinder can be found by mathematically locating elements along the stretch-out line, which is equal in length to the circumference of the right section.

Fig. 31.46 Development of an oblique cylinder

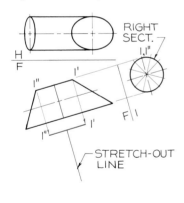

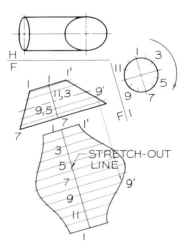

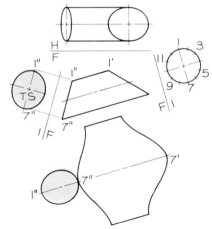

Step 1 The right section appears as an edge in the front view, in which it is perpendicular to the true-length axis. Construct an auxiliary view to determine the true size of the right section. Draw a stretch-out line parallel to the edge view of the right section. Locate element 1′–1″.

Step 2 Divide the true-size right section into equal divisions to locate point views of elements on the cylinder. Project these elements to the front view. Transfer measurements between the elements in the auxiliary view to the stretch-out line to locate the elements. Determine the lengths of the elements by projection to complete the development.

Step 3 The development of the end pieces will require auxiliary views that project these surfaces as ellipses, as shown for the left end. Attach this true-size ellipse to the pattern at a point. Note that the line of departure for the pattern was made along line 1″–1′, the shortest element, for economy.

2 to the right of point 1. This establishes the sequence of locating the elements.

The spacing between the elements in the top view can be conveniently done by drawing radial lines at 15° or 30° intervals. Using this technique, the elements will be equally spaced, making it convenient to lay them out along the stretch-out line. The lengths of the elements are found by projecting from the front view to complete the pattern.

An application of a developed cylinder with a beveled end is the air-conditioning duct from an automobile shown in Fig. 31.44. The development of a similar cylinder is shown in Fig. 31.45. The base of the front view is the edge view of the right section, which appears true size in the top view. Elements are located around the circumference of the right section in the top view. Two alternative methods are shown to illustrate how the elements are located at 30° intervals. One employs the 30°–60° triangle, and the other uses a compass with the radius equal to the radius of the right section.

The stretch-out line is drawn parallel to the right section in the front view. The total length is found mathematically by the formula $C = \pi D$. The stretch-out line is divided into the same number of divisions as there are elements, 12 in this case. This provides a high degree of accuracy in finding the circumference.

The pattern for the end of the cylinder is found by combining the partial top view and the auxiliary view. This is connected to the overall pattern to complete the solution.

31.17 DEVELOPMENT OF OBLIQUE CYLINDERS

The pattern for an oblique cylinder (Fig. 31.46) is found in the same manner as the previous examples, but with the addition of one preceding step: The right section must be found true-size in an auxiliary view. A series of equally spaced elements is located around the right section in the auxiliary view and is projected back to the true-length view, Step 1. The stretch-out line is drawn parallel to the edge of the right section in the front view.

In Step 2, the spacing between the elements is laid out along the stretch-out line, and the elements are drawn through these points perpendicular to the stretch-out line. The lengths of the elements are found by projecting from the front view.

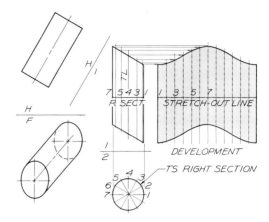

Fig. 31.47 The development of an oblique cylinder is found by constructing an auxiliary view in which the elements appear true length. The right section is found true size in a secondary auxiliary view.

The ends of the cylinder are found in Step 3 to complete the pattern. Only one end pattern is shown as an example.

The oblique cylinder in Fig. 31.47 is a more general case where the elements are not true length in the given views. A primary auxiliary view is used to find a view where the elements are true length, and a secondary auxiliary view is drawn to find the true-size view of the right section. The stretch-out line is drawn parallel to the edge view of the right section in the primary auxiliary view and the elements are located along this line by transferring their distances apart from the true-size right section.

The elements are drawn perpendicular to the stretch-out line. The length of each element is found by projecting from the primary auxiliary view. The endpoints are connected with a smooth curve to complete the pattern.

31.18 DEVELOPMENT OF PYRAMIDS

All lines used to draw a pattern must be true length. Pyramids have only a few lines that are true length in the given views; for this reason, the sloping corner lines must be found true-length at the outset.

The corner lines of a pyramid can be found by revolution, as shown in Fig. 31.48. Line 0–5 is revolved in the frontal position of 0–5′ in the top

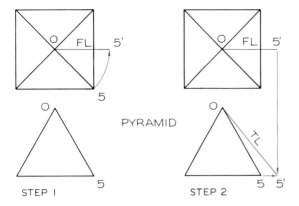

Fig. 31.48 True length by revolution

Step 1 Corner 0–5 of a pyramid is found true length by revolving it into the frontal plane in the top view, 0–5′.

Step 2 Point 5′ is projected to the front view where 0–5′ is true length.

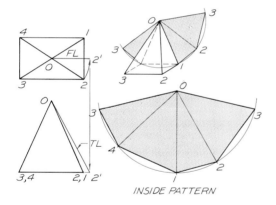

Fig. 31.49 Development of a right pyramid.

view, Step 1. Since 0–5′ is a frontal line, it will be true-length in the front view, Step 2.

The development of a pyramid is given in Fig. 31.49. Lines 0–1 and 0–2 are revolved into the frontal plane in the top view to find their true lengths in the front view. All bend lines are equal in length since this is a right pyramid. Line 0–1 is used as a radius to construct the base circle for drawing the development. Distance 1–2 is transferred from the base in the top view to the development, where it forms a chord on the base circle. Lines 2–3, 3–4, and 4–1 are found in the same manner and in sequence. The bend lines are drawn as thin lines from the base to the apex, point O.

A variation of this problem is given in Fig. 31.50, in which the pyramid has been truncated or cut at an angle to its axis. The development of the inside pattern is found in the same manner as the previous example; however, an additional step is required to establish the upper lines of the development. The true-length lines from the apex to points 1′, 2′, 3′, and 4′ are found by revolution. These distances are laid on their respective lines of the pattern to locate the upper limits of the pattern.

The mounting pads in Fig. 31.51 are sections of pyramids that intersect an engine body.

The development of an oblique pyramid is shown in Fig. 31.52. In Step 1, the corner lines are

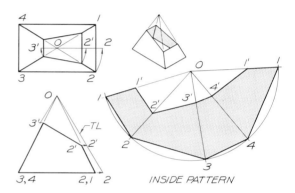

Fig. 31.50 The development of an inside pattern of a truncated pyramid. The corner lines are found true length by revolution.

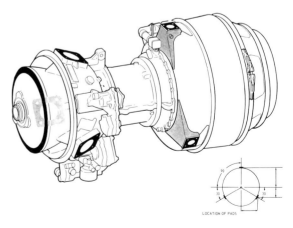

Fig. 31.51 Examples of pyramid shapes in the design of mounting pads for an engine. (Courtesy of Avco Lycoming.)

Fig. 31.52 Development of an oblique pyramid.

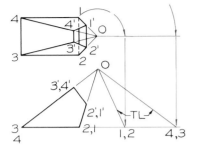
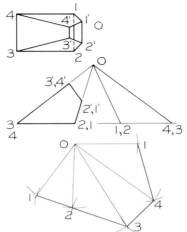
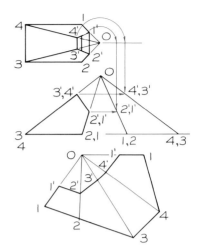

Step 1 Revolve each of the bend lines in the top view until they are parallel to the frontal plane. Project to the front view where the true-length views of the revolved lines can be found. Let point 0 remain stationary but project points 1, 2, 3, and 4 horizontally in the front view to the projectors from the top view.

Step 2 The base lines appear true length in the top view. Using these true-length lines from the top view and the revolved lines in the front view, draw the development triangles. All triangles have one side and point 0 in common. This gives a development of the surface, excluding the truncated section.

Step 3 The true lengths of the lines from point 0 to 1', 2', 3', and 4' are found by revolving these lines. These distances are laid off from point 0 along their respective lines to establish points along the upper edge of the pattern. The points are then sequentially connected by straight lines.

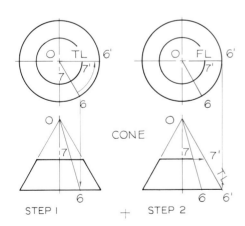

Fig 31.53 True length by revolution

Step 1 An element of a cone, 0–6, is revolved into a frontal plane in the top view.

Step 2 Point 6' is projected to the front view where it is the outside element of the cone and is true length. Line 0–7' is true length and is found by projecting to the outside element in the front view.

found true length by revolution. Using these true-length lines and those that are given in the principal views, the pattern is drawn by triangulation using a compass, Step 2. In Step 3, the upper limits of the pattern are found by measuring along the fold lines from point O.

31.19 DEVELOPMENT OF CONES

All elements of a right cone are equal in length as illustrated in Fig. 31.53 where O–6 is found true length by revolution. When revolved to O–6' position, it is a frontal line and is therefore true length in the front view where it is an outside element of the cone. Point 7' is found by projecting horizontally to element O–6'.

The right cone in Fig. 31.54 is developed by dividing the base into equally spaced elements in the top view and by projecting these to the base in the front view. These elements radiate to the apex at O. The outside elements in the front view, O–10 and O–4, are true length.

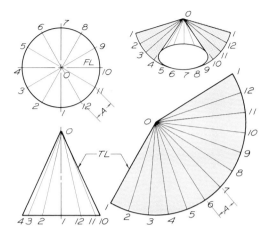

Fig. 31.54 The development of an inside pattern of a right cone. In this case, all elements are equal in length.

Using element O–10 as a radius, draw the base arc of the development. The elements are located along the base arc equal to the chordal distances between the points around the base in the top view. You can see that this is an inside pattern by inspecting the top view where point 2 is to the right of point 1 when viewed from the inside. Always label the points of a pattern in order to designate whether it is an inside or outside pattern.

The sheet metal conical vessel in Fig. 31.55 is an example of a large vessel that was designed using principles of a development.

The development of a truncated cone is shown in Fig. 31.56. The pattern is found by laying out the total cone ignoring the portions removed from it. This is found by following the steps of the previous example.

The hyperbolic sections through the front view of the cone can be found on their respective elements in the top and front views. Lines O–2' and O–3' are projected horizontally to the true length element O–1 in the front view, where they will appear true length. These distances are measured off along their respective elements in the development to establish a smooth curve.

31.20 DEVELOPMENT OF OBLIQUE CONES

The development of an oblique cone is found in Fig. 31.57. The elements of this cone are of varying lengths, but the pattern will be symmetrical about an axis.

Elements are located in the top view by dividing the base into equal arcs and drawing the elements to point O, the apex. These elements are

Fig. 31.55 An example of a conical shape that was formed from steel panels by applying principles of developments.

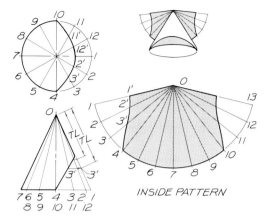

Fig. 31.56 The development of a conical surface with a side opening.

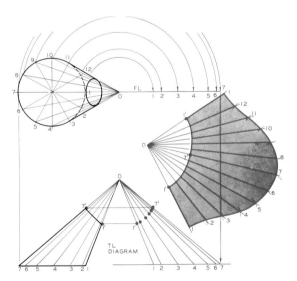

Fig. 31.57 The development of an oblique cone. All elements must be found true length by a true-length diagram before the pattern can be constructed.

projected to the frontal view. Each element is found true length by revolving it into a frontal plane in the top view. When projected to the front view, the revolved lines are found true length in a true-length diagram.

The development is begun by constructing a series of triangles using the true-length elements and the chordal distances found on the base in the top view. Line O–1 is chosen for the line of separation since it is the shortest element. The base of the pattern is drawn as a smooth curve.

The distances from point *O* to the upper cut through the cone are found by projecting to their respective elements in the true-length diagram from the front view. Line *O*–7′ is shown as an example. These shorter elements are located on their respective elements in the pattern to give the upper limits of the pattern. This will be an irregular curve rather than an arc because this is not a right cone.

31.21 DEVELOPMENT OF WARPED SURFACES

The geometric shape in Fig. 31.58 is an approximate cone with a warped surface; it is similar to the oblique cone in Fig. 31.57. The development of this surface will be an approximation, since a warped surface cannot be laid out on a flat surface.

The surface is divided into a series of triangles in the top and front views by dividing the upper and lower views into equal sectors. The true lengths of all lines are found in the true-length diagrams drawn on both sides of the front and by projecting horizontally from the ends of the lines. If necessary, review Fig. 31.57 to see how the true-length diagrams were found.

The chordal distances between the points on the base appear true-length in the top view since the base is horizontal. However, an auxiliary view

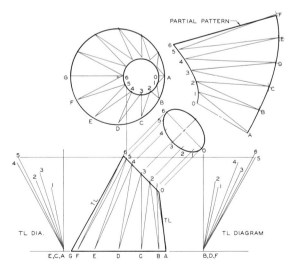

Fig. 31.58 The development of a partial pattern of a warped surface.

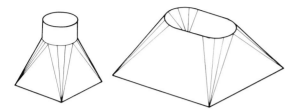

Fig. 31.59 Examples of transition pieces that join together parts that have different cross section.

is necessary to find the true distance about the upper surface of the object.

The developed surface is found by triangulation using true-length lines from (1) the true-length diagram, (2) the horizontal base in the top view, and (3) the primary auxiliary view. Each point should be carefully labeled as it is laid out to avoid confusion.

31.22 DEVELOPMENT OF TRANSITION PIECES

A transition piece is a form that transforms the section at one end to a different shape at the other (Fig. 31.59). Huge transition pieces can be seen in the industrial installation in Fig. 31.60.

The development of a transition piece is shown in Fig. 31.61. In Step 1, radial elements are

Fig. 31.60 Transition-piece developments are used to join a circular shape with a rectangular section. (Courtesy of Western Precipitation Group, Joy Manufacturing Company.)

drawn from each corner to the equally spaced points on the circular end of the piece. Each of these elements is found true length by revolution.

In Step 2, the true-length lines are used with the true-length chordal distance in the top view to lay out a series of adjacent triangles to form the pattern beginning with element A–2.

The triangles A–1–2 and G–3–4 are added at each end of the pattern to complete the development of the half pattern.

A similar transition piece is developed in Fig. 31.62 using the same techniques of construction. Elements are established at the corners of the given views and are found true length in the true-length diagrams by revolution. By triangulation, using the true-length lines, the full pattern is drawn to complete the development.

31.23 DEVELOPMENT OF SPHERES—ZONE METHOD

In Fig. 31.63, a development of a sphere is found by the zone method. A series of parallels, called latitudes in cartography, are drawn in the front view. Each is spaced an equal distance, D, apart along the surface in the front view. Distance D can be found mathematically to improve the accuracy of this step.

Cones are passed through the sphere's surface so that they pass through two parallels at the outer surface of the sphere. The largest cone with element R_1 is found by extending it through where the equator and the next parallel intersect on the sphere's surface in the front view until R_1 intersects the extended centerline of the sphere. Elements R_2, R_3, and R_4 are found by repeating this process.

The development is begun by laying out the largest zone, using R_1 as the radius, on the arc that represents the base of an imaginary cone. The breadth of the zone is found by laying off distance D from the front view to the development and drawing the upper portion of the zone with a radius equal to R_1–D, using the same center. No regard is given to finding the arc lengths at this point.

The next zone is drawn using the radius R_2 with its center located on a line through the center of arc R_1. The center of R_2 is positioned along this line such that the arc to be drawn will be tangent to the preceding arc, which was drawn with ra-

Fig. 31.61 Development of a transition piece

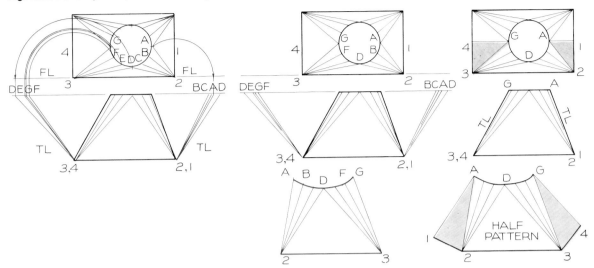

Step 1 Divide the circular edge of the surface into equal parts in the top view. Connect these points with bend lines to the corner points, 2 and 3. Find the true length of these lines by revolving them into a frontal plane and projecting them to the front view. These lines represent elements on the surface of an oblique cone.

Step 2 Using the TL lines found in the TL diagram and the lines on the circular edge in the top view, draw a series of triangles, which are joined together at common sides, to form the development. *Example:* Arcs 2D and 2C are drawn from point 2. Point C is found by drawing arc DC from point D to find point C. Line DC is TL in the top view.

Step 3 Construct the remaining planes, A–1–2 and G–3–4, by triangulation to complete the inside half-pattern of the transition piece. Draw the fold lines as thin lines at the places where the surface is to be bent slightly. The line of departure for the pattern is chosen along A–1, the shortest possible line, for economy.

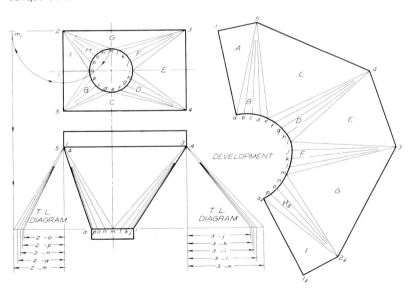

Fig. 31.62 The development of this transition piece is found by finding the conical elements at the corner's true-length in a true-length diagram. Line 2—m is found true-length as an example. The pattern is constructed by a series of triangulations.

dius R_1–D. The upper arc of this second zone is drawn with a radius of R_2–D. The remaining zones are constructed successively in this manner. The last zone will appear as a circle with R_4 as its radius.

The lengths of the arcs can be established by dividing the top view with vertical cutting planes that radiate through the poles. These lines, which lie on the surface of the sphere, are called longitudes in cartography. Arc distances S_1, S_2, S_3, and S_4 are found on each parallel in the top view.

These distances are measured off on the constructed arcs in the development. In this case there are 12 divisions, but smaller divisions would provide a more accurate measurement. A series of zones found in this manner can be joined to give an approximate development of a sphere.

31.24 DEVELOPMENT OF SPHERES—GORE METHOD

Figure 31.64 illustrates an alternative method of developing a flat pattern for a sphere. This method uses a series of spherical elements called gores. Equally spaced vertical cutting planes are passed through the poles in the top view. Parallels are located in the front view by dividing the surface into equal zones of dimension D.

A true-size view of one of the gores is projected from the top view. Dimensions can be checked mathematically for all points. A series of these gores is laid out in sequence to complete the pattern.

31.25 DEVELOPMENT OF ELBOWS

An elbow is a cylindrical shape that turns a 90° angle. An elbow that is also a transition piece is shown in Fig. 31.65.

The method of constructing the pattern for an elbow is illustrated in Fig. 31.66. In Step 1, the 90° arc is drawn and the cylinder is drawn to size about its centerline. Divide this arc into one divi-

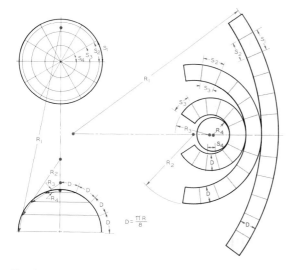

Fig. 31.63 The zone method of finding the inside development of a spherical pattern.

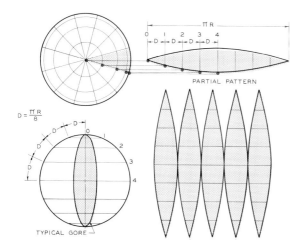

Fig. 31.64 The gore method of developing an inside pattern of a sphere.

Fig. 31.65 The radial bend of the cylinder was developed using the technique covered in Fig. 31.66.

Fig. 31.66 Construction of an elbow

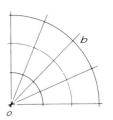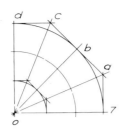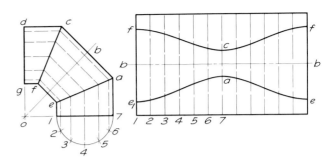

Step 1 Divide the arc into one division less than the number of pieces desired in the elbow (two in this case) along *Ob*. Bisect the two arcs.

Step 2 Draw a tangent line through point *b* to intersect tangents *dc* and *a-7*.

Step 3 Draw a semicircle using *1–7* as the diameter. This is half of the right section that will be used to space the elements along the stretch-out line.

Step 4 The three patterns can be cut from a rectangular piece of material whose height is equal to the sum of *a–7*, *e–f a*, and *d cd*. The curves for each cut are found by transferring the lengths of the elements to parallel element lines in the pattern.

sion less than the number of pieces that will be used in the elbow. Since three pieces are to be used in this example, the arc is divided into two along line *O–b*. The two arcs are then bisected.

In Step 2, perpendiculars are drawn tangent to the arc at the division lines and *O–b*. In Step 3, a half-circular view of the cylinder is drawn for locating equally spaced elements that are projected back to the given view parallel to the elements of the three pieces of the elbow. The elements will be true-length in this view.

In Step 4, the flat pattern is developed, with the patterns of the two short pieces of the elbow located at the top and bottom of the rectangular piece. The middle segment is located between these two smaller patterns so that there will be no wasted material, and only two cuts will be necessary.

31.26 DEVELOPMENT OF STRAPS

Figure 31.67 illustrates the steps of finding the development of a strap that has been bent to serve as a support bracket, between point *A* on a vertical surface and point *B* on a horizontal surface. In Step 1, the strap is drawn in the side view where the strap appears as an edge using the specified radius to show the bend. The bend is divided into equal arcs and is developed into a flat piece in the vertical plane through point *A*. Point *B'* is located in this view.

In Step 2, the location of the hole at *B'* is found in the front view by projection. The true-size development of the strap is drawn in this view and it is projected back to the side view to complete that view of the strap.

In Step 3, the projected view of the strap in its bent position in the front view can be found by using projectors from the side view and the true pattern of the strap.

31.27 INTERSECTIONS AND DEVELOPMENTS IN COMBINATION

Intersections between parts must be found before developments of each can be completed. An example of this type is shown in Fig. 31.68.

The intersection between the two prisms is found in the given top and front views. The pattern of the vertical prism is found to the left of the front view and the intersections are shown to indicate cuts that must be made in the pattern.

The pattern of the inclined prism cannot be found without the construction of an auxiliary view in which the fold lines appear true length. The true-size right section is found in a secondary auxiliary view. The fold lines of the pattern are laid off along a stretch-out line by transferring the distances around the right section to it.

The resulting two patterns can be cut out and folded to form shapes that will intersect as shown in the top and front views.

Fig. 31.67 Strap development

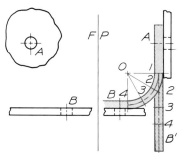

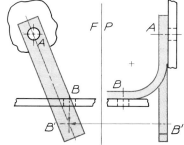

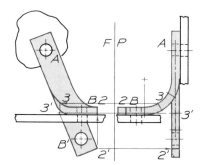

Step 1 Construct the edge view of the strap in the side view using the specified radius of bend. Locate 1, 2, 3, and 4 on the neutral axis at the bend. Revolve this portion of the strap into the vertical plane and measure the distances along this view of the neutral axis. Check the arc distances by mathematics. The hole is located at B' in this view.

Step 2 Construct the front view of B' by revolving point B parallel to the profile plane until it intersects the projector from B' in the side view. Draw the centerline of the true-size strap from A to B' in the front view. Add the outline of the strap around this centerline and the holes at each end, allowing enough material to provide sufficient strength.

Step 3 Determine the projection of the strap in the front view by projecting points from the given views. Points 3 and 2 are shown in the views to illustrate the system of projection used. The ends of the strap are drawn in each view to form true projections.

Fig. 31.68 Two prisms intersect in this example. Their intersections are found and the developments of each are found by using the principles covered in this chapter. The development of the inclined prism is found by using primary and secondary auxiliary views.

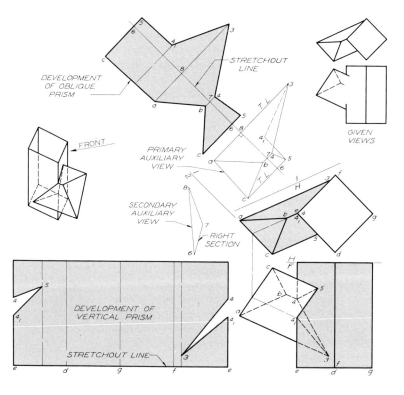

PROBLEMS

These problems are designed to fit on Size A and Size B sheets. Some problems can be grouped two or more on a single sheet. For example, two problems might be solved on a Size A sheet and four on a Size B sheet. The proper layout of a problem should be considered as an important part of its solution.

Intersections

1–5. Lay out two problems on Size A sheets. Use the given scales to transfer the dimensions from the views to the drawing. In Problems 1–3, the labeled points represent points on cutting planes that pass through the objects. Find their lines of intersection and show visibility.

6–7. For Problem 6, lay out the top and front views twice on a Size A sheet, one above the other on the sheet. Using the auxiliary sections assigned, A, B, C, D, or E, find the intersection in the front and top views. Lay out the views of Problem 7 in the same manner and use right sections A through E as assigned to find the intersections.

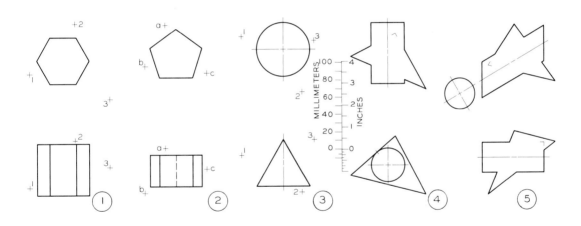

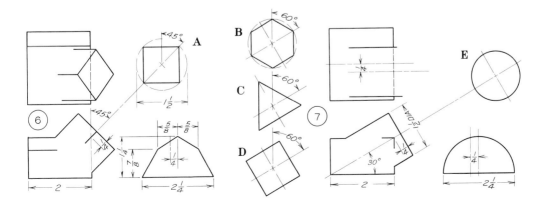

8–20. Lay out one problem per Size A sheet. Find the intersections and show the visibility for each problem.

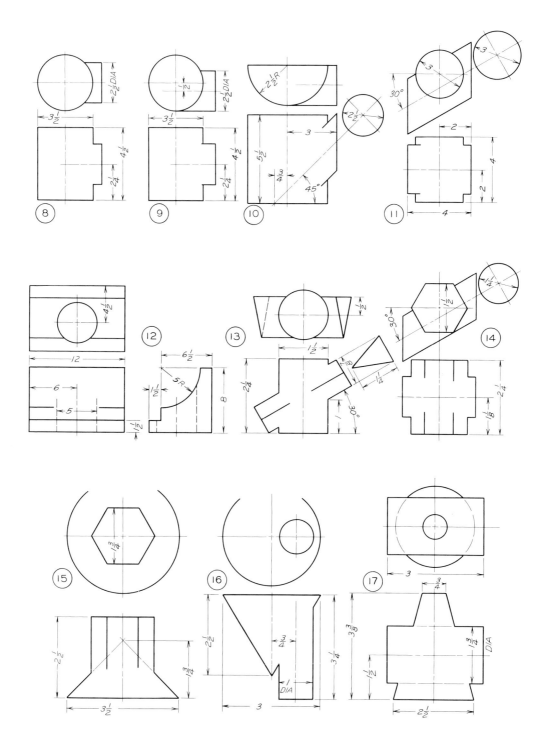

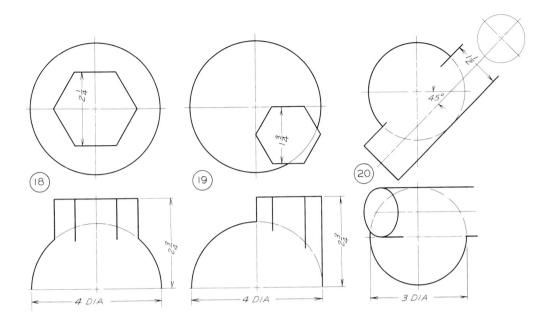

Developments

21–47. Lay out two problems per Size A sheet after dividing the sheet in half by a division line. Find the flat inside patterns of each and label the fold lines in all views.

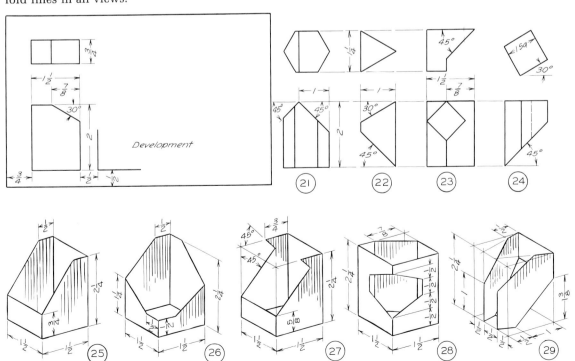

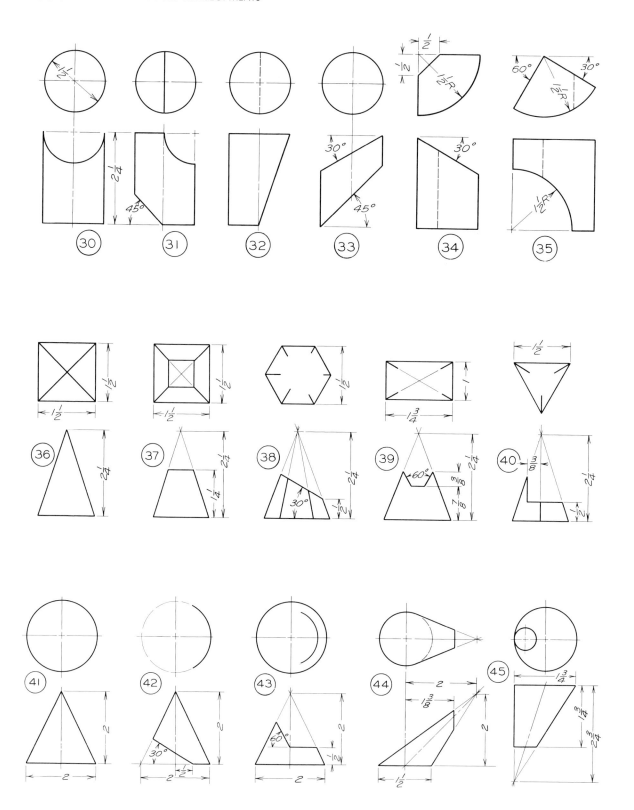

48–51. Lay out two problems per Size A sheet. Find the inside developments of each and remove the ends of the object that have been cut away by the cutting planes. Label the points in all views.

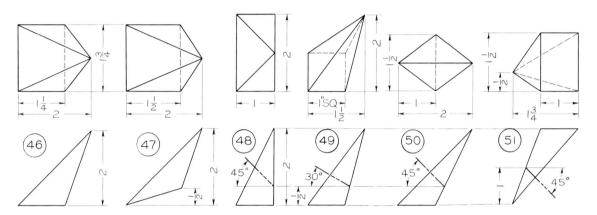

Combination Problems

52–63. Using Problems 6–17, lay them out with one per size B sheet. Find the intersections and inside developments of each. Label the points in all views.

GRAPHS

32.1 INTRODUCTION

Data and information expressed as numbers and words are difficult to analyze or evaluate unless they are transcribed into graphical form. Drawings of data shown graphically are called **graphs.** They are sometimes called **charts,** an acceptable term, but one that is more appropriate when referring to maps, which are specialized forms of graphs.

Graphs are helpful in the communication of data to others (Fig. 32.1); consequently, this is a popular means of briefing other people on trends that would otherwise be difficult to communicate. The trends of a plotted curve on a graph can be compared to the expressions on a person's face,

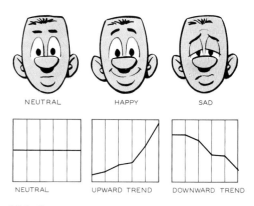

Fig. 32.2 Curves on a graph are similar to expressions on a face.

which is a graph of sorts that reveals a person's feelings (Fig. 32.2). For example, a flat curve shows no change while an upwardly inclined curve indicates a positive increase. A downward curve, on the other hand, represents a downward trend and a negative result.

This chapter will deal with the more commonly used graphs. The basic types are:

1. Pie graphs,
2. Bar graphs,
3. Linear coordinate graphs,
4. Logarithmic coordinate graphs,
5. Semilogarithmic coordinate graphs, and
6. Schematics and diagrams.

Fig. 32.1 Graphs are helpful in presenting technical data to one's associates.

32.2 SIZE PROPORTIONS OF GRAPHS

Graphs may be used to illustrate technical reports that are reproduced in quantity, and they may be used for projection by slide or overhead projectors. In all cases, the proportion of the graph must be determined so that it will match the proportion of the space or the format of the visual aid.

If a graph is to be photographed by a 35 mm camera (Fig. 32.3), the graph must conform to the standard size of the 35 mm film that is used. This proportion is approximately 2 × 3, as shown in Fig. 32.4.

The proportions of the area in which the graph is to be drawn can be enlarged or reduced by using the diagonal-line method, as illustrated in Fig. 32.4. The horizontal dimension of the slide opening is extended to the right and the left edge is extended upward. Any point on either of these extended lines is projected to the diagonal, and then to the other extended line, to give an area of an equal proportion.

Fig. 32.3 When graphs are drawn to be photographed, they must be laid out at a proportion that will match the proportion of the film in the camera.

32.3 PIE GRAPHS

Pie graphs compare the relationship of parts to a whole when there are only a few parts. Figure 32.5 shows the distribution of skilled workers employed in industry; this graph gives a good visual comparison of these groups.

The method of drawing a pie graph is shown in Fig. 32.6. Note that the given data does not give as good an impression of the comparisons as does the pie graph, even though the data is quite simple.

To facilitate lettering within narrow spaces, the thin sectors should be placed as nearly horizontal as possible. This provides more room for the label and the percentage. The actual percent-

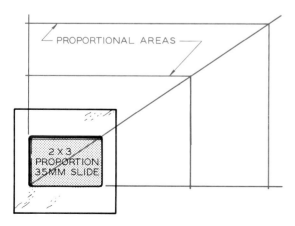

Fig. 32.4 This diagonal-line method can be used for constructing areas whose sides are proportional to those of a 35 mm slide.

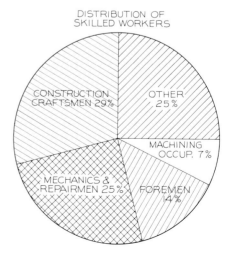

Fig. 32.5 A pie graph shows the relationship of the parts to a whole. It is effective only when there are a few parts.

Fig. 32.6 Drawing pie graphs

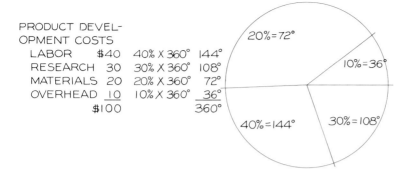

PRODUCT DEVEL-
OPMENT COSTS

LABOR	$40	40% X 360°	144°
RESEARCH	30	30% X 360°	108°
MATERIALS	20	20% X 360°	72°
OVERHEAD	10	10% X 360°	36°
	$100		360°

NEW PRODUCT DEVELOPMENT
COST - PER UNIT

Step 1 The total sum of the parts is found, and the percentage of each is found. Each percentage is multiplied times 360° to find the angle of each sector of the pie graph.

Step 2 The circle is drawn to represent the pie graph. Each sector is drawn using the degrees found in step 1. The smaller sectors should be placed as nearly horizontal as possible.

Step 3 The sectors are labeled with their proper names and percentages. In some cases it might be desirable to include the exact numbers in each sector as well.

age should be given in all cases, and it may be desirable to give the actual numbers or values as well in each sector.

32.4 BAR GRAPHS

Bar graphs are effective to compare values, especially since they are well understood by the general public. For example, the production of timber for various uses is compared in Fig. 32.7. In this

example, the bars not only show the overall production (the total lengths of the bars), but the portions of the total devoted to the three uses of the timber.

A bar graph can be composed of a single bar (Fig. 32.8). The total length of the bar is 100% and the bar is divided into lengths that are proportional to the percentages represented by each of the three parts of the bar.

The method of constructing a bar graph is given in Fig. 32.9. The title of the graph is placed inside the graph where space is available in this case. The title could have been placed under or over the graph.

For bar graphs, the data should be sorted by arranging the bars in ascending or descending order, since it is desirable to know how the data represented by the bars rank from category to category

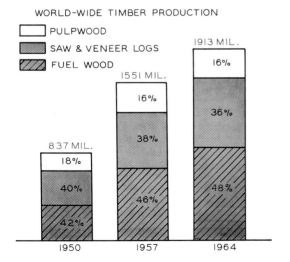

Fig. 32.7 In this example, each bar represents 100% of the total amount, and each bar represents different totals.

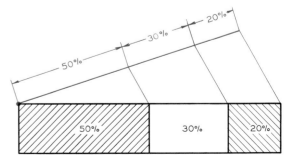

Fig. 32.8 The method of constructing a single bar where the sum of all of the parts will be 100%.

Fig. 32.9 Construction of a bar graph   669

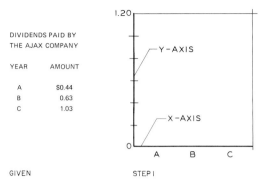

DIVIDENDS PAID BY
THE AJAX COMPANY

YEAR	AMOUNT
A	$0.44
B	0.63
C	1.03

GIVEN   STEP I

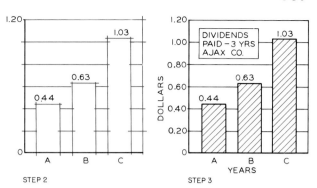

STEP 2   STEP 3

Given These data are to be plotted as a bar graph.

Step 1 Lay off the vertical and horizontal axes so that the data will fit on the grid. Make the bars begin at zero.

Step 2 Construct and label the bars. The width of the bars should be different from the space between them. Horizontal grid lines should not pass through the bars.

Step 3 Strengthen lines, title the graph, label the axes, and crosshatch the bars.

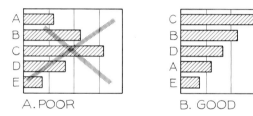

A. POOR   B. GOOD

Fig. 32.10 The bars at A are arranged alphabetically. The resulting graph is not as easy to evaluate as the one at B, where the bars have been sorted and arranged in descending order.

(Fig. 32.10). An arbitrary arrangement of bars, alphabetically or numerically, results in a graph that is more difficult to evaluate than the descending arrangement at B.

If the data are sequential and involve time, such as sales per month, it would be less effective to rank the data in ascending order because it is more important to see variations in the data as related to periods of time. The determination of the method of arranging the bars depends upon the purpose of the graph and will be left to the judgment of the person constructing the graph.

Bars in a bar graph may be horizontal, as shown in Fig. 32.11, or vertical as shown in Fig. 32.12. In both of these cases, the bars are arranged in order of length of the bars for ease of comparison and ranking of the data. It is desirable for the bars of a graph to begin at zero to show a true comparison in the data.

32.5 LINEAR COORDINATE GRAPHS

A typical coordinate graph is given in Fig. 32.13 with the accompanying notes that explain its important features. The divisions along an axis of the graph should be equal; in other words, the scale is linear. The other axis is also divided into equal units, and therefore, it is a linear scale also.

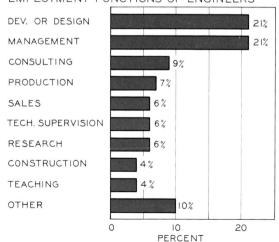

EMPLOYMENT FUNCTIONS OF ENGINEERS

Fig. 32.11 A horizontal bar graph that is arranged in descending order to show where engineers are employed.

TRANSPORTATION SYSTEM COMPARISON

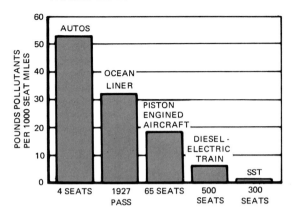

Fig. 32.12 This bar graph has the bars arranged in descending order to compare several sources of pollution. (Courtesy of Boeing Company.)

Points are plotted on the grid by using two measurements, called coordinates, made along each axis. The plotted points are indicated by using easy-to-draw symbols, such as circles, that can be drawn with a template.

The horizontal line of the graph is called the abscissa or x-axis, and the vertical scale at the left is called the ordinate or y-axis. In mathematics, data is plotted in terms of x- and y-coordinates on a graph.

Once the points have been plotted, the curve is drawn from point to point. (The line that is drawn to represent the plotted points is called a curve whether it is a smooth or broken line.) The curve should not close up the plotted points, but they should be left as open circles or symbols.

The curve must be drawn as a heavy prominent line, since it is the most important part of the graph and shows the data. In Fig. 32.13, there are two curves; therefore, it is helpful if they are drawn as different types of lines to distinguish between them. Each is labeled with a note and a leader.

The title of the graph is placed inside the graph in a box to explain the graph. It is a good rule to give enough information on a graph to make it understandable without additional text.

Units are given along the x- and y-axes with labels that designate the units that the graph is comparing.

Broken-Line Graphs

The steps of drawing a linear coordinate graph are shown in Fig. 32.14. In this case, the points are connected with a broken-line curve since the data points are ten years apart on the x-axis. Thus it is impossible to assume that the change in the data is a smooth, continual progression from point to point.

For the best appearance the plotted points should not be crossed by the curve or the grid lines of the graph (Fig. 32.15). Each circle or symbol used to plot points should be about ⅛" (5 mm) in diameter.

Different symbols can be used to plot points, along with distinctively different lines to represent the curves. Several approved symbols and lines are shown in Fig. 32.16.

Fig. 32.13 The basic linear coordinate graph with the important features identified.

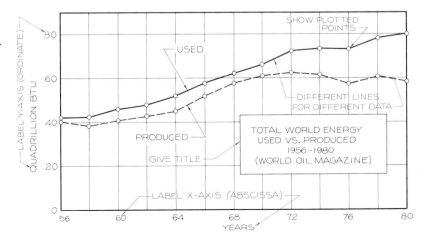

Fig. 32.14 Construction of a broken-line graph

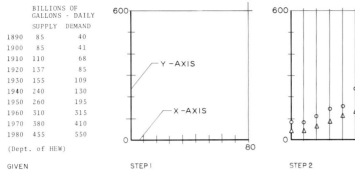

	BILLIONS OF GALLONS - DAILY	
	SUPPLY	DEMAND
1890	85	40
1900	85	41
1910	110	68
1920	137	85
1930	155	109
1940	240	130
1950	260	195
1960	310	315
1970	380	410
1980	455	550

(Dept. of HEW)

GIVEN   STEP 1   STEP 2   STEP 3

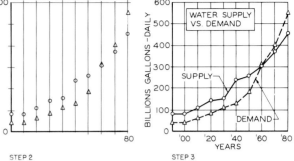

Given A record of water supply and water demand since 1890 plotted as a line graph.

Step 1 The vertical and horizontal axes are laid off to provide space for the largest values.

Step 2 The points are plotted directly over the respective years. Different symbols are used for each curve.

Step 3 The data points are connected with straight lines, the axes are labeled, the graph is titled, and the lines are strengthened.

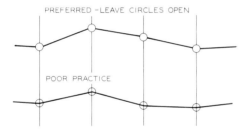

Fig. 32.15 The curve of a graph should be drawn from point to point, but it should not close up the symbols used to locate the plotted points.

Fig. 32.16 Any of these symbols or lines can be used effectively to represent different curves on a single graph. The symbols are about ⅛" (4 mm) in diameter.

Fig. 32.17 Title placement on a graph

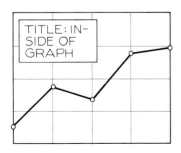

A. The title of a graph can be lettered inside a box placed within the area of the graph. The perimeter lines of the box should not coincide with grid lines within the graph.

B. The title can be placed above the graph. The title should be drawn in ⅛" letters, or slightly larger.

C. The title can be placed under a graph. It is good practice to be consistent when a series of graphs is used in the same report.

The title of a graph can be placed in any of the three positions shown in Fig. 32.17. The title should never be one as meaningless as "graph" or "coordinate graph." Instead, it should explain the graph by giving the important information such as company, date, source of the data, and the general comparisons being shown.

The example in Fig. 32.18A shows a properly labeled axis. You can see that the axis at part B has too many grid lines and too many units labeled along the axis; this clutters the graph without adding to its value. On the other hand, the units selected at C make it difficult to easily interpolate between the labeled values. For example, it is more difficult to locate a value such as 22 by eye on this scale than on the one at A.

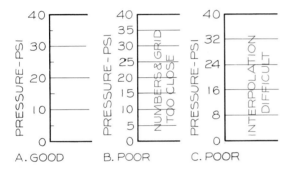

Fig. 32.18 The scale at A is the best. It has about the right number of grid lines and divisions, and the numbers are given in well-spaced, easy-to-interpolate form. The numbers at B are too close and there are too many grid lines. The units at C are given in units that make interpolation difficult by eye.

Smooth-Line Graphs

In order to determine whether the points on a graph will be connected with a broken-line curve as previously shown, or a smooth-line curve as shown in Fig. 32.19, you must have an understanding of the data. You instinctively understand that the strength of cement and its curing time will result in a smooth, continuous relationship that should be connected by a smooth curve. Even if the data points do not plot to lie on the curve, you know that the deviation of the points from this curve is due to errors of measurement or the methods used in collecting the data.

Similarly, the strength of clay tile, as related to its absorption characteristics, is an example of

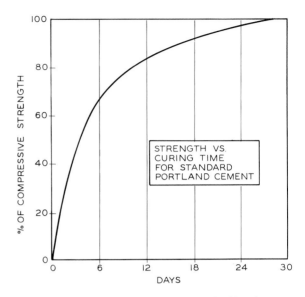

Fig. 32.19 When the process that is graphed involves gradual, continuous changes in relationships, the curve should be drawn as a smooth line.

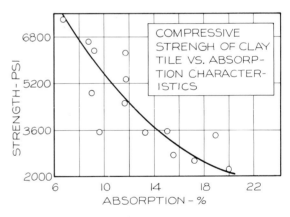

Fig. 32.20 If it is known that a relationship plotted on a graph should yield a smooth gradual curve, a smooth-line "best" curve is drawn to represent the average of the plotted points. You must use your judgment and knowledge of the data in cases of this type.

data that yields a smooth curve (Fig. 32.20). Note that the plotted data does not lie on the curve. Since you know that the relationship is a continuous one that should be connected with a smooth curve, the *best curve* is drawn to interpret the data to give an average representation of the points.

For the same reasons, you know that there is a smooth-line curve relationship between miles

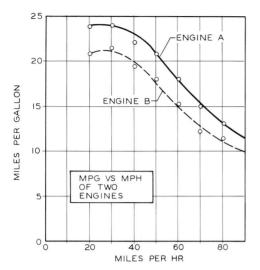

Fig. 32.21 These are "best" curves that approximate the data without necessarily passing through each data point. Inspection of the data tells you that this curve should be a smooth-line curve rather than a broken-line curve.

per gallon and the speed at which a car is driven. Two engines are compared in Fig. 32.21 with two smooth-line curves. The effect of speed on several automotive characteristics is compared in Fig. 32.22.

When a smooth-line curve is used to connect data points, there is the implication that you can *interpolate* between the plotted points to estimate

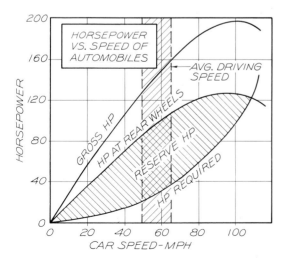

Fig. 32.22 A linear coordinate graph is used here to analyze data affecting the design of an automobile's power system.

other values. Points connected by a broken-line curve imply that you cannot interpolate between the plotted points.

Straight-Line Graphs

Some graphs have neither broken-line curves nor smooth-line curves, but straight-line curves as shown in the example in Fig. 32.23. You can determine a third value from the two given values using this graph. For example, if you are driving 70 miles per hour and it takes 5 seconds to react to apply your brakes, you will have traveled 500 feet in this interval of time.

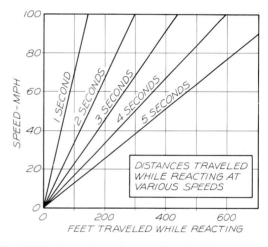

Fig. 32.23 A graph can be used to determine a third value when two variables are known. Taking this information from a graph is easier than computing each answer separately.

Two-Scale Coordinate Graphs

Graphs can be drawn with different scales in combination, such as the one shown in Fig. 32.24. The vertical scale at the left is in units of pounds, and the one at the right is expressed in units of degrees of temperature. Both curves are drawn using their respective y-axes, and each curve is labeled.

Care must be taken to avoid confusing the reader of a graph of this type. However, these graphs are effective when comparing related variables such as the drag force and air temperature of a type of automobile tire, as shown in this example.

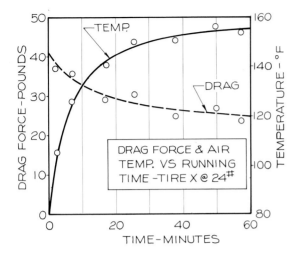

Fig. 32.24 This is a composite graph with different scales along each y-axis. The curves are labeled so reference can be made to the applicable scale.

Optimization Graphs

Graphs can be used effectively to optimize data. For example, the optimization of the depreciation of an automobile and its increase in maintenance costs is shown in Fig. 32.25. These two sets of data are plotted and the curves cross at an x-axis value of four years. At this time, the cost of maintenance is equal to the value of the car, which indicates

that this might be a desirable time to exchange it for a new one.

Another optimization graph is constructed in two steps in Fig. 32.26. The manufacturing cost per unit is reduced as more units are made, but the warehousing cost increases. A third curve is found in Step 2, by adding the two curves. Value A is shown to illustrate how this value is transferred with your dividers to add it to the manufacturing cost curve. The "total" curve tells you that the optimum number to manufacture at a time is about 11,000 units. When more or fewer are manufactured, the expense per unit is greater.

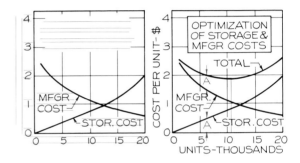

Fig. 32.26 Optimization graphs

Step 1 Lay out the graph and plot the given curves.

Step 2 Add the two curves to find a third curve. Distance A is shown transferred to locate a point on the third curve. The lowest point of the "total" curve is the optimum point of 11,000 units.

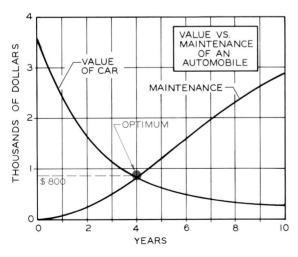

Fig. 32.25 This graph shows the optimum time to sell a car, based on the intersection of two curves that represent the depreciation of the car's value and its increasing maintenance costs.

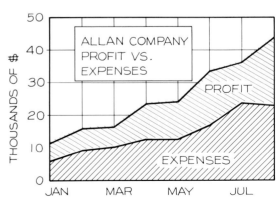

Fig. 32.27 This graph is a combination of a coordinate graph and an area graph. The upper curve represents the total of two values plotted, one above the other.

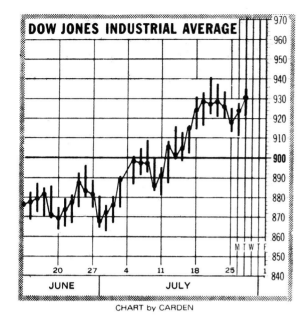

CHART by CARDEN

Fig. 32.28 This graph is a combination of a coordinate graph and a bar graph. The bars represent the ranges of selling during a day, and the broken-line curve connects the points at which the market closed each day.

Composite Graphs

The graph in Fig. 32.27 is a composite between an area graph and a coordinate graph. The lower curve is plotted first. The upper curve is found by adding the values to the lower curve so that the two areas represent the data. The upper curve is equal to the sum of the two y-values.

The graph in Fig. 32.28 is a combination of a coordinate graph and a bar graph that is used to show the Dow-Jones Industrial stock average. The bars represent the daily ranges in the index. The broken-line curve connects the points where the market closed for each day.

Break-Even Graphs

Coordinate graphs are helpful in evaluating marketing and manufacturing costs that are used to determine the selling cost for a product. The break-even graph in Fig. 32.29 is drawn to reveal that 10,000 units must be sold at $3.50 each to cover the manufacturing and development costs. Sales in excess of 10,000 result in profit.

A second type of break-even graph (Fig. 32.30) uses the cost of manufacturing per unit versus the number of units produced. In this example, the development costs must be incorporated into the unit costs. The manufacturer can determine how many units must be sold to break even at a given

Fig. 32.29 Break-even graph

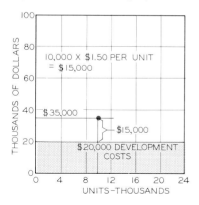

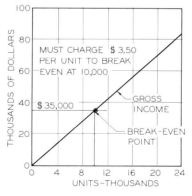

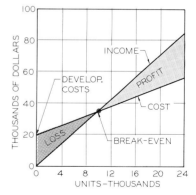

Step 1 The graph is drawn to show the cost ($20,000 in this case) of developing the product. It is determined that each unit would cost $1.50 to manufacture. This is a total investment of $35,000 for 10,000 units.

Step 2 In order for the manufacturer to break even at 10,000, the units must be sold for $3.50 each. Draw a line from zero through the break-even point for $35,000.

Step 3 The loss is $20,000 at zero units and becomes progressively less until the break-even point is reached. The profit is the difference between the cost and income to the right of the break-even point.

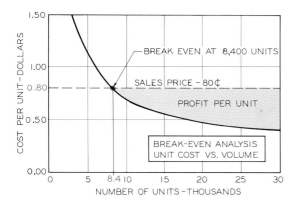

Fig. 32.30 The break-even point can be found on a graph that shows the relationship between the cost per unit, which includes the development cost, and the number of units produced. The sales price is a fixed price. The break-even point is reached when 8400 units have been sold at 80¢ each.

price, or the price per unit if a given number is selected. In this example, a sales price of 80¢ requires that 8400 units be sold to break even.

32.6 LOGARITHMIC COORDINATE GRAPHS

Both scales of a logarithmic grid are calibrated into divisions that are equal to the logarithms of the units represented. Commercially printed logarithmic grid paper is available in many variations that can be used for graphing data.

The graph in Fig. 32.31 has a logarithmic grid and shows the geometry of standard railroad cars as they relate to the tracks so that there will not be more than a maximum projection width of 12 feet around curves. Extremely large values can be shown on logarithmic grids since the lengths are considerably compressed.

32.7 SEMILOGARITHMIC COORDINATE GRAPHS

Semilogarithmic graphs are referred to as ratio graphs because they give graphical representations of ratios. One scale, usually the vertical scale, is logarithmic, and the other is linear (divided into equal divisions). Two curves that are parallel on a semilogarithmic graph have equal percentage increases.

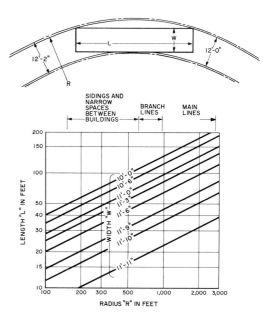

Fig. 32.31 This logarithmic graph shows the maximum load projection of 12 feet in relation to the length of a railroad car and the radius of the curve. (Courtesy of *Plant Engineering.*)

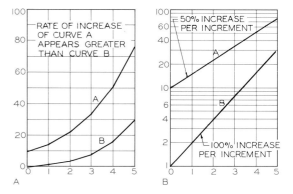

Fig. 32.32 When plotted on a standard grid, curve *A* appears to be increasing at a greater rate than curve *B*. However, the true rate of increase can be seen when the same data are plotted on a semilogarithmic graph in part B.

In Fig. 32.32, the same data is shown plotted on a linear grid and on a semilogarithmic grid. The semilogarithmic graph reveals that the percent of change from 0 to 5 is greater for Curve B than for Curve A, since Curve B is steeper. This comparison was not apparent in the plot on the linear grid.

The relationship between the linear scale and the logarithmic scale is shown in Fig. 32.33. Equal divisions along the linear scale have unequal ratios, and equal divisions along the log scale have equal ratios.

Log scales can be drawn to have one or many cycles. Each cycle increases by a factor of 10. For example, the scale in Fig. 32.34A is a three-cycle scale, and the one at B is a two-cycle scale. When these must be drawn to a special length, commercially printed log scales can be used to graphically

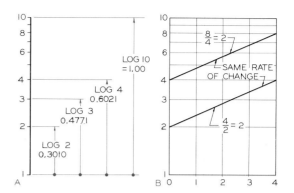

Fig. 32.35 A number's logarithm is used to locate its position on a log scale (A). This makes it possible to see the true rate of change at any location on a semilogarithmic graph (B).

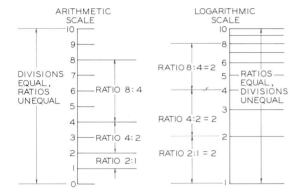

Fig. 32.33 The spacings on an arithmetic scale are equal, with unequal ratios between points. The spacings on logarithmic scales are unequal, but equal spaces represent equal ratios.

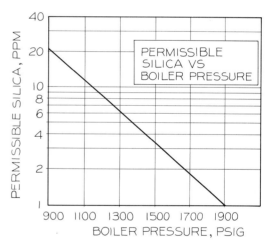

Fig. 32.36 A semilogarithmic graph is used to compare the permissible silica (parts per million) in relation to the boiler pressure.

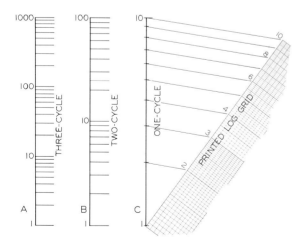

Fig. 32.34 Logarithmic paper can be purchased or drawn using several cycles. Three-, two-, and one-cycle scales are shown here. Calibrations can be drawn on a scale of any length by projecting from a printed scale as shown in part C.

transfer the calibrations to the scale being drawn (Fig. 32.34C).

In Fig. 32.35, the calibrations along the log scale are separated by the difference in their logarithms. The logarithms are laid off using a convenient scale that is calibrated in decimal divisions.

It can be seen in Fig. 32.35B that parallel straight-line curves yield equal ratios of increase. Figure 32.36 is an example of a semilogarithmic graph used to present industrial data.

Semilog graphs have several disadvantages. The most critical one is that they are misunderstood by many people who do not recognize them as being different from linear coordinate graphs. Also, zero values cannot be shown on log scales.

Percentage Graphs

The percent that one number is of another, or the percent increase of one number that is greater than the other, can be determined by using a semilogarithmic graph. Examples of both are shown in Fig. 32.37.

Data plotted in Step 1 are used to find the percent that 30 is of 60, two points on the curve. The vertical distance between the two points is equal to the difference of their logarithms. This distance is subtracted from the log of 100 at the right of the graph. This gives a value of 50% as a direct reading.

In Step 2, the percent of increase between two points is transferred from the grid to the lower end of the log scale and measured upward. It is measured upward since the increase is greater than zero. The percent of increase is measured directly from scale. These methods can be used to find percent increases or decreases of any sets of points on the grid.

32.8 POLAR GRAPHS

Polar graphs are drawn with a series of concentric circles with the origin at the center. Lines are drawn from the center toward the perimeter of the graph, where the data can be plotted through 360° by measuring values from the origin. The illumination of a lamp is shown in Fig. 32.38, where the maximum lighting of the lamp is 550 lumens at 35° from the vertical, for example.

This type of graph is used to plot the areas of illumination of all types of lighting fixtures. Polar graph paper is available commercially for drawing graphs of this type.

32.9 SCHEMATICS

Miscellanous types of graphs can be used to explain organizations, events, and relationships in a simplified manner. The block diagram in Fig. 32.39 shows the progressive steps of the completion of a construction project. Each step is blocked in and connected with arrows to explain the sequence of events.

Fig. 32.37 Percentage graphs

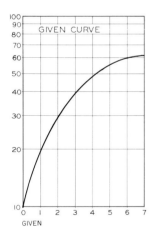

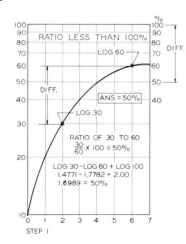

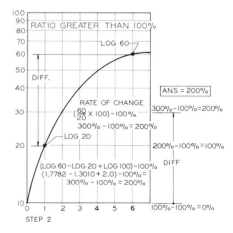

Given The data are plotted on a semilogarithmic graph to enable you to determine percentages and ratios in much the same manner that you use a slide rule.

Step 1 In finding the percent that a smaller number is of a larger number, you know that the percent will be less than 100%. The log of 30 is subtracted from the log of 60 with dividers and is transferred to the percent scale at the right, where 30 is found to be 50% of 60.

Step 2 To find the percent of increase, a smaller number is divided into a larger number to give a value greater than 100%. The difference between the logs of 60 and 20 is found with dividers, and is measured upward from 100% at the right, to find that the percent of increase is 200%.

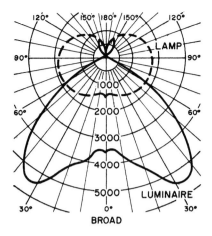

Fig. 32.38 A polar graph is used to show the illumination characteristics of luminaires.

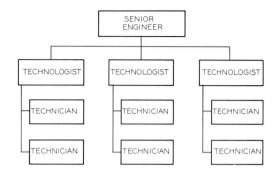

Fig. 32.40 This schematic shows the organization of a design team in a block diagram.

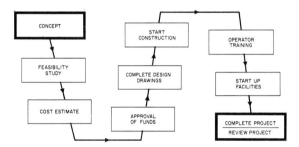

Fig. 32.39 This schematic shows a block diagram of the steps required to complete a project. (Courtesy of *Plant Engineering.*)

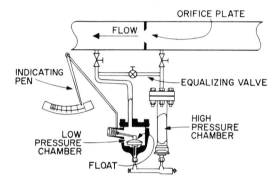

Fig. 32.41 A schematic showing the components of a gauge that measures the flow in a pipeline. (Courtesy of *Plant Engineering.*)

The organization of a company or a group of people can be depicted in an organizational chart of the type shown in Fig. 32.40. The offices represented by the blocks at the lower part of the graph are responsible to the blocks above them as they are connected by lines of authority. All blocks are finally connected with lines that converge at the top to the principal office in charge of all those below. The lines connecting the blocks also suggest the routes for communication from one to another in an upward or downward direction.

The schematic in Fig. 32.41 is not a graph, nor is it a true view of the apparatus. Instead, it is a schematic that effectively shows how the parts and their functions relate to each other.

Geographical graphs are used to combine maps and other relationships such as weather (Fig. 32.42). Different symbols are used to represent the annual rainfall that various areas of the nation receive. The graph in Fig. 32.43 shows the locations of reclamation dams by using circular symbols.

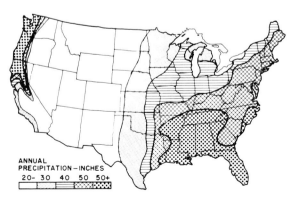

Fig. 32.42 A map chart that shows the weather characteristics of various geographical areas. (Courtesy of the *Structural Clay Products Institute.*)

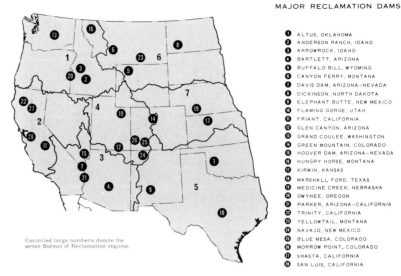

MAJOR RECLAMATION DAMS

❶ ALTUS, OKLAHOMA
❷ ANDERSON RANCH, IDAHO
❸ ARROWROCK, IDAHO
❹ BARTLETT, ARIZONA
❺ BUFFALO BILL, WYOMING
❻ CANYON FERRY, MONTANA
❼ DAVIS DAM, ARIZONA–NEVADA
❽ DICKINSON, NORTH DAKOTA
❾ ELEPHANT BUTTE, NEW MEXICO
❿ FLAMING GORGE, UTAH
⓫ FRIANT, CALIFORNIA
⓬ GLEN CANYON, ARIZONA
⓭ GRAND COULEE, WASHINGTON
⓮ GREEN MOUNTAIN, COLORADO
⓯ HOOVER DAM, ARIZONA–NEVADA
⓰ HUNGRY HORSE, MONTANA
⓱ KIRWIN, KANSAS
⓲ MARSHALL FORD, TEXAS
⓳ MEDICINE CREEK, NEBRASKA
⓴ OWYHEE, OREGON
㉑ PARKER, ARIZONA–CALIFORNIA
㉒ TRINITY, CALIFORNIA
㉓ YELLOWTAIL, MONTANA
㉔ NAVAJO, NEW MEXICO
㉕ BLUE MESA, COLORADO
㉖ MORROW POINT, COLORADO
㉗ SHASTA, CALIFORNIA
㉘ SAN LUIS, CALIFORNIA

Fig. 32.43 A chart that locates the various reclamation dams in the western portion of the United States. (Couresy of the *Bureau of Reclamation.)*

Uncircled large numbers denote the
seven Bureau of Reclamation regions.

32.10 HOW TO LIE WITH GRAPHS

Graphs can be used to distort data to the extent that the graph is actually lying. You should become familiar with these techniques so you will be able to properly interpret the data shown by a graph.

The three bar graphs in Fig. 32.44 present the same data, but a different impression is given by each. The upper graph gives the impression that Fuel B is almost five times better than Fuel A. This is because the bars do not begin at zero, thereby distorting a true comparison.

The center graph begins at zero, which gives a true comparison between the two bars. Fuel B is shown to be about 35% better than Fuel A.

The lower graph de-emphasizes the difference between the two fuels since the x-axis is much longer than is necessary. The implication of this horizontal scale is that a car might be expected to get over 90 miles per gallon. The bars appear insignificant on this graph, and even though they are drawn accurately, the difference in their lengths appears much less than in the other two graphs.

The width of bars and the colors used can give a misleading impression in a bar graph. Beware of graphs in which the bars run out of the bounds of the graph. These seldom give a true graphical picture.

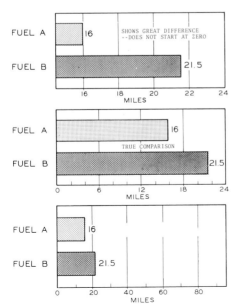

Fig. 32.44 All three graphs show the same data, but each gives a different impression of the data.

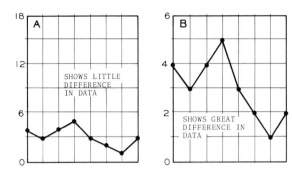

Fig. 32.45 Due to a variation in the scales, the change in the curve at B appears greater than the curve at A, although the data is the same in each.

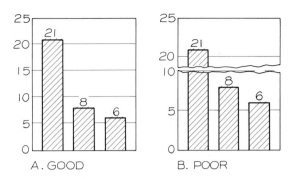

Fig. 32.47 This bar graph distorts the data at B because a portion of the graph has been removed, negating the pictorial value of the graph.

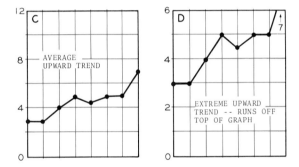

Fig. 32.46 The upward trend of the data at D appears greater than the curve at C. Both graphs show the same data.

Identical data are plotted on two graphs with different y-axes in Fig. 32.45. Graph A shows only a little variation in the data. Graph B gives a more dramatic effect because the expanded vertical scale emphasizes the difference in the data.

The two graphs in Fig. 32.46 give a different appearance even though the same data are plotted on both. The data in Graph D appear to have a more significant rate of increase because of the selection of the vertical scale. The curve of Graph D is drawn to run off the top of the graph, giving the impression that the increase is too great to be contained on a graph.

Another method of misrepresenting data is the removal of a portion of the graph (Fig. 32.47). This distorts the comparison between the bars and negates the value of the graph.

PROBLEMS

The problems below are to be drawn on Size A sheets (8½″ × 11″) in pencil or ink as specified. Follow the techniques that were covered in this chapter, and the examples that are given as the problems are being solved.

Pie Graphs

1. Draw a pie graph that compares the employment of male youth between the ages of 16 and 21: Operators—25%; Craftsmen—9%; Professionals, technicians, and managers—6%; Clerical and sales—17%; Service—11%; Farm workers—13%; and Laborers—19%.

2. Draw a pie graph that shows the relationship between the following members of the technological team: Engineers—985,000; Technicians—932,000; and Scientists—410,000.

3. Construct a pie graph of the following percentages of the employment status of graduates of two-year technician programs one year after graduation; Employed—63%; Continuing full-time study—23%; Considering job offers—6%; Military—6%; and Other—2%.

4. Construct a pie graph that shows the relationship between the types of degrees held by engineers in aeronautical engineering: Bachelor—65%; Master—29%; and Ph.D.—6%.

5. Draw a pie graph for the following average annual expenditures of a state on public roads. The approximate figures are: Construction—$13,600,000; Maintenance—$7,100,000; Purchase and upkeep of equipment—$2,900,000; Bonds—$5,600,000; Engineering and administration—$1,600,000.

6. Draw a pie that shows the data given in Problem 10.

7. Draw a pie that shows the data given in Problem 11.

Bar Graphs

8. Draw a bar graph that depicts the unemployment rate of high-school graduates and dropouts in various age categories. The age groups and the percent of unemployment of each group are given in the following table:

Ages	Percent of labor force	
	Graduates	Dropouts
16–17	18	22
18–19	12.5	17.5
20–21	8	13
22–24	5	9

9. Draw a single bar that represents 100% of a die casting alloy. The proportional parts of the alloy are: Tin—16%; Lead—24%; Zinc—38.8%; Aluminum—16.4%; and Copper—4.8%.

10. Draw a bar graph that compares the number of skilled workers employed in various occupations. Arrange the graph for ease of interpretation and comparison of occupations. Use the following data: Carpenters—82,000; All-round machinists—310,000; Plumbers—350,000; Bricklayers—200,000; Appliance servicers—185,000; Automotive mechanics—760,000; Electricians—380,000; and Painters—400,000.

11. Draw a bar graph that represents the flow of a river in cubic feet per second (cfs) as shown in the following table. Show bars that represent the data for ten days in the first month only. Omit the second month.

Day of month	Rate of flow in 1000 cfs	
	1st Month	2nd Month
1	19	19
2	130	70
3	228	79
4	129	33
5	65	19
6	32	14
7	17	15
8	13	11
9	22	19
10	32	27

12. Draw a bar graph that shows the airline distances in statute miles from New York to the cities listed in the table below. Arrange in ascending or descending order.

- Berlin.....................................3965
- Buenos Aires 5300
- Honolulu 4960
- London 3465
- Manila...................................8510
- Mexico City 2090
- Moscow 4665
- Paris.....................................3634
- Tokyo 6740

13. Draw a bar graph that compares the corrosion resistance of the materials listed in the table below:

	Loss in weight %	
	In atmosphere	In sea water
Common Steel	100	100
10% Nickel Steel	70	80
25% Nickel Steel	20	55

14. Draw a bar graph using the data in Problem 1.

15. Draw a bar graph using the data in Problem 2.

16. Draw a bar graph using the data in Problem 3.

17. Construct a rectangular grid graph to show the accident experience of Company A. Plot the numbers of disabling accidents per million

Table 32.1

	1890	1900	1910	1920	1930	1940	1950	1960	1970	1980
Supply	80	90	110	135	155	240	270	315	380	450
Demand	35	35	60	80	110	125	200	320	410	550

person-hours of work on the y-axis. Years will be plotted on the x-axis. Data: 1970—1.21; 1971—0.97; 1972—0.86; 1973—0.63; 1974—0.76; 1975—0.99; 1976—0.95; 1977—0.55; 1978—0.76; 1979—0.68; 1980—0.55; 1981—0.73; 1982—0.52; 1983—0.46.

Linear Coordinate Graphs

18. Using the data given in Table 32.1, draw a linear coordinate graph that compares the supply and demand of water in the United States from 1890 to 1980. Supply and demand are given in billions of gallons of water per day.

19. Present the data in Table 32.2 in a linear coordinate graph to decide which lamps should be selected to provide economical lighting for an industrial plant. The table gives the candlepower directly under the lamps (0°) and at various angles from the vertical when the lamps are mounted at a height of 25 feet.

20. Construct a linear coordinate graph that shows the relationship in energy costs (mills per kilowatt-hour) and the percent capacity of two types of power plants. Plot energy costs along the y-axis, and the capacity factor along the x-axis. The plotted curve will compare the costs of a nuclear plant with a gas- or oil-fired plant. Data for a gas-fired plant: 17 mills, 10%; 12 mills, 20%; 8 mills, 40%; 7 mills, 60%; 6 mills, 80%; 5.8 mills, 100%. Nuclear plant data: 24 mills, 10%; 14 mills, 20%; 7 mills, 40%; 5 mills, 60%; 4.2 mills, 80%; 3.7 mills, 100%.

21. Plot the data from Problem 17 as a linear coordinate graph.

22. Construct a linear coordinate graph to show the relationship between the transverse resilience in inch-pounds (y-axis) and the single-blow impact in foot-pounds (x-axis) of gray iron. Data: 21 fp, 375 ip; 22 fp, 350 ip; 23 fp, 380 ip; 30 fp, 400 ip; 32 fp 420 ip; 33 fp, 410 ip; 38 fp, 510 ip; 45 fp, 615 ip; 50 fp, 585 ip; 60 fp, 785 ip; 70 fp, 900 ip; 75 fp, 920 ip.

23. Draw a linear composite coordinate graph to compare the two sets of data in the following table: capacity vs. diameter, and capacity vs. weight of a brine cooler. The horizontal scale is to be tons of capacity, and the vertical scales are to be outside diameter on the left, and weight (cwt) on the right.

Tons refrigerating capacity	Outside diameter, inches	Weight, cwt
15	22	25
30	28	46
50	34	73
85	42	116
100	46	136
130	50	164
160	58	215
210	60	263

Use 20 × 20 graph paper 8½ × 11. Horizontal scale of 1″ = 40 tons. Vertical scales of 1″ = 10″ of outside DIA, and 1″ = 40 cwt.

24. Draw a linear coordinate graph that shows the voltage characteristics for a generator as given in the following table of values: Abscissa—arma-

Table 32.2

Angle with vertical	0	10	20	30	40	50	60	70	80	90
Candlepower (thous.) 2–400W	37	34	25	12	5.5	2.5	2	0.5	0.5	0.5
Candlepower (thous.) 1–1000W	22	21	19	16	12.3	7	3	2	0.5	0.5

ture current in amperes (I_a); ordinate—terminal voltage in volts (E_t).

I_a	E_t	I_a	E_t	I_a	E_t
0	288	31.1	181.8	41.5	68
5.4	275	35.4	156	40.5	42.5
11.8	257	39.7	108	39.5	26.5
15.6	247	40.5	97	37.8	16
22.2	224.5	40.7	90	13.0	0
26.2	217	41.4	77.5		

25. Draw a linear coordinate graph for the centrifugal pump test data in the table below. The units along the x-axis are to be gallons per minute. There will be four curves to represent the variables given.

Gallons per min.	Discharge pressure	Water HP	Electric HP	Efficiency %
0	19.0	0.00	1.36	0.00
75	17.5	0.72	2.25	32.0
115	15.0	1.00	2.54	39.4
154	10.0	1.00	2.74	36.5
185	5.0	0.74	2.80	26.5
200	3.0	0.63	2.83	2.22

26. Draw a linear coordinate graph that compares two of the values shown in the table below—ultimate strength and elastic limit—with degrees of temperature labeled along the x-axis.

°F	Ultimate strength	Elastic limit	Elongation %	Red of area %	Brinell hardness no.
400	257,500	208,000	10.8	31.3	500
500	247,000	224,500	12.5	39.5	483
600	232,500	214,000	13.3	42.0	453
700	207,500	193,500	15.0	47.5	410
800	180,500	169,000	17.0	52.5	358
900	159,500	146,500	18.5	56.5	313
1000	142,400	128,500	20.3	59.2	285
1100	126,500	114,000	23.0	60.8	255
1200	114,500	96,500	26.3	67.8	230
1300	108,000	85,500	25.8	58.3	235

27. Draw a linear coordinate graph that compares two of the values shown in the table in Problem 26—percent of elongation and percent of reduction of area of the cross section—with the degrees of temperature along the x-axis.

Break-Even Graphs

28. Construct a break-even graph that shows the earnings for a new product that has a development cost of $12,000. It will cost 50¢ each to manufacture, and you wish to break even at 8000. What would be the profit at a volume of 20,000 and at 25,000?

29. Same as Problem 28 except that the development costs are $80,000, the manufacturing cost is $2.30 each, and the desired break-even point is 10,000. What would be the profit at a volume of 20,000 and at 30,000? What sales price would be required to break even at 10,000 units?

30. A manufacturer has incorporated the manufacturing and development costs into a cost-per-unit estimate. He wishes to sell the product at $1.50 each. Construct a graph of the following data. On the y-axis plot cost per unit in dollars; on the x-axis, number of units in thousands. Data: 1000, $2.55; 2000, $2.01; 3000, $1.55; 4000, $1.20; 5000, $0.98; 6000, $0.81; 7000, $0.80; 8000, $0.75; 9000, $0.73; 10,000, $0.70. How many must be sold to break even? What will be the total profit when 9000 are sold?

31. The cost per unit to produce a product by a manufacturing plant is given below. Construct a break-even graph with the cost per unit plotted on the y-axis and and the number of units on the x-axis. Data: 1000, $5.90; 2000, $4.50; 3000, $3.80; 4000, $3.20; 5000, $2.85; 6000, $2.55; 7000, $2.30; 8000, $2.17; 9000, $2.00; 10,000, $1.95.

Logarithmic Graphs

32. Using the data given in Table 32.3, construct a logarithmic graph where the vibration amplitude (A) is plotted as the ordinate and vibration frequency (F) as the abscissa. The data for Curve 1 represent the maximum limits of machinery in good condition with no danger from vibration. The data for Curve 2 are the lower limits of machinery that is being vibrated excessively to the danger point. The vertical scale should be three cycles and the horizontal scale two cycles.

Table 32.3

F	100	200	500	1000	2000	5000	10,000
A(1)	0.0028	0.002	0.0015	0.001	0.0006	0.0003	0.00013
A(2)	0.06	0.05	0.04	0.03	0.018	0.005	0.001

33. Plot the data below on a two-cycle log graph to show the current in amperes (y-axis) versus the voltage in volts (x-axis) of precision temperature-sensing resistors. Data: 1 volt, 1.9 amps; 2 volts, 4 amps; 4 volts, 8 amps; 8 volts, 17 amps; 10 volts, 20 amps; 20 volts, 30 amps; 40 volts, 36 amps; 80 volts, 31 amps; 100 volts, 30 amps.

34. Plot the data from Problem 18 as a logarithmic graph.

35. Plot the data from Problem 24 as a logarithmic graph.

Semilogarithmic Graphs

36. Construct a semilogarithmic graph with the y-axis a two-cycle log scale from 1 to 100 and the x-axis a linear scale from 1 to 7. Plot the data below to show the survivability of a shelter at varying distances from a one-megaton air burst. The data consists of overpressure in psi along the y-axis, and distance from ground zero in miles along the x-axis. The data points represent an 80% chance of survival of the shelter. Data: 1 mile, 55 psi; 2 miles, 11 psi; 3 miles, 4.5 psi; 4 miles, 2.5 psi; 5 miles, 2.0 psi; 6 miles, 1.3 psi.

37. The growth of two divisions of a company, Division A and Division B, is given in the data below. Plot the data on a rectilinear graph and on a semilog graph. The semilog graph should have a one-cycle log scale on the y-axis for sales in thousands of dollars, and a linear scale on the x-axis showing years for a six-year period. Data in dollars: 1 yr, A = $11,700 and B = $44,000; 2 yr, A = $19,500 and B = $50,000; 3 yr, A = $25,000 and B = $55,000; 4 yr, A = $32,000 and B = $64,000; 5 yr, A = $42,000 and B = $66,000; 6 yr, A = $48,000 and B = $75,000. Which division has the better growth rate?

38. Draw a semilog chart showing probable engineering progress. Use the following indices: 40,000 B.C. = 21; 30,000 B.C. = 21.5; 20,000 B.C. = 22; 16,000 B.C. = 23; 10,000 B.C. = 27; 6000

B.C. = 34; 4000 B.C. = 39; 2000 B.C. = 49; 500 B.C. = 60; A.D. 1900 = 100. Horizontal scale 1″ = 10,000 years. Height of cycle = about 5″. Two-cycle printed paper may be used if available.

39. Plot the data from Problem 24 as a semilogarithmic graph.

40. Plot the data from Problem 26 as a semilogarithmic graph.

Percentage Graphs

41. Plot the data given in Problem 18 on a semilog graph. What is the percent of increase in the demand for water from 1890 to 1920? What percent of the demand is the supply for the following years: 1900, 1930, and 1970?

42. Using the graph plotted in Problem 37, determine the percent of increase of Division A and Division B from year 1 to year 4. Also, what percent of sales of Division A are the sales of Division B at the end of year 2? At the end of year 6?

43. Plot two values from Problem 26—water horsepower and electric horsepower—on semilog paper compared with gallons per minute along the x-axis. What is the percent that water horsepower is of the electric horsepower when 75 gallons per minute are being pumped? What is the percent increase of the electric horsepower from 0 to 185 gallons per minute?

Organizational Charts

44. Draw an organization chart for a city government organized as follows: The electorate elects school board, city council, and municipal court officers. The city council is responsible for the civil service commission, city manager, and city planning board. The city manager's duties cover finance, public safety, public works, health and welfare, and law.

45. Draw an organization chart for a manufacturing plant. The sales manager, chief engineer, treasurer, and general manager are responsible to the president. Each of these officers has a department force. The general manager has three department heads: master mechanic, plant superintendent, and purchasing agent. The plant superintendent has charge of the shop foremen, under whom are the working forces, and also has direct charge of the shipping, tool and die, inspection, order, and stores and supplies departments.

Polar Graphs

46. Construct a polar graph of the data given in Problem 19.

47. Construct a polar graph of the following illumination, in lumens at various angles, emitted from a luminaire. The zero-degrees position is vertically under the overhead lamp. Data: 0°, 12,000; 10°, 15,000; 20°, 10,000; 30°, 8000; 40°, 4200; 50°, 2500; 60°, 1000; 70°, 0. The illumination is symmetrical about the vertical.

NOMOGRAPHY

33.1 NOMOGRAPHY

An additional aid in analyzing data is a graphical computer called a **nomogram** or **nomograph.** Basically, a nomogram or "number chart" is any graphical arrangement of calibrated scales and lines that may be used to permit calculations, usually those of a repetitive nature.

The term "nomogram" is frequently used to denote a specific type of scale arrangement called an alignment graph. Typical examples of alignment graphs are shown in Fig. 33.1. Many other types are also used that have curved scales or other scale arrangements for more complex problems. The discussion of nomograms in this chapter will be limited to the simpler conversion, parallel-scale, and N-type graphs and their variations.

Using an Alignment Chart

An alignment graph is usually constructed to solve for one or more unknowns in a formula or empirical relationship between two or more quantities. For example, it can be used to convert degrees Celsius to degrees Fahrenheit, to find the size of a structural member to sustain a certain load, and so on. An alignment graph is read by placing a straightedge, or by drawing a line called

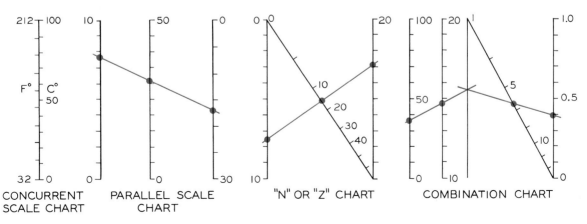

Fig. 33.1 Typical examples of types of alignment charts.

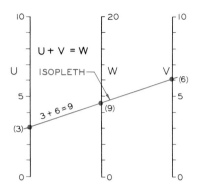

Fig. 33.2 Use of an isopleth to solve graphically for unknowns in the given equation.

an **isopleth,** across the scales of the chart and reading corresponding values from the scale on this line. The example in Fig. 33.2 shows readings for the formula $U + V = W$.

33.2 ALIGNMENT-GRAPH SCALES

To construct any alignment graph, you must first determine the graduations of the scales. Alignment-graph scales are called **functional scales.** A functional scale is one that is graduated according to values of some *function* of a variable, but *calibrated* with values of the variable. A functional scale for $F(U) = U^2$ is illustrated in Fig. 33.3. It can be seen in this example that if a value of $U = 2$ was substituted into the equation, the position of U on the functional scale would be 4 units from zero, or $2^2 = 4$. This procedure can be repeated with all values of U by substitution.

VALUES OF U

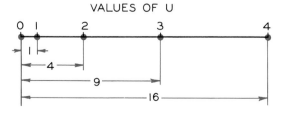

Fig. 33.3 Functional scale for units of measurement that are proportional to $F(U) = U^2$.

The Scale Modulus

Since the graduations on a functional scale are spaced in proportion to values of the function, a proportionality, or scaling factor, is needed. This constant of proportionality is called the **scale modulus** and it is given by the equation

$$m = \frac{L}{F(U_2) - F(U_1)} \tag{1}$$

where

$m =$ scale modulus, in inches per functional unit,
$L =$ desired length of the scale, in inches,
$F(U_2) =$ function value at the end of the scale,
$F(U_1) =$ function value at the start of the scale.

For example, suppose that we are to construct a functional scale for $F(U) = \sin U$, with $0° \leq U \leq 45°$ and a scale 6″ in length. Thus $L = 6″$, $F(U_2) = \sin 45° = 0.707$, $F(U_1) = \sin 0° = 0$. Therefore, Eq. (1) can be written in the following form by substitution:

$$m = \frac{6}{0.707 - 0} = 8.49 \text{ inches per (sine) unit.}$$

The Scale Equation

Graduation and calibration of a functional scale are made possible by a **scale equation.** The general form of this equation may be written as a variation of Eq. (1) in the following form:

$$X = m[F(U) - F(U_1)] \tag{2}$$

where

$X =$ distance from the measuring point of the scale to any graduation point,
$m =$ scale modulus,
$F(U) =$ functional value at the graduation point,
$F(U_1) =$ functional value at the measuring point of the scale.

For example, a functional scale is constructed for the previous equation, $F(U) = \sin U$ ($0° \leq U \leq 45°$). It has been determined that $m = 8.49$, $F(U) = \sin U$, and $F(U_1) = \sin 0° = 0$. Thus by substitution the scale equation, (2), becomes

$$X = 8.49 (\sin U - 0) = 8.49 \sin U.$$

Using the eqation, we can substitute values of U

Table 33.1

U	0°	5°	10°	15°	20°	25°	30°	35°	40°	45°
X	0	0.74	1.47	2.19	2.90	3.58	4.24	4.86	5.45	6.00

and construct a table of positions. In this case, the scale is calibrated at 5° intervals, as reflected in Table 33.1.

The values of X from the table give the positions, in inches, for the corresponding graduations, measured from the start of the scale ($U = 0°$); see Fig. 33.4. It should be noted that the measuring point does *not* need to be at one end of the scale, but it is usually the most convenient point, especially if the functional value is zero at that point.

A graphical method of locating the functional values along a scale can be found as shown in Fig. 33.5 by the proportional-line method. The sine

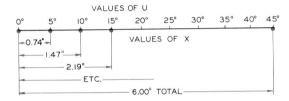

Fig. 33.4 Construction of a functional scale using values from Table 33.1, which were derived from the scale equation.

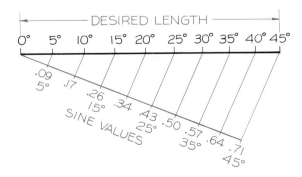

Fig. 33.5 A functional scale that shows the sine of the angles from 0° to 45° can be drawn graphically by the proportional-line method. The scale is drawn to a desired length and the sine values of angles at 5° intervals are laid off along a construction line that passes through the 0° end of the scale. These values are projected back to the scale.

functions are measured off along a line at 5° intervals, with the end of the line passing through the 0° end of the scale. The functions are transferred from the inclined line with parallel lines back to the scale where the functions are represented and labeled.

33.3 CONCURRENT SCALES

Concurrent scales are useful in the rapid conversion of one value into terms of a second system of measurement. Formulas of the type $F_1 = F_2$, which relate two variables, can be adapted to the concurrent-scale format. Typical examples might be the Fahrenheit–Celsius temperature relation.

$$°F = \frac{9}{5}°C + 32$$

or the area of a circle,

$$A = \pi r^2.$$

Design of a concurrent-scale chart involves the construction of a functional scale for each side of the mathematical formula in such a manner that the *position* and *lengths* of each scale coincide. For example, to design a conversion chart 5″ long that will give the areas of circles whose radii range from 1 to 10, we first write $F_1(A) = A$, $F_2(r) = \pi r^2$, and $r_1 = 1$, $r_2 = 10$. The scale modulus for r is

$$m_r = \frac{L}{F_2(r_2) - F_2(r_1)}$$

$$= \frac{5}{\pi(10)^2 - \pi(1)^2} = 0.0161.$$

Thus the scale equation for r becomes

$$
\begin{aligned}
X_r &= m_r[F_2(r) - F_2(r_1)] \\
&= 0.0161\,[\pi r^2 - \pi(1)^2] \\
&= 0.0161\,\pi(r^2 - 1) \\
&= 0.0505(r^2 - 1).
\end{aligned}
$$

A table of values for X_r and r may now be completed as shown in Table 33.2. The r-scale can be

Table 33.2

r	1	2	3	4	5	6	7	8	9	10
X_r	0	0.15	0.40	0.76	1.21	1.77	2.42	3.18	4.04	5.00

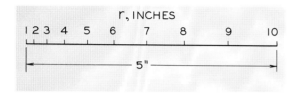

Fig. 33.6 Calibration of one scale of a concurrent scale graph using values from Table 33.2.

drawn from this table, as shown in Fig. 33.6. From the original formula, $A = \pi r^2$, the limits of A are found to be $A_1 = \pi = 3.14$ and $A_2 = 100\pi = 314$. The scale modulus for concurrent scales is always the same for equal-length scales; therefore $m_A = m_r = 0.0161$, and the scale equation for A becomes

$$X_A = m_A[F_1(A) - F_1(A_1)]$$
$$= 0.0161(A - 3.14).$$

The corresponding table of values is then computed for selected values of A, as shown in Table 33.3.

The A-scale is now superimposed on the r-scale; its calibrations have been placed on the

other side of the line to facilitate reading (Fig. 33.7). It may be desired to expand or contract one of the scales, in which case an alternative arrangement may be used, as shown in Fig. 33.8. The two scales are drawn parallel at any convenient distance, and calibrated in *opposite* directions. A different scale modulus and corresponding scale equation must be calculated for each scale if they are *not* the same length.

A graphical method can be used to construct concurrent scales as shown in Fig. 33.9 by using the proportional-line method. Since there are 101.6 mm in 4 inches, the units of millimeters can be located on the upper side of the inch scale by projecting to the scale with a series of parallel projectors.

33.4 CONSTRUCTION OF ALIGNMENT GRAPHS WITH THREE VARIABLES

For a formula of three functions (of one variable each), the general approach is to select the lengths and positions of *two* scales according to the range of variables and size of the graph desired. These are then calibrated by means of the scale equations, as shown in the preceding article. The position and calibration of the third scale will then depend upon these initial constructions. Although definite mathematical relationships exist that may be used to locate the third scale, graphical constructions are simpler and usually less subject to error. Examples of the various forms are presented in the following articles.

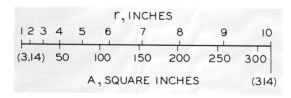

Fig. 33.7 The completed concurrent scale graph for the formula $A = \pi r^2$. Values for the A-scale are taken from Table 33.3.

Table 33.3

A	(3.14)	50	100	150	200	250	300	(314)
X_1	0	0.76	1.56	2.36	3.16	3.96	4.76	5.00

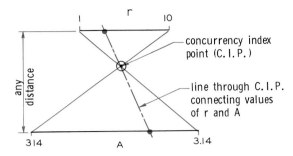

Fig. 33.8 Concurrent scale graph with unequal scales.

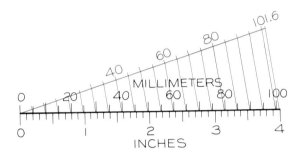

Fig. 33.9 The proportional-line method can be used to construct an alignment graph that converts inches to millimeters. This requires that the units at each end of the scales be known. For example, there are 101.6 millimeters in 4 inches.

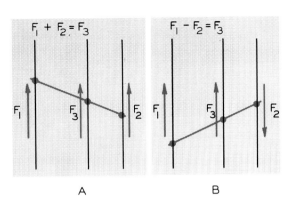

Fig. 33.10 Two common forms of parallel-scale alignment nomographs.

33.5 PARALLEL-SCALE NOMOGRAPHS

Many engineering relationships involve three variables that can be computed graphically on a repetitive basis. Any formula of the type $F_1 + F_2 = F_3$ may be represented as a parallel-scale alignment chart, as shown in Fig. 33.10A. Note that all scales increase (functionally) in the same direction and that the function of the middle scale represents the *sum* of the other two. Reversing the direction of any scale changes the sign of its func-

Fig. 33.11 Parallel-scale nomogram (linear)

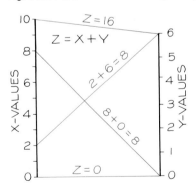

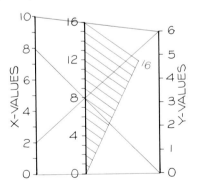

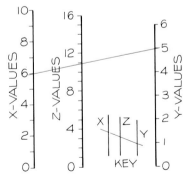

Step 1 Two parallel scales are drawn at any length and calibrated. The location of the parallel Z-scale is found by selecting two sets of values that will give the same value of Z, 8 in this case. The ends of the Z-scale will be 0 and 16, the sum of the end values of X and Y.

Step 2 The Z-scale is drawn through the point located in step 1 parallel to the other scales. The scale is calibrated from 0 to 16 by using the proportional-line method. Note that the two sets of X- and Y-values cross at 8, the sum of each set.

Step 3 The Z-scale is labeled and calibrated with easy-to-read units. A key is drawn to show how the nomograph is used. If the Y-scale were calibrated with 0 at the upper end instead of the bottom, a different Z-scale could be computed and the nomograph could be used for Z = X − Y.

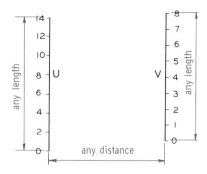

Fig. 33.12 Calibration of the outer scales for the formula $U + 2V = 3W$, where $0 \leqslant U \leqslant 14$ and $0 \leqslant V \leqslant 8$.

tion in the formula, as for $F_1 - F_2 = F_3$ in Fig. 33.10B.

To illustrate this type of alignment graph we will use the formula $Z = X + Y$ as illustrated in Fig. 33.11. The outer scales for X and Y are drawn and calibrated. They can be drawn to any length and positioned any distance apart as shown in Fig. 33.12. Two sets of data that yield a Z of 8 in Step 1 are used to locate the parallel Z-scale. In Step 2, the Z-scale is drawn and divided into 16 equal units. The finished nomograph in Step 3 can be used to add various values of X and Y to find their sums along the Z-scale.

A more complex alignment graph is illustrated in Fig. 33.13 where the formula $U + 2V = 3W$ is expressed in the form of a nomograph.

First it is necessary to determine and calibrate the two outer scales for U and V; we can make them any convenient length and any convenient distance apart, as shown in Fig. 33.12. These scales are used as the basis for the step-by-step construction shown in Fig. 33.13.

The limits of calibration for the middle scale are found by connecting the endpoints of the outer scales and substituting these values into the formula. Here, W is found to be 0 and 10 at the extreme ends (Step 1). Two pairs of corresponding values of U and V are selected that will give the same value of W. For example, values of $U = 0$ and $V = 7.5$ give a value of 5 for W. We also find that $W = 5$ when $U = 14$ and $V = 0.5$. This should be verified by substitution before continuing with construction. We connect these corresponding pairs of values with isopleths to locate their intersection and the position of the W-scale.

Since the W-scale is linear ($3W$ is a linear function), it may be subdivided into uniform intervals (Step 2). For a nonlinear scale, the scale modulus (and the scale equation) may be found in Step 2 by substituting its length and its two end values into Eq. (1) of Section 33.2. The nomograph

Fig. 33.13 Parallel-scale nomograph (linear)

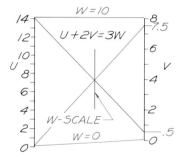

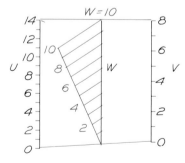

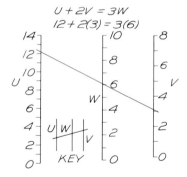

Step 1 Substitute the end values of the U- and V-scales into the formula to find the end values of the W-scale: $W = 10$ and $W = 0$. Select two sets of U and V that will give the same value of W. *Example:* When $U = 0$ and $V = 7.5$, W will equal 5, and when $U = 14$ and $V = 0.5$, W will equal 5. Connect these sets of values; the intersection of their lines locates the position of the W-scale.

Step 2 Draw the W-scale parallel to the outer scales; its length is controlled by the previously established lines of $W = 10$ and $W = 0$. Since this scale is 10 linear divisions long, divide it graphically into ten units as shown. This will be a linear scale constructed as shown in Fig. 33.11.

Step 3 The nomogram can be used as illustrated by selecting any two known variables and connecting them with an isopleth to determine the third unknown. A key is always included to illustrate how the nomogram is to be used. An example of $U = 12$ and $V = 3$ is shown to verify the accuracy of the graph.

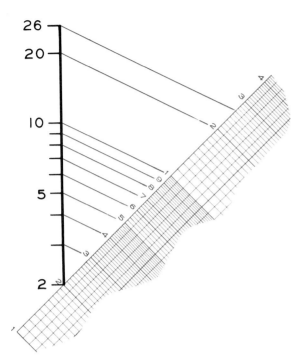

Fig. 33.14 Graphical calibration of a scale using logarithmic paper.

can be used to determine an infinite number of problem solutions when sets of two variables are known, as illustrated in Step 3.

Parallel-Scale Graph with Logarithmic Scales

Problems involving formulas of the type $F_1 \times F_2 = F_3$ can be solved in a manner very similar to the example given in Fig. 33.1 when logarithmic scales are used.

The first step in drawing a nomograph with logarithmic scales is learning how to transfer logarithmic functions to the scale. The graphical method is shown in Fig. 33.14, where units along the scale are found by projecting from a printed logarithmic scale with parallel lines. The mathematical method can also be used to locate these logarithmic spacings.

The formula $Z = XY$ is converted into a nomograph in Fig. 33.15. In Step 1, the X and Y scales are drawn within the desired limits from 1 to 10 on each. Sets of values of X and Y that yield the same value of Z, 10 in this case, are used to locate the Z-axis. The limits of the Z-axis are 1 and 100.

In Step 2, the Z-axis is drawn and calibrated as a two-cycle log scale. In Step 3, a key is given to explain how an isopleth is used to add the log-

Fig. 33.15 Parallel-scale nomogram (logarithmic)

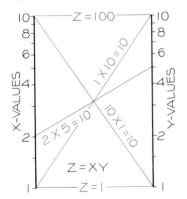 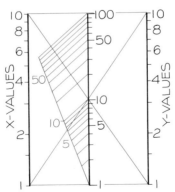 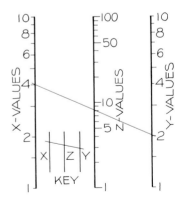

Step 1 To find scales that will solve the equation, $Z = XY$, parallel log scales are drawn. Sets of X and Y points that yield the same value of Z, 10 in this example, are drawn. Their intersection locates the Z-scale. The end values of the scale are 1 and 100.

Step 2 The Z-axis is graphically calibrated as a two-cycle logarithmic scale from 1 to 100. This scale is parallel to the X- and Y-scales.

Step 3 A key is drawn to explain how the nomograph is used. An example isopleth is drawn to show that $4 \times 2 = 8$. By reversing the numerical order of the Y-value scale and computing a different Z-scale, the nomogram could be used for the equation $Z = X/Y$.

Table 33.4

S	0.1	0.2	0.3	0.4	0.5	0.6	0.7	0.8	0.9	1.0
X_s	0	1.80	2.88	3.61	4.19	4.67	5.07	5.42	5.72	6.00

Table 33.5

T	1	2	4	6	8	10	20	40	60	80	100
X_T	0	0.91	1.80	2.33	2.71	3.00	3.91	4.81	5.33	5.77	6.00

arithms of X and Y to give the log of Z. When logarithms are added the result is multiplication. Had the Y-axis been calibrated in the opposite direction with 1 at the upper end and 10 at the lower end, a new Z-axis could have been calibrated and the nomograph used for the formula, $Z = Y/X$, since it would be subtracting logarithms.

A more advanced example of this type of problem is the formula $R = S\sqrt{T}$, for $0.1 \leqslant S \leqslant 1.0$ and $1 \leqslant T \leqslant 100$. Assume the scales to be 6" long. These scales need not be equal except for convenience. This formula may be converted into the required form by taking common logarithms of both sides, which gives

$$\log R = \log S + \tfrac{1}{2}\log T.$$

Thus we have

$$F_1(S) + F_2(T) = F_3(R), \qquad (1)$$

where $F_1(S) = \log S$, $F_2(T) = \tfrac{1}{2}\log T$, and $F_3(R) = \log R$. The scale modulus for $F_1(S)$ is, from Eq. (1),

$$m_s = \frac{6}{\log 1.0 - \log 0.1} = \frac{6}{0 - (-1)} = 6 \qquad (2)$$

Choosing the scale measuring point from $S = 0.1$, we find from Eq. (2) that the scale equation for $F_1(S)$ is

$$X_s = 6(\log S - \log 0.1) = 6(\log S + 1) \qquad (3)$$

Similarly, the scale modulus for $F_2(T)$ is

$$m_T = \frac{6}{\tfrac{1}{2}\log 100 - \tfrac{1}{2}\log 1} = \frac{6}{\tfrac{1}{2}(2) - \tfrac{1}{2}(0)} = 6 \qquad (4)$$

Thus, the scale equation, measuring from $T = 1$, is

$$X_T = 6(\tfrac{1}{2}\log T - \tfrac{1}{2}\log 1) = 3\log T \qquad (5)$$

The corresponding tables for the two scale equations may be computed as shown in Tables 33.4 and 33.5. We shall position the two scales 5" apart, as shown in Fig. 33.16. The logarithmic scales are graduated using the values in Tables 33.4 and 33.5. The step-by-step procedure for constructing the remainder of the nomogram is given in Fig. 33.17 using the two outer scales determined here.

The end values of the middle (R) scale are found from the formula $R = S\sqrt{T}$ to be $R = 1.0$ $\sqrt{100} = 10$ and $R = 0.1\sqrt{1} = 0.1$. Choosing a value of $R = 1.0$, we find pairs of S and T might be $S = 0.1$, $T = 100$, and $S = 1.0$, $T = 1.0$ that yield $R = 1$. We connect these pairs with isopleths in Step 1 and position the middle scale at their intersection. The R-scale is drawn parallel to the outer scales and is calibrated by deriving its

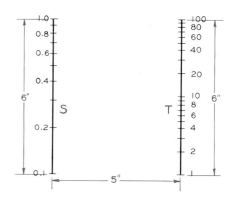

Fig. 33.16 Calibration of the outer scales for the formula $R = S\sqrt{T}$, where $0.1 \leqslant S \leqslant 1.0$ and $1 \leqslant T \leqslant 100$.

Fig. 33.17 Parallel-scale chart (logarithmic)

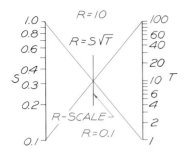 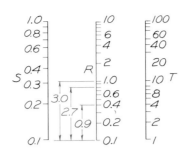 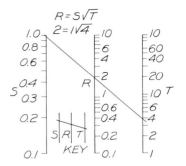

Step 1 Connect the end values of the outer scales to determine the extreme values of the R-scale, R = 10 and R = 0.1. Select corresponding values of S and T that will give the same value of R. Values of S = 0.1, T = 100 and S = 1.0, T = 1.0 give a value of T = 1.0. Connect the pairs to locate the position of the R-scale.

Step 2 Draw the R-scale to extend from 0.1 to 10. Calibrate it by substituting values determined from its scale equation. These values have been computed and tabulated in Table 33.6. The resulting tabulation is a logarithmic, two-cycle scale.

Step 3 Add labels to the finished nomogram and draw a key to indicate how it is to be used. An isopleth has been used to determine R when S = 1.0 and T = 4. The result of 2 is the same as that obtained mathematically, thus verifying the accuracy of the chart. Other combinations can be solved in this same manner.

scale modulus:

$$m_R = \frac{6}{\log 10 - \log 0.1} = \frac{6}{1 - (-1)} = 3.$$

Thus its scale equation (measuring from R = 0.1) is

$$X_R = 3(\log R - \log 0.1) = 3(\log R + 1.0)$$

Table 33.6 is computed to give the values for the scale. These values are applied to the R-scale as shown in Step 2. The finished nomogram can be used as illustrated in Step 3 to compute the unknown variables when two variables are given.

Note that this example illustrates a general method of creating a parallel-scale graph for all formulas of the type $F_1 + F_2 = F_3$ through the use of a table of values computed from the scale equation.

33.6 N- OR Z-GRAPHS

Whenever F_2 and F_3 are linear functions, we can partially avoid using logarithmic scales for for-

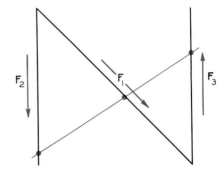

Fig. 33.18 An N-graph for solving an equation of the form $F_1 = F_2/F_3$.

mulas of the type

$$F_1 = \frac{F_2}{F_3}.$$

Instead, we use an N-graph, as shown in Fig. 33.18. The outer scales, or "legs" of the N are functional scales and will therefore be linear if F_2 and F_3 are linear, whereas if the same formula were drawn as a parallel-scale graph, all scales would have to be logarithmic.

Table 33.6

R	0.1	0.2	0.4	0.6	0.8	1.0	2.0	4.0	6.0	8.0	10.0
X_R	0	0.91	1.80	2.33	2.71	3.00	3.91	4.81	5.33	5.71	6.00

Fig. 33.19 An N-chart nomograph

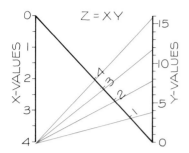

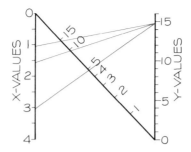

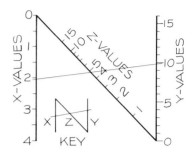

Step 1 An *N*-nomograph for the equation of *Z=Y/X* can be drawn with two parallel scales. The zero ends of each scale are connected with a diagonal scale. Isopleths are drawn to locate units along the diagonal.

Step 2 Additional isopleths are drawn to locate other units along the diagonal. It is important that the units labeled on the diagonal be whole units that are easy to interpolate between when the nomogram is used.

Step 3 The diagonal scale is labeled and a key is drawn to illustrate how the nomogram is used. An example isopleth is drawn to show that 10 ÷ 2 = 5. Note that the accuracy of the N-chart is greater at the 0 end of the diagonal. It approaches infinity at the other end.

Some main features of the N-chart are:

1. The outer scales are parallel functional scales of F_2 and F_3.

2. They increase (functionally) in *opposite* directions.

3. The diagonal scale connects the (functional) *zeros* of the outer scale.

4. In general, the diagonal scale is not a functional scale for the function F_1 and is generally nonlinear.

Construction of an N-graph is simplified by the fact that locating the middle (diagonal) scale is usually less of a problem than it is for a parallel-scale graph. Calibration of the diagonal scale is most easily accomplished by graphical methods.

The steps in constructing a basic N-graph of the equation $Z = Y/X$ are shown in Fig. 33.19. In Step 1, the diagonal is drawn to connect the zero ends of the scales. Whole values are located along the diagonal by using combinations of *X*- and *Y*-values. This process of locating values is continued in Step 2. It is important that the units located along the diagonal are whole values that are easy to interpolate between. For example, fractional units such as 1.36 and 2.25 would be complex values that would give a scale that would be difficult to use.

In Step 3, the diagonal is labeled and a key is given to explain how to use the nomograph. A

sample isopleth is given that verifies the correctness of the graphical relationship between the scales.

A more advanced N-graph is constructed (see Fig. 33.21) for the equation

$$A = \frac{B + 2}{C + 5}$$

where $0 \leqslant B \leqslant 8$ and $0 \leqslant C \leqslant 15$. This equation follows the form of

$$F_1 = \frac{F_2}{F_3}$$

where $F_1(A) = A$, $F_2(B) = B + 2$, and $F_3(C) = C$

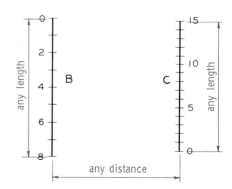

Fig. 33.20 Calibration of the outer scales of an N-graph for the equation $A = (B + 2)/(C + 5)$.

+ 5. Thus the outer scales will be for $B + 2$ and $C + 5$, and the diagonal scale will be for A.

Construction is begun in the same manner as for a parallel-scale graph by selecting the layout of the outer scales (Fig. 33.20). As before, the limits of the diagonal are determined by connecting the endpoints on the outer scales, giving $A = 0.1$ for $B = 0$, $C = 15$ and $A = 2.0$ for $B = 8$, $C = 0$, as shown in Step 1 of Fig. 33.21.

The diagonal scale is located by finding the *function zeros* of the outer scales, i.e., the points where $B + 2 = 0$ or $B = -2$, and $C + 5 = 0$ or $C = -5$. The diagonal scale is drawn by connecting these points as shown in Step 1. Calibration of the diagonal scale is most easily accomplished by substituting into the formula. Select the upper limit of an outer scale, for example, $B = 8$. This gives the formula

$$A = \frac{10}{C + 5}.$$

Solve this equation for the other scale variable:

$$C = \frac{10}{A} - 5.$$

Using this as a "scale equation," make a table of values for the desired values of A and corresponding values of C (up to the limit of C in the chart), as shown in Table 33.7. Connect isopleths from $B = 8$ to the tabulated values of C. Their intersections with the diagonal scale give the required calibrations for approximately half the diagonal scale, as shown in Step 2 of Fig. 33.21.

The remainder of the diagonal scale is calibrated by substituting the end value of the other outer scale $(C = 15)$ into the formula, giving

$$A = \frac{B + 2}{20}$$

Solving this for B yields

$$B = 20A - 2.$$

Table 33.7

A	2.0	1.5	1.0	0.9	0.8	0.7	0.6	0.5
C	0	1.67	5.0	6.11	7.50	9.28	11.7	15.0

Fig. 33.21 Construction of an N-graph

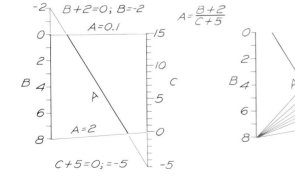

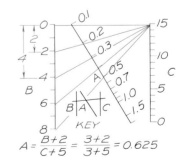

Step 1 Locate the diagonal scale by finding the functional zeros of the outer scales. This is done by setting $B + 2 = 0$ and $C + 5 = 0$, which gives a zero value for A when $B = -2$ and $C = -5$. Connect these points with diagonal scale A.

Step 2 Select the upper limit of one of the outer scales, $B = 8$ in this case, and substitute it into the given equation to find a series of values of C for the desired

values of A, as shown in Table 33.7. Draw isopleths from $B = 8$ to the values of C to calibrate the A-scale.

Step 3 Calibrate the remainder of the A-scale in the same manner by substituting $C = 15$ into the equation to determine a series of values on the B-scale for desired values on the A-scale, as listed in Table 33.8. Draw isopleths from $C = 15$ to calibrate the A-scale as shown. Draw a key to indicate how the nomogram is to be used.

Table 33.8

A	0.5	0.4	0.3	0.2	0.1
B	8.0	6.0	4.0	2.0	0

A table for the desired values of A can be constructed as shown in Table 33.8. Isopleths connecting $C = 15$ with the tabulated values of B will locate the remaining calibrations on the A-scale, as shown in Step 3.

33.7 COMBINATION FORMS OF ALIGNMENT GRAPHS

The types of graphs discussed above may be used in combination to handle different types of formulas. For example, formulas of the type $F_1/F_2 = F_3/F_4$ (four variables) may be represented as *two* N-charts by the insertion of a "dummy" function. To do this, let

$$\frac{F_1}{F_2} = S \quad \text{and then} \quad S = \frac{F_3}{F_4}.$$

Each of these may be represented as shown in Step 1 of Fig. 33.22, where one N-graph is inverted and rotated 90°. In this way, the charts may be superimposed as shown in Step 2 if the S-scales are of equal length. The S-scale, being a "dummy" scale, does not need to be calibrated; it is merely a "turning" scale for intermediate values of S that do not actually enter into the formula itself. The chart is read with *two* isopleths that connect the four variable values and cross on the S-scale as shown in Step 3. Nomograms of this form are commonly called *ratio graphs*.

Formulas of the type $F_1 + F_2 = F_3 F_4$ are handled similarly. As in the preceding example, a "dummy" function is used: $F_1 + F_2 = S$ and $S = F_3 F_4$. In order to apply the superimposition principle, a more equitable arrangement is obtained by rewriting the equations as $F_2 = S - F_1$ and $F_3 = S/F_4$. These two equations then take the form of a parallel-scale nomogram and an N-graph, respectively, as shown Step 1 in Fig. 33.23. Again the S-scales must be identical but need not be calibrated. The nomograms are superimposed in Step 2. The S-scale is used as a "turning" scale for the two isopleths, as shown in Step 3. Many other combinations are possible, limited only by the ingenuity of the nomographer in adapting formulas and scale arrangements to specific needs.

Fig. 33.22 Four-variable graph

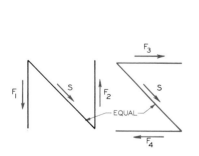 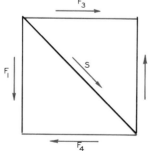 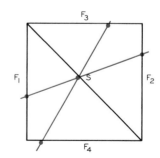

Step 1 A combination chart can be developed to handle four variables in the form $F_1/F_2 = F_3/F_4$ by developing two N-graphs in the forms $F_1/F_2 = S$ and $F_3/F_4 = S$, where S is a dummy scale of equal length in both charts.

Step 2 If equal-length scales are used in each of the N-graphs and if the S-scales are equal, then the charts can be overlapped so that each is common to the S-scale.

Step 3 Two lines (isopleths) are drawn to cross at a common point on the S-scale. Numerous combinations of the four variables can be read on the surrounding scales. The S-scale need not be calibrated, since no values are read from it.

Fig. 33.23 Combination parallel-scale chart and N-graph

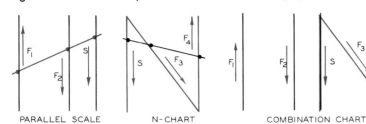

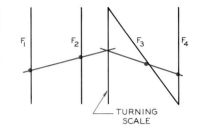

PARALLEL SCALE   N-CHART   COMBINATION CHART   TURNING SCALE

Step 1 Formulas of the type $F_1 + F_2 = F_3F_4$ can be combined into one nomogram by constructing a parallel-scale chart and an N-graph with an equal S-scale (a dummy scale).

Step 2 By superimposing the two equal S-scales, the two nomograms are combined into a combination chart. The S-scale need not be calibrated, since values are not read from it.

Step 3 The addition of the variables can be handled at the left of the chart. The S-scale is the turning scale from which the N-graph can be used to find the unknown variables.

PROBLEMS

The following problems are to be solved on Size A sheets ($8\frac{1}{2}'' \times 11''$) using the principles covered in this chapter. Problems that involve geometric construction and mathematical calculations should show the construction and calculations as part of the solutions. If the mathematical calculations are extensive, these should be included on a separate sheet.

Concurrent Scales

Construct concurrent scales for converting the following relationships of one type of unit to another. The range of units for the scales is given for each.

1. Kilometers and miles: 1.609 km = 1 mile, from 10 to 100 miles.

2. Liters and U.S. gallons: 1 liter = 0.2692 U.S. gallons, from 1 to 10 liters.

3. Knots and miles per hour: 1 knot = 1.15 miles per hour, from 0 to 45 knots.

4. Horsepower and British thermal units: 1 horsepower = 42.4 btu, from 0 to 1200 hp.

5. Centigrade and Fahrenheit: $°F = \frac{9}{5}°C + 32$, from $32°F$ to $212°F$.

6. Radius and area of a circle: Area = πr^2, from $r = 0$ to 10.

7. Inches and millimeters: 1 inch = 25.4 millimeters, from 0 to 5 inches.

8. Numbers and their logarithms: (use logarithm tables), numbers from 1 to 10.

Addition and Subtraction Nomographs

Construct parallel-scale nomographs to solve the following addition and subtraction problems.

9. $A = B + C$, where B = 0 to 10 and C = 0 to 5.

10. $Z = X + Y$, where X = 0 to 8 and Y = 0 to 12.

11. $Z = Y - X$, where X = 0 to 6 and Y = 0 to 24.

12. $A = C - B$, where C = 0 to 30 and B = 0 to 6.

13. $3W = 2V + U$, where U = 0 to 12 and V = 0 to 9.

14. $W = 3U + V$, where U = 0 to 10 and V = 0 to 10.

15. Electrical current at a circuit junction:

$$I = I_1 + I_2.$$

I = current entering the junction in amperes
I_1 = current leaving the junction, varying from 2 to 15 amps
I_2 = current leaving junction, varying from 7 to 36 amps.

16. Pressure change in fluid flowing in a pipe:

$$\Delta P = P_2 - P_1.$$

ΔP = pressure change between two points in pounds per square inch,

P_1 = pressure upstream, varying from 3 psi to 12 psi,

P_2 = pressure downstream, varying from 10 psi to 15 psi.

Multiplication and Division: Parallel Scales

Construct parallel-scale nomographs with logarithmic scales that will perform the following multiplication and division operations:

17. Area of a rectangle: Area = Height × Width, where H = 1 to 10 and W = 1 to 12.

18. Area of a triangle: A = Base × Height/2, where B = 1 to 10 and H = 1 to 5.

19. Electrical potential between terminals of a conductor: $E = IR$.

E = electrical potential in volts,

I = current, varying from 1 to 10 amperes,

R = resistance, varying from 5 to 30 ohms.

20. Pythagorean theorem: $C^2 = A^2 + B^2$.

C = hypotenuse of a right triangle in centimeters,

A = one leg of the right triangle, varying from 5 to 50 cm,

B = second leg of the right triangle, varying from 20 to 80 cm.

21. Allowable pressure on a shaft bearing: $P = ZN/100$.

P = pressure in pounds per square inch,

Z = viscosity of lubricant from 15 to 50 cp- (centipoises),

N = angular velocity of shaft from 10 to 1000 rpm.

22. Miles per gallon and automobile travels: mgp = miles/gallon. Miles vary from 1 to 500 and gallons from 1 to 24.

23. Cost per mile (cpm) of an automobile: cpm = cost/miles. Miles vary from 1 to 500 and cost varies from $1 to $28.

24. $R = S\sqrt{T}$ where S varies from 1 to 10 and T from 1 to 10.

25. Angular velocity of a rotating body: $W = V/R$.

W = angular velocity, in radians per second,

V = Peripheral velocity, varying from 1 to 100 meters per second,

R = radius, varying from 0.1 to 1 meter.

N-Graphs

Construct N-graphs that will solve the following equations.

26. Stress = P/A: where P varies from 0 to 1000 psi and A varies from 0 to 15 square inches.

27. Volume of a cylinder: $V = \pi r^2 h$

V = volume in cubic inches,

r = radius, varying from 5 to 10 feet,

h = height, varying from 2 to 20 inches.

28. Same as Problem 17.

29. Same as Problem 18.

30. Same as Problem 19.

31. Same as Problem 20.

32. Same as Problem 21.

33. Same as Problem 22.

34. Same as Problem 23.

35. Same as Problem 24.

36. Same as Problem 25.

Combination Nomographs

37. Construct a combination nomograph to express the law of sines: $a/\text{sine } A = b/\text{sine } B$. Assume that a and b vary from 0 to 10, and that A and B vary from 0° to 90°.

38. Construct a combination nomograph to determine the velocity of sound in a solid, using the formula

$$C = \sqrt{\frac{E + 4\mu/3}{p}}$$

where E varies from 10^6 to 10^7 psi, μ varies from 1×10^6 to 2×10^6 psi, and C varies from 1000 to 1500 fps. (*Hint*: rewrite the formula as $C^2 p = E + 4/3\ \mu$.)

EMPIRICAL EQUATIONS AND CALCULUS

34.1 EMPIRICAL DATA

Data gathered from laboratory experiments and tests of prototypes or from actual field tests are called **empirical data.** Often empirical data can be transformed to equation form by means of one of three types of equations to be covered here.

The analysis of empirical data begins with the plotting of the data on rectangular grids, logarithmic grids, and semilogarithmic grids. Curves are then sketched through each point to determine which of the grids renders a straight-line relationship (Fig. 34.1). When the data plots as a straight line, its equation may be determined. Each curve appears as a straight line in one of the graphs. We use this straight-line curve to write an equation for the data.

34.2 SELECTION OF POINTS ON A CURVE

Two methods of finding the equation of a curve are (1) the selected-points method and (2) the slope-intercept method. These are compared on a linear graph in Fig. 34.2.

Selected-Points Method

Two widely separated points, such as (1, 30) and (4, 60) can be selected on the curve. These points are substituted in the equation below:

$$\frac{Y - 30}{X - 1} = \frac{60 - 30}{4 - 1}$$

The resulting data for the equation is

$$Y = 10X + 20.$$

Slope-Intercept Method

To apply the slope-intercept method, the intercept on the y-axis where $X = 0$ must be known. If the x-axis is logarithmic, then the log of $X = 1$ is 0 and the intercept must be found above the value of $X = 1$.

In the slope-intercept method in Fig. 34.2, the data do not intercept the y-axis; therefore, the curve must be extended to find the intercept $B = 20$. The slope of the curve is found ($\Delta Y/\Delta X$) and substituted into the slope-intercept form to give the equation as $Y = 10X + 20$.

Other methods of converting data to equations are used, but the two methods illustrated here make the best use of the graphical process and are the most direct methods of introducing these concepts.

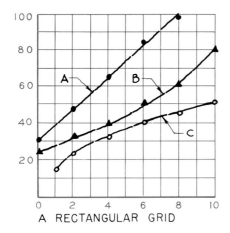

A RECTANGULAR GRID

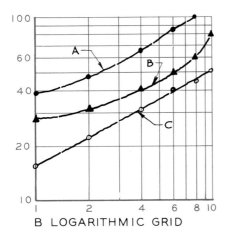

B LOGARITHMIC GRID

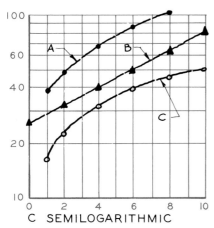

C SEMILOGARITHMIC

Fig. 34.1 Empirical data are plotted on each of these types of grids to determine which will render a straight-line plot. If the data can be plotted as a straight line on one of these grids, their equation can be found.

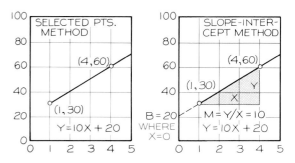

Fig. 34.2 The equation of a straight line on a grid can be determined by selecting any two points on the line. The slope-intercept method requires that the intercept be found where $X = 0$ on a semilog grid. This requires the extension of the curve to the Y-axis.

34.3 THE LINEAR EQUATION: $Y = MX + B$

The curve fitting the experimental data plotted in Fig. 34.3 is a straight line; therefore, we may assume that these data are linear, meaning that each measurement along the y-axis is directly proportional to x-axis units. We may use the slope-intercept form or the selected-points method to find the equation for the data.

In the slope-intercept method, two known points are selected along the curve. The vertical and horizontal differences between the coordinates of each of these points are determined to establish the right triangle shown in part A of the

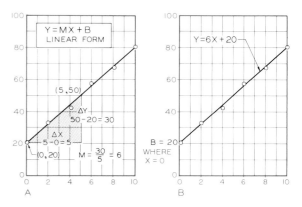

Fig. 34.3 (A) A straight line on an arithmetic grid will have an equation in the form $Y = MX + B$. The slope, M, is found to be 6. (B) The intercept, B, is found to be 20. The equation is written as $Y = 6X + 20$.

figure. In the slope-intercept equation, $Y = MX + B$, M is the tangent of the angle between the curve and the horizontal, B is the intercept of the curve with the y-axis where $X = 0$, and X and Y are variables. In this example $M = {}^{30}/_5 = 6$ and the intercept is 20.

If the curve has sloped downward to the right, the slope would have been negative. By substituting this information into the slope-intercept equation, we obtain $Y = 6X + 20$, from which we can determine values of Y by substituting any value of X into the equation.

The selected-points method could also have been used to arrive at the same equation if the intercept were not known. By selecting two widely separated points such as (2, 32) and (10, 80), one can write the equation in this form:

$$\frac{Y - 32}{X - 2} = \frac{80 - 32}{10 - 2}, \quad \therefore Y = 6X + 20,$$

which results in the same equation as was found by the slope-intercept method ($Y = MX + B$).

34.4 THE POWER EQUATION: $Y = BX^M$

Since the data shown plotted on a rectangular grid in Fig. 34.4 do not form a straight line, they cannot be expressed in the form of a linear equation.

However, when the data are plotted on a logarithmic grid, they form a straight line (Step 1). Therefore, we express the data in the form of a power equation in which Y is a function of X raised to a given power, or $Y = BX^M$. The equation of the data is obtained in much the same manner as was the linear equation, using the point where the curve intersects the Y-axis where $X = 0$, and letting M equal the slope of the curve. Two known points are selected on the curve to form the slope triangle. The engineers' scale can be used, when the cycles along the X- and Y-axes are equal, to measure the slope between the coordinates of the two points.

If the horizontal distance of the right triangle is drawn to be 1 or 10 or a multiple of 10, the vertical distance can be read directly. In Step 2, the slope M (tangent of the angle) is found to be 0.54. The intercept B is 7; thus the equation is $Y = 7X^{0.54}$, which can be evaluated for each value of Y by converting this power equation into the logarithmic form of $\log Y$:

$$\log Y = \log B + M \log X.$$
$$\log Y = \log 7 + 0.54 \log X.$$

Note that when the slope-intercept method is used, the intercept is found on the y-axis where $X = 1$. In Fig. 34.5, the y-axis at the left of the graph has an X-value of 0.1; consequently, the intercept is located midway across the graph where

Fig. 34.4 The power equation, $Y = BX^M$

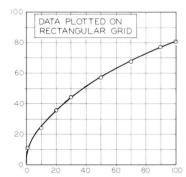

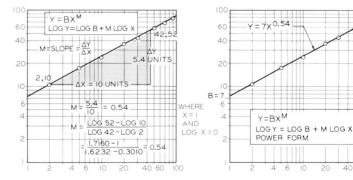

Given The data plotted on the rectangular grid give an approximation of a parabola. Since the data does not form a straight line on the rectangular grid, the equation will not be linear.

Step 1 The data forms a straight line on a logarithmic grid, making it possible to find its equation. The slope, M, can be found graphically with an engineer's scale, setting dX at 10 units and measuring the slope (dY) using the same scale.

Step 2 The intercept $B = 7$ is found on the y-axis where $X = 1$. The slope and intercept are substituted into the equation, which then becomes $Y = 7X^{0.54}$.

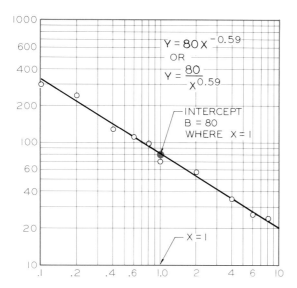

Fig. 34.5 When the slope-intercept equation is used, the intercept can be found only where $X = 1$. Therefore, in this example the intercept is found at the middle of the graph.

$X = 1$. This is analogous to the linear form of the equation, since the log of 1 is 0. The curve slopes downward to the right; thus the slope, M, is negative. The selected-points method can be applied to find the equation of the data as discussed in the previous article.

Base-10 logarithms are used in these examples, but natural logs could be used with e (2.718) as the base.

34.5 THE EXPONENTIAL EQUATION: $Y = BM^X$

The experimental data plotted in Fig. 34.6 form a curve, indicating that they are not linear. When the data are plotted on a semilogarithmic grid (Step 1) they approximate a straight line for which we can write the equation $Y = BM^X$, where B is the Y-intercept of the curve and M is the slope of the curve. The procedure for deriving the equation is shown in Step 2, in which two points are selected along the curve so that a right triangle can be drawn to represent the differences between the coordinates of the points selected. The slope of the curve is found to be

$$\log M = \frac{\log 40 - \log 6}{8 - 3} = 0.1648$$

Fig. 34.6 The exponential equation: $Y = BM^X$

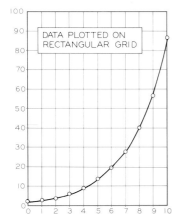

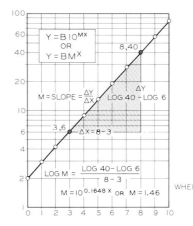

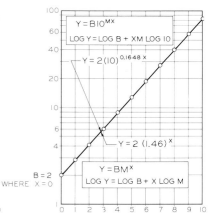

Given These data do not give a straight line on either a rectangular grid or a logarithmic grid. However, when plotted on a semilogarithmic grid, they give a straight line.

Step 1 The slope must be found by mathematical calculations; it cannot be found graphically. The slope may be written in either of the forms shown here.

Step 2 The intercept $B = 2$ is found where $X = 0$. The slope, M, and the intercept, B, are substituted into the equation to give $Y = 2(10)^{0.1648X}$ or $Y = 2(1.46)^X$.

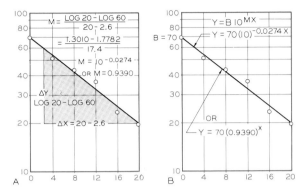

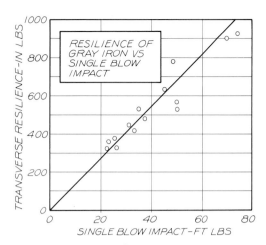

Fig. 34.7 When a curve slopes downward to the right, its slope is negative as calculated in part A. Two forms of a final equation are shown in part B by substitution.

or

$$M = (10)^{0.1648} = 1.46$$

The value of M can be substituted in the equation in the following manner:

$$Y = BM^X \quad \text{or} \quad Y = 2(1.46)^X,$$
$$Y = B(10)^{MX} \quad \text{or} \quad Y = 2(10)^{0.1648X},$$

where X is a variable that can be substituted into the equation to give an infinite number of values for Y. We can write this equation in its logarithmic form, which enables us to solve it for the unknown value of Y for any given value of X. The equation can be written as

$$\log Y = \log B + X \log M$$

or

$$\log Y = \log 2 + X \log 1.46.$$

The same methods are used to find the slope of a curve with a negative slope. The curve of the data in Fig. 34.7 slopes downward to the right; therefore, the slope is negative. Two points are selected in order to find which is the antilog of -0.0274. The intercept, 70, can be combined with the slope, M, to find the final equations as illustrated in Step 2 of Fig. 34.7.

34.6 APPLICATIONS OF EMPIRICAL GRAPHS

Figure 34.8 is an example of how empirical data can be plotted to compare the transverse strength and impact resistance of gray iron. Note that the data are somewhat scattered, but the best curve is drawn. Since the curve is a straight line on a linear graph, the equation of these data can be found by the equation

$$Y = MX + B.$$

Figure 34.9 is an example of how empirical data can be plotted to compare the specific weight (pounds per horsepower) of generators and hy-

Fig. 34.8 The relationship between the transverse strength of gray iron and impact resistance results in a straight line with an equation of the form $Y = MX + B$.

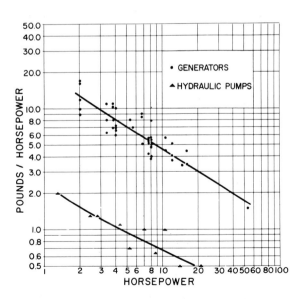

Fig. 34.9 Empirical data plotted on a logarithmic grid, showing the specific weight versus horsepower of electric generators and hydraulic pumps. The curve is the average of joints plotted. (Courtesy of General Motors Corporation, *Engineering Journal*.)

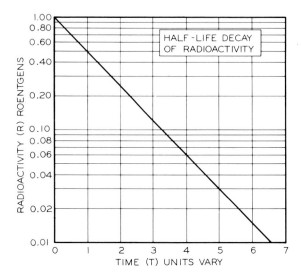

Fig. 34.10 The relative decay of radioactivity is plotted as a straight line on this semilog graph, making it possible for its equation to be found in the form $Y = BM^X$.

draulic pumps versus horsepower. Note that the weight of these units decreases linearly as the horsepower increases. Therefore, these data can be written in the form of the power equation

$$Y = BX^M$$

We obtain the equation of these data by applying the procedures covered in Section 34.3 and thus mathematically analyze these relationships.

The half-life decay of radioactivity is plotted in Fig. 34.10 to show the relationship of decay to time. Since the half-life of different isotopes varies, different units would have to be assigned to time along the X-axis; however, the curve would be a straight line for all isotopes. The exponential form of the equation discussed in Section 34.5 can be applied to find the equation for these data in the form of

$$Y = BM^X.$$

34.7 INTRODUCTION TO GRAPHICAL CALCULUS

Engineers, designers, and technicians must often deal with relationships between variables that must be solved using the principles of calculus. If the equation of the curve is known, traditional

methods of calculus will solve the problem. However, many experimental data cannot be converted to standard equations. In these cases, it is desirable to use the graphical method of calculus which provides relatively accurate solutions to irregular problems.

The two basic forms of calculus are (1) differential calculus, and (2) integral calculus. Differential calculus is used to determine the rate of change of one variable with respect to another. For example, the curve plotted in Fig. 34.11A represents the relationship between two variables. Note that the Y-variable increases as the X-variable increases. The rate of change at any instant along the curve is the slope of a line that is tangent to the curve at that particular point. This exact slope is often difficult to determine graphically; consequently, it can be approximated by constructing a chord at a given interval, as shown in Fig. 34.11A. The slope of this chord can be measured by finding the tangent of $\Delta Y / \Delta X$.

This slope can represent miles per hour, weight versus length, or a number of other meaningful rates that are important to the analysis of data.

Integral calculus is the reverse of differential calculus. Integration is the process of finding the area under a given curve, which can be thought of generally as the product of the two variables plotted on the x- and y-axes. The area under a curve is approximated by dividing one of the variables into a number of very small intervals, which become small rectangular areas at a particular zone under the curve, as shown in Fig. 34.11B. The bars are extended so that as much of the square end of

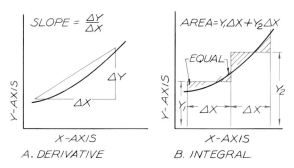

Fig. 34.11 The derivative of a curve is the change at any point that is the slope of the curve, Y/X. The integral of a curve is the cumulative area enclosed by the curve, which is the summation of the products of the areas.

the bar is under the curve as above the curve and the average height of the bar is, therefore, near its midpoint.

34.8 GRAPHICAL DIFFERENTIATION

Graphical differentiation is defined as the determination of the rate of change of two variables with respect to each other at any given point. Figure 34.12 illustrates the preliminary construction of the derivative scale that would be used to plot a continuous derivative curve from the given data.

Step 1   The original data are plotted graphically and the axes are labeled with the proper units of measurement. A chord can be constructed to estimate the maximum slope by inspection. In the given curve, the maximum slope is estimated to be 2.3. A vertical scale is constructed in excess of this to provide for the plotting of slopes that may exceed the estimate. This ordinate scale is drawn to a convenient scale to facilitate measurement.

Step 2   A known slope is plotted on the given data grid. This slope need not be related to the

curve in any way. In this case, the slope can be read directly as 1.

Step 3   The pole can be found by drawing a line from the ordinate of 1 (the known slope) on the derivative scale parallel to the slope line. These similar triangles are used to obtain the pole, which will be used in determining the derivative curve.

The steps in completing the graphical differentiation are given in Fig. 34.13. Note that the same horizontal intervals used in the given curve are projected directly beneath on the derivative scale. The maximum slope of the data curve is estimated to be slightly less than 12. A scale is selected that will provide an ordinate that will accommodate the maximum slope. A line is drawn from point 12 on the ordinate axis of the derivative grid that is parallel to the known slope on the given curve grid. The point of intersection of this line and the extension of the x-axis is the pole point.

A series of chords is constructed on the given curve. Lines are constructed parallel to these chords through point P and extended to the y-axis of the derivative grid to locate bars at each interval. Note that the interval between 0 and 1 was divided in half to provide a more accurate plot.

Fig. 34.12 Scales for graphical differentiation

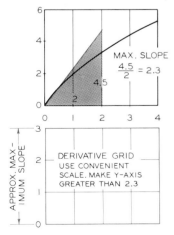

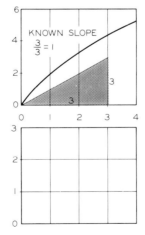

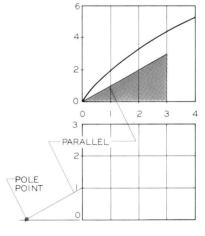

Step 1 The maximum slope of a curve is found by constructing a line tangent to the curve where it is steepest. The maximum slope, 2.3, is found and the derivative grid is laid off with a maximum ordinate of 3.0.

Step 2 A known slope is found on the given grid; this value is 1 in this example. The known slope has no relationship to the curve at this point.

Step 3 Construct a line from 1 on the Y-axis of the derivative grid that is parallel to the slope of the triangle in the given grid. This line locates the pole point where it crosses the extension of the X-axis.

Fig. 34.13 Graphical differentiation

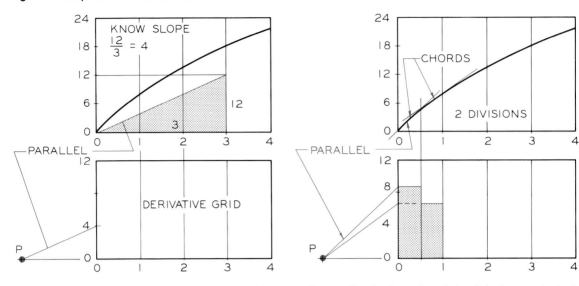

Required Find the derivative curve of the given data.

Step 1 Find the derivative grid and the pole point using the construction illustrated in Fig. 34.12.

Step 2 Construct a series of chords between selected intervals on the given curve and draw lines parallel to these chords through point P on the derivative grid. These lines locate the heights of bars in their respective intervals. The first interval is divided in two since the curve is changing sharply in this interval.

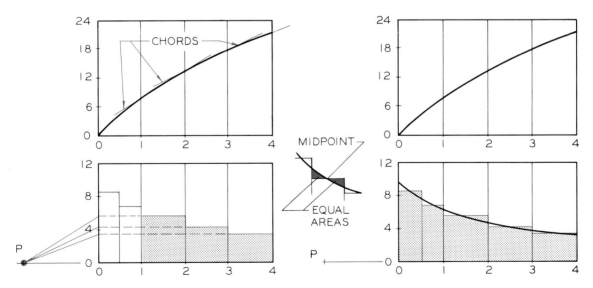

Step 3 Additional chords are drawn in the last two intervals. Lines parallel to these chords are drawn from the pole point to the Y-axis to find additional bars in their respective intervals.

Step 4 The vertical bars represent the different slopes of the given curve at different intervals. The derivative curve is drawn through the midpoints of the bars so that the areas under and above the bars are approximately equal.

The curve is the sharpest in this interval. A smooth curve is constructed through the top of these bars in such a manner that the area above the horizontal top of the bar is the same as that below it. This curve represents the derivative of the given data. The rate of change, $\Delta Y / \Delta X$, can be found at any interval of the variable X by reading directly from the graph at the value of X in question.

34.9 APPLICATIONS OF GRAPHICAL DIFFERENTIATION

The mechanical handling shuttle shown in Fig. 34.14 is used to convert rotational motion into controlled linear motion. A scale drawing of the linkage components is given so that graphical analysis can be applied to determine the motion resulting from this system.

The linkage is drawn to show the end positions of point P, which will be used as the zero point for plotting the travel versus the degrees of revolution. Since rotation is constant at one revolution per three seconds, the degrees of revolution can be converted to time, as shown in the data curve given at the top of Fig. 34.15. The drive crank, R_1, is revolved at 30° intervals, and the distance that point P travels from its end position is plotted on the graph, as shown in the given data. This gives the distance-versus-time relationship.

We determine the ordinate scale of the derivative grid by estimating the maximum slope of the given data curve, which is found to be a little less than 100 in./sec. A slope of 40 is drawn on the given data curve; this will be used in determining the location of pole P in the derivative grid. From point 40 on the derivative ordinate scale, we draw a line parallel to the known slope, which is found

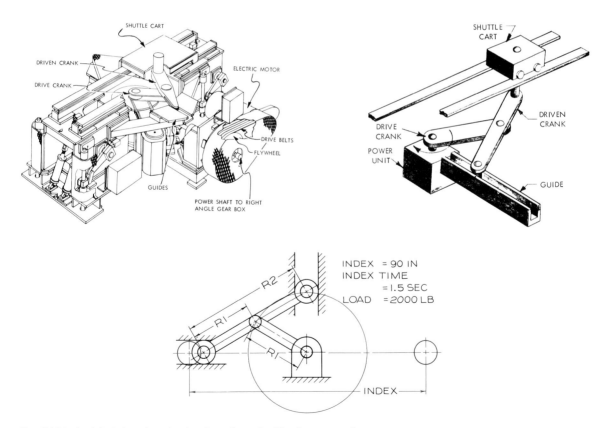

INDEX   = 90 IN
INDEX TIME
         = 1.5 SEC
LOAD   = 2000 LB

Fig. 34.14 A pictorial and scale drawing of an electrically powered mechanical handling shuttle used to move automobile parts on an assembly line. (Courtesy of General Motors Corporation.)

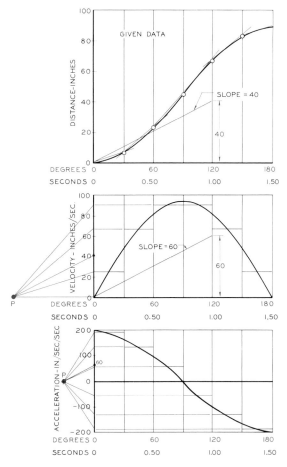

Fig. 34.15 Graphical determination of velocity and acceleration of the mechanical handling shuttle by differential calculus.

on the given grid. Point P is the point where this line intersects the extension of the x-axis.

A series of chords are drawn on the given curve to approximate the slope at various points. Lines are constructed through point P of the derivative scale parallel to the chord lines of the given curve and extended to the ordinate scale. The points thus obtained are then projected across to their respective intervals to form vertical bars. A smooth curve is drawn through the top of each of the bars to give an average of the bars. This curve can be used to find the velocity of the shuttle in inches per second at any time interval.

The construction of the second derivative curve, the acceleration, is very similar to that of the first derivative. By inspecting the first deriva-

tive, we estimate the maximum slope to be 200 in./sec/sec. An easily measured scale is established for the ordinate scale of the second derivative curve. Point P is found in the same manner as for the first derivative.

Chords are drawn at intervals on the first derivative curve. Lines are drawn parallel to· these chords from point P in the second derivative curve to the y-axis, where they are projected horizontally to their respective intervals to form a series of bars. A smooth curve is drawn through the tops of the bars to give a close approximation of the average areas of the bars. Note that a minus scale is given for the acceleration curve to indicate deceleration.

The maximum acceleration is found to be at the extreme endpoints and the minimum acceleration is at 90°, where the velocity is the maximum. It can be seen from the velocity and acceleration plots that the parts being handled by the shuttle are accelerated at a rapid rate until the maximum velocity is attained at 90°, at which time deceleration begins and continues until the parts come to rest.

34.10 GRAPHICAL INTEGRATION

Integration is the process of determining the area (product of two variables) under a given curve. For example, if the y-axis were pounds and the x-axis were feet, the integral curve would give the product of the variables, foot-pounds, at any interval of feet along the x-axis. Figure 34.16 depicts the method of constructing scales for graphical integration.

Step 1   It is customary to locate the integral curve above the given data curve, since the integral will be an equation raised to a higher power. A line is drawn through the given data curve to approximate the total area under the curve. The approximate area is 4 times 5, or 20, square units of area. The ordinate scale is drawn on the integral curve in excess of 20 units to provide a margin for any overage. The horizontal scale intervals are projected from the given curve to the integral grid.

Step 2   The ordinate at any point on the integral scale will have the same numerical value as the area under the curve as measured from the origin

Fig. 34.16 Scales for graphical integration

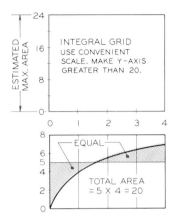

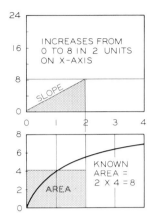

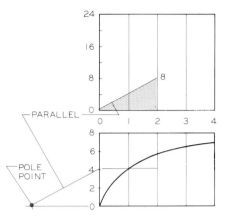

Step 1 To determine the maximum value on the Y-axis of the integral grid, a line is drawn to approximate the area under the curve. This is found to be 20, and the Y-axis is constructed with 24 as the maximum value.

Step 2 A known area, 8 in this case, is found in the given grid. A slope line from 0 to 8 is constructed in the integral grid directly above the known area, which establishes the integral for this model.

Step 3 A line is drawn from 4 on the Y-axis of the given grid parallel to the slope line in the integral grid. This line from 4 crosses the extension of the X-axis to locate the pole point.

to that point on the given data grid. The ordinate at point 2 on the x-axis directly above the rectangle must be equal to its area of 8. A slope is drawn from the origin to the ordinate of 8.

Step 3 Point P is found by drawing a line from point 4 on the given grid parallel to the slope established in the integral grid. This line intersects the extension of the x-axis at point P. This point will be used to find the integral curve.

The technique illustrated in Fig. 34.17 can be applied to most integration problems. The equation of the given curve is $Y = 2X^2$, which can also be integrated mathematically as a check.

From the given grid, the total area under the curve can be estimated to be less than 40 units. This value becomes the maximum height of the y-axis on the integral curve. A convenient scale is selected, units are assigned to the ordinate, and the pole point, P, is found.

A series of vertical bars is constructed to approximate the areas under the curve at these intervals. The narrower the bars, the more accurate will be the resulting calculations. Notice that the interval between 1 and 2 was divided in half to provide a more accurate plot. The top lines of the bars are extended horizontally to the y-axis, where the points are then connected by lines to point P.

Lines are drawn parallel to AP, BP, CP, DP, and EP in the integral grid to correspond to the respective intervals in the given grid. The intersection points of the chords are connected by a smooth curve—the integral curve. This curve gives the cumulative product of the X- and Y-variables at any value along the x-axis. For example, the area under the curve at $X = 3$ can be read directly as 18.

Mathematical integration gives the following result for the area under the curve from 0 to 3:

$$\text{Area } A = \int_0^3 Y\,dX, \quad \text{where } Y = 2X^2;$$

$$A = \int_0^3 2X^2 dX = \tfrac{2}{3}X^3 \Big|_0^3 = 18.$$

34.11 APPLICATIONS OF GRAPHICAL INTEGRATION

Integration is commonly used in the study of the strength of materials to determine shear, moments, and deflections of beams. An example problem of this type is shown in Fig. 34.18 in which a truck exerts a total force of 36,000 lb on a beam that is used to span a portion of a bridge. The first step is to determine the resultants supporting each end of the beam.

Fig. 34.17 Graphical integration

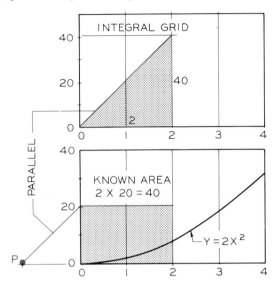

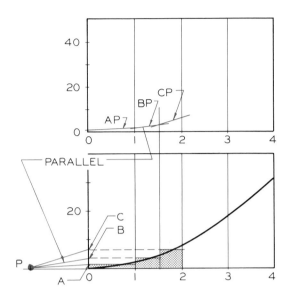

Required Plot the integral curve of the given data.

Step 1 Find the pole point, *P*, using the technique illustrated in Fig. 34.16.

Step 2 Construct bars at intervals to approximate the areas under the curve. The interval from 1 to 2 was divided in half to improve the accuracy of the approximation. The heights of the bars are projected to the *Y*-axis and lines are drawn to the pole point. Sloping lines *AP, BP,* and *CP* are drawn in their respective intervals and parallel to the lines drawn to the pole, *P*.

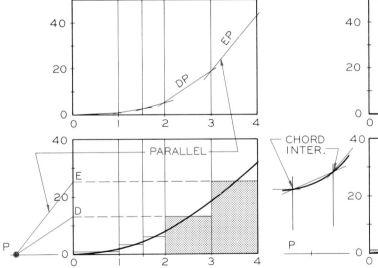

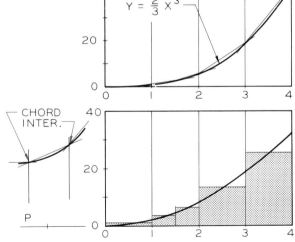

Step 3 Additional bars are drawn from 2 to 4 on the *X*-axis. The heights of the bars are projected to the *Y*-axis and rays are drawn to the pole point, *P*. Lines *DP* and *DE* are drawn in their respective intervals and parallel to their rays in the integral grid.

Step 4 The straight lines connected in the integral grid represent chords of the integral curve. Construct the integral curve to pass through the points where the chords intersect. Any ordinate value on the integral curve represents the cumulative area under the given curve from zero to that point on the *X*-axis.

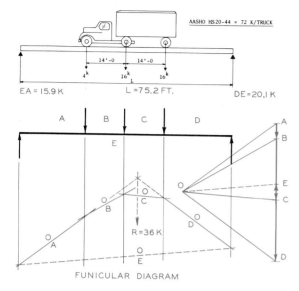

AASHO HS 20-44 = 72 K/TRUCK

14'-0   14'-0

4ᵏ   16ᵏ   16ᵏ

EA = 15.9 K   L = 75.2 FT.   DE = 20.1 K

A   B   C   D

E

O   O

B   C

R = 36 K

O   A

O   E

FUNICULAR DIAGRAM

A
B
E
C
D

Fig. 34.18 Determination of the forces on a beam of a bridge and its total resultant.

A scale drawing of the beam is made with the loads concentrated at their respective positions. A force diagram is drawn using Bow's notational system for laying out the vectors in sequence. Pole point O is located, and rays are drawn from the ends of each vector to point O. The lines of force in the load diagram at the top of the figure are extended to the funicular diagram.

Then lines are drawn parallel to the rays between the corresponding lines of force. For example, ray OA is drawn in the A-interval in the funicular diagram. The closing ray of the funicular diagram, OE, is transferred to the vector diagram by drawing a parallel through point O to locate point E. Vector DE is the right-end resultant of 20.1 kips (one kip equals 1000 lb) and EA is the left-hand resultant of 15.9 kips. The origin of the resultant force of 36 kips is found by extending OA and OD in the funicular diagram to their point of intersection.

From the load diagram shown in Fig. 34.19 we can, by integration, find the shear diagram, which indicates the points in the beam where failure is most critical. Since the applied loads are concentrated rather than uniformly applied, the shear diagram will be composed of straight-line segments. In the shear diagram, the left-end resultant of 15.9 kips is drawn to scale from the axis. The first load of 4 kips, acting in a downward di-

rection, is subtracted from this value directly over its point of application, which is projected from the load diagram. The second load of 16 kips also exerts a downward force and so is subtracted from the 11.9 kips (15.9 − 4). The third load of 16 kips is also subtracted, and the right-end resultant will bring the shear diagram back to the x-axis. It can be seen that the beam must withstand maximum shear at each support and minimum shear at the center.

The moment diagram is used to evaluate the bending characteristics of the applied loads in foot-pounds at any interval along the beam. The ordinate of any X-value in the moment diagram must represent the cumulative foot-pounds in the shear diagram as measured from either end of the beam.

Pole point P is located in the shear diagram by applying the method described in Fig. 34.16. A rectangular area of 200 ft-kips is found in the shear diagram. We estimate the total area to be less than 600 ft-kips; so we select a convenient scale that will allow an ordinate scale of 600 units for

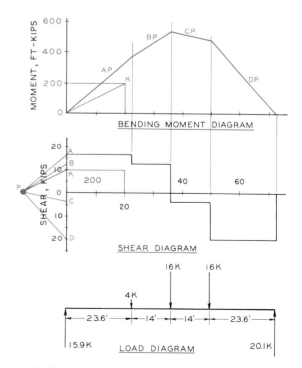

Fig. 34.19 Determination of shear and bending moment by graphical integration.

the moment diagram. We draw a known area of 200 (10 × 20) on the shear diagram. A diagonal line in the moment diagram is drawn that slopes upward from 0 to 200, where X = 20. The diagonal, OK, is transferred to the shear diagram, where it is drawn from the ordinate of the given rectangle to point P on the extension of the x-axis. Rays AP, BP, CP, and DP are found in the shear diagram by projecting horizontally from the various values of shear.

In the moment diagram, these rays are then drawn in their respective intervals to form a straight-line curve that represents the cumulative area of the shear diagram, which is in units of ft-kips. Maximum bending will occur at the center of the beam, where the shear is zero. The bending is scaled to be about 560 ft-kips. The beam selected for this span must be capable of withstanding a shear of 20.1 kips and a bending moment of 560 ft-kips.

PROBLEMS

The following problems are to be solved on Size A sheets (8½" × 11"). Solutions involving mathematical calculations should show these calculations on separate sheets if space is not available on the sheet where the graphical solution is drawn. Legible lettering practices and principles of good layout should be followed when constructing these solutions.

Empirical Equations—Logarithmic

1. Find the equation of the data shown in the following table. The empirical data compare input voltage (V) with input current in amperes (I) to a heat pump.

y-axis	V	0.8	1.3	1.75	1.85
x-axis	I	20	30	40	45

2. Find the equation of the data in the following table. The empirical data give the relationship between peak allowable current in amperes (I) with the overload operating time in cycles at 60 cycles per second (C).

y-axis	I	2000	1840	1640	1480	1300	1200	1000
x-axis	C	1	2	5	10	20	50	100

3. Find the equation of the data in the following table. The empirical data for a low-voltage circuit breaker used on a welding machine give the

maximum loading during weld in amperes (rms) for the percent of duty (pdc).

y-axis	rms	7500	5200	4400	3400	2300	1700
x-axis	pdc	3	6	9	15	30	60

4. Construct a three-cycle times three-cycle logarithmic graph to find the equation of a machine's vibration during operation. Plot vibration displacement in mills along the y-axis and vibration frequency in cycles per minute (cpm) along the x-axis. Data: 100 cpm, 0.80 mills; 400 cpm, 0.22 mills; 1000 cpm, 0.09 mills; 10,000 cpm, 0.009 mills; 50,000 cpm, 0.0017 mills.

5. Find the equation of the data in the following table that compares the velocities of air moving over a plane surface in feet per second (v) at different heights in inches (y) above the surface. Plot y-values on the y-axis.

y	v
0.1	18.8
0.2	21.0
0.3	22.6
0.4	24.1
0.6	26.0
0.8	27.3
1.2	29.2
1.6	30.6
2.4	32.4
3.2	33.7

6. Find the equation of the data in the following table that shows the distance traveled in feet

(s) at various times in seconds (t) of a test vehicle. Plot s on the y-axis and t on the x-axis.

t	s
1	15.8
2	63.3
3	146.0
4	264.0
5	420.0
6	580.0

Empirical Equations—Linear

7. Construct a linear graph to determine the equation for the yearly cost of a compressor in relationship to the compressor's size in horsepower. The yearly cost should be plotted on the y-axis and the compressor's size in horsepower on the x-axis. Data: 0 hp, $0; 50 hp, $2100; 100 hp, $4500; 150 hp, $6700; 200 hp, $9000; 250 hp, $11,400. What is the equation of these data?

8. Construct a linear graph to determine the equation for the cost of soil investigation by boring to determine the proper foundation design for varying sizes of buildings. Plot the cost of borings in dollars along the y-axis and the building area in sq ft along the x-axis. Data: 0 sq ft, $0; 25,000 sq ft, $35,000; 50,000 sq ft, $70,000; 750,000 sq ft, $100,000; 1,000,000 sq ft, $130,000.

9. Find the equation of the empirical data plotted in Fig. 34.8.

10. Plot the data in the table below on a linear graph and determine its equation. The empirical data show the deflection in centimeters of a spring (d) when it is loaded with different weights in kilograms (W). Plot W along the x-axis and d along the y-axis.

w	d
0	0.45
1	1.10
2	1.45
3	2.03
4	2.38
5	3.09

11. Plot the data in the table below on a linear graph and determine its equation. The empirical data show the temperatures that are read from a Fahrenheit thermometer at B and a centigrade thermometer at A. Plot the A-values along the x-axis and the B-values along the y-axis.

°A	°B
−6.8	20.0
6.0	43.0
16.0	60.8
32.2	90.0
52.0	125.8
76.0	169.0

Empirical Equations— Semilogarithmic

12. Construct a semilog graph of the following data to determine their equation. The y-axis should be a two-cycle log scale and the x-axis a 10-unit linear scale. Plot the voltage (E) along the y-axis and time (T) in sixteenths of a second along the x-axis to represent resistor voltage during capacitor charging. Data: 0, 10 volts; 2, 6 volts; 4, 3.6 volts; 6, 2.2 volts; 8, 1.4 volts; 10, 0.8 volts.

13. Find the equation of the data that is plotted in Fig. 34.10.

14. Construct a semilog graph of the following data to determine their equation. The y-axis should be a three-cycle log scale and the x-axis a linear scale from 0 to 250. These data give a comparison of the reduction factor, R (y-axis), with the mass thickness per square foot (x-axis) of a nuclear protection barrier. Data: 0, 1.0R; 100, 0.9R; 150, 0.028R; 200, 0.009R; 300, 0.0011R.

15. An engineering firm is considering its expansion by reviewing its past sales that are shown in the table below. Their years of operation are represented by x and N is their annual profit in tens of thousands. Plot x along the x-axis and N along the y-axis and determine the equation of their progress.

x	N
1	0.05
2	0.08
3	0.12
4	0.20
5	0.32
6	0.51
7	0.80
8	1.30
9	2.05
10	3.25

Empirical Equations—General Types

16–21. Plot the experimental data shown in Table 34.1 on the grid where the data will appear as straight-line curves. Determine the equations of the data.

Calculus—Differentiation

22. Plot the equation $Y = X^3/6$ as a rectangular graph. Graphically differentiate the curve to determine the first and second derivatives.

23. Plot the following data on a graph and find the derivative curve of the data on a graph placed below the first: $Y = 2X^2$.

24. Plot the following equation on a graph and find the derivative curve of the data on a graph placed below the first: $4Y = 8 - X^2$.

25. Plot the following data on a graph and find the derivative curve of the data on a graph placed below the first: $3Y = X^2 + 16$.

26. Plot the following data on a graph and find the derivative curve of the data on a graph placed below the first: $X = 3Y^2 - 5$.

Calculus—Integration

27. Plot the following equation on a graph and find the integral curve of the data on a graph placed above the first: $Y = X^2$.

28. Plot the following equation on a graph and find the integral curve of the data on a graph placed above the first: $Y = 9 - X^2$.

29. Plot the following equation on a graph and find the integral curve of the data on a graph placed above the first: $Y = X$.

30. Using graphical calculus, analyze a vertical strip 12″ wide on the inside face of the dam in Fig. 34.20. The force on this strip will be 52.0 lb/in. at the bottom of the dam. The first graph will be pounds per inch (ordinate) versus height in

Table 34.1

A	x	0	40	80	120	160	200	240	280		
	y	4.0	7.0	9.8	12.5	15.3	17.2	21.0	24.0		
B	x	1	2	5	10	20	50	100	200	500	1000
	y	1.5	2.4	3.3	6.0	9.2	15.0	23.0	24.0	60	85
C	x	1	5	10	50	100	500	1000			
	y	3	10	19	70	110	400	700			
D	x	2	4	6	8	12	14				
	y	6.5	14.0	32.0	75.0	320	710				
E	x	0	2	4	6	8	10	12	14		
	y	20	34	53	96	115	270	430	730		
F	x	0	1	2	3	4	5	6	7	8	9	10
	y	1.8	2.1	2.2	2.5	2.7	3.0	3.4	3.7	4.1	4.5	5.0

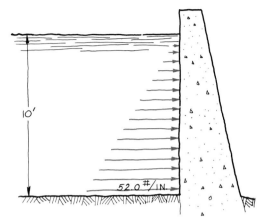

Fig. 34.20 Pressure on a 12"-wide section of a dam. (Problem 30.)

inches (abscissa). The second graph will be the integral of the first to give shear in pounds (ordinate) versus height in inches (abscissa). The third will be the integral of the second graph to give the moment in inch-pounds (ordinate) versus height in inches (abscissa). Convert these scales to give feet instead of inches.

31. A plot plan shows that a tract of land is bounded by a lake front (Fig. 34.21). By graphical integration, determine a graph that will represent the cumulative area of the land from point A to E. What is the total area? What is the area of each lot?

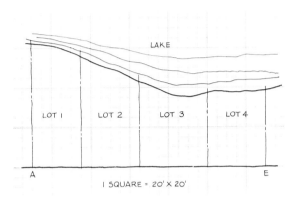

Fig. 34.21 Plot plan of a tract bounded by a lake front. (Problem 31.)

PIPE DRAFTING

35.1 INTRODUCTION

An understanding of pipe drafting begins with a familiarity of the types of pipe that are available. The commonly used types of pipe are (1) steel pipe, (2) cast-iron pipe, (3) copper, brass, and bronze pipe and tubing, and (4) plastic pipe.

The standards for the grades and weights for pipe and pipe fittings are specified by several organizations to ensure the uniformity of size and strength of the interchangeable components. Several of these organizations are: American National Standards Institute (ANSI), American Society for Testing and Materials (ASTM), American Petroleum Institute (API), and Manufacturers Standardization Society (MSS).

35.2 WELDED AND SEAMLESS STEEL PIPE

Traditionally, steel pipe has been specified in three weights—standard (STD), extra strong (XS), and double extra strong (XXS). These designations are still in use and their specifications are listed in the ANSI B 36.10-1979 standards. However, additional designations for pipe called *schedules* have been introduced to provide the pipe designer with a wider selection of pipe to cover more applications.

The ten schedules are: Schedule 10, Schedule 20, Schedule 30, Schedule 40, Schedule 60,

Schedule 80, Schedule 100, Schedule 120, Schedule 140, and Schedule 160. The wall thicknesses of the pipes vary from the thinnest, in Schedule 10, to the thickest in Schedule 160. The outside diameters are of a constant size for pipes of the same nominal size in all Schedules.

Schedule designations correspond to STD, XS, and XXS specifications in some cases, as is shown partially in Table 35.1. This table has been abbreviated from the ANSI B 36.10-1979 tables by omitting a number of the pipe sizes and Schedules. The most often used Schedules are 40, 80, and 120.

You will note in the table that pipes from the smallest size up to and including 12″ pipes are specified by their inside diameter (ID), which means that the outside diameter (OD) is larger than the specified size. The inside diameters are the same size as the nominal sizes of the pipe for Standard weight pipe. For XS and XXS pipe, the inside diameters are slightly different in size from the nominal size. Beginning with the 14-inch diameter pipes, the nominal sizes represent the outside diameters of the pipe.

The standard lengths for steel pipe are 20 feet and 40 feet. Seamless steel (SMLS STL) pipe is a smooth pipe with no weld seams along its length. Welded pipe is formed into a cylinder and is butt-welded (BW) at the seam, or it is joined with an electric resistance weld (ERW).

Welded and seamless pipe are the most widely used types of pipe for process and chemical refining plants. The type and grade of pipe with the proper characteristics must be selected to resist a wide range of pressures and corrosive agents.

35.3 CAST-IRON PIPE

Cast-iron pipe is used for the transportation of liquids, water, gas, and sewerage. When used as a sewerage pipe, cast-iron pipe is referred to as "soil pipe." Cast-iron pipe is available in diameter sizes from 3 inches to 60 inches.

The standard lengths of this type of pipe are 5 feet and 10 feet. Cast iron is more brittle and more subject to cracking when it is loaded than is steel pipe. Therefore, it should not be used where high pressures of weights will be applied to it.

35.4 COPPER, BRASS, AND BRONZE PIPING

Copper, brass, and bronze are used to manufacture piping and tubing for use in applications where there must be a high resistance to corrosive ele-

Table 35.1 DIMENSIONS AND WEIGHTS OF WELDED AND SEAMLESS STEEL PIPE (ANSI B 36.10–1979)

Inch Units				Identification		SI Units		
Inch Nominal size	O.D. Dia. (in.)	Wall Thk (in.)	Weight lbs/ft	*STD XS XXS	Sch. no.	O.D. Dia. (mm)	Wall Thk (mm)	Weight kg/m
½	0.84	0.11	0.85	STD	40	21.3	2.8	1.3
1	1.32	0.13	1.68	STD	40	33.4	3.4	2.5
1	1.3	0.18	2.17	XS	80	33.4	4.6	3.2
1	1.3	0.36	3.66	XXS		33.4	9.1	5.5
2	2.38	0.22	3.65	STD	40	60.3	3.9	5.4
2	2.38	0.22	5.02	XS	80	60.3	5.5	7.5
2	2.38	0.44	9.03	XXS		60.3	11.1	13.4
4	4.50	0.23	10.79	STD	40	114.3	6.0	16.1
4	4.50	0.34	14.98	XS	80	114.3	8.6	42.6
4	4.50	0.67	27.54	XXS		114.3	17.1	41.0
8	8.63	0.32	28.55	STD	40	219.1	8.2	42.6
8	8.63	0.50	43.39	XS	80	219.1	12.7	64.6
8	8.63	0.88	74.40	XXS		219.1	22.2	107.9
12	12.75	0.38	49.56	STD		323.0	9.5	67.9
12	12.75	0.50	65.42	XS		323.0	12.7	97.5
12	12.75	1.00	125.4	XXS	120	133.9	25.4	187.0
14	†14.00	0.38	54.57	STD	30	355.6	9.5	87.3
14	14.00	0.50	72.08	XS		355.6	12.7	107.4
18	18.00	0.38	70.59	STD		457	9.5	106.2
18	18.00	0.50	93.45	XS		457	12.7	139.2
24	24.00	0.38	94.62	STD	20	610	9.5	141.1
24	24.00	0.50	125.49	XS		610	12.7	187.1
30	30.00	0.38	118.65	STD		762	9.5	176.8
30	30.00	0.50	157.53	XS	20	762	12.7	234.7
40	40.00	0.38	158.70	STD		1016	9.5	236.5
40	40.00	0.50	210.90	XS		1016	12.7	314.2

*Standard (STD)
X-strong (XS)
XX-strong (XXS)

†Beginning with 14″ DIA pipe, the nominal size represents the outside diameter (O.D.)

This table has been compressed by omitting many of the available pipe sizes. The nominal sizes of pipes that are listed in the complete table are: ⅛″, ¼″,⅜″, ½″, ¾″, 1″, 1¼″, 1½″, 2″, 2½″, 3″, 3½″, 4″, 5″, 6″, 8″, 10″, 12″, 14″, 16″, 18″, 20″, 22″ . . . (at 2″ increments up to 60″).

ments, such as acidic soils and chemicals that are transmitted through the pipes. Copper pipe is used for gas and water transmission when the pipes are placed within or under concrete slab foundations of buildings. This ensures that the pipes will not have to be replaced because of corrosion. The standard length of pipes made of these nonferrous materials is 12 feet.

Tubing is a smaller-size pipe that can be easily bent when it is made of copper, brass, or bronze. An example of a system of tubing is shown in Fig. 35.1. The term piping is considered to apply to rigid pipes that are larger than tubes, usually in excess of 2″ in diameter.

35.5   MISCELLANEOUS PIPES

Other materials that are used to manufacture pipes are aluminum, asbestos-cement, concrete, polyvinyl chloride (PVC), and various other plastics.

Each of these materials has its special characteristics that make it desirable or economical for certain applications. The method of designing and detailing piping systems by the pipe drafter is essentially the same regardless of the piping material used.

35.6   PIPE JOINTS

The basic connection in a pipe system is the joint where two straight sections of pipe fit together. Three types of joints are illustrated in Fig. 35.2: *screwed*, *welded*, and *flanged*.

Screwed joints   are joined by pipe threads of the type covered in Chapter 17. A table of pipe-thread specifications is given in Appendix 10. Pipe threads are tapered at a ratio of 1 to 16 along the outside diameter (Fig. 35.3). As the pipes are screwed together, the threads bind to form a snug,

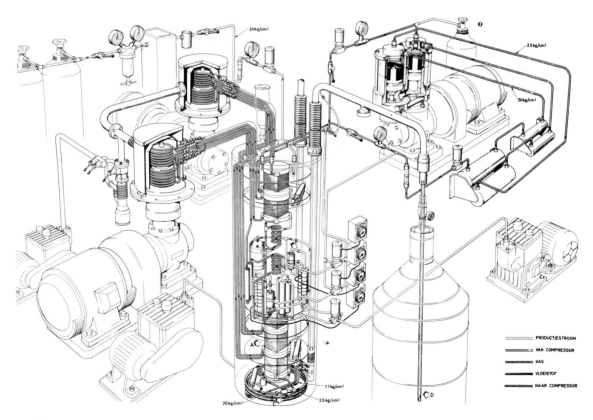

Fig. 35.1 The small pipes in this illustration are called tubes because they are less than 2″ in diameter and can be bent to form angles.

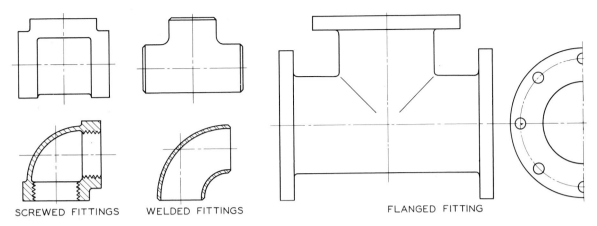

Fig. 35.2 The three basic types of joints are screwed, welded, and flanged joints.

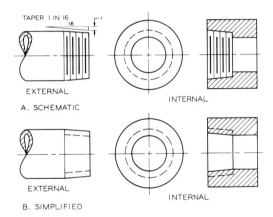

Fig. 35.3 A screwed pipe utilizes a type of threading called pipe threads that binds tightly as they are screwed together.

locking fit. A cementing compound is applied to the threads before joining, to improve the seal.

Flanged joints   shown in Fig. 35.4 are welded to the straight sections of pipe, which are then bolted together with a series of bolts around the perimeter of the flanges. Flanged joints form strong rigid joints that can withstand high pressure and permit disassembly of the joints when needed. Several types of flange faces are shown in Fig. 35.5 and in Appendix 13.

Welded joints   are joined by welded seams around the perimeter of the pipe to form butt welds. Welded joints are used extensively in "big inch" pipelines that are used for transporting petroleum products cross-country.

Bell and spigot (B&S)   joints are used to join cast-iron pipes, as illustrated in Fig. 35.6. The spigot is

WELDING NECK FLANGES   LAP JOINT FLANGES   SOCKET WELDING FLANGES   SLIP-ON WELDING FLANGES   THREADED FLANGES

Fig. 35.4 Types of flanged joints and the methods of attaching the flanges to the pipes. (Courtesy of Vogt Machine Co.)

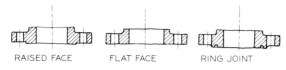

Fig. 35.5 Three types of flange faces are the raised face (RF), flat face (FF), and ring joint (RJ).

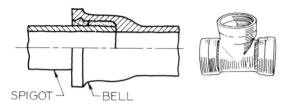

SPIGOT BELL

Fig. 35.6 A bell and spigot joint (B&S) is used to connect cast-iron pipes.

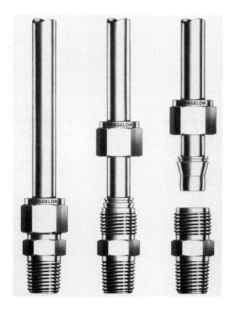

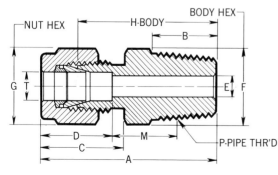

NUT HEX   H-BODY   BODY HEX

B

G T   E F

D   M

C   P-PIPE THR'D

A

Fig. 35.7 This fitting is used to attach small tubing. (Courtesy of Crawford Fitting Co.)

placed inside the bell and the two are sealed with molten lead or a commercially available sealing ring that snaps into position to form a sealed joint.

Soldering is used to connect smaller pipes and tubular connections. Soldering is usually limited to nonferrous tubing. However, screwed fittings are available to connect tubing, as shown in Fig. 35.7. In this example, the joint is sealed more firmly as the fitting is tightened.

35.7 SCREWED FITTINGS

A number of standard fittings used to assemble a piping system are shown in Fig. 35.8. The two types of graphical symbols that are used to represent fittings and pipe are *double-line* and *single-line symbols*.

Double-line symbols are pictorially more representative of the fittings and pipes since they are drawn to scale with double lines. Single-line symbols are more symbolic since the size of the pipe is drawn with a single line and the fittings are drawn as single-line schematic symbols. Either type of representation can be used to present a piping system in orthographic and pictorial views.

Fittings are available in three weights: standard (STD), extra strong (XS), and double extra strong (XXS). These weights match the standard weights of the pipes with which they will be connected. Other weights of fittings are available from manufacturers in addition to these, but these three weights are stocked by practically all suppliers.

A piping system of screwed fittings is shown in Fig. 35.9 with single-line symbols in a single-line system to call attention to them. These could just as well have been drawn using the double-line symbols. You should become familiar with the names and methods of specifying the various fittings by type and size.

The most common symbols for representing fittings are shown in Appendix 9. These have been extracted from ANSI Z 32.2.3 standards.

35.8 FLANGED FITTINGS

Flanges are used to connect fittings into a piping system when heavy loads are supported in large pipes and where pressures are great. Flanges are expensive and their use should be kept to a mini-

REDUCER   **HALF COUPLING**   **PIPE CAP**   **SQUARE HEAD PLUG**   **HEX. HD. PLUG**   **ROUND HEAD PLUG**   **HEXAGON BUSHING**   **FLUSH BUSHING**

DOUBLE-LINE SYMBOLS: SCREWED

SINGLE-LINE SYMBOLS: SCREWED

4 X 2 RED   3 X 3 HLF CPLG   1- CAP   2-SQ HD PLUG   3-HEX HD PLUG   4-RD HD PLUG   3 X 2 HEX BUSH.   2 X 1 FLUSH BUSH.

DESIGNATIONS

90° ELBOW   **TEE**   **45° ELBOW**   **CROSS**   **STREET ELBOW**   **LATERAL**   **COUPLING**

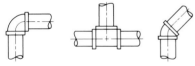 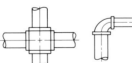  

DOUBLE-LINE SYMBOLS: SCREWED

 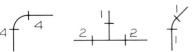 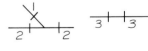

SINGLE-LINE SYMBOLS: SCREWED

4 X 90° ELL   2 X 2 X 1 TEE   1 X 45° ELL   3 X 3 X 2 X 1 CROSS   3 X 2 RED ELL   2 X 2 X 1 LAT   3 COUPL

DESIGNATIONS

Fig. 35.8 Standard fittings for screwed connections are shown in this example, along with the single-line and double-line symbols that are used to represent them. Nominal pipe sizes can be indicated by numbers placed near the joints. The major flow direction is labeled first with the branches labeled second. The large openings are labeled to precede the smaller openings.

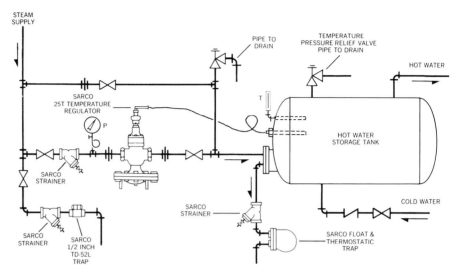

Fig. 35.9 This is a single-line piping system with the major valves represented as double-line symbols. (Courtesy of Sarco, Inc.)

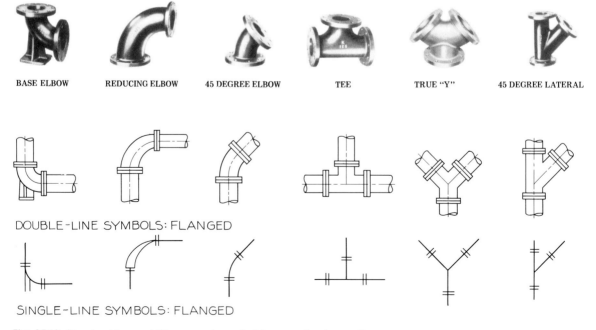

Fig. 35.10 Standard flanged fittings are shown in this example along with the single-line and double-line symbols that are used to represent them.

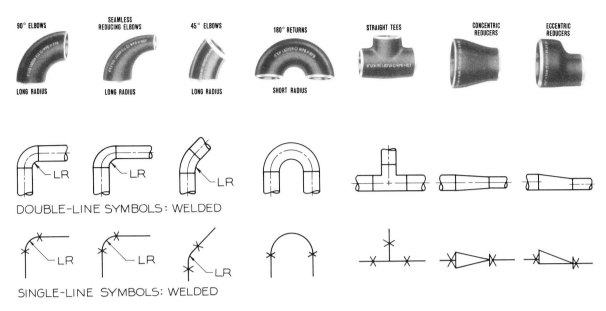

Fig. 35.11 Standard fittings, welded, are shown in this example along with single-line and double-line symbols that are used to represent them.

mum if other joining methods can be used. Flanges are connected to straight pipe sections by welding to match the flanges on the fittings so that they can be bolted together.

Examples of several fittings are drawn as double-line and single-line symbols in Fig. 35.10. The elbow is commonly referred to as an "ell" and it is available in angles of turn of 90° and 45° in both long and short radii. The long-radius ells have radii that are approximately 1.5 times the nominal diameter of the large end of the ell. The radius of a short-radius ell is equal to the diameter of the larger end. The long radius ells are noted as LR.

Tables of dimensions for 125 LB and 250 LB cast-iron fittings are given in Appendixes 11–13.

35.9 WELDED FITTINGS

Welding is a widely used method of joining pipes and fittings for permanent, pressure-resistant joints. Examples of double-line and single-line fittings connected by welding are shown in Fig. 35.11. Fittings are available with beveled edges that are prepared for welding.

A piping layout in Fig. 35.12 illustrates a series of welded joints with a double-line drawing. The location of the welded joints has been dimen-

sioned. Note that several flanged fittings have been welded into the system in order for the flanges to be used.

A comparison of flanged and welded joints is shown in Fig. 35.13. Double-line symbols are used in these comparisons.

35.10 VALVES

Valves are used to regulate the flow within a pipeline or to turn off the flow completely. Types of valves are gate, globe, angle, check, safety, diaphragm, float, and relief valves, to name a few. The three basic types—*gate, globe,* and *check valves*—are shown in Fig. 35.14 using single-line symbols.

Gate valves are used to turn the flow within a pipe on or off with the least restriction of flow through the valve. These valves are not meant to be used to regulate the degree of flow.

Globe valves are not only used to turn the flow on and off, but they are also used to regulate the flow to a desired level by adjusting the handwheel control.

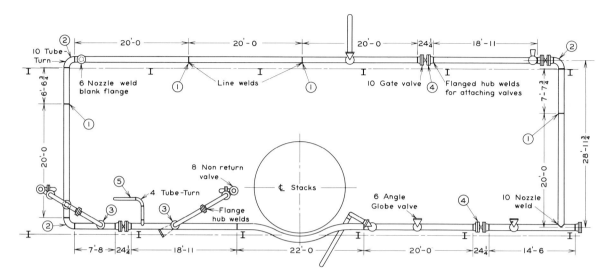

Fig. 35.12 A piping layout that uses a series of welded and flanged joints. This system is drawn using double-line symbols.

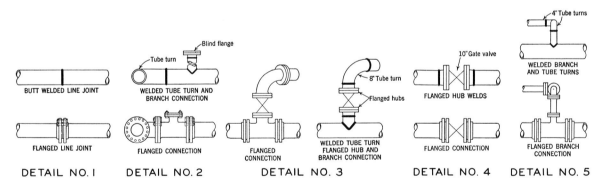

Fig. 35.13 A comparison of flanged and welded joints drawn using double-line symbols.

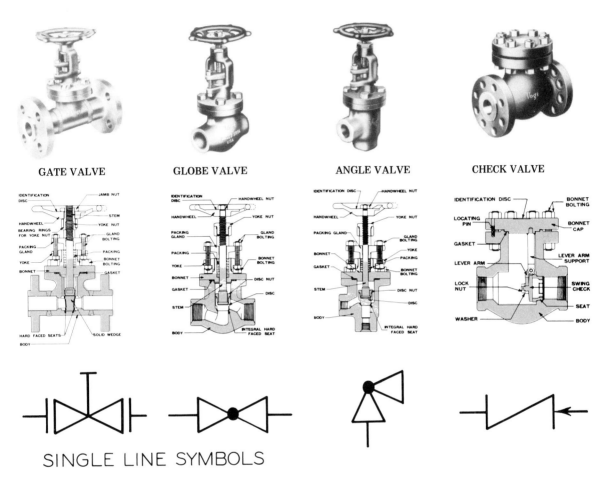

GATE VALVE   GLOBE VALVE   ANGLE VALVE   CHECK VALVE

SINGLE LINE SYMBOLS

Fig. 35.14 The three basic types of valves are gate, globe, and check valves.
(Photographs courtesy of Vogt Machine Co.)

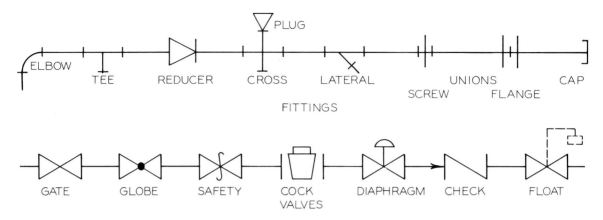

Fig. 35.15 Examples of types of valves and fittings.

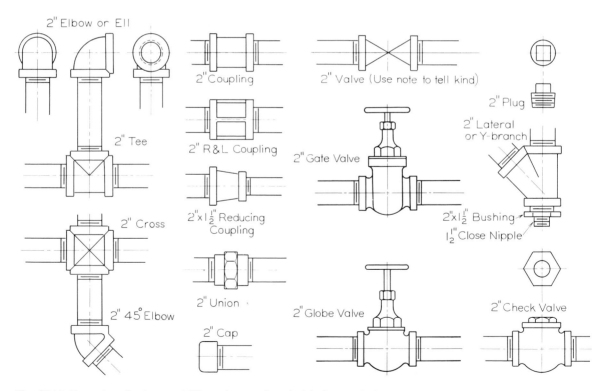

Fig. 35.16 Examples of valves and fittings drawn using double-line symbols.

Angle valves are types of globe valves that turn at 90° angles at bends in the piping system. They have the same controlling features as the straight globe valves.

Check valves permit the flow in the pipe in only one direction. A backward flow is checked by either a movable piston or a swinging washer which will be activated by a change of flow to close the opening (Fig. 35.14).

The symbols for the other types of valves can be found in Fig. 35.15. Examples of double-line symbols depicting valves and fittings are shown in Fig. 35.16.

35.11 FITTINGS IN ORTHOGRAPHIC VIEWS

Fittings and valves must be shown from any view in orthographic projection; consequently you must be familiar with the methods of representing different views of these components. Two and three views of typical fittings and valves are shown in Fig. 35.17. These fittings are drawn as single-line

screwed fittings, but the same general principles can be used to represent other types of joints as double-line drawings. Several views of fittings are given using double-line symbols in Fig. 35.16.

Observe the various views of the fittings and notice how the direction of an elbow can be shown by a slight variation in the different views. The same techniques are used to represent tees and laterals.

A piping system is shown in a single orthographic view in Fig. 35.18, where a combination of double-line and single-line symbols is drawn. Note that arrows are used to give the direction of flow in the system. Different joints are screwed, welded, and flanged. Horizontal elevation lines are given to dimension the heights of each horizontal pipe. Station $5 + 12 - 0\text{-}\frac{1}{4}$ represents a distance of 500 feet plus $12' - 0\text{-}\frac{1}{4}''$, or $512' - 0\text{-}\frac{1}{4}''$ from the beginning station point of $0 + 00$.

The dimensions in Fig. 35.18 are measured from the centerlines of the pipes and this is indicated by the CL symbols. In some cases, the elevations of the pipes are dimensioned to the bottom of the pipe, which is abbreviated as BOP. An example of this type can be seen dimensioned in Fig. 35.23.

35.12 PIPING SYSTEMS IN PICTORIAL

Isometric drawings of piping systems are very helpful in the representation of three-dimensional installations that would be difficult to interpret if drawn in orthographic projection. Isometric and axonometric drawings of piping systems are called "spool drawings." They can be drawn using either one-line or double-line symbols.

A three-dimensional piping system is drawn orthographically in Fig. 35.19 with top and front views. Although this is a relatively simple three-dimensional system, a thorough understanding of orthographic projection is required to read the drawing. Since a major role of pipe drafters is to clarify their drawings to be as understandable as possible, they must often rely on spool drawings as an aid.

In Fig. 35.20, the piping system is drawn with all of the pipes revolved into the same horizontal plane. You will notice that the vertical pipes and their fittings are drawn true size in the top view. This is called a **developed** pipe drawing. The fit-

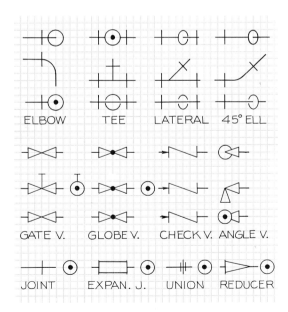

Fig. 35.17 Fittings and valves drawn here with single-line symbols are drawn differently in the various orthographic views. Observe these differences and how the various views are drawn.

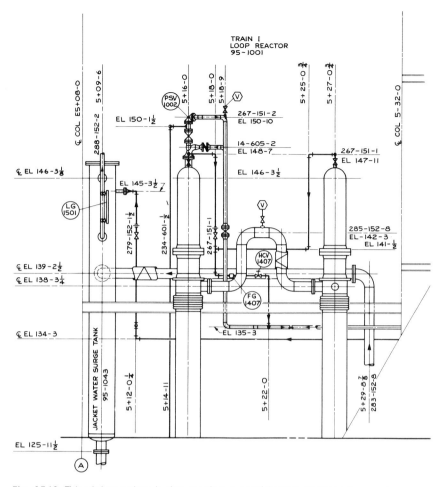

Fig. 35.18 This piping system is drawn using a combination of single-line and double-line symbols. The connections are shown as screwed, welded, and flanged. (Courtesy of Bechtel Corporation.)

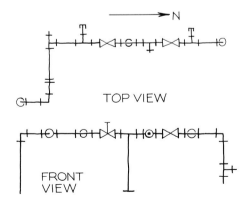

Fig. 35.19 A top and front view are used to represent this three-dimensional piping system using single-line symbols.

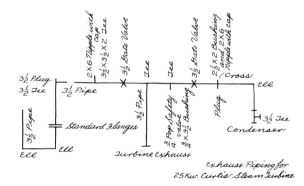

Fig. 35.20 The vertical pipes shown in Fig. 35.19 are revolved into the horizontal plane to form a developed drawing. The fittings and valves are noted on the sketch.

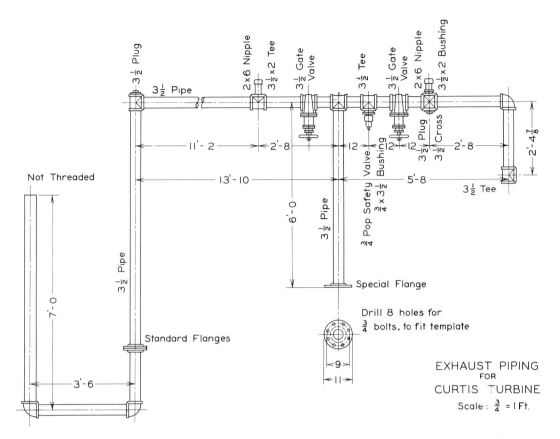

Fig. 35.21 A finished developed drawing that shows all of the components in the system true size with double-line symbols.

tings and pipe sizes are noted on this preliminary sketch from which the finished drawing will be made in Fig. 35.21.

An axonometric schematic is drawn in Fig. 35.22 to explain the three-dimensional relationship of the parts of the system. The rounded bends in the elbows in an isometric drawing can be constructed with ellipses using the isometric ellipse template, or the corners can be drawn square to reduce the effort and time required.

A north arrow is drawn on the plan view of the piping system in Fig. 35.19; this can be used to orient the isometric pictorial. This north direction is not necessarily related to compass north, but it is a direction that is selected parallel to a major set of pipes within the system. In the iso-

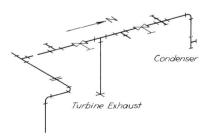

Fig. 35.22 An axonometric pictorial of the pipe system shown in Fig. 35.19 is drawn to give a three-dimensional pictorial of the system.

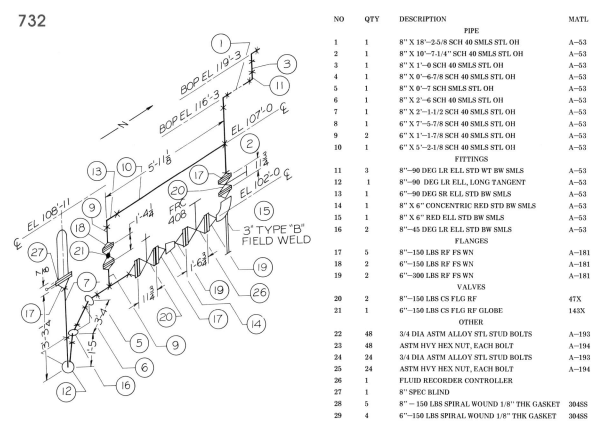

NO	QTY	DESCRIPTION	MATL
		PIPE	
1	1	8" X 18'–2-5/8 SCH 40 SMLS STL OH	A–53
2	1	8" X 10'–7-1/4" SCH 40 SMLS STL OH	A–53
3	1	8" X 1'–0 SCH 40 SMLS STL OH	A–53
4	1	8" X 0'–6-7/8 SCH 40 SMLS STL OH	A–53
5	1	8" X 0'–7 SCH SMLS STL OH	A–53
6	1	8" X 2'–6 SCH 40 SMLS STL OH	A–53
7	1	8" X 2'–1-1/2 SCH 40 SMLS STL OH	A–53
8	1	6" X 7'–5-7/8 SCH 40 SMLS STL OH	A–53
9	2	6" X 1'–1-7/8 SCH 40 SMLS STL OH	A–53
10	1	6" X 5'–2-1/8 SCH 40 SMLS STL OH	A–53
		FITTINGS	
11	3	8"–90 DEG LR ELL STD WT BW SMLS	A–53
12	1	8"–90 DEG LR ELL, LONG TANGENT	A–53
13	1	6"–90 DEG SR ELL STD BW SMLS	A–53
14	1	8" X 6" CONCENTRIC RED STD BW SMLS	A–53
15	1	8" X 6" RED ELL STD BW SMLS	A–53
16	2	8"–45 DEG LR ELL STD BW SMLS	A–53
		FLANGES	
17	5	8"–150 LBS RF FS WN	A–181
18	2	6"–150 LBS RF FS WN	A–181
19	2	6"–300 LBS RF FS WN	A–181
		VALVES	
20	2	8"–150 LBS CS FLG RF	47X
21	1	6"–150 LBS CS FLG RF GLOBE	143X
		OTHER	
22	48	3/4 DIA ASTM ALLOY STL STUD BOLTS	A–193
23	48	ASTM HVY HEX NUT, EACH BOLT	A–194
24	24	3/4 DIA ASTM ALLOY STL STUD BOLTS	A–193
25	24	ASTM HVY HEX NUT, EACH BOLT	A–194
26	1	FLUID RECORDER CONTROLLER	
27	1	8" SPEC BLIND	
28	5	8" – 150 LBS SPIRAL WOUND 1/8" THK GASKET	304SS
29	4	6"–150 LBS SPIRAL WOUND 1/8" THK GASKET	304SS

Fig. 35.23 This dimensioned isometric pictorial is called a "spool drawing." It is sufficiently complete that it can serve as a working drawing when used with the bill of materials.

Table 35.2 STANDARD ABREVIATIONS ASSOCIATED WITH PIPE SPECIFICATIONS

AVG	average	FS	forged steel	SPEC	specification
BC	bold circle	FSS	forged stainless steel	SR	short radius
BE	beveled ends	FW	field weld	SS	stainless steel
BF	blind flange	GALV	galvanized	STD	standard
BM	bill of materials	GR	grade	STL	steel
BOP	bottom of pipe	ID	inside diameter	STM	steam
B&S	bell & spigot	INS	insulate	SW	socketweld
BWG	Birmingham wire gage	IPS	iron pipe size	SWP	standard working pressure
CAS	cast alloy steel	LR	long radius	T&C	test connection
CI	cast iron	LW	lap weld	TE	threaded end
CO	clean out	MI	8.1malleable iron	TEMP	temperature
CONC.	concentric	MFG	manufacture	T&G	tongue & groove
CPLG	coupling	OD	outside diameter	TOS	top of steel
CS	carbon steel, cast steel	OH	open hearth	TYP	typical
DWG	drawing	PE	plain end–not beveled	VC	vitified clay
ECC	eccentric	PR	pair	WE	weld end
EF	electric furnace	RED	reducer	WN	weld neck
EFW	electric fusion weld	RF	raised face	WB	welded bonnet
ELEV	elevation	RTG or RJ	ring type joint	WT	weight
ERW	electric resistance weld	SCH	schedule	XS	extra strong
FF	flat face	SCRD	screwed	XXS	double extra strong
FLG	flange	SMLS	seamless		
FOB	flat on bottom	SO	slip-on		

metric drawing, it is preferred that the north arrow point to the upper left or upper right corner of the pictorial.

35.13 DIMENSIONED ISOMETRICS

An isometric drawing can be drawn as a fully dimensioned and specified drawing from which a piping system can be constructed. The spool drawing in Fig. 35.23 is an example where the specifications for the pipe, fittings, flanges, and valves are noted on the drawing and are itemized in the bill of materials.

A number of abbreviations are used to specify piping components and fittings by referring to the bill of materials. Many of the standard abbreviations that are associated with pipe drawings and specifications are given in Table 35.2. Part number 1, for example, is an 8″ diameter pipe of a Schedule 40 weight that is made of seamless steel by the open hearth (OH) process. Instead of OH, the ab-

breviation EF may be used, which is the abbreviation for electric furnace.

Under the column "materials," you will notice a code that begins with the letter A, such as A–53. The letter A is used to represent a grade of carbon steel that is listed in Table A of the ANSI B31.3: *Petroleum Refinery Piping Standards*. The codes for fittings, flanges, and valves are taken from the manufacturers' catalogs of these products.

A suggested format for spool drawings is given in Fig. 35.24. This format is used by the Bechtel Corporation, a major construction company, in designing and constructing pipelines and refineries.

35.14 VESSEL DETAILING

Vessels are containers, usually cylindrical in shape, that are used to contain petroleum products and other chemicals that are transported to them by pipelines. The cylinders can be installed in

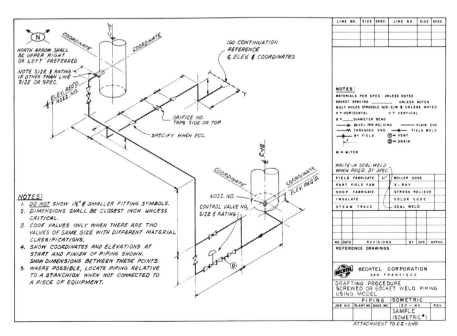

Fig. 35.24 A suggested format for preparing spool drawings by the Bechtel Corporation is shown here. (Courtesy of the Bechtel Corporation.)

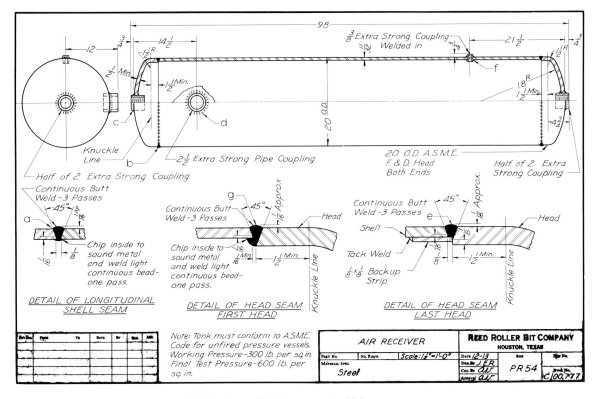

Fig. 35.25 A detail drawing of a cylindrical vessel that is connected into a pipe system.

vertical or horizontal positions. Vessels can also be spherical or ellipsoidal, although the cylindrical shape is the most common.

The pipe drafter must become familiar with vessels since they too are fittings of a sort within the overall piping system.

A detailed drawing of a cylindrical vessel is shown in Fig. 35.25. Its dimensions and specifications are given by following the general rules of working drawings that have been previously covered. In addition, the welded joints are specified to ensure that the vessel is properly fabricated to withstand the pressures and weights that it will be subjected to.

35.15   SUMMARY

The area of pipe drafting is a complex study in graphics and technology worthy of a sizable textbook on this field alone. The coverage here has been presented as an introduction to familiarize you with the basics of the field. The standards of pipe drafting vary to a notable degree from company to company.

PROBLEMS

1. On a Size A sheet, draw five orthographic views of the fittings listed below. The views should include the front view, the top view, the bottom view, and the left and right views. Draw two fittings per page. Refer to Fig. 35.17 and Appendix 9 as guides in making these drawings. Use single-line symbols to draw the following screwed fittings: 90° ell, 45° ell, tee, lateral, cap reducing ell, cross, concentric reducer, check valve, union, globe valve, gate valve, and bushing.

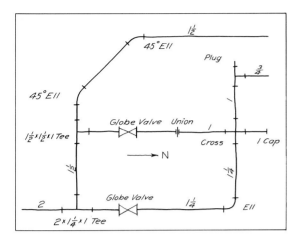

Fig. 35.26 Problem 7.

2. Same as Problem 1, but draw the fittings as flanged fittings.

3. Same as Problem 1, but draw the fittings as welded fittings.

4. Same as Problem 1, but draw the fittings as double-line screwed fittings.

5. Same as Problem 1, but draw the fittings as double-line flanged fittings.

6. Same as Problem 1, but draw the fittings as double-line welded fittings.

7. Convert the single-line sketch in Fig. 35.26 into a double-line system that will fit on a Size A sheet.

8. Convert the single-line pipe system in Fig. 35.27 into a double-line drawing that will fit on a Size A sheet, using the graphical scale given in the drawing to select the best scale for the system.

9. Convert the pipe system shown in isometric in Fig. 35.20 into a double-line isometric drawing that will fit on a Size B sheet.

10. Convert the pipe system given in Fig. 35.20 into a two-view orthographic drawing using single-line symbols.

11. Convert the isometric drawing of the pipe system in Fig. 35.24 into a two-view orthographic drawing that will fit on a Size B sheet. Take the measurements from the given drawing, and select a convenient scale.

12. Convert the isometric of the pipe system in Fig. 35.19 into a double-line isometric that will fit on a Size B sheet.

13. Convert the orthographic pipe system in Fig. 35.9 into a single-line isometric drawing that will fit on a Size B sheet. Estimate the dimensions.

14. Convert the orthographic pipe system in Fig. 35.9 into a double-line orthographic view that will fit on a Size B sheet.

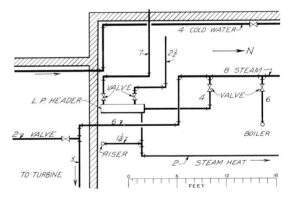

Fig. 35.27 Problem 8.

ELECTRIC/ ELECTRONICS DRAFTING

36.1 INTRODUCTION

Electric/electronics drafting is a specialty area in the field of drafting technology. Electrical drafting is related to the transmission of electrical power that is used in large quantities in the home and industry for lighting, heating, and equipment operation. Electronics drafting deals with circuits in which electronic tubes or transistors are used, where power is used in smaller quantities. Examples of electronic equipment are radios, televisions, computers, and similar products.

Electronics drafters are responsible for the preparation of drawings that will be used in fabricating the circuit, and thereby bringing the product into being. They will work from sketches and specifications developed by the engineer or electronics technologist. Although it is not necessary that they be knowledgeable in the design of a circuit, the drafters must be familiar with the standard methods of representing the various parts within the electrical/electronic circuit.

A major portion of the text in this chapter has been extracted from ANSI Y14.15, *Electrical and Electronics Diagrams,* the standard that regulates the drafting techniques used in this area. The symbols used were taken from ANSI Y32.2, *Graphic Symbols for Electrical and Electronics Diagrams.*

36.2 TYPES OF DIAGRAMS

Electronic circuits are classified and drawn in the format of one of the following types of diagrams:

1. Single-line diagrams,
2. Schematic diagrams, or
3. Connection diagrams.

The suggested line weights for drawing these diagrams are shown in Fig. 36.1.

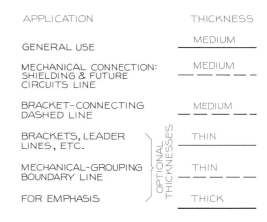

APPLICATION	THICKNESS
GENERAL USE	MEDIUM
MECHANICAL CONNECTION: SHIELDING & FUTURE CIRCUITS LINE	MEDIUM
BRACKET-CONNECTING DASHED LINE	MEDIUM
BRACKETS, LEADER LINES, ETC.	THIN
MECHANICAL-GROUPING BOUNDARY LINE	THIN
FOR EMPHASIS	THICK

Fig. 36.1 The recommended line weights for drawing electronics diagrams.

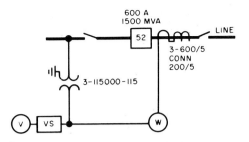

Fig. 36.2 A portion of a single-line diagram where heavy lines represent the primary circuits, and medium lines represent the connections to the current and potential sources. (Courtesy of ANSI.)

Single-Line Diagrams

Single-line diagrams use single lines and graphic symbols to show the course of an electric circuit or system of circuits and the parts and devices within it. A single-line is used to represent both AC and DC systems as illustrated in Fig. 36.2. An example of a single-line diagram of an audio system is shown in Fig. 36.3. Primary circuits are indicated by thick connecting lines, and medium

lines are used to represent connections to the current and potential sources.

Single-line diagrams show the connections of meters, major equipment, and instruments. In addition, ratings are often given to supplement the graphic symbols to provide such information as: kilowatt ratings, voltages, cycles and revolutions per minute, and generator ratings. An example of a diagram of this type with the ratings added is shown in Fig. 36.4.

Schematic Diagrams

Schematic diagrams are graphic symbols to show the electrical connections and functions of a specific circuit arrangement. Although the schematic diagram enables one to trace the circuit and its functions, the physical sizes, shapes, and loca-matic diagram is illustrated in Fig. 36.5 (see pages 740–741); this diagram should be referred to for applications of the principles that are covered in this chapter.

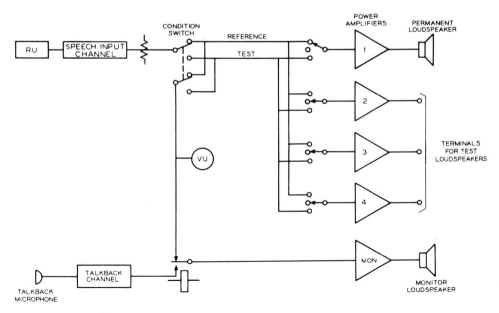

Fig. 36.3 A typical single-line diagram for illustrating electronics and communications circuits. (Courtesy of ANSI.)

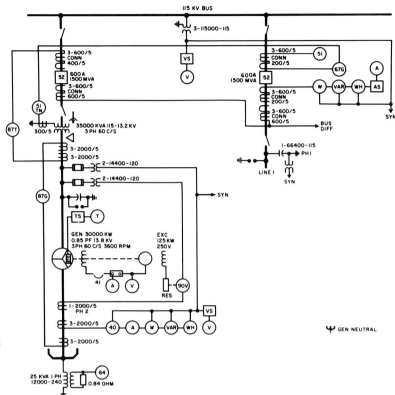

Fig. 36.4 A single-line diagram that is used to illustrate a power switchgear complete with the device designations noted on the diagram. (Courtesy of ANSI.)

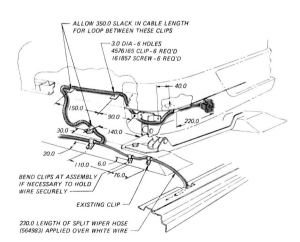

Fig. 36.6 This is a three-dimensional connection diagram that shows the circuit and its components with the necessary dimensions to explain how it is connected, or installed. (Courtesy of the General Motors Corporation.)

Connection Diagrams

Connection diagrams show the connections and installations of the parts and devices of the system. In addition to showing the internal connections, external connections, or both, it shows the physical arrangement of the parts. It can be described as an installation diagram such as the one shown in Fig. 36.6.

36.3 SCHEMATIC DIAGRAM CONNECTING SYMBOLS

Of the symbols used to communicate the composition of a circuit graphically, the most basic symbols are those that are used to represent connections of parts within the circuit. The use of dots to show connections is optional and it is preferable to omit them if clarity is not sacrificed by so doing (Fig. 36.7A). The dots are used to distinguish between connecting lines and those that

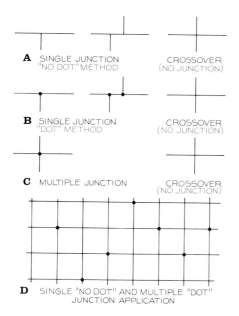

Fig. 36.7 Connections should be shown with single-point junctions as shown at (A). Dots may be used to call attention to connections as shown at (B); and dots *must* by used when there are multiples of the type shown at (C).

simply pass over each other (B). It is preferred that connecting wires have single junctions wherever possible.

When the layout of a circuit does not permit the use of single junctions, and lines within the circuit must cross, then dots must be used to distinguish between crossing and connecting lines (C, D).

Interrupted paths are breaks in lines within a schematic diagram that are interrupted to conserve space when this can be done without confusion. For example, the circuit in Fig. 36.8 has been interrupted since the lines do not connect the left and the right sides of the circuit. Instead, the ends of the lines are labeled to correspond to the matching notes at the other side of the interrupted circuit.

There will be occasions where sets of lines in a horizontal or vertical direction will be interrupted (Fig. 36.9). Brackets will be used to interrupt the circuit and notes will be placed outside the brackets to indicate the destinations of the wires or their connections.

In some cases, a dashed line is used to connect brackets that interrupt circuits (Fig. 36.10). The dashed line should be drawn so that it will not be mistaken as a continuation of one of the lines within the bracket.

Mechanical linkages that are closely related to electronic functions may be shown as part of a

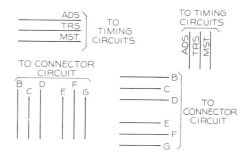

Fig. 36.9 Brackets and notes may be used to specify the destinations of interrupted circuits as shown in this illustration.

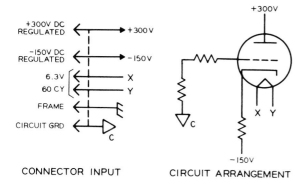

Fig. 36.8 Circuits may be interrupted and connections not shown by lines if they are properly labeled to clarify their relationship to the part of the circuit that has been removed. The connections above are labeled to match those on the left and right sides of the illustration. (Courtesy of ANSI.)

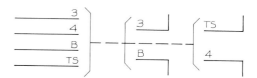

Fig. 36.10 The connections of interrupted circuits can be indicated by using brackets and a dashed line in addition to labeling the lines. The dashed line should not be drawn to appear as an extension of one of the lines in the circuit.

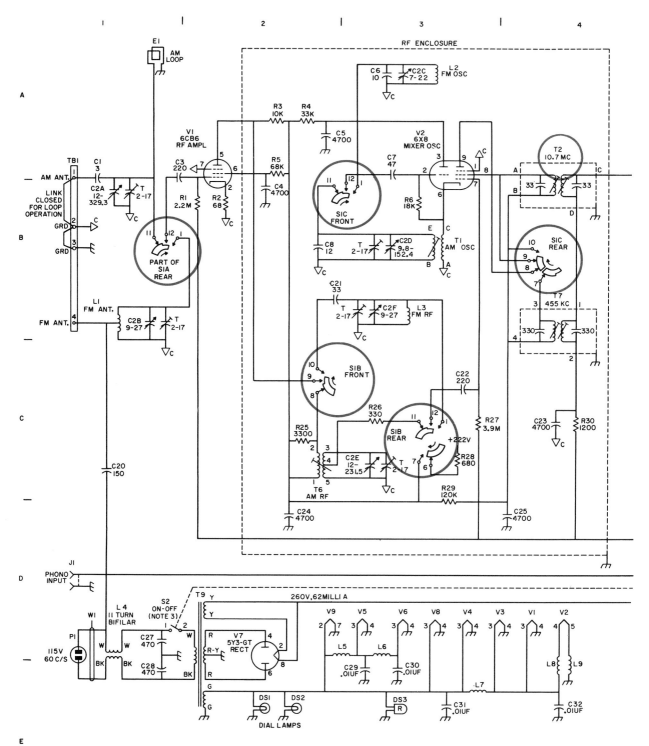

Fig. 36.5 A typical schematic diagram of an AM/FM radio circuit with all parts of the system labeled. The title of a to the type of circuit being presented. (Courtesy of ANSI.)

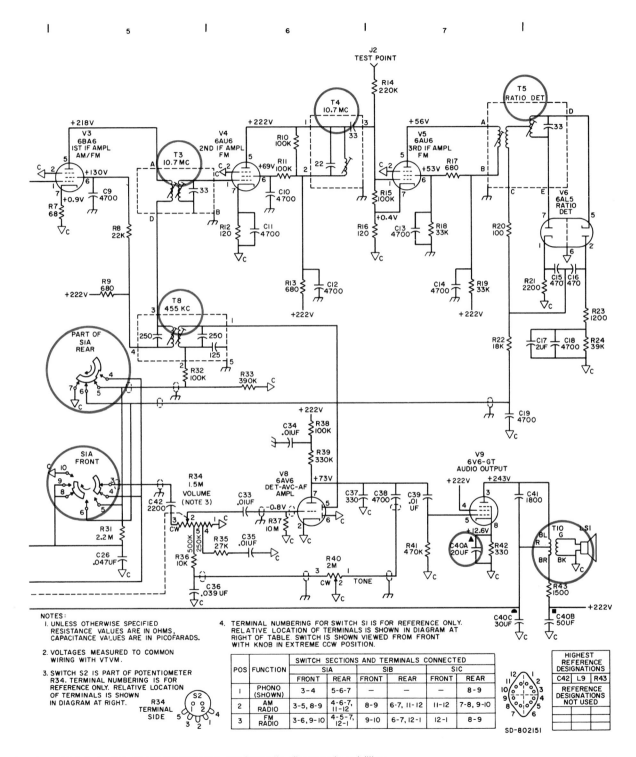

NOTES:
1. UNLESS OTHERWISE SPECIFIED
   RESISTANCE VALUES ARE IN OHMS,
   CAPACITANCE VALUES ARE IN PICOFARADS.

2. VOLTAGES MEASURED TO COMMON
   WIRING WITH VTVM.

3. SWITCH S2 IS PART OF POTENTIOMETER
   R34. TERMINAL NUMBERING IS FOR
   REFERENCE ONLY. RELATIVE LOCATION
   OF TERMINALS IS SHOWN
   IN DIAGRAM AT RIGHT.

4. TERMINAL NUMBERING FOR SWITCH S1 IS FOR REFERENCE ONLY.
   RELATIVE LOCATION OF TERMINALS IS SHOWN IN DIAGRAM AT
   RIGHT OF TABLE. SWITCH IS SHOWN VIEWED FROM FRONT
   WITH KNOB IN EXTREME CCW POSITION.

POS	FUNCTION	SWITCH SECTIONS AND TERMINALS CONNECTED					
		S1A		S1B		S1C	
		FRONT	REAR	FRONT	REAR	FRONT	REAR
1	PHONO (SHOWN)	3-4	5-6-7	—	—	—	8-9
2	AM RADIO	3-5, 8-9	4-6-7, 11-12	8-9	6-7, 11-12	11-12	7-8, 9-10
3	FM RADIO	3-6, 9-10	4-5-7, 12-1	9-10	6-7, 12-1	12-1	8-9

S2
R34 TERMINAL SIDE

HIGHEST REFERENCE DESIGNATIONS		
C42	L9	R43
REFERENCE DESIGNATIONS NOT USED		

SD-802151

drawing of this type should specify that it is a schematic diagram in addition

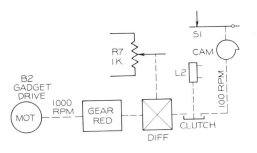

Fig. 36.11 If mechanical functions are closely related to electrical functions, it may be desirable to link the mechanical components within the schematic diagram.

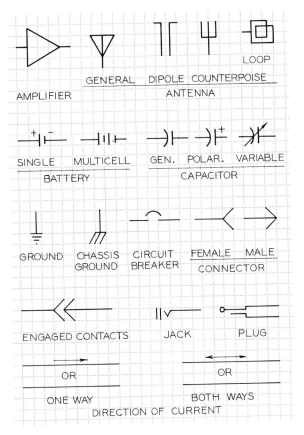

Fig. 36.12 These graphic symbols that are used to represent parts within a schematic diagram are drawn on a 5-mm (0.20-inch) grid. The suggested sizes of the symbols can be found by taking the dimensions from the grid to draw the symbols full size.

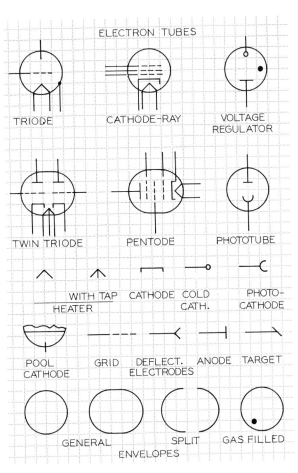

Fig. 36.13 The upper six symbols are used to represent often-used types of electron tubes. The interpretation of the parts that make up each symbol is given in the lower half of the figure.

schematic diagram (Fig. 36.11). An arrangement of this type helps clarify the relationship of the electronics circuit with the mechanical components.

36.4 GRAPHIC SYMBOLS

The symbols covered in this article are extracted from the ANSI Y32.2 standard, and they are adequate for practically all diagrams. However, when a highly specialized part needs to be shown and a symbol for it is not provided in these standards, it is permissible for the drafter to develop his or her own symbol provided it is properly labeled and the meaning is clearly understood.

The symbols that are presented in Figs. 36.12 through 36.16 are drawn on a grid composed of 5-

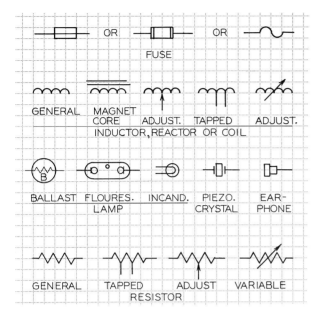

Fig. 36.14 Graphics symbols of standard circuit components.

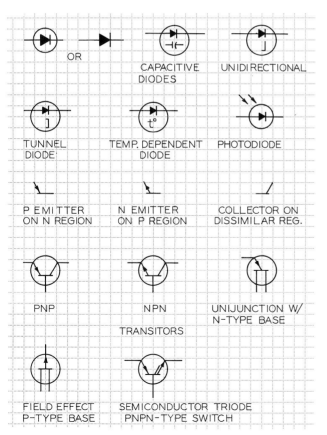

Fig. 36.15 Graphics symbols for semiconductor devices and transistors. The arrows in the middle of the figure illustrate the meanings of the arrows used in the transistor symbols shown below them.

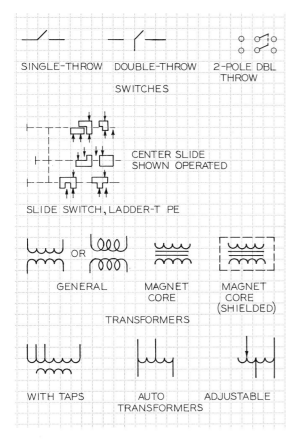

SINGLE-THROW   DOUBLE-THROW   2-POLE DBL THROW

SWITCHES

CENTER SLIDE
SHOWN OPERATED

SLIDE SWITCH, LADDER-T PE

OR

GENERAL   MAGNET CORE   MAGNET CORE (SHIELDED)

TRANSFORMERS

WITH TAPS   AUTO   ADJUSTABLE
TRANSFORMERS

Fig. 36.16 Graphic symbols for representing switches and transformers.

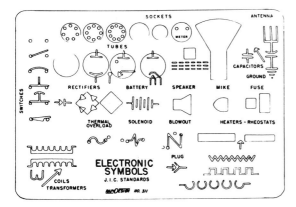

SOCKETS   ANTENNA
METER
TUBES
CAPACITORS
GROUND
SWITCHES
RECTIFIERS   BATTERY   SPEAKER   MIKE   FUSE
THERMAL
OVERLOAD   SOLENOID   BLOWOUT   HEATERS - RHEOSTATS
PLUG
ELECTRONIC
SYMBOLS
J.I.C. STANDARDS
COILS
TRANSFORMERS   NO. 311

Fig. 36.17 A number of various types of template are available for drawing the graphic symbols used in schematic diagrams. (Courtesy of Frederick Post Company.)

mm (0.20 inch) squares that have been reduced. The actual dimensions of the symbols can be approximated by using the grid and equating each square to its full-size measurement of 5 mm. Symbols may be drawn larger or smaller to fit the size of your layout provided the relative proportions of the symbols are kept about the same.

The symbols in Figs. 36.12 through 36.16 are but a few of the more commonly used symbols. There are between five and six hundred different symbols in the ANSI standard for variations of the basic electrical/electronics symbols.

Electronics symbols can be drawn with conventional drawing equipment and instruments by approximating the proportions of the symbols; drafting templates of electronics symbols are available if you wish to save time and effort. An example of an electronics-symbol template is shown in Fig. 36.17.

Some of the symbols presented as examples have been noted to provide designations of their sizes or ratings. The need for this additional information depends upon the requirements and the usage of the schematic diagrams in which the designations appear. Additional information on the preparation of identification information is covered in Section 36.7.

36.5 TERMINALS

Terminals are the ends of devices that are attached in a circuit with connecting wires. Examples of devices with terminals that are specified in circuit diagrams are switches, relays, and transformers. The graphic symbol for a terminal is an open circle that is the same size as the solid circle used to indicate a connection.

Switches are used to turn a circuit on or off, or else to actuate a certain part of it while turning another off. Examples of graphical techniques for labeling switches are shown in the schematic diagram in Fig. 36.5. In this case, a table is used with notes to clarify the switching connections of the terminals.

When a group of parts is enclosed or shielded (drawn enclosed with dashes lines) and the terminal circles have been omitted, the terminal markings should be placed immediately outside the enclosure, as shown in Fig. 36.5 at T2, T3, T4, T5 and T8. The terminal identifications should be

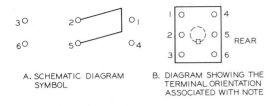

A. SCHEMATIC DIAGRAM
SYMBOL

B. DIAGRAM SHOWING THE
TERMINAL ORIENTATION
ASSOCIATED WITH NOTE

Fig. 36.18 An example of a method of labeling the terminals of a toggle switch on a schematic diagram (A), and a diagram that illustrates the toggle switch when it is viewed from its rear (B).

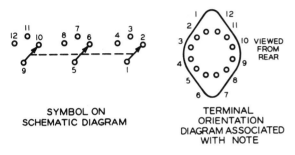

SYMBOL ON
SCHEMATIC DIAGRAM

TERMINAL
ORIENTATION
DIAGRAM ASSOCIATED
WITH NOTE

Fig. 36.19 An example of a rotary switch as it would appear on a schematic diagram (left), and a diagram that shows the numbered terminals of the switch when viewed from its rear. (Courtesy of ANSI.)

added to the graphic symbols that correspond to the actual physical markings that appear on or near the terminals of the part, such as 10.7 MC for transformer T3 in Fig. 36.5. When terminal parts are not marked with identifying numbers, notes should be assigned arbitrarily by the drafter or engineer. Several examples of notes and symbols that explain the parts of a diagram are shown in Figs. 36.18, 36.19, and 36.20.

Colored wires or symbols are often used to identify the various leads that connect to terminals. When colored wires need to be identified on a diagram, the colors are lettered on the drawing, as was done for the transformer T10 of Fig. 36.5. An example of the identification of a capacitor, C40, that is marked with geometric symbols is shown in Fig. 36.5.

Rotary terminals are used to regulate the resistance in some circuits, and the direction of rotation of the dial is indicated on the schematic diagram. The abbreviations CW (clockwise) or CCW (counterclockwise) are placed adjacent to the movable contact when it is in its extreme clockwise position as shown in Fig. 36.21A. The movable contact has an arrow at its end.

If the device terminals are not marked, numbers may be used with the resistor symbol and the

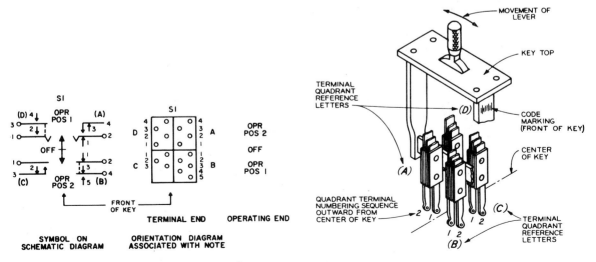

Fig. 36.20 An example of a typical lever switch as it would appear on a schematic diagram (left, part A) and an orientation diagram that shows the numbered terminals of the switch when viewed from its operating end (right, part A). A pictorial of the lever switch and its four quadrants is given in part B of the figure. (Courtesy of ANSI.)

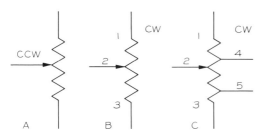

Fig. 36.21 To indicate the direction of rotation of rotary switches on a schematic diagram, the abbreviations CW (clockwise) and CCW (counterclockwise) are placed near the movable contact (part A). If the device terminals are not marked, numbers may be used with the resistor symbols and the number 2 assigned to the adjustable contact (part B). Additional contacts may be labeled as shown at C.

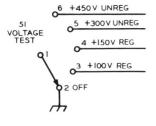

SI VOLTAGE TEST	
FUNCTION	TERM.
OFF	1-2
+100V REG	1-3
+150V REG	1-4
+300V UNREG	1-5
+450V UNREG	1-6

FUNCTIONS SHOWN AT SYMBOL   FUNCTIONS SHOWN IN TABULAR FORM

Fig. 36.22 For more complex switches, position-to-position function relations may be shown using symbols on the schematic diagram, or by a table of values located elsewhere on the diagram. (Courtesy of ANSI.)

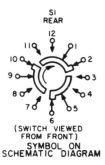

(SWITCH VIEWED FROM FRONT)
SYMBOL ON SCHEMATIC DIAGRAM

SI (REAR)		
POS	FUNCTION	TERM.
1	OFF(SHOWN)	1-2,5-6,9-10
2	STANDBY	1-3,5-7,9-11
3	OPERATE	1-4,5-8,9-12

FUNCTIONS SHOWN IN TABULAR FORM

Fig. 36.23 A rotary switch may be shown on a schematic diagram with its terminals labeled as shown at the left, or its functions can be given in a table placed elsewhere on the drawing as shown at the right. Dashes are used to indicate the linkage of the numbered terminals. For example, 1–2 means that terminals 1 and 2 are connected in the "off" position. (Courtesy of ANSI.)

number 2 assigned to the adjustable contact (Fig. 36.21B). Other fixed taps may be sequentially numbered and added as shown in part C.

The position of a switch as it relates to the function of a circuit should be indicated on the schematic diagram. A method of showing functions of a variable switch is shown in Fig. 36.22. The arrow represents the movable end of the switch that can be positioned to connect with several circuits. The different functional positions of the rotary switch are shown both by symbol and by table in this illustration.

Another method of representing a rotary switch is shown in Fig. 36.23 by symbol and by table. The tabular form is preferred due to the complexity of this particular switch. The dashes between the numbers in the table indicate that the numbers have been connected. For example, when the switch is in position 2 the following terminals are connected: 1 and 3, 5 and 7, and 9 and 11. A table of this type is used at the bottom of Fig. 36.5.

Electron tubes have pins that fit into sockets that have terminals connecting into circuits. Pins are labeled with numbers placed outside the symbol used to represent the tube as shown in Fig. 36.24, and are numbered in a clockwise direction with the tube viewed from its bottom.

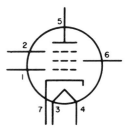

Fig. 36.24 Tube pin numbers should be placed outside the tube envelope and adjacent to the connecting lines. (Courtesy of ANSI.)

36.6 SEPARATION OF PARTS

In complex circuits, it is often advantageous to separate elements of a multi-element part with portions of the graphic symbols drawn in different locations on the drawing. An example of this method of separation is the switch labeled S1A, S1B, S1C, etc., in Fig. 36.5. The switch is labeled

S1 and the letters that follow, called suffixes, are used to designate different parts of the same switch. Suffix letters may also be used to label subdivisions of an enclosed unit that is made up of a series of internal parts, such as the crystal unit shown in Fig. 36.25. These crystals are referred to as Y1A and Y1B.

Rotary switches of the type shown in Fig. 36.26 are designated as S1A, S1B, etc. The suffix letters A, B, etc., are labeled in sequence begin-

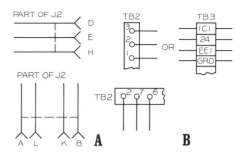

Fig. 36.27 The portions of connectors or terminal boards are functionally separated on a diagram; the words PART OF may precede the reference designation of the entire portion. Or conventional breaks can be used to indicate graphically that the part drawn is only a portion of the whole.

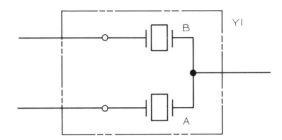

Fig. 36.25 As subdivisions within the complete part, crystals A and B are referred to as Y1A and Y1B.

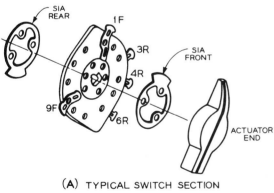

(A) TYPICAL SWITCH SECTION

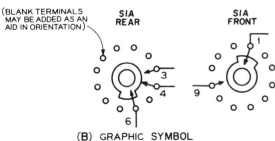

(BLANK TERMINALS MAY BE ADDED AS AN AID IN ORIENTATION)

(B) GRAPHIC SYMBOL

Fig. 36.26 Parts of rotary switches are designated with suffix letters A, B, C, etc., and are referred to as S1A, S1B, S1C, etc. The words FRONT and REAR are added to these designations when both sides of the switch are used. (Courtesy of ANSI.)

ning with the knob and working away from it. Each end of the various sections of the switch should be viewed from the same end. When the rear and front of the switches need to be used, the words FRONT and REAR are added to the designations.

Portions of items such as terminal boards, connectors, or rotary switches may be separated on a diagram provided this is noted for clarity. The words PART OF may precede the identification of the portion of the circuit of which it is a part, as shown in Fig. 36.27A. A second method of showing a part of a system is by using conventional break lines that make the note PART OF unnecessary.

36.7 REFERENCE DESIGNATIONS

A combination of letters and numbers that identify items on a schematic diagram are called reference designations. These designations are used to identify the components not only on the drawing, but in the related documents that refer to them. Reference designations should be placed close to the symbols that represent the replaceable items of a circuit on a drawing. Items that are not separately replaceable may be identified if this is considered necessary. Mounting devices for electron tubes, lamps, fuses, etc., are seldom identified on schematic diagrams.

It is standard practice to begin each reference designation with an uppercase letter that may be

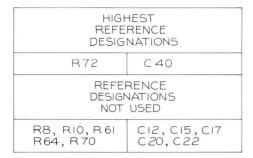

HIGHEST REFERENCE DESIGNATIONS	
R 72	C 40
REFERENCE DESIGNATIONS NOT USED	
R 8, R 10, R 61 R 64, R 70	C 12, C 15, C 17 C 20, C 22

Fig. 36.28 Reference designations are used to identify parts of a circuit. They are labeled in a numerical sequence from left to right beginning at the upper left of the diagram. If parts are later deleted from the system, the ones deleted should be listed in a table, along with the highest reference number designations.

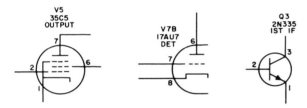

Fig. 36.29 Three lines of notes can be used with electron tubes to specify reference designations, type designation, and function. This information should be located adjacent to the symbol, and preferably above it. (Courtesy of ANSI.)

followed by a numeral with no hyphen between them. The number usually represents a portion of the part being represented. The lowest number of a designation should be assigned to begin at the upper left of the schematic diagram and proceed consecutively from left to right and top to bottom throughout the drawing.

Some of the standard abbreviations used to designate parts of an assembly are: amplifier—A, battery—BT, capacitor—C, connector—J, piezoelectric crystal—Y, fuse—F, electron tube—V, generator—G, rectifier—CR, resistor—R, transformer—T, and transistor—Q.

As the circuit is being designed, some of the numbered elements may be deleted from the circuit drawing. The numbered elements that remain should not be renumbered even though there is a missing element within the sequence of numbers used to label the parts. Instead, a table of the type shown in Fig. 36.28 can be used to list the parts that have been omitted from the circuit. The highest designations are also given in the table as a check to be sure that all parts are considered.

Electron tubes are labeled not only with reference designations but with type designation and circuit function as shown in Fig. 36.29. This information is labeled in three lines, such as V5/35C5/OUTPUT, which are located adjacent to the symbol.

Fig. 36.30 Multipliers should be used to reduce the number of zeros in a number (part A). Examples of expressing units of capacitance and resistance are shown at B and C. (Courtesy of ANSI.)

Multiplier	Prefix	Symbol	
		Method 1	Method 2
10^{12}	tera	T	T
10^9	giga	G	G
10^6 (1,000,000)	mega	M	M
10^3 (1000)	kilo	k	K
10^{-3} (.001)	milli	m	MILLI
10^{-6} (.000001)	micro	μ	U
10^{-9}	nano	n	N
10^{-12}	pico	p	P
10^{-15}	femto	f	F
10^{-18}	atto	a	A

A. MULTIPLIERS

Range in Ohms	Express as	Example
Less than 1000	ohms	.031 470
1000 to 99,999	ohms or kilohms	1800 15,853 10k 82k
100,000 to 999,999	kilohms or megohms	220k .22M
1,000,000 or more	megohms	3.3M

B. RESISTANCE

Range in Picofarads	Express as	Example
Less than 10,000	picofarads	152.4 pF 4700 pF
10,000 or more	microfarads	.015μF 30μF

C. CAPACITANCE

36.8 NUMERICAL UNITS OF FUNCTION

Functional units such as the values of resistance, capacitance, inductance, and voltage should be specified with the fewest number of zeros by using the multipliers in Fig. 36.30A as prefixes. Examples using this method of expression are shown in the B and C parts of Fig. 36.30, where units of resistance and capacitance are given. When four-digit numbers are given, the commas should be omitted. One thousand should be written as 1000, not as 1,000. You should recognize and use the lowercase or uppercase prefixes as indicated in the table of Fig. 36.30.

A general note can be used where certain units are repeated on a drawing to reduce time and effort. The following is recommended form for a note of this type:

UNLESS OTHERWISE SPECIFIED: RESIST-ANCE VALUES ARE IN OHMS. CAPACITANCE VALUES ARE IN MICROFARADS.

or

CAPACITANCE VALUES ARE IN PICOFARADS.

A note for specifying capacitance values is:

CAPACITANCE VALUES SHOWN AS NUMBERS EQUAL TO OR GREATER THAN UNITS ARE IN pF AND NUMBERS LESS THAN UNITY ARE IN µF.

Examples of the placement of the reference designations and the numerical values of resistors are shown in Fig. 36.31.

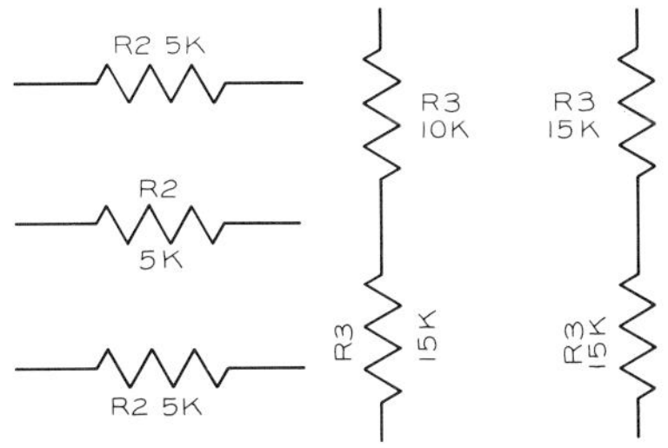

Fig. 36.31 Methods of labeling the units of resistance on a schematic diagram.

36.9 FUNCTIONAL IDENTIFICATION OF PARTS

The readability of a circuit is improved if parts are labeled to indicate their functions. Test points are labeled on drawings with the letters "Tp" and their suffix numbers. The sequence of the suffix numbers should be the same as the sequence of troubleshooting the circuit when it is defective. As an alternative, the test function can be indicated on the diagram below the reference designation.

Additional information may be included on a schematic diagram to aid in the maintenance of the system. Examples of additional information are:

- DC resistance of windings and coils.
- Critical input and output impedance values.
- Wave shapes (voltage or current) at significant points.
- Wiring requirements for critical ground points, shielding, pairing, etc.
- Power or voltage ratings of parts.
- Caution notation for electrical hazards at maintenance points.
- Circuit voltage values at significant points (tube pins, test points, terminal boards, etc.).
- Zones (grid system) on complex schematics.
- Signal flow direction in main signal paths shall be emphasized.

36.10 PRINTED CIRCIUTS

Printed circuits are universally used for miniature electronic components and computer systems. The degree of miniaturization that has occurred in the electronics industry can be seen in Fig. 36.32 where the function of a vacuum tube has been replaced by a chip transistor the size of a dot.

In this method, the drawings of the circuits are drawn up to four times or more the size that the circuit will ultimately be printed. The drawings for the printed circuit are usually drawn in black India ink on acetate film and are then photographically reduced to the desired size. The circuit is "printed" onto an insulated board made of plastic or ceramics, and the devices within the circuit are connected and soldered (Fig. 36.33).

The steps of fabricating a solid state logic technology (SLT) module that was used in the

Fig. 36.32 The result of the miniaturization of electronic devices can be seen here where a solid state logic technology (SLT) chip the size of a dot gives the same function as its predecessors, the transistor and the vacuum tube. (Courtesy of IBM.)

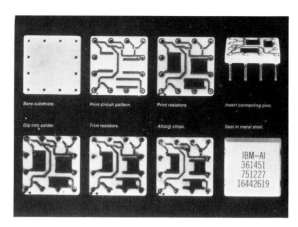

Fig. 36.34 The steps of fabricating a printed circuit are shown here where the circuit and its parts are printed on a ceramic substrate base, and the pins are connected. Next, it is dipped into solder to connect the parts of the circuit and build up the printed circuit, the resistors are trimmed and the chips are attached to prepare the module for sealing in a metal shield. (Courtesy of IBM.)

Fig. 36.33 This is a magnified view of a printed circuit where the circuit has actually been printed and etched on a circuit board, and the devices are then soldered into position. (Courtesy of Bishop Industries Corporation.)

IBM 360 computer is shown in Fig. 36.34. The tiny module (about one-half inch square) had its chip transistor positioned and attached in the last step of assembly before it was encased in its protective metal shell (Fig. 36.35).

Some printed circuits are printed on both sides of the circuit board; this requires two photographic negatives, as shown in Fig. 36.36, that were made from positive drawings (black lines on

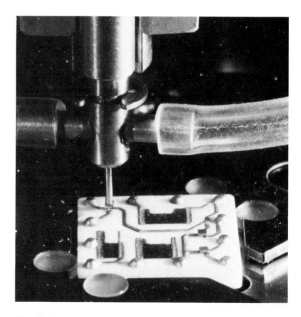

Fig. 36.35 A chip is shown being attached to the one-half inch square SLT module at a rate of better than one per second. (Courtesy of IBM.)

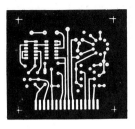

Fig. 36.36 A printed circuit that is attached to both sides of the circuit board requires two circuit drawings, one for each side, that are photographically converted to reduced negatives for printing. (Courtesy of Bishop Industries Corporation.)

a white background). Each drawing for each side can be made on separate sheets of acetate that are laid over each other when the second diagram is drawn. However, a more efficient method uses red and blue tape that can be used for making a single drawing (Fig. 36.37) from which two negatives are photographically made. Filters are used on the process camera to drop out the red for one negative and a different filter for dropping out the blue for the second negative. Register marks are used for aligning the negatives when the circuit is printed.

Printed circuits are usually coated with silicone varnish to prevent malfunction because of the collection of moisture or dust on the surface. They may also be enclosed in protective shells.

36.11 SHORTCUT SYMBOLS

Several manufacturers produce preprinted symbols that can be used for "drawing" high quality electronic circuits and printed circuits. The symbols are available on sheets or on tapes that can be burnished onto the surface of the drawing to form a permanent schematic diagram. Examples of symbols that are used for printed circuits are shown in Fig. 36.38. They are made in a variety of sizes so that they fill practically all needs.

The symbols can be connected with a matching tape to represent wires between them instead of drawing the lines. Schematic diagrams made with these materials are of a very high quality that holds up well when they are reduced in size for publication in specifications or technical journals.

36.12 INSTALLATION DRAWINGS

Many types of electrical/electronics drawings are used to produce the finished installation, from the designer who visualized the system at the outset of the project to the contractor who builds it. Drawings are used to design the circuit, detail its parts for fabrication, specify the arrangement of the devices within the system, and instruct the contractor how to install the project.

A combination arrangement and wiring diagram drawing is shown in Fig. 36.39 where the system is shown in a front and right-side view. The wiring diagram explains how the wires and components within the system are connected for the metal-encased switchgear. Bus bars are conductors for the primary circuits.

A power substation is detailed in Fig. 36.40 with sufficient details and dimensions to allow the contractor to install the parts of the system. The contractor on a job of this type will not be involved in manufacturing the devices in the system, but will be responsible for connecting the parts together in accordance with the details given in the drawing and the accompanying written specifications.

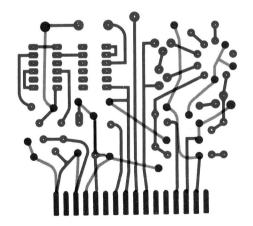

Fig. 36.37 By using two colors, such as blue and red, one circuit drawing can be made, and two negatives made from the same drawing by using camera filters that screen out one of the colors with each shot. The circuits are then printed on each side of the board. (Courtesy of Bishop Industries Corporation.)

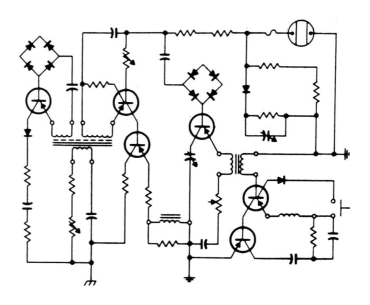

SYMBOLS SHOWN ACTUAL SIZE

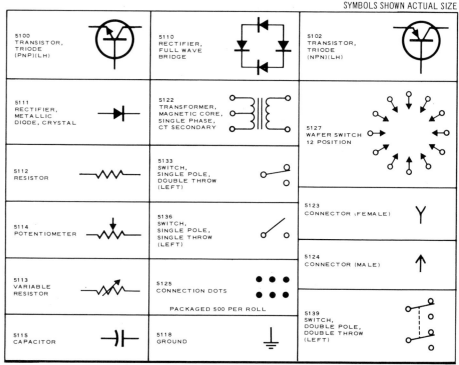

SYMBOLS ABOVE ARE PRICED AT $5.80 PER ROLL OF 250.

Fig. 36.38 Stick-on symbols are available for laying out schematic diagrams rather than drawing them. The resulting layouts have a higher contrast and sharpness that improves their reproducibility and reduction for publication. (Courtesy of Bishop Industries Corporation.)

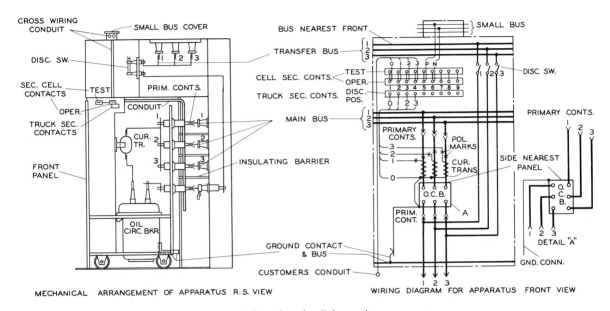

CROSS WIRING CONDUIT — SMALL BUS COVER

BUS NEAREST FRONT — SMALL BUS

DISC. SW.

TRANSFER BUS

DISC. SW.

SEC. CELL CONTACTS — TEST

PRIM. CONTS.

CELL SEC. CONTS. — TEST / OPER.

TRUCK SEC. CONTS. — DISC. POS.

OPER.

TRUCK SEC. CONTACTS

CONDUIT

MAIN BUS

PRIMARY CONTS.

FRONT PANEL

CUR. TR.

PRIMARY CONTS.

POL. MARKS

CUR. TRANS.

SIDE NEAREST PANEL

INSULATING BARRIER

O.C.B.

PRIM. CONT.

A

O. C. B.

DETAIL "A"

OIL CIRC. BKR

GROUND CONTACT & BUS

CUSTOMERS CONDUIT

GND. CONN.

MECHANICAL ARRANGEMENT OF APPARATUS R.S. VIEW

WIRING DIAGRAM FOR APPARATUS FRONT VIEW

Fig. 36.39 This drawing shows views of a metal-enclosed switchgear to describe the arrangement of the apparatus and also gives the wiring diagram for the unit.

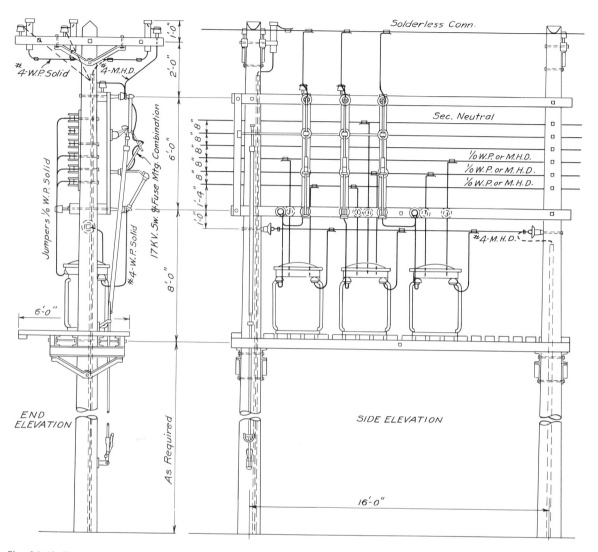

Fig. 36.40 This installation drawing of an electrical substation is sufficient to explain its installation to the contractor who must assemble and construct it. (Courtesy of Railway and Industrial Engineering Co., Greensburg, Pennsylvania.)

PROBLEMS

1. On a Size A sheet, make a schematic diagram of the circuit shown in Fig. 36.41.

2. On a Size A sheet, make a schematic diagram of the circuit shown in Fig. 36.42.

3. On a Size A sheet, make a schematic diagram of the circuit shown in Fig. 36.43.

4. On a Size A sheet, make a schematic diagram of the circuit shown in Fig. 36.44.

5. On a Size A sheet, make a schematic diagram of the circuit shown in Fig. 36.45.

6. On a Size B sheet, make a schematic diagram of the circuit shown in Fig. 36.46.

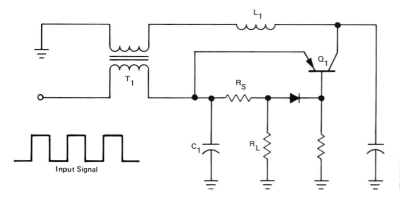

Fig. 36.41 Problem 1: A low-pass inductive-input filter. (Courtesy of NASA).

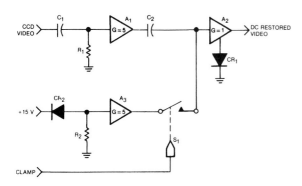

Fig. 36.42 Problem 2: A quadruple-sampling processor. (Courtesy of NASA.)

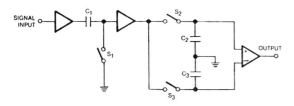

Fig. 36.43 Problem 3: A temperature-compensating DC restorer circuit designed to condition the video circuit of a Fairchild area-image sensor. (Courtesy of NASA.)

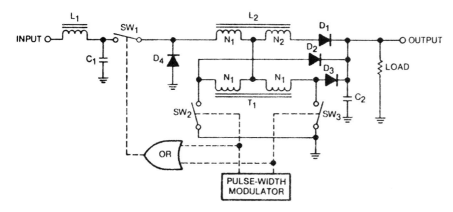

Fig. 36.44 Problem 4: A "buck/boost" voltage regulator. (Courtesy of NASA.)

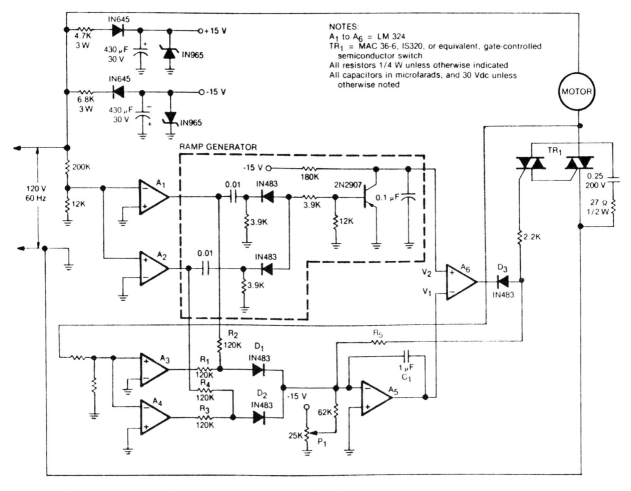

Fig. 36.45 Problem 5: An improved power-factor controller (Courtesy of NASA.)

Fig. 36.46 Problem 6: A magnetic-amplified DC transducer. (Courtesy of NASA.)

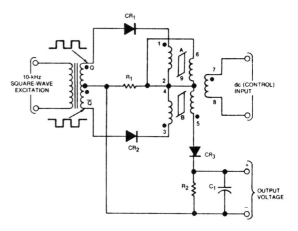

7. On a Size C sheet, make a schematic diagram of the circuit shown in Fig. 36.47.

8. On a Size C sheet, make a schematic diagram of the circuit shown in Fig. 36.48 and give the parts list.

9. On a Size C sheet, make a schematic diagram of the circuit shown in Fig. 36.49 and give the parts list.

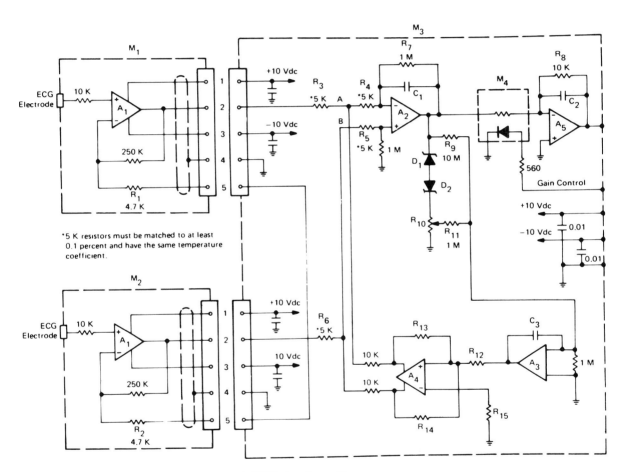

Fig. 36.47 Problem 7: An electrocardiograph (EKG) signal conditioner. (Courtesy of NASA.)

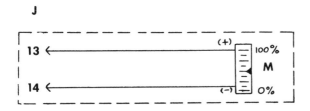

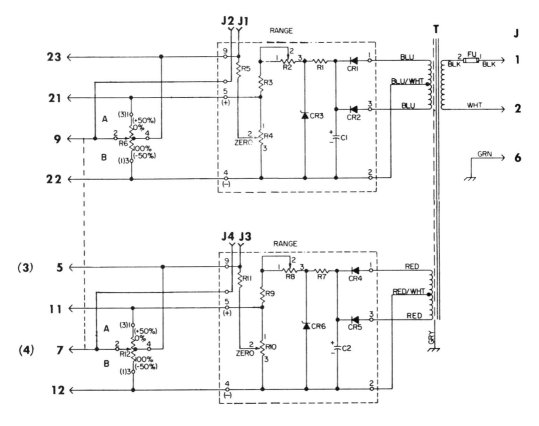

COMPONENT	DESCRIPTION	
CR1, CR2, CR3, CR4	DIODE	1N3193
CR3 & CR6	ZENER DIODE	24V 3W ±.5V TOL
R1 & R7	RESISTOR	560 Ω 5W 5%
R3 & R9	RESISTOR	56 Ω
R5 & R11	RESISTOR	2.7K 1/2W 5%
R2 & R8	TRIM POT	50 Ω
R4 & R10	TRIM POT	5K
R6 & R12	POT	500 Ω 2W DUAL
C1 & C2	CAPACITOR	60 MFD 60V
T	TRANSFORMER	
FU	FUSE	1 AMP
M	EDGEWISE METER	
J	CONNECTOR	
J1, J2, J3, J4	TEST JACK	

TYPE	R6 SCALE		M SCALE	
	A	B	BOTTOM	TOP
RS1100C	0%	100%	0%	100%
RS1110C	0%	100%	PER ENGINEERING DATA	
RS3100C	+50%	-50%	-50%	+50%
RS4100C	0%	100%	0%	100%
RS2120C	100%	0%	OMIT	
RS2100C	100%	0%	0%	100%

NOTE: OUTPUT WIRED TO TERMINALS 3 AND 4 ON
TYPES RS1100C, RS1110C, AND RS3100C.

OUTPUT WIRED TO TERMINALS 5 AND 7 ON
TYPES RS2100C, RS2120C, AND RS4100C.

Fig. 36.48 Problem 8: A schematic diagram of a dual output manual
station and its parts list. (Courtesy of NASA.)

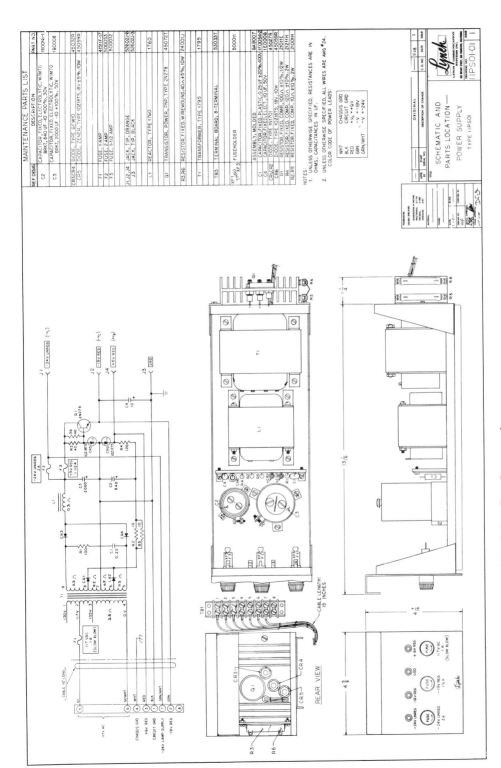

Fig. 36.49 Problem 9: A schematic and parts location diagram of a power supply. (Courtesy of Lynch Communication Systems, Inc.)

37

COMPUTER GRAPHICS

WILLIAM A. ZAGGLE
Texas A&M University

37.1 INTRODUCTION

Computers, in conjunction with graphical plotters, have already been adopted in many drafting departments. This trend is expected to increase as the price of computer equipment declines and the state of the art advances.

The first application of computer graphics where a drawing was made came about when a pen was used to replace the cutting tool in an automated milling machine. This was done to determine economically the path of the machine tool prior to cutting metal. This system soon evolved into a method of computer graphics whereby a drawing could be made by a plotter driven by a computer program.

Computer-aided manufacturing (CAM) involves computerized production machinery. Initially, this was confined to metal-cutting machines such as lathes, drills, and milling machines. Modern CAM systems make use of and control complicated robots used to assemble delicate printed circuit boards as well as heavy steel auto parts. These robots can have as many as ten arms that may weld, drill holes and tighten bolts in a sequence of steps.

Computer-aided design (CAD) is the computer-aided process of solving design problems in all areas of engineering. Computer-aided design equipment enables the designer to analyze and design a part in an accurate and rapid manner. The specifications of the design can also be stored, and

then recalled for further modification and evaluation at a later date. The graphic display of the CAD system aids designers in viewing and studying their designs as they are being developed.

Computer-aided design drafting (CADD) is used to produce the final working drawings once the design has been finalized. The designer can display the drawings on a screen, called a cathode-ray tube (CRT), before the final copies of the drawings are drawn on a sheet by the plotter. Some industries use programs similar to those written for making the drawings, to drive the automated machinery in the shop and make the finished part, thereby eliminating the need for drawings made on paper.

This chapter will touch only briefly on the many applications in which the various phases of computer graphics are involved. A general introduction to programming for CAD applications will be presented to familiarize you with some of the fundamental uses of computer graphics.

37.2 APPLICATIONS OF COMPUTER GRAPHICS

A natural application of computer graphics is the design and drawing of printed circuits of the type shown in Fig. 37.1. Printed circuits are drawn up to five times full size, and are then reduced photographically. A computer-driven plotter will

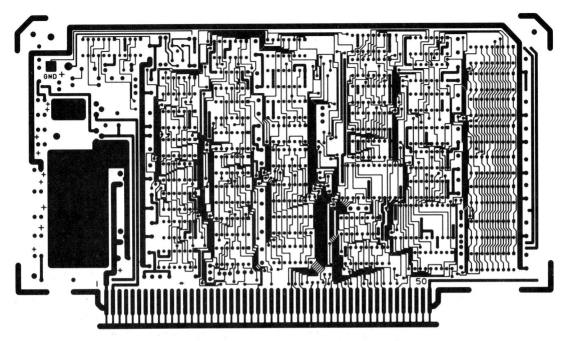

Fig. 37.1 An often-used application of computer graphics is the drawing of printed circuits for electronic systems. They can be drawn very accurately, photographically reduced, and then fabricated. (Courtesy of Computel Engineering, Inc.)

Fig. 37.2 The designer is using computer graphics to plot a piping system in color on the screen of a video display. (Courtesy of Applicon.)

draw this circuit within accuracy of approximately 0.001 in.

Piping systems can be designed and represented in both orthographic and pictorial views with all of the components shown. Once the system has been completed, the design and its specifications can be stored for rapid recall when the system needs to be reexamined. An example piping system is shown in Fig. 37.2.

Computer graphics is also used advantageously as an aid in finite element analysis, where a series of elements is used to represent an irregular three-dimensional shape. In Fig. 37.3, the designer is digitizing (assigning numerical coordinates to) the elements of a gun mount by working from a multiview drawing at his right. The mount is then displayed by the computer system as an isometric on the screen of the CRT.

Clothing patterns for a wide selection of graduated sizes (Fig. 37.4) can be drawn and cut by an application of computer graphics. The automatic cutter follows the computer-generated path to cut the most economical patterns from a section of material.

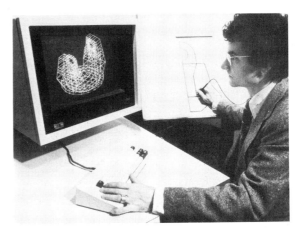

Fig. 37.3 This engineer has constructed a 3-D finite element model of a gun mount on the Applicon display by digitizing a multiview engineering drawing. (Courtesy of Applicon.)

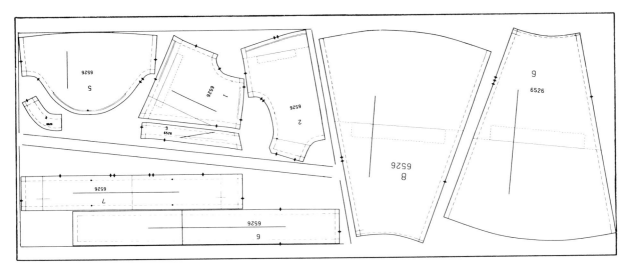

Fig. 37.4 This clothing pattern was designed, plotted and cut out by utilizing computer graphics on a Versatec plotter. Programs have been written that will plot this same pattern for graduated sizes of clothing. (Courtesy of Versatec.)

Fig. 37.5 This CalComp Graphic 7 system is a typical microcomputer system that is comprised of the keyboard, CRT, and computer. (Courtesy of California Computer Products, Inc.)

Pictorials drawn by the computer are very beneficial to the technical as well as the nontechnical person in observing a three-dimensional design. Most 3-D computer graphics systems permit the observer to view a design from any angle as it is revolved on the screen.

37.3 HARDWARE SYSTEMS

A basic computer graphics system consists of a *computer, terminal, plotter,* and *printer.* Additional devices such as *digitizers* and *light pens* may be used for direct input of graphics information (Fig. 37.5).

Computer

Computers are the devices that receive the input of the programmer, execute the programs, and then produce some form of useful output. The largest computers, called *mainframes,* are big, fast, powerful, and expensive. The smallest computers, called *microcomputers,* have recently come into widespread use for personal and small business applications. Microcomputers are much cheaper than mainframes, and they are excellent for graphical applications where massive data storage and high speed is not essential, yet low price and small size is important.

The *minicomputer* is a medium-sized computer that lies between the mainframe and the microcomputer in performance and cost.

Terminal

The terminal is the device by which the user communicates with the computer. It usually consists of a keyboard with some type of output device such as a typewriter or television screen (Fig. 37.6). Most graphics terminals use one of three types of television screens or CRTs (cathode-ray tubes). One type is **raster scanned,** which means that the picture display is being refreshed or scanned from left to right and top to bottom at a rate of about 60 times per second.

A second type is the **storage tube** where the image is drawn on the screen much like a drawing on an erasable blackboard.

The third and most powerful type of screen is **vector refreshed,** which means that each line in the picture is being continuously redrawn by the

Fig. 37.6 An engineer is shown working at a computer graphics system's terminal where he has access to a keyboard and a display screen. (Courtesy of Computervision.)

computer. The vector refresh type requires more computer power than is available from the smaller minicomputers and the microcomputers.

Plotters

The plotter is the machine that is directed by the computer and the program to make a drawing. The two basic types of plotters are **flatbed** and **drum.** The flatbed plotter is a large flat bed to which the drawing paper is attached. A pen is moved about over the paper in a raised or lowered position to complete the drawing (Fig. 37.7). This type of plotter is well suited to plotting on a variety of types of surfaces in addition to paper. The drum plotter uses a special type of paper that is held on a spool and rolled over a rotating drum (Fig. 37.8). The drum rotates in two directions as the pen that is suspended above the surface of the paper moves left or right along the drum.

Printers

The printer can be thought of as a typewriter that is operated by the program and computer. Many varieties are available in a wide range of prices. The speed and type quality offered by the printer are what usually determine the purchase price. One of the most important applications of the printer in a computer system is the printing of a copy of a computer program that can be used for

Fig. 37.7 A close-up of the flat-bed plotter that uses four pens of different colors to plot the output from the computer. (Courtesy of Computervision.)

review and modification by the programmer in "hard copy" form.

Digitizer

The digitizer is a device that inputs the coordinates of points of a drawing into the computer by tracing the drawing located on a digitizer board. A pen-like stylus, connected to the computer, is often used for this purpose. Some systems using the

Fig. 37.8 A drum plotter used for plotting the output from the computer. (Courtesy of California Computer Products, Inc.)

refreshed or scanned types of CRT terminals use a *light pen* that can change or establish points on the screen. These points can be either graphical data or commands used by the computer. The light pen is often considered to be a digitizing device.

37.4 USE OF THE SYSTEM

The following examples and explanations deal specifically with the computer system at Texas A&M University and may not be literally applicable to another hardware or software configuration (Fig. 37.9). However, the general principles covered throughout the remainder of this section are applicable to most systems.

The system referred to above consists of the following components.

HARDWARE:

- CPU (Central Processing Unit)—Northstar Advantage (Z-80 based) microcomputer with 64K RAM; Two 5.25″ floppy discs; 240 × 640 points raster-scan resolution 9″ × 7″ screen).
- Plotter—Houston Instrument DMP-3 HIPLOT single pen, 11″ × 8.5″ flatbed.
- Digitizer—Houston Instrument HIPAD 11″ × 8.5″ resolution to 0.005 inch.

SOFTWARE: Microsoft's MBASIC and CP/M operating system. Hardware drivers written in Z-80 machine code. (MBASIC is the trademark of Microsoft Inc., CP/M is the trademark of Digital Research Inc.)

Basic Functions

The basic functions of this system are discussed below.

TYPING IN THE PROGRAM: All typing, done on the keyboard while the computer is not executing a program, is displayed on the CRT screen by the computer, so that the user can see what is being received by the computer. One concept that should be remembered is that a line of program is accepted only if it begins with a number; otherwise it is considered to be a command and is executed immediately. (This concept will be discussed further in Section 37.5.) As illustrated in

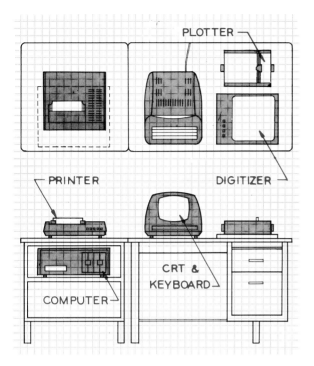

Fig. 37.9 A typical computer graphics work station used at Texas A&M University. This system consists of microcomputer, a cathode-ray tube, a plotter, and digitizer. (Courtesy of Alan Kent.)

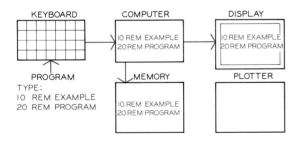

Fig. 37.10 Typing on the keyboard of a terminal feeds the program from the keyboard to the computer, and into its memory. At the same time, the input program is displayed on the screen as it is put in.

Fig. 37.10 a line entered with a line number is displayed on the screen and is also retained in the computer memory.

LISTING TO THE DISPLAY: Typing the command LIST on the keyboard (remember that commands such as this one do not have line numbers) will cause the computer to display the lines of program

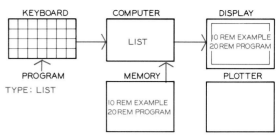

Fig. 37.11 A listing of a program on the display can be called from memory by typing in the command, LIST, on the keyboard. The program is called from memory by the computer and is then listed on the display.

it has in memory on the display screen. This concept is illustrated by Fig. 37.11.

LISTING TO THE PRINTER: Figure 37.12 shows the use of the command LLIST to cause the program listing to be made on the printer instead of on the display screen. As with the LIST command, only the lines of the program currently in memory are displayed. The extra L before the LIST command tells the computer that the output is to be sent to the printer.

GRAPHICS TO THE DISPLAY AND PLOTTER: Graphics can be drawn by the computer, either on the display or on the plotter. The plotting program determines which of these two is to receive the generated graphics, by the value of the variable P.OUT%. If the value of P.OUT% = 0, the graphics are displayed on the screen. If the value of P.OUT% = 1 the graphics are plotted on the plotter (Fig. 37.13 and Fig. 37.14).

After the RUN command is given and before the program is run by the computer, all variables

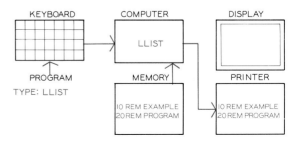

Fig. 37.12 A print of the program can be obtained by typing the command, LLIST, into the keyboard, which causes the computer to call the program from memory and then prints it out at the printer.

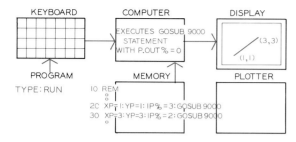

Fig. 37.13 A graphics display can be obtained at the display screen when a program has the statement P.OUT% = 0. The command RUN is typed into the keyboard, and the computer commands the display to plot the graphical output from the program.

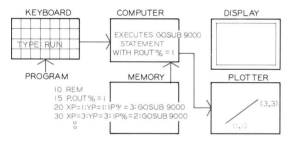

Fig. 37.14 A graphics plot can be obtained at the plotter when the program has the statement P.OUT% = 1. The command RUN is typed into the keyboard, and the computer directs the plotter to plot the graphical output.

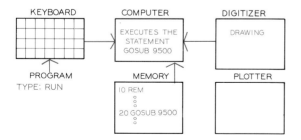

Fig. 37.15 Data from a drawing can be fed into the computer by means of a digitizer board by using a pen to digitize points on the drawing. The digitized points are transferred to the computer and this data is input into the existing program by the statement GOSUB 9500. The command RUN, input at the keyboard, runs the program.

are assigned a value of zero. This means that unless a program specifically assigns P.OUT% a value other than zero, all graphics are displayed on the screen. Once P.OUT% is set equal to one to send graphics to the plotter, it must be reassigned to the value of zero to display graphics on the screen.

The convention of using the variable, P.OUT%, to control the graphics output is dependent upon the subroutine used at TAMU, the Engineering Design Graphics Depart of Texas A&M University. This variable will not be compatible with all subroutines and systems.

GETTING DATA FROM THE DIGITIZER: Data is obtained from the digitizer by executing the subroutine, GOSUB 9500, which defines the variables DIGI.X, DIGI.Y, and DIGI.P%. The values of DIGI.X and DIGI.Y are the coordinates of the digitizer pen, and DIGI.P% has the value of either 2 or 3 for pen down and up commands. As shown in Fig. 37.15, the RUN command causes the computer to execute the program statements. Executing the subroutine, GOSUB 9500, causes the coordinates and pen value to be accepted from the digitizer.

37.5 BASIC PROGRAMMING RULES

The following example programs were written in a language called BASIC, which is different from but has many similarities to the FORTRAN language. Some elementary aspects of BASIC programming are covered in this article.

A *program* is a sequence of instructions and specifications that will be received by a computer, causing it to perform the operations desired by the programmer. Each statement must be given a number that is an integer between 0 and 65535. For example, a statement may be written as 30 X = X + 1, where 30 is the statement number and X = X + 1 is the statement.

Any statement that does not have a line number is interpreted as a BASIC command to be performed immediately rather than a BASIC statement to become part of a program and performed later, when the program is run.

The following rules apply to a statement that is entered after a line number:

1. If a line number of a newly entered statement is the same as a previously entered statement,

then the new line will replace the old line with the same line number.

2. If a line containing only a line number and no statement is entered, then any previously entered statement with the same line number will be deleted from the program.

3. If the line number of a newly entered statement is different from any other line in the program, the statement will be added to the program. When the statements are listed by the computer, they will be arranged in ascending order (i.e., 10, 20, 30, etc.).

Some BASIC *commands* should not be given line numbers and will be executed immediately by BASIC. These commands cause BASIC to operate on the current program or to perform some system function. Listed below are some of the commonly used commands and their functions:

```
LIST—CAUSES THE CURRENT PROGRAM
TO BE LISTED ON THE TERMINAL
DEVICE (USUALLY A CRT).

NEW—CAUSES THE CURRENT PROGRAM
TO BE ERASED.

SAVE ''PROGRAM NAME HERE''—
SAVES THE PROGRAM UNDER THE GIVEN
NAME THAT IS ENCLOSED IN
QUOTATIONS. EXAMPLE: SAVE
''PROG1''

LOAD ''PROGRAM NAME HERE''—
TRANSFERS A PROGRAM BY NAME, IN
QUOTATIONS, FROM THE DISC INTO
MEMORY AND MAKES IT THE CURRENT
PROGRAM. EXAMPLE: LOAD
''PROG1''

KILL ''PROGRAM NAME HERE''—
REMOVES THE NAMED PROGRAM, IN
QUOTATIONS, FROM THE DISC.
EXAMPLE: KILL ''PROG1''

FILES—PRODUCES A LIST OF THE
PROGRAMS THAT ARE STORED ON A
PARTICULAR DISC.

RUN—BEGINS THE EXECUTION OF THE
CURRENT PROGRAM.
```

All BASIC programs are executed (set into operation) by the command RUN, which is typed using the keyboard of the terminal. The program will continue to run until one of the following events occurs:

1. The statement STOP is executed,

2. An invalid statement is executed, or

3. The program runs out of sequential line numbers to execute.

Variable Names Alphabetic or numeric characters up to 40 characters long, provided there are no spaces, can be used for variable names. Periods can be used effectively to separate words such as: PROG.ONE. Variable names cannot be identical to commands such as LIST, RUN, etc.

Arrays Subscripted variables can be assigned any valid variable name followed by a number in parentheses. The letter, or letter and number combination, is the name of the array and the number of parentheses locates the element within the array and is called a subscript. For example, the tenth element in array B would be B(10) and the sixth element in array F3 would be F3(6).

To allocate space in the computer's memory for an array, the DIM (dimensioning) statement is used at the beginning of a BASIC program. A statement such as 10 DIM (50), Y(50), X9(10), P(2,10), would allocate 50 elements for the array of X, 50 elements for the array Y, 10 elements for the variable X9, and 2 groups of 10 elements each for the variable P.

Remark statements are used to document a program within the program for future reference, but they have no effect on the function of the program when it is run. The REMARK statement can be shortened to REM when written in the program. You will note in example programs in the following sections that REM statements are essential to understanding the logic of the programmer as he or she writes the program.

Examples of REM statements are given below:

```
10 REM THIS IS AN EXAMPLE OF
20 REM BASIC REMARK STATEMENTS
30 REM THAT ARE IGNORED DURING
40 REM EXECUTION
50 REM OF THE BASIC PROGRAM
```

Mathematical Functions The version of BASIC that is presented in this chapter has eight mathematical functions that can be used in numerical expressions. These functions are listed below:

- ABS (numerical values here)—Returns the absolute value of the numerical expression. *Example:* ABS(3) = 3, ABS(−3) = 3, and ABS(0) = 0.
- SGN (numerical expression here)—Returns 1, 0, or −1, which indicates whether the numerical expression is positive, zero-valued, or negative, respectively. *Example:* SGN (10) = 1, SGN (0) = 0, and SGN (−3.2) = −1.
- INT (numerical expression here)—Returns the greatest integer value that is less than or equal to the value of the numerical expression. *Example:* INT (3) = 3, INT (3.9) = 3, and INT (−3.5) = −4.
- LOG (numerical expression here)—Returns an approximation of the natural logarithm of the value of the numerical expression. If LOG is called with an argument value less than or equal to zero, a program error will occur. *Example:* LOG (1) = 0, LOG (7) = 1.945901, and LOG (0.1) = −2.3025851.
- EXP (numerical expression here)—Returns an approximation to the value of the base e raised to the power of the numerical expression. *Example:* EXP (0) = 1, EXP (2) =

7.3890562, EXP (−2.3025851) = 0.1, and EXP (1) = 2.7182817.

- SQR (numerical expression here)—Returns an approximation of the positive square root of the numerical expression. A program error will occur if this function is called with a negative argument. *Example:* SQR (0) = 0, SQR (10) = 3.1622776, and SQR (0.3) = 0.54772256.
- SIN (numerical expression here)— Computes an approximation of the trigonometric sine of the value of the numerical expression. The answers must be given in radians rather than degrees. (Note that $2 \times$ Pi radians = 360 degrees). *Example:* SIN (0) = 0, and SIN (3.1415926/2) = 1.
- COS (numerical expression here)—COS computes an approximation of the trigonometric cosine of the value of the numerical expression, which must be specified in radians. *Example:* COS (0) = 1, and COS (3.1415926/2) = 0.
- ATN (numerical expression here)—Computes an approximation of the trigonometric arctangent function. The angular value that is returned is expressed in radians. *Example:* ATN (5) = 1.3734007, and ATN (1.7) = 1.0390722.

The BASIC language has some similarities to FORTRAN. Table 37.1 lists some of the similarities.

Table 37.1 SIMILARITIES OF BASIC TO FORTRAN

Statement type	Basic	Fortran
Unconditional Branch	GOTO line #	GO TO statement #
Conditional Branch	IF X>Y THEN statement ELSE statement	IF (X.GT.Y) statement
Subroutine call	GOSUB line #	CALL program name (arguments)
Subroutines	line #statements statements line # RETURN	SUBROUTINE name (parameters) statements RETURN
Loops	line # FOR J = 0 TO 100 STEP 2 statements line # NEXT J	DO line # J = 0,100,2 statements line # CONTINUE

37.6 THE BASIC PLOT STATEMENT

To command the plotter to draw or move to a given point, a statement of the format shown in Fig. 37.16 is used. The X- and Y-coordinates of a point are given as XP and YP. The pen position, IP%, and the plotting subroutine, GOSUB 9000, complete the plotting commands. This statement communicates with a device, either a plotter or a CRT, in order for a point to be located and a line drawn to it.

The coordinates of XP and YP can be given at the right of the equal sign as numbers or as defined variables. For example, XP can be expressed as XP = 4.1 or as XP = A + 2. The coordinates XP = 0 and YP = 0 are customarily located at the lower left corner of the plotting area. The colons between the values of XP, YP, IP%, and GOSUB 9000 permit you to list these four statements on a single line. Colons can be used throughout any program to write more than one statement per line.

100 XP=3:YP=4:IP%=2:GOSUB 9000

Fig. 37.16 This is the plot statement that is used in a program to command the plotter either on the screen or on the plotter. The *line number* is required on all BASIC program statements. *XP* and *YP* are the desired X- and Y-coordinates to which the pen is to move. IP% gives the desired pen position: 2 for down and 3 for up. GOSUB 9000 is the *plotting subroutine* that commands the movement of the pen.

Pen positions can be assigned IP% of 0, 2, or 3. When P = 2, the pen is lowered to the surface and a line is drawn. When P = 3, the pen is raised and a line is not drawn. When P = 0, the origin of XP = 0, YP = 0 is reestablished and the pen is moved to the lower left corner of the plotting surface and clears the screen.

Pen commands lift and lower the pen in much the same manner as you lift your pencil when making a drawing.

The plotting area of a 11″ × 8.5″ flatbed printer is 10″ × 7″, while the same proportional area on the CRT screen is approximately 7.14″ ×

5.00″. A scale factor of 0.71 (71% of full size) is used to represent measurements on the screen that is only 71 percent of the area of the plotting surface. The scale factor of 0.71 is included within the plotting subroutine, GOSUB 9000, used in the examples that are illustrated in the following sections.

37.7 PLOTTING POINTS

An example of a point plotted on a screen is shown in Fig. 37.17. The rectangular outline represents the 7 × 5 inch screen of the CRT.

The first two statements, numbers 10 and 20, are REM (remark) statements that are used to document the program, but these statements are not executed when the program is run by the computer. The first two executable statements, lines 30 and 40, set the X- and Y-coordinates at the lower left corner of the screen with the pen in the up position.

Statement 50 moves the pen to X = 7 and Y = 5 with the pen up. Statement 60 lowers the pen by using IP% = 2, which marks the point.

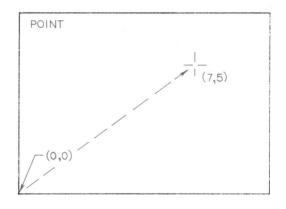

```
10 REM PROGRAM TO PLOT A POINT
20 REM POINT A ---> (X=7,Y=5)
30 IP%=0 : GOSUB 9000
40 XP=0 : YP=0 : IP%=3 : GOSUB 9000
50 XP=7 : YP=5 : IP%=3 : GOSUB 9000
60 XP=7 : YP=5 : IP%=2 : GOSUB 9000
70 XP=0 : YP=0 : IP%=0 : GOSUB 9000
80 STOP
90 END
```

Fig. 37.17 A program for plotting a point on the screen of the CRT. The information that appears in color will not appear on the screen, but it has been added for clarity.

Statement 70 moves the pen back to the origin with the pen up, and the last two statements stop and end the program.

37.8 PLOTTING LINES

A program has been written in Fig. 37.18 for drawing a line from A to B when the coordinates of each point are known. The coordinates are given in the remark statements, 100, 110, and 120.

The pen is initialized at 0,0 by statements 130 and 140. Statement 150 directs the pen to point A(6,5) with the pen up (IP% = 3). Statement 160 lowers the pen and moves it to point B at coordinates (2,3) and statement 170 moves the pen back to the point of origin.

A program has been written in Fig. 37.19 to illustrate how the line CD can be drawn on the screen, and how line AB can be drawn by the plotter. By inserting the statement in line 190, P.OUT% = 1, the program commands the computer to output the line on the plotter. When AB has been plotted, line 280 P.OUT% = 0 commands the computer to output the line CD on the display screen of the CRT. This technique will

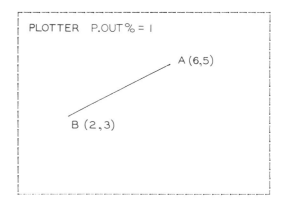

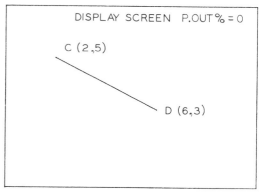

```
100 REM PORGRAM TO DRAW A LINE FROM A TO B
110 REM ON THE PLOTTER AND DRAW A LINE FROM
120 REM C TO D ON THE SCREEN.
130 REM POINT A ---> (X=6,Y=5)
140 REM POINT B ---> (X=2,Y=3)
150 REM POINT C ---> (X=2,Y=5)
160 REM POINT D ---> (X=6,Y=3)
170 REM --------------------
180 REM SELECT PLOTTER BY DEFINING P.OUT%=1.
190 P.OUT%=1
200 REM NOW PLOT LINE
210 IP%=0 : GOSUB 9000
220 XP=0 : YP=0 : IP%=3 : GOSUB 9000
230 XP=6 : YP=5 : IP%=3 : GOSUB 9000
240 XP=2 : YP=3 : IP%=2 : GOSUB 9000
250 XP=0 : YP=0 : IP%=3 : GOSUB 9000
260 REM NOW SELECT DISPLAY SCREEN
270 REM BY REDEFINING P.OUT%=0
280 P.OUT%=0
290 IP%=0 : GOSUB 9000
300 XP=0 : YP=0 : IP%=3 : GOSUB 9000
310 XP=2 : YP=5 : IP%=3 : GOSUB 9000
320 XP=6 : YP=3 : IP%=2 : GOSUB 9000
330 XP=0 : YP=0 : IP%=3 : GOSUB 9000
340 STOP
350 END
```

Fig. 37.19 The program commands the computer to output the drawing in "hard-copy" form by using the command, 190 P.OUT%=1. The program then directs the output back to the display screen by the command, 280 P.OUT%=0.

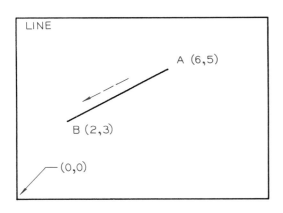

```
100 REM PORGRAM TO DRAW A LINE FROM A TO B
110 REM POINT A ---> (X=6,Y=5)
120 REM POINT B ---> (X=2,Y=3)
130 IP%=0 : GOSUB 9000
140 XP=0 : YP=0 : IP%=3 : GOSUB 9000
150 XP=6 : YP=5 : IP%=3 : GOSUB 9000
160 XP=2 : YP=3 : IP%=2 : GOSUB 9000
170 XP=0 : YP=0 : IP%=3 : GOSUB 9000
180 STOP
190 END
```

Fig. 37.18 A program and the resulting plot on the screen for drawing a line from point A to B.

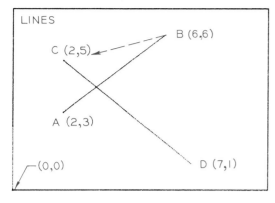

```
100 REM PROGRAM TO DRAW TWO LINES AB & CD
110 REM POINT A & B ---> XA,YA & XB,YB
120 REM POINT A & B ---> XC,YC & XD,YD
130 XA=2 : YA=3
140 XB=6 : YB=6
150 XC=2 : YC=5
160 XD=7 : YD=1
170 IP%=0 : GOSUB 9000
180 XP=0 : YP=0 : IP%=3 : GOSUB 9000
190 REM PLOT LINE AB
200 XP=XA : YP=YA : IP%=3 : GOSUB 9000
210 XP=XB : YP=YB : IP%=2 : GOSUB 9000
220 REM PLOT LINE CD
230 XP=XC : YP=YC : IP%=3 : GOSUB 9000
240 XP=XD : YP=YD : IP%=2 : GOSUB 9000
250 REM RETURN TO ORGIN
260 XP=0 : YP=0 : IP%=3 : GOSUB 9000
270 STOP
280 END
```

Fig. 37.20 A program for drawing lines AB and CD on the display screen.

work on any problem by using the two commands P.OUT% = 0 and P.OUT% = 1.

The same principles are used in the program shown in Fig. 37.20 where two lines are drawn: AB and CD. Note that instead of using numerical coordinates for the ends of the lines, variables (XA, YA, XB, etc.) are used with numerical values assigned to them as shown in statements 130 through 160. Variables XA and YB are used in the plot statement in line 200. This technique of using variable names enables you to define the variables at the beginning of the program without changing the variables in the plot statement when other values are substituted for different coordinates.

A line can be drawn using trigonometric functions when the following are given: length, angle of inclination, and the coordinates of one end.

In Fig. 37.21, end A is located at XA, YA, where XA = 2 and YA = 3 (statements 180 and 190). The 35-degree angle of inclination is converted to radians in statement 230.

The end at B is found by computing the XB-value as the cosine of the angle times the length of the line plus the X-component of point A(XA) (statement 250). The YB component is the sine of the angle times the length of the line plus the Y component of point A, (YA). These trigonometric values are substituted into the plot statements in lines 300 and 310.

37.9 PLOTTING RECTANGLES

The plotting of a rectangle is done by a program that is written in the same format as was used to plot a series of lines in Fig. 37.22. The example in Fig. 37.22 gives the coordinates of each of the four

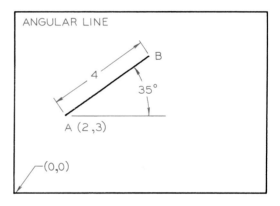

```
100 REM PROGRAM TO DRAW A LINE
110 REM GIVEN ONE END POINT, ITS LENGTH,
120 REM AND ITS ANGLE OF ROTATION
130 REM ================================
140 REM POINT A -------------> XA,YA
150 REM POINT B -------------> (COMPUTED)
160 REM LENGTH --------------> LENGTH (INCHES)
170 REM ANGLE OF ROTATION ---> ANGLE (DEGREES)
180 XA = 2
190 YA = 3
200 LENGTH = 4
210 ANGLE = 35
220 REM CONVERT ANGLE TO RADIANS
230 ANGLE = ANGLE * 3.1416 / 180
240 REM COMPUTE THE COORDINATES OF POINT B
250 XB = XA + LENGTH * COS(ANGLE)
260 YB = YA + LENGTH * SIN(ANGLE)
270 REM PLOT THE LINE FROM A TO B
280 IP%=0 : GOSUB 9000
290 XP=0 : YP=0 : IP%=3 : GOSUB 9000
300 XP=XA : YP=YA : IP%=3 : GOSUB 9000
310 XP=XB : YP=YB : IP%=2 : GOSUB 9000
320 XP=0 : YP=0 : IP%=3 : GOSUB 9000
330 STOP
340 END
```

Fig. 37.21 A line can be drawn on the display screen by using its trigonometric functions when its length, angle of inclination, and one point on it are known.

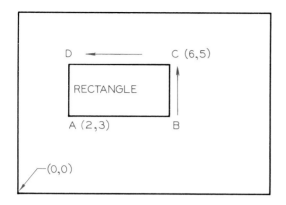

```
100 REM PROGRAM TO DRAW A RECTANGLE
110 REM GIVEN COORDINATES OF TWO CORNERS
120 REM POINT A AND POINT C
130 REM ===============================
140 REM POINT A ---> XA,YA
150 REM POINT B ---> XC,YA
160 REM POINT C ---> XC,YC
170 REM POINT D ---> XA,YC
180 XA = 2 : YA = 3
190 XC = 6 : YC = 5
200 IP%=0 : GOSUB 9000
210 XP=0 : YP=0 : IP%=3 : GOSUB 9000
220 REM PLOT TO POINT A
230 XP=XA : YP=YA : IP%=3 : GOSUB 9000
240 REM PLOT TO POINT B
250 XP=XC : YP=YA : IP%=2 : GOSUB 9000
260 REM PLOT TO POINT C
270 XP=XC : YP=YC : IP%=2 : GOSUB 9000
280 REM PLOT TO POINT D
290 XP=XA : YP=YC : IP%=2 : GOSUB 9000
300 REM PLOT TO POINT A
310 XP=XA : YP=YA : IP%=2 : GOSUB 9000
320 REM RETURN TO ORIGIN
330 XP=0 : YP=0 : IP%=3 : GOSUB 9000
340 STOP
350 END
```

Fig. 37.22 A program and a display of its output for drawing a rectangle when the coordinates of each corner are known.

corners of a rectangle in terms of variables XA, YA, XC, and YC in lines 180 and 190.

After the pen is initialized at 0, 0, it is directed to point A (XA, YA) with the pen up in line 230. Statements 250, 270, 290, and 310 successively move the pen to points B, C, D, and back to A. You can see that a variety of different rectangles can be drawn by changing the numerical values of the variables in statements 180 and 190.

Inclined rectangles can be constructed when the following are known: The coordinates of one corner, the height and width, and the angle of inclination from 0 to 360 degrees. In the example in Fig. 37.23, these known values are given in statements 190 through 220.

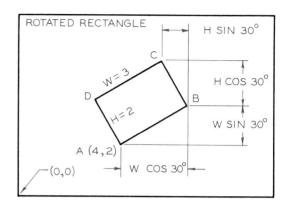

```
100 REM PROGRAM TO DRAW A RECTANGLE
110 REM GIVEN THE COORDINATE OF THE POINT A,
120 REM THE HEIGHT, THE WIDTH, AND THE ROTATION
130 REM ABOUT THE POINT A IN DEGREES
140 REM =======================================
150 REM POINT A --------------------> XA,YA
160 REM HEIGHT --------------------> R.HEIGHT
170 REM WIDTH ---------------------> R.WIDTH
180 REM ROTATION ABOUT POINT A -----> R.ANGLE
190 XA = 4 : YA = 2
200 R.HEIGHT = 2
210 R.WIDTH = 3
220 R.ANGLE = 30
230 REM CONVERT R.ANGLE TO RADIANS
240 R.ANGLE = R.ANGLE * 3.1416 / 180
250 REM COMPUTE POINT B
260 XB = XA + R.WIDTH * COS(R.ANGLE)
270 YB = YA + R.WIDTH * SIN(R.ANGLE)
280 REM COMPUTE POINT C
290 XC = XB - R.HEIGHT * SIN(R.ANGLE)
300 YC = YB + R.HEIGHT * COS(R.ANGLE)
310 REM COMPUTE POINT D
320 XD = XA - R.HEIGHT * SIN(R.ANGLE)
330 YD = YA + R.HEIGHT * COS(R.ANGLE)
340 IP%=0 : GOSUB 9000
350 XP=0 : YP=0 : IP%=3 : GOSUB 9000
360 REM PLOT TO POINT A
370 XP=XA : YP=YA : IP%=3 : GOSUB 9000
380 REM PLOT TO POINT B
390 XP=XB : YP=YB : IP%=2 : GOSUB 9000
400 REM PLOT TO POINT C
410 XP=XC : YP=YC : IP%=2 : GOSUB 9000
420 REM PLOT TO POINT D
430 XP=XD : YP=YD : IP%=2 : GOSUB 9000
440 REM PLOT TO POINT A
450 XP=XA : YP=YA : IP%=2 : GOSUB 9000
460 REM RETURN TO ORIGIN
470 XP=0 : YP=0 : IP%=3 : GOSUB 9000
480 STOP
490 END
```

Fig. 37.23 A program and a display of its output for drawing a rectangle by expressing its dimensions in variables that can be easily changed for other variations of size and rotation.

The angle is converted to radians in line 240. The coordinates at corner B are XB and YB, and these are computed in lines 260 and 270 as the cosine and sine of the 30-degree angle multiplied by the width of the rectangle. Similarly, the coordinates of corner C are found using the sine and cosine of the height of the rectangle in statements 290 and 300. The coordinates of corner D are defined in lines 320 and 330.

The variables for each corner are sequentially inserted into the plot statement in statements 370 through 450 to drive the pen to each of the four corners. The use of variables makes it possible to change any or all of the variables in statements 190 through 220 and thereby obtain a different rectangle with a minimum of programming modification.

37.10 PLOTTING POLYGONS

A regular polygon, having a number of sides each of the same length, can be constructed using the program shown in Fig. 37.24 when the following are given: The coordinates of the center, the distance from the center to each corner (RADIUS), and the number of sides (NSIDES). This information is given in lines 170 through 190.

The coordinates of corner 1 are found in statements 260 and 270 where distance RADIUS is multiplied by the sine and cosine of the angle between each corner point and each is added to the coordinates of the center of the polygon. An IF statement is used in line 290 that tells the plotter to lift the pen when the angle is at zero degrees (horizontal) and to move to the first calculated point. A FOR NEXT loop is used between statements 220 and 310 where the steps of angle ANGLE are specified to be 360/NSIDES, which is 30° in this example.

The values of each set of X- and Y-coordinates are computed and plotted for each successive 30° angle through 360° and back to the point of beginning where the pen is then lifted and the loop is ended. The pen is then directed by statement 320 to return to the origin, 0, 0.

37.11 CIRCLES

The program for drawing a circle is identical to the one used to draw regular polygons. The only exception is the smallness of the size of the steps

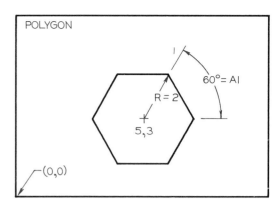

```
100 REM PROGRAM TO DRAW A POLYGON
110 REM GIVEN CENTER, RADIUS,
120 REM AND NUMBER OF SIDES.
130 REM =========================
140 REM CENTER ---> XCENTER , YCENTER
150 REM RADIUS ---> RADIUS
160 REM #SIDES ---> NSIDES
170 XCENTER=5 : YCENTER=3
180 RADIUS = 2
190 NSIDES = 6
200 IP%=0 : GOSUB 9000
210 XP=0 : YP=0 : IP%=3 : GOSUB 9000
220 FOR ANGLE = 0 TO 360 STEP 360/NSIDES
230    REM CONVERT ANGLE TO RADIANS
240    ANGLE.R = ANGLE * 3.1416 / 180
250    REM COMPUTE COORDINATES OF POINT X,Y
260    XP = XCENTER + RADIUS * COS(ANGLE.R)
270    YP = YCENTER + RADIUS * SIN(ANGLE.R)
280    REM MOVE PEN UP IF ANGLE = 0
290    IF ANGLE = 0 THEN IP%=3 ELSE IP%=2
300    GOSUB 9000
310 NEXT ANGLE
320 XP=0 : YP=0 : IP%=3 : GOSUB 9000
330 STOP
340 END
```

Fig. 37.24 A program and its resulting display for drawing regular polygons.

of the angle from corner to corner of the polygon. The circle is really no more than a polygon with many small sides.

Like the polygon, the following must be given: The coordinates of the center (XCENTER and YCENTER) and the radius R, as given in statements 170 and 180 of the program in Fig. 37.25. The angular steps are given in the FOR NEXT loop as 5° increments and the resulting polygon approximates a circle.

As the circle is drawn larger, the angular steps can be specified in smaller increments. Small circles can be drawn with slightly larger increments, and therefore plotting time can be reduced without affecting the appearance of the plotted circle.

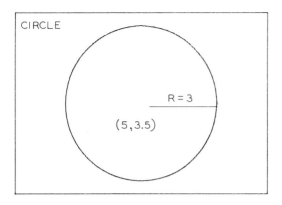

```
100 REM PROGRAM TO DRAW A CIRCLE
110 REM GIVEN THE COORDINATES OF THE
120 REM CENTER AND THE LENGTH OF THE
125 REM RADIUS
130 REM =========================
140 REM CENTER ---> XCENTER , YCENTER
150 REM RADIUS ---> RADIUS
170 XCENTER=5 : YCENTER=3.5
180 RADIUS = 3
190 IP%=0 : GOSUB 9000
200 FOR ANGLE = 0 TO 360 STEP 5
210    REM CONVERT ANGLE TO RADIANS
220    ANGLE.R = ANGLE * 3.1416 / 180
230    REM COMPUTE COORDINATES OF POINT X,Y
240    XP = XCENTER + RADIUS * COS(ANGLE.R)
250    YP = YCENTER + RADIUS * SIN(ANGLE.R)
260    REM MOVE PEN UP IF ANGLE = 0
270    IF ANGLE = 0 THEN IP%=3 ELSE IP%=2
280    GOSUB 9000
290 NEXT ANGLE
300 XP=0 : YP=0 : IP%=3 : GOSUB 9000
310 STOP
320 END
```

Fig. 37.25 A program for drawing a circle is the same one used to draw regular polygons with the only difference being the number of sides that are plotted.

37.12 PROGRAMS WITH DATA STATEMENTS

An alternative method of programming the coordinates of points is the uses of DATA statements as given in lines 160 through 200 of Fig. 37.26. The three numbers in each DATA statement represent X, Y, and pen values for each point of a triangle, and the coordinates are listed in this order for each point.

A READ statement is given in lines 260, 280, and 290. Once read, the values are transferred to the plot statement in line 310 where the values are executed. Statement 320 sends the program back to line 260 where the second set of data is read. This loop continues until the data in line 210 is encountered and the program is then ended.

37.13 SUBROUTINES

A main program can be written that will recall previously written programs, called **subroutines.** Subroutines are loaded from the disc and appended to the main program, using the MERGE, rather than the LOAD command.

An example main program and subroutine 600 is used to draw a rectangle in Fig. 37.27. The main program gives the following specifics: The initial point, the coordinates of a corner (XA and YA), the height and width (R.HEIGHT and R.WIDTH), and the angle of inclination (R.ANGLE). The subroutine is called by statement 220 GOSUB 600.

The variable values given in the main program are transferred to the subroutine where the

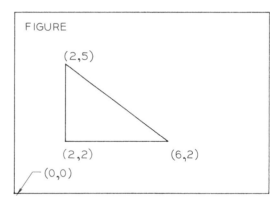

```
100 REM PROGRAM TO DRAW A FIGURE
110 REM GIVEN X,Y AND PEN COORDINATES
120 REM IN "DATA" STATEMENTS
130 REM ===========================
140 REM LAST DATA VALUE IS ALWAYS 999
150 REM DEFINE DATA
160 DATA 2,2,3
170 DATA 6,2,2
180 DATA 2,5,2
190 DATA 2,2,2
200 DATA 0,0,3
210 DATA 999
220 REM PLOT DATA
230 IP%=0 : GOSUB 9000
240 XP=0 : YP=0 : IP%=3 : GOSUB 9000
250 REM ----- READ AND PLOT LOOP -----
260 READ XP
270 IF XP=999 THEN STOP
280 READ YP
290 READ IP%
300 REM PLOT POINT
310 GOSUB 9000
320 GOTO 260
330 REM ----- END OF LOOP -----
340 END
```

Fig. 37.26 This program utilizes points of a triangle that are fed to the program by DATA statements as shown in lines 160 through 210.

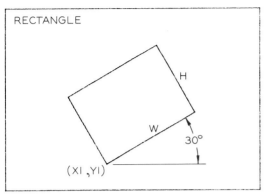

```
100 REM --- MAIN ROUTINE ---
110 REM DRAWS A RECTANGLE BY CALLING THE
120 REM RECTANGLE SUBROUTINE AT LINE 600
130 REM ==== STEP 1 ====
140 IP%=0 : GOSUB 9000
150 XP=0 : YP=0 : IP%=3 : GOSUB 9000
160 REM ==== STEP 2 ====
170 REM DEFINE VARIABLES NEEDED BY THE
180 REM RECTANGLE SUBROUTINE
190 XA = 4 : YA = 1 : R.HEIGHT = 3
200 R.WIDTH = 4 : R.ANGLE = 30
210 REM CALL THE RECTANGLE SUBROUTINE
220 GOSUB 600
230 STOP
240 END


600 REM PROGRAM TO DRAW A RECTANGLE
610 REM GIVEN THE COORDINATE OF THE POINT A,
620 REM THE HEIGHT, THE WIDTH, AND THE ROTATION
630 REM ABOUT THE POINT A IN DEGREES
640 REM ====================================
650 REM POINT A ---------------------> XA,YA
660 REM HEIGHT --------------------> R.HEIGHT
670 REM WIDTH ---------------------> R.WIDTH
680 REM ROTATION ABOUT POINT A -----> R.ANGLE
690 REM CONVERT R.ANGLE TO RADIANS
700 R.ANGLE = R.ANGLE * 3.1416 / 180
710 REM COMPUTE POINT B
720 XB = XA + R.WIDTH * COS(R.ANGLE)
730 YB = YA + R.WIDTH * SIN(R.ANGLE)
740 REM COMPUTE POINT C
750 XC = XB - R.HEIGHT * SIN(R.ANGLE)
760 YC = YB + R.HEIGHT * COS(R.ANGLE)
770 REM COMPUTE POINT D
780 XD = XA - R.HEIGHT * SIN(R.ANGLE)
790 YD = YA + R.HEIGHT * COS(R.ANGLE)
800 REM PLOT TO POINT A
810 XP=XA : YP=YA : IP%=3 : GOSUB 9000
820 REM PLOT TO POINT B
830 XP=XB : YP=YB : IP%=2 : GOSUB 9000
840 REM PLOT TO POINT C
850 XP=XC : YP=YC : IP%=2 : GOSUB 9000
860 REM PLOT TO POINT D
870 XP=XD : YP=YD : IP%=2 : GOSUB 9000
880 REM PLOT TO POINT A
890 XP=XA : YP=YA : IP%=2 : GOSUB 9000
900 REM RETURN TO ORIGIN
910 RETURN
920 END
```

Fig. 37.27 A subroutine is used in conjunction with a main program for drawing this rectangle. The subroutine 600 is appended to the main program, and control is then returned to the main program once the subroutine has been utilized.

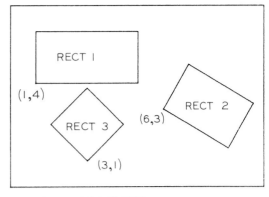

```
100 REM --- MAIN ROUTINE ---
110 REM DRAWS 3 RECTANGLES BY CALLING THE
120 REM RECTANGLE SUBROUTINE AT LINE 600
130 IP%=0 : GOSUB 9000
140 XP=0 : YP=0 : IP%=3 : GOSUB 9000
150 REM DRAW RECTANGLE 1
160 XA=1 : YA=4 : R.HEIGHT=2 : R.WIDTH=4
170 R.ANGLE=0 : GOSUB 600
180 REM DRAW RECTANGLE 2
190 XA=6 : YA=3 : R.HEIGHT=2 : R.WIDTH=3
200 R.ANGLE=-30 : GOSUB 600
210 REM DRAW RECTANGLE 3
220 XA=3 : YA=1 : R.HEIGHT=2 : R.WIDTH=2
230 R.ANGLE=45 : GOSUB 600
240 REM RETURN TO ORIGIN
250 XP=0 : YP=0 : IP%=3 : GOSUB 9000
260 END


600 REM PROGRAM TO DRAW A RECTANGLE
610 REM GIVEN THE COORDINATE OF THE POINT A,
620 REM THE HEIGHT, THE WIDTH, AND THE ROTATION
630 REM ABOUT THE POINT A IN DEGREES
640 REM ====================================
650 REM POINT A ---------------------> XA,YA
660 REM HEIGHT --------------------> R.HEIGHT
670 REM WIDTH ---------------------> R.WIDTH
680 REM ROTATION ABOUT POINT A -----> R.ANGLE
690 REM CONVERT R.ANGLE TO RADIANS
700 R.ANGLE = R.ANGLE * 3.1416 / 180
710 REM COMPUTE POINT B
720 XB = XA + R.WIDTH * COS(R.ANGLE)
730 YB = YA + R.WIDTH * SIN(R.ANGLE)
740 REM COMPUTE POINT C
750 XC = XB - R.HEIGHT * SIN(R.ANGLE)
760 YC = YB + R.HEIGHT * COS(R.ANGLE)
770 REM COMPUTE POINT D
780 XD = XA - R.HEIGHT * SIN(R.ANGLE)
790 YD = YA + R.HEIGHT * COS(R.ANGLE)
800 REM PLOT TO POINT A
810 XP=XA : YP=YA : IP%=3 : GOSUB 9000
820 REM PLOT TO POINT B
830 XP=XB : YP=YB : IP%=2 : GOSUB 9000
840 REM PLOT TO POINT C
850 XP=XC : YP=YC : IP%=2 : GOSUB 9000
860 REM PLOT TO POINT D
870 XP=XD : YP=YD : IP%=2 : GOSUB 9000
880 REM PLOT TO POINT A
890 XP=XA : YP=YA : IP%=2 : GOSUB 9000
900 REM RETURN TO ORIGIN
910 RETURN
920 END
```

Fig. 37.28 In this plot, you can see that the main program has called subroutine 600 (GOSUB 600) three times for drawing three rectangles. The dimensions of the rectangles are given in the main program.

other values of the rectangle are calculated and plotted. When the subroutine has been executed, it returns to the main program (line 910) where the main program resumes its control over the next steps of the program.

A subroutine can be called several times and in combination with other subroutines by the main program. Fig. 37.28 is an example of subrou-

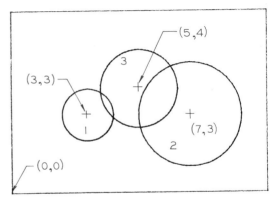

```
100 REM --- MAIN PROGRAM ---
110 REM PROGRAM TO DRAW 3 CIRCLES BY
120 REM CALLING THE CIRCLE SUBROUTINE
130 REM AT LINE 800.
140 IP%=0 : GOSUB 9000
150 REM DRAW CIRCLE 1
160 XCENTER=3 : YCENTER=3
170 RADIUS=1 : GOSUB 800
180 REM DRAW CIRCLE 2
190 XCENTER=7 : YCENTER=3
200 RADIUS=2 : GOSUB 800
210 REM DRAW CIRCLE 3
220 XCENTER=5 : YCENTER=4
230 RADIUS=1.5 : GOSUB 800
240 REM RETURN TO ORIGIN
250 XP=0 : YP=0 : IP%=3 : GOSUB 9000
260 END
```

```
800 REM SUBROUTINE TO DRAW A CIRCLE
810 REM GIVEN THE COORDINATES OF THE
820 REM CENTER AND THE LENGTH OF THE
830 REM RADIUS
840 REM =========================
850 REM CENTER ---> XCENTER , YCENTER
860 REM RADIUS ---> RADIUS
870 FOR ANGLE = 0 TO 360 STEP 5
880    REM CONVERT ANGLE TO RADIANS
890    ANGLE.R = ANGLE * 3.1416 / 180
900    REM COMPUTE COORDINATES OF POINT X,Y
910    XP = XCENTER + RADIUS * COS(ANGLE.R)
920    YP = YCENTER + RADIUS * SIN(ANGLE.R)
930    IF ANGLE = 0 THEN IP%=3 ELSE IP%=2
940    GOSUB 9000
950 NEXT ANGLE
960 XP=0 : YP=0 : IP%=3 : GOSUB 9000
970 RETURN
980 END
```

Fig. 37.30 Subroutine 800 (circle) is called three times by the main program to draw the different circles.

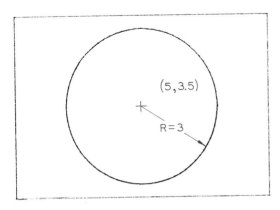

```
100 REM --- MAIN PROGRAM ---
110 REM PROGRAM TO DRAW A CIRCLE BY
120 REM CALLING THE CIRCLE SUBROUTINE
130 REM AT LINE 800.
140 REM ==== STEP 1 ====
150 REM INITIALIZE
160 IP%=0 : GOSUB 9000
170 REM ==== STEP 2 ====
180 REM DEFINE THE VARIABLES NEEDED BY
190 REM THE CIRCLE SUBROUTINE
200 XCENTER=5 : YCENTER=3.5 : RADIUS=3
210 REM ==== STEP 3 ====
220 REM CALL THE CIRCLE SUBROUTINE
230 GOSUB 800
240 STOP
250 END
```

```
800 REM SUBROUTINE TO DRAW A CIRCLE
810 REM GIVEN THE COORDINATES OF THE
820 REM CENTER AND THE LENGTH OF THE
830 REM RADIUS
840 REM =========================
850 REM CENTER ---> XCENTER , YCENTER
860 REM RADIUS ---> RADIUS
870 FOR ANGLE = 0 TO 360 STEP 5
880    REM CONVERT ANGLE TO RADIANS
890    ANGLE.R = ANGLE * 3.1416 / 180
900    REM COMPUTE COORDINATES OF POINT X,Y
910    XP = XCENTER + RADIUS * COS(ANGLE.R)
920    YP = YCENTER + RADIUS * SIN(ANGLE.R)
930    IF ANGLE = 0 THEN IP%=3 ELSE IP%=2
940    GOSUB 9000
950 NEXT ANGLE
960 XP=0 : YP=0 : IP%=3 : GOSUB 9000
970 RETURN
980 END
```

SUBROUTINE

Fig. 37.29 Subroutine 800 is called by the main program to plot a circle. Control is then returned to the main program.

tine 600 that is called three times for drawing different rectangles, 1, 2, and 3.

Rectangle 1 is drawn by using subroutine 600 and the command statement in lines 160 and 170 of the main program. The subroutine returns to the main program where new data is given and the subroutine is called once more. Finally, rectangle 3 is drawn using the data given in lines 220 and 230 of the main program and subroutine 600.

Circle subroutines can be called in the same manner as rectangular subroutines. In Fig. 37.29 a

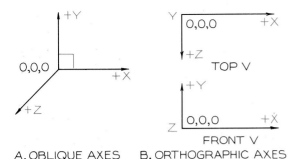

Fig. 37.31 The axes of an oblique drawing are shown in part A. The same axes are shown orthographically in part B where (0, 0, 0) is the origin.

main program is used to define the values of the coordinates of the center and the radius of the circle (line 200). Line 230 calls the subroutine by the statement GOSUB 800.

The circle subroutine is called three times in Fig. 37.30 to draw three different circles. The main program gives the data for each circle and calls the subroutine in lines 160 through 230. Circles 1, 2, and 3 are drawn each time the subroutine 800 is called. The subroutine returns its control to the main program after the third circle has been drawn.

Subroutines of various types for drawing different geometric shapes can be accessed. However, they must be listed in ascending order where the subroutine with the largest number is listed last.

37.14 OBLIQUES

The program for plotting an oblique pictorial can be written by using X-, Y-, and Z-axes as shown in Fig. 37.31, with the origin at the rear of the position of the oblique. Three points are shown plot-

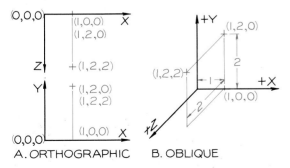

Fig. 37.32 The orthographic coordinates of three points are shown on the X-, X-, and Z-axes at A. The points shown at A are shown plotted in oblique at B.

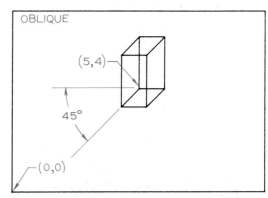

```
100 REM PROGRAM TO DRAW A FIGURE GIVEN THE
110 REM COORDINATES FOR ITS POINTS ON THE
120 REM X,Y, & Z AXIS.  THESE VALUES ARE
130 REM TRANSLATED ONTO AN OBLIQUE AXIS
140 REM BEFORE THEY ARE PLOTTED.  THE
150 REM COORDINATES ARE DEFINED WITH "DATA"
160 REM STATEMENTS CONTAINING X,Y,Z & PEN
170 REM VALUES.  X.ORG AND Y.ORG ARE THE
180 REM X & Y LOCATION OF THE ORIGIN OF THE
190 REM FIGURE.  Z.SCALE IS THE Z-AXIS
200 REM SCALE FACTOR.  SCALE IS THE OVERALL
210 REM SCALE FACTOR. ANGLE IS THE Z-AXIS
220 REM ANGLE. THE LAST DATA VALUE IS
230 REM ALWAYS 999.
240 REM DEFINE ORIGIN, SCALE AND ROTATION
250 X.ORG=5 : Y.ORG=4
260 Z.SCALE=.5 : SCALE=1 : ANGLE = 45
270 DATA 0,0,0,3,1,0,0,2,1,2,0,2,1,2,2,2
280 DATA 1,0,2,2,0,0,2,2,0,0,0,2,0,2,0,2
290 DATA 0,2,2,2,0,0,2,2,1,0,2,3,1,0,0,2
300 DATA 0,2,2,3,1,2,2,2,0,2,0,3,1,2,0,2
310 DATA 0,0,0,3,999
320 IP%=0 : GOSUB 9000
330 REM CONVERT ANGLE TO RADIANS
340 ANGLE = ANGLE * 3.1416 /180
350 REM ------ READ LOOP ------
360 READ XP
370 IF XP = 999 THEN STOP
380 READ YP
390 READ ZP
400 READ IP%
410 XP=X.ORG+((XP-(ZP*COS(ANGLE)*Z.SCALE)))*SCALE
420 YP=Y.ORG+((YP-(ZP*SIN(ANGLE)*Z.SCALE)))*SCALE
430 REM PLOT XP,YP & IP%
440 GOSUB 9000
450 GOTO 360
460 REM ------ END OF LOOP ------
470 END
```

Fig. 37.33 A program and its plot that converts given data points into an oblique drawing.

ted on these axes in Fig. 37.32 with their coordinates given at part A. If these views were drawn in a top and front view the points would appear orthographically as shown in Fig. 37.32B.

The program in Fig. 37.33 gives the coordinates of the origin of the oblique (X.ORG and Y.ORG), the scale factor (Z.SCALE) of the Z-axis, the overall scale factor (SCALE), and the angle of the receding axis (ANGLE). If Z.SCALE = 1, the oblique will be a cavalier oblique; if S = 0.5, the oblique will be a cabinet oblique. Scale factors

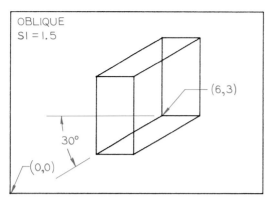

```
100 REM PROGRAM TO DRAW A FIGURE GIVEN THE
110 REM COORDINATES FOR ITS POINTS ON THE
120 REM X,Y, & Z AXIS.   THESE VALUES ARE
130 REM TRANSLATED ONTO AN OBLIQUE AXIS
140 REM BEFORE THEY ARE PLOTTED.   THE
150 REM COORDINATES ARE DEFINED WITH "DATA"
160 REM STATEMENTS CONTAINING X,Y,Z & PEN
170 REM VALUES.   X.ORG AND Y.ORG ARE THE
180 REM X & Y LOCATION OF THE ORIGIN OF THE
190 REM FIGURE.   Z.SCALE IS THE Z-AXIS
200 REM SCALE FACTOR.   SCALE IS THE OVERALL
210 REM SCALE FACTOR. ANGLE IS THE Z-AXIS
220 REM ANGLE. THE LAST DATA VALUE IS
230 REM ALWAYS 999.
240 REM DEFINE ORIGIN, SCALE AND ROTATION
250 X.ORG=6 : Y.ORG=3
260 Z.SCALE=1 : SCALE=1.5 : ANGLE = 30
270 DATA 0,0,0,3,1,0,0,2,1,2,0,2,1,2,2,2
280 DATA 1,0,2,2,0,0,2,2,0,0,0,2,0,2,0,2
290 DATA 0,2,2,2,0,0,2,2,1,0,2,3,1,0,0,2
300 DATA 0,2,2,3,1,2,2,2,0,2,0,3,1,2,0,2
310 DATA 0,0,0,3,999
320 IP%=0 : GOSUB 9000
330 REM CONVERT ANGLE TO RADIANS
340 ANGLE = ANGLE * 3.1416 /180
350 REM ------ READ LOOP ------
360 READ XP
370 IF XP = 999 THEN STOP
380 READ YP
390 READ ZP
400 READ IP%
410 XP=X.ORG+((XP-(ZP*COS(ANGLE)*Z.SCALE)))*SCALE
420 YP=Y.ORG+((YP-(ZP*SIN(ANGLE)*Z.SCALE)))*SCALE
430 REM PLOT XP,YP & IP%
440 GOSUB 9000
450 GOTO 360
460 REM ------ END OF LOOP ------
470 END
```

Fig. 37.34 A cavalier oblique has been plotted by the accompanying program. Note that the drawing has been enlarged since a scale factor (SCALE) of 1.5 was given.

(SCALE) are used to reduce the size of a figure so it will fit on the screen's area.

Data statements are used in lines 270 through 310 to give the X-, Y-, Z-, and P-values of each point of the oblique. The data statements are read in statements 360, 380, 390, and 400; calculations are made in statements 410 and 420; the points are plotted in statements 440; and then the program

returns to statement 360 for the next values. This is repeated until all of the points have been read and plotted.

A cavalier oblique is drawn by the program in Fig. 37.34. The scale factor of the receding axis, Z.SCALE = 1, means it will be full size, which is the characteristic of the cavalier oblique. The same data points that were given in the previous example are drawn in this oblique and the angle of the receding axis is changed from 45° to 30°.

Note the method of converting Z-values into X- and Y-coordinates in statements 410 and 420. Since points on a sheet or on a screen are located using only X- and Y-coordinates, the Z-coordinates must be converted by these formulas. A loop is used between lines 360 and 450 for reading, calculating, and plotting the points.

37.15 AXONOMETRICS AND ISOMETRICS

An axonometric pictorial can be programmed and plotted by revolving the views of the part about the Y-axis and then the X-axis by the desired angles as shown orthographically in Fig. 37.35. The resulting front view is the axonometric pictorial of the object. Notice the assignments of positive and negative directions of rotation in each view.

A program and the plot of an axonometric that has been revolved − 30° about the Y-axis in a negative direction, and revolved 30° in a positive di-

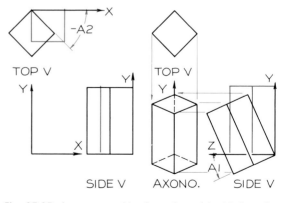

Fig. 37.35 An axonometric view of an object is found by revolving the top view about the Y-axis and the side view about the X-axis. The resulting front view is an axonometric projection.

rection about the X-axis are shown in Fig. 37.36. In line 270, the coordinates of the origin of the figure (X.ORG, Y.ORG), the two angles of rotation (X.ANGLE and Y.ANGLE), and the scale factor (SCALE) are given.

The data statements, lines 270 through 310, give the coordinates of the points of the pictorial that are to be drawn. The equations in lines 410,

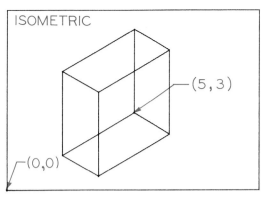

```
100 REM PROGRAM TO DRAW A FIGURE GIVEN THE
110 REM COORDINATES FOR ITS POINTS ON THE
120 REM X,Y, & Z AXIS.   THESE VALUES ARE
130 REM TRANSLATED ONTO AN ISOMETRIC AXIS
140 REM BEFORE THEY ARE PLOTTED.   THE
150 REM COORDINATES ARE DEFINED WITH "DATA"
160 REM STATEMENTS CONTAINING X,Y,Z & PEN
170 REM VALUES.   X.ORG AND Y.ORG ARE THE
180 REM X & Y LOCATION OF THE ORIGIN OF THE
190 REM FIGURE.   SCALE IS THE OVERALL SCALE
200 REM FACTOR.   X.ANGLE IS THE ROTATION
210 REM ABOUT THE X AXIS(-45).   Y.ANGLE IS
220 REM THE ROTATION ABOUT THE Y AXIS(35.3).
230 REM ====================================
240 REM DEFINE ORIGIN, SCALE AND ROTATION
250 X.ORG=5 : Y.ORG=3 : SCALE=2
260 X.ANGLE=35.3 : Y.ANGLE=-45
270 DATA 0,0,0,3,1,0,0,2,1,2,0,2,1,2,2,2
280 DATA 1,0,2,2,0,0,2,2,0,0,0,2,0,2,0,2
290 DATA 0,2,2,2,0,0,2,2,1,0,2,3,1,0,0,2
300 DATA 0,2,2,3,1,2,2,2,0,2,0,3,1,2,0,2
310 DATA 0,0,0,3,999
320 REM PLOT DATA
330 IP%=0 : GOSUB 9000
340 REM CONVERT ANGLE TO RADIANS
350 X.ANGLE = X.ANGLE * 3.1416 / 180
360 Y.ANGLE = Y.ANGLE * 3.1416 / 180
370 REM ------ READ LOOP ------
380 READ XP
390 IF XP = 999 THEN STOP
400 READ YP,ZP,IP%
410 YP=XP*SIN(Y.ANGLE)*SIN(X.ANGLE)+YP*COS(X.ANGLE)
420 YP=(YP-ZP*COS(Y.ANGLE)*SIN(X.ANGLE))*SCALE+Y.ORG
430 XP=(XP*COS(Y.ANGLE)+ZP*SIN(Y.ANGLE))*SCALE+X.ORG
440 REM PLOT XP,YP & IP%
450 GOSUB 9000
460 GOTO 380
470 REM ------ END OF LOOP ------
480 END
```

Fig. 37.37 The axonometric program can be used to draw an isometric pictorial by revolving the top view and the side view so the point view of a cube's diagonal will be found in the axonometric view.

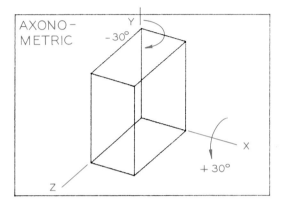

```
100 REM PROGRAM TO DRAW A FIGURE GIVEN THE
110 REM COORDINATES FOR ITS POINTS ON THE
120 REM X,Y, & Z AXIS.   THESE VALUES ARE
130 REM TRANSLATED ONTO AN AXONOMETRIC AXIS
140 REM BEFORE THEY ARE PLOTTED.   THE
150 REM COORDINATES ARE DEFINED WITH "DATA"
160 REM STATEMENTS CONTAINING X,Y,Z & PEN
170 REM VALUES.   X.ORG AND Y.ORG ARE THE
180 REM X & Y LOCATION OF THE ORIGIN OF THE
190 REM FIGURE.   SCALE IS THE OVERALL SCALE
200 REM FACTOR.   X.ANGLE IS THE ROTATION
210 REM ABOUT THE X AXIS.   Y.ANGLE IS THE
220 REM ROTATION ABOUT THE Y AXIS.
230 REM ====================================
240 REM DEFINE ORIGIN, SCALE AND ROTATION
250 X.ORG=5 : Y.ORG=3 : SCALE=2
260 X.ANGLE=30 : Y.ANGLE=-30
270 DATA 0,0,0,3,1,0,0,2,1,2,0,2,1,2,2,2
280 DATA 1,0,2,2,0,0,2,2,0,0,0,2,0,2,0,2
290 DATA 0,2,2,2,0,0,2,2,1,0,2,3,1,0,0,2
300 DATA 0,2,2,3,1,2,2,2,0,2,0,3,1,2,0,2
310 DATA 0,0,0,3,999
320 REM PLOT DATA
330 IP%=0 : GOSUB 9000
340 REM CONVERT ANGLE TO RADIANS
350 X.ANGLE = X.ANGLE * 3.1416 / 180
360 Y.ANGLE = Y.ANGLE * 3.1416 / 180
370 REM ------ READ LOOP ------
380 READ XP
390 IF XP = 999 THEN STOP
400 READ YP,ZP,IP%
410 YP=XP*SIN(Y.ANGLE)*SIN(X.ANGLE)+YP*COS(X.ANGLE)
420 YP=(YP-ZP*COS(Y.ANGLE)*SIN(X.ANGLE))*SCALE+Y.ORG
430 XP=(XP*COS(Y.ANGLE)+ZP*SIN(Y.ANGLE))*SCALE+X.ORG
440 REM PLOT XP,YP & IP%
450 GOSUB 9000
460 GOTO 380
470 REM ------ END OF LOOP ------
480 END
```

Fig. 37.36 This program rotates the top and side views by 30° to give this axonometric pictorial.

420, and 430 determine the axonometric coordinates of the points that are to be plotted. Each of the data sets is looped through these equations until all have been calculated and plotted.

Since the three types of axonometrics are isometrics, dimetrics, and trimetrics, you can see that this program can be used for a variety of axonometric pictorials. An isometric projection is shown plotted in Fig. 37.37. The program that was used for this pictorial was the one used to draw the dimetric in Fig. 37.36. The only difference was the two angles of rotation.

37.16 THE BASIC CHARACTER-PLOTTING FUNCTION

Much like the basic plotting subroutine located at line number 9000 and called using GOSUB 9000, characters are plotted using a subroutine located at 9300 that is called using GOSUB 9300 (Fig. 37.38). However, before the character subroutine can be called, an initialization subroutine must be called using GOSUB 9400.

The initialization subroutine performs certain functions needed before the character subroutine can work properly. The initializing subroutine, called GOSUB 9400, should be executed only once per program.

Similar to the plotting subroutine, certain key variables must be assigned values before calling the character-plotting subroutine. These are CHAR$, which should be assigned the string of characters to be plotted; CHAR.X and CHAR.Y, which should be assigned the X- and Y-coordinate

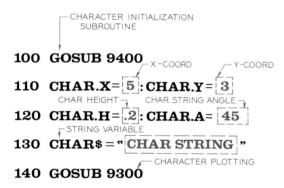

Fig. 37.38 This plot statement is used for writing text or characters. GOSUB 9400 is the *character initialization subroutine,* which loads the characters and is called only once per program. CHAR.X and CHAR.Y are the X- and Y-coordinates of the lower left corner of a string of characters that are to be plotted. CHAR.H is the desired height of the characters in inches. CHAR.A is the angle of rotation of a string of characters in a counterclockwise direction from the horizontal. CHAR$ is assigned the string of characters that are to be plotted. The character plotting subroutine is called with GOSUB 9300.

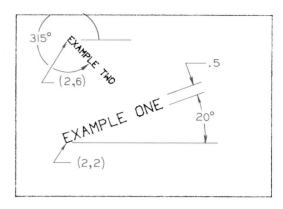

```
100 REM PROGRAM TO SHOW HOW CHARACTERS ARE
110 REM PLOTTED USING THE CHARACTER PLOTTING
120 REM SUBROUTINE.
130 REM =====================================
140 REM INITIALIZE
150 IP%=0 : GOSUB 9000 : GOSUB 9400
160 REM NOW DEFINE THE CHARACTER STRING TO BE
170 REM PLOTTED, IT'S ORIGIN, HEIGHT
180 REM AND ROTATION.
190 CHAR$="EXAMPLE ONE"
200 REM DEFINE ORIGIN
210 CHAR.X=2 : CHAR.Y=2
220 REM DEFINE DEFINE ROTATION AND HEIGHT.
230 CHAR.A=20 : CHAR.H=.5
240 REM NOW PLOT THE STRING
250 GOSUB 9300
260 REM NOW PLOT EXAMPLE STRING 2
270 CHAR.X=2 : CHAR.Y=6 : CHAR.H=.3
280 CHAR.A=315 : CHAR$="EXAMPLE TWO"
290 GOSUB 9300
300 REM RETURN TO ORIGIN
310 XP=0 : YP=0 : IP%=3 : GOSUB 9000
320 STOP
330 END
```

Fig. 37.39 An example program and its output on a display screen is shown where the words, "EXAMPLE ONE," are plotted.

locations of the lower left corner of the first character string; CHAR.H, which should be assigned a value representing the height of the characters in inches; and CHAR.A, which should be assigned the degrees of rotation of the string about the lower left corner of the first character of the string (Fig. 37.39).

37.17 ADVANCED APPLICATIONS

Using the concepts of computer graphics learned thus far, one might imagine how the following more advanced applications were created. All of these plotted examples are the output of programs using the ideas of circles, rectangles, or simple lines and points.

Bar Graph   The bar graph in Fig. 37.40 was generated from information about the number of bars, the maximum bar value, X-axis and graph titles. Looking at the graph in parts rather than as a whole, it can be seen that the bars are just simple

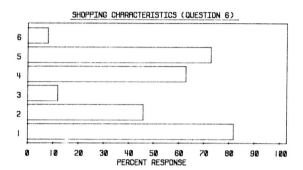

SHOPPING CHARACTERISTICS (QUESTION 6)

PERCENT RESPONSE

1. IS FOR DISCOUNT STORES.
2. IS FOR T.V. SPECIALITY STORES.
3. IS FOR WHOLESALE OUTLETS.
4. IS FOR DEPARTMENT STORES.
5. IS FOR RETAIL OUTLETS.
6. IS FOR MAIL ORDER CATALOGUE OR ADVERTISEMENTS.

Fig. 37.40 A bar graph is the result of using a number of plot statements and rectangle subroutines in addition to the text plotting statement.

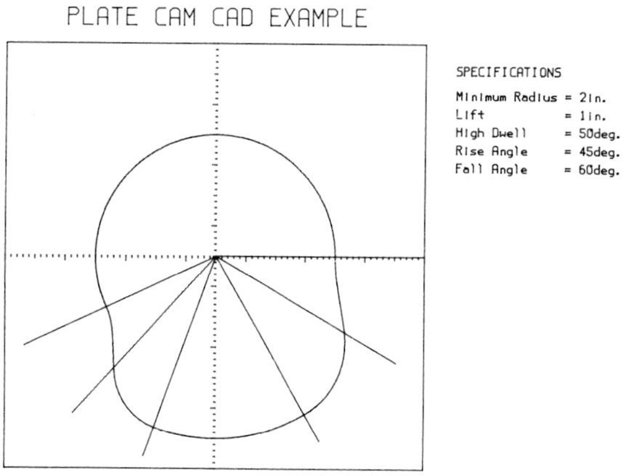

PLATE CAM CAD EXAMPLE

SPECIFICATIONS

Minimum Radius = 2in.
Lift   = 1in.
High Dwell   = 50deg.
Rise Angle   = 45deg.
Fall Angle   = 60deg.

Fig. 37.41 This cam profile is an example of an advanced computer graphics plotting problem.

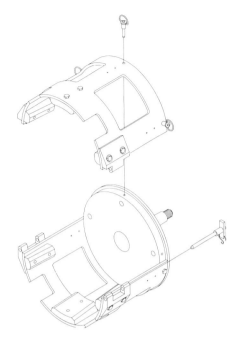

Fig. 37.42 A computer-drawn pictorial of an assembly of several parts. (Courtesy of Texas Instruments.)

rectangles. The tick marks along the X-axis are single lines and the text was plotted using the FNC character function.

Putting these parts together in the right places created the bar graph in Fig. 37.40.

Plate Cam   The plate cam example in Fig. 37.41 is a very good example of computer-aided design of different shapes of cams. The one shown, designed for a knife-edge follower, was developed from the given specifications. This cam design was plotted by a computer program that used only the specifications shown.

The program is based upon a circle routine with the radius of the circle defined according to the step angle and a parabolic function of the lift and the rise and fall angles. This program used an axis-drawing subroutine to plot the boarder and tick-marks and the FNC character plotting functions to draw the text.

A more advanced example of computer graphics is the pictorial in Fig. 37.42 that illustrates an assembly of parts. This drawing was developed by Texas Instruments.

PROBLEMS

The problems below can be assigned to be written, programmed, and run on a computer graphics system. When a system is not available, the programs can be written and then plotted by hand on a 7.5" × 10" grid to simulate a display screen. When plotted by hand, it is suggested that the programs and plots be done on a Size AV sheet or sheets.

1. Write a program that will plot the lines given below. Obtain a printout of your program and a plot of the lines when your program has run.

Problem	Line	
A	A (2,2)	B (6,6)
B	B (1,6)	C(10,2)
C	E (3,5)	F (8,3)
D	G (8,5)	H (3,3)

2. Write a program that will plot the horizontally positioned rectangles from the given information. Obtain a printout of your program and a plot of the rectangles when your program has run.

Problem	Lower Left Corner	Height	Width
A	(2,2)	4"	6"
B	(1,1)	6"	5"
C	(3,1)	2"	3"
D	(2,1)	5"	4"

3. Write a program that will plot the angular lines below by using trigonometric functions of the angle of inclination, and the length of the line. Obtain a printout of your program and plot the lines when your program has run.

Problem	Lower Left Point	Length	Angle with Horizontal
A	(1,1)	5"	30°
B	(3,1)	4"	45°
C	(2,2)	3"	135°
D	(5,2)	4"	195°

4. Write a program that will plot the rectangles that incline with the horizontal at the angles given in the table below. Use variables similar to those illustrated in Fig. 37.23. Obtain a printout of your program and plot the rectangles when your program has run.

Problem	Lower Left Corner	Weight	Width	Angle
A	(2,2)	4"	5"	15°
B	(1,1)	5"	4"	35°
C	(3,1)	2"	3.4"	40°
D	(2,1)	4.5"	4"	28°

5. Write a program that will plot regular polygons using the specifications given in the table below for each. Refer to Fig. 37.24. Obtain a printout of your program and plot the solutions when your program has run.

Problem	No. of Sides	Center	Radius
A	4	(5,4)	2.5"
B	6	(5,4)	2.75"
C	8	(5,4)	2.00"
D	10	(5,4)	2.25"

6. Write a program that will plot circles using the specifications given in Problem 5 by increasing the number of sides of the polygons to approximate the circles. Obtain a printout of your program and plot the solutions when your program has run.

7. Write a program that will use data points to plot the front views of any of the assigned objects shown in Fig. 37.43. Place the lower left corner at (2,2). Obtain a printout of your program and plot the solutions when your program has run.

8. Write a main program and a subroutine for plotting the rectangles assigned in Problem 4. Refer to Fig. 37.27. Obtain a printout of your program and a plot of the solutions when the program has run.

9. Write a main program and a subroutine for plotting the circles assigned in Problem 6. Refer to Fig. 37.29. Obtain a printout of your program and a plot of the solutions when the program has run.

10. By referring to Fig. 37.33, write a similar program to plot one of the oblique pictorials below using the orthographic views of the objects in Fig. 37.43 and the supplementary data in Table 37.2.

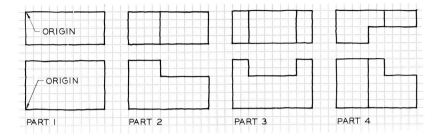

Fig. 37.43 Orthographic views of problems where each square is equal to 0.50 inches. The same corner point in the top and front views is the origin for each problem.

Obtain a printout of your program and a plot of the solutions when the program has run.

Table 37.2

Part No.	Origin	Overall Scale Factor	Z-axis Scale Factor	Angle
1	(5,3)	0.5	0.5	30°
2	(5,3)	0.65	1.00	45°
3	(5,3)	0.40	0.75	36°
4	(5,3)	0.70	0.60	25°

11. By referring to Fig. 37.36, write a similar program to plot one of the axonometric pictorials using the orthographic views of the objects in Fig. 37.43 and the supplementary data in Table 37.3.

Obtain a printout of your program and a plot of the solutions when the program has run.

Table 37.3

Part	Origin	Overall Scale Factor	Rotation about Y-axis	Rotation about X-axis
1	(5,3)	0.40	−20°	40°
2	(5,3)	0.50	−45°	35.5°
3	(5,3)	0.60	−50°	15°
4	(5,3)	0.37	−15°	20°

12. Write a program similar to the one given in Fig. 37.38 that will plot three lines of text (of your choice) that will be plotted at different angles and will not overlap.

APPENDIXES

APPENDIXES CONTENTS

APPENDIX 1 ABBREVIATIONS (ANSI Z 32.13)

Word	Abbreviation	Word	Abbreviation	Word	Abbreviation
Abbreviate	ABBR	Bench mark	BM	Centigrade	C
Absolute	ABS	Between	BET.	Centigram	CG
Account	ACCT	Between centers	BC	Centimeter	cm
Actual	ACT.	Between		Chain	CH
Adapter	ADPT	perpendiculars	BP	Chamfer	CHAM
Addendum	ADD.	Bevel	BEV	Change notice	CN
Adjust	ADJ	Bill of material	B/M	Change order	CO
Advance	ADV	Birmingham wire gage	BWG	Channel	CHAN
After	AFT.	Blueprint	BP	Check	CHK
Aggregate	AGGR	Board	BD	Check valve	CV
Air condition	AIR COND	Boiler	BLR	Chemical	CHEM
Airport	AP	Bolt circle	BC	Chord	CHD
Airplane	APL	Both sides	BS	Circle	CIR
Allowance	ALLOW	Bottom	BOT	Circuit	CKT
Alloy	ALY	Bottom chord	BC	Circular	CIR
Alteration	ALT	Boundary	BDY	Circular pitch	CP
Alternate	ALT	Bracket	BRKT	Circumference	CIRC
Alternating current	AC	Brake horsepower	BHP	Clockwise	CW
Altitude	ALT	Brass	BRS	Coated	CTD
Aluminum	AL	Brazing	BRZG	Cold drawn	CD
American National		Break	BRK	Cold drawn steel	CDS
Standard	AMER NATL STD	Breaker	BKR	Cold finish	CF
American wire gage	AWG	Bridge	BRDG	Cold punched	CP
Ammeter	AM	Brinell hardness	BH	Cold rolled	CR
Amount	AMT	British Standard	BR STD	Cold rolled steel	CRS
Ampere	AMP	British Thermal Units	BTU	Column	COL
Anneal	ANL	Broach	BRO	Combination	COMB.
Antenna	ANT.	Bronze	BRZ	Combustion	COMB
Apparatus	APP	Brown & Sharp (Wire gage,		Commutator	COMM
Appendix	APPX	same as AWG)	B&S	Company	CO
Approved	APPD	Building	BLDG	Concentric	CONC
Approximate	APPROX	Bulkhead	BHD	Concrete	CONC
Arc weld	ARC/W	Bureau	BU	Condition	COND
Area	A	Bureau of Standards	BU STD	Connect	CONN
Armature	ARM.	Bushing	BUSH.	Contact	CONT
Asbestos	ASB	Button	BUT.	Cord	CD
Asphalt	ASPH	Buzzer	BUZ	Corporation	CORP
Assembly	ASSY	By-pass	BYP	Corrugate	CORR
Association	ASSN			Cotter	COT
Atomic	AT	Cabinet	CAB.	Counterclockwise	CCW
Authorized	AUTH	Cadmium plate	CD PL	Counterbore	CBORE
Auxiliary	AUX	Calculate	CALC	Counterdrill	CDRILL
Avenue	AVE	Calibrate	CAL	Counterpunch	CPUNCH
Average	AVG	Calorie	CAL	Countersink	CSK
Avoirdupois	AVDP	Capacitor	CAP	Coupling	CPLG
Azimuth	AZ	Cap screw	CAP SCR	Crank	CRK
		Case harden	CH	Cross section	XSECT
Babbitt	BAB	Cast iron	CI	Cubic	CU
Back pressure	BP	Cast steel	CS	Cubic centimeter	cc
Balance	BAL	Casting	CSTG	Cubic feet per minute	CFM
Ball bearing	BB	Castle nut	CAS NUT	Cubic feet per second	CFS
Barometer	BAR	Catalogue	CAT.	Cubic foot	CU FT
Barrel	BBL	Cement	CEM	Cubic inch	CU IN.
Base line	BL	Center	CTR	Cubic meter	CU M
Base plate	BP	Centerline	CL	Cubic yard	CU YD
Battery	BAT.	Center of gravity	CG	Current	CUR
Bearing	BRG	Center to center	C to C	Cylinder	CYL

Cont.

APPENDIX 1 ABBREVIATIONS (ANSI Z 32.13) (Cont.)

Word	Abbreviation	Word	Abbreviation	Word	Abbreviation
Decimal	DEC	Fillister	FIL	Illustrate	ILLUS
Dedendum	DED	Filter	FLT	Inch	(") IN.
Degree	(°) DEG	Finish	FIN.	Inches per second	IPS
Department	DEPT	Finish all over	FAO	Include	INCL
Design	DSGN	Flange	FLG	Industrial	IND
Detail	DET	Flat head	FH	Information	INFO
Develop	DEV	Fluid	FL	Inside diameter	ID
Diagonal	DIAG	Focus	FOC	Instrument	INST
Diagram	DIAG	Foot	(') FT	Insulate	INS
Diameter	DIA	Forging	FORG	Interior	INT
Diametrical pitch	DP	Forward	FWD	Internal	INT
Dimension	DIM.	Foundation	FDN	Intersect	INT
Direct current	DC	Foundry	FDRY	Iron	I
Discharge	DISCH	Frequency	FREQ	Irregular	IRREG
Distance	DIST	Front	FR		
District	DIST			Jack	J
Ditto	DO.			Joint	JT
Dovetail	DVTL	Gage	GA	Junction	JCT
Dowel	DWL	Gallon	GAL	Junction box	JB
Down	DN	Galvanize	GALV		
Dozen	DOZ	Galvanized iron	GI	Key	K
Drafting	DFTG	Galvanized steel	GS	Keyseat	KST
Draftsman	DFTSMN	Gasket	GSKT	Keyway	KWY
Drawing	DWG	General	GEN	Kiln-dried	KD
Drill	DR	Government	GOVT	Kip (1000 lb)	K
Drive	DR	Governor	GOV	Knots	KN
Drop forge	DF	Grade	GR		
		Grade line	GL	Laboratory	LAB
Each	EA	Gram	G	Lateral	LAT
East	E	Gravity	G	Latitude	LAT
Eccentric	ECC	Grind	GRD	Left	L
Effective	EFF	Groove	GRV	Left hand	LH
Elbow	ELL	Ground	GRD	Length	LG
Electric	ELEC	Gypsum	GYP	Letter	LTR
Elevation	ELEV			Light	LT
Engineer	ENGR	Half-round	½ RD	Line	L
Equal	EQ	Handle	HDL	Logarithm	LOG.
Equipment	EQUIP.	Hanger	HGR	Lubricate	LUB
Equivalent	EQUIV	Hard	H	Lumber	LBR
Estimate	EST	Hard-drawn	HD		
Exterior	EXT	Harden	HDN	Machine	MACH
Extra heavy	X HVY	Hardware	HDW	Malleable	MALL
Extra strong	X STR	Head	HD	Malleable iron	MI
		Headless	HDLS	Manhole	MH
		Headquarters	HQ	Manual	MAN.
Fabricate	FAB	Heat	HT	Manufacture	MFR
Face to face	F to F	Heat treat	HT TR	Material	MATL
Fahrenheit	F	Hexagon	HEX	Maximum	MAX
Fairing	FAIR.	High-pressure	HP	Mechanical	MECH
Far side	FS	High-speed	HS	Mechanism	MECH
Federal	FED.	Horizontal	HOR	Median	MED
Feet	(') FT	Horsepower	HP	Metal	MET.
Feet per minute	FPM	Hot rolled	HR	Meter (Instrument or	
Feet per second	FPS	Hot rolled steel	HRS	measure of length)	M
Field	FLD	Hour	HR	Miles	MI
Figure	FIG.	Hundredweight	CWT	Miles per gallon	MPG
Fillet	FIL	Hydraulic	HYD	Miles per hour	MPH

APPENDIX 1 ABBREVIATIONS (ANSI Z 32.13) (Cont.)

Word	Abbreviation	Word	Abbreviation	Word	Abbreviation
Millimeter	MM	Precast	PRCST	Shaft	SFT
Minimum	MIN	Prefabricated	PREFAB	Sketch	SK
Minute	(') MIN	Preferred	PFD	Sleeve	SLV
Miscellaneous	MISC	Prepare	PREP	Slide	SL
Mixture	MIX.	Pressure	PRESS.	Slotted	SLOT.
Model	MOD	Pressure angle	PA.	Socket	SOC
Month	MO	Process	PROC	Solder	SLD
Morse taper	MOR T	Production	PROD	South	S
Multiple	MULT	Profile	PF	Space	SP
		Project	PROJ	Special	SPL
National	NATL	Proof	PRF	Specific gravity	SP GR
Near side	NS			Spherical	SPHER
Negative	NEG	Quadrant	QUAD	Spot faced	SF
Neutral	NEUT	Quart	QT	Spring	SPG
Nipple	NIP.	Quarter	QTR	Square	SQ
Nominal	NOM	Quarter-round	¼ RD	Stainless	STN
Normal	NOR			Stainless steel	SST
North	N	Radial	RAD	Standard	STD
Not to scale	NTS	Radius	R	Station	STA
Number	NO.	Railroad	RR	Steel	STL
		Ream	RM	Stock	STK
Obsolete	OBS	Received	RECD	Straight	STR
Octagon	OCT	Record	REC	Structural	STR
Ohm	Ω	Rectangle	RECT	Substitute	SUB
On center	OC	Reference	REF	Summary	SUM.
Opposite	OPP	Reference line	REF L	Supply	SUP
Optical	OPT	Relief	REL	Surface	SUR
Original	ORIG	Remove	REM	Symbol	SYM
Ounce	OZ	Require	REQ	Symmetrical	SYM
Outlet	OUT.	Required	REQD	System	SYS
Outside diameter	OD	Return	RET.		
Outside face	OF	Revise	REV	Tangent	TAN.
Outside radius	OR	Revolution	REV	Taper	TPR
Overall	OA	Revolutions per minute	RPM	Technical	TECH
		Rheostat	RHEO	Temperature	TEMP
Pack	PK	Right	R	Template	TEMP
Packing	PKG	Right hand	RH	Tensile strength	TS
Parallel	PAR.	Rivet	RIV	Tension	TENS.
Part	PT	Rockwell hardness	RH	Thick	THK
Patent	PAT.	Roller bearing	RB	Thousand	M
Permanent	PERM	Room	RM	Thousand pound	KIP
Perpendicular	PERP	Root diameter	RD	Thread	THD
Photograph	PHOTO	Root mean square	RMS	Tolerance	TOL
Piece	PC	Rough	RGH	Tongue & groove	T&G
Pint	PT	Round	RD	Tool steel	TS
Pitch	P	Rubber	RUB.	Tooth	T
Pitch circle	PC			Total	TOT
Pitch diameter	PD	Safety	SAF	Transfer	TRANS
Plastic	PLSTC	Sand blast	SD BL	Typical	TYP
Plate	PL	Schedule	SCH		
Point	PT	Screen	SCRN	Ultimate	ULT
Polish	POL	Screw	SCR	Unit	U
Position	POS	Sea level	SL	Universal	UNIV
Positive	POS	Second	SEC		
Pound	LB	Section	SECT	Vacuum	VAC
Pounds per square inch	PSI	Separate	SEP	Valve	V
Power	PWR	Set screw	SS	Variable	VAR
					Cont.

APPENDIX 1 ABBREVIATIONS (ANSI Z 32.13) (Cont.)

Word	Abbreviation	Word	Abbreviation	ABBREVIATIONS FOR COLORS	
Vertical	VERT	West	W	Amber	AMB
Volt	V	Width	W	Black	BLK
Voltmeter	VM	Wood	WD	Blue	BLU
Volume	VOL	Woodruff	WDF	Brown	BRN
		Wrought iron	WI	Green	GRN
Washer	WASH.			Orange	ORN
Watt	W	Yard	YD	White	WHT
Weight	WT	Year	YR	Yellow	YEL

Appendix 2 CONVERSION TABLES

Length conversions		
Angstrom units	$\times\ 1 \times 10^{-10}$	= meters
	$\times\ 1 \times 10^{-4}$	= microns
	$\times\ 1.650\ 763\ 73 \times 10^{-4}$	= wavelengths of orange-red line of krypton 86
Cables	$\times\ 120$	= fathoms
	$\times\ 720$	= feet
	$\times\ 219.456$	= meters
Fathoms	$\times\ 6$	= feet
	$\times\ 1.828\ 8$	= meters
Feet	$\times\ 12$	= inches
	$\times\ 0.3048$	= meters
Furlongs	$\times\ 660$	= feet
	$\times\ 201.168$	= meters
	$\times\ 220$	= yards
Inches	$\times\ 2.54 \times 10^{8}$	= Angstroms
	$\times\ 25.4$	= millimeters
	$\times\ 8.333\ 33 \times 10^{-2}$	= feet
Kilometers	$\times\ 3.280\ 839 \times 10^{3}$	= feet
	$\times\ 0.62$	= miles
	$\times\ 0.539\ 956$	= nautical miles
	$\times\ 0.621\ 371$	= statute miles
	$\times\ 1.093\ 613 \times 10^{3}$	= yards
Light-years	$\times\ 9.460\ 55 \times 10^{12}$	= kilometers
	$\times\ 5.878\ 51 \times 10^{12}$	= statute miles
Meters	$\times\ 1 \times 10^{10}$	= Angstroms
	$\times\ 3.280\ 839\ 9$	= feet
	$\times\ 39.370\ 079$	= inches
	$\times\ 1.093\ 61$	= yards
Microns	$\times\ 10^{4}$	= Angstroms
	$\times\ 10^{-4}$	= centimeters
	$\times\ 10^{-6}$	= meters
Nautical Miles (International)	$\times\ 8.439\ 049$	= cables
	$\times\ 6.076\ 115\ 49 \times 10^{3}$	= feet
	$\times\ 1.852 \times 10^{3}$	= meters
	$\times\ 1.150\ 77$	= statute miles

Appendix 2 CONVERSION TABLES (Cont.)

Length conversions

Statute Miles	$\times\ 5.280 \times 10^3$	= feet
	$\times\ 8$	= furlongs
	$\times\ 6.336\ 0 \times 10^4$	= inches
	$\times\ 1.609\ 34$	= kilometers
	$\times\ 8.689\ 7 \times 10^{-1}$	= nautical miles
Miles	$\times\ 10^{-3}$	= inches
	$\times\ 2.54 \times 10^{-2}$	= millimeters
	$\times\ 25.4$	= micrometers
	$\times\ 0.61$	= kilometers
Yards	$\times\ 3$	= feet
	$\times\ 9.144 \times 10^{-1}$	= meters
Feet/hour	$\times\ 3.048 \times 10^{-4}$	= kilometers/hour
	$\times\ 1.645\ 788 \times 10^{-4}$	= knots
Feet/minute	$\times\ 0.3048$	= meters/minute
	$\times\ 5.08 \times 10^{-3}$	= meters/second
Feet/second	$\times\ 1.097\ 28$	= kilometers/hour
	$\times\ 18.288$	= meters/minute
Kilometers/hour	$\times\ 3.280\ 839 \times 10^3$	= feet/hour
	$\times\ 54.680\ 66$	= feet/minute
	$\times\ 0.277\ 777$	= meters/second
	$\times\ 0.621\ 371$	= miles/hour
Kilometers/minute	$\times\ 3.280\ 839 \times 10^3$	= feet/minute
	$\times\ 37.282\ 27$	= miles/hour
Knots	$\times\ 6.076\ 115 \times 10^3$	= feet/hour
	$\times\ 101.268\ 5$	= feet/minute
	$\times\ 1.687\ 809$	= feet/second
	$\times\ 1.852$	= kilometers/hour
	$\times\ 30.866$	= meters/minute
	$\times\ 0.514\ 4$	= meters/second
	$\times\ 1.150\ 77$	= statute miles/hour
Meters/hour	$\times\ 3.280\ 839$	= feet/hour
	$\times\ 88$	= feet/minute
	$\times\ 1.466$	= feet/second
	$\times\ 1 \times 10^{-3}$	= kilometers/hour
	$\times\ 1.667 \times 10^{-2}$	= meters/minute
Feet/second^2	$\times\ 1.097\ 28$	= kilometers/hour/second
	$\times\ 0.304\ 8$	= meters/second^2

Area conversions

Acres	$\times\ 4.046\ 85 \times 10^{-3}$	= square kilometers
	$\times\ 4.046\ 856 \times 10^3$	= square meters
	$\times\ 4.356\ 0 \times 10^4$	= square feet
Ares	$\times\ 2.471\ 053\ 8 \times 10^{-2}$	= acres
	$\times\ 1$	= square dekameters
	$\times\ 10^2$	= square meters
Barns	$\times\ 1 \times 10^{-28}$	= square meters
Circular mils	$\times\ 1 \times 10^{-6}$	= circular inches
	$\times\ 5.067\ 074\ 8 \times 10^{-4}$	= square millimeters
	$\times\ 0.785\ 398\ 1$	= square mils

Cont.

Appendix 2 CONVERSION TABLES (Cont.)

Area conversions

Hectares	$\times$ 2.471 05	= acres
	$\times 10^2$	= ares
	$\times 10^4$	= square meters
Square feet	$\times$ 2.295 684 $\times 10^{-5}$	= acres
	$\times$ 9.290 3 $\times 10^{-4}$	= ares
	$\times$ 144	= square inches
	$\times$ 9.290 304 $\times 10^{-2}$	= square meters
Square inches	$\times$ 1.273 239 5 $\times 10^6$	= circular mils
	$\times$ 6.944 4 $\times 10^{-3}$	= square feet
	$\times$ 6.451 6 $\times 10^{-4}$	= square meters
Square kilometers	$\times$ 247.105 38	= acres
	$\times$ 1.076 391 0 $\times 10^7$	= square feet
	$\times$ 1.000	= cubic meters
	$\times$ 1.307 950 6	= cubic yards
	$\times$ 219.969	= imperial gallons
Liters	$\times 10^3$	= cubic centimeters
	$\times$ 1.000 $\times 10^6$	= cubic millimeters
	$\times$ 1.000 $\times 10^{-3}$	= cubic meters
	$\times$ 61.023 74	= cubic inches
	$\times$ 3.531 5 $\times 10^{-2}$	= cubic feet
	$\times$ 1.307 95 $\times 10^{-3}$	= cubic yards
	$\times$ 0.22	= gallons
	$\times$ 0.219 969	= imperial gallons
	$\times$ 0.879 877	= imperial quarts
Imperial pints	$\times$ 0.125	= imperial gallons
	$\times$ 0.568 261	= liters
	$\times$ 20	= imperial fluid ounces
	$\times$ 0.5	= imperial quarts
	$\times$ 568.260 9	= cubic centimeters
Imperial quarts	$\times$ 1.136 52 $\times 10^3$	= cubic centimeters
	$\times$ 69.354 8	= cubic inches
	$\times$ 1.136 522 8	= liters

Power conversions

British Thermal Units/hour	$\times$ 2.928 7 $\times 10^{-4}$	= kilowatts
	$\times$ 0.292 875	= watts
BTU/minute	$\times$ 1.757 25 $\times 10^{-2}$	= kilowatts
BTU/pound	$\times$ 2.324 4	= joules/gram
BTU/second	$\times$ 1.413 91	= horsepower
	$\times$ 107.514	= kilogrammeters/second
	$\times$ 1.054 35	= kilowatts
	$\times$ 1.054 35 $\times 10^3$	= watts
Foot-pound-force/hour	$\times$ 5.050 $\times 10^{-7}$	= horsepower
	$\times$ 3.766 16 $\times 10^{-7}$	= kilowatts
Foot-pound-force/ minute	$\times$ 3.030 303 $\times 10^{-5}$	= horsepower
	$\times$ 2.259 70 $\times 10^{-2}$	= joules/second
	$\times$ 2.259 70 $\times 10^{-5}$	= kilowatts
Horsepower	$\times$ 42.435 6	= BTU/minute
	$\times$ 550	= footpounds/second
	$\times$ 0.746	= kilowatts
	$\times$ 746	= joules/second

Appendix 2 CONVERSION TABLES (Cont.)

Power conversions

Kilogrammeters/second	$\times$ 9.806 65	= watts
Kilowatts	$\times$ 3.414 43 $\times 10^3$	= BTU/hour
	$\times$ 2.655 22 $\times 10^6$	= footpounds/hour
	$\times$ 4.425 37 $\times 10^4$	= footpounds/minute
	$\times$ 737.562	= footpounds/second
	$\times$ 1.019 726 $\times 10^7$	= gramcentimeters/second
	$\times$ 1.341 02	= horsepower
	$\times$ 3.6 $\times 10^6$	= joules/hour
	$\times 10^3$	= joules/second
	$\times$ 3.671 01 $\times 10^5$	= kilogrammeters/hour
	$\times$ 999.835	= international watt
Watts	$\times$ 44.253 7	= footpounds/minute
	$\times$ 1.341 02 $\times 10^{-3}$	= horsepower
	$\times$ 1	= joules/second

Time conversions

(No attempt has been made in this brief treatment to correlate solar, mean solar, sidereal, and mean sidereal days.)

Mean solar days	$\times$ 24	= mean solar hours
Mean solar hours	$\times$ 3.600 $\times 10^3$	= mean solar seconds
	$\times$ 60	= mean solar minutes

Angle conversions

Degrees	$\times$ 60	= minutes
	$\times$ 1.745 329 3 $\times 10^{-2}$	= radians
Degrees/foot	$\times$ 5.726 145 $\times 10^{-4}$	= radians/centimeter
Degrees/minute	$\times$ 2.908 8 $\times 10^{-4}$	= radians/second
	$\times$ 4.629 629 $\times 10^{-5}$	= revolutions/second
Degrees/second	$\times$ 1.745 329 3 $\times 10^{-2}$	= radians/second
	$\times$ 0.166	= revolutions/minute
	$\times$ 2.77 $\times 10^{-3}$	= revolutions/second
Minutes	$\times$ 1.667 $\times 10^{-2}$	= degrees
	$\times$ 2.908 8 $\times 10^{-4}$	= radians
	$\times$ 60	= seconds
Radians	$\times$ 0.159 154	= circumferences
	$\times$ 57.295 77	= degrees
	$\times$ 3.437 746 $\times 10^3$	= minutes
Seconds	$\times$ 2.777 $\times 10^{-4}$	= degrees
	$\times$ 1.667 $\times 10^{-2}$	= minutes
	$\times$ 4.848 136 8 $\times 10^{-6}$	= radians
Steradians	$\times$ 0.159 154 9	= hemispheres
	$\times$ 7.957 74 $\times 10^{-2}$	= spheres
	$\times$ 0.636 619 7	= spherical right angles

Mass conversions

Grains	$\times$ 6.479 8 $\times 10^{-2}$	= grams
	$\times$ 2.285 71 $\times 10^{-3}$	= avoirdupois ounces

Cont.

Appendix 2 CONVERSION TABLES (Cont.)

Mass conversions

Grams	$\times$ 15.432 358	= grains
	$\times$ 3.527 396 $\times 10^{-2}$	= avoirdupois ounces
	$\times$ 2.204 62 $\times 10^{-3}$	= avoirdupois pounds
Kilograms	$\times$ 564.383 4	= avoirdupois drams
	$\times$ 2.204 622 6	= avoirdupois pounds
	$\times$ 2.2	= pounds
	$\times$ 9.842 065 $\times 10^{-4}$	= long tons
	$\times 10^{-3}$	= metric tons
	$\times$ 1.102 31 $\times 10^{-3}$	= short tons
Avoirdupois ounces	$\times$ 28.349 5	= grams
	$\times$ 6.25 $\times 10^{-2}$	= avoirdupois pounds
	$\times$ 0.911 458	= troy ounces
Avoirdupois pounds	$\times$ 256	= drams
	$\times$ 4.535 923 7 $\times 10^2$	= grams
	$\times$ 0.453 592 4	= kilograms
	$\times$ 16	= ounces
Long tons	$\times$ 2.24 $\times 10^3$	= avoirdupois pounds
	$\times$ 1.106 046 9	= metric tons
	$\times$ 1.12	= short tons
Metric tons	$\times 10^3$	= kilograms
	$\times$ 2.204 622 $\times 10^3$	= avoirdupois pounds
Short tons	$\times$ 2 $\times 10^3$	= avoirdupois pounds
	$\times$ 907.184 74	= kilograms

Force conversions

Dynes	$\times 10^{-5}$	= newtons
Newtons	$\times 10^5$	= dynes
	$\times$ 0.224 808	= pounds-force
Pounds	$\times$ 4.448 22	= newtons

Energy conversions

British Thermal Units (thermochemical)	$\times$ 1.054 35 $\times 10^3$	= joules
	$\times$ 2.928 27 $\times 10^{-4}$	= kilowatthours
	$\times$ 1.054 35 $\times 10^3$	= wattseconds
Foot-pound-force	$\times$ 1.355 818 0	= joules
	$\times$ 0.138 255	= kilogramforce-meters
	$\times$ 3.766 16 $\times 10^{-7}$	= kilowatthours
	$\times$ 1.355 818 0	= newtonmeters
Joules	$\times$ 9.484 5 $\times 10^{-4}$	= British Thermal Units
	$\times$ 0.737 562	= foot-pounds-force
	$\times$ 0.101 971 6	= kilogramforce-meters
	$\times$ 2.777 7 $\times 10^{-7}$	= kilowatthours
	$\times$ 1	= wattseconds
Kilogramforce-meters	$\times$ 9.287 7 $\times 10^{-3}$	= British Thermal Units
	$\times$ 7.233 01	= foot-pounds-force
	$\times$ 9.806 65	= joules
	$\times$ 9.806 65	= newtonmeters
	$\times$ 2.724 0 $\times 10^{-3}$	= watthours

Appendix 2 CONVERSION TABLES (Cont.)

Energy conversions

Kilowatthours	$\times\ 3.409\ 52 \times 10^3$	= British Thermal Units
	$\times\ 2.655\ 22 \times 10^6$	= foot-pounds-force
	$\times\ 1.341\ 02$	= horsepowerhours
	$\times\ 3.6 \times 10^6$	= joules
	$\times\ 3.670\ 98 \times 10^5$	= kilogramforce-meters
Newtonmeters	$\times\ 0.101\ 971$	= kilogramforce-meters
	$\times\ 0.737\ 562$	= poundforce-feet
Watthours	$\times\ 3.414\ 43$	= British Thermal Units
	$\times\ 2.655\ 22 \times 10^3$	= foot-pounds-force
	$\times\ 3.6 \times 10^3$	= joules
	$\times\ 3.670\ 98 \times 10^2$	= kilogramforce-meters

Pressure conversions

Atmospheres	$\times\ 1.013\ 25$	= bars
	$\times\ 1.033\ 23 \times 10^3$	= grams/square centimeter
	$\times\ 1.033\ 23 \times 10^7$	= grams/square meter
	$\times\ 14.696\ 0$	= pounds/square inch
	$\times\ 760$	= torrs
	$\times\ 101$	= kilopascals
Bars	$\times\ 0.986\ 923$	= atmospheres
	$\times\ 10^6$	= baryes
	$\times\ 1.019\ 716 \times 10^7$	= grams/square meter
	$\times\ 1.019\ 716 \times 10^4$	= kilogramsforce/square meter
	$\times\ 14.503\ 8$	= poundsforce/square inch
Baryes	$\times\ 10^{-6}$	= bars
Inches of mercury	$\times\ 3.386\ 4 \times 10^{-2}$	= bars
	$\times\ 345.316$	= kilogramsforce/square meter
	$\times\ 70.726\ 2$	= poundsforce/square foot
Pascal	$\times\ 1$	= newton/square meter

Appendix 3 LOGARITHMS OF NUMBERS

N	0	1	2	3	4	5	6	7	8	9
1.0	.0000	.0043	.0086	.0128	.0170	.0212	.0253	.0294	.0334	.0374
1.1	.0414	.0453	.0492	.0531	.0569	.0607	.0645	.0682	.0719	.0755
1.2	.0792	.0828	.0864	.0899	.0934	.0969	.1004	.1038	.1072	.1106
1.3	.1139	.1173	.1206	.1239	.1271	.1303	.1335	.1367	.1399	.1430
1.4	.1461	.1492	.1523	.1553	.1584	.1614	.1644	.1673	.1703	.1732
1.5	.1761	.1790	.1818	.1847	.1875	.1903	.1931	.1959	.1987	.2014
1.6	.2041	.2068	.2095	.2122	.2148	.2175	.2201	.2227	.2253	.2279
1.7	.2304	.2330	.2355	.2380	.2405	.2430	.2455	.2480	.2504	.2529
1.8	.2553	.2577	.2601	.2625	.2648	.2672	.2695	.2718	.2742	.2765
1.9	.2788	.2810	.2833	.2856	.2878	.2900	.2923	.2945	.2967	.2989
2.0	.3010	.3032	.3054	.3075	.3096	.3118	.3139	.3160	.3181	.3201
2.1	.3222	.3243	.3263	.3284	.3304	.3324	.3345	.3365	.3385	.3404
2.2	.3424	.3444	.3464	.3483	.3502	.3522	.3541	.3560	.3579	.3598
2.3	.3617	.3636	.3655	.3674	.3692	.3711	.3729	.3747	.3766	.3784
2.4	.3802	.3820	.3838	.3856	.3874	.3892	.3909	.3927	.3945	.3962
2.5	.3979	.3997	.4014	.4031	.4048	.4065	.4082	.4099	.4116	.4133
2.6	.4150	.4166	.4183	.4200	.4216	.4232	.4249	.4265	.4281	.4298
2.7	.4314	.4330	.4346	.4362	.4378	.4393	.4409	.4425	.4440	.4456
2.8	.4472	.4487	.4502	.4518	.4533	.4548	.4564	.4579	.4594	.4609
2.9	.4624	.4639	.4654	.4669	.4683	.4698	.4713	.4728	.4742	.4757
3.0	.4771	.4786	.4800	.4814	.4829	.4843	.4857	.4871	.4886	.4900
3.1	.4914	.4928	.4942	.4955	.4969	.4983	.4997	.5011	.5024	.5038
3.2	.5051	.5065	.5079	.5092	.5105	.5119	.5132	.5145	.5159	.5172
3.3	.5185	.5198	.5211	.5224	.5237	.5250	.5263	.5276	.5289	.5302
3.4	.5315	.5328	.5340	.5353	.5366	.5378	.5391	.5403	.5416	.5428
3.5	.5441	.5453	.5465	.5478	.5490	.5502	.5514	.5527	.5539	.5551
3.6	.5563	.5575	.5587	.5599	.5611	.5623	.5635	.5647	.5658	.5670
3.7	.5682	.5694	.5705	.5717	.5729	.5740	.5752	.5763	.5775	.5786
3.8	.5798	.5809	.5821	.5832	.5843	.5855	.5866	.5877	.5888	.5899
3.9	.5911	.5922	.5933	.5944	.5955	.5966	.5977	.5988	.5999	.6010
4.0	.6021	.6031	.6042	.6053	.6064	.6075	.6085	.6096	.6107	.6117
4.1	.6128	.6138	.6149	.6160	.6170	.6180	.6191	.6201	.6212	.6222
4.2	.6232	.6243	.6253	.6263	.6274	.6284	.6294	.6304	.6314	.6325
4.3	.6335	.6345	.6355	.6365	.6375	.6385	.6395	.6405	.6415	.6425
4.4	.6435	.6444	.6454	.6464	.6474	.6484	.6493	.6503	.6513	.6522
4.5	.6532	.6542	.6551	.6561	.6571	.6580	.6590	.6599	.6609	.6618
4.6	.6628	.6637	.6646	.6656	.6665	.6675	.6684	.6693	.6702	.6712
4.7	.6721	.6730	.6739	.6749	.6758	.6767	.6776	.6785	.6794	.6803
4.8	.6812	.6821	.6830	.6839	.6848	.6857	.6866	.6875	.6884	.6893
4.9	.6902	.6911	.6920	.6928	.6937	.6946	.6955	.6964	.6972	.6981
5.0	.6990	.6998	.7007	.7016	.7024	.7033	.7042	.7050	.7059	.7067
5.1	.7076	.7084	.7093	.7101	.7110	.7118	.7126	.7135	.7143	.7152
5.2	.7160	.7168	.7177	.7185	.7193	.7202	.7210	.7218	.7226	.7235
5.3	.7243	.7251	.7259	.7267	.7275	.7284	.7292	.7300	.7308	7316
5.4	.7324	.7332	.7340	.7348	.7356	.7364	.7372	.7380	.7388	.7396
N	0	1	2	3	4	5	6	7	8	9

Appendix 3 LOGARITHMS OF NUMBERS (Cont.)

N	0	1	2	3	4	5	6	7	8	9
5.5	.7404	.7412	.7419	.7427	.7435	.7443	.7451	.7459	.7466	.7474
5.6	.7482	.7490	.7497	.7505	.7513	.7520	.7528	.7536	.7543	.7551
5.7	.7559	.7566	.7574	.7582	.7589	.7597	.7604	.7612	.7619	.7627
5.8	.7634	.7642	.7649	.7657	.7664	.7672	.7679	.7686	.7694	.7701
5.9	.7709	.7716	.7723	.7731	.7738	.7745	.7752	.7760	.7767	.7774
6.0	.7782	.7789	.7796	.7803	.7810	.7818	.7825	.7832	.7839	.7846
6.1	.7853	.7860	.7868	.7875	.7882	.7889	.7896	.7903	.7910	.7917
6.2	.7924	.7931	.7938	.7945	.7952	.7959	.7966	.7973	.7980	.7987
6.3	.7993	.8000	.8007	.8014	.8021	.8028	.8035	.8041	.8048	.8055
6.4	.8062	.8069	.8075	.8082	.8089	.8096	.8102	.8109	.8116	.8122
6.5	.8129	.8136	.8142	.8149	.8156	.8162	.8169	.8176	.8182	.8189
6.6	.8195	.8202	.8209	.8215	.8222	.8228	.8235	.8241	.8248	.8254
6.7	.8261	.8267	.8274	.8280	.8287	.8293	.8299	.8306	.8312	.8319
6.8	.8325	.8331	.8338	.8344	.8351	.8357	.8363	.8370	.8376	.8382
6.9	.8388	.8395	.8401	.8407	.8414	.8420	.8426	.8432	.8439	.8445
7.0	.8451	.8457	.8463	.8470	.8476	.8482	.8488	.8494	.8500	.8506
7.1	.8513	.8519	.8525	.8531	.8537	.8543	.8549	.8555	.8561	.8567
7.2	.8573	.8579	.8585	.8591	.8597	.8603	.8609	.8615	.8621	.8627
7.3	.8633	.8639	.8645	.8651	.8657	.8663	.8669	.8675	.8681	.8686
7.4	.8692	.8698	.8704	.8710	.8716	.8722	.8727	.8733	.8739	.8745
7.5	.8751	.8756	.8762	.8768	.8774	.8779	.8785	.8791	.8797	.8802
7.6	.8808	.8814	.8820	.8825	.8831	.8837	.8842	.8848	.8854	.8859
7.7	.8865	.8871	.8876	.8882	.8887	.8893	.8899	.8904	.8910	.8915
7.8	.8921	.8927	.8932	.8938	.8943	.8949	.8954	.8960	.8965	.8971
7.9	.8976	.8982	.8987	.8993	.8998	.9004	.9009	.9015	.9020	.9025
8.0	.9031	.9036	.9042	.9047	.9053	.9058	.9063	.9069	.9074	.9079
8.1	.9085	.9090	.9096	.9101	.9106	.9112	.9117	.9122	.9128	.9133
8.2	.9138	.9143	.9149	.9154	.9159	.9165	.9170	.9175	.9180	.9186
8.3	.9191	.9196	.9201	.9206	.9212	.9217	.9222	.9227	.9232	.9238
8.4	.9243	.9248	.9253	.9258	.9263	.9269	.9274	.9279	.9284	.9289
8.5	.9294	.9299	.9304	.9309	.9315	.9320	.9325	.9330	.9335	.9340
8.6	.9345	.9350	.9355	.9360	.9365	.9370	.9375	.9380	.9385	.9390
8.7	.9395	.9400	.9405	.9410	.9415	.9420	.9425	.9430	.9435	.9440
8.8	.9445	.9450	.9455	.9460	.9465	.9469	.9474	.9479	.9484	.9489
8.9	.9494	.9499	.9504	.9509	.9513	.9518	.9523	.9528	.9533	.9538
9.0	.9542	.9547	.9552	.9557	.9562	.9566	.9571	.9576	.9581	.9586
9.1	.9590	.9595	.9600	.9605	.9609	.9614	.9619	.9624	.9628	.9633
9.2	.9638	.9643	.9647	.9652	.9657	.9661	.9666	.9671	.9675	.9680
9.3	.9685	.9689	.9694	.9699	.9703	.9708	.9713	.9717	.9722	.9727
9.4	.9731	.9736	.9741	.9745	.9750	.9754	.9759	.9763	.9768	.9773
9.5	.9777	.9782	.9786	.9791	.9795	.9800	.9805	.9809	.9814	.9818
9.6	.9823	.9827	.9832	.9836	.9841	.9845	.9850	.9854	.9859	.9863
9.7	.9868	.9872	.9877	.9881	.9886	.9890	.9894	.9899	.9903	.9908
9.8	.9912	.9917	.9921	.9926	.9930	.9934	.9939	.9943	.9948	.9952
9.9	.9956	.9961	.9965	.9969	.9974	.9978	.9983	.9987	.9991	.9996
N	0	1	2	3	4	5	6	7	8	9

Appendix 4 VALUES OF TRIGONOMETRIC FUNCTIONS

Degrees	Radians	Sine	Tangent	Cotangent	Cosine		
0° 00′	.0000	.0000	.0000		1.0000	1.5708	90° 00′
10′	.0029	.0029	.0029	343.77	1.0000	1.5679	50′
20′	.0058	.0058	.0058	171.89	1.0000	1.5650	40′
30′	.0087	.0087	.0087	114.59	1.0000	1.5621	30′
40′	.0116	.0116	.0116	85.940	.9999	1.5592	20′
50′	.0145	.0145	.0145	68.750	.9999	1.5563	10′
1° 00′	.0175	.0175	.0175	57.290	.9998	1.5533	89° 00′
10′	.0204	.0204	.0204	49.104	.9998	1.5504	50′
20′	.0233	.0233	.0233	42.964	.9997	1.5475	40′
30′	.0262	.0262	.0262	38.188	.9997	1.5446	30′
40′	.0291	.0291	.0291	34.368	.9996	1.5417	20′
50′	.0320	.0320	.0320	31.242	.9995	1.5388	10′
2° 00′	.0349	.0349	.0349	28.636	.9994	1.5359	88° 00′
10′	.0378	.0378	.0378	26.432	.9993	1.5330	50′
20′	.0407	.0407	.0407	24.542	.9992	1.5301	40′
30′	.0436	.0436	.0437	22.904	.9990	1.5272	30′
40′	.0465	.0465	.0466	21.470	.9989	1.5243	20′
50′	.0495	.0494	.0495	20.206	.9988	1.5213	10′
3° 00′	.0524	.0523	.0524	19.081	.9986	1.5184	87° 00′
10′	.0553	.0552	.0553	18.075	.9985	1.5155	50′
20′	.0582	.0581	.0582	17.169	.9983	1.5126	40′
30′	.0611	.0610	.0612	16.350	.9981	1.5097	30′
40′	.0640	.0640	.0641	15.605	.9980	1.5068	20′
50′	.0669	.0669	.0670	14.924	.9978	1.5039	10′
4° 00′	.0698	.0698	.0699	14.301	.9976	1.5010	86° 00′
10′	.0727	.0727	.0729	13.727	.9974	1.4981	50′
20′	.0756	.0756	.0758	13.197	.9971	1.4952	40′
30′	.0785	.0785	.0787	12.706	.9969	1.4923	30′
40′	.0814	.0814	.0816	12.251	.9967	1.4893	20′
50′	.0844	.0843	.0846	11.826	.9964	1.4864	10′
5° 00′	.0873	.0872	.0875	11.430	.9962	1.4835	85° 00′
10′	.0902	.0901	.0904	11.059	.9959	1.4806	50′
20′	.0931	.0929	.0934	10.712	.9957	1.4777	40′
30′	.0960	.0958	.0963	10.385	.9954	1.4748	30′
40′	.0989	.0987	.0992	10.078	.9951	1.4719	20′
50′	.1018	.1016	.1022	9.7882	.9948	1.4690	10′
6° 00′	.1047	.1045	.1051	9.5144	.9945	1.4661	84° 00′
10′	.1076	.1074	.1080	9.2553	.9942	1.4632	50′
20′	.1105	.1103	.1110	9.0098	.9939	1.4603	40′
30′	.1134	.1132	.1139	8.7769	.9936	1.4573	30′
40′	.1164	.1161	.1169	8.5555	.9932	1.4544	20′
50′	.1193	.1190	.1198	8.3450	.9929	1.4515	10′
7° 00′	.1222	.1219	.1228	8.1443	.9925	1.4486	83° 00′
10′	.1251	.1248	.1257	7.9530	.9922	1.4457	50′
20′	.1280	.1276	.1287	7.7704	.9918	1.4428	40′
30′	.1309	.1305	.1317	7.5958	.9914	1.4399	30′
40′	.1338	.1334	.1346	7.4287	.9911	1.4370	20′
50′	.1367	.1363	.1376	7.2687	.9907	1.4341	10′
8° 00′	.1396	.1392	.1405	7.1154	.9903	1.4312	82° 00′
10′	.1425	.1421	.1435	6.9682	.9899	1.4283	50′
20′	.1454	.1449	.1465	6.8269	.9894	1.4254	40′
30′	.1484	.1478	.1495	6.6912	.9890	1.4224	30′
40′	.1513	.1507	.1524	6.5606	.9886	1.4195	20′
50′	.1542	.1536	.1554	6.4348	.9881	1.4166	10′
9° 00′	.1571	.1564	.1584	6.3138	.9877	1.4137	81° 00′
		Cosine	Cotangent	Tangent	Sine	Radians	Degrees

Appendix 4 VALUES OF TRIGONOMETRIC FUNCTIONS (Cont.)

Degrees	Radians	Sine	Tangent	Cotangent	Cosine		
9° 00′	.1571	.1564	.1584	6.3138	.9877	1.4137	81° 00′
10′	.1600	.1593	.1614	6.1970	.9872	1.4108	50′
20′	.1629	.1622	.1644	6.0844	.9868	1.4079	40′
30′	.1658	.1650	.1673	5.9758	.9863	1.4050	30′
40′	.1687	.1679	.1703	5.8708	.9858	1.4021	20′
50′	.1716	.1708	.1733	5.7694	.9853	1.3992	10′
10° 00′	.1745	.1736	.1763	5.6713	.9848	1.3963	80° 00′
10′	.1774	.1765	.1793	5.5764	.9843	1.3934	50′
20′	.1804	.1794	.1823	5.4845	.9838	1.3904	40′
30′	.1833	.1822	.1853	5.3955	.9833	1.3875	30′
40′	.1862	.1851	.1883	5.3093	.9827	1.3846	20′
50′	.1891	.1880	.1914	5.2257	.9822	1.3817	10′
11° 00′	.1920	.1908	.1944	5.1446	.9816	1.3788	79° 00′
10′	.1949	.1937	.1974	5.0658	.9811	1.3759	50′
20′	.1978	.1965	.2004	4.9894	.9805	1.3730	40′
30′	.2007	.1994	.2035	4.9152	.9799	1.3701	30′
40′	.2036	.2022	.2065	4.8430	.9793	1.3672	20′
50′	.2065	.2051	.2095	4.7729	.9787	1.3643	10′
12° 00′	.2094	.2079	.2126	4.7046	.9781	1.3614	78° 00′
10′	.2123	.2108	.2156	4.6382	.9775	1.3584	50′
20′	.2153	.2136	.2186	4.5736	.9769	1.3555	40′
30′	.2182	.2164	.2217	4.5107	.9763	1.3526	30′
40′	.2211	.2193	.2247	4.4494	.9757	1.3497	20′
50′	.2240	.2221	.2278	4.3897	.9750	1.3468	10′
13° 00′	.2269	.2250	.2309	4.3315	.9744	1.3439	77° 00′
10′	.2298	.2278	.2339	4.2747	.9737	1.3410	50′
20′	.2327	.2306	.2370	4.2193	.9730	1.3381	40′
30′	.2356	.2334	.2401	4.1653	.9724	1.3352	30′
40′	.2385	.2363	.2432	4.1126	.9717	1.3323	20′
50′	.2414	.2391	.2462	4.0611	.9710	1.3294	10′
14° 00′	.2443	.2419	.2493	4.0108	.9703	1.3265	76° 00′
10′	.2473	.2447	.2524	3.9617	.9696	1.3235	50′
20′	.2502	.2476	.2555	3.9136	.9689	1.3206	40′
30′	.2531	.2504	.2586	3.8667	.9681	1.3177	30′
40′	.2560	.2532	.2617	3.8208	.9674	1.3148	20′
50′	.2589	.2560	.2648	3.7760	.9667	1.3119	10′
15° 00′	.2618	.2588	.2679	3.7321	.9659	1.3090	75° 00′
10′	.2647	.2616	.2711	3.6891	.9652	1.3061	50′
20′	.2676	.2644	.2742	3.6470	.9644	1.3032	40′
30′	.2705	.2672	.2773	3.6059	.9636	1.3003	30′
40′	.2734	.2700	.2805	3.5656	.9628	1.2974	20′
50′	.2763	.2728	.2836	3.5261	.9621	1.2945	10′
16° 00′	.2793	.2756	.2867	3.4874	.9613	1.2915	74° 00′
10′	.2822	.2784	.2899	3.4495	.9605	1.2886	50′
20′	.2851	.2812	.2931	3.4124	.9596	1.2857	40′
30′	.2880	.2840	.2962	3.3759	.9588	1.2828	30′
40′	.2909	.2868	.2994	3.3402	.9580	1.2799	20′
50′	.2938	.2896	.3026	3.3052	.9572	1.2770	10′
17° 00′	.2967	.2924	.3057	3.2709	.9563	1.2741	73° 00′
10′	.2996	.2952	.3089	3.2371	.9555	1.2712	50′
20′	.3025	.2979	.3121	3.2041	.9546	1.2683	40′
30′	.3054	.3007	.3153	3.1716	.9537	1.2654	30′
40′	.3083	.3035	.3185	3.1397	.9528	1.2625	20′
50′	.3113	.3062	.3217	3.1084	.9520	1.2595	10′
18° 00′	.3142	.3090	.3249	3.0777	.9511	1.2566	72° 00′
		Cosine	Cotangent	Tangent	Sine	Radians	Degrees

Cont.

Appendix 4 VALUES OF TRIGONOMETRIC FUNCTIONS (Cont.)

Degrees	Radians	Sine	Tangent	Cotangent	Cosine		
18° 00'	.3142	.3090	.3249	3.0777	.9511	1.2566	72° 00'
10'	.3171	.3118	.3281	3.0475	.9502	1.2537	50'
20'	.3200	.3145	.3314	3.0178	.9492	1.2508	40'
30'	.3229	.3173	.3346	2.9887	.9483	1.2479	30'
40'	.3258	.3201	.3378	2.9600	.9474	1.2450	20'
50'	.3287	.3228	.3411	2.9319	.9465	1.2421	10'
19° 00'	.3316	.3256	.3443	2.9042	.9455	1.2392	71° 00'
10'	.3345	.3283	.3476	2.8770	.9446	1.2363	50'
20'	.3374	.3311	.3508	2.8502	.9436	1.2334	40'
30'	.3403	.3338	.3541	2.8239	.9426	1.2305	30'
40'	.3432	.3365	.3574	2.7980	.9417	1.2275	20'
50'	.3462	.3393	.3607	2.7725	.9407	1.2246	10'
20° 00'	.3491	.3420	.3640	2.7475	.9397	1.2217	70° 00'
10'	.3520	.3448	.3673	2.7228	.9387	1.2188	50'
20'	.3549	.3475	.3706	2.6985	.9377	1.2159	40'
30'	.3578	.3502	.3739	2.6746	.9367	1.2130	30'
40'	.3607	.3529	.3772	2.6511	.9356	1.2101	20'
50'	.3636	.3557	.3805	2.6279	.9346	1.2072	10'
21° 00'	.3665	.3584	.3839	2.6051	.9336	1.2043	69° 00'
10'	.3694	.3611	.3872	2.5826	.9325	1.2014	50'
20'	.3723	.3638	.3906	2.5605	.9315	1.1985	40'
30'	.3752	.3665	.3939	2.5386	.9304	1.1956	30'
40'	.3782	.3692	.3973	2.5172	.9293	1.1926	20'
50'	.3811	.3719	.4006	2.4960	.9283	1.1897	10'
22° 00'	.3840	.3746	.4040	2.4751	.9272	1.1868	68° 00'
10'	.3869	.3773	.4074	2.4545	.9261	1.1839	50'
20'	.3898	.3800	.4108	2.4342	.9250	1.1810	40'
30'	.3927	.3827	.4142	2.4142	.9239	1.1781	30'
40'	.3956	.3854	.4176	2.3945	.9228	1.1752	20'
50'	.3985	.3881	.4210	2.3750	.9216	1.1723	10'
23° 00'	.4014	.3907	.4245	2.3559	.9205	1.1694	67° 00'
10'	.4043	.3934	.4279	2.3369	.9194	1.1665	50'
20'	.4072	.3961	.4314	2.3183	.9182	1.1636	40'
30'	.4102	.3987	.4348	2.2998	.9171	1.1606	30'
40'	.4131	.4014	.4383	2.2817	.9159	1.1577	20'
50'	.4160	.4041	.4417	2.2637	.9147	1.1548	10'
24° 00'	.4189	.4067	.4452	2.2460	.9135	1.1519	66° 00'
10'	.4218	.4094	.4487	2.2286	.9124	1.1490	50'
20'	.4247	.4120	.4522	2.2113	.9112	1.1461	40'
30'	.4276	.4147	.4557	2.1943	.9100	1.1432	30'
40'	.4305	.4173	.4592	2.1775	.9088	1.1403	20'
50'	.4334	.4200	.4628	2.1609	.9075	1.1374	10'
25° 00'	.4363	.4226	.4663	2.1445	.9063	1.1345	65° 00'
10'	.4392	.4253	.4699	2.1283	.9051	1.1316	50'
20'	.4422	.4279	.4734	2.1123	.9038	1.1286	40'
30'	.4451	.4305	.4770	2.0965	.9026	1.1257	30'
40'	.4480	.4331	.4806	2.0809	.9013	1.1228	20'
50'	.4509	.4358	.4841	2.0655	.9001	1.1199	10'
26° 00'	.4538	.4384	.4877	2.0503	.8988	1.1170	64° 00'
10'	.4567	.4410	.4913	2.0353	.8975	1.1141	50'
20'	.4596	.4436	.4950	2.0204	.8962	1.1112	40'
30'	.4625	.4462	.4986	2.0057	.8949	1.1083	30'
40'	.4654	.4488	.5022	1.9912	.8936	1.1054	20'
50'	.4683	.4514	.5059	1.9768	.8923	1.1025	10'
27° 00'	.4712	.4540	.5095	1.9626	.8910	1.0996	63° 00'
		Cosine	Cotangent	Tangent	Sine	Radians	Degrees

Appendix 4 VALUES OF TRIGONOMETRIC FUNCTIONS (Cont.)

Degrees	Radians	Sine	Tangent	Cotangent	Cosine		
27° 00′	.4712	.4540	.5095	1.9626	.8910	1.0996	63° 00′
10′	.4741	.4566	.5132	1.9486	.8897	1.0966	50′
20′	.4771	.4592	.5169	1.9347	.8884	1.0937	40′
30′	.4800	.4617	.5206	1.9210	.8870	1.0908	30′
40′	.4829	.4643	.5243	1.9074	.8857	1.0879	20′
50′	.4858	.4669	.5280	1.8940	.8843	1.0850	10′
28° 00′	.4887	.4695	.5317	1.8807	.8829	1.0821	62° 00′
10′	.4916	.4720	.5354	1.8676	.8816	1.0792	50′
20′	.4945	.4746	.5392	1.8546	.8802	1.0763	40′
30′	.4974	.4772	.5430	1.8418	.8788	1.0734	30′
40′	.5003	.4797	.5467	1.8291	.8774	1.0705	20′
50′	.5032	.4823	.5505	1.8165	.8760	1.0676	10′
29° 00′	.5061	.4848	.5543	1.8040	.8746	1.0647	61° 00′
10′	.5091	.4874	.5581	1.7917	.8732	1.0617	50′
20′	.5120	.4899	.5619	1.7796	.8718	1.0588	40′
30′	.5149	.4924	.5658	1.7675	.8704	1.0559	30′
40′	.5178	.4950	.5696	1.7556	.8689	1.0530	20′
50′	.5207	.4975	.5735	1.7437	.8675	1.0501	10′
30° 00′	.5236	.5000	.5774	1.7321	.8660	1.0472	60° 00′
10′	.5265	.5025	.5812	1.7205	.8646	1.0443	50′
20′	.5294	.5050	.5851	1.7090	.8631	1.0414	40′
30′	.5323	.5075	.5890	1.6977	.8616	1.0385	30′
40′	.5352	.5100	.5930	1.6864	.8601	1.0356	20′
50′	.5381	.5125	.5969	1.6753	.8587	1.0327	10′
31° 00′	.5411	.5150	.6009	1.6643	.8572	1.0297	59° 00′
10′	.5440	.5175	.6048	1.6534	.8557	1.0268	50′
20′	.5469	.5200	.6088	1.6426	.8542	1.0239	40′
30′	.5498	.5225	.6128	1.6319	.8526	1.0210	30′
40′	.5527	.5250	.6168	1.6212	.8511	1.0181	20′
50′	.5556	.5275	.6208	1.6107	.8496	1.0152	10′
32° 00′	.5585	.5299	.6249	1.6003	.8480	1.0123	58° 00′
10′	.5614	.5324	.6289	1.5900	.8465	1.0094	50′
20′	.5643	.5348	.6330	1.5798	.8450	1.0065	40′
30′	.5672	.5373	.6371	1.5697	.8434	1.0036	30′
40′	.5701	.5398	.6412	1.5597	.8418	1.0007	20′
50′	.5730	.5422	.6453	1.5497	.8403	.9977	10′
33° 00′	.5760	.5446	.6494	1.5399	.8387	.9948	57° 00′
10′	.5789	.5471	.6536	1.5301	.8371	.9919	50′
20′	.5818	.5495	.6577	1.5204	.8355	.9890	40′
30′	.5847	.5519	.6619	1.5108	.8339	.9861	30′
40′	.5876	.5544	.6661	1.5013	.8323	.9832	20′
50′	.5905	.5568	.6703	1.4919	.8307	.9803	10′
34° 00′	.5934	.5592	.6745	1.4826	.8290	.9774	56° 00′
10′	.5963	.5616	.6787	1.4733	.8274	.9745	50′
20′	.5992	.5640	.6830	1.4641	.8258	.9716	40′
30′	.6021	.5664	.6873	1.4550	.8241	.9687	30′
40′	.6050	.5688	.6916	1.4460	.8225	.9657	20′
50′	.6080	.5712	.6959	1.4370	.8208	.9628	10′
35° 00′	.6109	.5736	.7002	1.4281	.8192	.9599	55° 00′
10′	.6138	.5760	.7046	1.4193	.8175	.9570	50′
20′	.6167	.5783	.7089	1.4106	.8158	.9541	40′
30′	.6196	.5807	.7133	1.4019	.8141	.9512	30′
40′	.6225	.5831	.7177	1.3934	.8124	.9483	20′
50′	.6254	.5854	.7221	1.3848	.8107	.9454	10′
36° 00′	.6283	.5878	.7265	1.3764	.8090	.9425	54° 00′
		Cosine	Cotangent	Tangent	Sine	Radians	Degrees

Cont.

Appendix 4 VALUES OF TRIGONOMETRIC FUNCTIONS (Cont.)

Degrees	Radians	Sine	Tangent	Cotangent	Cosine		
36° 00′	.6283	.5878	.7265	1.3764	.8090	.9425	54° 00′
10′	.6312	.5901	.7310	1.3680	.8073	.9396	50′
20′	.6341	.5925	.7355	1.3597	.8056	.9367	40′
30′	.6370	.5948	.7400	1.3514	.8039	.9338	30′
40′	.6400	.5972	.7445	1.3432	.8021	.9308	20′
50′	.6429	.5995	.7490	1.3351	.8004	.9279	10′
37° 00′	.6458	.6018	.7536	1.3270	.7986	.9250	53° 00′
10′	.6487	.6041	.7581	1.3190	.7969	.9221	50′
20′	.6516	.6065	.7627	1.3111	.7951	.9192	40′
30′	.6545	.6088	.7673	1.3032	.7934	.9163	30′
40′	.6574	.6111	.7720	1.2954	.7916	.9134	20′
50′	.6603	.6134	.7766	1.2876	.7898	.9105	10′
38° 00′	.6632	.6157	.7813	1.2799	.7880	.9076	52° 00′
10′	.6661	.6180	.7860	1.2723	.7862	.9047	50′
20′	.6690	.6202	.7907	1.2647	.7844	.9018	40′
30′	.6720	.6225	.7954	1.2572	.7826	.8988	30′
40′	.6749	.6248	.8002	1.2497	.7808	.8959	20′
50′	.6778	.6271	.8050	1.2423	.7790	.8930	10′
39° 00′	.6807	.6293	.8098	1.2349	.7771	.8901	51° 00′
10′	.6836	.6316	.8146	1.2276	.7753	.8872	50′
20′	.6865	.6338	.8195	1.2203	.7735	.8843	40′
30′	.6894	.6361	.8243	1.2131	.7716	.8814	30′
40′	.6923	.6383	.8292	1.2059	.7698	.8785	20′
50′	.6952	.6406	.8342	1.1988	.7679	.8756	10′
40° 00′	.6981	.6428	.8391	1.1918	.7660	.8727	50° 00′
10′	.7010	.6450	.8441	1.1847	.7642	.8698	50′
20′	.7039	.6472	.8491	1.1778	.7623	.8668	40′
30′	.7069	.6494	.8541	1.1708	.7604	.8639	30′
40′	.7098	.6517	.8591	1.1640	.7585	.8610	20′
50′	.7127	.6539	.8642	1.1571	.7566	.8581	10′
41° 00′	.7156	.6561	.8693	1.1504	.7547	.8552	49° 00′
10′	.7185	.6583	.8744	1.1436	.7528	.8523	50′
20′	.7214	.6604	.8796	1.1369	.7509	.8494	40′
30′	.7243	.6626	.8847	1.1303	.7490	.8465	30′
40′	.7272	.6648	.8899	1.1237	.7470	.8436	20′
50′	.7301	.6670	.8952	1.1171	.7451	.8407	10′
42° 00′	.7330	.6691	.9004	1.1106	.7431	.8378	48° 00′
10′	.7359	.6713	.9057	1.1041	.7412	.8348	50′
20′	.7389	.6734	.9110	1.0977	.7392	.8319	40′
30′	.7418	.6756	.9163	1.0913	.7373	.8290	30′
40′	.7447	.6777	.9217	1.0850	.7353	.8261	20′
50′	.7476	.6799	.9271	1.0786	.7333	.8232	10′
43° 00′	.7505	.6820	.9325	1.0724	.7314	.8203	47° 00′
10′	.7534	.6841	.9380	1.0661	.7294	.8174	50′
20′	.7563	.6862	.9435	1.0599	.7274	.8145	40′
30′	.7592	.6884	.9490	1.0538	.7254	.8116	30′
40′	.7621	.6905	.9545	1.0477	.7234	.8087	20′
50′	.7650	.6926	.9601	1.0416	.7214	.8058	10′
44° 00′	.7679	.6947	.9657	1.0355	.7193	.8029	46° 00′
10′	.7709	.6967	.9713	1.0295	.7173	.7999	50′
20′	.7738	.6988	.9770	1.0235	.7153	.7970	40′
30′	.7767	.7009	.9827	1.0176	.7133	.7941	30′
40′	.7796	.7030	.9884	1.0117	.7112	.7912	20′
50′	.7825	.7050	.9942	1.0058	.7092	.7883	10′
45° 00′	.7854	.7071	1.0000	1.0000	.7071	.7854	45° 00′
		Cosine	Cotangent	Tangent	Sine	Radians	Degrees

Appendix 5 WEIGHTS AND MEASURES

UNITED STATES SYSTEM

LINEAR MEASURE

Inches	Feet	Yards	Rods	Furlongs	Miles
1.0 =	.08333 =	.02778 =	.00012626 =	.00001578
12.0 =	1.0 =	.33333 =	.0606061 =	.00151515 =	.00018939
36.0 =	3.0 =	1.0 =	.1818182 =	.00454545 =	.00056818
198.0 =	16.5 =	5.5 =	1.0 =	.025 =	.003125
7920.0 =	660.0 =	220.0 =	40.0 =	1.0 =	.125
63360.0 =	5280.0 =	1760.0 =	320.0 =	8.0 =	1.0

SQUARE AND LAND MEASURE

Sq. Inches	Square Feet	Square Yards	Sq. Rods	Acres	Sq. Miles
1.0 =	.006944 =	.000772			
144.0 =	1.0 =	.111111			
1296.0 =	9.0 =	1.0 =	.03306 =	.000207	
39204.0 =	272.25 =	30.25 =	1.0 =	.00625 =	.0000098
	43560.0 =	4840.0 =	160.0 =	1.0 =	.0015625
		3097600.0 =	102400.0 =	640.0 =	1.0

AVOIRDUPOIS WEIGHTS

Grains	Drams	Ounces	Pounds	Tons
1.0 =	.03657 =	.002286 =	.000143 =	.0000000714
27.34375 =	1.0 =	.0625 =	.003906 =	.00000195
437.5 =	16.0 =	1.0 =	.0625 =	.00003125
7000.0 =	256.0 =	16.0 =	1.0 =	.0005
14000000.0 =	512000.0 =	32000.0 =	2000.0 =	1.0

DRY MEASURE

Pints	Quarts	Pecks	Cubic Feet	Bushels
1.0 =	.5 =	.0625 =	.01945 =	.01563
2.0 =	1.0 =	.125 =	.03891 =	.03125
16.0 =	8.0 =	1.0 =	.31112 =	.25
51.42627 =	25.71314 =	3.21414 =	1.0 =	.80354
64.0 =	32.0 =	4.0 =	1.2445 =	1.0

LIQUID MEASURE

Gills	Pints	Quarts	U. S. Gallons	Cubic Feet
1.0 =	.25 =	.125 =	.03125 =	.00418
4.0 =	1.0 =	.5 =	.125 =	.01671
8.0 =	2.0 =	1.0 =	.250 =	.03342
32.0 =	8.0 =	4.0 =	1.0 =	.1337
			7.48052 =	1.0

METRIC SYSTEM

UNITS

Length—Meter : Mass—Gram : Capacity—Liter

for pure water at 4°C. (39.2°F.)

1 cubic decimeter or 1 liter = 1 kilogram

1000 Milli $\begin{cases} meters \text{ (mm)} \\ grams \text{ (mg)} \\ liters \text{ (ml)} \end{cases}$ = 100 Centi $\begin{cases} meters \text{ (cm)} \\ grams \text{ (cg)} \\ liters \text{ (cl)} \end{cases}$ = 10 Deci $\begin{cases} meters \text{ (dm)} \\ grams \text{ (dg)} \\ liters \text{ (dl)} \end{cases}$ = 1 $\begin{cases} meter \\ gram \\ liter \end{cases}$

1000 $\begin{cases} meters \\ grams \\ liters \end{cases}$ = 100 Deka $\begin{cases} meters \text{ (dkm)} \\ grams \text{ (dkg)} \\ liters \text{ (dkl)} \end{cases}$ = 10 Hecto $\begin{cases} meters \text{ (hm)} \\ grams \text{ (hg)} \\ liters \text{ (hl)} \end{cases}$ = 1 Kilo $\begin{cases} meter \text{ (km)} \\ gram \text{ (kg)} \\ liter \text{ (kl)} \end{cases}$

1 Metric Ton	= 1000 Kilograms
100 Square Meters	= 1 Are
100 Ares	= 1 Hectare
100 Hectares	= 1 Square Kilometer

Appendix 6 DECIMAL EQUIVALENTS AND TEMPERATURE CONVERSION

DECIMAL EQUIVALENTS — INCH-MILLIMETER CONVERSION TABLE

1/2	1/4	1/8	1/16	1/32	1/64	Decimals	Millimeters
					1	.015625	.396875
				1		.031250	.793750
					3	.046875	1.190625
			1			.062500	1.587500
					5	.078125	1.984375
				3		.093750	2.381250
					7	.109375	2.778125
		1				.125000	3.175000
					9	.140625	3.571875
				5		.156250	3.968750
					11	.171875	4.365625
			3			.187500	4.762500
					13	.203125	5.159375
				7		.218750	5.556250
					15	.234375	5.953125
	1					.250000	6.350000
					17	.265625	6.746875
				9		.281250	7.143750
					19	.296875	7.540625
			5			.312500	7.937500
					21	.328125	8.334375
				11		.343750	8.731250
					23	.359375	9.128125
		3				.375000	9.525000
					25	.390625	9.921875
				13		.406250	10.318750
					27	.421875	10.715625
			7			.437500	11.112500
					29	.453125	11.509375
				15		.468750	11.906250
					31	.484375	12.303125
1						.500000	12.700000

1/2	1/4	1/8	1/16	1/32	1/64	Decimals	Millimeters
					33	.515625	13.096875
				17		.531250	13.493750
					35	.546875	13.890625
			9			.562500	14.287500
					37	.578125	14.684375
				19		.593750	15.081250
					39	.609375	15.478125
		5				.625000	15.875000
					41	.640625	16.271875
				21		.656250	16.668750
					43	.671875	17.065625
			11			.687500	17.462500
					45	.703125	17.859375
				23		.718750	18.256250
					47	.734375	18.653125
	3					.750000	19.050000
					49	.765625	19.446875
				25		.781250	19.843750
					51	.796875	20.240625
			13			.812500	20.637500
					53	.828125	21.034375
				27		.843750	21.481250
					55	.859375	21.828125
		7				.875000	22.225000
					57	.890625	22.621875
				29		.906250	23.018750
					59	.921875	23.415625
			15			.937500	23.812500
					61	.953125	24.209375
				31		.968750	24.606250
					63	.984375	25.003125
2	4	8	16	32	64	1.000000	25.400000

Appendix 6 DECIMAL EQUIVALENTS AND TEMPERATURE CONVERSION (Cont.)

TEMPERATURE CONVERSION

−210 to 0

C.	C. or F.	F.
−134	−210	−346
−129	−200	−328
−123	−190	−310
−118	−180	−292
−112	−170	−274
−107	−160	−256
−101	−150	−238
−95.6	−140	−220
−90.0	−130	−202
−84.4	−120	−184
−78.9	−110	−166
−73.3	−100	−148
−67.8	−90	−130
−62.2	−80	−112
−56.7	−70	−94
−51.1	−60	−76
−45.6	−50	−58
−40.0	−40	−40
−34.4	−30	−22
−28.9	−20	−4
−23.3	−10	14
−17.8	0	32

1 to 25

C.	C. or F.	F.
−17.2	1	33.8
−16.7	2	35.6
−16.1	3	37.4
−15.6	4	39.2
−15.0	5	41.0
−14.4	6	42.8
−13.9	7	44.6
−13.3	8	46.4
−12.8	9	48.2
−12.2	10	50.0
−11.7	11	51.8
−11.1	12	53.6
−10.6	13	55.4
−10.0	14	57.2
−9.44	15	59.0
−8.89	16	60.8
−8.33	17	62.6
−7.78	18	64.4
−7.22	19	66.2
−6.67	20	68.0
−6.11	21	69.8
−5.56	22	71.6
−5.00	23	73.4
−4.44	24	75.2
−3.89	25	77.0

26 to 50

C.	C. or F.	F.
−3.33	26	78.8
−2.78	27	80.6
−2.22	28	82.4
−1.67	29	84.2
−1.11	30	86.0
−0.56	31	87.8
0	32	89.6
0.56	33	91.4
1.11	34	93.2
1.67	35	95.0
2.22	36	96.8
2.78	37	98.6
3.33	38	100.4
3.89	39	102.2
4.44	40	104.0
5.00	41	105.8
5.56	42	107.6
6.11	43	109.4
6.67	44	111.2
7.22	45	113.0
7.78	46	114.8
8.33	47	116.6
8.89	48	118.4
9.44	49	120.2
10.0	50	122.0

51 to 75

C.	C. or F.	F.
10.6	51	123.8
11.1	52	125.6
11.7	53	127.4
12.2	54	129.2
12.8	55	131.0
13.3	56	132.8
13.9	57	134.6
14.4	58	136.4
15.0	59	138.2
15.6	60	140.0
16.1	61	141.8
16.7	62	143.6
17.2	63	145.4
17.8	64	147.2
18.3	65	149.0
18.9	66	150.8
19.4	67	152.6
20.0	68	154.4
20.6	69	156.2
21.1	70	158.0
21.7	71	159.8
22.2	72	161.6
22.8	73	163.4
23.3	74	165.2
23.9	75	167.0

76 to 100

C.	C. or F.	F.
24.4	76	168.8
25.0	77	170.6
25.6	78	172.4
26.1	79	174.2
26.7	80	176.0
27.2	81	177.8
27.8	82	179.6
28.3	83	181.4
28.9	84	183.2
29.4	85	185.0
30.0	86	186.8
30.6	87	188.6
31.1	88	190.4
31.7	89	192.2
32.2	90	194.0
32.8	91	195.8
33.3	92	197.6
33.9	93	199.4
34.4	94	201.2
35.0	95	203.0
35.6	96	204.8
36.1	97	206.6
36.7	98	208.4
37.2	99	210.2
37.8	100	212.0

101 to 340

C.	C. or F.	F.
43	110	230
49	120	248
54	130	266
60	140	284
66	150	302
71	160	320
77	170	338
82	180	356
88	190	374
93	200	392
99	210	410
100	212	413
104	220	428
110	230	446
116	240	464
121	250	482
127	260	500
132	270	518
138	280	536
143	290	554
149	300	572
154	310	590
160	320	608
166	330	626
171	340	644

341 to 490

C.	C. or F.	F.
177	350	662
182	360	680
188	370	698
193	380	716
199	390	734
204	400	752
210	410	770
216	420	788
221	430	806
227	440	824
232	450	842
238	460	860
243	470	878
249	480	896
254	490	914

491 to 750

C.	C. or F.	F.
260	500	932
266	510	950
271	520	968
277	530	986
282	540	1004
288	550	1022
293	560	1040
299	570	1058
304	580	1076
310	590	1094
316	600	1112
321	610	1130
327	620	1148
332	630	1166
338	640	1184
343	650	1202
349	660	1220
354	670	1238
360	680	1256
366	690	1274
371	700	1292
377	710	1310
382	720	1328
388	730	1346
393	740	1364
399	750	1382

$$°F = \frac{9}{5}(°C) + 32$$

$$°C = \frac{5}{9}(°F - 32)$$

INTERPOLATION FACTORS

C.	F.	F.	C.	F.	F.
0.56	1	1.8	3.33	6	10.8
1.11	2	3.6	3.89	7	12.6
1.67	3	5.4	4.44	8	14.4
2.22	4	7.2	5.00	9	16.2
2.78	5	9.0	5.56	10	18.0

NOTE:—The numbers in bold face type refer to the temperature either in degrees Centigrade or Fahrenheit which it is desired to convert into the other scale. If converting from Fahrenheit degrees to Centigrade degrees the equivalent temperature will be found in the left column, while if converting from degrees Centigrade to degrees Fahrenheit, the answer will be found in the column on the right.

Appendix 7 WEIGHTS AND SPECIFIC GRAVITIES

Substance	Weight Lb. per Cu. Ft.	Specific Gravity	Substance	Weight Lb. per Cu. Ft.	Specific Gravity
METALS, ALLOYS, ORES			**TIMBER, U. S. SEASONED**		
Aluminum, cast,			**Moisture Content by**		
hammered	165	2.55-2.75	**Weight:**		
Brass, cast, rolled	534	8.4-8.7	Seasoned timber 15 to 20%		
Bronze, 7.9 to 14% Sn	509	7.4-8.9	Green timber up to 50%		
Bronze, aluminum	481	7.7	Ash, white, red	40	0.62-0.65
Copper, cast, rolled	556	8.8-9.0	Cedar, white, red	22	0.32-0.38
Copper ore, pyrites	262	4.1-4.3	Chestnut	41	0.66
Gold, cast, hammered	1205	19.25-19.3	Cypress	30	0.48
Iron, cast, pig	450	7.2	Fir, Douglas spruce	32	0.51
Iron, wrought	485	7.6-7.9	Fir, eastern	25	0.40
Iron, spiegel-eisen	468	7.5	Elm, white	45	0.72
Iron, ferro-silicon	437	6.7-7.3	Hemlock	29	0.42-0.52
Iron ore, hematite	325	5.2	Hickory	49	0.74-0.84
Iron ore, hematite in bank	160-180	....	Locust	46	0.73
Iron ore, hematite loose	130-160	....	Maple, hard	43	0.68
Iron ore, limonite	237	3.6-4.0	Maple, white	33	0.53
Iron ore, magnetite	315	4.9-5.2	Oak, chestnut	54	0.86
Iron slag	172	2.5-3.0	Oak, live	59	0.95
Lead	710	11.37	Oak, red, black	41	0.65
Lead ore, galena	465	7.3-7.6	Oak, white	46	0.74
Magnesium, alloys	112	1.74-1.83	Pine, Oregon	32	0.51
Manganese	475	7.2-8.0	Pine, red	30	0.48
Manganese ore, pyrolusite	259	3.7-4.6	Pine, white	26	0.41
Mercury	849	13.6	Pine, yellow, long-leaf	44	0.70
Monel Metal	556	8.8-9.0	Pine, yellow, short-leaf	38	0.61
Nickel	565	8.9-9.2	Poplar	30	0.48
Platinum, cast, hammered	1330	21.1-21.5	Redwood, California	26	0.42
Silver, cast, hammered	656	10.4-10.6	Spruce, white, black	27	0.40-0.46
Steel, rolled	490	7.85	Walnut, black	38	0.61
Tin, cast, hammered	459	7.2-7.5			
Tin ore, cassiterite	418	6.4-7.0			
Zinc, cast, rolled	440	6.9-7.2			
Zinc ore, blende	253	3.9-4.2	**VARIOUS LIQUIDS**		
			Alcohol, 100%	49	0.79
			Acids, muriatic 40%	75	1.20
			Acids, nitric 91%	94	1.50
VARIOUS SOLIDS			Acids, sulphuric 87%	112	1.80
Cereals, oats........bulk	32	....	Lye, soda 66%	106	1.70
Cereals, barley........bulk	39	....	Oils, vegetable	58	0.91-0.94
Cereals, corn, rye........bulk	48	....	Oils, mineral, lubricants	57	0.90-0.93
Cereals, wheat........bulk	48	....	Water, 4°C. max. density	62.428	1.0
Hay and Straw........bales	20		Water, 100°C.	59.830	0.9584
Cotton, Flax, Hemp	93	1.47-1.50	Water, ice	56	0.88-0.92
Fats	58	0.90-0.97	Water, snow, fresh fallen	8	.125
Flour, loose	28	0.40-0.50	Water, sea water	64	1.02-1.03
Flour, pressed	47	0.70-0.80			
Glass, common	156	2.40-2.60			
Glass, plate or crown	161	2.45-2.72	**GASES**		
Glass, crystal	184	2.90-3.00			
Leather	59	0.86-1.02	Air, 0°C. 760 mm.	.08071	1.0
Paper	58	0.70-1.15	Ammonia	.0478	0.5920
Potatoes, piled	42	....	Carbon dioxide	.1234	1.5291
Rubber, caoutchouc	59	0.92-0.96	Carbon monoxide	.0781	0.9673
Rubber goods	94	1.0-2.0	Gas, illuminating	.028-.036	0.35-0.45
Salt, granulated, piled	48	....	Gas, natural	.038-.039	0.47-0.48
Saltpeter	67	....	Hydrogen	.00559	0.0693
Starch	96	1.53	Nitrogen	.0784	0.9714
Sulphur	125	1.93-2.07	Oxygen	.0892	1.1056
Wool.	82	1.32			

The specific gravities of solids and liquids refer to water at 4°C., those of gases to air at 0°C. and 760 mm. pressure. The weights per cubic foot are derived from average specific gravities, except where stated that weights are for bulk, heaped or loose material, etc.

(Courtesy of the American Institute of Steel Construction.)

Appendix 7 WEIGHTS AND SPECIFIC GRAVITIES (Cont.)

Substance	Weight Lb. per Cu. Ft.	Specific Gravity	Substance	Weight Lb. per Cu. Ft.	Specific Gravity
ASHLAR MASONRY			**MINERALS**		
Granite, syenite, gneiss	165	2.3-3.0	Asbestos	153	2.1-2.8
Limestone, marble	160	2.3-2.8	Barytes	281	4.50
Sandstone, bluestone	140	2.1-2.4	Basalt	184	2.7-3.2
			Bauxite	159	2.55
MORTAR RUBBLE			Borax	109	1.7-1.8
MASONRY			Chalk	137	1.8-2.6
Granite, syenite, gneiss	155	2.2-2.8	Clay, marl	137	1.8-2.6
Limestone, marble	150	2.2-2.6	Dolomite	181	2.9
Sandstone, bluestone	130	2.0-2.2	Feldspar, orthoclase	159	2.5-2.6
			Gneiss, serpentine	159	2.4-2.7
DRY RUBBLE MASONRY			Granite, syenite	175	2.5-3.1
Granite, syenite, gneiss	130	1.9-2.3	Greenstone, trap	187	2.8-3.2
Limestone, marble	125	1.9-2.1	Gypsum, alabaster	159	2.3-2.8
Sandstone, bluestone	110	1.8-1.9	Hornblende	187	3.0
			Limestone, marble	165	2.5-2.8
BRICK MASONRY			Magnesite	187	3.0
Pressed brick	140	2.2-2.3	Phosphate rock, apatite	200	3.2
Common brick	120	1.8-2.0	Porphyry	172	2.6-2.9
Soft brick	100	1.5-1.7	Pumice, natural	40	0.37-0.90
			Quartz, flint	165	2.5-2.8
CONCRETE MASONRY			Sandstone, bluestone	147	2.2-2.5
Cement, stone, sand	144	2.2-2.4	Shale, slate	175	2.7-2.9
Cement, slag, etc.	130	1.9-2.3	Soapstone, talc	169	2.6-2.8
Cement, cinder, etc.	100	1.5-1.7			
VARIOUS BUILDING					
MATERIALS			**STONE, QUARRIED, PILED**		
Ashes, cinders	40-45	....	Basalt, granite, gneiss	96	....
Cement, portland, loose	90	....	Limestone, marble, quartz	95	....
Cement, portland, set	183	2.7-3.2	Sandstone	82	....
Lime, gypsum, loose	53-64	....	Shale	92	....
Mortar, set	103	1.4-1.9	Greenstone, hornblende	107	....
Slags, bank slag	67-72	....			
Slags, bank screenings	98-117	....			
Slags, machine slag	96	....			
Slags, slag sand	49-55	....	**BITUMINOUS SUBSTANCES**		
			Asphaltum	81	1.1-1.5
EARTH, ETC., EXCAVATED			Coal, anthracite	97	1.4-1.7
Clay, dry	63	....	Coal, bituminous	84	1.2-1.5
Clay, damp, plastic	110	....	Coal, lignite	78	1.1-1.4
Clay and gravel, dry	100	....	Coal, peat, turf, dry	47	0.65-0.85
Earth, dry, loose	76	....	Coal, charcoal, pine	23	0.28-0.44
Earth, dry, packed	95	....	Coal, charcoal, oak	33	0.47-0.57
Earth, moist, loose	78	....	Coal, coke	75	1.0-1.4
Earth, moist, packed	96	....	Graphite	131	1.9-2.3
Earth, mud, flowing	108	....	Paraffine	56	0.87-0.91
Earth, mud, packed	115	....	Petroleum	54	0.87
Riprap, limestone	80-85	....	Petroleum, refined	50	0.79-0.82
Riprap, sandstone	90	....	Petroleum, benzine	46	0.73-0.75
Riprap, shale	105	....	Petroleum, gasoline	42	0.66-0.69
Sand, gravel, dry, loose	90-105	....	Pitch	69	1.07-1.15
Sand, gravel, dry, packed	100-120	....	Tar, bituminous	75	1.20
Sand, gravel, dry, wet	118-120	....			
EXCAVATIONS IN WATER					
Sand or gravel	60	....	**COAL AND COKE, PILED**		
Sand or gravel and clay	65	....	Coal, anthracite	47-58	....
Clay	80	....	Coal, bituminous, lignite	40-54	....
River mud	90	....	Coal, peat, turf	20-26	....
Soil	70	....	Coal, charcoal	10-14	....
Stone riprap	65	....	Coal, coke	23-32	....

The specific gravities of solids and liquids refer to water at 4°C., those of gases to air at 0°C. and 760 mm. pressure. The weights per cubic foot are derived from average specific gravities, except where stated that weights are for bulk, heaped or loose material, etc.

Appendix 8 WIRE AND SHEET METAL GAGES

WIRE AND SHEET METAL GAGES
IN DECIMALS OF AN INCH

Name of Gage	United States Standard Gage*		The United States Steel Wire Gage	American or Brown & Sharpe Wire Gage	New Birmingham Standard Sheet & Hoop Gage	British Imperial or English Legal Standard Wire Gage	Birmingham or Stubs Iron Wire Gage	Name of Gage
Principal Use	Uncoated Steel Sheets and Light Plates		Steel Wire except Music Wire	Non-Ferrous Sheets and Wire	Iron and Steel Sheets and Hoops	Wire	Strips, Bands, Hoops and Wire	Principal Use
Gage No.	Weight Oz. per Sq. Ft.	Approx. Thickness Inches	Thickness, Inches					Gage No.
7/0's			.4900		.6666	.500		7/0's
6/0's			.4615	.5800	.625	.464		6/0's
5/0's			.4305	.5165	.5883	.432	.500	5/0's
4/0's			.3938	.4600	.5416	.400	.454	4/0's
3/0's			.3625	.4096	.500	.372	.425	3/0's
2/0's			.3310	.3648	.4452	.348	.380	2/0's
0			.3065	.3249	.3964	.324	.340	0
1			.2830	.2893	.3532	.300	.300	1
2			.2625	.2576	.3147	.276	.284	2
3	160	.2391	.2437	.2294	.2804	.252	.259	3
4	150	.2242	.2253	.2043	.250	.232	.238	4
5	140	.2092	.2070	.1819	.2225	.212	.220	5
6	130	.1943	.1920	.1620	.1981	.192	.203	6
7	120	.1793	.1770	.1443	.1764	.176	.180	7
8	110	.1644	.1620	.1285	.1570	.160	.165	8
9	100	.1495	.1483	.1144	.1398	.144	.148	9
10	90	.1345	.1350	.1019	.1250	.128	.134	10
11	80	.1196	.1205	.0907	.1113	.116	.120	11
12	70	.1046	.1055	.0808	.0991	.104	.109	12
13	60	.0897	.0915	.0720	.0882	.092	.095	13
14	50	.0747	.0800	.0641	.0785	.080	.083	14
15	45	.0673	.0720	.0571	.0699	.072	.072	15
16	40	.0598	.0625	.0508	.0625	.064	.065	16
17	36	.0538	.0540	.0453	.0556	.056	.058	17
18	32	.0478	.0475	.0403	.0495	.048	.049	18
19	28	.0418	.0410	.0359	.0440	.040	.042	19
20	24	.0359	.0348	.0320	.0392	.036	.035	20
21	22	.0329	.0318	.0285	.0349	.032	.032	21
22	20	.0299	.0286	.0253	.0313	.028	.028	22
23	18	.0269	.0258	.0226	.0278	.024	.025	23
24	16	.0239	.0230	.0201	.0248	.022	.022	24
25	14	.0209	.0204	.0179	.0220	.020	.020	25
26	12	.0179	.0181	.0159	.0196	.018	.018	26
27	11	.0164	.0173	.0142	.0175	.0164	.016	27
28	10	.0149	.0162	.0126	.0156	.0148	.014	28
29	9	.0135	.0150	.0113	.0139	.0136	.013	29
30	8	.0120	.0140	.0100	.0123	.0124	.012	30
31	7	.0105	.0132	.0089	.0110	.0116	.010	31
32	6.5	.0097	.0128	.0080	.0098	.0108	.009	32
33	6	.0090	.0118	.0071	.0087	.0100	.008	33
34	5.5	.0082	.0104	.0063	.0077	.0092	.007	34
35	5	.0075	.0095	.0056	.0069	.0084	.005	35
36	4.5	.0067	.0090	.0050	.0061	.0076	.004	36
37	4.25	.0064	.0085	.0045	.0054	.0068		37
38	4	.0060	.0080	.0040	.0048	.0060		38
39			.0075	.0035	.0043	.0052		39
40			.0070	.0031	.0039	.0048		40

* U. S. Standard Gage is officially a weight gage, in oz. per sq. ft. as tabulated. The Approx. Thickness shown is the "Manufacturers' Standard" of the American Iron and Steel Institute, based on steel as weighing 501.81 lbs. per cu. ft. (489.6 true weight plus 2.5 percent for average over-run in area and thickness). The A.I.S.I. standard nomenclature for flat rolled carbon steel is as follows:

Widths, Inches	Thicknesses, Inch						
	0.2500 and thicker	0.2499 to 0.2031	0.2030 to 0.1875	0.1874 to 0.0568	0.0567 to 0.0344	0.0343 to 0.0255	0.0254 to 0.0142	0.0141 and thinner
To 3½ incl............................	Bar	Bar	Strip	Strip	Strip	Strip	Sheet	Sheet
Over 3½ to 6 incl...........	Bar	Bar	Strip	Strip	Strip	Sheet	Sheet	Sheet
" 6 to 12 "	Plate	Strip	Strip	Strip	Sheet	Sheet	Sheet	Sheet
" 12 to 32 "	Plate	Sheet	Sheet	Sheet	Sheet	Sheet	Sheet	Black Plate
" 32 to 48 "	Plate	Sheet	Sheet	Sheet	Sheet	Sheet	Sheet	Sheet
" 48	Plate	Plate	Plate	Sheet	Sheet	Sheet	Sheet	——

Appendix 9 PIPING SYMBOLS

TYPE OF FITTING		DOUBLE LINE CONVENTION					SINGLE LINE CONVENTION					FLOW DIAGRAM
		FLANGED	SCREWED	B & S	WELDED	SOLDERED	FLANGED	SCREWED	B & S	WELDED	SOLDERED	
1	Joint											
2	Joint - Expansion											
3	Union											
4	Sleeve											
5	Reducer											
6	Reducer - Eccentric											
7	Reducing Flange											
8	Bushing											
9	Elbow - 45°											
10	Elbow - 90°											
11	Elbow - Long radius											
12	Elbow - (turned up)											
13	Elbow - (turned down)											
14	Elbow - Side outlet (outlet up)											
15	Elbow - Side outlet (outlet down)											
16	Elbow - Base											
17	Elbow - Double branch											
18	Elbow - Reducing											
19	Lateral											
20	Tee											
21	Tee - Single sweep											

Cont.

Appendix 9 PIPING SYMBOLS (Cont.)

TYPE OF FITTING		DOUBLE LINE CONVENTION					SINGLE LINE CONVENTION					FLOW DIAGRAM
		FLANGED	SCREWED	B & S	WELDED	SOLDERED	FLANGED	SCREWED	B & S	WELDED	SOLDERED	
22	Tee-Double sweep											
23	Tee-(outlet up)											
24	Tee-(outlet down)											
25	Tee-Side outlet (outlet up)											
26	Tee-Side outlet (outlet down)											
27	Cross											
28	Valve-Globe											
29	Valve-Angle											
30	Valve-Motor operated globe											Motor operated
31	Valve-Gate											
32	Valve-Angle gate											
33	Valve-Motor operated gate											Motor operated
34	Valve-Check											
35	Valve-Angle check											
36	Valve-Safety											
37	Valve-Angle safety											
38	Valve-Quick opening											
39	Valve-Float operating											
40	Stop Cock											

Appendix 10 AMERICAN STANDARD TAPER PIPE THREADS, NPT[1]

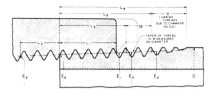

1	2	3	4	5	6	7	8	9	10	11
				Pitch Diameter at Beginning of External Thread E_0	Hand-Tight Engagement			Effective Thread, External		
	Outside Diameter of Pipe D	Threads per Inch n	Pitch of Thread p		Length[2] L_1		Dia E_1	Length L_2		Dia E_2
Nominal Pipe Size					In.	Thds		In.	Thds	In.
$\frac{1}{16}$	0.3125	27	0.03704	0.27118	0.160	4.32	0.28118	0.2611	7.05	0.28750
$\frac{1}{8}$	0.405	27	0.03704	0.36351	0.180	4.86	0.37476	0.2639	7.12	0.38000
$\frac{1}{4}$	0.540	18	0.05556	0.47739	0.200	3.60	0.48989	0.4018	7.23	0.50250
$\frac{3}{8}$	0.675	18	0.05556	0.61201	0.240	4.32	0.62701	0.4078	7.34	0.63750
$\frac{1}{2}$	0.840	14	0.07143	0.75843	0.320	4.48	0.77843	0.5337	7.47	0.79179
$\frac{3}{4}$	1.050	14	0.07143	0.96768	0.339	4.75	0.98887	0.5457	7.64	1.00179
1	1.315	$11\frac{1}{2}$	0.08696	1.21363	0.400	4.60	1.23863	0.6828	7.85	1.25630
$1\frac{1}{4}$	1.660	$11\frac{1}{2}$	0.08696	1.55713	0.420	4.83	1.58338	0.7068	8.13	1.60130
$1\frac{1}{2}$	1.900	$11\frac{1}{2}$	0.08696	1.79609	0.420	4.83	1.82234	0.7235	8.32	1.84130
2	2.375	$11\frac{1}{2}$	0.08696	2.26902	0.436	5.01	2.29627	0.7565	8.70	2.31630
$2\frac{1}{2}$	2.875	8	0.12500	2.71953	0.682	5.46	2.76216	1.1375	9.10	2.79062
3	3.500	8	0.12500	3.34062	0.766	6.13	3.38850	1.2000	9.60	3.41562
$3\frac{1}{2}$	4.000	8	0.12500	3.83750	0.821	6.57	3.88881	1.2500	10.00	3.91562
4	4.500	8	0.12500	4.33438	0.844	6.75	4.38712	1.3000	10.40	4.41562
5	5.563	8	0.12500	5.39073	0.937	7.50	5.44929	1.4063	11.25	5.47862
6	6.625	8	0.12500	6.44609	0.958	7.66	6.50597	1.5125	12.10	6.54062
8	8.625	8	0.12500	8.43359	1.063	8.50	8.50003	1.7125	13.70	8.54062
10	10.750	8	0.12500	10.54531	1.210	9.68	10.62094	1.9250	15.40	10.66562
12	12.750	8	0.12500	12.53281	1.360	10.88	12.61781	2.1250	17.00	12.66562
14 OD	14.000	8	0.12500	13.77500	1.562	12.50	13.87262	2.2500	18.90	13.91562
16 OD	16.000	8	0.12500	15.76250	1.812	14.50	15.87575	2.4500	19.60	15.91562
18 OD	18.000	8	0.12500	17.75000	2.000	16.00	17.87500	2.6500	21.20	17.91562
20 OD	20.000	8	0.12500	19.73750	2.125	17.00	19.87031	2.8500	22.80	19.91562
24 OD	24.000	8	0.12500	23.71250	2.375	19.00	23.86094	3.2500	26.00	23.91562

All dimensions are given in inches.

[1] The basic dimensions of the American Standard Taper Pipe Thread are given in inches to four or five decimal places. While this implies a greater degree of precision than is ordinarily attained, these dimensions are the basis of gage dimensions and are so expressed for the purpose of eliminating errors in computations.

[2] Also length of thin ring gage and length from gaging notch to small end of plug gage.

(Courtesy of ANSI; B2.1–1960.)

Appendix 11 AMERICAN STANDARD 250-LB CAST IRON FLANGED FITTINGS

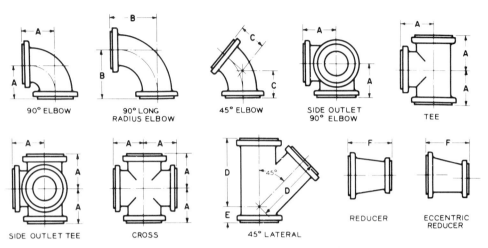

90° ELBOW 90° LONG RADIUS ELBOW 45° ELBOW SIDE OUTLET 90° ELBOW TEE

SIDE OUTLET TEE CROSS 45° LATERAL REDUCER ECCENTRIC REDUCER

Dimensions of 250-lb Cast Iron Flanged Fittings

Nominal Pipe Size	Flanges			Fittings		Straight					
	Dia of Flange	Thickness of Flange (Min)	Dia of Raised Face	Inside Dia of Fittings (Min)	Wall Thickness	Center to Face 90 Deg Elbow Tees, Crosses and True "Y" A	Center to Face 90 Deg Long Radius Elbow B	Center to Face 45 Deg Elbow C	Center to Face Lateral D	Short Center to Face True "Y" and Lateral E	Face to Face Reducer F
1	$4\frac{7}{8}$	$\frac{11}{16}$	$2\frac{11}{16}$	1	$\frac{7}{16}$	4	5	2	$6\frac{1}{2}$	2	....
$1\frac{1}{4}$	$5\frac{1}{4}$	$\frac{3}{4}$	$3\frac{1}{16}$	$1\frac{1}{4}$	$\frac{7}{16}$	$4\frac{1}{4}$	$5\frac{1}{2}$	$2\frac{1}{2}$	$7\frac{1}{4}$	$2\frac{1}{4}$	....
$1\frac{1}{2}$	$6\frac{1}{8}$	$\frac{13}{16}$	$3\frac{9}{16}$	$1\frac{1}{2}$	$\frac{7}{16}$	$4\frac{1}{2}$	6	$2\frac{3}{4}$	$8\frac{1}{2}$	$2\frac{1}{2}$	....
2	$6\frac{1}{2}$	$\frac{7}{8}$	$4\frac{3}{16}$	2	$\frac{7}{16}$	5	$6\frac{1}{2}$	3	9	$2\frac{1}{2}$	5
$2\frac{1}{2}$	$7\frac{1}{2}$	1	$4\frac{15}{16}$	$2\frac{1}{2}$	$\frac{1}{2}$	$5\frac{1}{2}$	7	$3\frac{1}{2}$	$10\frac{1}{2}$	$2\frac{1}{2}$	$5\frac{1}{2}$
3	$8\frac{1}{4}$	$1\frac{1}{8}$	$5\frac{11}{16}$	3	$\frac{9}{16}$	6	$7\frac{3}{4}$	$3\frac{1}{2}$	11	3	6
$3\frac{1}{2}$	9	$1\frac{3}{16}$	$6\frac{5}{16}$	$3\frac{1}{2}$	$\frac{9}{16}$	$6\frac{1}{2}$	$8\frac{1}{2}$	4	$12\frac{1}{2}$	3	$6\frac{1}{2}$
4	10	$1\frac{1}{4}$	$6\frac{15}{16}$	4	$\frac{5}{8}$	7	9	$4\frac{1}{2}$	$13\frac{1}{2}$	3	7
5	11	$1\frac{3}{8}$	$8\frac{5}{16}$	5	$\frac{11}{16}$	8	$10\frac{1}{4}$	5	15	$3\frac{1}{2}$	8
6	$12\frac{1}{2}$	$1\frac{7}{16}$	$9\frac{11}{16}$	6	$\frac{3}{4}$	$8\frac{1}{2}$	$11\frac{1}{2}$	$5\frac{1}{2}$	$17\frac{1}{2}$	4	9
8	15	$1\frac{5}{8}$	$11\frac{15}{16}$	8	$\frac{13}{16}$	10	14	6	$20\frac{1}{2}$	5	11
10	$17\frac{1}{2}$	$1\frac{7}{8}$	$14\frac{1}{16}$	10	$\frac{15}{16}$	$11\frac{1}{2}$	$16\frac{1}{2}$	7	24	$5\frac{1}{2}$	12
12	$20\frac{1}{2}$	2	$16\frac{7}{16}$	12	1	13	19	8	$27\frac{1}{2}$	6	14
14	23	$2\frac{1}{8}$	$18\frac{15}{16}$	$13\frac{1}{4}$	$1\frac{1}{8}$	15	$21\frac{1}{2}$	$8\frac{1}{2}$	31	$6\frac{1}{2}$	16
16	$25\frac{1}{2}$	$2\frac{1}{4}$	$21\frac{1}{16}$	$15\frac{1}{4}$	$1\frac{1}{4}$	$16\frac{1}{2}$	24	$9\frac{1}{2}$	$34\frac{1}{2}$	$7\frac{1}{2}$	18
18	28	$2\frac{3}{8}$	$23\frac{5}{16}$	17	$1\frac{3}{8}$	18	$26\frac{1}{2}$	10	$37\frac{1}{2}$	8	19
20	$30\frac{1}{2}$	$2\frac{1}{2}$	$25\frac{9}{16}$	19	$1\frac{1}{2}$	$19\frac{1}{2}$	29	$10\frac{1}{2}$	$40\frac{1}{2}$	$8\frac{1}{2}$	20
24	36	$2\frac{3}{4}$	$30\frac{5}{16}$	23	$1\frac{5}{8}$	$22\frac{1}{2}$	34	12	$47\frac{1}{2}$	10	24
30	43	3	$37\frac{3}{16}$	29	2	$27\frac{1}{2}$	$41\frac{1}{2}$	15	....	....	30

All dimensions are given in inches.
(Courtesy of ANSI; B16.1–1967.)

Appendix 12 AMERICAN STANDARD 125-LB CAST IRON FLANGED FITTINGS

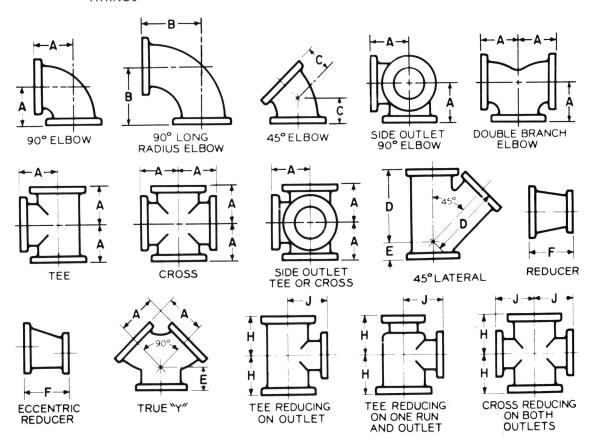

90° ELBOW

90° LONG RADIUS ELBOW

45° ELBOW

SIDE OUTLET 90° ELBOW

DOUBLE BRANCH ELBOW

TEE

CROSS

SIDE OUTLET TEE OR CROSS

45° LATERAL

REDUCER

ECCENTRIC REDUCER

TRUE "Y"

TEE REDUCING ON OUTLET

TEE REDUCING ON ONE RUN AND OUTLET

CROSS REDUCING ON BOTH OUTLETS

Appendix 12 AMERICAN STANDARD 125-LB CAST IRON FLANGED FITTINGS (Cont.)

Nominal Pipe Size	Flanges — Dia of Flange	Flanges — Thickness of Flange (Min)	General — Inside Dia of Flange Fittings	General — Wall Thickness	Straight Fittings — Center to Face 90 deg Elbow Tees, Crosses, True "Y" and Double Branch Elbow (A)	Straight Fittings — Center to Face 90 deg Long Radius Elbow (B)	Straight Fittings — Center to Face 45 deg Elbow (C)	Straight Fittings — Center to Face Lateral (D)	Straight Fittings — Short Center to Face True "Y" and Lateral (E)	Straight Fittings — Face to Face Reducer (F)	Reducing Fittings (Short Body Patterns) Tees and Crosses — Size of Outlet and Smaller	Reducing Fittings Tees and Crosses — Center to Face Run (H)	Reducing Fittings — Center to Face Outlet or Side Outlet (J)
1	4¼	7/16	1	5/16	3½	5	1¾	5¾	1¾	· · · ·			
1¼	4⅝	½	1¼	5/16	3¾	5½	2	6¼	1¾	· · · ·			
1½	5	9/16	1½	5/16	4	6	2¼	7	2	· · · ·			
2	6	5/8	2	5/16	4½	6½	2½	8	2½	5			
2½	7	11/16	2½	5/16	5	7	3	9½	2½	5½			
3	7½	¾	3	3/8	5½	7¾	3	10	3	6			
3½	8½	13/16	3½	7/16	6	8½	3½	11½	3	6½			
4	9	15/16	4	½	6½	9	4	12	3	7			
5	10	15/16	5	½	7½	10¼	4½	13½	3½	8			
6	11	1	6	9/16	8	11½	5	14½	3½	9			
8	13½	1⅛	8	5/8	9	14	5½	17½	4½	11			
10	16	1 3/16	10	¾	11	16½	6½	20½	5	12			
12	19	1¼	12	13/16	12	19	7½	24½	5½	14			
14	21	1 3/8	14	7/8	14	21½	7½	27	6	16			
16	23½	1 7/16	16	1	15	24	8	30	6½	18			
18	25	1 9/16	18	1 1/16	16½	26½	8½	32	7	19	12	13	15½
20	27½	1 11/16	20	1⅛	18	29	9½	35	8	20	14	14	17
24	32	1 7/8	24	1¼	22	34	11	40½	9	24	16	15	19
30	38¾	2⅛	30	1 7/16	25	41½	15	49	10	30	20	18	23
36	46	2⅜	36	1 5/8	28*	49	18	· · · ·	· · · ·	36	24	20	26
42	53	2⅝	42	1 13/16	31*	56½	21	· · · ·	· · · ·	42	24	23	30
48	59½	2¾	48	2	34*	64	24	· · · ·	· · · ·	48	30	26	34

All reducing tees and crosses, sizes 16 in. and smaller, shall have same center to face dimensions as straight size fittings, corresponding to the size of the largest opening.

All dimensions are given in inches.
(Courtesy of ANSI: B16.1–1967.)

Appendix 13 AMERICAN STANDARD 125-LB CAST IRON FLANGES*

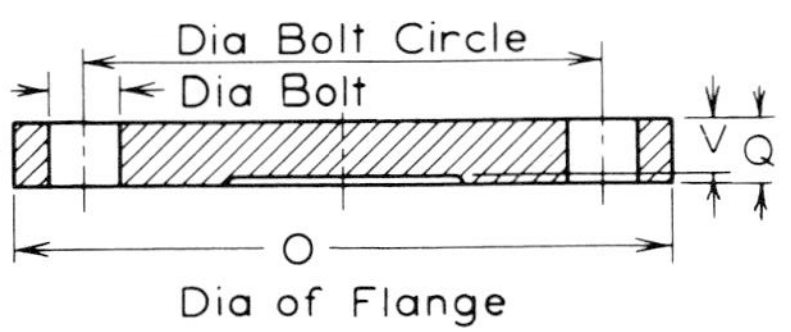

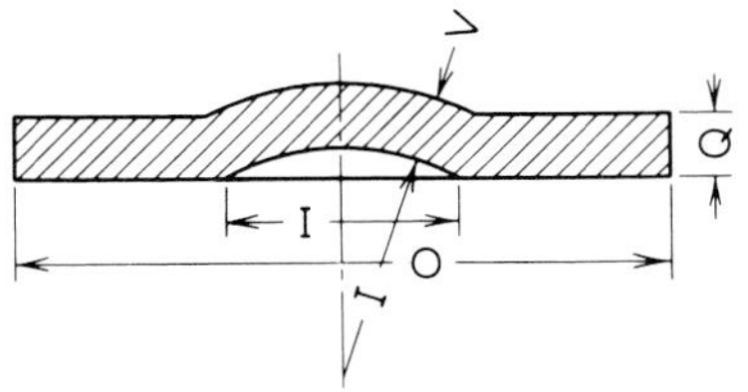

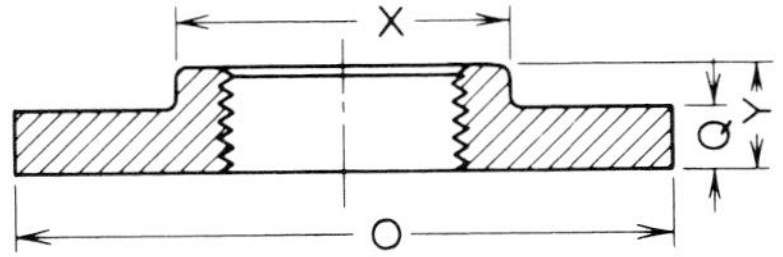

Size *I*	*O*	*Q*	*V*	*X*	*Y*	Dia. Bolt Circle	No. of Bolts	Dia. Bolts	Dia. Bolt Holes	Length of Bolts
1	4¼	7/16	—	1 15/16	0.68	3⅛	4	½	⅝	1¾
1¼	4⅝	½	—	2 5/16	0.76	3½	4	½	⅝	2
1½	5	9/16	—	2 9/16	0.87	3⅞	4	½	⅝	2
2	6	⅝	—	3 1/16	1.00	4¾	4	⅝	¾	2¼
2½	7	11/16	—	3 9/16	1.14	5½	4	⅝	¾	2½
3	7½	¾	—	4¼	1.20	6	4	⅝	¾	2½
3½	8½	13/16	—	4 13/16	1.25	7	8	⅝	¾	2¾
4	9	15/16	—	5 5/16	1.30	7½	8	⅝	¾	3
5	10	15/16	—	6 7/16	1.41	8½	8	¾	⅞	3¼
6	11	1	—	7 9/16	1.51	9½	8	¾	⅞	3¼
8	13½	1⅛	—	9 11/16	1.71	11¾	8	¾	⅞	3½
10	16	1 3/16	—	11 15/16	1.93	14¼	12	⅞	1	3¾
12	19	1¼	12/16	14 1/16	2.13	17	12	⅞	1	3¾
14 O.D.	21	1⅜	⅞	15⅜	2.25	18¾	12	1	1⅛	4¼
16 O.D.	23½	1 7/16	1	17½	2.45	21¼	16	1⅛	1⅛	4½
18 O.D.	25	1 9/16	1 1/16	19⅝	2.65	22¾	16	1⅛	1¼	4¾

All dimensions in inches.

* Extracted from American Standards, "Cast-Iron Pipe Flanges and Flanged Fittings" (ANSI B16.1), with the permission of the publisher, The American Society of Mechanical Engineers.

Appendix 14 AMERICAN NATIONAL STANDARD 125-LB CAST IRON SCREWED FITTINGS*

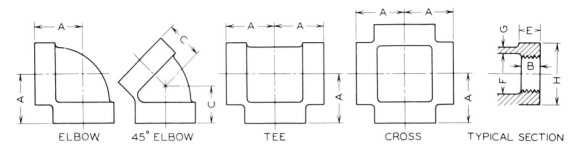

ELBOW   45° ELBOW   TEE   CROSS   TYPICAL SECTION

Nominal Pipe Size	A	C	B Min	E Min	F Min	F Max	G Min	H Min
¼	0.81	0.73	0.32	0.38	0.540	0.584	0.110	0.93
⅜	0.95	0.80	0.36	0.44	0.675	0.719	0.120	1.12
½	1.12	0.88	0.43	0.50	0.840	0.897	0.130	1.34
¾	1.31	0.98	0.50	0.56	1.050	1.107	0.155	1.63
1	1.50	1.12	0.58	0.62	1.315	1.385	0.170	1.95
1¼	1.75	1.29	0.67	0.69	1.660	1.730	0.185	2.39
1½	1.94	1.43	0.70	0.75	1.900	1.970	0.200	2.68
2	2.25	1.68	0.75	0.84	2.375	2.445	0.220	3.28
2½	2.70	1.95	0.92	0.94	2.875	2.975	0.240	3.86
3	3.08	2.17	0.98	1.00	3.500	3.600	0.260	4.62
3½	3.42	2.39	1.03	1.06	4.000	4.100	0.280	5.20
4	3.79	2.61	1.08	1.12	4.500	4.600	0.310	5.79
5	4.50	3.05	1.18	1.18	5.563	5.663	0.380	7.05
6	5.13	3.46	1.28	1.28	6.625	6.725	0.430	8.28
8	6.56	4.28	1.47	1.47	8.625	8.725	0.550	10.63
10	8.08	5.16	1.68	1.68	10.750	10.850	0.690	13.12
12	9.50	5.97	1.88	1.88	12.750	12.850	0.800	15.47
14 O.D.	10.40	—	2.00	2.00	14.000	14.100	0.880	16.94
16 O.D.	11.82	—	2.20	2.20	16.000	16.100	1.000	19.30

All dimensions in inches.
 * Extracted from American National Standards, "Cast-Iron Screwed Fittings, 125- and 250-lb" (ANSI B16.4), with the permission of the publisher, The American Society of Mechanical Engineers.

Appendix 15 AMERICAN NATIONAL STANDARD UNIFIED INCH SCREW THREADS (UN AND UNR THREAD FORM)*

Sizes Primary	Sizes Secondary	Basic Major Diameter	Coarse UNC	Fine UNF	Extra Fine UNEF	4UN	6UN	8UN	12UN	16UN	20UN	28UN	32UN	Sizes
0		0.0600	—	80	—	—	—	—	—	—	—	—	—	0
	1	0.0730	64	72	—	—	—	—	—	—	—	—	—	1
2		0.0860	56	64	—	—	—	—	—	—	—	—	—	2
	3	0.0990	48	56	—	—	—	—	—	—	—	—	—	3
4		0.1120	40	48	—	—	—	—	—	—	—	—	—	4
5		0.1250	40	44	—	—	—	—	—	—	—	—	—	5
6		0.1380	32	40	—	—	—	—	—	—	—	—	UNC	6
8		0.1640	32	36	—	—	—	—	—	—	—	—	UNC	8
10		0.1900	24	32	—	—	—	—	—	—	—	—	UNF	10
	12	0.2160	24	28	32	—	—	—	—	—	—	UNF	UNEF	12
1/4		0.2500	20	28	32	—	—	—	—	—	UNC	UNF	UNEF	1/4
5/16		0.3125	18	24	32	—	—	—	—	—	20	28	UNEF	5/16
3/8		0.3750	16	24	32	—	—	—	—	UNC	20	28	UNEF	3/8
7/16		0.4375	14	20	28	—	—	—	—	16	UNF	UNEF	32	7/16
1/2		0.5000	13	20	28	—	—	—	—	16	UNF	UNEF	32	1/2
9/16		0.5625	12	18	24	—	—	—	UNC	16	20	28	32	9/16
5/8		0.6250	11	18	24	—	—	—	12	16	20	28	32	5/8
	11/16	0.6875	—	—	24	—	—	—	12	16	20	28	32	11/16
3/4		0.7500	10	16	20	—	—	—	12	UNF	UNEF	28	32	3/4
	13/16	0.8125	—	—	20	—	—	—	12	16	UNEF	28	32	13/16
7/8		0.8750	9	14	20	—	—	—	12	16	UNEF	28	32	7/8
	15/16	0.9375	—	—	20	—	—	—	12	16	UNEF	28	32	15/16
1		1.0000	8	12	20	—	—	UNC	UNF	16	UNEF	28	32	1
	1 1/16	1.0625	—	—	18	—	—	8	12	16	20	28	—	1 1/16
1 1/8		1.1250	7	12	18	—	—	8	UNF	16	20	28	—	1 1/8
	1 3/16	1.1875	—	—	18	—	—	8	12	16	20	28	—	1 3/16
1 1/4		1.2500	7	12	18	—	—	8	UNF	16	20	28	—	1 1/4
	1 5/16	1.3125	—	—	18	—	—	8	12	16	20	28	—	1 5/16
1 3/8		1.3750	6	12	18	—	UNC	8	UNF	16	20	28	—	1 3/8
	1 7/16	1.4375	—	—	18	—	6	8	12	16	20	28	—	1 7/16
1 1/2		1.5000	6	12	18	—	UNC	8	UNF	16	20	28	—	1 1/2
	1 9/16	1.5625	—	—	18	—	6	8	12	16	20	—	—	1 9/16
1 5/8		1.6250	—	—	18	—	6	8	12	16	20	—	—	1 5/8
	1 11/16	1.6875	—	—	18	—	6	8	12	16	20	—	—	1 11/16
1 3/4		1.7500	5	—	—	—	6	8	12	16	20	—	—	1 3/4
	1 13/16	1.8125	—	—	—	—	6	8	12	16	20	—	—	1 13/16
1 7/8		1.8750	—	—	—	—	6	8	12	16	20	—	—	1 7/8
	1 15/16	1.9375	—	—	—	—	6	8	12	16	20	—	—	1 15/16
2		2.0000	4½	—	—	—	6	8	12	16	20	—	—	2
	2 1/8	2.1250	—	—	—	—	6	8	12	16	20	—	—	2 1/8
2 1/4		2.2500	4½	—	—	—	6	8	12	16	20	—	—	2 1/4
	2 3/8	2.3750	—	—	—	—	6	8	12	16	20	—	—	2 3/8
2 1/2		2.5000	4	—	—	UNC	6	8	12	16	20	—	—	2 1/2
	2 5/8	2.6250	—	—	—	4	6	8	12	16	20	—	—	2 5/8
2 3/4		2.7500	4	—	—	UNC	6	8	12	16	20	—	—	2 3/4
	2 7/8	2.8750	—	—	—	4	6	8	12	16	20	—	—	2 7/8

*Series designation shown indicates the UN thread form; however, the UNR thread form may be specified by substituting UNR in place of UN in all designations for external use only.

Cont.

Appendix 15 AMERICAN NATIONAL STANDARD UNIFIED INCH
SCREW THREADS (UN AND UNR THREAD FORM)*
(Cont.)

Sizes		Basic Major Diameter	Threads per Inch											Sizes
			Series with Graded Pitches			Series with Constant Pitches								
Primary	Secondary		Coarse UNC	Fine UNF	Extra Fine UNEF	4UN	6UN	8UN	12UN	16UN	20UN	28UN	32UN	
3		3.0000	4	—	—	UNC	6	8	12	16	20	—	—	3
	$3\frac{1}{8}$	3.1250	—	—	—	4	6	8	12	16	—	—	—	$3\frac{1}{8}$
$3\frac{1}{4}$		3.2500	4	—	—	UNC	6	8	12	16	—	—	—	$3\frac{1}{4}$
	$3\frac{3}{8}$	3.3750	—	—	—	4	6	8	12	16	—	—	—	$3\frac{3}{8}$
$3\frac{1}{2}$		3.5000	4	—	—	UNC	6	8	12	16	—	—	—	$3\frac{1}{2}$
	$3\frac{5}{8}$	3.6250	—	—	—	4	6	8	12	16	—	—	—	$3\frac{5}{8}$
$3\frac{3}{4}$		3.7500	4	—	—	UNC	6	8	12	16	—	—	—	$3\frac{3}{4}$
	$3\frac{7}{8}$	3.8750	—	—	—	4	6	8	12	16	—	—	—	$3\frac{7}{8}$
4		4.0000	4	—	—	UNC	6	8	12	16	—	—	—	4
	$4\frac{1}{8}$	4.1250	—	—	—	4	6	8	12	16	—	—	—	$4\frac{1}{8}$
$4\frac{1}{4}$		4.2500	—	—	—	4	6	8	12	16	—	—	—	$4\frac{1}{4}$
	$4\frac{3}{8}$	4.3750	—	—	—	4	6	8	12	16	—	—	—	$4\frac{3}{8}$
$4\frac{1}{2}$		4.5000	—	—	—	4	6	8	12	16	—	—	—	$4\frac{1}{2}$
	$4\frac{5}{8}$	4.6250	—	—	—	4	6	8	12	16	—	—	—	$4\frac{5}{8}$
$4\frac{3}{4}$		4.7500	—	—	—	4	6	8	12	16	—	—	—	$4\frac{3}{4}$
	$4\frac{7}{8}$	4.8750	—	—	—	4	6	8	12	16	—	—	—	$4\frac{7}{8}$
5		5.0000	—	—	—	4	6	8	12	16	—	—	—	5
	$5\frac{1}{8}$	5.1250	—	—	—	4	6	8	12	16	—	—	—	$5\frac{1}{8}$
$5\frac{1}{4}$		5.2500	—	—	—	4	6	8	12	16	—	—	—	$5\frac{1}{4}$
	$5\frac{3}{8}$	5.3750	—	—	—	4	6	8	12	16	—	—	—	$5\frac{3}{8}$
$5\frac{1}{2}$		5.5000	—	—	—	4	6	8	12	16	—	—	—	$5\frac{1}{2}$
	$5\frac{5}{8}$	5.6250	—	—	—	4	6	8	12	16	—	—	—	$5\frac{5}{8}$
$5\frac{3}{4}$		5.7500	—	—	—	4	6	8	12	16	—	—	—	$5\frac{3}{4}$
	$5\frac{7}{8}$	5.8750	—	—	—	4	6	8	12	16	—	—	—	$5\frac{7}{8}$
6		6.0000	—	—	—	4	6	8	12	16	—	—	—	6

(Courtesy of ANSI; B1.1–1974.)

Appendix 16 TAP DRILL SIZES FOR AMERICAN NATIONAL AND
UNIFIED COARSE AND FINE THREADS

$$p = \text{pitch} = \frac{1}{\text{No. thrd. per in.}}$$

$$d = \text{depth} = p \times .649519$$

$$f = \text{flat} = \frac{p}{8}$$

$$\text{pitch diameter} = d - \frac{.6495}{N}$$

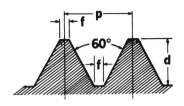

For Nos. 575 and 585 Screw Thread Micrometers

Size	Threads per inch NC UNC	Threads per inch NF UNF	Outside Diameter Inches	Pitch Diameter Inches	Root Diameter Inches	Tap Drill Approx. 75% Full Thread	Decimal Equiv. of Tap Drill
0	..	80	.0600	.0519	.0438	3/64	.0469
1	64	..	.0730	.0629	.0527	53	.0595
1	..	72	.0730	.0640	.0550	53	.0595
2	56	..	.0860	.0744	.0628	50	.0700
2	..	64	.0860	.0759	.0657	50	.0700
3	48	..	.0990	.0855	.0719	47	.0785
3	..	56	.0990	.0874	.0758	46	.0810
4	40	..	.1120	.0958	.0795	43	.0890
4	..	48	.1120	.0985	.0849	42	.0935
5	40	..	.1250	.1088	.0925	38	.1015
5	..	44	.1250	.1102	.0955	37	.1040
6	32	..	.1380	.1177	.0974	36	.1065
6	..	40	.1380	.1218	.1055	33	.1130
8	32	..	.1640	.1437	.1234	29	.1360
8	..	36	.1640	.1460	.1279	29	.1360
10	24	..	.1900	.1629	.1359	26	.1470
10	..	32	.1900	.1697	.1494	21	.1590
12	24	..	.2160	.1889	.1619	16	.1770
12	..	28	.2160	.1928	.1696	15	.1800
1/4	20	..	.2500	.2175	.1850	7	.2010
1/4	..	28	.2500	.2268	.2036	3	.2130
5/16	18	..	.3125	.2764	.2403	F	.2570
5/16	..	24	.3125	.2854	.2584	I	.2720
3/8	16	..	.3750	.3344	.2938	5/16	.3125
3/8	..	24	.3750	.3479	.3209	Q	.3320
7/16	14	..	.4375	.3911	.3447	U	.3680
7/16	..	20	.4375	.4050	.3726	25/64	.3906
1/2	13	..	.5000	.4500	.4001	27/64	.4219
1/2	..	20	.5000	.4675	.4351	29/64	.4531
9/16	12	..	.5625	.5084	.4542	31/64	.4844
9/16	..	18	.5625	.5264	.4903	33/64	.5156
5/8	11	..	.6250	.5660	.5069	17/32	.5312
5/8	..	18	.6250	.5889	.5528	37/64	.5781
3/4	10	..	.7500	.6850	.6201	21/32	.6562
3/4	..	16	.7500	.7094	.6688	11/16	.6875
7/8	9	..	.8750	.8028	.7307	49/64	.7656
7/8	..	14	.8750	.8286	.7822	13/16	.8125

Appendix 16 TAP DRILL SIZES FOR AMERICAN NATIONAL AND UNIFIED COARSE AND FINE THREADS (Cont.)

Size	Threads per inch NC UNC	Threads per inch NF UNF	Outside Diameter Inches	Pitch Diameter Inches	Root Diameter Inches	Tap Drill Approx. 75% Full Thread	Decimal Equiv. of Tap Drill
1	8	..	1.0000	.9188	.8376	$\frac{7}{8}$	.8750
1	..	12	1.0000	.9459	.8917	$\frac{59}{64}$	.9219
1⅛	7	..	1.1250	1.0322	.9394	$\frac{63}{64}$	.9844
1⅛	..	12	1.1250	1.0709	1.0168	$1\frac{3}{64}$	1.0469
1¼	7	..	1.2500	1.1572	1.0644	$1\frac{7}{64}$	1.1094
1¼	..	12	1.2500	1.1959	1.1418	$1\frac{11}{64}$	1.1719
1⅜	6	..	1.3750	1.2667	1.1585	$1\frac{7}{32}$	1.2187
1⅜	..	12	1.3750	1.3209	1.2668	$1\frac{19}{64}$	1.2969
1½	6	..	1.5000	1.3917	1.2835	$1\frac{11}{32}$	1.3437
1½	..	12	1.5000	1.4459	1.3918	$1\frac{27}{64}$	1.4219
1¾	5	..	1.7500	1.6201	1.4902	$1\frac{9}{16}$	1.5625
2	4½	..	2.0000	1.8557	1.7113	$1\frac{25}{32}$	1.7812
2¼	4½	..	2.2500	2.1057	1.9613	$2\frac{1}{32}$	2.0313
2½	4	..	2.5000	2.3376	2.1752	2¼	2.2500
2¾	4	..	2.7500	2.5876	2.4252	2½	2.5000
3	4	..	3.0000	3.8376	2.6752	2¾	2.7500
3¼	4	..	3.2500	3.0876	2.9252	3	3.0000
3½	4	..	3.5000	3.3376	3.1752	3¼	3.2500
3¾	4	..	3.7500	3.5876	3.4252	3½	3.5000
4	4	..	4.0000	3.3786	3.6752	3¾	3.7500

(Courtesy of the L. S. Starrett Company.)

Appendix 17 LENGTH OF THREAD ENGAGEMENT GROUPS

Nominal Size Diam.		Pitch P	Length of Thread Engagement				Nominal Size Diam.		Pitch P	Length of Thread Engagement			
			Group S	Group N		Group L				Group S	Group N		Group L
Over	To and Incl		To and Incl	Over	To and Incl	Over	Over	To and Incl		To and Incl	Over	To and Incl	Over
1.5	2.8	0.2	0.5	0.5	1.5	1.5	22.4	45	1	4	4	12	12
		0.25	0.6	0.6	1.9	1.9			1.5	6.3	6.3	19	19
		0.35	0.8	0.8	2.6	2.6			2	8.5	8.5	25	25
		0.4	1	1	3	3			3	12	12	36	36
		0.45	1.3	1.3	3.8	3.8			3.5	15	15	45	45
									4	18	18	53	53
2.8	5.6	0.35	1	1	3	3			4.5	21	21	63	63
		0.5	1.5	1.5	4.5	4.5	45	90	1.5	7.5	7.5	22	22
		0.6	1.7	1.7	5	5			2	9.5	9.5	28	28
		0.7	2	2	6	6			3	15	15	45	45
		0.75	2.2	2.2	6.7	6.7			4	19	19	56	56
		0.8	2.5	2.5	7.5	7.5			5	24	24	71	71
5.6	11.2	0.75	2.4	2.4	7.1	7.1			5.5	28	28	85	85
		1	3	3	9	9			6	32	32	95	95
		1.25	4	4	12	12	90	180	2	12	12	36	36
		1.5	5	5	15	15			3	18	18	53	53
									4	24	24	71	71
11.2	22.4	1	3.8	3.8	11	11			6	36	36	106	106
		1.25	4.5	4.5	13	13	180	355	3	20	20	60	60
		1.5	5.6	5.6	16	16			4	26	26	80	80
		1.75	6	6	18	18			6	40	40	118	118
		2	8	8	24	24							
		2.5	10	10	30	30							

All dimensions are given in millimeters. (Courtesy of ISO Standards.)

Appendix 18 ISO METRIC SCREW THREAD STANDARD SERIES

Nominal Size Dia. (mm) Column[a]			Pitches (mm) — Series with Graded Pitches		Pitches (mm) — Series with Constant Pitches												Nominal Size Dia. (mm)
1	2	3	Coarse	Fine	6	4	3	2	1.5	1.25	1	0.75	0.5	0.35	0.25	0.2	(mm)
0.25			0.075	—	—	—	—	—	—	—	—	—	—	—	—	—	0.25
0.3			0.08	—	—	—	—	—	—	—	—	—	—	—	—	—	0.3
	0.35		0.09	—	—	—	—	—	—	—	—	—	—	—	—	—	0.35
0.4			0.1	—	—	—	—	—	—	—	—	—	—	—	—	—	0.4
	0.45		0.1	—	—	—	—	—	—	—	—	—	—	—	—	—	0.45
0.5			0.125	—	—	—	—	—	—	—	—	—	—	—	—	—	0.5
	0.55		0.125	—	—	—	—	—	—	—	—	—	—	—	—	—	0.55
0.6			0.15	—	—	—	—	—	—	—	—	—	—	—	—	—	0.6
	0.7		0.175	—	—	—	—	—	—	—	—	—	—	—	—	—	0.7
0.8			0.2	—	—	—	—	—	—	—	—	—	—	—	—	—	0.8
	0.9		0.225	—	—	—	—	—	—	—	—	—	—	—	—	—	0.9
			0.25	—	—	—	—	—	—	—	—	—	—	—	—	0.2	1
	1.1		0.25	—	—	—	—	—	—	—	—	—	—	—	—	0.2	1.1
1.2			0.25	—	—	—	—	—	—	—	—	—	—	—	—	0.2	1.2
	1.4		0.3	—	—	—	—	—	—	—	—	—	—	—	—	0.2	1.4
1.6			0.35	—	—	—	—	—	—	—	—	—	—	—	—	0.2	1.6
	1.8		0.35	—	—	—	—	—	—	—	—	—	—	—	—	0.2	1.8
2			0.4	—	—	—	—	—	—	—	—	—	—	—	0.25	—	2
	2.2		0.45	—	—	—	—	—	—	—	—	—	—	—	0.25	—	2.2
2.5			0.45	—	—	—	—	—	—	—	—	—	—	0.35	—	—	2.5
3			0.5	—	—	—	—	—	—	—	—	—	—	0.35	—	—	3
	3.5		0.6	—	—	—	—	—	—	—	—	—	—	0.35	—	—	3.5
4			0.7	—	—	—	—	—	—	—	—	—	0.5	—	—	—	4
	4.5		0.75	—	—	—	—	—	—	—	—	—	0.5	—	—	—	4.5
5			0.8	—	—	—	—	—	—	—	—	—	0.5	—	—	—	5
		5.5	—	—	—	—	—	—	—	—	—	—	0.5	—	—	—	5.5
6			1	—	—	—	—	—	—	—	—	0.75	—	—	—	—	6
		7	1	—	—	—	—	—	—	—	—	0.75	—	—	—	—	7
8			1.25	1	—	—	—	—	—	—	1	0.75	—	—	—	—	8
		9	1.25	—	—	—	—	—	—	—	1	0.75	—	—	—	—	9
10			1.5	1.25	—	—	—	—	—	1.25	1	0.75	—	—	—	—	10
		11	1.5	—	—	—	—	—	—	—	1	0.75	—	—	—	—	11
12			1.75	1.25	—	—	—	—	1.5	1.25	1	—	—	—	—	—	12
	14		2	1.5	—	—	—	—	1.5	1.25[b]	1	—	—	—	—	—	14
		15	—	1.5	—	—	—	—	1.5	—	1	—	—	—	—	—	15
16			2	1.5	—	—	—	—	1.5	—	1	—	—	—	—	—	16
		17	—	—	—	—	—	—	1.5	—	1	—	—	—	—	—	17
	18		2.5	1.5	—	—	—	2	1.5	—	1	—	—	—	—	—	18
20			2.5	1.5	—	—	—	2	1.5	—	1	—	—	—	—	—	20
	22		2.5	1.5	—	—	—	2	1.5	—	1	—	—	—	—	—	22

[a] Thread diameter should be selected from columns 1, 2 or 3, with preference being in that order.

[b] Pitch 1.25 mm in combination with diameter 14 mm has been included for sparkplug applications.

[c] Diameter 35 mm has been included for bearing locknut applications.

The use of pitches shown in parentheses should be avoided wherever possible.

The pitches enclosed in the bold frame, together with the corresponding nominal diameters in columns 1 and 2, are those combinations which have been established by ISO Recommendations as a selected "coarse" and "fine" series for commercial fasteners.

Appendix 18 ISO METRIC SCREW THREAD STANDARD SERIES (Cont.)

Nominal Size Dia. (mm)			Pitches (mm)													Nominal Size Dia. (mm)
Column[a]			Series with Graded Pitches		Series with Constant Pitches											
1	2	3	Coarse	Fine	6	4	3	2	1.5	1.25	1	0.75	0.5	0.35	0.25	0.2	
24			3	2	—	—	—	2	1.5	—	1	—	—	—	—	—	24
		25	—	—	—	—	—	2	1.5	—	1	—	—	—	—	—	25
		26	—	—	—	—	—	—	1.5	—	1	—	—	—	—	—	26
	27		3	2	—	—	—	2	1.5	—	1	—	—	—	—	—	27
		28	—	—	—	—	—	2	1.5	—	1	—	—	—	—	—	28
30			3.5	2	—	—	(3)	2	1.5	—	1	—	—	—	—	—	30
		32	—	—	—	—	—	2	1.5	—	—	—	—	—	—	—	32
	33		3.5	2	—	—	(3)	2	1.5	—	—	—	—	—	—	—	33
		35[c]	—	—	—	—	—	—	1.5	—	—	—	—	—	—	—	35[c]
36			4	3	—	—	—	2	1.5	—	—	—	—	—	—	—	36
		38	—	—	—	—	—	—	1.5	—	—	—	—	—	—	—	38
	39		4	3	—	—	—	2	1.5	—	—	—	—	—	—	—	39
		40	—	—	—	—	3	2	1.5	—	—	—	—	—	—	—	40
42			4.5	3	—	4	3	2	1.5	—	—	—	—	—	—	—	42
	45		4.5	3	—	4	3	2	1.5	—	—	—	—	—	—	—	45
48			5	3	—	4	3	2	1.5	—	—	—	—	—	—	—	48
		50	—	—	—	—	3	2	1.5	—	—	—	—	—	—	—	50
	52		5	3	—	4	3	2	1.5	—	—	—	—	—	—	—	52
		55	—	—	—	4	3	2	1.5	—	—	—	—	—	—	—	55
56			5.5	4	—	4	3	2	1.5	—	—	—	—	—	—	—	56
		58	—	—	—	4	3	2	1.5	—	—	—	—	—	—	—	58
	60		5.5	4	—	4	3	2	1.5	—	—	—	—	—	—	—	60
		62	—	—	—	4	3	2	1.5	—	—	—	—	—	—	—	62
64			6	4	—	4	3	2	1.5	—	—	—	—	—	—	—	64
		65	—	—	—	4	3	2	1.5	—	—	—	—	—	—	—	65
	68		6	4	—	4	3	2	1.5	—	—	—	—	—	—	—	68
		70	—	—	6	4	3	2	1.5	—	—	—	—	—	—	—	70
72			—	—	6	4	3	2	1.5	—	—	—	—	—	—	—	72
		75	—	—	—	4	3	2	1.5	—	—	—	—	—	—	—	75
	76		—	—	6	4	3	2	1.5	—	—	—	—	—	—	—	76
		78	—	—	—	—	—	2	—	—	—	—	—	—	—	—	78
80			—	—	6	4	3	2	1.5	—	—	—	—	—	—	—	80
		82	—	—	—	—	—	2	—	—	—	—	—	—	—	—	82
	85		—	—	6	4	3	2	—	—	—	—	—	—	—	—	85
90			—	—	6	4	3	2	—	—	—	—	—	—	—	—	90

Cont.

Appendix 18 ISO METRIC SCREW THREAD STANDARD SERIES (Cont.)

Nominal Size Dia. (mm) Column[a]			Series with Graded Pitches		Pitches (mm) Series with Constant Pitches												Nominal Size Dia. (mm)
1	2	3	Coarse	Fine	6	4	3	2	1.5	1.25	1	0.75	0.5	0.35	0.25	0.2	
	95		—	—	6	4	3	2	—	—	—	—	—	—	—	—	95
100			—	—	6	4	3	2	—	—	—	—	—	—	—	—	100
	105		—	—	6	4	3	2	—	—	—	—	—	—	—	—	105
110			—	—	6	4	3	2	—	—	—	—	—	—	—	—	110
	115		—	—	6	4	3	2	—	—	—	—	—	—	—	—	115
	120		—	—	6	4	3	2	—	—	—	—	—	—	—	—	120
125			—	—	6	4	3	2	—	—	—	—	—	—	—	—	125
	130		—	—	6	4	3	2	—	—	—	—	—	—	—	—	130
		135	—	—	6	4	3	2	—	—	—	—	—	—	—	—	135
140			—	—	6	4	3	2	—	—	—	—	—	—	—	—	140
		145	—	—	6	4	3	2	—	—	—	—	—	—	—	—	145
	150		—	—	6	4	3	2	—	—	—	—	—	—	—	—	150
		155	—	—	6	4	3	—	—	—	—	—	—	—	—	—	155
160			—	—	6	4	3	—	—	—	—	—	—	—	—	—	160
		165	—	—	6	4	3	—	—	—	—	—	—	—	—	—	165
	170		—	—	6	4	3	—	—	—	—	—	—	—	—	—	170
		175	—	—	6	4	3	—	—	—	—	—	—	—	—	—	175
180			—	—	6	4	3	—	—	—	—	—	—	—	—	—	180
		185	—	—	6	4	3	—	—	—	—	—	—	—	—	—	185
	190		—	—	6	4	3	—	—	—	—	—	—	—	—	—	190
		195	—	—	6	4	3	—	—	—	—	—	—	—	—	—	195
200			—	—	6	4	3	—	—	—	—	—	—	—	—	—	200
		205	—	—	6	4	3	—	—	—	—	—	—	—	—	—	205
	210		—	—	6	4	3	—	—	—	—	—	—	—	—	—	210
220			—	—	6	4	3	—	—	—	—	—	—	—	—	—	220
		225	—	—	6	4	3	—	—	—	—	—	—	—	—	—	225
		230	—	—	6	4	3	—	—	—	—	—	—	—	—	—	230
		235	—	—	6	4	3	—	—	—	—	—	—	—	—	—	235
	240		—	—	6	4	3	—	—	—	—	—	—	—	—	—	240
		245	—	—	6	4	3	—	—	—	—	—	—	—	—	—	245
250			—	—	6	4	3	—	—	—	—	—	—	—	—	—	250
		255	—	—	6	4	—	—	—	—	—	—	—	—	—	—	255
	260		—	—	6	4	—	—	—	—	—	—	—	—	—	—	260
		265	—	—	6	4	—	—	—	—	—	—	—	—	—	—	265
		270	—	—	6	4	—	—	—	—	—	—	—	—	—	—	270
		275	—	—	6	4	—	—	—	—	—	—	—	—	—	—	275
280			—	—	6	4	—	—	—	—	—	—	—	—	—	—	280
		285	—	—	6	4	—	—	—	—	—	—	—	—	—	—	285
		290	—	—	6	4	—	—	—	—	—	—	—	—	—	—	290
		295	—	—	6	4	—	—	—	—	—	—	—	—	—	—	295
	300		—	—	6	4	—	—	—	—	—	—	—	—	—	—	300

[a] Thread diameter should be selected from columns 1, 2, or 3; with preference being in that order.

Appendix 19 SQUARE AND ACME THREADS

Size	Threads per Inch	Size	Threads per Inch
$\frac{3}{8}$	12	2	$2\frac{1}{2}$
$\frac{7}{16}$	10	$2\frac{1}{4}$	2
$\frac{1}{2}$	10	$2\frac{1}{2}$	2
$\frac{9}{16}$	8	$2\frac{3}{4}$	2
$\frac{5}{8}$	8	3	$1\frac{1}{2}$
$\frac{3}{4}$	6	$3\frac{1}{4}$	$1\frac{1}{2}$
$\frac{7}{8}$	5	$3\frac{1}{2}$	$1\frac{1}{3}$
1	5	$3\frac{3}{4}$	$1\frac{1}{3}$
$1\frac{1}{8}$	4	4	$1\frac{1}{3}$
$1\frac{1}{4}$	4	$4\frac{1}{4}$	$1\frac{1}{3}$
$1\frac{1}{2}$	3	$4\frac{1}{2}$	1
$1\frac{3}{4}$	$2\frac{1}{2}$	over $4\frac{1}{2}$	1

Appendix 20 AMERICAN STANDARD SQUARE BOLTS AND NUTS

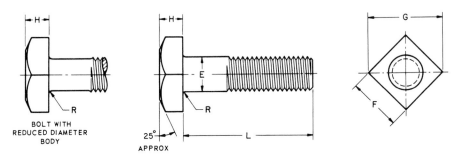

BOLT WITH
REDUCED DIAMETER
BODY

25°
APPROX

Dimensions of Square Bolts

Nominal Size or Basic Product Dia		Body Dia E	Width Across Flats F			Width Across Corners G		Height H			Radius of Fillet R
		Max	Basic	Max	Min	Max	Min	Basic	Max	Min	Max
1/4	0.2500	0.260	3/8	0.3750	0.362	0.530	0.498	11/64	0.188	0.156	0.031
5/16	0.3125	0.324	1/2	0.5000	0.484	0.707	0.665	13/64	0.220	0.186	0.031
3/8	0.3750	0.388	9/16	0.5625	0.544	0.795	0.747	1/4	0.268	0.232	0.031
7/16	0.4375	0.452	5/8	0.6250	0.603	0.884	0.828	19/64	0.316	0.278	0.031
1/2	0.5000	0.515	3/4	0.7500	0.725	1.061	0.995	21/64	0.348	0.308	0.031
5/8	0.6250	0.642	15/16	0.9375	0.906	1.326	1.244	27/64	0.444	0.400	0.062
3/4	0.7500	0.768	1 1/8	1.1250	1.088	1.591	1.494	1/2	0.524	0.476	0.062
7/8	0.8750	0.895	1 5/16	1.3125	1.269	1.856	1.742	19/32	0.620	0.568	0.062
1	1.0000	1.022	1 1/2	1.5000	1.450	2.121	1.991	21/32	0.684	0.628	0.093
1 1/8	1.1250	1.149	1 11/16	1.6875	1.631	2.386	2.239	3/4	0.780	0.720	0.093
1 1/4	1.2500	1.277	1 7/8	1.8750	1.812	2.652	2.489	27/32	0.876	0.812	0.093
1 3/8	1.3750	1.404	2 1/16	2.0625	1.994	2.917	2.738	29/32	0.940	0.872	0.093
1 1/2	1.5000	1.531	2 1/4	2.2500	2.175	3.182	2.986	1	1.036	0.964	0.093

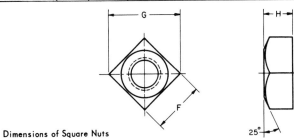

25°

Dimensions of Square Nuts

Nominal Size or Basic Major Dia of Thread		Width Across Flats F			Width Across Corners G		Thickness H		
		Basic	Max	Min	Max	Min	Basic	Max	Min
1/4	0.2500	7/16	0.4375	0.425	0.619	0.584	7/32	0.235	0.203
5/16	0.3125	9/16	0.5625	0.547	0.795	0.751	17/64	0.283	0.249
3/8	0.3750	5/8	0.6250	0.606	0.884	0.832	21/64	0.346	0.310
7/16	0.4375	3/4	0.7500	0.728	1.061	1.000	3/8	0.394	0.356
1/2	0.5000	13/16	0.8125	0.788	1.149	1.082	7/16	0.458	0.418
5/8	0.6250	1	1.0000	0.969	1.414	1.330	35/64	0.569	0.525
3/4	0.7500	1 1/8	1.1250	1.088	1.591	1.494	21/32	0.680	0.632
7/8	0.8750	1 5/16	1.3125	1.269	1.856	1.742	49/64	0.792	0.740
1	1.0000	1 1/2	1.5000	1.450	2.121	1.991	7/8	0.903	0.847
1 1/8	1.1250	1 11/16	1.6875	1.631	2.386	2.239	1	1.030	0.970
1 1/4	1.2500	1 7/8	1.8750	1.812	2.652	2.489	1 3/32	1.126	1.062
1 3/8	1.3750	2 1/16	2.0625	1.994	2.917	2.738	1 13/64	1.237	1.169
1 1/2	1.5000	2 1/4	2.2500	2.175	3.182	2.986	1 5/16	1.348	1.276

(Courtesy of ANSI; B18.2.1–1965 and ANSI; B18.2.2–1965.)

Appendix 21 AMERICAN STANDARD HEXAGON HEAD BOLTS AND NUTS

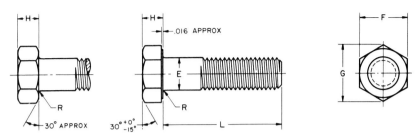

Dimensions of Hex Cap Screws (Finished Hex Bolts)

Nominal Size or Basic Product Dia		Body Dia E		Width Across Flats F			Width Across Corners G		Height H			Radius of Fillet R	
		Max	Min	Basic	Max	Min	Max	Min	Basic	Max	Min	Max	Min
1/4	0.2500	0.2500	0.2450	7/16	0.4375	0.428	0.505	0.488	5/32	0.163	0.150	0.025	0.015
5/16	0.3125	0.3125	0.3065	1/2	0.5000	0.489	0.577	0.557	13/64	0.211	0.195	0.025	0.015
3/8	0.3750	0.3750	0.3690	9/16	0.5625	0.551	0.650	0.628	15/64	0.243	0.226	0.025	0.015
7/16	0.4375	0.4375	0.4305	5/8	0.6250	0.612	0.722	0.698	9/32	0.291	0.272	0.025	0.015
1/2	0.5000	0.5000	0.4930	3/4	0.7500	0.736	0.866	0.840	5/16	0.323	0.302	0.025	0.015
9/16	0.5625	0.5625	0.5545	13/16	0.8125	0.798	0.938	0.910	23/64	0.371	0.348	0.045	0.020
5/8	0.6250	0.6250	0.6170	15/16	0.9375	0.922	1.083	1.051	25/64	0.403	0.378	0.045	0.020
3/4	0.7500	0.7500	0.7410	1 1/8	1.1250	1.100	1.299	1.254	15/32	0.483	0.455	0.045	0.020
7/8	0.8750	0.8750	0.8660	1 5/16	1.3125	1.285	1.516	1.465	35/64	0.563	0.531	0.065	0.040
1	1.0000	1.0000	0.9900	1 1/2	1.5000	1.469	1.732	1.675	39/64	0.627	0.591	0.095	0.060
1 1/8	1.1250	1.1250	1.1140	1 11/16	1.6875	1.631	1.949	1.859	11/16	0.718	0.658	0.095	0.060
1 1/4	1.2500	1.2500	1.2390	1 7/8	1.8750	1.812	2.165	2.066	25/32	0.813	0.749	0.095	0.060
1 3/8	1.3750	1.3750	1.3630	2 1/16	2.0625	1.994	2.382	2.273	27/32	0.878	0.810	0.095	0.060
1 1/2	1.5000	1.5000	1.4880	2 1/4	2.2500	2.175	2.598	2.480	15/16	0.974	0.902	0.095	0.060
1 3/4	1.7500	1.7500	1.7380	2 5/8	2.6250	2.538	3.031	2.893	1 3/32	1.134	1.054	0.095	0.060
2	2.0000	2.0000	1.9880	3	3.0000	2.900	3.464	3.306	1 7/32	1.263	1.175	0.095	0.060
2 1/4	2.2500	2.2500	2.2380	3 3/8	3.3750	3.262	3.897	3.719	1 3/8	1.423	1.327	0.095	0.060
2 1/2	2.5000	2.5000	2.4880	3 3/4	3.7500	3.625	4.330	4.133	1 17/32	1.583	1.479	0.095	0.060
2 3/4	2.7500	2.7500	2.7380	4 1/8	4.1250	3.988	4.763	4.546	1 11/16	1.744	1.632	0.095	0.060
3	3.0000	3.0000	2.9880	4 1/2	4.5000	4.350	5.196	4.959	1 7/8	1.935	1.815	0.095	0.060

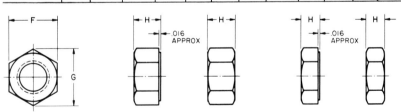

Dimensions of Hex Nuts and Hex Jam Nuts

Nominal Size or Basic Major Dia of Thread		Width Across Flats F			Width Across Corners G		Thickness Hex Nuts H			Thickness Hex Jam Nuts H		
		Basic	Max	Min	Max	Min	Basic	Max	Min	Basic	Max	Min
1/4	0.2500	7/16	0.4375	0.428	0.505	0.488	7/32	0.226	0.212	5/32	0.163	0.150
5/16	0.3125	1/2	0.5000	0.489	0.577	0.557	17/64	0.273	0.258	3/16	0.195	0.180
3/8	0.3750	9/16	0.5625	0.551	0.650	0.628	21/64	0.337	0.320	7/32	0.227	0.210
7/16	0.4375	11/16	0.6875	0.675	0.794	0.768	3/8	0.385	0.365	1/4	0.260	0.240
1/2	0.5000	3/4	0.7500	0.736	0.866	0.840	7/16	0.448	0.427	5/16	0.323	0.302
9/16	0.5625	7/8	0.8750	0.861	1.010	0.982	31/64	0.496	0.473	5/16	0.324	0.301
5/8	0.6250	15/16	0.9375	0.922	1.083	1.051	35/64	0.559	0.535	3/8	0.387	0.363
3/4	0.7500	1 1/8	1.1250	1.088	1.299	1.240	41/64	0.665	0.617	27/64	0.446	0.398
7/8	0.8750	1 5/16	1.3125	1.269	1.516	1.447	3/4	0.776	0.724	31/64	0.510	0.458
1	1.0000	1 1/2	1.5000	1.450	1.732	1.653	55/64	0.887	0.831	35/64	0.575	0.519
1 1/8	1.1250	1 11/16	1.6875	1.631	1.949	1.859	31/32	0.999	0.939	39/64	0.639	0.579
1 1/4	1.2500	1 7/8	1.8750	1.812	2.165	2.066	1 1/16	1.094	1.030	23/32	0.751	0.687
1 3/8	1.3750	2 1/16	2.0625	1.994	2.382	2.273	1 11/64	1.206	1.138	25/32	0.815	0.747
1 1/2	1.5000	2 1/4	2.2500	2.175	2.598	2.480	1 9/32	1.317	1.245	27/32	0.880	0.808

(Courtesy of ANSI; B18.2.1–1965 and ANSI; B18.2.2–1965.)

Appendix 22 FILLISTER HEAD AND ROUND HEAD CAP SCREWS

Fillister Head Cap Screws

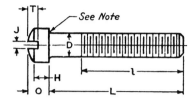

Nominal Size	D Body Diameter		A Head Diameter		H Height of Head		O Total Height of Head		J Width of Slot		T Depth of Slot	
	Max	Min	Max	Min	Max	Min	Max	Min	Max	Min	Max	Min
1/4	0.250	0.245	0.375	0.363	0.172	0.157	0.216	0.194	0.075	0.064	0.097	0.077
5/16	0.3125	0.307	0.437	0.424	0.203	0.186	0.253	0.230	0.084	0.072	0.115	0.090
3/8	0.375	0.369	0.562	0.547	0.250	0.229	0.314	0.284	0.094	0.081	0.142	0.112
7/16	0.4375	0.431	0.625	0.608	0.297	0.274	0.368	0.336	0.094	0.081	0.168	0.133
1/2	0.500	0.493	0.750	0.731	0.328	0.301	0.413	0.376	0.106	0.091	0.193	0.153
9/16	0.5625	0.555	0.812	0.792	0.375	0.346	0.467	0.427	0.118	0.102	0.213	0.168
5/8	0.625	0.617	0.875	0.853	0.422	0.391	0.521	0.478	0.133	0.116	0.239	0.189
3/4	0.750	0.742	1.000	0.976	0.500	0.466	0.612	0.566	0.149	0.131	0.283	0.223
7/8	0.875	0.866	1.125	1.098	0.594	0.556	0.720	0.668	0.167	0.147	0.334	0.264
1	1.000	0.990	1.312	1.282	0.656	0.612	0.803	0.743	0.188	0.166	0.371	0.291

All dimensions are given in inches.

The radius of the fillet at the base of the head:
    For sizes 1/4 to 3/8 in. incl. is 0.016 min and 0.031 max,
        7/16 to 9/16 in. incl. is 0.016 min and 0.047 max,
        5/8 to 1 in. incl. is 0.031 min and 0.062 max.

Round Head Cap Screws

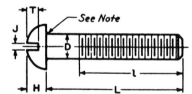

Nominal Size	D Body Diameter		A Head Diameter		H Height of Head		J Width of Slot		T Depth of Slot	
	Max	Min	Max	Min	Max	Min	Max	Min	Max	Min
1/4	0.250	0.245	0.437	0.418	0.191	0.175	0.075	0.064	0.117	0.097
5/16	0.3125	0.307	0.562	0.540	0.245	0.226	0.084	0.072	0.151	0.126
3/8	0.375	0.369	0.625	0.603	0.273	0.252	0.094	0.081	0.168	0.138
7/16	0.4375	0.431	0.750	0.725	0.328	0.302	0.094	0.081	0.202	0.167
1/2	0.500	0.493	0.812	0.786	0.354	0.327	0.106	0.091	0.218	0.178
9/16	0.5625	0.555	0.937	0.909	0.409	0.378	0.118	0.102	0.252	0.207
5/8	0.625	0.617	1.000	0.970	0.437	0.405	0.133	0.116	0.270	0.220
3/4	0.750	0.742	1.250	1.215	0.546	0.507	0.149	0.131	0.338	0.278

All dimensions are given in inches.

Radius of the fillet at the base of the head:
    For sizes 1/4 to 3/8 in. incl. is 0.016 min and 0.031 max,
        7/16 to 9/16 in. incl. is 0.016 min and 0.047 max,
        5/8 to 1 in. incl. is 0.031 min and 0.062 max.

(Courtesy of ANSI; B18.6.2–1956.)

Appendix 23 FLAT HEAD CAP SCREWS

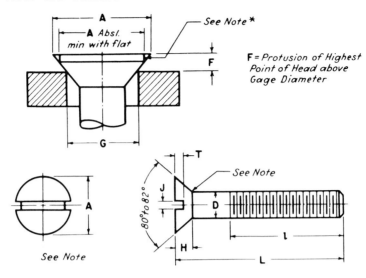

	D		A			G	H	J		T		F	
	Body Diameter		Head Diameter			Gaging Diameter	Height of Head	Width of Slot		Depth of Slot		Protrusion Above Gaging Diameter	
Nominal Size	Max	Min	Max	Min	Absolute Min with Flat		Average	Max	Min	Max	Min	Max	Min
1/4	0.250	0.245	0.500	0.477	0.452	0.4245	0.140	0.075	0.064	0.068	0.045	0.0452	0.0307
5/16	0.3125	0.307	0.625	0.598	0.567	0.5376	0.177	0.084	0.072	0.086	0.057	0.0523	0.0354
3/8	0.375	0.369	0.750	0.720	0.682	0.6507	0.210	0.094	0.081	0.103	0.068	0.0594	0.0401
7/16	0.4375	0.431	0.8125	0.780	0.736	0.7229	0.210	0.094	0.081	0.103	0.068	0.0649	0.0448
1/2	0.500	0.493	0.875	0.841	0.791	0.7560	0.210	0.106	0.091	0.103	0.068	0.0705	0.0495
9/16	0.5625	0.555	1.000	0.962	0.906	0.8691	0.244	0.118	0.102	0.120	0.080	0.0775	0.0542
5/8	0.625	0.617	1.125	1.083	1.020	0.9822	0.281	0.133	0.116	0.137	0.091	0.0846	0.0588
3/4	0.750	0.742	1.375	1.326	1.251	1.2085	0.352	0.149	0.131	0.171	0.115	0.0987	0.0682
7/8	0.875	0.866	1.625	1.568	1.480	1.4347	0.423	0.167	0.147	0.206	0.138	0.1128	0.0776
1	1.000	0.990	1.875	1.811	1.711	1.6610	0.494	0.188	0.166	0.240	0.162	0.1270	0.0870
1 1/8	1.125	1.114	2.062	1.992	1.880	1.8262	0.529	0.196	0.178	0.257	0.173	0.1401	0.0964
1 1/4	1.250	1.239	2.312	2.235	2.110	2.0525	0.600	0.211	0.193	0.291	0.197	0.1542	0.1056
1 3/8	1.375	1.363	2.562	2.477	2.340	2.2787	0.665	0.226	0.208	0.326	0.220	0.1684	0.1151
1 1/2	1.500	1.488	2.812	2.720	2.570	2.5050	0.742	0.258	0.240	0.360	0.244	0.1825	0.1245

All dimensions are given in inches.

The maximum and minimum head diameters, A, are extended to the theoretical sharp corners.

The radius of the fillet at the base of the head shall not exceed 0.4 Max. D.

*Edge of head may be flat as shown or slightly rounded.

(Courtesy of ANSI; B18.6.2–1956.)

Appendix 24 MACHINE SCREWS

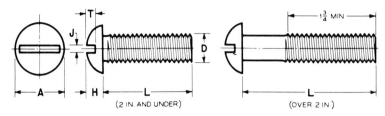

Dimensions of Slotted Round Head Machine Screws

Nom-inal Size	D Diameter of Screw	A Head Diameter		H Head Height		J Width of Slot		T Depth of Slot	
	Basic	Max	Min	Max	Min	Max	Min	Max	Min
0	0.0600	0.113	0.099	0.053	0.043	0.023	0.016	0.039	0.029
1	0.0730	0.138	0.122	0.061	0.051	0.026	0.019	0.044	0.033
2	0.0860	0.162	0.146	0.069	0.059	0.031	0.023	0.048	0.037
3	0.0990	0.187	0.169	0.078	0.067	0.035	0.027	0.053	0.040
4	0.1120	0.211	0.193	0.086	0.075	0.039	0.031	0.058	0.044
5	0.1250	0.236	0.217	0.095	0.083	0.043	0.035	0.063	0.047
6	0.1380	0.260	0.240	0.103	0.091	0.048	0.039	0.068	0.051
8	0.1640	0.309	0.287	0.120	0.107	0.054	0.045	0.077	0.058
10	0.1900	0.359	0.334	0.137	0.123	0.060	0.050	0.087	0.065
12	0.2160	0.408	0.382	0.153	0.139	0.067	0.056	0.096	0.073
1/4	0.2500	0.472	0.443	0.175	0.160	0.075	0.064	0.109	0.082
5/16	0.3125	0.590	0.557	0.216	0.198	0.084	0.072	0.132	0.099
3/8	0.3750	0.708	0.670	0.256	0.237	0.094	0.081	0.155	0.117
7/16	0.4375	0.750	0.707	0.328	0.307	0.094	0.081	0.196	0.148
1/2	0.5000	0.813	0.766	0.355	0.332	0.106	0.091	0.211	0.159
9/16	0.5625	0.938	0.887	0.410	0.385	0.118	0.102	0.242	0.183
5/8	0.6250	1.000	0.944	0.438	0.411	0.133	0.116	0.258	0.195
3/4	0.7500	1.250	1.185	0.547	0.516	0.149	0.131	0.320	0.242

All dimensions are given in inches.

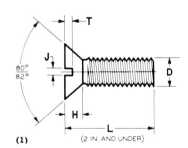

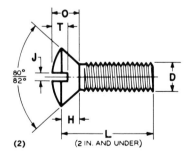

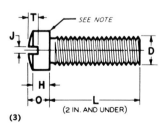

(1)   (2)   (3)

Three other common forms of machine screws are shown above: (1) flat head, (2) oval head, and (3) fillister head. Although dimension tables are not given for these three types of machine screws in this text, their general dimensions are closely related to those shown in the table above. Additional information on these screws can be obtained from ANSI; B18.6.3–1962.

(Courtesy of ANSI; B18.6.3–1962.)

Appendix 25 AMERICAN STANDARD MACHINE SCREWS

(The proportions of the screws can be found by multiplying the major diameter, D, by the factors given below.)

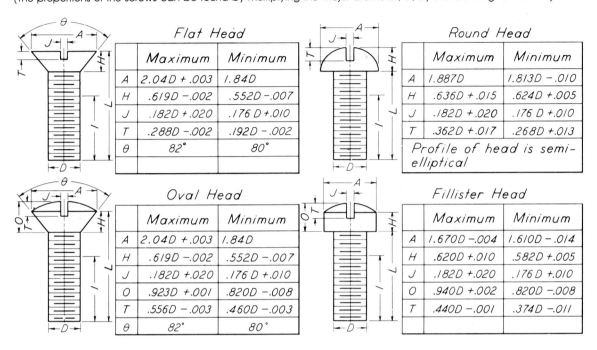

Flat Head

	Maximum	Minimum
A	2.04D +.003	1.84D
H	.619D −.002	.552D −.007
J	.182D +.020	.176D +.010
T	.288D −.002	.192D −.002
θ	82°	80°

Round Head

	Maximum	Minimum
A	1.887D	1.813D −.010
H	.636D +.015	.624D +.005
J	.182D +.020	.176D +.010
T	.362D +.017	.268D +.013

Profile of head is semi-elliptical

Oval Head

	Maximum	Minimum
A	2.04D +.003	1.84D
H	.619D −.002	.552D −.007
J	.182D +.020	.176D +.010
O	.923D +.001	.820D −.008
T	.556D −.003	.460D −.003
θ	82°	80°

Fillister Head

	Maximum	Minimum
A	1.670D −.004	1.610D −.014
H	.620D +.010	.582D +.005
J	.182D +.020	.176D +.010
O	.940D +.002	.820D −.008
T	.440D −.001	.374D −.011

Appendix 26 AMERICAN STANDARD MACHINE TAPERS*

No. of Taper	Taper per Foot (Basic)	Origin of Series	No. of Taper	Taper per Foot (Basic)	Origin of Series	No. of Taper	Taper per Foot (Basic)	Origin of Series	No. of Taper	Taper per Foot (Basic)	Origin of Series
0.239	0.50200	Brown & Sharpe	*	0.62326	Morse	250	0.750	¾ in. per ft.	600	0.750	¾ in. per ft.
.299	.50200	Brown & Sharpe	4½	.62400	Morse	300	.750	¾ in. per ft.	800	0.750	¾ in. per ft.
.375	.50200	Brown & Sharpe	5	.63151	Morse	350	.750	¾ in. per ft.	1000	0.750	¾ in. per ft.
1	.59858	Morse	6	.62565	Morse	400	.750	¾ in. per ft.	1200	0.750	¾ in. per ft.
2	.59941	Morse	7	.62400	Morse	450	.750	¾ in. per ft.			
3	.60235	Morse	200	.750	¾ in. per ft.	500	.750	¾ in. per ft.			

All dimensions in inches.

* Extracted from American Standards, "Machine Tapers, Self-Holding and Steep Taper Series" (ASA B5,10-1960), with the permission of the publisher, The American Society of Mechanical Engineers.

Appendix 27 AMERICAN NATIONAL STANDARD SQUARE HEAD SET
SCREWS (ANSI B18.6.2)

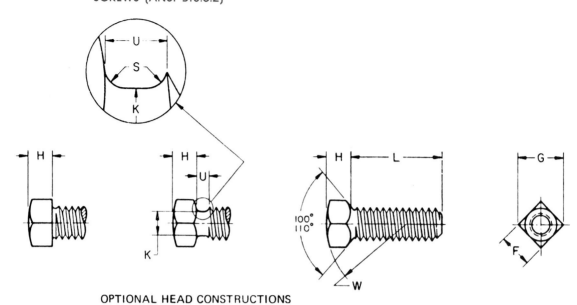

OPTIONAL HEAD CONSTRUCTIONS

Nominal Size[1] or Basic Screw Diameter		F Width Across Flats		G Width Across Corners		H Head Height		K Neck Relief Diameter		S Neck Relief Fillet Radius	U Neck Relief Width	W Head Radius
		Max	Min	Max	Min	Max	Min	Max	Min	Max	Min	Min
10	0.1900	0.188	0.180	0.265	0.247	0.148	0.134	0.145	0.140	0.027	0.083	0.48
1/4	0.2500	0.250	0.241	0.354	0.331	0.196	0.178	0.185	0.170	0.032	0.100	0.62
5/16	0.3125	0.312	0.302	0.442	0.415	0.245	0.224	0.240	0.225	0.036	0.111	0.78
3/8	0.3750	0.375	0.362	0.530	0.497	0.293	0.270	0.294	0.279	0.041	0.125	0.94
7/16	0.4375	0.438	0.423	0.619	0.581	0.341	0.315	0.345	0.330	0.046	0.143	1.09
1/2	0.5000	0.500	0.484	0.707	0.665	0.389	0.361	0.400	0.385	0.050	0.154	1.25
9/16	0.5625	0.562	0.545	0.795	0.748	0.437	0.407	0.454	0.439	0.054	0.167	1.41
5/8	0.6250	0.625	0.606	0.884	0.833	0.485	0.452	0.507	0.492	0.059	0.182	1.56
3/4	0.7500	0.750	0.729	1.060	1.001	0.582	0.544	0.620	0.605	0.065	0.200	1.88
7/8	0.8750	0.875	0.852	1.237	1.170	0.678	0.635	0.731	0.716	0.072	0.222	2.19
1	1.0000	1.000	0.974	1.414	1.337	0.774	0.726	0.838	0.823	0.081	0.250	2.50
1 1/8	1.1250	1.125	1.096	1.591	1.505	0.870	0.817	0.939	0.914	0.092	0.283	2.81
1 1/4	1.2500	1.250	1.219	1.768	1.674	0.966	0.908	1.064	1.039	0.092	0.283	3.12
1 3/8	1.3750	1.375	1.342	1.945	1.843	1.063	1.000	1.159	1.134	0.109	0.333	3.44
1 1/2	1.5000	1.500	1.464	2.121	2.010	1.159	1.091	1.284	1.259	0.109	0.333	3.75

[1] Where specifying nominal size in decimals, zeros preceding decimal and in the fourth decimal place shall be omitted.

Appendix 28 AMERICAN NATIONAL STANDARD POINTS FOR
SQUARE HEAD SET SCREWS (ANSI B18.6.2)

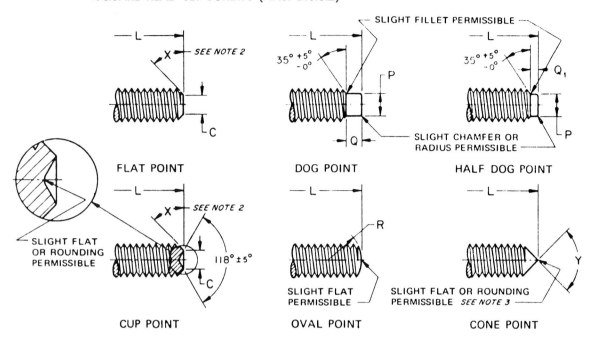

FLAT POINT | DOG POINT | HALF DOG POINT

CUP POINT | OVAL POINT | CONE POINT

Nominal Size[1] or Basic Screw Diameter		C		P		Q		Q₁		R	Y
		Cup and Flat Point Diameters		Dog and Half Dog Point Diameters		Point Length				Oval Point Radius	Cone Point Angle 90° ±2° For These Nominal Lengths or Longer; 118° ±2° For Shorter Screws
						Dog		Half Dog		+0.031 −0.000	
		Max	Min	Max	Min	Max	Min	Max	Min		
10	0.1900	0.102	0.088	0.127	0.120	0.095	0.085	0.050	0.040	0.142	1/4
1/4	0.2500	0.132	0.118	0.156	0.149	0.130	0.120	0.068	0.058	0.188	5/16
5/16	0.3125	0.172	0.156	0.203	0.195	0.161	0.151	0.083	0.073	0.234	3/8
3/8	0.3750	0.212	0.194	0.250	0.241	0.193	0.183	0.099	0.089	0.281	7/16
7/16	0.4375	0.252	0.232	0.297	0.287	0.224	0.214	0.114	0.104	0.328	1/2
1/2	0.5000	0.291	0.270	0.344	0.334	0.255	0.245	0.130	0.120	0.375	9/16
9/16	0.5625	0.332	0.309	0.391	0.379	0.287	0.275	0.146	0.134	0.422	5/8
5/8	0.6250	0.371	0.347	0.469	0.456	0.321	0.305	0.164	0.148	0.469	3/4
3/4	0.7500	0.450	0.425	0.562	0.549	0.383	0.367	0.196	0.180	0.562	7/8
7/8	0.8750	0.530	0.502	0.656	0.642	0.446	0.430	0.227	0.211	0.656	1
1	1.0000	0.609	0.579	0.750	0.734	0.510	0.490	0.260	0.240	0.750	1 1/8
1 1/8	1.1250	0.689	0.655	0.844	0.826	0.572	0.552	0.291	0.271	0.844	1 1/4
1 1/4	1.2500	0.767	0.733	0.938	0.920	0.635	0.615	0.323	0.303	0.938	1 1/2
1 3/8	1.3750	0.848	0.808	1.031	1.011	0.698	0.678	0.354	0.334	1.031	1 5/8
1 1/2	1.5000	0.926	0.886	1.125	1.105	0.760	0.740	0.385	0.365	1.125	1 3/4

[1] Where specifying nominal size in decimals, zeros preceding decimal and in the fourth decimal place shall be omitted.

[2] Point angle X shall be 45° plus 5°, minus 0°, for screws of nominal lengths equal to or longer than those listed in Column Y, and 30° minimum for screws of shorter nominal lengths.

[3] The extent of rounding or flat at apex of cone point shall not exceed an amount equivalent to 10 per cent of the basic screw diameter.

Appendix 29 AMERICAN NATIONAL STANDARD SLOTTED HEADLESS SET SCREWS (ANSI B18.6.2)

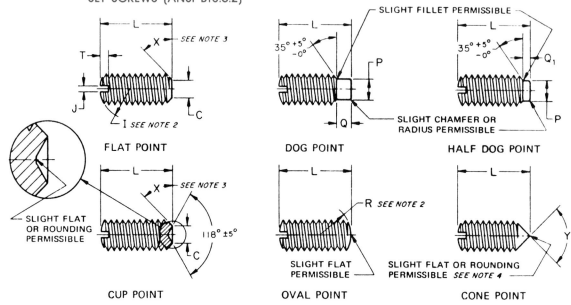

FLAT POINT   DOG POINT   HALF DOG POINT

CUP POINT   OVAL POINT   CONE POINT

Nominal Size[1] or Basic Screw Diameter		I[2] Crown Radius	J Slot Width		T Slot Depth		C Cup and Flat Point Diameters		P Dog Point Diameters		Q Point Length Dog		Q_1 Point Length Half Dog		R[2] Oval Point Radius	Y Cone Point Angle 90° ±2° For These Nominal Lengths or Longer; 118° ±2° For Shorter Screws
		Basic	Max	Min	Max	Min	Max	Min	Max	Min	Max	Min	Max	Min	Basic	
0	0.0600	0.060	0.014	0.010	0.020	0.016	0.033	0.027	0.040	0.037	0.032	0.028	0.017	0.013	0.045	5/64
1	0.0730	0.073	0.016	0.012	0.020	0.016	0.040	0.033	0.049	0.045	0.040	0.036	0.021	0.017	0.055	3/32
2	0.0860	0.086	0.018	0.014	0.025	0.019	0.047	0.039	0.057	0.053	0.046	0.042	0.024	0.020	0.064	7/64
3	0.0990	0.099	0.020	0.016	0.028	0.022	0.054	0.045	0.066	0.062	0.052	0.048	0.027	0.023	0.074	1/8
4	0.1120	0.112	0.024	0.018	0.031	0.025	0.061	0.051	0.075	0.070	0.058	0.054	0.030	0.026	0.084	5/32
5	0.1250	0.125	0.026	0.020	0.036	0.026	0.067	0.057	0.083	0.078	0.063	0.057	0.033	0.027	0.094	3/16
6	0.1380	0.138	0.028	0.022	0.040	0.030	0.074	0.064	0.092	0.087	0.073	0.067	0.038	0.032	0.104	3/16
8	0.1640	0.164	0.032	0.026	0.046	0.036	0.087	0.076	0.109	0.103	0.083	0.077	0.043	0.037	0.123	1/4
10	0.1900	0.190	0.035	0.029	0.053	0.043	0.102	0.088	0.127	0.120	0.095	0.085	0.050	0.040	0.142	1/4
12	0.2160	0.216	0.042	0.035	0.061	0.051	0.115	0.101	0.144	0.137	0.115	0.105	0.060	0.050	0.162	5/16
1/4	0.2500	0.250	0.049	0.041	0.068	0.058	0.132	0.118	0.156	0.149	0.130	0.120	0.068	0.058	0.188	5/16
5/16	0.3125	0.312	0.055	0.047	0.083	0.073	0.172	0.156	0.203	0.195	0.161	0.151	0.083	0.073	0.234	3/8
3/8	0.3750	0.375	0.068	0.060	0.099	0.089	0.212	0.194	0.250	0.241	0.193	0.183	0.099	0.089	0.281	7/16
7/16	0.4375	0.438	0.076	0.068	0.114	0.104	0.252	0.232	0.297	0.287	0.224	0.214	0.114	0.104	0.328	1/2
1/2	0.5000	0.500	0.086	0.076	0.130	0.120	0.291	0.270	0.344	0.334	0.255	0.245	0.130	0.120	0.375	9/16
9/16	0.5625	0.562	0.096	0.086	0.146	0.136	0.332	0.309	0.391	0.379	0.287	0.275	0.146	0.134	0.422	5/8
5/8	0.6250	0.625	0.107	0.097	0.161	0.151	0.371	0.347	0.469	0.456	0.321	0.305	0.164	0.148	0.469	3/4
3/4	0.7500	0.750	0.134	0.124	0.193	0.183	0.450	0.425	0.562	0.549	0.383	0.367	0.196	0.180	0.562	7/8

[1] Where specifying nominal size in decimals, zeros preceding decimal and in the fourth decimal place shall be omitted.

[2] Tolerance on radius for nominal sizes up to and including 5 (0.125 in.) shall be plus 0.015 in. and minus 0.000, and for larger sizes, plus 0.031 in. and minus 0.000. Slotted ends on screws may be flat at option of manufacturer.

[3] Point angle X shall be 45° plus 5°, minus 0°, for screws of nominal lengths equal to or longer than those listed in Column Y, and 30° minimum for screws of shorter nominal lengths.

[4] The extent of rounding or flat at apex of cone point shall not exceed an amount equivalent to 10 per cent of the basic screw diameter.

Appendix 30 TWIST DRILL SIZES

Number Size Drills

Size	Inches	mm	Size	Inches	mm	Size	Inches	mm	Size	Inches	mm
	Drill Diameter			Drill Diameter			Drill Diameter			Drill Diameter	
1	0.2280	5.7912	21	0.1590	4.0386	41	0.0960	2.4384	61	0.0390	0.9906
2	0.2210	5.6134	22	0.1570	3.9878	42	0.0935	2.3622	62	0.0380	0.9652
3	0.2130	5.4102	23	0.1540	3.9116	43	0.0890	2.2606	63	0.0370	0.9398
4	0.2090	5.3086	24	0.1520	3.8608	44	0.0860	2.1844	64	0.0360	0.9144
5	0.2055	5.2197	25	0.1495	3.7973	45	0.0820	2.0828	65	0.0350	0.8890
6	0.2040	5.1816	26	0.1470	3.7338	46	0.0810	2.0574	66	0.0330	0.8382
7	0.2010	5.1054	27	0.1440	3.6576	47	0.0785	1.9812	67	0.0320	0.8128
8	0.1990	5.0800	28	0.1405	3.5560	48	0.0760	1.9304	68	0.0310	0.7874
9	0.1960	4.9784	29	0.1360	3.4544	49	0.0730	1.8542	69	0.0292	0.7417
10	0.1935	4.9149	30	0.1285	3.2639	50	0.0700	1.7780	70	0.0280	0.7112
11	0.1910	4.8514	31	0.1200	3.0480	51	0.0670	1.7018	71	0.0260	0.6604
12	0.1890	4.8006	32	0.1160	2.9464	52	0.0635	1.6129	72	0.0250	0.6350
13	0.1850	4.6990	33	0.1130	2.8702	53	0.0595	1.5113	73	0.0240	0.6096
14	0.1820	4.6228	34	0.1110	2.8194	54	0.0550	1.3970	74	0.0225	0.5715
15	0.1800	4.5720	35	0.1100	2.7940	55	0.0520	1.3208	75	0.0210	0.5334
16	0.1770	4.4958	36	0.1065	2.7051	56	0.0465	1.1684	76	0.0200	0.5080
17	0.1730	4.3942	37	0.1040	2.6416	57	0.0430	1.0922	77	0.0180	0.4572
18	0.1695	4.3053	38	0.1015	2.5781	58	0.0420	1.0668	78	0.0160	0.4064
19	0.1660	4.2164	39	0.0995	2.5273	59	0.0410	1.0414	79	0.0145	0.3638
20	0.1610	4.0894	40	0.0980	2.4892	60	0.0400	1.0160	80	0.0135	0.3429

Letter Size Drills

Size	Inches	mm	Size	Inches	mm	Size	Inches	mm	Size	Inches	mm
	Drill Diameter			Drill Diameter			Drill Diameter			Drill Diameter	
A	0.234	5.944	H	0.266	6.756	O	0.316	8.026	V	0.377	9.576
B	0.238	6.045	I	0.272	6.909	P	0.323	8.204	W	0.386	9.804
C	0.242	6.147	J	0.277	7.036	Q	0.332	8.433	X	0.397	10.084
D	0.246	6.248	K	0.281	7.137	R	0.339	8.611	Y	0.404	10.262
E	0.250	6.350	L	0.290	7.366	S	0.348	8.839	Z	0.413	10.490
F	0.257	6.528	M	0.295	7.493	T	0.358	9.093			
G	0.261	6.629	N	0.302	7.601	U	0.368	9.347			

Appendix 31 STANDARD KEYS AND KEYWAYS

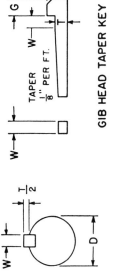

PARALLEL KEY

TAPER $\frac{1''}{8}$ PER FT.

TAPER KEY

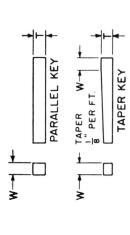

TAPER $\frac{1''}{8}$ PER FT.

GIB HEAD TAPER KEY

SPROCKET BORE (= SHAFT DIAM.) INCHES D	KEYWAY DIMENSIONS — INCHES				KEY DIMENSIONS — INCHES					GIB HEAD DIMENSIONS — INCHES				KEY TOLERANCES TAPER AND GIB HEAD	
	FOR SQUARE KEY		FOR FLAT KEY		SQUARE		FLAT		TOLERANCE ON W AND T (−)	SQUARE KEY		FLAT KEY		W (−)	T (+)
	WIDTH W	DEPTH T/2	WIDTH W	DEPTH T/2	WIDTH W	HEIGHT T	WIDTH W	HEIGHT T		H	G	H	G		
1/2 — 9/16	1/8	1/16	1/8	3/64	1/8	1/8	1/8	3/32	0.002	1/4	7/32	3/16	1/8	0.002	0.002
5/8 — 7/8	3/16	3/32	3/16	1/16	3/16	3/16	3/16	1/8	0.002	5/16	9/32	1/4	3/16	0.002	0.002
15/16 — 1 1/4	1/4	1/8	1/4	3/32	1/4	1/4	1/4	3/16	0.002	7/16	11/32	5/16	1/4	0.002	0.002
1 5/16 — 1 3/8	5/16	5/32	5/16	1/8	5/16	5/16	5/16	1/4	0.002	9/16	13/32	3/8	5/16	0.002	0.002
1 7/16 — 1 3/4	3/8	3/16	3/8	1/8	3/8	3/8	3/8	1/4	0.002	11/16	15/32	7/16	3/8	0.002	0.002
1 13/16 — 2 1/4	1/2	1/4	1/2	3/16	1/2	1/2	1/2	3/8	0.0025	7/8	19/32	5/8	1/2	0.0025	0.0025
2 5/16 — 2 3/4	5/8	5/16	5/8	7/32	5/8	5/8	5/8	7/16	0.0025	1 1/16	23/32	3/4	5/8	0.0025	0.0025
2 7/8 — 3 1/4	3/4	3/8	3/4	1/4	3/4	3/4	3/4	1/2	0.0025	1 1/4	7/8	7/8	3/4	0.0025	0.0025
3 3/8 — 3 3/4	7/8	7/16	7/8	5/16	7/8	7/8	7/8	5/8	0.003	1 1/2	1	1 1/16	7/8	0.003	0.003
3 7/8 — 4 1/2	1	1/2	1	3/8	1	1	1	3/4	0.003	1 3/4	1 3/16	1 1/4	1	0.003	0.003
4 3/4 — 5 1/2	1 1/4	5/8	1 1/4	7/16	1 1/4	1 1/4	1 1/4	7/8	0.003	2	1 7/16	1 1/2	1 1/4	0.003	0.003
5 3/4 — 7 7/8	1 1/2	3/4	1 1/2	1/2	1 1/2	1 1/2	1 1/2	1	0.003	2 1/2	1 3/4	1 3/4	1 1/2	0.003	0.003
7 1/2 — 9 7/8	1 3/4	7/8	. .	. .	1 3/4	1 3/4	. .	. .	0.004	3	2	. .	. .	0.004	0.004
10 — 12 1/2	2	1	. .	. .	2	2	. .	. .	0.004	3 1/2	2 3/8	. .	. .	0.004	0.004

Standard Keyway Tolerances: Straight Keyway — Width (W) $+\ .005\ \ -\ .000$ Depth (T/2) $+\ .010\ \ -\ .000$

Taper Keyway — Width (W) $+\ .005\ \ -\ .000$ Depth (T/2) $+\ .000\ \ -\ .010$

Appendix 32 WOODRUFF KEYS AND KEYSEATS

USA STANDARD

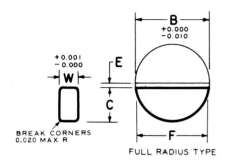

FULL RADIUS TYPE

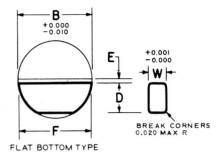

FLAT BOTTOM TYPE

Woodruff Keys

Key No.	Nominal Key Size W × B	Actual Length F +0.000-0.010	Height of Key				Distance Below Center E
			C		D		
			Max	Min	Max	Min	
202	1/16 × 1/4	0.248	0.109	0.104	0.109	0.104	1/64
202.5	1/16 × 5/16	0.311	0.140	0.135	0.140	0.135	1/64
302.5	3/32 × 5/16	0.311	0.140	0.135	0.140	0.135	1/64
203	1/16 × 3/8	0.374	0.172	0.167	0.172	0.167	1/64
303	3/32 × 3/8	0.374	0.172	0.167	0.172	0.167	1/64
403	1/8 × 3/8	0.374	0.172	0.167	0.172	0.167	1/64
204	1/16 × 1/2	0.491	0.203	0.198	0.194	0.188	3/64
304	3/32 × 1/2	0.491	0.203	0.198	0.194	0.188	3/64
404	1/8 × 1/2	0.491	0.203	0.198	0.194	0.188	3/64
305	3/32 × 5/8	0.612	0.250	0.245	0.240	0.234	1/16
405	1/8 × 5/8	0.612	0.250	0.245	0.240	0.234	1/16
505	5/32 × 5/8	0.612	0.250	0.245	0.240	0.234	1/16
605	3/16 × 5/8	0.612	0.250	0.245	0.240	0.234	1/16
406	1/8 × 3/4	0.740	0.313	0.308	0.303	0.297	1/16
506	5/32 × 3/4	0.740	0.313	0.308	0.303	0.297	1/16
606	3/16 × 3/4	0.740	0.313	0.308	0.303	0.297	1/16
806	1/4 × 3/4	0.740	0.313	0.308	0.303	0.297	1/16
507	5/32 × 7/8	0.866	0.375	0.370	0.365	0.359	1/16
607	3/16 × 7/8	0.866	0.375	0.370	0.365	0.359	1/16
707	7/32 × 7/8	0.866	0.375	0.370	0.365	0.359	1/16
807	1/4 × 7/8	0.866	0.375	0.370	0.365	0.359	1/16
608	3/16 × 1	0.992	0.438	0.433	0.428	0.422	1/16
708	7/32 × 1	0.992	0.438	0.433	0.428	0.422	1/16
808	1/4 × 1	0.992	0.438	0.433	0.428	0.422	1/16
1008	5/16 × 1	0.992	0.438	0.433	0.428	0.422	1/16
1208	3/8 × 1	0.992	0.438	0.433	0.428	0.422	1/16
609	3/16 × 1 1/8	1.114	0.484	0.479	0.475	0.469	5/64
709	7/32 × 1 1/8	1.114	0.484	0.479	0.475	0.469	5/64
809	1/4 × 1 1/8	1.114	0.484	0.479	0.475	0.469	5/64
1009	5/16 × 1 1/8	1.114	0.484	0.479	0.475	0.469	5/64

(Courtesy of ANSI; B17.2-1967.)

Cont.

Appendix 33 WOODRUFF KEYS AND KEYSEATS (Cont.)

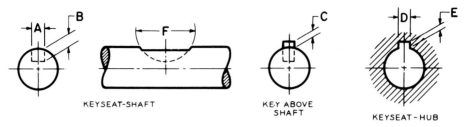

KEYSEAT-SHAFT   KEY ABOVE SHAFT   KEYSEAT-HUB

Keyseat Dimensions

Key Number	Nominal Size Key	Keyseat — Shaft					Key Above Shaft	Keyseat — Hub	
		Width A*		Depth B	Diameter F		Height C	Width D	Depth E
		Min	Max	+0.005 −0.000	Min	Max	+0.005 −0.005	+0.002 −0.000	+0.005 −0.000
202	1/16 × 1/4	0.0615	0.0630	0.0728	0.250	0.268	0.0312	0.0635	0.0372
202.5	1/16 × 5/16	0.0615	0.0630	0.1038	0.312	0.330	0.0312	0.0635	0.0372
302.5	3/32 × 5/16	0.0928	0.0943	0.0882	0.312	0.330	0.0469	0.0948	0.0529
203	1/16 × 3/8	0.0615	0.0630	0.1358	0.375	0.393	0.0312	0.0635	0.0372
303	3/32 × 3/8	0.0928	0.0943	0.1202	0.375	0.393	0.0469	0.0948	0.0529
403	1/8 × 3/8	0.1240	0.1255	0.1045	0.375	0.393	0.0625	0.1260	0.0685
204	1/16 × 1/2	0.0615	0.0630	0.1668	0.500	0.518	0.0312	0.0635	0.0372
304	3/32 × 1/2	0.0928	0.0943	0.1511	0.500	0.518	0.0469	0.0948	0.0529
404	1/8 × 1/2	0.1240	0.1255	0.1355	0.500	0.518	0.0625	0.1260	0.0685
305	3/32 × 5/8	0.0928	0.0943	0.1981	0.625	0.643	0.0469	0.0948	0.0529
405	1/8 × 5/8	0.1240	0.1255	0.1825	0.625	0.643	0.0625	0.1260	0.0685
505	5/32 × 5/8	0.1553	0.1568	0.1669	0.625	0.643	0.0781	0.1573	0.0841
605	3/16 × 5/8	0.1863	0.1880	0.1513	0.625	0.643	0.0937	0.1885	0.0997
406	1/8 × 3/4	0.1240	0.1255	0.2455	0.750	0.768	0.0625	0.1260	0.0685
506	5/32 × 3/4	0.1553	0.1568	0.2299	0.750	0.768	0.0781	0.1573	0.0841
606	3/16 × 3/4	0.1863	0.1880	0.2143	0.750	0.768	0.0937	0.1885	0.0997
806	1/4 × 3/4	0.2487	0.2505	0.1830	0.750	0.768	0.1250	0.2510	0.1310
507	5/32 × 7/8	0.1553	0.1568	0.2919	0.875	0.895	0.0781	0.1573	0.0841
607	3/16 × 7/8	0.1863	0.1880	0.2763	0.875	0.895	0.0937	0.1885	0.0997
707	7/32 × 7/8	0.2175	0.2193	0.2607	0.875	0.895	0.1093	0.2198	0.1153
807	1/4 × 7/8	0.2487	0.2505	0.2450	0.875	0.895	0.1250	0.2510	0.1310
608	3/16 × 1	0.1863	0.1880	0.3393	1.000	1.020	0.0937	0.1885	0.0997
708	7/32 × 1	0.2175	0.2193	0.3237	1.000	1.020	0.1093	0.2198	0.1153
808	1/4 × 1	0.2487	0.2505	0.3080	1.000	1.020	0.1250	0.2510	0.1310
1008	5/16 × 1	0.3111	0.3130	0.2768	1.000	1.020	0.1562	0.3135	0.1622
1208	3/8 × 1	0.3735	0.3755	0.2455	1.000	1.020	0.1875	0.3760	0.1935
609	3/16 × 1 1/8	0.1863	0.1880	0.3853	1.125	1.145	0.0937	0.1885	0.0997
709	7/32 × 1 1/8	0.2175	0.2193	0.3697	1.125	1.145	0.1093	0.2198	0.1153
809	1/4 × 1 1/8	0.2487	0.2505	0.3540	1.125	1.145	0.1250	0.2510	0.1310
1009	5/16 × 1 1/8	0.3111	0.3130	0.3228	1.125	1.145	0.1562	0.3135	0.1622

(Courtesy of ANSI; B17.2–1967.)

Appendix 34 STRAIGHT PINS

CHAMFERED

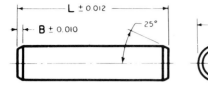

SQUARE END

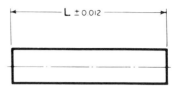

| Nominal Diameter | Diameter A | | Chamfer B |
	Max	Min	
0.062	0.0625	0.0605	0.015
0.094	0.0937	0.0917	0.015
0.109	0.1094	0.1074	0.015
0.125	0.1250	0.1230	0.015
0.156	0.1562	0.1542	0.015
0.188	0.1875	0.1855	0.015
0.219	0.2187	0.2167	0.015
0.250	0.2500	0.2480	0.015
0.312	0.3125	0.3095	0.030
0.375	0.3750	0.3720	0.030
0.438	0.4375	0.4345	0.030
0.500	0.500	0.4970	0.030

All dimensions are given in inches.

These pins must be straight and free from burrs or any other defects that will affect their serviceability.

(Courtesy of ANSI; B5.20–1958.)

Appendix 35 TAPER PINS

All dimensions are given in inches.

Standard reamers are available for pins given above the line.

Pins Nos. 11 (size 0.8600), 12 (size 1.032), 13 (size 1.241), and 14 (1.523) are special sizes—hence their lengths are special.

To find small diameter of pin, multiply the length by 0.2083 and subtract the result from the large diameter.

(Courtesy of ANSI; B5.20–1958.)

Number	7/0	6/0	5/0	4/0	3/0	2/0	0	1	2	3	4	5	6	7	8	9	10
Size (large end)	0.0625	0.0780	0.0940	0.1090	0.1250	0.1410	0.1560	0.1720	0.1930	0.2190	0.2500	0.2890	0.3410	0.4090	0.4920	0.5910	0.7060
Length, L																	
0.375	X	X															
0.500	X	X	X														
0.625	X	X	X	X	X	X	X										
0.750		X	X	X	X	X	X	X									
0.875			X	X	X	X	X	X	X								
1.000				X	X	X	X	X	X	X	X						
1.250					X	X	X	X	X	X	X	X					
1.500						X	X	X	X	X	X	X	X				
1.750							X	X	X	X	X	X	X	X			
2.000								X	X	X	X	X	X	X	X		
2.250									X	X	X	X	X	X	X		
2.500										X	X	X	X	X	X		
2.750										X	X	X	X	X	X		
3.000											X	X	X	X	X		
3.250												X	X	X	X	X	
3.500													X	X	X	X	X
3.750													X	X	X	X	X
4.000													X	X	X	X	X
4.250														X	X	X	X
4.500															X	X	X
4.750															X	X	X
5.000																X	X
5.250																X	X
5.500																X	X
5.750																X	X
6.000																X	X

Appendix 36 PLAIN WASHERS

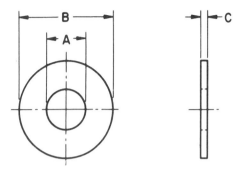

Dimensions of Preferred Sizes of Type A Plain Washers[a]

When specifying washers on drawings or in notes, give the inside diameter, outside diameter, and the thickness. *Example:* 0.938 × 1.750 × 0.134 TYPE A PLAIN WASHER.

Nominal Washer Size[b]			Inside Diameter A			Outside Diameter B			Thickness C		
				Tolerance			Tolerance				
			Basic	Plus	Minus	Basic	Plus	Minus	Basic	Max	Min
—	—		0.078	0.000	0.005	0.188	0.000	0.005	0.020	0.025	0.016
—	—		0.094	0.000	0.005	0.250	0.000	0.005	0.020	0.025	0.016
—	—		0.125	0.008	0.005	0.312	0.008	0.005	0.032	0.040	0.025
No. 6	0.138		0.156	0.008	0.005	0.375	0.015	0.005	0.049	0.065	0.036
No. 8	0.164		0.188	0.008	0.005	0.438	0.015	0.005	0.049	0.065	0.036
No. 10	0.190		0.219	0.008	0.005	0.500	0.015	0.005	0.049	0.065	0.036
$\frac{3}{16}$	0.188		0.250	0.015	0.005	0.562	0.015	0.005	0.049	0.065	0.036
No. 12	0.216		0.250	0.015	0.005	0.562	0.015	0.005	0.065	0.080	0.051
$\frac{1}{4}$	0.250	N	0.281	0.015	0.005	0.625	0.015	0.005	0.065	0.080	0.051
$\frac{1}{4}$	0.250	W	0.312	0.015	0.005	0.734[c]	0.015	0.007	0.065	0.080	0.051
$\frac{5}{16}$	0.312	N	0.344	0.015	0.005	0.688	0.015	0.007	0.065	0.080	0.051
$\frac{5}{16}$	0.312	W	0.375	0.015	0.005	0.875	0.030	0.007	0.083	0.104	0.064
$\frac{3}{8}$	0.375	N	0.406	0.015	0.005	0.812	0.015	0.007	0.065	0.080	0.051
$\frac{3}{8}$	0.375	W	0.438	0.015	0.005	1.000	0.030	0.007	0.083	0.104	0.064
$\frac{7}{16}$	0.438	N	0.469	0.015	0.005	0.922	0.015	0.007	0.065	0.080	0.051
$\frac{7}{16}$	0.438	W	0.500	0.015	0.005	1.250	0.030	0.007	0.083	0.104	0.064
$\frac{1}{2}$	0.500	N	0.531	0.015	0.005	1.062	0.030	0.007	0.095	0.121	0.074
$\frac{1}{2}$	0.500	W	0.562	0.015	0.005	1.375	0.030	0.007	0.109	0.132	0.086

Preferred sizes are for the most part from series previously designated "Standard Plate" and "SAE." Where common sizes existed in the two series, the SAE size is designated "N" (narrow) and the Standard Plate "W" (wide). These sizes as well as all other sizes of Type A Plain Washers are to be ordered by ID, OD, and thickness dimensions.

[b] Nominal washer sizes are intended for use with comparable nominal screw or bolt sizes.

[c] The 0.734 in., 1.156 in., and 1.469 in. outside diameters avoid washers which could be used in coin-operated devices.

Cont.

Appendix 36 PLAIN WASHERS (Cont.)

Nominal Washer Size[b]			Inside Diameter A			Outside Diameter B			Thickness C		
			Basic	Tolerance		Basic	Tolerance		Basic	Max	Min
				Plus	Minus		Plus	Minus			
$\frac{9}{16}$	0.562	N	0.594	0.015	0.005	1.156[c]	0.030	0.007	0.095	0.121	0.074
$\frac{9}{16}$	0.562	W	0.625	0.015	0.005	1.469[c]	0.030	0.007	0.109	0.132	0.086
$\frac{5}{8}$	0.625	N	0.656	0.030	0.007	1.312	0.030	0.007	0.095	0.121	0.074
$\frac{5}{8}$	0.625	W	0.688	0.030	0.007	1.750	0.030	0.007	0.134	0.160	0.108
$\frac{3}{4}$	0.750	N	0.812	0.030	0.007	1.469	0.030	0.007	0.134	0.160	0.108
$\frac{3}{4}$	0.750	W	0.812	0.030	0.007	2.000	0.030	0.007	0.148	0.177	0.122
$\frac{7}{8}$	0.875	N	0.938	0.030	0.007	1.750	0.030	0.007	0.134	0.160	0.108
$\frac{7}{8}$	0.875	W	0.938	0.030	0.007	2.250	0.030	0.007	0.165	0.192	0.136
1	1.000	N	1.062	0.030	0.007	2.000	0.030	0.007	0.134	0.160	0.108
1	1.000	W	1.062	0.030	0.007	2.500	0.030	0.007	0.165	0.192	0.136
$1\frac{1}{8}$	1.125	N	1.250	0.030	0.007	2.250	0.030	0.007	0.134	0.160	0.108
$1\frac{1}{8}$	1.125	W	1.250	0.030	0.007	2.750	0.030	0.007	0.165	0.192	0.136
$1\frac{1}{4}$	1.250	N	1.375	0.030	0.007	2.500	0.030	0.007	0.165	0.192	0.136
$1\frac{1}{4}$	1.250	W	1.375	0.030	0.007	3.000	0.030	0.007	0.165	0.192	0.136
$1\frac{3}{8}$	1.375	N	1.500	0.030	0.007	2.750	0.030	0.007	0.165	0.192	0.136
$1\frac{3}{8}$	1.375	W	1.500	0.045	0.010	3.250	0.045	0.010	0.180	0.213	0.153
$1\frac{1}{2}$	1.500	N	1.625	0.030	0.007	3.000	0.030	0.007	0.165	0.192	0.136
$1\frac{1}{2}$	1.500	W	1.625	0.045	0.010	3.500	0.045	0.010	0.180	0.213	0.153
$1\frac{5}{8}$	1.625		1.750	0.045	0.010	3.750	0.045	0.010	0.180	0.213	0.153
$1\frac{3}{4}$	1.750		1.875	0.045	0.010	4.000	0.045	0.010	0.180	0.213	0.153
$1\frac{7}{8}$	1.875		2.000	0.045	0.010	4.250	0.045	0.010	0.180	0.213	0.153
2	2.000		2.125	0.045	0.010	4.500	0.045	0.010	0.180	0.213	0.153
$2\frac{1}{4}$	2.250		2.375	0.045	0.010	4.750	0.045	0.010	0.220	0.248	0.193
$2\frac{1}{2}$	2.500		2.625	0.045	0.010	5.000	0.045	0.010	0.238	0.280	0.210
$2\frac{3}{4}$	2.750		2.875	0.065	0.010	5.250	0.065	0.010	0.259	0.310	0.228
3	3.000		3.125	0.065	0.010	5.500	0.065	0.010	0.284	0.327	0.249

(Courtesy of ANSI; B27.2–1965.)

Appendix 37 LOCK WASHERS (ANSI B27.1)

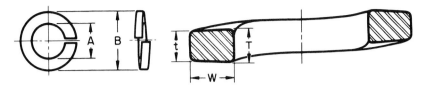

Dimensions of Regular* Helical Spring Lock Washers

Nominal Washer Size		Inside Diameter A		Outside Diameter B	Washer Section Width W	Washer Section Thickness $\frac{T+t}{2}$
		Min	Max	Max**	Min	Min
No. 2	0.086	0.088	0.094	0.172	0.035	0.020
No. 3	0.099	0.101	0.107	0.195	0.040	0.025
No. 4	0.112	0.115	0.121	0.209	0.040	0.025
No. 5	0.125	0.128	0.134	0.236	0.047	0.031
No. 6	0.138	0.141	0.148	0.250	0.047	0.031
No. 8	0.164	0.168	0.175	0.293	0.055	0.040
No. 10	0.190	0.194	0.202	0.334	0.062	0.047
No. 12	0.216	0.221	0.229	0.377	0.070	0.056
¼	0.250	0.255	0.263	0.489	0.109	0.062
⁵⁄₁₆	0.312	0.318	0.328	0.586	0.125	0.078
³⁄₈	0.375	0.382	0.393	0.683	0.141	0.094
⁷⁄₁₆	0.438	0.446	0.459	0.779	0.156	0.109
½	0.500	0.509	0.523	0.873	0.171	0.125
⁹⁄₁₆	0.562	0.572	0.587	0.971	0.188	0.141
⁵⁄₈	0.625	0.636	0.653	1.079	0.203	0.156
¹¹⁄₁₆	0.688	0.700	0.718	1.176	0.219	0.172
¾	0.750	0.763	0.783	1.271	0.234	0.188
¹³⁄₁₆	0.812	0.826	0.847	1.367	0.250	0.203
⁷⁄₈	0.875	0.890	0.912	1.464	0.266	0.219
¹⁵⁄₁₆	0.938	0.954	0.978	1.560	0.281	0.234
1	1.000	1.017	1.042	1.661	0.297	0.250
1¹⁄₁₆	1.062	1.080	1.107	1.756	0.312	0.266
1⅛	1.125	1.144	1.172	1.853	0.328	0.281
1³⁄₁₆	1.188	1.208	1.237	1.950	0.344	0.297
1¼	1.250	1.271	1.302	2.045	0.359	0.312
1⁵⁄₁₆	1.312	1.334	1.366	2.141	0.375	0.328
1⅜	1.375	1.398	1.432	2.239	0.391	0.344
1⁷⁄₁₆	1.438	1.462	1.497	2.334	0.406	0.359
1½	1.500	1.525	1.561	2.430	0.422	0.375

*Formerly designated Medium Helical Spring Lock Washers.

**The maximum outside diameters specified allow for the commercial tolerances on cold drawn wire.

Appendix 38 COTTER PINS

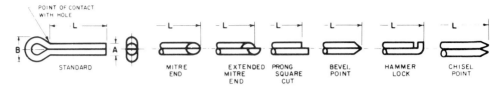

POINT OF CONTACT WITH HOLE

STANDARD | MITRE END | EXTENDED MITRE END | PRONG SQUARE CUT | BEVEL POINT | HAMMER LOCK | CHISEL POINT

Nominal Diameter	Diameter A		Outside Eye Diameter B	Hole Sizes Recommended
	Max	Min	Min	
0.031	0.032	0.028	$\frac{1}{16}$	$\frac{3}{64}$
0.047	0.048	0.044	$\frac{3}{32}$	$\frac{1}{16}$
0.062	0.060	0.056	$\frac{1}{8}$	$\frac{5}{64}$
0.078	0.076	0.072	$\frac{5}{32}$	$\frac{3}{32}$
0.094	0.090	0.086	$\frac{3}{16}$	$\frac{7}{64}$
0.109	0.104	0.100	$\frac{7}{32}$	$\frac{1}{8}$
0.125	0.120	0.116	$\frac{1}{4}$	$\frac{9}{64}$
0.141	0.134	0.130	$\frac{9}{32}$	$\frac{5}{32}$
0.156	0.150	0.146	$\frac{5}{16}$	$\frac{11}{64}$
0.188	0.176	0.172	$\frac{3}{8}$	$\frac{13}{64}$
0.219	0.207	0.202	$\frac{7}{16}$	$\frac{15}{64}$
0.250	0.225	0.220	$\frac{1}{2}$	$\frac{17}{64}$
0.312	0.280	0.275	$\frac{5}{8}$	$\frac{5}{16}$
0.375	0.335	0.329	$\frac{3}{4}$	$\frac{3}{8}$
0.438	0.406	0.400	$\frac{7}{8}$	$\frac{7}{16}$
0.500	0.473	0.467	1	$\frac{1}{2}$
0.625	0.598	0.590	$1\frac{1}{4}$	$\frac{5}{8}$
0.750	0.723	0.715	$1\frac{1}{2}$	$\frac{3}{4}$

All dimensions are given in inches.

A certain amount of leeway is permitted in the design of the head; however, the outside diameters given should be adhered to.

Prongs are to be parallel; ends shall not be open.

Points may be blunt, bevel, extended prong, mitre, etc., and purchaser may specify type required.

Lengths shall be measured as shown on the above illustration (L-dimension).

Cotter pins shall be free from burrs or any defects that will affect their serviceability.

(Courtesy of ANSI; B5.20–1958.)

Appendix 39 AMERICAN STANDARD RUNNING AND SLIDING FITS

Limits are in thousandths of an inch.
Limits for hole and shaft are applied algebraically to the basic size to obtain the limits of size for the parts.
Data in bold face are in accordance with ABC agreements.
Symbols H5, g5, etc., are Hole and Shaft designations used in ABC System.

Nominal Size Range Inches Over – To	Class RC 1 Limits of Clearance	Standard Limits Hole H5	Standard Limits Shaft g4	Class RC 2 Limits of Clearance	Standard Limits Hole H6	Standard Limits Shaft g5	Class RC 3 Limits of Clearance	Standard Limits Hole H7	Standard Limits Shaft f6	Class RC 4 Limits of Clearance	Standard Limits Hole H8	Standard Limits Shaft f7
0 – 0.12	0.1 / 0.45	+ 0.2 / 0	− 0.1 / − 0.25	0.1 / 0.55	+ 0.25 / 0	− 0.1 / − 0.3	0.3 / 0.95	+ 0.4 / 0	− 0.3 / − 0.55	0.3 / 1.3	+ 0.6 / 0	− 0.3 / − 0.7
0.12 – 0.24	0.15 / 0.5	+ 0.2 / 0	− 0.15 / − 0.3	0.15 / 0.65	+ 0.3 / 0	− 0.15 / − 0.35	0.4 / 1.12	+ 0.5 / 0	− 0.4 / − 0.7	0.4 / 1.6	+ 0.7 / 0	− 0.4 / − 0.9
0.24 – 0.40	0.2 / 0.6	0.25 / 0	− 0.2 / − 0.35	0.2 / 0.85	+ 0.4 / 0	− 0.2 / − 0.45	0.5 / 1.5	+ 0.6 / 0	− 0.5 / − 0.9	0.5 / 2.0	+ 0.9 / 0	− 0.5 / − 1.1
0.40 – 0.71	0.25 / 0.75	+ 0.3 / 0	− 0.25 / − 0.45	0.25 / 0.95	+ 0.4 / 0	− 0.25 / − 0.55	0.6 / 1.7	+ 0.7 / 0	− 0.6 / − 1.0	0.6 / 2.3	+ 1.0 / 0	− 0.6 / − 1.3
0.71 – 1.19	0.3 / 0.95	+ 0.4 / 0	− 0.3 / − 0.55	0.3 / 1.2	+ 0.5 / 0	− 0.3 / − 0.7	0.8 / 2.1	+ 0.8 / 0	− 0.8 / − 1.3	0.8 / 2.8	+ 1.2 / 0	− 0.8 / − 1.6
1.19 – 1.97	0.4 / 1.1	+ 0.4 / 0	− 0.4 / − 0.7	0.4 / 1.4	+ 0.6 / 0	− 0.4 / − 0.8	1.0 / 2.6	+ 1.0 / 0	− 1.0 / − 1.6	1.0 / 3.6	+ 1.6 / 0	− 1.0 / − 2.0
1.97 – 3.15	0.4 / 1.2	+ 0.5 / 0	− 0.4 / − 0.7	0.4 / 1.6	+ 0.7 / 0	− 0.4 / − 0.9	1.2 / 3.1	+ 1.2 / 0	− 1.2 / − 1.9	1.2 / 4.2	+ 1.8 / 0	− 1.2 / − 2.4
3.15 – 4.73	0.5 / 1.5	+ 0.6 / 0	− 0.5 / − 0.9	0.5 / 2.0	+ 0.9 / 0	− 0.5 / − 1.1	1.4 / 3.7	+ 1.4 / 0	− 1.4 / − 2.3	1.4 / 5.0	+ 2.2 / 0	− 1.4 / − 2.8
4.73 – 7.09	0.6 / 1.8	+ 0.7 / 0	− 0.6 / − 1.1	0.6 / 2.3	+ 1.0 / 0	− 0.6 / − 1.3	1.6 / 4.2	+ 1.6 / 0	− 1.6 / − 2.6	1.6 / 5.7	+ 2.5 / 0	− 1.6 / − 3.2
7.09 – 9.85	0.6 / 2.0	+ 0.8 / 0	− 0.6 / − 1.2	0.6 / 2.6	+ 1.2 / 0	− 0.6 / − 1.4	2.0 / 5.0	+ 1.8 / 0	− 2.0 / − 3.2	2.0 / 6.6	+ 2.8 / 0	− 2.0 / − 3.8
9.85 – 12.41	0.8 / 2.3	+ 0.9 / 0	− 0.8 / − 1.4	0.8 / 2.9	+ 1.2 / 0	− 0.8 / − 1.7	2.5 / 5.7	+ 2.0 / 0	− 2.5 / − 3.7	2.5 / 7.5	+ 3.0 / 0	− 2.5 / − 4.5
12.41 – 15.75	1.0 / 2.7	+ 1.0 / 0	− 1.0 / − 1.7	1.0 / 3.4	+ 1.4 / 0	− 1.0 / − 2.0	3.0 / 6.6	+ / 0	− 3.0 / − 4.4	3.0 / 8.7	+ 3.5 / 0	− 3.0 / − 5.2
15.75 – 19.69	1.2 / 3.0	+ 1.0 / 0	− 1.2 / − 2.0	1.2 / 3.8	+ 1.6 / 0	− 1.2 / − 2.2	4.0 / 8.1	+ 1.6 / 0	− 4.0 / − 5.6	4.0 / 10.5	+ 4.0 / 0	− 4.0 / − 6.5
19.69 – 30.09	1.6 / 3.7	+ 1.2 / 0	− 1.6 / − 2.5	1.6 / 4.8	+ 2.0 / 0	− 1.6 / − 2.8	5.0 / 10.0	+ 3.0 / 0	− 5.0 / − 7.0	5.0 / 13.0	+ 5.0 / 0	− 5.0 / − 8.0
30.09 – 41.49	2.0 / 4.6	+ 1.6 / 0	− 2.0 / − 3.0	2.0 / 6.1	+ 2.5 / 0	− 2.0 / − 3.6	6.0 / 12.5	+ 4.0 / 0	− 6.0 / − 8.5	6.0 / 16.0	+ 6.0 / 0	− 6.0 / −10.0
41.49 – 56.19	2.5 / 5.7	+ 2.0 / 0	− 2.5 / − 3.7	2.5 / 7.5	+ 3.0 / 0	− 2.5 / − 4.5	8.0 / 16.0	+ 5.0 / 0	− 8.0 / −11.0	8.0 / 21.0	+ 8.0 / 0	− 8.0 / −13.0
56.19 – 76.39	3.0 / 7.1	+ 2.5 / 0	− 3.0 / − 4.6	3.0 / 9.5	+ 4.0 / 0	− 3.0 / − 5.5	10.0 / 20.0	+ 6.0 / 0	−10.0 / −14.0	10.0 / 26.0	+10.0 / 0	−10.0 / −16.0
76.39 – 100.9	4.0 / 9.0	+ 3.0 / 0	− 4.0 / − 6.0	4.0 / 12.0	+ 5.0 / 0	− 4.0 / − 7.0	12.0 / 25.0	+ 8.0 / 0	−12.0 / −17.0	12.0 / 32.0	+12.0 / 0	−12.0 / −20.0
100.9 – 131.9	5.0 / 11.5	+ 4.0 / 0	− 5.0 / − 7.5	5.0 / 15.0	+ 6.0 / 0	− 5.0 / − 9.0	16.0 / 32.0	+10.0 / 0	−16.0 / −22.0	16.0 / 36.0	+16.0 / 0	−16.0 / −26.0
131.9 – 171.9	6.0 / 14.0	+ 5.0 / 0	− 6.0 / − 9.0	6.0 / 19.0	+ 8.0 / 0	− 6.0 / −11.0	18.0 / 38.0	+ 8.0 / 0	−18.0 / −26.0	18.0 / 50.0	+20.0 / 0	−18.0 / −30.0
171.9 – 200	8.0 / 18.0	+ 6.0 / 0	− 8.0 / −12.0	8.0 / 22.0	+10.0 / 0	− 8.0 / −12.0	22.0 / 48.0	+16.0 / 0	−22.0 / −32.0	22.0 / 63.0	+25.0 / 0	−22.0 / −38.0

(Courtesy of USASI; B4.1–1955.)

Cont.

Appendix 39 AMERICAN STANDARD RUNNING AND SLIDING FITS (Cont.)

Class RC 5 Limits of Clearance	Class RC 5 Hole H8	Class RC 5 Shaft e7	Class RC 6 Limits of Clearance	Class RC 6 Hole H9	Class RC 6 Shaft e8	Class RC 7 Limits of Clearance	Class RC 7 Hole H9	Class RC 7 Shaft d8	Class RC 8 Limits of Clearance	Class RC 8 Hole H10	Class RC 8 Shaft c9	Class RC 9 Limits of Clearance	Class RC 9 Hole H11	Class RC 9 Shaft	Nominal Size Range Inches Over	To
0.6 / 1.6	+ 0.6 / − 0	− 0.6 / − 1.0	0.6 / 2.2	+ 1.0 / − 0	− 0.6 / − 1.2	1.0 / 2.6	+ 1.0 / 0	− 1.0 / − 1.6	2.5 / 5.1	+ 1.6 / 0	− 2.5 / − 3.5	4.0 / 8.1	+ 2.5 / 0	− 4.0 / − 5.6	0 −	0.12
0.8 / 2.0	+ 0.7 / − 0	− 0.8 / − 1.3	0.8 / 2.7	+ 1.2 / − 0	− 0.8 / − 1.5	1.2 / 3.1	+ 1.2 / 0	− 1.2 / − 1.9	2.8 / 5.8	+ 1.8 / 0	− 2.8 / − 4.0	4.5 / 9.0	+ 3.0 / 0	− 4.5 / − 6.0	0.12−	0.24
1.0 / 2.5	+ 0.9 / − 0	− 1.0 / − 1.6	1.0 / 3.3	+ 1.4 / − 0	− 1.0 / − 1.9	1.6 / 3.9	+ 1.4 / 0	− 1.6 / − 2.5	3.0 / 6.6	+ 2.2 / 0	− 3.0 / − 4.4	5.0 / 10.7	+ 3.5 / 0	− 5.0 / − 7.2	0.24−	0.40
1.2 / 2.9	+ 1.0 / − 0	− 1.2 / − 1.9	1.2 / 3.8	+ 1.6 / − 0	− 1.2 / − 2.2	2.0 / 4.6	+ 1.6 / 0	− 2.0 / − 3.0	3.5 / 7.9	+ 2.8 / 0	− 3.5 / − 5.1	6.0 / 12.8	+ 4.0 / − 0	− 6.0 / − 8.8	0.40−	0.71
1.6 / 3.6	+ 1.2 / − 0	− 1.6 / − 2.4	1.6 / 4.8	+ 2.0 / − 0	− 1.6 / − 2.8	2.5 / 5.7	+ 2.0 / 0	− 2.5 / − 3.7	4.5 / 10.0	+ 3.5 / 0	− 4.5 / − 6.5	7.0 / 15.5	+ 5.0 / 0	− 7.0 / − 10.5	0.71−	1.19
2.0 / 4.6	+ 1.6 / − 0	− 2.0 / − 3.0	2.0 / 6.1	+ 2.5 / − 0	− 2.0 / − 3.6	3.0 / 7.1	+ 2.5 / 0	− 3.0 / − 4.6	5.0 / 11.5	+ 4.0 / 0	− 5.0 / − 7.5	8.0 / 18.0	+ 6.0 / 0	− 8.0 / − 12.0	1.19−	1.97
2.5 / 5.5	+ 1.8 / − 0	− 2.5 / − 3.7	2.5 / 7.3	+ 3.0 / − 0	− 2.5 / − 4.3	4.0 / 8.8	+ 3.0 / 0	− 4.0 / − 5.8	6.0 / 13.5	+ 4.5 / 0	− 6.0 / − 9.0	9.0 / 20.5	+ 7.0 / 0	− 9.0 / − 13.5	1.97−	3.15
3.0 / 6.6	+ 2.2 / − 0	− 3.0 / − 4.4	3.0 / 8.7	+ 3.5 / − 0	− 3.0 / − 5.2	5.0 / 10.7	+ 3.5 / 0	− 5.0 / − 7.2	7.0 / 15.5	+ 5.0 / 0	− 7.0 / − 10.5	10.0 / 24.0	+ 9.0 / 0	− 10.0 / − 15.0	3.15−	4.73
3.5 / 7.6	+ 2.5 / − 0	− 3.5 / − 5.1	3.5 / 10.0	+ 4.0 / − 0	− 3.5 / − 6.0	6.0 / 12.5	+ 4.0 / 0	− 6.0 / − 8.5	8.0 / 18.0	+ 6.0 / 0	− 8.0 / − 12.0	12.0 / 28.0	+ 10.0 / 0	− 12.0 / − 18.0	4.73−	7.09
4.0 / 8.6	+ 2.8 / − 0	− 4.0 / − 5.8	4.0 / 11.3	+ 4.5 / 0	− 4.0 / − 6.8	7.0 / 14.3	+ 4.5 / 0	− 7.0 / − 9.8	10.0 / 21.5	+ 7.0 / 0	− 10.0 / − 14.5	15.0 / 34.0	+ 12.0 / 0	− 15.0 / − 22.0	7.09−	9.85
5.0 / 10.0	+ 3.0 / 0	− 5.0 / − 7.0	5.0 / 13.0	+ 5.0 / 0	− 5.0 / − 8.0	8.0 / 16.0	+ 5.0 / 0	− 8.0 / − 11.0	12.0 / 25.0	+ 8.0 / 0	− 12.0 / − 17.0	18.0 / 38.0	+ 12.0 / 0	− 18.0 / − 26.0	9.85−	12.41
6.0 / 11.7	+ 3.5 / 0	− 6.0 / − 8.2	6.0 / 15.5	+ 6.0 / 0	− 6.0 / − 9.5	10.0 / 19.5	+ 6.0 / 0	− 10.0 / − 13.5	14.0 / 29.0	+ 9.0 / 0	− 14.0 / − 20.0	22.0 / 45.0	+ 14.0 / 0	− 22.0 / − 31.0	12.41−	15.75
8.0 / 14.5	+ 4.0 / 0	− 8.0 / −10.5	8.0 / 18.0	+ 6.0 / 0	− 8.0 / −12.0	12.0 / 22.0	+ 6.0 / 0	− 12.0 / − 16.0	16.0 / 32.0	+10.0 / 0	− 16.0 / − 22.0	25.0 / 51.0	+ 16.0 / 0	− 25.0 / − 35.0	15.75−	19.69
10.0 / 18.0	+ 5.0 / 0	−10.0 / −13.0	10.0 / 23.0	+ 8.0 / 0	−10.0 / −15.0	16.0 / 29.0	+ 8.0 / 0	− 16.0 / − 21.0	20.0 / 40.0	+12.0 / 0	− 20.0 / − 28.0	30.0 / 62.0	+ 20.0 / 0	− 30.0 / − 42.0	19.69−	30.09
12.0 / 22.0	+ 6.0 / 0	−12.0 / −16.0	12.0 / 28.0	+10.0 / 0	−12.0 / −18.0	20.0 / 36.0	+10.0 / 0	− 20.0 / − 26.0	25.0 / 51.0	+16.0 / 0	− 25.0 / − 35.0	40.0 / 81.0	+ 25.0 / 0	− 40.0 / − 56.0	30.09−	41.49
16.0 / 29.0	+ 8.0 / 0	−16.0 / −21.0	16.0 / 36.0	+12.0 / 0	−16.0 / −24.0	25.0 / 45.0	+12.0 / 0	− 25.0 / − 33.0	30.0 / 62.0	+20.0 / 0	− 30.0 / − 42.0	50.0 / 100	+ 30.0 / 0	− 50.0 / − 70.0	41.49−	56.19
20.0 / 36.0	+10.0 / 0	−20.0 / −26.0	20.0 / 46.0	+16.0 / 0	−20.0 / −30.0	30.0 / 56.0	+16.0 / 0	− 30.0 / − 40.0	40.0 / 81.0	+25.0 / 0	− 40.0 / − 56.0	60.0 / 125	+ 40.0 / 0	− 60.0 / − 85.0	56.19−	76.39
25.0 / 45.0	+12.0 / 0	−25.0 / −33.0	25.0 / 57.0	+20.0 / 0	−25.0 / −37.0	40.0 / 72.0	+20.0 / 0	− 40.0 / − 52.0	50.0 / 100	+30.0 / 0	− 50.0 / − 70.0	80.0 / 160	+ 50.0 / 0	− 80.0 / −110	76.39−100.9	
30.0 / 56.0	+16.0 / 0	−30.0 / −40.0	30.0 / 71.0	+25.0 / 0	−30.0 / −46.0	50.0 / 91.0	+25.0 / 0	− 50.0 / − 66.0	60.0 / 125	+40.0 / 0	− 60.0 / − 85.0	100 / 200	+ 60.0 / 0	−100 / −140	100.9 −131.9	
35.0 / 67.0	+20.0 / 0	−35.0 / −47.0	35.0 / 85.0	+30.0 / 0	−35.0 / −55.0	60.0 / 110	+30.0 / 0	− 60.0 / − 80.0	80.0 / 160	+50.0 / 0	− 80.0 / −110	130 / 260	+ 80.0 / 0	−130 / −180	131.9 −171.9	
45.0 / 86.0	+25.0 / 0	−45.0 / −61.0	45.0 / 110	+40.0 / 0	−45.0 / −70.0	80.0 / 145.0	+40.0 / 0	− 80.0 / −105.0	100 / 200	+60.0 / 0	−100 / −140	150 / 310	+100 / 0	−150 / −210	171.9 −200	

(Courtesy of ANSI; B4.1–1955.)

Appendix 40 AMERICAN STANDARD CLEARANCE LOCATIONAL FITS

Limits are in thousandths of an inch.
Limits for hole and shaft are applied algebraically to the basic size to obtain the limits of size for the parts.
Data in bold face are in accordance with ABC agreements.
Symbols H9,f8, etc., are Hole and Shaft designations used in ABC System.

Nominal Size Range Inches Over	To	LC 1 Limits of Clearance	LC 1 Hole H6	LC 1 Shaft h5	LC 2 Limits of Clearance	LC 2 Hole H7	LC 2 Shaft h6	LC 3 Limits of Clearance	LC 3 Hole H8	LC 3 Shaft h7	LC 4 Limits of Clearance	LC 4 Hole H10	LC 4 Shaft h9	LC 5 Limits of Clearance	LC 5 Hole H7	LC 5 Shaft g6
0 —	0.12	0 0.45	+0.25 −0	+0 −0.2	0 0.65	+0.4 −0	+0 −0.25	0 1	+0.6 −0	+0 −0.4	0 2.6	+1.6 −0	+0 −1.0	0.1 0.75	+0.4 −0	−0.1 −0.35
0.12—	0.24	0 0.5	+0.3 −0	+0 −0.2	0 0.8	+0.5 −0	+0 −0.3	0 1.2	+0.7 −0	+0 −0.5	0 3.0	+1.8 −0	+0 −1.2	0.15 0.95	+0.5 −0	−0.15 −0.45
0.24—	0.40	0 0.65	+0.4 −0	+0 −0.25	0 1.0	+0.6 −0	+0 −0.4	0 1.5	+0.9 −0	+0 −0.6	0 3.6	+2.2 −0	+0 −1.4	0.2 1.2	+0.6 −0	−0.2 −0.6
0.40—	0.71	0 0.7	+0.4 −0	+0 −0.3	0 1.1	+0.7 −0	+0 −0.4	0 1.7	+1.0 −0	+0 −0.7	0 4.4	+2.8 −0	+0 −1.6	0.25 1.35	+0.7 −0	−0.25 −0.65
0.71—	1.19	0 0.9	+0.5 −0	+0 −0.4	0 1.3	+0.8 −0	+0 −0.5	0 2	+1.2 −0	+0 −0.8	0 5.5	+3.5 −0	+0 −2.0	0.3 1.6	+0.8 −0	−0.3 −0.8
1.19—	1.97	0 1.0	+0.6 −0	+0 −0.4	0 1.6	+1.0 −0	+0 −0.6	0 2.6	+1.6 −0	+0 −1	0 6.5	+4.0 −0	+0 −2.5	0.4 2.0	+1.0 −0	−0.4 −1.0
1.97—	3.15	0 1.2	+0.7 −0	+0 −0.5	0 1.9	+1.2 −0	+0 −0.7	0 3	+1.8 −0	+0 −1.2	0 7.5	+4.5 −0	+0 −3	0.4 2.3	+1.2 −0	−0.4 −1.1
3.15—	4.73	0 1.5	+0.9 −0	+0 −0.6	0 2.3	+1.4 −0	+0 −0.9	0 3.6	+2.2 −0	+0 −1.4	0 8.5	+5.0 −0	+0 −3.5	0.5 2.8	+1.4 −0	−0.5 −1.4
4.73—	7.09	0 1.7	+1.0 −0	+0 −0.7	0 2.6	+1.6 −0	+0 −1.0	0 4.1	+2.5 −0	+0 −1.6	0 10	+6.0 −0	+0 −4	0.6 3.2	+1.6 −0	−0.6 −1.6
7.09—	9.85	0 2.0	+1.2 −0	+0 −0.8	0 3.0	+1.8 −0	+0 −1.2	0 4.6	+2.8 −0	+0 −1.8	0 11.5	+7.0 −0	+0 −4.5	0.6 3.6	+1.8 −0	−0.6 −1.8
9.85—	12.41	0 2.1	+1.2 −0	+0 −0.9	0 3.2	+2.0 −0	+0 −1.2	0 5	+3.0 −0	+0 −2.0	0 13	+8.0 −0	+0 −5	0.7 3.9	+2.0 −0	−0.7 −1.9
12.41—	15.75	0 2.4	+1.4 −0	+0 −1.0	0 3.6	+2.2 −0	+0 −1.4	0 5.7	+3.5 −0	+0 −2.2	0 15	+9.0 −0	+0 −6	0.7 4.3	+2.2 −0	−0.7 −2.1
15.75—	19.69	0 2.6	+1.6 −0	+0 −1.0	0 4.1	+2.5 −0	+0 −1.6	0 6.5	+4 −0	+0 −2.5	0 16	+10.0 −0	+0 −6	0.8 4.9	+2.5 −0	−0.8 −2.4
19.69—	30.09	0 3.2	+2.0 −0	+0 −1.2	0 5.0	+3 −0	+0 −2	0 8	+5 −0	+0 −3	0 20	+12.0 −0	+0 −8	0.9 5.9	+3.0 −0	−0.9 −2.9
30.09—	41.49	0 4.1	+2.5 −0	+0 −1.6	0 6.5	+4 −0	+0 −2.5	0 10	+6 −0	+0 −4	0 26	+16.0 −0	+0 −10	1.0 7.5	+4.0 −0	−1.0 −3.5
41.49—	56.19	0 5.0	+3.0 −0	+0 −2.0	0 8.0	+5 −0	+0 −3	0 13	+8 −0	+0 −5	0 32	+20.0 −0	+0 −12	1.2 9.2	+5.0 −0	−1.2 −4.2
56.19—	76.39	0 6.5	+4.0 −0	+0 −2.5	0 10	+6 −0	+0 −4	0 16	+10 −0	+0 −6	0 41	+25.0 −0	+0 −16	1.2 11.2	+6.0 −0	−1.2 −5.2
76.39—	100.9	0 8.0	+5.0 −0	+0 −3.0	0 13	+8 −0	+0 −5	0 20	+12 −0	+0 −8	0 50	+30.0 −0	+0 −20	1.4 14.4	+8.0 −0	−1.4 −6.4
100.9 —	131.9	0 10.0	+6.0 −0	+0 −4.0	0 16	+10 −0	+0 −6	0 26	+16 −0	+0 −10	0 65	+40.0 −0	+0 −25	1.6 17.6	+10.0 −0	−1.6 −7.6
131.9 —	171.9	0 13.0	+8.0 −0	+0 −5.0	0 20	+12 −0	+0 −8	0 32	+20 −0	+0 −12	0 8	+50.0 −0	+0 −30	1.8 21.8	+12.0 −0	−1.8 −9.8
171.9 —	200	0 16.0	+10.0 −0	+0 −6.0	0 26	+16 −0	+0 −10	0 41	+25 −0	+0 −16	0 100	+60.0 −0	+0 −40	1.8 27.8	+16.0 −0	−1.8 −11.8

(Courtesy of USASI; B4.1-1955.)

Cont.

Appendix 40 AMERICAN STANDARD CLEARANCE LOCATIONAL FITS (Cont.)

Class LC 6			Class LC 7			Class LC 8			Class LC 9			Class LC 10			Class LC 11			Nominal Size Range Inches	
Limits of Clearance	Hole H9	Shaft f8	Limits of Clearance	Hole H10	Shaft e9	Limits of Clearance	Hole H10	Shaft d9	Limits of Clearance	Hole H11	Shaft c10	Limits of Clearance	Hole H12	Shaft	Limits of Clearance	Hole H13	Shaft	Over	To
0.3 / 1.9	+1.0 / 0	−0.3 / −0.9	0.6 / 3.2	+1.6 / 0	−0.6 / −1.6	1.0 / 3.6	+1.6 / −0	−1.0 / −2.0	2.5 / 6.6	+2.5 / −0	−2.5 / −4.1	4 / 12	+4 / −0	−4 / −8	5 / 17	+6 / −0	−5 / −11	0	0.12
0.4 / 2.3	+1.2 / 0	−0.4 / −1.1	0.8 / 3.8	+1.8 / 0	−0.8 / −2.0	1.2 / 4.2	+1.8 / −0	−1.2 / −2.4	2.8 / 7.6	+3.0 / −0	−2.8 / −4.6	4.5 / 14.5	+5 / −0	−4.5 / −9.5	6 / 20	+7 / −0	−6 / −13	0.12	0.24
0.5 / 2.8	+1.4 / 0	−0.5 / −1.4	1.0 / 4.6	+2.2 / 0	−1.0 / −2.4	1.6 / 5.2	+2.2 / −0	−1.6 / −3.0	3.0 / 8.7	+3.5 / −0	−3.0 / −5.2	5 / 17	+6 / −0	−5 / −11	7 / 25	+9 / −0	−7 / −16	0.24	0.40
0.6 / 3.2	+1.6 / 0	−0.6 / −1.6	1.2 / 5.6	+2.8 / 0	−1.2 / −2.8	2.0 / 6.4	+2.8 / −0	−2.0 / −3.6	3.5 / 10.3	+4.0 / −0	−3.5 / −6.3	6 / 20	+7 / −0	−6 / −13	8 / 28	+10 / −0	−8 / −18	0.40	0.71
0.8 / 4.0	+2.0 / 0	−0.8 / −2.0	1.6 / 7.1	+3.5 / 0	−1.6 / −3.6	2.5 / 8.0	+3.5 / −0	−2.5 / −4.5	4.5 / 13.0	+5.0 / −0	−4.5 / −8.0	7 / 23	+8 / −0	−7 / −15	10 / 34	+12 / −0	−10 / −22	0.71	1.19
1.0 / 5.1	+2.5 / 0	−1.0 / −2.6	2.0 / 8.5	+4.0 / 0	−2.0 / −4.5	3.0 / 9.5	+4.0 / −0	−3.0 / −5.5	5 / 15	+6 / −0	−5 / −9	8 / 28	+10 / −0	−8 / −18	12 / 44	+16 / −0	−12 / −28	1.19	1.97
1.2 / 6.0	+3.0 / 0	−1.2 / −3.0	2.5 / 10.0	+4.5 / 0	−2.5 / −5.5	4.0 / 11.5	+4.5 / −0	−4.0 / −7.0	6 / 17.5	+7 / −0	−6 / −10.5	10 / 34	+12 / −0	−10 / −22	14 / 50	+18 / −0	−14 / −32	1.97	3.15
1.4 / 7.1	+3.5 / 0	−1.4 / −3.6	3.0 / 11.5	+5.0 / 0	−3.0 / −6.5	5.0 / 13.5	+5.0 / −0	−5.0 / −8.5	7 / 21	+9 / −0	−7 / −12	11 / 39	+14 / −0	−11 / −25	16 / 60	+22 / −0	−16 / −38	3.15	4.73
1.6 / 8.1	+4.0 / 0	−1.6 / −4.1	3.5 / 13.5	+6.0 / 0	−3.5 / −7.5	6 / 16	+6 / −0	−6 / −10	8 / 24	+10 / −0	−8 / −14	12 / 44	+16 / −0	−12 / −28	18 / 68	+25 / −0	−18 / −43	4.73	7.09
2.0 / 9.3	+4.5 / 0	−2.0 / −4.8	4.0 / 15.5	+7.0 / 0	−4.0 / −8.5	7 / 18.5	+7 / −0	−7 / −11.5	10 / 29	+12 / −0	−10 / −17	16 / 52	+18 / −0	−16 / −34	22 / 78	+28 / −0	−22 / −50	7.09	9.85
2.2 / 10.2	+5.0 / 0	−2.2 / −5.2	4.5 / 17.5	+8.0 / 0	−4.5 / −9.5	7 / 20	+8 / −0	−7 / −12	12 / 32	+12 / −0	−12 / −20	20 / 60	+20 / −0	−20 / −40	28 / 88	+30 / −0	−28 / −50	9.85	12.41
2.5 / 12.0	+6.0 / 0	−2.5 / −6.0	5.0 / 20.0	+9.0 / 0	−5 / −11	8 / 23	+9 / −0	−8 / −14	14 / 37	+14 / −0	−14 / −23	22 / 66	+22 / −0	−22 / −44	30 / 100	+35 / −0	−30 / −65	12.41	15.75
2.8 / 12.8	+6.0 / 0	−2.8 / −6.8	5.0 / 21.0	+10.0 / 0	−5 / −11	9 / 25	+10 / −0	−9 / −15	16 / 42	+16 / −0	−16 / −26	25 / 75	+25 / −0	−25 / −50	35 / 115	+40 / −0	−35 / −75	15.75	19.69
3.0 / 16.0	+8.0 / 0	−3.0 / −8.0	6.0 / 26.0	+12.0 / −0	−6 / −14	10 / 30	+12 / −0	−10 / −18	18 / 50	+20 / −0	−18 / −30	28 / 88	+30 / −0	−28 / −58	40 / 140	+50 / −0	−40 / −90	19.69	30.09
3.5 / 19.5	+10.0 / 0	−3.5 / −9.5	7.0 / 33.0	+16.0 / −0	−7 / −17	12 / 38	+16 / −0	−12 / −22	20 / 61	+25 / −0	−20 / −36	30 / 110	+40 / −0	−30 / −70	45 / 165	+60 / −0	−45 / −105	30.09	41.49
4.0 / 24.0	+12.0 / 0	−4.0 / −12.0	8.0 / 40.0	+20.0 / −0	−8 / −20	14 / 46	+20 / −0	−14 / −26	25 / 75	+30 / −0	−25 / −45	40 / 140	+50 / −0	−40 / −90	60 / 220	+80 / −0	−60 / −140	41.49	56.19
4.5 / 30.5	+16.0 / 0	−4.5 / −14.5	9.0 / 50.0	+25.0 / −0	−9 / −25	16 / 57	+25 / −0	−16 / −32	30 / 95	+40 / −0	−30 / −55	50 / 170	+60 / −0	−50 / −110	70 / 270	+100 / −0	−70 / −170	56.19	76.39
5.0 / 37.0	+20.0 / 0	−5 / −17	10.0 / 60.0	+30.0 / −0	−10 / −30	18 / 68	+30 / −0	−18 / −38	35 / 115	+50 / −0	−35 / −65	50 / 210	+80 / −0	−50 / −130	80 / 330	+125 / −0	−80 / −205	76.39	100.9
6.0 / 47.0	+25.0 / 0	−6 / −22	12.0 / 67.0	+40.0 / −0	−12 / −27	20 / 85	+40 / −0	−20 / −45	40 / 140	+60 / −0	−40 / −80	60 / 260	+100 / −0	−60 / −160	90 / 410	+160 / −0	−90 / −250	100.9	131.9
7.0 / 57.0	+30.0 / 0	−7 / −27	14.0 / 94.0	+50.0 / −0	−14 / −44	25 / 105	+50 / −0	−25 / −55	50 / 180	+80 / −0	−50 / −100	80 / 330	+125 / −0	−80 / −205	100 / 500	+200 / −0	−100 / −300	131.9	171.9
7.0 / 72.0	+40.0 / 0	−7 / −32	14.0 / 114.0	+60.0 / −0	−14 / −54	25 / 125	+60 / −0	−25 / −65	50 / 210	+100 / −0	−50 / −110	90 / 410	+160 / −0	−90 / −250	125 / 625	+250 / −0	−125 / −375	171.9	200

(Courtesy of ANSI; B4.1–1955.)

Appendix 41 AMERICAN STANDARD TRANSITION LOCATIONAL FITS

Limits are in thousandths of an inch.

Limits for hole and shaft are applied algebraically to the basic size to obtain the limits of size for the mating parts.

Data in bold face are in accordance with ABC agreements.

"Fit" represents the maximum interference (minus values) and the maximum clearance (plus values).

Symbols H7, js6, etc., are Hole and Shaft designations used in ABC System.

Nominal Size Range Inches Over	To	Class LT 1 Fit	Hole H7	Shaft js6	Class LT 2 Fit	Hole H8	Shaft js7	Class LT 3 Fit	Hole H7	Shaft k6	Class LT 4 Fit	Hole H8	Shaft k7	Class LT 5 Fit	Hole H7	Shaft n6	Class LT 6 Fit	Hole H7	Shaft n7
0	0.12	−0.10 / +0.50	+0.4 / −0	+0.10 / −0.10	−0.2 / +0.8	+0.6 / −0	+0.2 / −0.2							−0.5 / +0.15	+0.4 / −0	+0.5 / +0.25	−0.65 / +0.15	+0.4 / −0	−0.65 / +0.25
0.12	0.24	−0.15 / +0.65	+0.5 / −0	+0.15 / −0.15	−0.25 / +0.95	+0.7 / −0	+0.25 / −0.25							−0.6 / +0.2	+0.5 / −0	+0.6 / +0.3	−0.8 / +0.2	+0.5 / −0	+0.8 / +0.3
0.24	0.40	−0.2 / +0.8	+0.6 / −0	+0.2 / −0.2	−0.3 / +1.2	+0.9 / −0	+0.3 / −0.3	−0.5 / +0.5	+0.6 / −0	+0.5 / +0.1	−0.7 / +0.8	+0.9 / −0	+0.7 / +0.1	−0.8 / +0.2	+0.6 / −0	+0.8 / +0.4	−1.0 / +0.2	+0.6 / −0	+1.0 / +0.4
0.40	0.71	−0.2 / +0.9	+0.7 / −0	+0.2 / −0.2	−0.35 / +1.35	+1.0 / −0	+0.35 / −0.35	−0.5 / +0.6	+0.7 / −0	+0.5 / +0.1	−0.8 / +0.9	+1.0 / −0	+0.8 / +0.1	−0.9 / +0.2	+0.7 / −0	+0.9 / +0.5	−1.2 / +0.2	+0.7 / −0	+1.2 / +0.5
0.71	1.19	−0.25 / +1.05	+0.8 / −0	+0.25 / −0.25	−0.4 / +1.6	+1.2 / −0	+0.4 / −0.4	−0.6 / +0.7	+0.8 / −0	+0.6 / +0.1	−0.9 / +1.1	+1.2 / −0	+0.9 / +0.1	−1.1 / +0.2	+0.8 / −0	+1.1 / +0.6	−1.4 / +0.2	+0.8 / −0	+1.4 / +0.6
1.19	1.97	−0.3 / +1.3	+1.0 / −0	+0.3 / −0.3	−0.5 / +2.1	+1.6 / −0	+0.5 / −0.5	−0.7 / +0.9	+1.0 / −0	+0.7 / +0.1	−1.1 / +1.5	+1.6 / −0	+1.1 / +0.1	−1.3 / +0.3	+1.0 / −0	+1.3 / +0.7	−1.7 / +0.3	+1.0 / −0	+1.7 / +0.7
1.97	3.15	−0.3 / +1.5	+1.2 / −0	+0.3 / −0.3	−0.6 / +2.4	+1.8 / −0	+0.6 / −0.6	−0.8 / +1.1	+1.2 / −0	+0.8 / +0.1	−1.3 / +1.7	+1.8 / −0	+1.3 / +0.1	−1.5 / +0.4	+1.2 / −0	+1.5 / +0.8	−2.0 / +0.4	+1.2 / −0	+2.0 / +0.8
3.15	4.73	−0.4 / +1.8	+1.4 / −0	+0.4 / −0.4	−0.7 / +2.9	+2.2 / −0	+0.7 / −0.7	−1.0 / +1.3	+1.4 / −0	+1.0 / +0.1	−1.5 / +2.1	+2.2 / −0	+1.5 / +0.1	−1.9 / +0.4	+1.4 / −0	+1.9 / +1.0	−2.4 / +0.4	+1.4 / −0	+2.4 / +1.0
4.73	7.09	−0.5 / +2.1	+1.6 / −0	+0.5 / −0.5	−0.8 / +3.3	+2.5 / −0	+0.8 / −0.8	−1.1 / +1.5	+1.6 / −0	+1.1 / +0.1	−1.7 / +2.4	+2.5 / −0	+1.7 / +0.1	−2.2 / +0.4	+1.6 / −0	+2.2 / +1.2	−2.8 / +0.4	+1.6 / −0	+2.8 / +1.2
7.09	9.85	−0.6 / +2.4	+1.8 / −0	+0.6 / −0.6	−0.9 / +3.7	+2.8 / −0	+0.9 / −0.9	−1.4 / +1.6	+1.8 / −0	+1.4 / +0.2	−2.0 / +2.6	+2.8 / −0	+2.0 / +0.2	−2.6 / +0.4	+1.8 / −0	+2.6 / +1.4	−3.2 / +0.4	+1.8 / −0	+3.2 / +1.4
9.85	12.41	−0.6 / +2.6	+2.0 / −0	+0.6 / −0.6	−1.0 / +4.0	+3.0 / −0	+1.0 / −1.0	−1.4 / +1.8	+2.0 / −0	+1.4 / +0.2	−2.2 / +2.8	+3.0 / −0	+2.2 / +0.2	−2.6 / +0.6	+2.0 / −0	+2.6 / +1.4	−3.4 / +0.6	+2.0 / −0	+3.4 / +1.4
12.41	15.75	−0.7 / +2.9	+2.2 / −0	+0.7 / −0.7	−1.0 / +4.5	+3.5 / −0	+1.0 / −1.0	−1.6 / +2.0	+2.2 / −0	+1.6 / +0.2	−2.4 / +3.3	+3.5 / −0	+2.4 / +0.2	−3.0 / +0.6	+2.2 / −0	+3.0 / +1.6	−3.8 / +0.6	+2.2 / −0	+3.8 / +1.6
15.75	19.69	−0.8 / +3.3	+2.5 / −0	+0.8 / −0.8	−1.2 / +5.2	+4.0 / −0	+1.2 / −1.2	−1.8 / +2.3	+2.5 / −0	+1.8 / +0.2	−2.7 / +3.8	+4.0 / −0	+2.7 / +0.2	−3.4 / +0.7	+2.5 / −0	+3.4 / +1.8	−4.3 / +0.7	+2.5 / −0	+4.3 / +1.8

(Courtesy of ANSI; B4.1–1955.)

Appendix 42 AMERICAN STANDARD INTERFERENCE LOCATIONAL FITS

Limits are in thousandths of an inch.
Limits for hole and shaft are applied algebraically to the
basic size to obtain the limits of size for the parts.
Data in bold face are in accordance with ABC agreements.
Symbols H7, p6, etc., are Hole and Shaft designations
used in ABC System.

Nominal Size Range Inches Over — To	Class LN 1 Limits of Interference	Class LN 1 Standard Limits Hole H6	Class LN 1 Standard Limits Shaft n5	Class LN 2 Limits of Interference	Class LN 2 Standard Limits Hole H7	Class LN 2 Standard Limits Shaft p6	Class LN 3 Limits of Interference	Class LN 3 Standard Limits Hole H7	Class LN 3 Standard Limits Shaft r6
0 — 0.12	0 / 0.45	+ 0.25 / − 0	+0.45 / +0.25	0 / 0.65	+ 0.4 / − 0	+ 0.65 / + 0.4	0.1 / 0.75	+ 0.4 / − 0	+ 0.75 / + 0.5
0.12 — 0.24	0 / 0.5	+ 0.3 / − 0	+0.5 / +0.3	0 / 0.8	+ 0.5 / − 0	+ 0.8 / + 0.5	0.1 / 0.9	+ 0.5 / − 0	+ 0.9 / + 0.6
0.24 — 0.40	0 / 0.65	+ 0.4 / − 0	+0.65 / +0.4	0 / 1.0	+ 0.6 / − 0	+ 1.0 / + 0.6	0.2 / 1.2	+ 0.6 / − 0	+ 1.2 / + 0.8
0.40 — 0.71	0 / 0.8	+ 0.4 / − 0	+0.8 / +0.4	0 / 1.1	+ 0.7 / − 0	+ 1.1 / + 0.7	0.3 / 1.4	+ 0.7 / − 0	+ 1.4 / + 1.0
0.71 — 1.19	0 / 1.0	+ 0.5 / − 0	+1.0 / +0.5	0 / 1.3	+ 0.8 / − 0	+ 1.3 / + 0.8	0.4 / 1.7	+ 0.8 / − 0	+ 1.7 / + 1.2
1.19 — 1.97	0 / 1.1	+ 0.6 / − 0	+1.1 / +0.6	0 / 1.6	+ 1.0 / − 0	+ 1.6 / + 1.0	0.4 / 2.0	+ 1.0 / − 0	+ 2.0 / + 1.4
1.97 — 3.15	0.1 / 1.3	+ 0.7 / − 0	+1.3 / +0.7	0.2 / 2.1	+ 1.2 / − 0	+ 2.1 / + 1.4	0.4 / 2.3	+ 1.2 / − 0	+ 2.3 / + 1.6
3.15 — 4.73	0.1 / 1.6	+ 0.9 / − 0	+1.6 / +1.0	0.2 / 2.5	+ 1.4 / − 0	+ 2.5 / + 1.6	0.6 / 2.9	+ 1.4 / − 0	+ 2.9 / + 2.0
4.73 — 7.09	0.2 / 1.9	+ 1.0 / − 0	+1.9 / +1.2	0.2 / 2.8	+ 1.6 / − 0	+ 2.8 / + 1.8	0.9 / 3.5	+ 1.6 / − 0	+ 3.5 / + 2.5
7.09 — 9.85	0.2 / 2.2	+ 1.2 / − 0	+2.2 / +1.4	0.2 / 3.2	+ 1.8 / − 0	+ 3.2 / + 2.0	1.2 / 4.2	+ 1.8 / − 0	+ 4.2 / + 3.0
9.85 — 12.41	0.2 / 2.3	+ 1.2 / − 0	+2.3 / +1.4	0.2 / 3.4	+ 2.0 / − 0	+ 3.4 / + 2.2	1.5 / 4.7	+ 2.0 / − 0	+ 4.7 / + 3.5
12.41 — 15.75	0.2 / 2.6	+ 1.4 / − 0	+2.6 / +1.6	0.3 / 3.9	+ 2.2 / − 0	+ 3.9 / + 2.5	2.3 / 5.9	+ 2.2 / − 0	+ 5.9 / + 4.5
15.75 — 19.69	0.2 / 2.8	+ 1.6 / − 0	+2.8 / +1.8	0.3 / 4.4	+ 2.5 / − 0	+ 4.4 / + 2.8	2.5 / 6.6	+ 2.5 / − 0	+ 6.6 / + 5.0
19.69 — 30.09		+ 2.0 / − 0		0.5 / 5.5	+ 3 / − 0	+ 5.5 / + 3.5	4 / 9	+ 3 / − 0	+ 9 / + 7
30.09 — 41.49		+ 2.5 / − 0		0.5 / 7.0	+ 4 / − 0	+ 7.0 / + 4.5	5 / 11.5	+ 4 / − 0	+11.5 / + 9
41.49 — 56.19		+ 3.0 / − 0		1 / 9	+ 5 / − 0	+ 9 / + 6	7 / 15	+ 5 / − 0	+15 / +12
56.19 — 76.39		+ 4.0 / − 0		1 / 11	+ 6 / − 0	+11 / + 7	10 / 20	+ 6 / − 0	+20 / +16
76.39 — 100.9		+ 5.0 / − 0		1 / 14	+ 8 / − 0	+14 / + 9	12 / 25	+ 8 / − 0	+25 / +20
100.9 — 131.9		+ 6.0 / − 0		2 / 18	+10 / − 0	+18 / +12	15 / 31	+10 / − 0	+31 / +25
131.9 — 171.9		+ 8.0 / − 0		4 / 24	+12 / − 0	+24 / +16	18 / 38	+12 / − 0	+38 / +30
171.9 — 200		+10.0 / − 0		4 / 30	+16 / − 0	+30 / +20	24 / 50	+16 / − 0	+50 / +40

(Courtesy of ANSI; B4.1–1955.)

Appendix 43 AMERICAN STANDARD FORCE AND SHRINK FITS

Limits are in thousandths of an inch.
Limits for hole and shaft are applied algebraically to the basic size to obtain the limits of size for the parts.
Data in bold face are in accordance with ABC agreements.
Symbols H7, s6, etc., are Hole and Shaft designations used in ABC System.

Nominal Size Range Inches Over — To	Class FN 1 Limits of Interference	Class FN 1 Hole H6	Class FN 1 Shaft	Class FN 2 Limits of Interference	Class FN 2 Hole H7	Class FN 2 Shaft s6	Class FN 3 Limits of Interference	Class FN 3 Hole H7	Class FN 3 Shaft t6	Class FN 4 Limits of Interference	Class FN 4 Hole H7	Class FN 4 Shaft u6	Class FN 5 Limits of Interference	Class FN 5 Hole H8	Class FN 5 Shaft x7
0 — 0.12	0.05	+0.25	+0.5	0.2	+0.4	+0.85				0.3	+0.4	+0.95	0.3	+0.6	+1.3
	0.5	−0	+0.3	0.85	−0	+0.6				0.95	−0	+0.7	1.3	−0	+0.9
0.12 — 0.24	0.1	+0.3	+0.6	0.2	+0.5	+1.0				0.4	+0.5	+1.2	0.5	+0.7	+1.7
	0.6	−0	+0.4	1.0	−0	+0.7				1.2	−0	+0.9	1.7	−0	+1.2
0.24 — 0.40	0.1	+0.4	+0.75	0.4	+0.6	+1.4				0.6	+0.6	+1.6	0.5	+0.9	+2.0
	0.75	−0	+0.5	1.4	−0	+1.0				1.6	−0	+1.2	2.0	−0	+1.4
0.40 — 0.56	0.1	−0.4	+0.8	0.5	+0.7	+1.6				0.7	+0.7	+1.8	0.6	+1.0	+2.3
	0.8	−0	+0.5	1.6	−0	+1.2				1.8	−0	+1.4	2.3	−0	+1.6
0.56 — 0.71	0.2	+0.4	+0.9	0.5	+0.7	+1.6				0.7	+0.7	+1.8	0.8	+1.0	+2.5
	0.9	−0	+0.6	1.6	−0	+1.2				1.8	−0	+1.4	2.5	−0	+1.8
0.71 — 0.95	0.2	+0.5	+1.1	0.6	+0.8	+1.9				0.8	+0.8	+2.1	1.0	+1.2	+3.0
	1.1	−0	+0.7	1.9	−0	+1.4				2.1	−0	+1.6	3.0	−0	+2.2
0.95 — 1.19	0.3	+0.5	+1.2	0.6	+0.8	+1.9	0.8	+0.8	+2.1	1.0	+0.8	+2.3	1.3	+1.2	+3.3
	1.2	−0	+0.8	1.9	−0	+1.4	2.1	−0	+1.6	2.3	−0	+1.8	3.3	−0	+2.5
1.19 — 1.58	0.3	+0.6	+1.3	0.8	+1.0	+2.4	1.0	+1.0	+2.6	1.5	+1.0	+3.1	1.4	+1.6	+4.0
	1.3	−0	+0.9	2.4	−0	+1.8	2.6	−0	+2.0	3.1	−0	+2.5	4.0	−0	+3.0
1.58 — 1.97	0.4	+0.6	+1.4	0.8	+1.0	+2.4	1.2	+1.0	+2.8	1.8	+1.0	+3.4	2.4	+1.6	+5.0
	1.4	−0	+1.0	2.4	−0	+1.8	2.8	−0	+2.2	3.4	−0	+2.8	5.0	−0	+4.0
1.97 — 2.56	0.6	+0.7	+1.8	0.8	+1.2	+2.7	1.3	+1.2	+3.2	2.3	+1.2	+4.2	3.2	+1.8	+6.2
	1.8	−0	+1.3	2.7	−0	+2.0	3.2	−0	+2.5	4.2	−0	+3.5	6.2	−0	+5.0
2.56 — 3.15	0.7	+0.7	+1.9	1.0	+1.2	+2.9	1.8	+1.2	+3.7	2.8	+1.2	+4.7	4.2	+1.8	+7.2
	1.9	−0	+1.4	2.9	−0	+2.2	3.7	−0	+3.0	4.7	−0	+4.0	7.2	−0	+6.0
3.15 — 3.94	0.9	+0.9	+2.4	1.4	+1.4	+3.7	2.1	+1.4	+4.4	3.6	+1.4	+5.9	4.8	+2.2	+8.4
	2.4	−0	+1.8	3.7	−0	+2.8	4.4	−0	+3.5	5.9	−0	+5.0	8.4	−0	+7.0
3.94 — 4.73	1.1	+0.9	+2.6	1.6	+1.4	+3.9	2.6	+1.4	+4.9	4.6	+1.4	+6.9	5.8	+2.2	+9.4
	2.6	−0	+2.0	3.9	−0	+3.0	4.9	−0	+4.0	6.9	−0	+6.0	9.4	−0	+8.0
4.73 — 5.52	1.2	+1.0	+2.9	1.9	+1.6	+4.5	3.4	+1.6	+6.0	5.4	+1.6	+8.0	7.5	+2.5	+11.6
	2.9	−0	+2.2	4.5	−0	+3.5	6.0	−0	+5.0	8.0	−0	+7.0	11.6	−0	+10.0
5.52 — 6.30	1.5	+1.0	+3.2	2.4	+1.6	+5.0	3.4	+1.6	+6.0	5.4	+1.6	+8.0	9.5	+2.5	+13.6
	3.2	−0	+2.5	5.0	−0	+4.0	6.0	−0	+5.0	8.0	−0	+7.0	13.6	−0	+12.0
6.30 — 7.09	1.8	+1.0	+3.5	2.9	+1.6	+5.5	4.4	+1.6	+7.0	6.4	+1.6	+9.0	9.5	+2.5	+13.6
	3.5	−0	+2.8	5.5	−0	+4.5	7.0	−0	+6.0	9.0	−0	+8.0	13.6	−0	+12.0
7.09 — 7.88	1.8	+1.2	+3.8	3.2	+1.8	+6.2	5.2	+1.8	+8.2	7.2	+1.8	+10.2	11.2	+2.8	+15.8
	3.8	−0	+3.0	6.2	−0	+5.0	8.2	−0	+7.0	10.2	−0	+9.0	15.8	−0	+14.0
7.88 — 8.86	2.3	+1.2	+4.3	3.2	+1.8	+6.2	5.2	+1.8	+8.2	8.2	+1.8	+11.2	13.2	+2.8	+17.8
	4.3	−0	+3.5	6.2	−0	+5.0	8.2	−0	+7.0	11.2	−0	+10.0	17.8	−0	+16.0
8.86 — 9.85	2.3	+1.2	+4.3	4.2	+1.8	+7.2	6.2	+1.8	+9.2	10.2	+1.8	+13.2	13.2	+2.8	+17.8
	4.3	−0	+3.5	7.2	−0	+6.0	9.2	−0	+8.0	13.2	−0	+12.0	17.8	−0	+16.0
9.85 — 11.03	2.8	+1.2	+4.9	4.0	+2.0	+7.2	7.0	+2.0	+10.2	10.0	+2.0	+13.2	15.0	+3.0	+20.0
	4.9	−0	+4.0	7.2	−0	+6.0	10.2	−0	+9.0	13.2	−0	+12.0	20.0	−0	+18.0
11.03 — 12.41	2.8	+1.2	+4.9	5.0	+2.0	+8.2	7.0	+2.0	+10.2	12.0	+2.0	+15.2	17.0	+3.0	+22.0
	4.9	−0	+4.0	8.2	−0	+7.0	10.2	−0	+9.0	15.2	−0	+14.0	22.0	−0	+20.0
12.41 — 13.98	3.1	+1.4	+5.5	5.8	+2.2	+9.4	7.8	+2.2	+11.4	13.8	+2.2	+17.4	18.5	+3.5	+24.2
	5.5	−0	+4.5	9.4	−0	+8.0	11.4	−0	+10.0	17.4	−0	+16.0	24.2	+0	+22.0
13.98 — 15.75	3.6	+1.4	+6.1	5.8	+2.2	+9.4	9.8	+2.2	+13.4	15.8	+2.2	+19.4	21.5	+3.5	+27.2
	6.1	−0	+5.0	9.4	−0	+8.0	13.4	−0	+12.0	19.4	−0	+18.0	27.2	−0	+25.0
15.75 — 17.72	4.4	+1.6	+7.0	6.5	+2.5	+10.6	9.5	+2.5	+13.6	17.5	+2.5	+21.6	24.0	+4.0	+30.5
	7.0	−0	+6.0	10.6	−0	+9.0	13.6	−0	+12.0	21.6	−0	+20.0	30.5	−0	+28.0
17.72 — 19.69	4.4	+1.6	+7.0	7.5	+2.5	+11.6	11.5	+2.5	+15.6	19.5	+2.5	+23.6	26.0	+4.0	+32.5
	7.0	−0	+6.0	11.6	−0	+10.0	15.6	−0	+14.0	23.6	−0	+22.0	32.5	−0	+30.0

(Courtesy of ANSI; B4.1–1955.)

Appendix 44 THE INTERNATIONAL TOLERANCE GRADES (ANSI B4.2)

Dimensions are in mm.

Basic sizes Over	Up to and including	IT01	IT0	IT1	IT2	IT3	IT4	IT5	IT6	IT7	IT8	IT9	IT10	IT11	IT12	IT13	IT14	IT15	IT16
0	3	0.0003	0.0005	0.0008	0.0012	0.002	0.003	0.004	0.006	0.010	0.014	0.025	0.040	0.060	0.100	0.140	0.250	0.400	0.600
3	6	0.0004	0.0006	0.001	0.0015	0.0025	0.004	0.005	0.008	0.012	0.018	0.030	0.048	0.075	0.120	0.180	0.300	0.480	0.750
6	10	0.0004	0.0006	0.001	0.0015	0.0025	0.004	0.006	0.009	0.015	0.022	0.036	0.058	0.090	0.150	0.220	0.360	0.580	0.900
10	18	0.0005	0.0008	0.0012	0.002	0.003	0.005	0.008	0.011	0.018	0.027	0.043	0.070	0.110	0.180	0.270	0.430	0.700	1.100
18	30	0.0006	0.001	0.0015	0.0025	0.004	0.006	0.009	0.013	0.021	0.033	0.052	0.084	0.130	0.210	0.330	0.520	0.840	1.300
30	50	0.0006	0.001	0.0015	0.0025	0.004	0.007	0.011	0.016	0.025	0.039	0.062	0.100	0.160	0.250	0.390	0.620	1.000	1.600
50	80	0.0008	0.0012	0.002	0.003	0.005	0.008	0.013	0.019	0.030	0.046	0.074	0.120	0.190	0.300	0.460	0.740	1.200	1.900
80	120	0.001	0.0015	0.0025	0.004	0.006	0.010	0.015	0.022	0.035	0.054	0.087	0.140	0.220	0.350	0.540	0.870	1.400	2.200
120	180	0.0012	0.002	0.0035	0.005	0.008	0.012	0.018	0.025	0.040	0.063	0.100	0.160	0.250	0.400	0.630	1.000	1.600	2.500
180	250	0.002	0.003	0.0045	0.007	0.010	0.014	0.020	0.029	0.046	0.072	0.115	0.185	0.290	0.460	0.720	1.150	1.850	2.900
250	315	0.0025	0.004	0.006	0.008	0.012	0.016	0.023	0.032	0.052	0.081	0.130	0.210	0.320	0.520	0.810	1.300	2.100	3.200
315	400	0.003	0.005	0.007	0.009	0.013	0.018	0.025	0.036	0.057	0.089	0.140	0.230	0.360	0.570	0.890	1.400	2.300	3.600
400	500	0.004	0.006	0.008	0.010	0.015	0.020	0.027	0.040	0.063	0.097	0.155	0.250	0.400	0.630	0.970	1.550	2.500	4.000
500	630	0.0045	0.006	0.009	0.011	0.016	0.022	0.030	0.044	0.070	0.110	0.175	0.280	0.440	0.700	1.100	1.750	2.800	4.400
630	800	0.005	0.007	0.010	0.013	0.018	0.025	0.035	0.050	0.080	0.125	0.200	0.320	0.500	0.800	1.250	2.000	3.200	5.000
800	1000	0.0055	0.008	0.011	0.015	0.021	0.029	0.040	0.056	0.090	0.140	0.230	0.360	0.560	0.900	1.400	2.300	3.600	5.600
1000	1250	0.0065	0.009	0.013	0.018	0.024	0.034	0.046	0.066	0.105	0.165	0.260	0.420	0.660	1.050	1.650	2.600	4.200	6.600
1250	1600	0.008	0.011	0.015	0.021	0.029	0.040	0.054	0.078	0.125	0.195	0.310	0.500	0.780	1.250	1.950	3.100	5.000	7.800
1600	2000	0.009	0.013	0.018	0.025	0.035	0.048	0.065	0.092	0.150	0.230	0.370	0.600	0.920	1.500	2.300	3.700	6.000	9.200
2000	2500	0.011	0.015	0.022	0.030	0.041	0.057	0.077	0.110	0.175	0.280	0.440	0.700	1.100	1.750	2.800	4.400	7.000	11.000
2500	3150	0.013	0.018	0.026	0.036	0.050	0.069	0.093	0.135	0.210	0.330	0.540	0.860	1.350	2.100	3.300	5.400	8.600	13.500

Tolerance grades[3]

[3] IT Values for tolerance grades larger than IT16 can be calculated by using the following formulas: IT17 = IT12 × 10; IT18 = IT13 × 10; etc.

Appendix 45 PREFERRED HOLE BASIS CLEARANCE FITS— CYLINDRICAL FITS (ANSI B4.2)

**AMERICAN NATIONAL STANDARD
PREFERRED METRIC LIMITS AND FITS**

ANSI B4.2-1978

Dimensions in mm.

BASIC SIZE		LOOSE RUNNING Hole H11	Shaft c11	Fit	FREE RUNNING Hole H9	Shaft d9	Fit	CLOSE RUNNING Hole H8	Shaft f7	Fit	SLIDING Hole H7	Shaft g6	Fit	LOCATIONAL CLEARANCE Hole H7	Shaft h6	Fit
1	MAX	1.060	0.940	0.180	1.025	0.980	0.070	1.014	0.994	0.030	1.010	0.998	0.018	1.010	1.000	0.016
	MIN	1.000	0.880	0.060	1.000	0.955	0.020	1.000	0.984	0.006	1.000	0.992	0.002	1.000	0.994	0.000
1.2	MAX	1.260	1.140	0.180	1.225	1.180	0.070	1.214	1.194	0.030	1.210	1.198	0.018	1.210	1.200	0.016
	MIN	1.200	1.080	0.060	1.200	1.155	0.020	1.200	1.184	0.006	1.200	1.192	0.002	1.200	1.194	0.000
1.6	MAX	1.660	1.540	0.180	1.625	1.580	0.070	1.614	1.594	0.030	1.610	1.598	0.018	1.610	1.600	0.016
	MIN	1.600	1.480	0.060	1.600	1.555	0.020	1.600	1.584	0.006	1.600	1.592	0.002	1.600	1.594	0.000
2	MAX	2.060	1.940	0.180	2.025	1.980	0.070	2.014	1.994	0.030	2.010	1.998	0.018	2.010	2.000	0.016
	MIN	2.000	1.880	0.060	2.000	1.955	0.020	2.000	1.984	0.006	2.000	1.992	0.002	2.000	1.994	0.000
2.5	MAX	2.560	2.440	0.180	2.525	2.480	0.070	2.514	2.494	0.030	2.510	2.498	0.018	2.510	2.500	0.016
	MIN	2.500	2.380	0.060	2.500	2.455	0.020	2.500	2.484	0.006	2.500	2.492	0.002	2.500	2.494	0.000
3	MAX	3.060	2.940	0.180	3.025	2.980	0.070	3.014	2.994	0.030	3.010	2.998	0.018	3.010	3.000	0.016
	MIN	3.000	2.880	0.060	3.000	2.955	0.020	3.000	2.984	0.006	3.000	2.992	0.002	3.000	2.994	0.000
4	MAX	4.075	3.930	0.220	4.030	3.970	0.090	4.018	3.990	0.040	4.012	3.996	0.024	4.012	4.000	0.020
	MIN	4.000	3.855	0.070	4.000	3.940	0.030	4.000	3.978	0.010	4.000	3.988	0.004	4.000	3.992	0.000
5	MAX	5.075	4.930	0.220	5.030	4.970	0.090	5.018	4.990	0.040	5.012	4.996	0.024	5.012	5.000	0.020
	MIN	5.000	4.855	0.070	5.000	4.940	0.030	5.000	4.978	0.010	5.000	4.988	0.004	5.000	4.992	0.000
6	MAX	6.075	5.930	0.220	6.030	5.970	0.090	6.018	5.990	0.040	6.012	5.996	0.024	6.012	6.000	0.020
	MIN	6.000	5.855	0.070	6.000	5.940	0.030	6.000	5.978	0.010	6.000	5.988	0.004	6.000	5.992	0.000
8	MAX	8.090	7.920	0.260	8.036	7.960	0.112	8.022	7.987	0.050	8.015	7.995	0.029	8.015	8.000	0.024
	MIN	8.000	7.830	0.080	8.000	7.924	0.040	8.000	7.972	0.013	8.000	7.986	0.005	8.000	7.991	0.000
10	MAX	10.090	9.920	0.260	10.036	9.960	0.112	10.022	9.987	0.050	10.015	9.995	0.029	10.015	10.000	0.024
	MIN	10.000	9.830	0.080	10.000	9.924	0.040	10.000	9.972	0.013	10.000	9.986	0.005	10.000	9.991	0.000
12	MAX	12.110	11.905	0.315	12.043	11.950	0.136	12.027	11.984	0.061	12.018	11.994	0.035	12.018	12.000	0.029
	MIN	12.000	11.795	0.095	12.000	11.907	0.050	12.000	11.966	0.016	12.000	11.983	0.006	12.000	11.989	0.000
16	MAX	16.110	15.905	0.315	16.043	15.950	0.136	16.027	15.984	0.061	16.018	15.994	0.035	16.018	16.000	0.029
	MIN	16.000	15.795	0.095	16.000	15.907	0.050	16.000	15.966	0.016	16.000	15.983	0.006	16.000	15.989	0.000
20	MAX	20.130	19.890	0.370	20.052	19.935	0.169	20.033	19.980	0.074	20.021	19.993	0.041	20.021	20.000	0.034
	MIN	20.000	19.760	0.110	20.000	19.883	0.065	20.000	19.959	0.020	20.000	19.980	0.007	20.000	19.987	0.000
25	MAX	25.130	24.890	0.370	25.052	24.935	0.169	25.033	24.980	0.074	25.021	24.993	0.041	25.021	25.000	0.034
	MIN	25.000	24.760	0.110	25.000	24.883	0.065	25.000	24.959	0.020	25.000	24.980	0.007	25.000	24.987	0.000
30	MAX	30.130	29.890	0.370	30.052	29.935	0.169	30.033	29.980	0.074	30.021	29.993	0.041	30.021	30.000	0.034
	MIN	30.000	29.760	0.110	30.000	29.883	0.065	30.000	29.959	0.020	30.000	29.980	0.007	30.000	29.987	0.000

Cont.

AMERICAN NATIONAL STANDARD
PREFERRED METRIC LIMITS AND FITS

ANSI B4.2-1978

Dimensions in mm.

BASIC SIZE		LOOSE RUNNING			FREE RUNNING			CLOSE RUNNING			SLIDING			LOCATIONAL CLEARANCE		
		Hole H11	Shaft c11	Fit	Hole H9	Shaft d9	Fit	Hole H8	Shaft f7	Fit	Hole H7	Shaft g6	Fit	Hole H7	Shaft h6	Fit
40	MAX	40.160	39.880	0.440	40.062	39.920	0.204	40.039	39.975	0.089	40.025	39.991	0.050	40.025	40.000	0.041
	MIN	40.000	39.720	0.120	40.000	39.858	0.080	40.000	39.950	0.025	40.000	39.975	0.009	40.000	39.984	0.000
50	MAX	50.160	49.870	0.450	50.062	49.920	0.204	50.039	49.975	0.089	50.025	49.991	0.050	50.025	50.000	0.041
	MIN	50.000	49.710	0.130	50.000	49.858	0.080	50.000	49.950	0.025	50.000	49.975	0.009	50.000	49.984	0.000
60	MAX	60.190	59.860	0.520	60.074	59.900	0.248	60.046	59.970	0.106	60.030	59.990	0.059	60.030	60.000	0.049
	MIN	60.000	59.670	0.140	60.000	59.826	0.100	60.000	59.940	0.030	60.000	59.971	0.010	60.000	59.981	0.000
80	MAX	80.190	79.850	0.530	80.074	79.900	0.248	80.046	79.970	0.106	80.030	79.990	0.059	80.030	80.000	0.049
	MIN	80.000	79.660	0.150	80.000	79.826	0.100	80.000	79.940	0.030	80.000	79.971	0.010	80.000	79.981	0.000
100	MAX	100.220	99.830	0.610	100.087	99.880	0.294	100.054	99.964	0.125	100.035	99.988	0.069	100.035	100.000	0.057
	MIN	100.000	99.610	0.170	100.000	99.793	0.120	100.000	99.929	0.036	100.000	99.966	0.012	100.000	99.978	0.000
120	MAX	120.220	119.820	0.620	120.087	119.880	0.294	120.054	119.964	0.125	120.035	119.988	0.069	120.035	120.000	0.057
	MIN	120.000	119.600	0.180	120.000	119.793	0.120	120.000	119.929	0.036	120.000	119.966	0.012	120.000	119.978	0.000
160	MAX	160.250	159.790	0.710	160.100	159.855	0.345	160.063	159.957	0.146	160.040	159.986	0.079	160.040	160.000	0.065
	MIN	160.000	159.540	0.210	160.000	159.755	0.145	160.000	159.917	0.043	160.000	159.961	0.014	160.000	159.975	0.000
200	MAX	200.290	199.760	0.820	200.115	199.830	0.400	200.072	199.950	0.168	200.046	199.985	0.090	200.046	200.000	0.075
	MIN	200.000	199.470	0.240	200.000	199.715	0.170	200.000	199.904	0.050	200.000	199.956	0.015	200.000	199.971	0.000
250	MAX	250.290	249.720	0.860	250.115	249.830	0.400	250.072	249.950	0.168	250.046	249.985	0.090	250.046	250.000	0.075
	MIN	250.000	249.430	0.280	250.000	249.715	0.170	250.000	249.904	0.050	250.000	249.956	0.015	250.000	249.971	0.000
300	MAX	300.320	299.670	0.970	300.130	299.810	0.450	300.081	299.944	0.189	300.052	299.983	0.101	300.052	300.000	0.084
	MIN	300.000	299.350	0.330	300.000	299.680	0.190	300.000	299.892	0.056	300.000	299.951	0.017	300.000	299.968	0.000
400	MAX	400.360	399.600	1.120	400.140	399.790	0.490	400.089	399.938	0.208	400.057	399.982	0.111	400.057	400.000	0.093
	MIN	400.000	399.240	0.400	400.000	399.650	0.210	400.000	399.881	0.062	400.000	399.946	0.018	400.000	399.964	0.000
500	MAX	500.400	499.520	1.280	500.155	499.770	0.540	500.097	499.932	0.228	500.063	499.980	0.123	500.063	500.000	0.103
	MIN	500.000	499.120	0.480	500.000	499.615	0.230	500.000	499.869	0.068	500.000	499.940	0.020	500.000	499.960	0.000

AMERICAN NATIONAL STANDARD
PREFERRED METRIC LIMITS AND FITS

ANSI B4.2-1978

Dimensions in mm.

BASIC SIZE	MAX/MIN	LOCATIONAL TRANSN. Hole H7	Shaft k6	Fit	LOCATIONAL TRANSN. Hole H7	Shaft n6	Fit	LOCATIONAL INTERF. Hole H7	Shaft p6	Fit	MEDIUM DRIVE Hole H7	Shaft s6	Fit	FORCE Hole H7	Shaft u6	Fit
1	MAX	1.010	1.006	0.010	1.010	1.010	0.006	1.010	1.012	0.004	1.010	1.020	-0.004	1.010	1.024	-0.008
	MIN	1.000	1.000	-0.006	1.000	1.004	-0.010	1.000	1.006	-0.012	1.000	1.014	-0.020	1.000	1.018	-0.024
1.2	MAX	1.210	1.206	0.010	1.210	1.210	0.006	1.210	1.212	0.004	1.210	1.220	-0.004	1.210	1.224	-0.008
	MIN	1.200	1.200	-0.006	1.200	1.204	-0.010	1.200	1.206	-0.012	1.200	1.214	-0.020	1.200	1.218	-0.024
1.6	MAX	1.610	1.606	0.010	1.610	1.610	0.006	1.610	1.612	0.004	1.610	1.620	-0.004	1.610	1.624	-0.008
	MIN	1.600	1.600	-0.006	1.600	1.604	-0.010	1.600	1.606	-0.012	1.600	1.614	-0.020	1.600	1.618	-0.024
2	MAX	2.010	2.006	0.010	2.010	2.010	0.006	2.010	2.012	0.004	2.010	2.020	-0.004	2.010	2.024	-0.008
	MIN	2.000	2.000	-0.006	2.000	2.004	-0.010	2.000	2.006	-0.012	2.000	2.014	-0.020	2.000	2.018	-0.024
2.5	MAX	2.510	2.506	0.010	2.510	2.510	0.006	2.510	2.512	0.004	2.510	2.520	-0.004	2.510	2.524	-0.008
	MIN	2.500	2.500	-0.006	2.500	2.504	-0.010	2.500	2.506	-0.012	2.500	2.514	-0.020	2.500	2.518	-0.024
3	MAX	3.010	3.006	0.010	3.010	3.010	0.006	3.010	3.012	0.004	3.010	3.020	-0.004	3.010	3.024	-0.008
	MIN	3.000	3.000	-0.006	3.000	3.004	-0.010	3.000	3.006	-0.012	3.000	3.014	-0.020	3.000	3.018	-0.024
4	MAX	4.012	4.009	0.011	4.012	4.016	0.004	4.012	4.020	0.000	4.012	4.027	-0.007	4.012	4.031	-0.011
	MIN	4.000	4.001	-0.009	4.000	4.008	-0.016	4.000	4.012	-0.020	4.000	4.019	-0.027	4.000	4.023	-0.031
5	MAX	5.012	5.009	0.011	5.012	5.016	0.004	5.012	5.020	0.000	5.012	5.027	-0.007	5.012	5.031	-0.011
	MIN	5.000	5.001	-0.009	5.000	5.008	-0.016	5.000	5.012	-0.020	5.000	5.019	-0.027	5.000	5.023	-0.031
6	MAX	6.012	6.009	0.011	6.012	6.016	0.004	6.012	6.020	0.000	6.012	6.027	-0.007	6.012	6.031	-0.011
	MIN	6.000	6.001	-0.009	6.000	6.008	-0.016	6.000	6.012	-0.020	6.000	6.019	-0.027	6.000	6.023	-0.031
8	MAX	8.015	8.010	0.014	8.015	8.019	0.005	8.015	8.024	0.000	8.015	8.032	-0.008	8.015	8.037	-0.013
	MIN	8.000	8.001	-0.010	8.000	8.010	-0.019	8.000	8.015	-0.024	8.000	8.023	-0.032	8.000	8.028	-0.037
10	MAX	10.015	10.010	0.014	10.015	10.019	0.005	10.015	10.024	0.000	10.015	10.032	-0.008	10.015	10.037	-0.013
	MIN	10.000	10.001	-0.010	10.000	10.010	-0.019	10.000	10.015	-0.024	10.000	10.023	-0.032	10.000	10.028	-0.037
12	MAX	12.018	12.012	0.017	12.018	12.023	0.006	12.018	12.029	0.000	12.018	12.039	-0.010	12.018	12.044	-0.015
	MIN	12.000	12.001	-0.012	12.000	12.012	-0.023	12.000	12.018	-0.029	12.000	12.028	-0.039	12.000	12.033	-0.044
16	MAX	16.018	16.012	0.017	16.018	16.023	0.006	16.018	16.029	0.000	16.018	16.039	-0.010	16.018	16.044	-0.015
	MIN	16.000	16.001	-0.012	16.000	16.012	-0.023	16.000	16.018	-0.029	16.000	16.028	-0.039	16.000	16.033	-0.044
20	MAX	20.021	20.015	0.019	20.021	20.028	0.006	20.021	20.035	-0.001	20.021	20.048	-0.014	20.021	20.054	-0.020
	MIN	20.000	20.002	-0.015	20.000	20.015	-0.028	20.000	20.022	-0.035	20.000	20.035	-0.048	20.000	20.041	-0.054
25	MAX	25.021	25.015	0.019	25.021	25.028	0.006	25.021	25.035	-0.001	25.021	25.048	-0.014	25.021	25.061	-0.027
	MIN	25.000	25.002	-0.015	25.000	25.015	-0.028	25.000	25.022	-0.035	25.000	25.035	-0.048	25.000	25.048	-0.061
30	MAX	30.021	30.015	0.019	30.021	30.028	0.006	30.021	30.035	-0.001	30.021	30.048	-0.014	30.021	30.061	-0.027
	MIN	30.000	30.002	-0.015	30.000	30.015	-0.028	30.000	30.022	-0.035	30.000	30.035	-0.048	30.000	30.048	-0.061

Cont.

AMERICAN NATIONAL STANDARD
PREFERRED METRIC LIMITS AND FITS

ANSI B4.2-1978

Dimensions in mm.

BASIC SIZE		LOCATIONAL TRANSN. Hole H7	Shaft k6	Fit	LOCATIONAL TRANSN. Hole H7	Shaft n6	Fit	LOCATIONAL INTERF. Hole H7	Shaft p6	Fit	MEDIUM DRIVE Hole H7	Shaft s6	Fit	FORCE Hole H7	Shaft u6	Fit
40	MAX	40.025	40.018	0.023	40.025	40.033	0.008	40.025	40.042	-0.001	40.025	40.059	-0.018	40.025	40.076	-0.035
	MIN	40.000	40.002	-0.018	40.000	40.017	-0.033	40.000	40.026	-0.042	40.000	40.043	-0.059	40.000	40.060	-0.076
50	MAX	50.025	50.018	0.023	50.025	50.033	0.008	50.025	50.042	-0.001	50.025	50.059	-0.018	50.025	50.086	-0.045
	MIN	50.000	50.002	-0.018	50.000	50.017	-0.033	50.000	50.026	-0.042	50.000	50.043	-0.059	50.000	50.070	-0.086
60	MAX	60.030	60.021	0.028	60.030	60.039	0.010	60.030	60.051	-0.002	60.030	60.072	-0.023	60.030	60.106	-0.057
	MIN	60.000	60.002	-0.021	60.000	60.020	-0.039	60.000	60.032	-0.051	60.000	60.053	-0.072	60.000	60.087	-0.106
80	MAX	80.030	80.021	0.028	80.030	80.039	0.010	80.030	80.051	-0.002	80.030	80.078	-0.029	80.030	80.121	-0.072
	MIN	80.000	80.002	-0.021	80.000	80.020	-0.039	80.000	80.032	-0.051	80.000	80.059	-0.078	80.000	80.102	-0.121
100	MAX	100.035	100.025	0.032	100.035	100.045	0.012	100.035	100.059	-0.002	100.035	100.093	-0.036	100.035	100.146	-0.089
	MIN	100.000	100.003	-0.025	100.000	100.023	-0.045	100.000	100.037	-0.059	100.000	100.071	-0.093	100.000	100.124	-0.146
120	MAX	120.035	120.025	0.032	120.035	120.045	0.012	120.035	120.059	-0.002	120.035	120.101	-0.044	120.035	120.166	-0.109
	MIN	120.000	120.003	-0.025	120.000	120.023	-0.045	120.000	120.037	-0.059	120.000	120.079	-0.101	120.000	120.144	-0.166
160	MAX	160.040	160.028	0.037	160.040	160.052	0.013	160.040	160.068	-0.003	160.040	160.125	-0.060	160.040	160.215	-0.150
	MIN	160.000	160.003	-0.028	160.000	160.027	-0.052	160.000	160.043	-0.068	160.000	160.100	-0.125	160.000	160.190	-0.215
200	MAX	200.046	200.033	0.042	200.046	200.060	0.015	200.046	200.079	-0.004	200.046	200.151	-0.076	200.046	200.265	-0.190
	MIN	200.000	200.004	-0.033	200.000	200.031	-0.060	200.000	200.050	-0.079	200.000	200.122	-0.151	200.000	200.236	-0.265
250	MAX	250.046	250.033	0.042	250.046	250.060	0.015	250.046	250.079	-0.004	250.046	250.169	-0.094	250.046	250.313	-0.238
	MIN	250.000	250.004	-0.033	250.000	250.031	-0.060	250.000	250.050	-0.079	250.000	250.140	-0.169	250.000	250.284	-0.313
300	MAX	300.052	300.036	0.048	300.052	300.066	0.018	300.052	300.088	-0.004	300.052	300.202	-0.118	300.052	300.382	-0.298
	MIN	300.000	300.004	-0.036	300.000	300.034	-0.066	300.000	300.056	-0.088	300.000	300.170	-0.202	300.000	300.350	-0.382
400	MAX	400.057	400.040	0.053	400.057	400.073	0.020	400.057	400.098	-0.005	400.057	400.244	-0.151	400.057	400.471	-0.378
	MIN	400.000	400.004	-0.040	400.000	400.037	-0.073	400.000	400.062	-0.098	400.000	400.208	-0.244	400.000	400.435	-0.471
500	MAX	500.063	500.045	0.058	500.063	500.080	0.023	500.063	500.108	-0.005	500.063	500.292	-0.189	500.063	500.580	-0.477
	MIN	500.000	500.005	-0.045	500.000	500.040	-0.080	500.000	500.068	-0.108	500.000	500.252	-0.292	500.000	500.540	-0.580

AMERICAN NATIONAL STANDARD
PREFERRED METRIC LIMITS AND FITS

ANSI B4.2-1978

Dimensions in mm.

BASIC SIZE		LOOSE RUNNING			FREE RUNNING			CLOSE RUNNING			SLIDING			LOCATIONAL CLEARANCE		
		Hole C11	Shaft h11	Fit	Hole D9	Shaft h9	Fit	Hole F8	Shaft h7	Fit	Hole G7	Shaft h6	Fit	Hole H7	Shaft h6	Fit
1	MAX	1.120	1.000	0.180	1.045	1.000	0.070	1.020	1.000	0.030	1.012	1.000	0.018	1.010	1.000	0.016
	MIN	1.060	0.940	0.060	1.020	0.975	0.020	1.006	0.990	0.006	1.002	0.994	0.002	1.000	0.994	0.000
1.2	MAX	1.320	1.200	0.180	1.245	1.200	0.070	1.220	1.200	0.030	1.212	1.200	0.018	1.210	1.200	0.016
	MIN	1.260	1.140	0.060	1.220	1.175	0.020	1.206	1.190	0.006	1.202	1.194	0.002	1.200	1.194	0.000
1.6	MAX	1.720	1.600	0.180	1.645	1.600	0.070	1.620	1.600	0.030	1.612	1.600	0.018	1.610	1.600	0.016
	MIN	1.660	1.540	0.060	1.620	1.575	0.020	1.606	1.590	0.006	1.602	1.594	0.002	1.600	1.594	0.000
2	MAX	2.120	2.000	0.180	2.045	2.000	0.070	2.020	2.000	0.030	2.012	2.000	0.018	2.010	2.000	0.016
	MIN	2.060	1.940	0.060	2.020	1.975	0.020	2.006	1.990	0.006	2.002	1.994	0.002	2.000	1.994	0.000
2.5	MAX	2.620	2.500	0.180	2.545	2.500	0.070	2.520	2.500	0.030	2.512	2.500	0.018	2.510	2.500	0.016
	MIN	2.560	2.440	0.060	2.520	2.475	0.020	2.506	2.490	0.006	2.502	2.494	0.002	2.500	2.494	0.000
3	MAX	3.120	3.000	0.180	3.045	3.000	0.070	3.020	3.000	0.030	3.012	3.000	0.018	3.010	3.000	0.016
	MIN	3.060	2.940	0.060	3.020	2.975	0.020	3.006	2.990	0.006	3.002	2.994	0.002	3.000	2.994	0.000
4	MAX	4.145	4.000	0.220	4.060	4.000	0.090	4.028	4.000	0.040	4.016	4.000	0.024	4.012	4.000	0.020
	MIN	4.070	3.925	0.070	4.030	3.970	0.030	4.010	3.988	0.010	4.004	3.992	0.004	4.000	3.992	0.000
5	MAX	5.145	5.000	0.220	5.060	5.000	0.090	5.028	5.000	0.040	5.016	5.000	0.024	5.012	5.000	0.020
	MIN	5.070	4.925	0.070	5.030	4.970	0.030	5.010	4.988	0.010	5.004	4.992	0.004	5.000	4.992	0.000
6	MAX	6.145	6.000	0.220	6.060	6.000	0.090	6.028	6.000	0.040	6.016	6.000	0.024	6.012	6.000	0.020
	MIN	6.070	5.925	0.070	6.030	5.970	0.030	6.010	5.988	0.010	6.004	5.992	0.004	6.000	5.992	0.000
8	MAX	8.170	8.000	0.260	8.076	8.000	0.112	8.035	8.000	0.050	8.020	8.000	0.029	8.015	8.000	0.024
	MIN	8.080	7.910	0.080	8.040	7.964	0.040	8.013	7.985	0.013	8.005	7.991	0.005	8.000	7.991	0.000
10	MAX	10.170	10.000	0.260	10.076	10.000	0.112	10.035	10.000	0.050	10.020	10.000	0.029	10.015	10.000	0.024
	MIN	10.080	9.910	0.080	10.040	9.964	0.040	10.013	9.985	0.013	10.005	9.991	0.005	10.000	9.991	0.000
12	MAX	12.205	12.000	0.315	12.093	12.000	0.136	12.043	12.000	0.061	12.024	12.000	0.035	12.018	12.000	0.029
	MIN	12.095	11.890	0.095	12.050	11.957	0.050	12.016	11.982	0.016	12.006	11.989	0.006	12.000	11.989	0.000
16	MAX	16.205	16.000	0.315	16.093	16.000	0.136	16.043	16.000	0.061	16.024	16.000	0.035	16.018	16.000	0.029
	MIN	16.095	15.890	0.095	16.050	15.957	0.050	16.016	15.982	0.016	16.006	15.989	0.006	16.000	15.989	0.000
20	MAX	20.240	20.000	0.370	20.117	20.000	0.169	20.053	20.000	0.074	20.028	20.000	0.041	20.021	20.000	0.034
	MIN	20.110	19.870	0.110	20.065	19.948	0.065	20.020	19.979	0.020	20.007	19.987	0.007	20.000	19.987	0.000
25	MAX	25.240	25.000	0.370	25.117	25.000	0.169	25.053	25.000	0.074	25.028	25.000	0.041	25.021	25.000	0.034
	MIN	25.110	24.870	0.110	25.065	24.948	0.065	25.020	24.979	0.020	25.007	24.987	0.007	25.000	24.987	0.000
30	MAX	30.240	30.000	0.370	30.117	30.000	0.169	30.053	30.000	0.074	30.028	30.000	0.041	30.021	30.000	0.034
	MIN	30.110	29.870	0.110	30.065	29.948	0.065	30.020	29.979	0.020	30.007	29.987	0.007	30.000	29.987	0.000

Cont.

AMERICAN NATIONAL STANDARD
PREFERRED METRIC LIMITS AND FITS

ANSI B4.2-1978

Dimensions in mm.

BASIC SIZE		LOOSE RUNNING Hole C11	Shaft h11	Fit	FREE RUNNING Hole D9	Shaft h9	Fit	CLOSE RUNNING Hole F8	Shaft h7	Fit	SLIDING Hole G7	Shaft h6	Fit	LOCATIONAL CLEARANCE Hole H7	Shaft h6	Fit
40	MAX	40.280	40.000	0.440	40.142	40.000	0.204	40.064	40.000	0.089	40.034	40.000	0.050	40.025	40.000	0.041
	MIN	40.120	39.840	0.120	40.080	39.938	0.080	40.025	39.975	0.025	40.009	39.984	0.009	40.000	39.984	0.000
50	MAX	50.290	50.000	0.450	50.142	50.000	0.204	50.064	50.000	0.089	50.034	50.000	0.050	50.025	50.000	0.041
	MIN	50.130	49.840	0.130	50.080	49.938	0.080	50.025	49.975	0.025	50.009	49.984	0.009	50.000	49.984	0.000
60	MAX	60.330	60.000	0.520	60.174	60.000	0.248	60.076	60.000	0.106	60.040	60.000	0.059	60.030	60.000	0.049
	MIN	60.140	59.810	0.140	60.100	59.926	0.100	60.030	59.970	0.030	60.010	59.981	0.010	60.000	59.981	0.000
80	MAX	80.340	80.000	0.530	80.174	80.000	0.248	80.076	80.000	0.106	80.040	80.000	0.059	80.030	80.000	0.049
	MIN	80.150	79.810	0.150	80.100	79.926	0.100	80.030	79.970	0.030	80.010	79.981	0.010	80.000	79.981	0.000
100	MAX	100.390	100.000	0.610	100.207	100.000	0.294	100.090	100.000	0.125	100.047	100.000	0.069	100.035	100.000	0.057
	MIN	100.170	99.780	0.170	100.120	99.913	0.120	100.036	99.965	0.036	100.012	99.978	0.012	100.000	99.978	0.000
120	MAX	120.400	120.000	0.620	120.207	120.000	0.294	120.090	120.000	0.125	120.047	120.000	0.069	120.035	120.000	0.057
	MIN	120.180	119.780	0.180	120.120	119.913	0.120	120.036	119.965	0.036	120.012	119.978	0.012	120.000	119.978	0.000
160	MAX	160.460	160.000	0.710	160.245	160.000	0.345	160.106	160.000	0.146	160.054	160.000	0.079	160.040	160.000	0.065
	MIN	160.210	159.750	0.210	160.145	159.900	0.145	160.043	159.960	0.043	160.014	159.975	0.014	160.000	159.975	0.000
200	MAX	200.530	200.000	0.820	200.285	200.000	0.400	200.122	200.000	0.168	200.061	200.000	0.090	200.046	200.000	0.075
	MIN	200.240	199.710	0.240	200.170	199.885	0.170	200.050	199.954	0.050	200.015	199.971	0.015	200.000	199.971	0.000
250	MAX	250.570	250.000	0.860	250.285	250.000	0.400	250.122	250.000	0.168	250.061	250.000	0.090	250.046	250.000	0.075
	MIN	250.280	249.710	0.280	250.170	249.885	0.170	250.050	249.954	0.050	250.015	249.971	0.015	250.000	249.971	0.000
300	MAX	300.650	300.000	0.970	300.320	300.000	0.450	300.137	300.000	0.189	300.069	300.000	0.101	300.052	300.000	0.084
	MIN	300.330	299.680	0.330	300.190	299.870	0.190	300.056	299.948	0.056	300.017	299.968	0.017	300.000	299.968	0.000
400	MAX	400.760	400.000	1.120	400.350	400.000	0.490	400.151	400.000	0.208	400.075	400.000	0.111	400.057	400.000	0.093
	MIN	400.400	399.640	0.400	400.210	399.860	0.210	400.062	399.943	0.062	400.018	399.964	0.018	400.000	399.964	0.000
500	MAX	500.880	500.000	1.280	500.385	500.000	0.540	500.165	500.000	0.228	500.083	500.000	0.123	500.063	500.000	0.103
	MIN	500.480	499.600	0.480	500.230	499.845	0.230	500.068	499.937	0.068	500.020	499.960	0.020	500.000	499.960	0.000

AMERICAN NATIONAL STANDARD
PREFERRED METRIC LIMITS AND FITS

ANSI B4.2-1978

Dimensions in mm.

BASIC SIZE	MAX/MIN	LOCATIONAL TRANSN. Hole K7	Shaft h6	Fit	LOCATIONAL TRANSN. Hole N7	Shaft h6	Fit	LOCATIONAL INTERF. Hole P7	Shaft h6	Fit	MEDIUM DRIVE Hole S7	Shaft h6	Fit	FORCE Hole U7	Shaft h6	Fit
1	MAX	1.000	1.000	0.006	0.996	1.000	0.002	0.994	1.000	0.000	0.986	1.000	-0.008	0.982	1.000	-0.012
	MIN	0.990	0.994	-0.010	0.986	0.994	-0.014	0.984	0.994	-0.016	0.976	0.994	-0.024	0.972	0.994	-0.028
1.2	MAX	1.200	1.200	0.006	1.196	1.200	0.002	1.194	1.200	0.000	1.186	1.200	-0.008	1.182	1.200	-0.012
	MIN	1.190	1.194	-0.010	1.186	1.194	-0.014	1.184	1.194	-0.016	1.176	1.194	-0.024	1.172	1.194	-0.028
1.6	MAX	1.600	1.600	0.006	1.596	1.600	0.002	1.594	1.600	0.000	1.586	1.600	-0.008	1.582	1.600	-0.012
	MIN	1.590	1.594	-0.010	1.586	1.594	-0.014	1.584	1.594	-0.016	1.576	1.594	-0.024	1.572	1.594	-0.028
2	MAX	2.000	2.000	0.006	1.996	2.000	0.002	1.994	2.000	0.000	1.986	2.000	-0.008	1.982	2.000	-0.012
	MIN	1.990	1.994	-0.010	1.986	1.994	-0.014	1.984	1.994	-0.016	1.976	1.994	-0.024	1.972	1.994	-0.028
2.5	MAX	2.500	2.500	0.006	2.496	2.500	0.002	2.494	2.500	0.000	2.486	2.500	-0.008	2.482	2.500	-0.012
	MIN	2.490	2.494	-0.010	2.486	2.494	-0.014	2.484	2.494	-0.016	2.476	2.494	-0.024	2.472	2.494	-0.028
3	MAX	3.000	3.000	0.006	2.996	3.000	0.002	2.994	3.000	0.000	2.986	3.000	-0.008	2.982	3.000	-0.012
	MIN	2.990	2.994	-0.010	2.986	2.994	-0.014	2.984	2.994	-0.016	2.976	2.994	-0.024	2.972	2.994	-0.028
4	MAX	4.003	4.000	0.011	3.996	4.000	0.004	3.992	4.000	0.000	3.985	4.000	-0.007	3.981	4.000	-0.011
	MIN	3.991	3.992	-0.009	3.984	3.992	-0.016	3.980	3.992	-0.020	3.973	3.992	-0.027	3.969	3.992	-0.031
5	MAX	5.003	5.000	0.011	4.996	5.000	0.004	4.992	5.000	0.000	4.985	5.000	-0.007	4.981	5.000	-0.011
	MIN	4.991	4.992	-0.009	4.984	4.992	-0.016	4.980	4.992	-0.020	4.973	4.992	-0.027	4.969	4.992	-0.031
6	MAX	6.003	6.000	0.011	5.996	6.000	0.004	5.992	6.000	0.000	5.985	6.000	-0.007	5.981	6.000	-0.011
	MIN	5.991	5.992	-0.009	5.984	5.992	-0.016	5.980	5.992	-0.020	5.973	5.992	-0.027	5.969	5.992	-0.031
8	MAX	8.005	8.000	0.014	7.996	8.000	0.005	7.991	8.000	0.000	7.983	8.000	-0.008	7.978	8.000	-0.013
	MIN	7.990	7.991	-0.010	7.981	7.991	-0.019	7.976	7.991	-0.024	7.968	7.991	-0.032	7.963	7.991	-0.037
10	MAX	10.005	10.000	0.014	9.996	10.000	0.005	9.991	10.000	0.000	9.983	10.000	-0.008	9.978	10.000	-0.013
	MIN	9.990	9.991	-0.010	9.981	9.991	-0.019	9.976	9.991	-0.024	9.968	9.991	-0.032	9.963	9.991	-0.037
12	MAX	12.006	12.000	0.017	11.995	12.000	0.006	11.989	12.000	0.000	11.979	12.000	-0.010	11.974	12.000	-0.015
	MIN	11.988	11.989	-0.012	11.977	11.989	-0.023	11.971	11.989	-0.029	11.961	11.989	-0.039	11.956	11.989	-0.044
16	MAX	16.006	16.000	0.017	15.995	16.000	0.006	15.989	16.000	0.000	15.979	16.000	-0.010	15.974	16.000	-0.015
	MIN	15.988	15.989	-0.012	15.977	15.989	-0.023	15.971	15.989	-0.029	15.961	15.989	-0.039	15.956	15.989	-0.040
20	MAX	20.006	20.000	0.019	19.993	20.000	0.006	19.986	20.000	0.000	19.973	20.000	-0.014	19.967	20.000	-0.020
	MIN	19.985	19.987	-0.015	19.972	19.987	-0.028	19.965	19.987	-0.035	19.952	19.987	-0.048	19.946	19.987	-0.054
25	MAX	25.006	25.000	0.019	24.993	25.000	0.006	24.986	25.000	0.000	24.973	25.000	-0.014	24.960	25.000	-0.027
	MIN	24.985	24.987	-0.015	24.972	24.987	-0.028	24.965	24.987	-0.035	24.952	24.987	-0.048	24.939	24.987	-0.061
30	MAX	30.006	30.000	0.019	29.993	30.000	0.006	29.986	30.000	0.000	29.973	30.000	-0.014	29.960	30.000	-0.027
	MIN	29.985	29.987	-0.015	29.972	29.987	-0.028	29.965	29.987	-0.035	29.952	29.987	-0.048	29.939	29.987	-0.061

AMERICAN NATIONAL STANDARD
PREFERRED METRIC LIMITS AND FITS

ANSI B4.2-1978

Dimensions in mm.

BASIC SIZE		LOCATIONAL TRANSN. Hole K7	Shaft h6	Fit	LOCATIONAL TRANSN. Hole N7	Shaft h6	Fit	LOCATIONAL INTERF. Hole P7	Shaft h6	Fit	MEDIUM DRIVE Hole S7	Shaft h6	Fit	FORCE Hole U7	Shaft h6	Fit
40	MAX	40.007	40.000	0.023	39.992	40.000	0.008	39.983	40.000	-0.001	39.966	40.000	-0.018	39.949	40.000	-0.035
	MIN	39.982	39.984	-0.018	39.967	39.984	-0.033	39.958	39.984	-0.042	39.941	39.984	-0.059	39.924	39.984	-0.076
50	MAX	50.007	50.000	0.023	49.992	50.000	0.008	49.983	50.000	-0.001	49.966	50.000	-0.018	49.939	50.000	-0.045
	MIN	49.982	49.984	-0.018	49.967	49.984	-0.033	49.958	49.984	-0.042	49.941	49.984	-0.059	49.914	49.984	-0.086
60	MAX	60.009	60.000	0.028	59.991	60.000	0.010	59.979	60.000	-0.002	59.958	60.000	-0.023	59.924	60.000	-0.057
	MIN	59.979	59.981	-0.021	59.961	59.981	-0.039	59.949	59.981	-0.051	59.928	59.981	-0.072	59.894	59.981	-0.106
80	MAX	80.009	80.000	0.028	79.991	80.000	0.010	79.979	80.000	-0.002	79.952	80.000	-0.029	79.909	80.000	-0.072
	MIN	79.979	79.981	-0.021	79.961	79.981	-0.039	79.949	79.981	-0.051	79.922	79.981	-0.078	79.879	79.981	-0.121
100	MAX	100.010	100.000	0.032	99.990	100.000	0.012	99.976	100.000	-0.002	99.942	100.000	-0.036	99.889	100.000	-0.089
	MIN	99.975	99.978	-0.025	99.955	99.978	-0.045	99.941	99.978	-0.059	99.907	99.978	-0.093	99.854	99.978	-0.146
120	MAX	120.010	120.000	0.032	119.990	120.000	0.012	119.976	120.000	-0.002	119.934	120.000	-0.044	119.869	120.000	-0.109
	MIN	119.975	119.978	-0.025	119.955	119.978	-0.045	119.941	119.978	-0.059	119.899	119.978	-0.101	119.834	119.978	-0.166
160	MAX	160.012	160.000	0.037	159.988	160.000	0.013	159.972	160.000	-0.003	159.915	160.000	-0.060	159.825	160.000	-0.150
	MIN	159.972	159.975	-0.028	159.948	159.975	-0.052	159.932	159.975	-0.068	159.875	159.975	-0.125	159.785	159.975	-0.215
200	MAX	200.013	200.000	0.042	199.986	200.000	0.015	199.967	200.000	-0.004	199.895	200.000	-0.076	199.781	200.000	-0.190
	MIN	199.967	199.971	-0.033	199.940	199.971	-0.060	199.921	199.971	-0.079	199.849	199.971	-0.151	199.735	199.971	-0.265
250	MAX	250.013	250.000	0.042	249.986	250.000	0.015	249.967	250.000	-0.004	249.877	250.000	-0.094	249.733	250.000	-0.238
	MIN	249.967	249.971	-0.033	249.940	249.971	-0.060	249.921	249.971	-0.079	249.831	249.971	-0.169	249.687	249.971	-0.313
300	MAX	300.016	300.000	0.048	299.986	300.000	0.018	299.964	300.000	-0.004	299.850	300.000	-0.118	299.670	300.000	-0.298
	MIN	299.964	299.968	-0.036	299.934	299.968	-0.066	299.912	299.968	-0.088	299.798	299.968	-0.202	299.618	299.968	-0.382
400	MAX	400.017	400.000	0.053	399.984	400.000	0.020	399.959	400.000	-0.005	399.813	400.000	-0.151	399.586	400.000	-0.378
	MIN	399.960	399.964	-0.040	399.927	399.964	-0.073	399.902	399.964	-0.098	399.756	399.964	-0.244	399.529	399.964	-0.471
500	MAX	500.018	500.000	0.058	499.983	500.000	0.023	499.955	500.000	-0.005	499.771	500.000	-0.189	499.483	500.000	-0.477
	MIN	499.955	499.960	-0.045	499.920	499.960	-0.080	499.892	499.960	-0.108	499.708	499.960	-0.292	499.420	499.960	-0.580

Appendix 49 ENGINEERING FORMULAS

Motion

S = distance (inches, feet, miles)
t = time (seconds, minutes, hours)
v = average velocity (feet per second, miles per hour, etc.
v_1 = initial velocity
v_2 = final velocity
a = acceleration (feet per second per second)

(1) $S = vt$

(2) $V(\text{avg.}) = \dfrac{v_2 + v_1}{2}$

(3) $S = \left(\dfrac{v_1 + v_2}{2}\right) \dfrac{\text{ft}}{\text{sec}} (t \text{ sec})$

(4) $a = \dfrac{v_2 - v_1}{t}$

(5) $S = v_1 t + \frac{1}{2} a t^2$

Angular Motion

V = linear velocity
N = number of revolutions per min
θ = angular distance in radians
1 radian = $360°/2\pi = 57.3°$
$\omega(\text{omega})$ = average angular velocity
$\quad\quad = \theta/t$ (rad per sec, rev per min)
ω_1 = initial velocity
ω_2 = final velocity
$\alpha(\text{alpha})$ = angular acceleration = rad per sec^2
S = length of arc
r = radius of arc
D = diameter

(6) $\theta = (\text{avg. } \omega)t$

(7) $\omega(\text{avg.}) = \dfrac{\omega_2 + \omega_1}{2}$

(8) $\alpha = \dfrac{\omega_2 - \omega_1}{t}$

(9) $\omega_2 = \omega_1 + \alpha t$

(10) $\theta = \omega_1 t + \dfrac{\alpha t^2}{2}$

(11) $V = \pi D N$ or $V = r\omega$ (ft per sec, ft per min, etc.)

(12) $\omega = \dfrac{2\pi r n}{r} = 2\pi N$

Force and Acceleration

F = force (pounds)
M = mass
a = acceleration (ft per sec^2)
g = gravitational acceleration = 32.2 ft/sec^2
W = weight (pounds)
$M = F/a = W/g$ (units of mass in slugs)

Work

W = work (ft · lb)
F = force (lb)
d = distance
$W = Fd$

Power

W = work
t = time

1 horsepower = $550 \dfrac{\text{ft} \cdot \text{lb}}{\text{sec}}$

Avg. power = $\dfrac{W}{t} = \dfrac{\text{ft} \cdot \text{lb}}{\text{sec}}$ or $\dfrac{\text{ft} \cdot \text{lb}}{\text{min}}$ etc.

Kinetic Energy

W = weight
V = velocity
g = 32.2 ft per sec^2
K.E. = $WV^2/2g$

Appendix 50 GRADING GRAPH

This graph can be used to determine the individual grades of members of a team and to compute grade averages for those who do extra assignments.

The percent participation of each team member should be determined by the team as a whole (see Chapter 18 problems).

Example: written or oral report grades

Overall team grade: 82

Team members N = 5	Contribution C = %	F = CN	Grade (graph)
J. Doe	20%	100	82.0
H. Brown	16%	80	76.4
L. Smith	24%	120	86.0
R. Black	20%	100	82.0
T. Jones	20%	100	82.0
	100%		

Example: quiz or problem sheet grades

Number assigned: 30
Number extra: 6

Total 36

Average grade for total (36): 82

$$F = \frac{\text{No. completed} \times 100}{\text{No. assigned}} = \frac{36 \times 100}{30} = 120$$

Final grade (from graph): 86.0

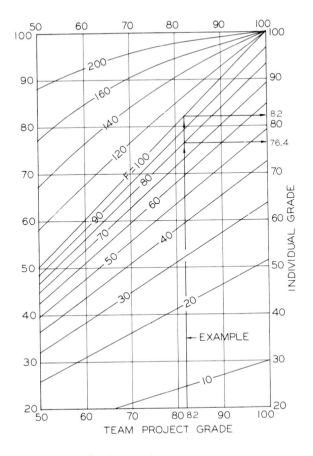

Fig. A50.—1. Grading graph.

INDEX

INDEX